CW00997033

Bayesian Speech and Language Processing

With this comprehensive guide you will learn how to apply Bayesian machine learning techniques systematically to solve various problems in speech and language processing.

A range of statistical models is detailed, from hidden Markov models to Gaussian mixture models, *n*-gram models, and latent topic models, along with applications including automatic speech recognition, speaker verification, and information retrieval. Approximate Bayesian inferences based on MAP, Evidence, Asymptotic, VB, and MCMC approximations are provided as well as full derivations of calculations, useful notations, formulas, and rules.

The authors address the difficulties of straightforward applications and provide detailed examples and case studies to demonstrate how you can successfully use practical Bayesian inference methods to improve the performance of information systems.

This is an invaluable resource for students, researchers, and industry practitioners working in machine learning, signal processing, and speech and language processing.

Shinji Watanabe received his Ph.D. from Waseda University in 2006. He has been a research scientist at NTT Communication Science Laboratories, a visiting scholar at Georgia Institute of Technology and a senior principal member at Mitsubishi Electric Research Laboratories (MERL), as well as having been an associate editor of the *IEEE Transactions on Audio Speech and Language Processing*, and an elected member of the IEEE Speech and Language Processing Technical Committee. He has published more than 100 papers in journals and conferences, and received several awards including the Best Paper Award from IEICE in 2003.

Jen-Tzung Chien is with the Department of Electrical and Computer Engineering and the Department of Computer Science at the National Chiao Tung University, Taiwan, where he is now the University Chair Professor. He received the Distinguished Research Award from the Ministry of Science and Technology, Taiwan, and the Best Paper Award of the 2011 IEEE Automatic Speech Recognition and Understanding Workshop. He serves currently as an elected member of the IEEE Machine Learning for Signal Processing Technical Committee.

"This book provides an overview of a wide range of fundamental theories of Bayesian learning, inference, and prediction for uncertainty modeling in speech and language processing. The uncertainty modeling is crucial in increasing the robustness of practical systems based on statistical modeling under real environment, such as automatic speech recognition systems under noise, and question answering systems based on limited size of training data. This is the most advanced and comprehensive book for learning fundamental Bayesian approaches and practical techniques."

Sadaoki Furui, Tokyo Institute of Technology

Bayesian Speech and Language Processing

SHINJI WATANABE
Mitsubishi Electric Research Laboratories

JEN-TZUNG CHIEN
National Chiao Tung University

CAMBRIDGE
UNIVERSITY PRESS

University Printing House, Cambridge CB2 8BS, United Kingdom

Cambridge University Press is part of the University of Cambridge.

It furthers the University's mission by disseminating knowledge in the pursuit of education, learning and research at the highest international levels of excellence.

www.cambridge.org
Information on this title: www.cambridge.org/9781107055575

First published 2015

Printed in the United Kingdom by Clays, St Ives plc

A catalog record for this publication is available from the British Library

Library of Congress Cataloging in Publication data
Watanabe, Shinji (Communications engineer) author.
Bayesian speech and language processing / Shinji Watanabe, Mitsubishi Electric Research Laboratories; Jen-Tzung Chien, National Chiao Tung University.
pages cm
ISBN 978-1-107-05557-5 (hardback)
1. Language and languages – Study and teaching – Statistical methods. 2. Bayesian statistical decision theory. I. Title.
P53.815.W38 2015
410.1′51–dc23
2014050265

ISBN 978-1-107-05557-5 Hardback

Contents

Preface

In general, speech and language processing involves extensive knowledge of statistical models. The acoustic model using hidden Markov models and the language model using n-grams are mainly introduced here. Both acoustic and language models are important parts of modern speech recognition systems where the learned models from real-world data are full of complexity, ambiguity, and uncertainty. The uncertainty modeling is crucial to tackle the lack of robustness for speech and language processing.

This book addresses fundamental theories of Bayesian learning, inference, and prediction for the uncertainty modeling. Uniquely, compared with standard textbooks for dealing with the fundamental Bayesian approaches, this book focuses on the practical methods of the approaches to make them applicable to actual speech and language problems. We (the authors) have been studying these topics for a long time with a strong belief that the Bayesian approaches could solve "robustness" issues in speech and language processing, which are the most difficult problem and most serious shortcoming of real systems based on speech and language processing. In our experience, the most difficult issue in applying Bayesian approaches is how to appropriately choose a specific technique among the many Bayesian techniques proposed in statistics and machine learning so far. One of our answers to this question is to provide the approximated Bayesian inference methods rather than focusing on covering the whole Bayesian techniques. We categorize the Bayesian approaches into five categories: the maximum a-posteriori estimation; evidence approximation; asymptotic approximation; variational Bayes; and Markov chain Monte Carlo. We also describe the speech and language processing applications within this categorization so that readers can appropriately choose the approximated Bayesian techniques for their problems.

This book is part of our long-term cooperative efforts to promote the Bayesian approaches in speech and language processing. We have been pursuing this goal for more than ten years, and part of our efforts was to organize a tutorial lecture with this theme at the 37th International Conference on Acoustics, Speech, and Signal Processing (ICASSP) in Kyoto, Japan, March 2012. The success of this tutorial lecture prompted the idea of writing a textbook with this theme. We strongly believe in the importance of the Bayesian approaches, and we sincerely encourage the researchers who work with Bayesian speech and language processing.

Acknowledgments

First we want to thank all of our colleagues and research friends, especially members of NTT Communication Science Laboratories, Mitsubishi Electric Research Laboratories (MERL), National Cheng Kung University, IBM T. J. Watson Research Center, and National Chiao Tung University (NCTU). Some of the studies in this book were actually conducted when the authors were working in these institutes. We also would like to thank many people for reading a draft and giving us valuable comments which greatly improved this book, including Tawara Naohiro, Yotaro Kubo, Seong-Jun Hahm, Yu Tsao, and all of the students from the Machine Learning Laboratory at NCTU. We are very grateful for support from Anthony Vetro, John R. Hershey, and Jonathan Le Roux at MERL, and Sin-Horng Chen, Hsueh-Ming Hang, Yu-Chee Tseng, and Li-Chun Wang at NCTU. The great efforts of the editors of Cambridge University Press, Phil Meyler, Sarah Marsh, and Heather Brolly, are also appreciated. Finally, we would like to thank our families for supporting our whole research lives.

Shinji Watanabe
Jen-Tzung Chien

Notation and abbreviations

General notation

This book observes the following general mathematical notation to avoid any confusion arising from notation:

$\mathbb{B} = \{\text{true, false}\}$

Set of boolean values

$\mathbb{Z}^+ = \{1, 2, \cdots\}$

Set of positive integers

$\mathbb{R}$

Set of real numbers

$\mathbb{R}_{>0}$

Set of positive real numbers

$\mathbb{R}^D$

Set of D dimensional real numbers

Σ^*

Set of all possible strings composed of letters

$\emptyset$

Empty set

a

Scalar variable

$\mathbf{a}$

Vector variable

$$\mathbf{a} = [a_1 \quad \cdots \quad a_N]^\mathsf{T} = \begin{bmatrix} a_1 \\ \vdots \\ a_N \end{bmatrix}$$

Elements of a vector, which can be described with the square brackets $[\cdots]$, $^\mathsf{T}$ denotes the transpose operation

$\mathbf{A}$

Matrix variable

$$\mathbf{A} = \begin{bmatrix} a & b \\ c & d \end{bmatrix}$$

Elements of a matrix, which can be described with the square brackets $[\cdots]$

$\mathbf{I}_D$

$D \times D$ identity matrix

$|\mathbf{A}|$

Determinant of square matrix

$\mathrm{tr}[\mathbf{A}]$

Trace of square matrix

$A, \mathcal{A}$

Set or sequential variable

$A = \{a_1, \cdots, a_N\} = \{a_n\}_{n=1}^N$

Elements in a set, which can be described with the curly brackets $\{\cdots\}$

$A = \{a_n\}$

Elements in a set, where the range of index n is omitted for simplicity

$a_{n:n'} = \{a_n, \cdots, a_{n'}\} \quad n' > n$

A set of sequential variables, which explicitly describes the range of elements from n to n' by using : in the subscript

$|A|$

The number of elements in a set A. For example $|\{a_n\}_{n=1}^N| = N$

$f(x)$ or f_x

Function of x

$p(x)$ or $q(x)$

Probabilistic distribution function of x

Functional of f. Note that a functional uses the square brackets $[\cdot]$ while a function uses the bracket $(\cdot)$.

$\mathbb{E}_{p(x|y)}[f(x)|y] = \int f(x)p(x|y)dx$

The expectation of $f(x)$ with respect to probability distribution $p(x|y)$

$\mathbb{E}_{(x)}[f(x)|y] = \int f(x)p(x|y)dx$ or $\mathbb{E}_{(x)}[f(x)] = \int f(x)p(x|y)dx$

Another form of the expectation of $f(x)$, where the subscript with the probability distribution and/or the conditional variable is omitted, when it is trivial.

$$\delta(a, a') = \begin{cases} 1 & a = a' \\ 0 & \text{Otherwise} \end{cases}$$

Kronecker delta function for discrete variables a and a'

$\delta(x - x')$

Dirac delta function for continuous variables x and x'

$A^{\mathrm{ML}}, A^{\mathrm{ML2}}, A^{\mathrm{MAP}}, A^{\mathrm{DT}}, \ldots$

The variables estimated by a specific criterion (e.g., Maximum Likelihood (ML)) are represented with the superscript of the abbreviation of the criterion.

Basic notation used for speech and language processing

We also list the notation specific for speech and language processing. This book tries to maintain consistency by using the same notation, while it also tries to use commonly used notation in each application. Therefore, some of the same characters are used to denote different variables, since this book needs to introduce many variables.

Common notation

Set of model parameters

Model variable including types of models, structure, hyperparameters, etc.

Set of hyperparameters

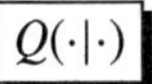

Auxiliary function used in the EM algorithm

Hessian matrix

Acoustic modeling

$T \in \mathbb{Z}^+$

Number of speech frames

$t \in \{1, \cdots, T\}$

Speech frame index

$\mathbf{o}_t \in \mathbb{R}^D$

D dimensional feature vector at time t

$\mathbf{O} = \{\mathbf{o}_t | t = 1, \cdots, T\}$

Sequence of T feature vectors

$J \in \mathbb{Z}^+$

Number of unique HMM states in an HMM

$s_t \in \{1, \cdots, J\}$

HMM state at time t

$S = \{s_t | t = 1, \cdots, T\}$

Sequence of HMM states for T speech frames

$K \in \mathbb{Z}^+$

Number of unique mixture components in a GMM

$v_t \in \{1, \cdots, K\}$

Latent mixture variable at time t

$V = \{v_t | t = 1, \cdots, T\}$

Sequence of latent mixture variables for T speech frames

$\alpha_t(j) \in [0, 1]$

Forward probability of the partial observations $\{\mathbf{o}_1, \cdots, \mathbf{o}_t\}$ until time t and state j at time t

$\beta_t(j) \in [0, 1]$

Backward probability of the partial observations $\{\mathbf{o}_{t+1}, \cdots, \mathbf{o}_T\}$ from $t + 1$ to the end given state j at time t

$\delta_t(j) \in [0, 1]$

The highest probability along a single path, at time t which accounts for previous observations $\{\mathbf{o}_1, \cdots, \mathbf{o}_t\}$ and ends in state j at time t

$\xi_t(i, j) \in [0, 1]$

Posterior probability of staying state i at time t and state j at time $t + 1$

$\gamma_t(j, k) \in [0, 1]$

Posterior probability of staying at state j and mixture component k at time t

$\pi_j \in [0, 1]$

Initial state probability of state j at time $t = 1$

$a_{ij} \in [0, 1]$

State transition probability from state $s_{t-1} = i$ to state $s_t = j$

$\omega_{jk} \in [0, 1]$

Gaussian mixture weight at component k of state j

$\boldsymbol{\mu}_{jk} \in \mathbb{R}^D$

Gaussian mean vector at component k of state j

$\boldsymbol{\Sigma}_{jk} \in \mathbb{R}^{D \times D}$

Gaussian covariance matrix at component k of state j. Symmetric matrix

$\mathbf{R}_{jk} \in \mathbb{R}^{D \times D}$

Gaussian precision matrix at component k of state j. Symmetric matrix, and the inverse of covariance matrix $\boldsymbol{\Sigma}_{jk}$

Language modeling

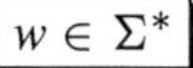

$w \in \Sigma^*$

Category (e.g., word in most cases, phoneme sometimes). The element is represented by a string in Σ^* (e.g., "I" and "apple" for words and /a/ and /k/ for phonemes) or a natural number in $\mathbb{Z}^+$ when the elements of categories are numbered.

$\mathcal{V} \subset \Sigma^*$

Vocabulary (dictionary), i.e., a set of distinct words, which is a subset of Σ^*

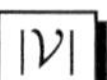

$|\mathcal{V}|$

Vocabulary size

$v \in \{1, \cdots, |\mathcal{V}|\}$

Ordered index number of distinct words in vocabulary $\mathcal{V}$

$w_{(v)} \in \mathcal{V}$

Word pointed by an ordered index v

$\{w_{(v)} | v = 1, \cdots, |\mathcal{V}|\} = \mathcal{V}$

A set of distinct words, which is equivalent to vocabulary $\mathcal{V}$

$J \in \mathbb{Z}^+$

Number of categories in a chunk (e.g., number of words in a sentence or number of phonemes or HMM states in a speech segment)

$i \in \{1, \cdots, J\}$

ith position of category (e.g., word or phoneme)

$w_i \in \mathcal{V}$

Word at ith position

$W = \{w_i | i = 1, \cdots, J\}$

Word sequence from 1 to J

$w_{i-n+1}^{i} = \{w_{i-n+1} \cdots w_i\}$

Word sequence from $i - n + 1$ to i

$p(w_i | w_{i-n+1}^{i-1}) \in [0, 1]$

n-gram probability, which considers $n - 1$ order Markov model

$c(w_{i-n+1}^{i-1}) \in \mathbb{Z}^+$

Number of occurrences of word sequence w_{i-n+1}^{i-1} in a training corpus

$\lambda_{w_{i-n+1}^{i-1}}$

Interpolation weight for each w_{i-n+1}^{i-1}

$M \in \mathbb{Z}^+$

Number of documents

$m \in \{1, \cdots, M\}$

Document index

mth document, which would be represented by a string or positive integer

$c(w_{(v)}, d_m) \in \mathbb{Z}^+$

Number of co-occurrences of word $w_{(v)}$ in document d_m

$K \in \mathbb{Z}^+$

Number of unique latent topics

$z_i \in \{1, \cdots, K\}$

ith latent topic variable for word w_i

$Z = \{z_j | j = 1, \cdots, J\}$

Sequence of latent topic variables for J words

Abbreviations

AIC: Akaike Information Criterion (page 217)
AM: Acoustic Model (page 3)
ARD: Automatic Relevance Determination (page 194)
ASR: Automatic Speech Recognition (page 58)
BIC: Bayesian Information Criterion (page 8)
BNP: Bayesian Nonparametrics (pages 337, 345)
BPC: Bayesian Predictive Classification (page 218)
CDHMM: Continuous Density Hidden Markov Model (page 157)
CRP: Chinese Restaurant Process (page 350)
CSR: Continuous Speech Recognition (page 334)
DCLM: Dirichlet Class Language Model (page 326)

DHMM: Discrete Hidden Markov Model (page 62)
DNN: Deep Neural Network (page 224)
DP: Dirichlet Process (page 348)
EM: Expectation Maximization (page 9)
fMLLR: feature-space MLLR (page 204)
GMM: Gaussian Mixture Model (page 63)
HDP: Hierarchical Dirichlet Process (page 337)
HMM: Hidden Markov Model (page 59)
HPY: Hierarchical Pitman–Yor Process (page 383)
HPYLM: Hierarchical Pitman–Yor Language Model (page 384)
iid: Independently, identically distributed (page 216)
KL: Kullback–Leibler (page 79)
KN: Kneser–Ney (page 102)
LDA: Latent Dirichlet Allocation (page 318)
LM: Language Model (page 3)
LSA: Latent Semantic Analysis (page 113)
LVCSR: Large Vocabulary Continuous Speech Recognition (page 97)
MAP: Maximum A-Posteriori (page 7)
MAPLR: Maximum A-Posteriori Linear Regression (page 287)
MBR: Minimum Bayes Risk (page 56)
MCE: Minimum Classification Error (page 59)
MCMC: Markov Chain Monte Carlo (page 337)
MDL: Minimum Description Length (page 9)
MFCC: Mel-Frequency Cepstrum Coefficients (page 249)
MKN: Modified Kneser–Ney (page 111)
ML: Maximum Likelihood (page 77)
ML2: Type-2 Maximum Likelihood (page 188)
MLLR: Maximum Likelihood Linear Regression (page 200)
MLP: MultiLayer Perceptron (page 326)
MMI: Maximum Mutual Information (page 167)
MMSE: Minimum Mean Square Error (page 139)
MPE: Minimum Phone Error (page 167)
nCRP: nested Chinese Restaurant Process (page 360)
NDP: Nested Dirichlet Process (page 360)
NMF: Non-negative Matrix Factorization (page 124)
pdf: probability density function (page 63)
PLP: Perceptual Linear Prediction (page 54)
PLSA: Probabilistic Latent Semantic Analysis (page 113)
PY: Pitman–Yor Process (page 379)
QB: Quasi-Bayes (page 180)
RHS: Right-Hand Side (page 199)
RLS: Regularized Least-Squares (page 188)
RVM: Relevance Vector Machine (page 192)
SBL: Sparse Bayesian Learning (page 194)

SBP: Stick Breaking Process (page 348)
SMAP: Structural Maximum A-Posteriori (page 288)
SMAPLR: Structural Maximum A-Posteriori Linear Regression (page 288)
SVD: Singular Value Decomposition (page 114)
SVM: Support Vector Machine (page 188)
tf–idf: term frequency – inverse document frequency (page 113)
UBM: Universal Background Model (page 172)
VB: Variational Bayes (page 7)
VC: Vapnik–Chervonenkis (page 191)
VQ: Vector Quantization (page 62)
WB: Witten–Bell (page 102)
WER: Word Error Rate (page 56)
WFST: Weighted Finite State Transducer (page 60)
WSJ: Wall Street Journal (page 108)

Part I

General discussion

1 Introduction

1.1 Machine learning and speech and language processing

Speech and language processing is one of the most successful examples of applying machine learning techniques to real problems. Current speech and language techniques embody our real-world information processing, automatically including information extraction, question answering, summarization, dialog, conversational agent, and machine translation (Jurafsky & Martin 2000). Among these, one of the most exciting applications of speech and language processing is speech recognition based voice search technologies (by Google, Nuance) and conversational agent technologies (by Apple) (Schalkwyk, Beeferman, Beaufays *et al.* 2010). These successful applications started to make people in general casually use speech interface rather than text interface in mobile devices, and the applications of speech and language processing are widely expanding.

One of the core technologies of speech and language processing is automatic speech recognition (ASR) and related techniques. Surprisingly, these techniques are fully based on statistical approaches by using large amounts of data. The machine learning techniques are applied to utilize these data. For example, the main components of ASR are acoustic and language models. The acoustic model (AM) provides a statistical model of each phoneme/word unit, and it is represented by a hidden Markov model (HMM). The HMM is one of the most typical examples of dealing with sequential data based on machine learning techniques (Bishop 2006), and machine learning techniques provide an efficient method of computing a maximum likelihood value for the HMM and an efficient training algorithm of the HMM parameters. The language model (LM) also provides an n-gram based statistical model for word sequences, which is also trained by using the large amount of data based on machine learning techniques. These statistical models and their variants are used for the other speech and language applications, including speaker verification and information retrieval, and thus, machine learning is a core component of speech and language processing.

Machine learning covers a wide range of applications in addition to speech and language processing, including bioinformatics, data mining, and computer vision. Machine learning also covers various theoretical fields including pattern recognition, information theory, statistics, control theory, and applied mathematics. Therefore, many people are studying and developing machine learning techniques, and the progress of machine learning is rather fast. By following the rapid progress of machine learning, researchers in speech and language processing interact positively with the machine learning community or communities in the machine learning application field by importing (and sometimes exporting) advanced machine learning techniques. For example, the recent great improvement of ASR comes from this interaction for discriminative approaches (recent progress summaries for discriminative speech recognition techniques are found in Gales, Watanabe & Fossler-Lussier (2012), Heigold, Ney, Schluter *et al.* (2012), Saon & Chien (2012*b*), Hinton, Deng, Yu *et al.* (2012). The discriminative training of HMM parameters has been mainly studied in speech recognition research since the 1990s, and became a standard technique around the 2000s. In addition, the deep neural network replaces the emission probability of the HMM from the Gaussian mixture model (GMM) (or is used as feature extraction (Hermansky, Ellis & Sharma 2000, Grézl, Karafiát, Kontár *et al.* 2007) for the GMM) and achieves further improvement on the discriminative training based ASR performance. Actually, current successful applications of speech and language processing are highly supported by these breakthroughs based on the discriminative techniques developed through the interaction with the machine learning community. By following the successful experience, researchers in speech and language processing try to collaborate with the machine learning community further to find new technologies.

1.2 Bayesian approach

This book also follows the trend of tight interaction with the machine learning community, but focuses on another active research topic in machine learning, called the *Bayesian approach*. The Bayesian approach is a major probabilistic theory that represents a causal relationship of data. By dealing with variables introduced in a model as probabilistic variables, we can consider uncertainties included in these variables based on the probabilistic theory.

As a simple example of uncertainty, we think of statistically modeling several data (x_1, $x_2, \cdots, x_N$) by a Gaussian distribution $\mathcal{N}(x|\mu, \Sigma)$ with mean and variance parameters μ and Σ, as shown in Figure 1.1, i.e.,

$$p(x) \approx \mathcal{N}(x|\mu, \Sigma). \tag{1.1}$$

Now, we consider the Bayesian approach, where the mean parameter is uncertain, and is distributed by a probabilistic function $p(\mu)$. Since μ is uncertain, we can consider the several possible μs instead of one fixed μ, and the Bayesian approach considers representing a distribution of x by several Gaussians with possible mean parameters (μ_1, μ_2, and μ_3 in the example of Figure 1.1),

$$p(x) \approx \frac{1}{N} \sum_{\mu=\{\mu_1,\mu_2,\cdots,\mu_N\}} \mathcal{N}(x|\mu, \Sigma), \tag{1.2}$$

where μ_1, μ_2, $\cdots$ are generated from the distribution $p(\mu)$. The extreme case of this uncertainty consideration is to represent the distribution of x, which is represented by

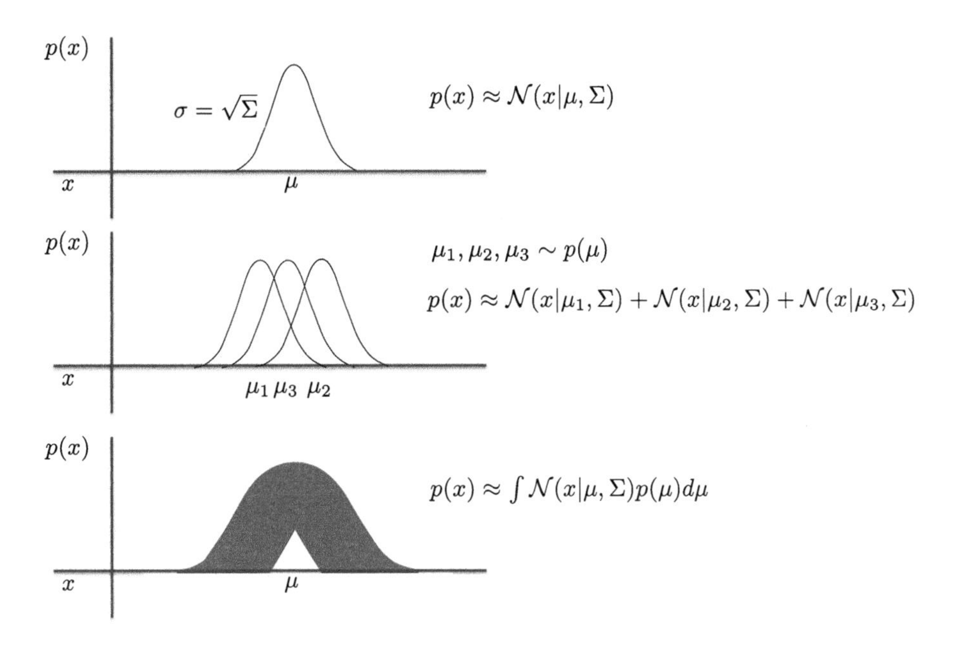

Figure 1.1 The uncertainty of the mean parameter of a Gaussian distribution.

all possible Gaussian distributions, with *all possible* mean parameters weighted by the probability $p(\mu)$, which is represented as the following integral equation:

$$p(x) \approx \int \mathcal{N}(x|\mu, \Sigma)p(\mu)d\mu. \tag{1.3}$$

This expectation over uncertain variables is called *marginalization*. The grayed band in Figure 1.1 provides an image of the expected distribution, where the mean parameter is marginalized over all possible infinite mean values. This is a unique aspect of the Bayesian approach that represents variables in a model by probabilistic distributions, and holds their *uncertainties*.

This uncertainty consideration often improves the generalization capability of a model that yields to mitigate the mismatch between training and unseen data and avoid over-fitting problems by the effect of regularization and marginalization. For example, Figure 1.2 points out the over-fitting problem. In real applications, we often face phenomena where observed data cannot cover all possible unobserved data (unseen data) and their distributions are mismatched. The Gaussian distribution with a fixed mean parameter that is estimated from observed data can well represent the observed data, but it cannot represent unseen data properly. This is a well-known over-fitting problem, that the model overly fits its parameters to represent observed data. However, since the Bayesian approach considers all possible mean parameters by the marginalization, some of these parameters would properly model unseen data more accurately than the fixed mean parameter case. The effect of this more powerful representation ability for unseen data leads to improved generalization capability, which is a famous advantage of the Bayesian approach. In addition, the example can also be viewed as

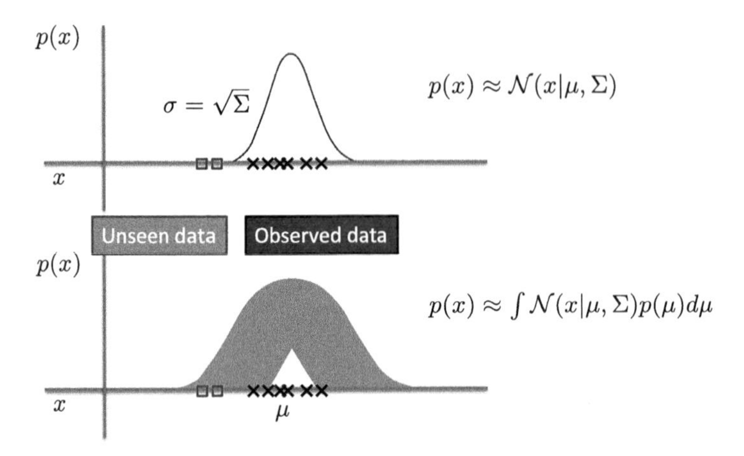

Figure 1.2 The Bayesian approach that holds uncertainty of the mean parameter has an ability to describe unseen data.

showing that the mean parameter is *regularized* not to overly fit the observed data to $p(\mu)$. The effect of setting the constraints for variables via their probabilistic distributions is a typical example of the *regularization*, and the regularization also leads to improved generalization capability.

Furthermore, the Bayesian theory provides a straightforward mathematical way to infer (predict) unobservable probabilistic variables by using the basic rules (including the Bayes theorem) within the basic probabilistic theory. The beauties of this mathematical treatment based on probabilistic theory and the expected robustness by considering uncertainties attract many machine learning researchers to study the Bayesian approach. Actually, Bayesian machine learning has also rapidly grown similarly to discriminative techniques, and various interesting approaches have been proposed (Bishop 2006, Barber 2012).

However, compared with the successful examples of the discriminative techniques, the applications of the Bayesian approach in speech and language processing are rather limited despite its many advantages. One of the most difficult problems is that the exact Bayesian approach cannot be applied in our speech and language processing without some approximations. For example, the Bayesian approach is based on the conditional distribution given datum x (that is called posterior distribution, $p(a|x)$). However, it is generally difficult to obtain the posterior distribution analytically, since we cannot solve the equation analytically to obtain the posterior distribution. The computation is often performed numerically, which limits the practical applications of the Bayesian approach, especially for speech and language processing that deals with large amounts of data. Thus, *how to make the Bayesian approach more practical for speech and language processing* is the most critical problem in Bayesian speech and language processing.

This book is aimed to guide readers in machine learning or speech and language processing to apply the Bayesian approach to speech and language processing in a systematic way. In other words, this book aims to bridge the gap between the machine learning community and the speech and language community by removing

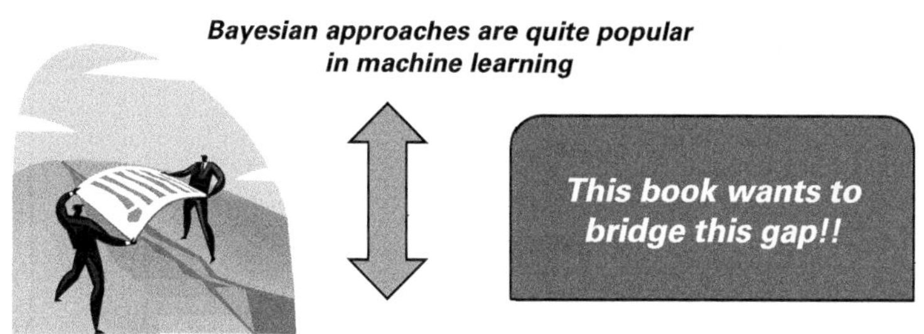

Figure 1.3 The aim of this book.

their preconceived ideas of the difficulty in these applications (Figure 1.3) (Watanabe & Chien 2012). The key idea for this guidance is *how to approximate the Bayesian approach* for specific speech and language applications (Ghahramani 2004). There are several approximations developed mainly in the machine learning and Bayesian statistics fields to deal with the problems. This book mainly deals with the following approximations:

- Chapter 4: Maximum a-posteriori approximation (MAP);
- Chapter 5: Evidence approximation;
- Chapter 6: Asymptotic approximation;
- Chapter 7: Variational Bayes (VB);
- Chapter 8: Markov chain Monte Carlo (MCMC).

Note that there are some other interesting approximations (e.g., loopy belief propagation (Murphy, Weiss & Jordan 1999, Yedidia, Freeman & Weiss 2003) and expectation propagation (Minka 2001)), but our book focuses on the above approximations. These approximations are described in the corresponding chapters in detail. We organize the chapters in Part II categorized by these approximation techniques, unlike application-oriented categorization (e.g., speech recognition and speech synthesis) as has appeared in many other speech and language books (Jurafsky & Martin 2000, Huang, Acero & Hon 2001), to emphasize how to approximate the Bayesian approach for practical applications. For example, the first sections in each chapter in Part II describe the introduction of the corresponding approximation, and provide the recipe for how to use the approximation in machine learning problems in general. The following sections in the chapter provide the approximated solutions of statistical models used in specific speech and language applications by following the recipe. This book mainly deals with popular statistical models including HMM, GMM, neural network, factor analysis, n-gram, and latent topic models. Table 1.1 summarizes approximated Bayesian inferences for statistical models discussed in this book. The applications covered by this book are typical topics in speech and language processing based on these statistical models with the approximated Bayesian treatment, and mainly related to automatic speech recognition.

Table 1.1 Approximated Bayesian inference for statistical models discussed in this book.

Approximation	HMM	GMM	Neural network	Factor analysis	n-gram	Latent topic model
MAP	4.3, 4.5	4.6			4.7	4.8
Evidence	5.2				5.3	
Asymptotic	6.3, 6.5	6.6	6.4			
VB	7.3			7.5		7.6
MCMC	8.3, 8.4				8.5	

1.3 History of Bayesian speech and language processing

There have been various studies to apply the Bayesian approach in speech and language processing. Although one of the aims of this book is to summarize these studies in a more systematic way in terms of an approximated Bayesian inference view, this section briefly reviews the history of these studies, according to time, by categorizing them within four trends.

The major earliest trend of using the Bayesian approach to speech and language processing started with *the statistical modeling* of automatic speech recognition in the 1980s. Furui (2010) reviews a historical perspective of automatic speech recognition and calls the technologies developed with statistical modeling around the 1980s *the third generation technology*. In a Bayesian perspective, this statistical modeling movement corresponded to the first introduction of *probabilistic variables* for speech recognition outputs (e.g., word sequences). As a result, the statistical modeling of automatic speech recognition formulates the speech recognition process as Bayes decision theory, and the noisy channel model is provided to solve the decision problem of determining most probable word sequences by considering the product of acoustic and language model distributions based on the Bayes theory (Jelinek 1976). The language model is used to provide a prior distribution of word sequences.

The second trend in the 1990s was to expand the Bayesian perspective from the speech recognition outputs to *model parameters* by regarding these as probabilistic variables. Maximum a-posteriori (MAP) estimation of HMM parameters is known as the most successful application of the Bayesian approach (Lee, Lin & Juang 1991, Gauvain & Lee 1994). This approach was used for speaker adaptation, where the prior distribution of HMM parameters is estimated from a large amount of speaker-independent data and the MAP estimation is used to estimate target speaker's HMM parameters with a small amount of target speaker data. The prior distribution regularizes the target speaker's HMM parameters to guarantee the performance of speaker independent HMMs instead of overly tuning to the model. Another example of the expansion was to treat *model structure* as a probabilistic variable in the late 1990s. This treatment enables *model selection* by selecting a most probable model structure from the posterior distribution of the model structure given training data (e.g., the numbers of HMM states and Gaussians). The Bayesian information criterion (BIC) and the minimum description length (MDL) criterion are successful examples (Shinoda & Watanabe 1996, Chen & Gopinath 1999, Chou & Reichl 1999, Zhou & Hansen 2000) that cover wide applications of speech and language processing (e.g., acoustic model selection, speaker clustering, and speaker segmentation).

This second trend made the importance of the Bayesian approach come alive within the speech and language communities. From the late 1990s to 2000s, many other Bayesian studies have been applied to speech and language processing, which are classified as the third trend. The Bayesian techniques (MAP and BIC) used in the second trend miss the important Bayesian concept, *marginalization*, which makes the full treatment of the Bayesian approach difficult. By following the progress of VB, MCMC, Evidence approximation, and graphical model techniques in machine learning in the 1990s, people in speech and language processing started to apply more exact Bayesian approaches by *fully incorporating marginalization*. These studies covered almost all statistical models in speech and language processing (Bilmes & Zweig 2002, Watanabe, Minami, Nakamura & Ueda 2002, Blei, Ng & Jordan 2003, Saon & Chien 2011). The most successful approach in the third trend is latent Dirichlet allocation (LDA (Blei *et al.* 2003)), which provides a VB solution of a latent topic model from a probabilistic latent semantic analysis (Hofmann 1999*b*). LDA has been mainly developed in machine learning and natural language processing by incorporating a Gibbs sampling solution (Griffiths & Steyvers 2004), and structured topic models (Wallach 2006), and LDA was extended to incorporate *Bayesian nonparametrics* (Teh, Jordan, Beal & Blei 2006), which became the fourth trend in Bayesian speech and language processing.

The fourth trend is a still ongoing trend that tries to fully incorporate Bayesian nonparametrics with speech and language processing. Due to their computational costs and algorithmic difference from the standard expectation maximization (EM) type algorithms (Dempster, Laird & Rubin 1976), the applications were limited to latent topic models and related studies. However, recent computational progress (e.g., many core processing and GPU processing) has enabled broadening of the fourth trend in statistical models in speech and language processing other than extended latent topic models (Teh 2006, Goldwater 2007, Fox, Sudderth, Jordan *et al.* 2008, Ding & Ou 2010, Lee & Glass 2012).

This book covers all four trends and categorizes these popular techniques with the approximated Bayesian inference techniques.

1.4 Applications

This book aims to describe the following target applications:

- Automatic speech recognition (ASR, Figure 1.4)
  This is a main application in this book, that converts human speech to texts. The main techniques required in ASR are based on speech enhancement for noise reduction, speech feature extraction, acoustic modeling, language modeling, pronunciation

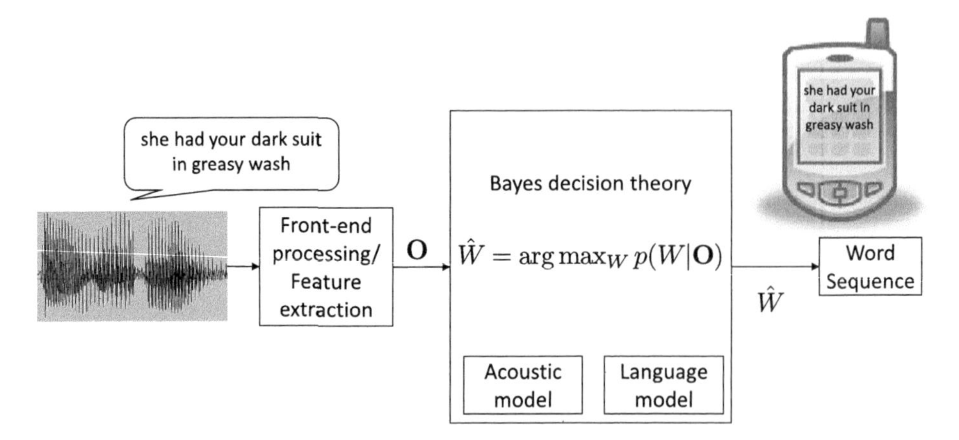

Figure 1.4 Automatic speech recognition.

lexicon modeling, and search. The book mainly deals with acoustic models represented by HMM and neural networks, and language models represented by n-gram and latent topic models.

- Speaker verification/segmentation/clustering (Figure 1.5)
  Speaker recognition techniques automatically provide a speaker identity given speech data. This often requires extracting speech segments uttered by target speakers from the recorded utterances (speaker segmentation). In addition, some of the real applications cannot have speaker labels in advance, and clustering speech segments to a specific speaker cluster (speaker clustering) is also another important direction. Gaussian or GMMs are used as statistical models. However, state-of-the-art speaker verification systems use GMMs as preprocessing, namely GMM parameters estimated from a speech segment are used as speaker features. The estimated features are further processed by a factor analysis to remove speaker-independent feature characteristics statistically.
- (Spoken) Information retrieval (IR, Figure 1.6)
  Information retrieval via document classification given a text (or spoken) query is another application of speech and language processing. The document can be represented by various units including newspaper articles, web pages, dialog conversations, sentences/utterances; which is used depends on applications. The most successful application for information retrieval is a search engine that uses a web page as a document unit. Voice search is an instance of a search engine based on spoken information retrieval where spoken terms are converted to text-form terms by using ASR and these terms are used as a query. The approach is based on the vector space model, which represents documents and queries as vectors in a vector space. The vector is simply represented by count or weighted count of unique words in a vocabulary. The approach often uses n-gram or latent topic models to provide more informative vector representation of documents.

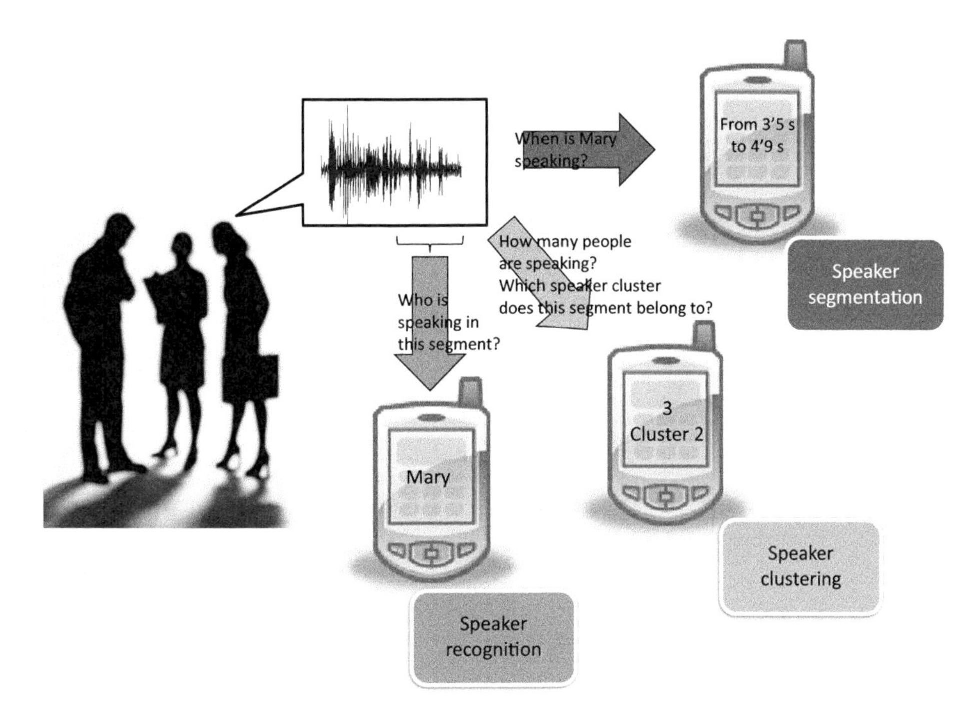

Figure 1.5 Speaker verification/segmentation/clustering.

1.5 Organization of this book

This book is divided into two parts. Part I starts by describing the Bayesian approach in general in Chapter 2, but the topics discussed in this book are focused on practical machine learning problems regarding speech and language processing rather than on general Bayesian statistics. Chapter 3 also provides general discussion of speech and language processing by providing some important statistical models. The discussion is based on the conventional style based on the maximum likelihood approach. These models are re-formulated in the latter sections in a Bayesian manner with an appropriated approximation. The final section in Chapter 3 also describes the application of the Bayesian approach to statistical models in automatic speech recognition, and discusses the difficulty of straightforward application.

As described before, Part II provides approximated Bayesian inferences based on MAP, Evidence, Asymptotic, VB, and MCMC approximations, and relevant sections describe statistical models and their applications, as shown in Table 1.1. We also provide in the appendices some useful formulas and rules used.

For readers

Our book assumes three types of readers: 1) undergraduate and graduate students; 2) those who have some knowledge in machine learning; and 3) those who have some

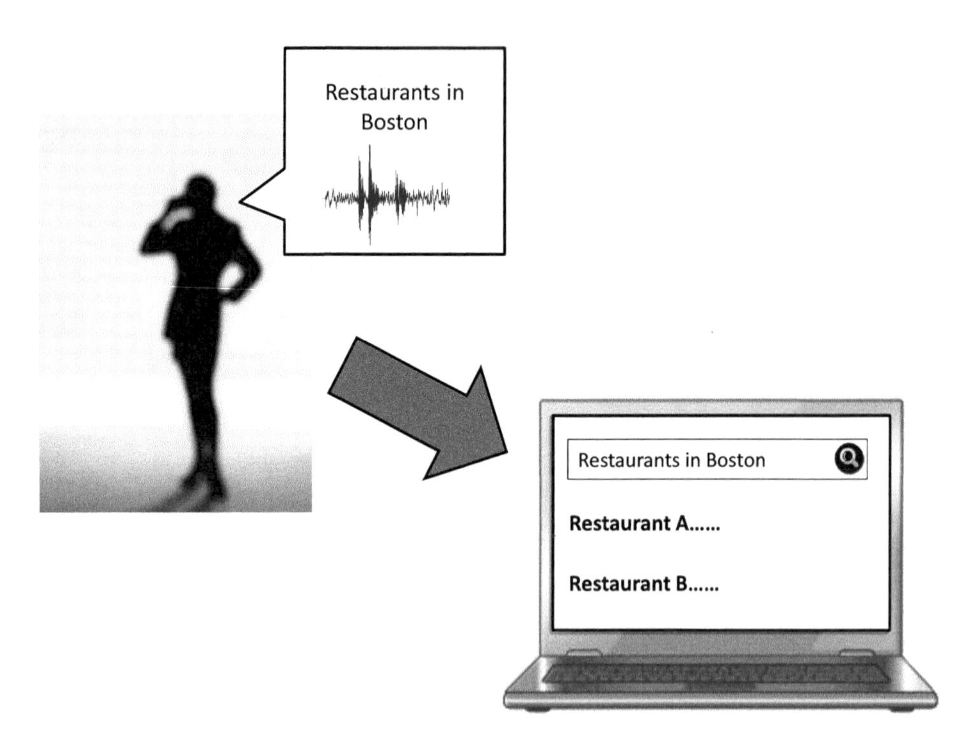

Figure 1.6 (Spoken) Information retrieval.

knowledge in speech and language processing. We recommend undergraduate and graduate students to read through this book, as we have made it self-consistent, especially for deriving equations. Readers who have knowledge in machine learning can skip the beginning part of Chapter 2 about the "Bayesian approach" in general. Similarly, readers who have knowledge in speech and language processing may skip Chapter 3 about statistical models in "speech and language processing". However, compared with conference and journal papers in this field, the chapter provides comprehensive understanding of standard statistical models with full derivations of equations. Therefore, we highly recommend readers familiar with these models to read through the whole book.

2 Bayesian approach

This chapter describes a general concept and statistics of the Bayesian approach. The Bayesian approach covers wide areas of statistics (Bernardo & Smith 2009, Gelman, Carlin, Stern *et al.* 2013), pattern recognition (Fukunaga 1990), machine learning (Bishop 2006, Barber 2012), and applications of these approaches. In this chapter, we start the discussion from the basic probabilistic theory, and mainly describe the Bayesian approach by aiming to follow a machine learning fashion of constructing and refining statistical models from data. The role of the Bayesian approach in machine learning is very important since the Bayesian approach provides a systematic way to infer unobserved variables (e.g., classification category, model parameters, latent variables, model structure) given data. This chapter limits the discussions considering the speech and language problems in the latter chapters, by providing simple probabilistic rules, and prior and posterior distributions in Section 2.1. The section also provides analytical solutions of posterior distributions of simple models. Based on the basic introduction, Section 2.2 introduces a useful representation of the relationship of probabilistic variables in the Bayesian approach, called the *Graphical model*. The graphical model representation gives us an intuitive view of statistical models even when they have complicated relationships between their variables. Section 2.3 explains the difference between Bayesian and maximum likelihood (ML) approaches. The following chapters extend the general Bayesian approach described in this chapter to deal with statistical models in speech and language processing.

2.1 Bayesian probabilities

This section describes the basic Bayesian framework based on probabilistic theory. Although some of the definitions, equations, and concepts are trivial, this section reviews the basics to assist readers to fully understand the Bayesian approach.

In the Bayesian approach, all the variables that are introduced when models are parameterized, such as model parameters and latent variables, are regarded as probabilistic variables. Thus, let a be a discrete valuable, then the Bayesian approach deals with a as a probabilistic variable, and aims to obtain $p(a)$:

$$a \rightarrow p(a). \tag{2.1}$$

Hereinafter, we assume that a is a discrete variable, and the expectation is performed by the summation over a for simplicity. Since $p(a)$ is a probabilistic distribution, $p(a)$ always satisfies the following condition:

$$\sum_a p(a) = 1, \quad p(a) \geq 0 \quad \forall a. \tag{2.2}$$

These properties help us to solve some calculations appearing in the following sections. In the continuous variable case, the summation $\sum$ is replaced with the integral $\int$.

2.1.1 Sum and product rules

Since the Bayesian approach treats all variables as probabilistic variables, the probabilistic theory gives us the two important probabilistic rules to govern the relationship between the variables. Let a and b be arbitrary probabilistic variables,

- Sum rule

$$p(b) = \sum_a p(a, b); \tag{2.3}$$

- Product rule

$$p(a, b) = p(a|b)p(b) = p(b|a)p(a). \tag{2.4}$$

Here, $p(a, b)$ is a joint probability that represents a probability of all possible joint events of a and b. $p(a|b)$ or $p(b|a)$ is a conditional probability. These are generalized to N probabilistic variables, for example, if we have $a_1, \cdots, a_N$ probabilistic variables, there rules are represented as:

- Sum rule

$$p(a_i) = \sum_{a_1} \cdots \sum_{a_{i-1}} \sum_{a_{i+1}} \cdots \sum_{a_N} p(a_1, \cdots, a_N); \tag{2.5}$$

- Product rule

$$\begin{aligned} p(a_1, \cdots, a_N) &= p(a_1|a_2, \cdots, a_N)p(a_2, \cdots, a_N) = \cdots \\ &= p(a_N) \prod_{n=1}^{N-1} p(a_n|a_{n+1}, \cdots, a_N). \end{aligned} \tag{2.6}$$

In a Bayesian manner, we formulate the probability distributions based on these rules. For example, the famous Bayes theorem can be derived by reforming the product rule in Eq. (2.4), as follows:

$$p(a|b) = \frac{p(a, b)}{p(b)} = \frac{p(b|a)p(a)}{p(b)} \tag{2.7}$$

$$= \frac{p(b|a)p(a)}{\sum_a p(b|a)p(a)}. \tag{2.8}$$

To derive Eq. (2.8), we use the sum and product rules for $p(a)$. The following discussion provides more practical examples based on this discussion.

2.1.2 Prior and posterior distributions

The above Bayes theorem has an interesting meaning if we consider a conditional probability distribution of a given an observation x. The conditional distribution $p(a|x)$ is called *posterior distribution*, and the main purpose of the Bayesian approach is to infer the posterior distribution of various valuables. Based on the Bayes theorem in Eq. (2.7), the posterior distribution is decomposed in Eq. (2.9) to the following three distributions:

$$p(a|x) = \frac{p(x|a)p(a)}{p(x)} \tag{2.9}$$

$$= \frac{p(x|a)p(a)}{\sum_a p(x|a)p(a)}, \tag{2.10}$$

where $p(x|a)$ is a likelihood function of x and $p(a)$ is a distribution without considering any observation, and called *prior distribution*. $p(x)$ is a distribution of an observation, and can be computed by using $p(x|a)$ and $p(a)$ based on Eq. (2.10). In most speech processing applications, it is difficult to estimate the posterior distribution directly (Section 3.8 describes it in detail). Therefore, the posterior distribution $p(a|x)$ is indirectly estimated via this Bayes theorem, which is derived from the sum and product rules, which are equivalence equations without approximation.

Since the posterior distribution provides a probability of a given data x, this matches one of the machine learning goals of refining information of a from data x. Therefore, the posterior distribution plays an important role in machine learning, and obtaining an appropriate posterior distribution for our problems in speech and language processing is a main goal of this book.

Once we obtain the posterior distribution $p(a|x)$, we can obtain the values of a via:

- Maximum a-posteriori (MAP) procedure:

$$a^{\text{MAP}} = \arg\max_a p(a|x); \tag{2.11}$$

- Expectation with respect to the posterior distribution:

$$a^{\text{EXP}} = \mathbb{E}_{(a)}[a|x] \triangleq \sum_a a \cdot p(a|x). \tag{2.12}$$

The MAP and expectation are typical ways to obtain meaningful information about a given x in terms of the probabilistic theory. From Eq. (2.10), $p(x)$ is disregarded in the MAP procedure as a constant factor that is independent of a, MAP, and expectation, which makes the calculation simple. The MAP and expectation are generalized to obtain meaningful information $f(a)$ given the posterior distribution $p(a|x)$. More specifically, if we consider a likelihood function of unseen data y given a, i.e., $p(y|a)$, these procedures are rewritten as:

- Maximum a-posteriori (MAP) procedure:

$$p^{\text{MAP}}(y|a) = p(y|\arg\max_a p(a|x)) = p(y|a^{\text{MAP}}); \tag{2.13}$$

- Expectation with respect to the posterior distribution:

$$p^{\text{EXP}}(y) = \mathbb{E}_{(a)}[p(y|a)|x] \triangleq \sum_a p(y|a)p(a|x). \tag{2.14}$$

Thus, we can predict y by using these procedures. Note that the MAP procedure decides a deterministic value of a, while the expectation procedure keeps possible a for the expectation. Therefore, the MAP procedure is called *hard decision* and the expectation procedure is called *soft decision*.

The expectation is a more general operation than MAP in terms of considering the distribution shape. For example, if we approximate $p(a|x)$ with a specific Kronecker delta function $\delta(a, a^{\text{MAP}})$ where $a^{\text{MAP}} = \arg\max_a p(a|x)$, Eq. (2.14) is represented as

$$\begin{aligned} p^{\text{EXP}}(y) &= \sum_a p(y|a)\delta(a, a^{\text{MAP}}) = p(y|a^{\text{MAP}}) \\ &= p(y|\arg\max_a p(a|x)) \\ &= p^{\text{MAP}}(y|a), \end{aligned} \tag{2.15}$$

where

$$\delta(a, a') = \begin{cases} 1 & a = a' \\ 0 & \text{otherwise.} \end{cases} \tag{2.16}$$

Thus, the MAP value is obtained from the specific case of the expectation value without considering the distribution shape of $p(a|x)$. However, in many cases, the MAP value is also often used since the expectation needs a complex computation due to the summation over a. Note that the above derivation via a Kronecker delta function (or Dirac delta function when we consider continuous variables) is often used to provide the relationship of the MAP and expectation values.

2.1.3 Exponential family distributions

The previous section introduces the posterior distribution. This section focuses on a specific problem of posterior distributions that consider the model parameter Θ given a set of D dimensional observation vectors, i.e., $\mathbf{X} = \{\mathbf{x}_n \in \mathbb{R}^D | n = 1, \cdots, N\}$. The problem here is to obtain the posterior distribution $p(\Theta|\mathbf{X})$, i.e., it is a general estimation problem of obtaining the distribution of Θ from data $\mathbf{X}$. Once we obtain $p(\Theta|\mathbf{X})$, for example, we can estimate Θ^{MAP} or compute some expectation values, as we discussed in Section 2.1.2.

Then, the Bayes theorem, which provides the relationship between prior and posterior distributions in Eq. (2.9) or (2.10), can be represented as follows:

$$p(\Theta|\mathbf{X}) = \frac{p(\mathbf{X}|\Theta)p(\Theta)}{p(\mathbf{X})} \tag{2.17}$$

$$= \frac{p(\mathbf{X}|\Theta)p(\Theta)}{\int p(\mathbf{X}|\Theta)p(\Theta)d\Theta}. \tag{2.18}$$

Here, we use the integral $\int$ rather than $\sum$ in Eq. (2.18), since model parameters are often represented by continuous variables (e.g., mean and variance parameters in a Gaussian distribution). In this particular case, the Bayes theorem has the more practical meaning that the posterior distribution $p(\Theta|\mathbf{X})$ is represented by the likelihood function $p(\mathbf{X}|\Theta)$, and the prior distribution of the model parameters $p(\Theta)$. Thus

$$p(\mathbf{X}) = \int p(\mathbf{X}|\Theta)p(\Theta)d\Theta, \tag{2.19}$$

which is also called *evidence* function or *marginal likelihood* . The evidence plays an important role in Bayesian inference, which is described in Chapter 5 in detail.

Basically, we can set any distributions (e.g., Gaussian, gamma, Dirichlet, Laplace, Rayleigh distributions, etc.) to prior and posterior distributions. However, a particular family of distributions called *conjugate distribution* makes analytical derivation simpler. Before we describe the conjugate distributions, the following section explains *exponential family* distributions, which are required to explain conjugate distributions.

Exponential family

The *exponential family* is a general distribution family, which contains standard distributions including Gaussian distribution, gamma distribution, and multinomial distribution. Let $\boldsymbol{\theta}$ be a vector form of model parameters. A distribution of a set of observation vectors $\mathbf{X} = \{\mathbf{x}_1, \cdots, \mathbf{x}_N\}$ given $\boldsymbol{\theta}$ (likelihood function), which belongs to the exponential family, is represented by the following exponential form:

$$p(\mathbf{X}|\boldsymbol{\theta}) \triangleq h(\mathbf{X}) \exp\left(\boldsymbol{\gamma}(\boldsymbol{\theta})^{\mathsf{T}}\mathbf{t}(\mathbf{X}) - g(\boldsymbol{\gamma})\right), \tag{2.20}$$

where $\mathbf{t}(\mathbf{X})$ is a sufficient statistics vector obtained from observation vector $\mathbf{X}$, $g(\boldsymbol{\gamma})$ is a logarithmic normalization factor. $\boldsymbol{\gamma}$ is a transformed vector of $\boldsymbol{\theta}$, and is called a *natural parameter* vector. If $\boldsymbol{\gamma}(\boldsymbol{\theta}) = \boldsymbol{\theta}$, it is called the *canonical form*, that simplifies Eq. (2.20) as follows:

$$p(\mathbf{X}|\boldsymbol{\theta}) = h(\mathbf{X}) \exp\left(\boldsymbol{\theta}^{\mathsf{T}}\mathbf{t}(\mathbf{X}) - g(\boldsymbol{\theta})\right). \tag{2.21}$$

The canonical form makes the calculation of posterior distributions simple.

If we have J multiple parameter vectors, we can represent the exponential form as the factorized form:

$$p(\mathbf{X}|\boldsymbol{\theta}_1, \cdots, \boldsymbol{\theta}_J) \triangleq h(\mathbf{X}) \prod_{i=1}^{J} \exp\left(\boldsymbol{\gamma}_i(\boldsymbol{\theta}_i)^{\mathsf{T}}\mathbf{t}_i(\mathbf{x}) - g_i(\boldsymbol{\gamma}_i)\right). \tag{2.22}$$

When the transformed model parameters are composed of a matrix or a vector, we can also define the exponential family distribution. For example, a multivariate Gaussian distribution $\mathcal{N}(\cdot|\boldsymbol{\mu}, \boldsymbol{\Sigma})$ is parameterized by a mean vector $\boldsymbol{\mu}$ and a covariance matrix $\boldsymbol{\Sigma}$, and the corresponding transformed parameters are also represented by vector $\boldsymbol{\gamma}_1$ and matrix $\boldsymbol{\Gamma}_2$. Then, the exponential family distribution for $\Theta = \{\boldsymbol{\theta}_1, \boldsymbol{\Theta}_2\}$ is defined as follows:

$$p(\mathbf{X}|\Theta) \triangleq h(\mathbf{X}) \exp\left(\boldsymbol{\gamma}_1^{\mathsf{T}}\mathbf{t}_1(\mathbf{X}) + \text{tr}[\boldsymbol{\Gamma}_2^{\mathsf{T}}\mathbf{T}_2(\mathbf{x})] - g(\boldsymbol{\gamma}_1, \boldsymbol{\Gamma}_2)\right). \tag{2.23}$$

Here, $\mathbf{t}_1$ and $\mathbf{T}_2$ are vector and matrix representations of sufficient statistics, respectively. The rest of this section provides examples of $h(\cdot)$, $g(\cdot)$, $\boldsymbol{\gamma}(\cdot)$, and $\mathbf{t}(\cdot)$ for standard distributions (Gaussian, multivariate Gaussian, and multinomial distributions).

Example 2.1 Gaussian (unknown mean):
We focus on the exponential family form of the Gaussian distribution for scalar observation $X = \{x_n \in \mathbb{R} | n = 1, \cdots, N\}$. As a simple example, we only focus on the Gaussian mean as a model parameter, and regard the precision parameter r as a constant value, i.e., $\mathcal{N}(x_n|\mu; r^{-1})$ where the variables located right after the semicolon; means that these are not treated as probabilistic variables, but specific values. That is $\theta = \mu$ in Eq. (2.20). We use the precision parameter r instead of the variance parameter Σ,[1] which makes the solution simple. Based on the definition in Appendix C.5, the standard form of the Gaussian distribution is represented as

$$\prod_{n=1}^{N} \mathcal{N}(x_n|\mu; r^{-1}) = \left(\frac{2\pi}{r}\right)^{-\frac{N}{2}} \exp\left(-\sum_{n=1}^{N} \frac{r}{2}(x_n - \mu)^2\right). \tag{2.24}$$

We assume that $x_1, \cdots, x_N$ are independent and identically distributed random variables from the Gaussian. The standard form of the Gaussian distribution is rewritten as the following exponential form:

$$\begin{aligned}
\prod_{n=1}^{N} \mathcal{N}(x_n|\mu; r^{-1}) &= \left(\frac{2\pi}{r}\right)^{-\frac{N}{2}} \exp\left(-\frac{r}{2}\sum_{n=1}^{N} x_n^2\right) \exp\left(r\mu\sum_{n=1}^{N} x_n - \frac{N\mu^2 r}{2}\right) \\
&= \underbrace{\left(\frac{2\pi}{r}\right)^{-\frac{N}{2}} \exp\left(-\frac{r}{2}\sum_{n=1}^{N} x_n^2\right)}_{=h(X)} \exp\left(\underbrace{r\sum_{n=1}^{N} x_n}_{=t(X)} \underbrace{\mu}_{=\gamma} - \underbrace{\frac{N\mu^2 r}{2}}_{=g(\gamma)}\right).
\end{aligned} \tag{2.25}$$

Thus, the Gaussian distribution is represented by the following exponential form in Eq. (2.20):

$$\begin{cases}
t(X) = r\sum_{n=1}^{N} x_n \\
h(X) = \left(\frac{2\pi}{r}\right)^{-\frac{N}{2}} \exp\left(-\frac{r\sum_{n=1}^{N} x_n^2}{2}\right) \\
\gamma(\mu) = \mu \\
g(\gamma) = \frac{N\gamma^2 r}{2}.
\end{cases} \tag{2.26}$$

[1] This book regards Σ as the variance parameter (not the standard deviation, which is represented as σ), as shown in Appendix C.5, to make the notation consistent with the covariance matrix $\boldsymbol{\Sigma}$.

Since $\gamma(\mu) = \mu$ in Eq. (2.26), it is regarded as a canonical form, as discussed in Eq. (2.21). Note that the parameterization of $\gamma(\mu)$ and $t(X)$ is not unique. For example, we can obtain the following parameterization from:

$$
\begin{cases}
t(X) = \sum_{n=1}^{N} x_n \\
h(X) = \left(\frac{2\pi}{r}\right)^{-\frac{N}{2}} \exp\left(-\frac{r \sum_{n=1}^{N} x_n^2}{2}\right) \\
\gamma(\mu) = \mu r \\
g(\gamma) = \frac{N\mu^2 r}{2} = \frac{N\gamma^2}{2r}.
\end{cases}
\tag{2.27}
$$

This is also another exponential form of the Gaussian distribution with unknown mean.

Example 2.2 Gaussian (unknown mean and precision):
Similarly to Example 2.1, we focus on the exponential family form of the Gaussian distribution for scalar observation X, but regard r as also unknown. Therefore, $\boldsymbol{\theta} = [\mu, r]^\mathsf{T}$. Thus, unlike the scalar forms of the natural parameter γ and sufficient statistics t in Eq. (2.24), the Gaussian distribution is represented by the vector form of these as

$$
\begin{aligned}
\prod_{n=1}^{N} \mathcal{N}(x_n|\mu, r^{-1}) &= \left(\frac{2\pi}{r}\right)^{-\frac{N}{2}} \exp\left(-\frac{N\mu^2 r}{2}\right) \exp\left(-\frac{r}{2}\sum_{n=1}^{N} x_n^2 + \mu r \sum_{n=1}^{N} x_n\right) \\
&= \exp\left(\underbrace{\begin{bmatrix} \mu r \\ r \end{bmatrix}^\mathsf{T}}_{=\boldsymbol{\gamma}(\boldsymbol{\theta})} \underbrace{\begin{bmatrix} \sum_{n=1}^{N} x_n \\ -\frac{\sum_{n=1}^{N} x_n^2}{2} \end{bmatrix}}_{\mathbf{t}(X)} - \underbrace{\left(\frac{N}{2}\log\left(\frac{2\pi}{r}\right) + \frac{N\mu^2 r}{2}\right)}_{=g(\boldsymbol{\gamma})} \right).
\end{aligned}
\tag{2.28}
$$

Therefore,

$$
\begin{cases}
\mathbf{t}(X) = \begin{bmatrix} \sum_{n=1}^{N} x_n \\ -\frac{\sum_{n=1}^{N} x_n^2}{2} \end{bmatrix} \\
h(X) = 1 \\
\boldsymbol{\gamma}(\boldsymbol{\theta}) = \begin{bmatrix} \mu r \\ r \end{bmatrix} \\
g(\boldsymbol{\gamma}) = \frac{N}{2}\left(\log\frac{2\pi}{r} + \mu^2 r\right) \\
\qquad = \frac{N}{2}\left(\log\frac{2\pi}{\gamma_2} + \frac{\gamma_1^2}{\gamma_2}\right).
\end{cases}
\tag{2.29}
$$

Again, the parameterization of $\boldsymbol{\gamma}(\boldsymbol{\theta})$ and $\mathbf{t}(x)$ is not unique and the parameterization of $\boldsymbol{\gamma}(\boldsymbol{\theta}) = \left[\mu r, -\frac{r}{2}\right]^\mathsf{T}$ and $\mathbf{t}(X) = [\sum_{n=1}^{N} x_n, \sum_{n=1}^{N} x_n^2]^\mathsf{T}$ is also possible.

Example 2.3 Multivariate Gaussian (unknown mean and precision):
The next example is to derive an exponential form of the multivariate Gaussian distribution with D dimensional mean vector $\boldsymbol{\mu}$ and $D \times D$ precision matrix $\mathbf{R}$ (we use precision matrix $\mathbf{R}$ instead of covariance matrix $\boldsymbol{\Sigma}$ to make the solution simple). A set of the parameters is $\Theta = \{\boldsymbol{\mu}, \mathbf{R}\}$. This is the most important example in this book, since statistical models in speech and language processing are often represented by multivariate Gaussian distributions, as discussed in Chapter 3. Let $\mathbf{X} = \{\mathbf{x}_n \in \mathbb{R}^D | n = 1, \cdots, N\}$ be independent and identically distributed random variables from the multivariate Gaussian distribution. Again, based on the definition in Appendix C.6, the standard form of the Gaussian distribution is represented as

$$\begin{aligned}\prod_{n=1}^{N} \mathcal{N}(\mathbf{x}_n|\boldsymbol{\mu}, \mathbf{R}^{-1}) &= \prod_{n=1}^{N} (2\pi)^{-\frac{D}{2}} |\mathbf{R}|^{\frac{1}{2}} \exp\left(-\frac{1}{2}(\mathbf{x}_n - \boldsymbol{\mu})^{\mathsf{T}}\mathbf{R}(\mathbf{x}_n - \boldsymbol{\mu})\right) \\ &= (2\pi)^{-\frac{ND}{2}} |\mathbf{R}|^{\frac{N}{2}} \exp\left(-\frac{1}{2}\sum_{n=1}^{N}(\mathbf{x}_n - \boldsymbol{\mu})^{\mathsf{T}}\mathbf{R}(\mathbf{x}_n - \boldsymbol{\mu})\right). \end{aligned} \tag{2.30}$$

Now we focus on the exponential part in Eq. (2.30), which is rewritten as follows:

$$\begin{aligned}&\sum_{n=1}^{N}(\mathbf{x}_n - \boldsymbol{\mu})^{\mathsf{T}}\mathbf{R}(\mathbf{x}_n - \boldsymbol{\mu}) \\ &= -\boldsymbol{\mu}^{\mathsf{T}}\mathbf{R}\sum_{n=1}^{N}\mathbf{x}_n - \left(\sum_{n=1}^{N}\mathbf{x}_n^{\mathsf{T}}\right)\mathbf{R}\boldsymbol{\mu} + \sum_{n=1}^{N}\mathbf{x}_n^{\mathsf{T}}\mathbf{R}\mathbf{x}_n + N\boldsymbol{\mu}^{\mathsf{T}}\mathbf{R}\boldsymbol{\mu}. \end{aligned} \tag{2.31}$$

To make the observation vector and parameter the inner product form, we first use the trace representation of the quadratic term of $\mathbf{x}_n$ as

$$\begin{aligned}\sum_{n=1}^{N}\mathbf{x}_n^{\mathsf{T}}\mathbf{R}\mathbf{x}_n &= \mathrm{tr}\left[\sum_{n=1}^{N}\mathbf{x}_n^{\mathsf{T}}\mathbf{R}\mathbf{x}_n\right] \\ &= \mathrm{tr}\left[\sum_{n=1}^{N}\mathbf{R}\mathbf{x}_n\mathbf{x}_n^{\mathsf{T}}\right] \\ &= \mathrm{tr}\left[\mathbf{R}\sum_{n=1}^{N}\mathbf{x}_n\mathbf{x}_n^{\mathsf{T}}\right], \end{aligned} \tag{2.32}$$

where we use the fact that the trace of the scalar value is equal to the original scalar value, the cyclic property, and the distributive property of the trace as in Appendix B:

$$a = \mathrm{tr}[a], \tag{2.33}$$

$$\mathrm{tr}[\mathbf{ABC}] = \mathrm{tr}[\mathbf{BCA}], \tag{2.34}$$

$$\mathrm{tr}[\mathbf{A}(\mathbf{B} + \mathbf{C})] = \mathrm{tr}[\mathbf{AB} + \mathbf{AC}]. \tag{2.35}$$

In addition, we can also use the following equation:

$$\boldsymbol{\mu}^{\mathsf{T}}\mathbf{R}\sum_{n=1}^{N}\mathbf{x}_n + \left(\sum_{n=1}^{N}\mathbf{x}_n^{\mathsf{T}}\right)\mathbf{R}\boldsymbol{\mu} = 2\boldsymbol{\mu}^{\mathsf{T}}\mathbf{R}\sum_{n=1}^{N}\mathbf{x}_n. \tag{2.36}$$

Here, since these values are scalar values, we use the following equation to derive Eq. (2.36).

$$\begin{aligned}\left(\sum_{n=1}^{N}\mathbf{x}_n^{\mathsf{T}}\right)\mathbf{R}\boldsymbol{\mu} &= \left(\left(\sum_{n=1}^{N}\mathbf{x}_n^{\mathsf{T}}\right)\mathbf{R}\boldsymbol{\mu}\right)^{\mathsf{T}} \\ &= \boldsymbol{\mu}^{\mathsf{T}}\mathbf{R}^{\mathsf{T}}\left(\sum_{n=1}^{N}\mathbf{x}_n^{\mathsf{T}}\right)^{\mathsf{T}} = \boldsymbol{\mu}^{\mathsf{T}}\mathbf{R}\sum_{n=1}^{N}\mathbf{x}_n,\end{aligned} \tag{2.37}$$

since the transpose of the scalar value is the same as the original scalar value ($a^{\mathsf{T}} = a$) and $\mathbf{R}$ is a symmetric matrix ($\mathbf{R}^{\mathsf{T}} = \mathbf{R}$). Thus, by substituting Eqs. (2.32) and (2.36) into Eq. (2.31), Eq. (2.31) is rewritten as

$$\begin{aligned}&\sum_{n=1}^{N}(\mathbf{x}_n - \boldsymbol{\mu})^{\mathsf{T}}\mathbf{R}(\mathbf{x}_n - \boldsymbol{\mu}) \\ &\quad = -2\boldsymbol{\mu}^{\mathsf{T}}\mathbf{R}\sum_{n=1}^{N}\mathbf{x}_n + \mathrm{tr}\left[\mathbf{R}\sum_{n=1}^{N}\mathbf{x}_n\mathbf{x}_n^{\mathsf{T}}\right] + N\boldsymbol{\mu}^{\mathsf{T}}\mathbf{R}\boldsymbol{\mu}.\end{aligned} \tag{2.38}$$

Note that Eq. (2.38) is a useful form, and it is used in the following sections to calculate the various equations for the multivariate Gaussian distribution.

Therefore, by substituting Eq. (2.38) into Eq. (2.30), we can obtain the exponential form of the multivariate Gaussian distribution as follows:

$$\begin{aligned}&\prod_{n=1}^{N}\mathcal{N}(\mathbf{x}_n|\boldsymbol{\mu}, \mathbf{R}^{-1}) \\ &= (2\pi)^{-\frac{ND}{2}}|\mathbf{R}|^{\frac{N}{2}}\exp\left(\boldsymbol{\mu}^{\mathsf{T}}\mathbf{R}\sum_{n=1}^{N}\mathbf{x}_n - \frac{1}{2}\mathrm{tr}\left[\mathbf{R}\sum_{n=1}^{N}\mathbf{x}_n\mathbf{x}_n^{\mathsf{T}}\right] - \frac{N}{2}\boldsymbol{\mu}^{\mathsf{T}}\mathbf{R}\boldsymbol{\mu}\right) \\ &= \exp\left(\boldsymbol{\mu}^{\mathsf{T}}\mathbf{R}\sum_{n=1}^{N}\mathbf{x}_n - \frac{1}{2}\mathrm{tr}\left[\mathbf{R}\sum_{n=1}^{N}\mathbf{x}_n\mathbf{x}_n^{\mathsf{T}}\right] - \frac{N}{2}\left(\log((2\pi)^{D}|\mathbf{R}|^{-1}) + \boldsymbol{\mu}^{\mathsf{T}}\mathbf{R}\boldsymbol{\mu}\right)\right).\end{aligned} \tag{2.39}$$

Thus, by comparing with Eq. (2.23), we obtain the following parameterization for the multivariate Gaussian distribution:

$$
\begin{cases}
\mathbf{t}_1(\mathbf{X}) = \sum_{n=1}^{N} \mathbf{x}_n \\
\mathbf{T}_2(\mathbf{X}) = -\frac{1}{2} \sum_{n=1}^{N} \mathbf{x}_n \mathbf{x}_n^{\mathsf{T}} \\
h(x) = 1 \\
\boldsymbol{\gamma}_1(\Theta) = \mathbf{R}\boldsymbol{\mu} \\
\boldsymbol{\Gamma}_2(\Theta) = \mathbf{R} \\
g(\boldsymbol{\gamma}_1, \boldsymbol{\Gamma}_2) = \frac{N}{2} \left(\log((2\pi)^D |\mathbf{R}|^{-1}) + \boldsymbol{\mu}^{\mathsf{T}} \mathbf{R} \boldsymbol{\mu} \right) \\
\qquad = \frac{N}{2} \left(\log((2\pi)^D |\boldsymbol{\Gamma}_2|^{-1}) + \boldsymbol{\gamma}_1^{\mathsf{T}} \boldsymbol{\Gamma}_2^{-1} \boldsymbol{\gamma}_1 \right).
\end{cases}
\tag{2.40}
$$

Note that if $D \to 1$, we have $\mathbf{x}_n \to x_n$, $\boldsymbol{\mu} \to \mu$, $\mathbf{R} \to r$, and Eq. (2.40) is equivalent to Eq. (2.29).

Example 2.4 Multinomial distribution:
The standard form of the multinomial distribution (Eq. (C.2)) is represented as follows:

$$
\mathrm{Mult}(x_1, \cdots, x_J | \omega_1, \cdots, \omega_J) \triangleq \frac{N!}{\prod_{j=1}^{J} x_j!} \prod_{j=1}^{J} \omega_j^{x_j}, \tag{2.41}
$$

where x_j is a non-negative integer, and

$$
\sum_{j=1}^{J} x_j = N. \tag{2.42}
$$

The parameter $\{\omega_1, \cdots, \omega_J\}$ has the following constraint:

$$
\sum_{j=1}^{J} \omega_j = 1, \quad 0 \leq \omega_j \leq 1 \quad \forall j. \tag{2.43}
$$

Therefore, the number of the free parameters is $J-1$. To deal with the constraint, we first consider the $\{\omega_1, \cdots, \omega_{J-1}\}$ as the target vector parameters, i.e., $\boldsymbol{\theta} \triangleq [\omega_1, \cdots, \omega_{J-1}]^{\mathsf{T}}$. ω_J is represented by

$$
\omega_J = 1 - \sum_{j=1}^{J-1} \omega_j. \tag{2.44}
$$

Similarly to the previous Gaussian-based distributions, the multinomial distribution is also represented as the exponential form as follows:

$$\begin{aligned}\mathrm{Mult}(x_1,\cdots,x_J|\omega_1,\cdots,\omega_J) &= \frac{N!}{\prod_{j=1}^{J} x_j!}\exp\left(\log\left(\prod_{j=1}^{J}\omega_j^{x_j}\right)\right)\\ &= \frac{N!}{\prod_{j=1}^{J} x_j!}\exp\left(\sum_{j=1}^{J} x_j\log\omega_j\right).\end{aligned} \tag{2.45}$$

By using Eqs. (2.42) and (2.44) for x_J and ω_J, respectively, the exponential part of Eq. (2.45) is rewritten as

$$\begin{aligned}&\mathrm{Mult}(x_1,\cdots,x_J|\omega_1,\cdots,\omega_J)\\ &\propto \exp\left(\sum_{j=1}^{J-1} x_j\log\omega_j + \left(N-\sum_{j=1}^{J-1}x_j\right)\log\left(1-\sum_{j=1}^{J-1}\omega_j\right)\right)\\ &= \exp\left(\sum_{j=1}^{J-1} x_j\log\omega_j - \sum_{j=1}^{J-1}x_j\log\left(1-\sum_{j=1}^{J-1}\omega_j\right) + N\log\left(1-\sum_{j=1}^{J-1}\omega_j\right)\right)\\ &= \exp\left(\underbrace{\sum_{j=1}^{J-1} x_j\log\frac{\omega_j}{1-\sum_{j=1}^{J-1}\omega_j}}_{\triangleq \mathbf{x}^{\mathsf{T}}\boldsymbol{\gamma}} + N\log\left(1-\sum_{j'=1}^{J-1}\omega_{j'}\right)\right),\end{aligned} \tag{2.46}$$

where $\propto$ denotes the proportional relation between left- and right-hand-side equations. Since the probabilistic function has the normalization factor, which can be neglected for most of the calculations, $\propto$ is often used to omit the normalization constant from the equations. Thus, we can derive the linear relationship between x_j and γ_j, which is defined with $\{\omega_j\}_{j=1}^{J-1}$ as follows:

$$\gamma_j \triangleq \log\frac{\omega_j}{1-\sum_{j'=1}^{J-1}\omega_{j'}}. \tag{2.47}$$

Note that ω_j is represented by γ_j by using the following equation:

$$\omega_j = \frac{\exp(\gamma_j)}{1+\sum_{j'=1}^{J-1}\exp(\gamma_{j'})}. \tag{2.48}$$

This is confirmed by substituting Eq. (2.47) into Eq. (2.48) as

$$\begin{aligned}\frac{\exp(\gamma_j)}{1+\sum_{j'=1}^{J-1}\gamma_{j'}} &= \frac{\frac{\omega_j}{1-\sum_{j'=1}^{J-1}\omega_{j'}}}{1+\sum_{j'=1}^{J-1}\frac{\omega_{j'}}{1-\sum_{j''=1}^{J-1}\omega_{j''}}}\\ &= \frac{\omega_j}{1-\sum_{j'=1}^{J-1}\omega_{j'}+\sum_{j'=1}^{J-1}\omega_{j'}} = \omega_j.\end{aligned} \tag{2.49}$$

Therefore, the canonical form of the multinomial distribution with the parameter $\boldsymbol{\theta} = [\omega_1, \cdots, \omega_{J-1}]^\mathsf{T}$ for $\mathbf{x} = [x_1, \cdots, x_{J-1}]^\mathsf{T}$ is represented as

$$\begin{cases} \mathbf{t}(\mathbf{x}) = \mathbf{x} \\ h(x) = \dfrac{N!}{\prod_{j=1}^{J} x_j!} \\ \boldsymbol{\gamma}(\boldsymbol{\theta}) = \left[\log \frac{\omega_1}{1-\sum_{j=1}^{J-1}\omega_j}, \cdots, \log \frac{\omega_{J-1}}{1-\sum_{j=1}^{J-1}\omega_j}\right]^\mathsf{T} \\ g(\boldsymbol{\gamma}) = -N \log \left(1 - \displaystyle\sum_{j=1}^{J-1} \omega_j\right) \\ \qquad = -N \log \left(1 - \displaystyle\sum_{j=1}^{J-1} \frac{\exp(\gamma_j)}{1 + \sum_{j'=1}^{J-1} \exp(\gamma_{j'})}\right) \\ \qquad = N \log \left(1 + \displaystyle\sum_{j=1}^{J-1} \exp(\gamma_j)\right). \end{cases} \tag{2.50}$$

Note that since the multinomial distribution has constraints for the observation x_j in Eq. (2.42) and the parameter ω_j in Eq. (2.44), the obtained canonical form of the multinomial distribution involves these constraints with $J-1$ variables for sufficient statistics $\mathbf{t}$ and the transformed vector $\boldsymbol{\gamma}$.

The obtained exponential family forms for Gaussian, multivariate Gaussian, and multinomial distributions are often used in the Bayesian treatment of statistical models in speech and language processing.

2.1.4 Conjugate distributions

The previous section introduces the exponential family distributions and provides some examples of these distributions. Based on the exponential family distributions, this section explains how to obtain the posterior distributions when we use the exponential family distributions as the likelihood functions. For such a distribution, we can find a nice property to obtain the posterior distribution analytically if we set a particular type of distribution.

Let $p(\mathbf{X}|\boldsymbol{\theta})$ be a likelihood function for a set of observation vectors $\mathbf{X} = \{\mathbf{x}_1, \cdots, \mathbf{x}_N\}$. We first start the discussion from the simple case that the parameters are represented as a vector form, i.e., $\boldsymbol{\theta}$. An exponential family distribution of $p(\mathbf{X}|\boldsymbol{\theta})$ is defined in Eq. (2.20) as

$$p(\mathbf{X}|\boldsymbol{\theta}) = h(\mathbf{X}) \exp\left(\boldsymbol{\gamma}^\mathsf{T}\mathbf{t}(\mathbf{X}) - g(\boldsymbol{\gamma})\right). \tag{2.51}$$

Here use $\boldsymbol{\gamma}(\boldsymbol{\theta}) \to \boldsymbol{\gamma}$ for simplicity. Then, we use the following Bayes theorem for $\boldsymbol{\theta}$ based on Eq. (2.18) to calculate the posterior distribution $p(\boldsymbol{\theta}|\mathbf{X})$:

$$p(\boldsymbol{\theta}|\mathbf{X}) \propto p(\mathbf{X}|\boldsymbol{\theta})p(\boldsymbol{\theta}), \tag{2.52}$$

where we disregard the normalization factor $p(\mathbf{X})$. For this calculation, we need to prepare a prior distribution $p(\boldsymbol{\theta})$. Instead of considering the prior distribution of $p(\boldsymbol{\theta})$, we consider the prior distribution of $p(\boldsymbol{\gamma})$. We set a prior distribution for $p(\boldsymbol{\gamma})$, which is parameterized with additional variables $\boldsymbol{\nu}$ and ϕ, where the parameters of prior and posterior distributions are called *hyperparameters*. The hyperparameter appearing in this book is often used as the parameter of prior or posterior distributions. Then, the prior distribution is proportional to the following function form:

$$p(\boldsymbol{\theta}) \to p(\boldsymbol{\gamma}|\boldsymbol{\nu},\phi) \propto \exp\left(\boldsymbol{\gamma}^{\mathsf{T}}\boldsymbol{\nu} - \phi g(\boldsymbol{\gamma})\right). \tag{2.53}$$

Here, $g(\boldsymbol{\gamma})$ is introduced in Eqs. (2.20) and (2.51) as a logarithmic normalization factor of the likelihood function. This form of prior distribution is called *conjugate prior* distribution.

We can calculate the posterior distribution of $p(\boldsymbol{\theta}|\mathbf{X})$ via $\boldsymbol{\gamma}$ by substituting Eqs. (2.51) and (2.53) into Eq. (2.52):

$$\begin{aligned}
p(\boldsymbol{\theta}|\mathbf{X}) &\to p(\mathbf{X}|\boldsymbol{\theta})p(\boldsymbol{\gamma}|\boldsymbol{\nu},\phi) \\
&= h(\mathbf{X})\exp\left(\boldsymbol{\gamma}^{\mathsf{T}}\mathbf{t}(\mathbf{X}) - g(\boldsymbol{\gamma})\right)\exp\left(\boldsymbol{\gamma}^{\mathsf{T}}\boldsymbol{\nu} - \phi g(\boldsymbol{\gamma})\right) \\
&\propto \exp\left(\boldsymbol{\gamma}^{\mathsf{T}}(\boldsymbol{\nu} + \mathbf{t}(\mathbf{X})) - (\phi + 1)g(\boldsymbol{\gamma})\right) \\
&= p(\boldsymbol{\gamma}|\boldsymbol{\nu} + \mathbf{t}(\mathbf{X}), \phi + 1),
\end{aligned} \tag{2.54}$$

where we use the definition used in the conjugate prior distribution (Eq. (2.53)). This solution means that the conjugate posterior distribution is analytically obtained with the same distribution function as the conjugate prior distribution by just using the simple rule of changing hyperparameters from ($\boldsymbol{\nu}$, ϕ) to ($\boldsymbol{\nu} + \mathbf{t}(\mathbf{X})$, $\phi + 1$).

Note that the setting of ϕ is not unique. We consider the case that $g(\boldsymbol{\gamma})$ is decomposed into M functions, i.e.,

$$g(\boldsymbol{\gamma}) \triangleq \sum_{m=1}^{M} g_m(\boldsymbol{\gamma}). \tag{2.55}$$

Then, similarly to Eq. (2.53), we can provide M hyperparameters for a prior distribution as follows:

$$p(\boldsymbol{\theta}) \to p(\boldsymbol{\gamma}|\boldsymbol{\nu},\phi) \propto \exp\left(\boldsymbol{\gamma}^{\mathsf{T}}\boldsymbol{\nu} - \sum_{m=1}^{M}\phi_m g_m(\boldsymbol{\gamma})\right). \tag{2.56}$$

The corresponding posterior distribution is similarly derived by substituting Eqs. (2.51), (2.55), and (2.56) into Eq. (2.52) as:

$$\begin{aligned}
p(\boldsymbol{\theta}|\mathbf{X}) &\to p(\mathbf{X}|\boldsymbol{\theta})p(\boldsymbol{\gamma}|\boldsymbol{\nu},\{\phi_m\}_{m=1}^{M}) \\
&= h(\mathbf{X})\exp\left(\boldsymbol{\gamma}^{\mathsf{T}}\mathbf{t}(\mathbf{X}) - \sum_{m=1}^{M} g_m(\boldsymbol{\gamma})\right)\exp\left(\boldsymbol{\gamma}^{\mathsf{T}}\boldsymbol{\nu} - \sum_{m=1}^{M}\phi_m g_m(\boldsymbol{\gamma})\right) \\
&\propto \exp\left(\boldsymbol{\gamma}^{\mathsf{T}}(\boldsymbol{\nu} + \mathbf{t}(\mathbf{X})) - \sum_{m=1}^{M}(\phi_m + 1)g_m(\boldsymbol{\gamma})\right) \\
&= p(\boldsymbol{\gamma}|\boldsymbol{\nu} + \mathbf{t}(\mathbf{X}), \{\phi_m + 1\}_{m=1}^{M}).
\end{aligned} \tag{2.57}$$

Thus, we can derive the posterior distribution with M hyperparameters. The setting of $\{\phi_m\}$ is an additional flexibility of the prior distribution. If we use many $\{\phi_m\}$, we could precisely represent a prior distribution. However, by using a few $\{\phi_m\}$, we can easily control the shape of a prior distribution with a few free parameters.

If the transformed model parameters are composed of a vector $\boldsymbol{\gamma}_1$ and matrix $\boldsymbol{\Gamma}_2$, as discussed in Eq. (2.23), we also have similar result. A likelihood function of this exponential family distribution is represented by the following general form:

$$p(\mathbf{X}|\Theta) \triangleq h(\mathbf{X}) \exp\left(\boldsymbol{\gamma}_1^\top \mathbf{t}_1(\mathbf{X}) + \mathrm{tr}[\boldsymbol{\Gamma}_2^\top \mathbf{T}_2(\mathbf{x})] - \sum_{m=1}^{M} g_m(\boldsymbol{\gamma}_1, \boldsymbol{\Gamma}_2)\right). \tag{2.58}$$

Here, similarly to Eq. (2.55), we use the following equation for the $g(\cdot)$ function:

$$g(\boldsymbol{\gamma}_1, \boldsymbol{\Gamma}_2) \triangleq \sum_{m=1}^{M} g_m(\boldsymbol{\gamma}_1, \boldsymbol{\Gamma}_2). \tag{2.59}$$

Therefore, by providing the following prior distribution form as a conjugate prior with hyperparameters $\boldsymbol{\nu}_1, \mathbf{N}_2$ and $\{\phi_m\}_{m=1}^M$:

$$p(\boldsymbol{\gamma}_1, \boldsymbol{\Gamma}_2|\boldsymbol{\nu}_1, \mathbf{N}_2, \{\phi\}_{m=1}^M) \propto \exp\left(\boldsymbol{\gamma}_1^\top \boldsymbol{\nu}_1 + \mathrm{tr}[\boldsymbol{\Gamma}_2^\top \mathbf{N}_2] - \sum_{m=1}^{M} \phi_m g_m(\boldsymbol{\gamma}_1, \boldsymbol{\Gamma}_2)\right). \tag{2.60}$$

We can calculate the posterior distribution by substituting Eqs. (2.60), (2.58), and (2.59) into Eq. (2.52):

$$\begin{aligned} &p(\Theta|\mathbf{X}) \\ &\quad \to p(\mathbf{X}|\Theta)p(\boldsymbol{\gamma}_1, \boldsymbol{\Gamma}_2|\mathbf{N}, \{\phi_m\}_{m=1}^M) \\ &\quad \propto \exp\left(\boldsymbol{\gamma}^\top(\boldsymbol{\nu}_1 + \mathbf{t}_1(\mathbf{X})) + \mathrm{tr}[\boldsymbol{\Gamma}_2^\top(\mathbf{N}_2 + \mathbf{T}_2(\mathbf{X}))] - \sum_{m=1}^{M}(\phi_m + 1)g_m(\boldsymbol{\gamma}_1, \boldsymbol{\Gamma}_2)\right) \\ &\quad = p(\boldsymbol{\gamma}_1, \boldsymbol{\Gamma}_2|\boldsymbol{\nu}_1 + \mathbf{t}_1(\mathbf{X}), \mathbf{N}_2 + \mathbf{T}_2(\mathbf{X}), \{\phi_m + 1\}_{m=1}^M). \end{aligned} \tag{2.61}$$

Here we use the distributive property of the trace in Appendix B that:

$$\mathrm{tr}[\mathbf{AB}] + \mathrm{tr}[\mathbf{AC}] = \mathrm{tr}[\mathbf{A}(\mathbf{B} + \mathbf{C})]. \tag{2.62}$$

Now, we summarize the conjugate prior and posterior distributions. The exponential family distributions with the vector form parameters $\boldsymbol{\theta}$ have the following relationship:

$$\begin{cases} \text{Prior: } p(\boldsymbol{\gamma}|\boldsymbol{\nu}, \{\phi_m\}_{m=1}^M) \\ \text{Posterior: } p(\boldsymbol{\gamma}|\boldsymbol{\nu} + \mathbf{t}(\mathbf{X}), \{\phi_m + 1\}_{m=1}^M). \end{cases} \tag{2.63}$$

When the distribution has vector and matrix parameters, we have the following relationship:

$$\begin{cases} \text{Prior: } p(\boldsymbol{\gamma}_1, \boldsymbol{\Gamma}_2|\boldsymbol{\nu}_1, \mathbf{N}_2, \{\phi_m\}_{m=1}^M) \\ \text{Posterior: } p(\boldsymbol{\gamma}_1, \boldsymbol{\Gamma}_2|\boldsymbol{\nu}_1 + \mathbf{t}_1(\mathbf{X}), \mathbf{N}_2 + \mathbf{T}_2(\mathbf{X}), \{\phi_m + 1\}_{m=1}^M). \end{cases} \tag{2.64}$$

Therefore, the posterior distribution of the natural parameters $(\boldsymbol{\gamma}, \boldsymbol{\gamma}_1, \boldsymbol{\Gamma}_2)$ is analytically obtained by using Eqs. (2.63) and (2.64) as a rule. The posterior distribution of the original parameters $p(\Theta|\mathbf{X})$ is obtained by transforming the posterior distribution of the natural parameters.

The rest of this section provides examples of the conjugate prior and posterior distributions for some exponential family distributions.

Example 2.5 Conjugate distributions for Gaussian (unknown mean):
We first describe the case that we only consider a Gaussian mean parameter μ, and the precision parameter $r = \Sigma^{-1}$ is regarded as a constant value. Based on the discussion in Example 2.1, the canonical form of the Gaussian distribution is represented as follows:

$$\prod_{n=1}^{N} \mathcal{N}(x_n|\mu; r^{-1}) = h(X) \exp\left(\gamma t(X) - g(\gamma)\right), \tag{2.65}$$

where

$$\begin{cases} t(X) = \sum_{n=1}^{N} x_n \\ h(X) = \left(\frac{2\pi}{r}\right)^{-\frac{N}{2}} \exp\left(-\frac{r\sum_{n=1}^{N} x_n^2}{2}\right) \\ \gamma = \mu r \\ g(\gamma) = \frac{N\gamma^2}{2r}. \end{cases} \tag{2.66}$$

Therefore, by substituting γ and $g(\gamma)$ in Eq. (2.66) into the general form of the conjugate distribution in Eq. (2.53), we can derive the function of mean μ as follows:

$$\begin{aligned} p(\gamma|\nu, \phi) &\propto \exp\left(\gamma\nu - \phi g(\gamma)\right) = \exp\left(\mu r\nu - \phi\frac{N\mu^2 r}{2}\right) \\ &\propto \exp\left(-\frac{N\phi r}{2}\left(\mu - \frac{\nu}{N\phi}\right)^2\right). \end{aligned} \tag{2.67}$$

Thus, the prior distribution of μ is represented by a Gaussian distribution with $\frac{\nu}{N\phi}$ and $N\phi r$ as the mean and precision parameters, respectively:

$$p(\mu) \propto \mathcal{N}\left(\mu \middle| \frac{\nu}{N\phi}, (N\phi r)^{-1}\right). \tag{2.68}$$

Based on the conjugate distribution rule (Eq. (2.63)), the posterior distribution is easily solved by just replacing $\nu \to \nu + t(X)$ and $\phi \to \phi + 1$ in Eq. (2.68) without complex calculations:

$$p(\gamma|\nu + t(X), \phi + 1) \propto \exp\left(-\frac{N(\phi+1)r}{2}\left(\mu - \frac{\nu + \sum_{n=1}^{N} x_n}{N(\phi+1)}\right)^2\right)$$
$$\to \mathcal{N}\left(\mu \left| \frac{\nu + \sum_{n=1}^{N} x_n}{N(\phi+1)}, (N(\phi+1)r)^{-1}\right.\right). \tag{2.69}$$

Therefore, similarly to the prior distribution, the posterior distribution of μ is represented by a Gaussian distribution with $\frac{\nu+\sum_{n=1}^{N} x_n}{N(\phi+1)}$ and $(N(\phi + 1)r)^{-1}$ as the mean and variance parameters, respectively:

$$p(\mu|X) \propto \mathcal{N}\left(\mu \left| \frac{\nu + \sum_{n=1}^{N} x_n}{N(\phi+1)}, (N(\phi+1)r)^{-1}\right.\right). \tag{2.70}$$

Thus, both prior and posterior distributions are represented in the same form as a Gaussian distribution with different parameters.

Now we consider the meaning of the solution of Eqs. (2.68) and (2.70). We parameterize the ϕ and ν by newly introducing the following parameters:

$$\phi \triangleq \frac{\phi^{\mu}}{N}$$
$$\nu \triangleq \phi^{\mu}\mu^{0}. \tag{2.71}$$

Then, the prior and posterior distributions of μ in Eqs. (2.68) and (2.70) are rewritten as:

$$\begin{cases} p(\mu) = \mathcal{N}\left(\mu \left| \mu^0, (\phi^0 r)^{-1}\right.\right) \\ p(\mu|X) = \mathcal{N}\left(\mu \left| \hat{\mu}, (\hat{\phi}^{\mu} r)^{-1}\right.\right). \end{cases} \tag{2.72}$$

where

$$\hat{\phi}^{\mu} \triangleq \phi^{\mu} + N$$
$$\hat{\mu} \triangleq \frac{\phi^{\mu}\mu^0 + \sum_{n=1}^{N} x_n}{\phi^{\mu} + N}. \tag{2.73}$$

These are famous Bayesian solutions of the posterior distribution of the Gaussian mean. We can consider the two extreme cases that the amount of data is zero or very large. Then, the posterior distribution is represented as:

- $N \to 0$

$$\lim_{N\to 0} p(\mu|X) = \mathcal{N}\left(\mu \left| \mu^0, (\phi^{\mu} r)^{-1}\right.\right) = p(\mu). \tag{2.74}$$

This solution means that we only use the prior information when we don't have data.

- $N \gg 1$

$$\lim_{N\to\infty} p(\mu|X) \approx \lim_{N\to\infty} \mathcal{N}\left(\mu \left| \frac{\sum_{n=1}^{N} x_n}{N}, \frac{1}{Nr}\right.\right) \to \delta(\mu - \mu^{\mathrm{ML}}), \tag{2.75}$$

where μ^{ML} is the ML estimate of μ, and the posterior distribution is close to the ML value with small standard deviation, which is similar to the delta function that has a peak value at the ML estimate.

Thus, the solution of Eq. (2.72) approaches the delta function with the ML estimate when the amount of data is very large and approaches the prior distribution when the amount of data is very small. The mean parameter of the posterior distribution,

$$\frac{\phi^{\mu}\mu^{0} + \sum_{n=1}^{N} x_n}{\phi^{\mu} + N}, \tag{2.76}$$

is interpolated by the prior mean parameter μ^0 and the ML estimate, and ϕ^{μ} can control an interpolation ratio.

Example 2.6 Conjugate distributions for Gaussian (unknown mean and precision): Similarly to Example 2.5, we first rewrite a Gaussian distribution. In this situation, the set of the parameters is $\theta = \{\mu, r\}$. From Eq. (2.29), the Gaussian distribution with precision r has the following exponential form:

$$\begin{aligned}\prod_{n=1}^{N} \mathcal{N}(x_n|\mu, r^{-1}) &= h(X) \exp\left(\boldsymbol{\gamma} \mathbf{t}(X)^{\mathsf{T}} - g(\boldsymbol{\gamma})\right) \\ &= h(X) \exp\left(\boldsymbol{\gamma} \mathbf{t}(X)^{\mathsf{T}} - \phi_1 g_1(\boldsymbol{\gamma}) - \phi_2 g_2(\boldsymbol{\gamma})\right),\end{aligned} \tag{2.77}$$

where we introduce ϕ_1 and ϕ_2 that are discussed in Eq. (2.55). The variables in the above equations are represented as follows:

$$\begin{cases} \mathbf{t}(X) = \begin{bmatrix} \sum_{n=1}^{N} x_n \\ -\frac{\sum_{n=1}^{N} x_n^2}{2} \end{bmatrix} \\ h(X) = 1 \\ \boldsymbol{\gamma} = \begin{bmatrix} \mu r \\ r \end{bmatrix} \\ g_1(\boldsymbol{\gamma}) = \frac{N}{2}\left(\frac{\gamma_1^2}{\gamma_2}\right) \\ g_2(\boldsymbol{\gamma}) = \frac{N}{2}\left(\log \frac{2\pi}{\gamma_2}\right). \end{cases} \tag{2.78}$$

Therefore, by substituting $\boldsymbol{\gamma}$, $g_1(\boldsymbol{\gamma})$, and $g_2(\boldsymbol{\gamma})$ in Eq. (2.78) into the general form of the conjugate distribution in Eq. (2.56), we can derive the function of mean μ and precision r as follows:

$$\begin{aligned} p(\boldsymbol{\gamma}|\boldsymbol{\nu}, \phi_1, \phi_2) &\propto \exp\left(\boldsymbol{\gamma}^{\mathsf{T}}\boldsymbol{\nu} - \phi_1 g_1(\boldsymbol{\gamma}) - \phi_2 g_2(\boldsymbol{\gamma})\right) \\ &\propto \exp\left([\mu r, r]\begin{bmatrix}\nu_1 \\ \nu_2\end{bmatrix} - \frac{N\phi_1}{2} r\mu^2 - \frac{N\phi_2}{2}\log\left(\frac{2\pi}{r}\right)\right) \\ &\propto r^{\frac{N\phi_2}{2}} \exp\left(\nu_1 r\mu - N\phi_1 \frac{r\mu^2}{2} + r\nu_2\right), \end{aligned} \tag{2.79}$$

where we omit the factor that does not depend on r and μ. By making a complete square form of μ, we can obtain a Gaussian distribution of μ with mean $\frac{\nu_1}{N\phi_1}$ and precision $N\phi_1 r$:

$$
\begin{aligned}
p(\mathbf{y}|\mathbf{v}, \phi_1, \phi_2) & \\
&\propto r^{\frac{N\phi_2}{2}} \exp\left(-\frac{N\phi_1 r}{2}\left(\mu - \frac{\nu_1}{N\phi_1}\right)^2 + \frac{r\nu_1^2}{2N\phi_1} + r\nu_2\right) \\
&= r^{\frac{N\phi_2}{2}} \left(\frac{2\pi}{N\phi_1 r}\right)^{\frac{1}{2}} \left(\frac{2\pi}{N\phi_1 r}\right)^{-\frac{1}{2}} \exp\left(-\frac{N\phi_1 r}{2}\left(\mu - \frac{\nu_1}{N\phi_1}\right)^2 + \frac{r\nu_1^2}{2N\phi_1} + r\nu_2\right) \\
&\propto \mathcal{N}\left(\mu \middle| \frac{\nu_1}{N\phi_1}, (N\phi_1 r)^{-1}\right) \underbrace{r^{\frac{N\phi_2}{2}} r^{-\frac{1}{2}} \exp\left(r\frac{\nu_1^2}{2N\phi_1} + r\nu_2\right)}_{\triangleq (*1)}.
\end{aligned}
\tag{2.80}
$$

Now we consider the rest of the exponential factor $(*1)$. By focusing on r and using the definition of a gamma distribution (Appendix C.11), the factor is rewritten as follows:

$$
\begin{aligned}
(*1) &\propto r^{\frac{N\phi_2+1}{2}-1} \exp\left(-\left(-\frac{\nu_1^2}{2N\phi_1} - \nu_2\right) r\right) \\
&\propto \mathrm{Gam}\left(r \middle| \frac{N\phi_2 + 1}{2}, -\frac{\nu_1^2}{2N\phi_1} - \nu_2\right),
\end{aligned}
\tag{2.81}
$$

where the definition of a gamma distribution is as follows:

$$
\mathrm{Gam}(r|\alpha, \beta) \triangleq \frac{1}{\Gamma(\alpha)}\beta^\alpha r^{\alpha-1} \exp\left(-\beta r\right),
\tag{2.82}
$$

where $\Gamma(\cdot)$ is a Gamma function (Appendix A.4). Thus, precision $r = \frac{1}{\Sigma}$ is represented by a gamma distribution with $\frac{N\phi_2+1}{2}$ and $-\frac{\nu_1^2}{2N\phi_1} - \nu_2$ as parameters.

This representation can be simplified by using the following definition for the other definition of the gamma distribution $\mathrm{Gam}_2(y|\phi, r^0)$ described in Eq. (C.81) instead of the original gamma distribution defined in Eq. (C.74):

$$
\begin{aligned}
\mathrm{Gam}_2(y|\phi, r^0) &\triangleq \mathrm{Gam}\left(y \middle| \frac{\phi}{2}, \frac{r^0}{2}\right) \\
&\propto y^{\frac{\phi}{2}-1} \exp\left(-\frac{r^0 y}{2}\right).
\end{aligned}
\tag{2.83}
$$

Equation (2.81) is rewritten as

$$
(*1) \propto \mathrm{Gam}_2\left(r \middle| N\phi_2 + 1, -\frac{\nu_1^2}{N\phi_1} - 2\nu_2\right).
\tag{2.84}
$$

Thus, the conjugate prior distribution is represented as the product form of the following Gaussian and gamma distributions:

$$p(\mu, r) = \mathcal{N}\left(\mu \left| \frac{\nu_1}{N\phi_1}, (N\phi_1 r)^{-1}\right.\right) \mathrm{Gam}_2\left(r \left| N\phi_2 + 1, -\frac{\nu_1^2}{N\phi_1} - 2\nu_2\right.\right). \quad (2.85)$$

This can be also represented as a Gaussian-gamma distribution (or so-called normal-gamma) defined in Appendix C.13, as follows:

$$p(\mu, r) = \mathcal{N}\mathrm{Gam}\left(\mu, r \left| \frac{\nu_1}{N\phi_1}, (N\phi_1 r)^{-1}, -\frac{\nu_1^2}{N\phi_1} - 2\nu_2, N\phi_2 + 1\right.\right). \quad (2.86)$$

The Gaussian-gamma distribution is a conjugate prior distribution of the joint variable μ and r.

Similarly to the previous example, we introduce the following new parameters:

$$\begin{cases} \phi^\mu \triangleq N\phi_1 \\ \mu^0 \triangleq \dfrac{\nu_1}{N\phi_1} \\ \phi^r \triangleq N\phi_2 + 1 \\ r^0 \triangleq -\dfrac{\nu_1^2}{N\phi_1} - 2\nu_2. \end{cases} \quad (2.87)$$

By using Eq. (2.87), the conjugate prior distribution of Eq. (2.85) is rewritten by using these new parameters as follows:

$$p(\mu, r) = \mathcal{N}\left(\mu \left| \mu^0, (\phi^\mu r)^{-1}\right.\right) \mathrm{Gam}_2\left(r \left| \phi^r, r^0\right.\right). \quad (2.88)$$

Note that we can also use Gaussian-gamma distribution as:

$$\begin{aligned} p(\mu, r) &= \mathcal{N}(\mu|\mu^0, (r\phi^\mu)^{-1})\, \mathrm{Gam}_2\left(r \left| \phi^r, r^0\right.\right) \\ &= \mathcal{N}\mathrm{Gam}(\mu, r|\mu^0, \phi^\mu, r^0, \phi^r). \end{aligned} \quad (2.89)$$

Thus, we can derive the prior distribution of joint variable μ and r as the product of the Gaussian and gamma distributions in Eq. (2.88), or the single Gaussian-gamma distribution in Eq. (2.89).

Now, we focus on the posterior distribution of μ and r. Based on the conjugate distribution theory, the posterior distribution is represented as the same form of the Gaussian-gamma distribution as the prior distribution (2.89) with hyperparameters $\hat{\phi}^\mu, \hat{\mu}, \hat{\phi}^r$, and $\hat{r}$ as follows:

$$p(\mu, r|X) = \mathcal{N}\mathrm{Gam}(\mu, r|\hat{\mu}^0, \hat{\phi}^\mu, \hat{r}^0, \hat{\phi}^r). \quad (2.90)$$

Based on the conjugate distribution rule (Eq. (2.63)), the hyperparameters of the posterior distribution are easily solved by just replacing $\boldsymbol{\nu} \to \boldsymbol{\nu} + \mathbf{t}(X)$ and $\phi_m \to \phi_m + 1$ in Eq. (2.87) without complex calculations, as follows:

$$
\begin{aligned}
\hat{\phi}^{\mu} &= N(\phi_1 + 1) = \phi^{\mu} + N \\
\hat{\mu} &= \frac{\nu_1 + \sum_{n=1}^{N} x_n}{N(\phi_1 + 1)} = \frac{\phi^{\mu}\mu^0 + \sum_{n=1}^{N} x_n}{\phi^{\mu} + N} \\
\hat{\phi}^{r} &= N(\phi_2 + 1 + 1) + 1 = N(\phi_2 + 1) + 1 + N \\
&= \phi^{r} + N \\
\hat{r} &= -\frac{\left(\nu_1 + \sum_{n=1}^{N} x_n\right)^2}{N(\phi_1 + 1)} - \left(2\nu_2 - \sum_{n=1}^{N} x_n^2\right) \\
&= -\frac{\left(\phi^{\mu}\mu^0 + \sum_{n=1}^{N} x_n\right)^2}{\phi^{\mu} + N} - \left(-\frac{\nu_1^2}{N\phi_1} - r^0 - \sum_{n=1}^{N} x_n^2\right) \\
&= -\frac{\left(\phi^{\mu}\mu^0 + \sum_{n=1}^{N} x_n\right)^2}{\phi^{\mu} + N} + \phi^{\mu}(\mu^0)^2 + r^0 + \sum_{n=1}^{N} x_n^2.
\end{aligned}
\tag{2.91}
$$

Thus, we summarize the result of the hyperparameters of the conjugate posterior distribution as

$$
\begin{cases}
\hat{\phi}^{\mu} = \phi^{\mu} + N \\
\hat{\mu} = \dfrac{\phi^{\mu}\mu^0 + \sum_{n=1}^{N} x_n}{\phi^{\mu} + N} \\
\hat{\phi}^{r} = \phi^{r} + N \\
\hat{r} = -\hat{\phi}^{\mu}(\hat{\mu})^2 + \phi^{\mu}(\mu^0)^2 + r^0 + \sum_{n=1}^{N} x_n^2.
\end{cases}
\tag{2.92}
$$

Note that in this representation, the posterior distribution parameters of $\hat{\phi}^{\mu}$ and $\hat{\phi}^{r}$ are obtained by simply adding the number of observations N to the prior distribution parameters of ϕ^{μ} and ϕ^{r}, respectively.

Finally, we summarize the result. The prior and posterior distributions of μ and r in Eqs. (2.89) and (2.90) are also summarized as:

$$
\begin{cases}
p(\mu, r) = \mathcal{N}\mathrm{Gam}(\mu, r|\mu^0, \phi^{\mu}, r^0, \phi^{r}) \\
p(\mu, r|X) = \mathcal{N}\mathrm{Gam}(\mu, r|\hat{\mu}, \hat{\phi}^{\mu}, \hat{r}, \hat{\phi}^{r}),
\end{cases}
\tag{2.93}
$$

or

$$
\begin{cases}
p(\mu, r) = p(\mu|r)p(r) = \mathcal{N}(\mu|\mu^0, (\phi^{\mu} r)^{-1})\,\mathrm{Gam}_2(r|\phi^{r}, r^0) \\
p(\mu, r|X) = p(\mu|r, X)p(r|X) = \mathcal{N}(\mu|\hat{\mu}\,(\hat{\phi}^{\mu} r)^{-1})\,\mathrm{Gam}_2(r|\hat{\phi}^{r}, \hat{r}).
\end{cases}
\tag{2.94}
$$

Similarly to the discussion about the mean parameter μ in Example 2.5, we can consider the two extreme cases, that the amount of data is zero or very large, for the behavior of the precision parameter solution r. The posterior distribution of r is represented as:

- $N \to 0$

$$
\lim_{N \to 0} p(r|X) = \mathrm{Gam}_2(r|\phi^{r}, r^0) = p(r).
\tag{2.95}
$$

This solution means that we only use the prior information when we don't have data.

- $N \gg 1$

$$\lim_{N \gg 1} p(r|X) \approx \text{Gam}_2\left(r \left| N, \sum_{n=1}^{N} x_n^2 - \frac{(\sum_{n=1}^{N} x_n)^2}{N}\right.\right). \tag{2.96}$$

Since the mean of the gamma distribution r^{Mean} (with $\frac{1}{2}$ factor) is defined in Eq. (C.83), the mean of r in this limit is represented as:

$$\begin{aligned} r^{\text{Mean}} &= \frac{N}{\sum_{n=1}^{N} x_n^2 - \frac{(\sum_{n=1}^{N} x_n)^2}{N}} \\ &= \left(\frac{\sum_{n=1}^{N} x_n^2}{N} - \left(\frac{\sum_{n=1}^{N} x_n}{N}\right)^2\right)^{-1}. \end{aligned} \tag{2.97}$$

This is equivalent to the maximum likelihood estimation of r^{ML} represented as follows:

$$r^{\text{ML}} = \left(\text{Mean}[x^2] - (\text{Mean}[x])^2\right)^{-1} = r^{\text{Mean}}. \tag{2.98}$$

Thus, the mean of the posterior distribution approaches the ML estimate of r when the amount of data is large.

Similarly, based on the definition of the variance of the gamma distribution (with $\frac{1}{2}$ factor) in Eq. (C.84), the variance is also represented as

$$\begin{aligned} r^{\text{Variance}} &= \frac{2N}{\left(\sum_{n=1}^{N} x_n^2 - \frac{(\sum_{n=1}^{N} x_n)^2}{N}\right)^2} \\ &= \frac{\frac{2}{N}}{\left(\frac{\sum_{n=1}^{N} x_n^2}{N} - \left(\frac{\sum_{n=1}^{N} x_n}{N}\right)^2\right)^2} \\ &= \frac{2\left(r^{\text{ML}}\right)^2}{N} \approx 0. \end{aligned} \tag{2.99}$$

Note that the order of r^{ML} in Eq. (2.98) is a constant for N, and the variance of the precision parameter r^{Variance} approaches 0, i.e., the posterior distribution of r has a strong peak at r^{ML} with a very small variance. Therefore, the posterior distribution of precision parameter $p(r|X)$ in the case of a large amount of data can be approximated as the following Dirac delta function with the ML estimate:

$$\lim_{N \gg 1} p(r|X) \approx \delta(r - r^{\text{ML}}). \tag{2.100}$$

This conclusion is similar to the case of the large amount limitation of the posterior distribution of mean parameter $p(\mu|X)$ in Eq. (2.75).

This Gaussian-gamma distribution is used to model the prior and posterior distributions of Gaussian parameters (μ and r) for scalar continuous observations, or can be used for vector continuous observations when we use a diagonal covariance matrix.

Example 2.7 Conjugate distributions for multivariate Gaussian (unknown mean vector and precision matrix):
Based on the discussion of Eq. (2.39) in Example 2.3, the canonical form of the multivariate Gaussian distribution is represented as follows:

$$\prod_{n=1}^{N} \mathcal{N}(\mathbf{x}_n|\boldsymbol{\mu}, \mathbf{R}^{-1})$$
$$\propto \exp\left(\boldsymbol{\mu}^\mathsf{T}\mathbf{R}\sum_{n=1}^{N}\mathbf{x}_n - \frac{1}{2}\mathrm{tr}\left[\mathbf{R}\sum_{n=1}^{N}\mathbf{x}_n\mathbf{x}_n^\mathsf{T}\right] - \frac{N}{2}\left(\log(2\pi|\mathbf{R}|^{-1}) + \boldsymbol{\mu}^\mathsf{T}\mathbf{R}\boldsymbol{\mu}\right)\right), \tag{2.101}$$

where

$$\begin{cases} \mathbf{t}_1(\mathbf{X}) = \sum_{n=1}^{N}\mathbf{x}_n \\ \mathbf{T}_2(\mathbf{X}) = -\frac{1}{2}\sum_{n=1}^{N}\mathbf{x}_n\mathbf{x}_n^\mathsf{T} \\ h(x) = 1 \\ \boldsymbol{\gamma}_1(\Theta) = \mathbf{R}\boldsymbol{\mu} \\ \boldsymbol{\Gamma}_2(\Theta) = \mathbf{R} \\ g_1(\boldsymbol{\gamma}_1, \boldsymbol{\Gamma}_2) = \frac{N}{2}\boldsymbol{\gamma}_1^\mathsf{T}\boldsymbol{\Gamma}_2^{-1}\boldsymbol{\gamma}_1 \\ g_2(\boldsymbol{\gamma}_1, \boldsymbol{\Gamma}_2) = \frac{N}{2}\log(2\pi|\boldsymbol{\Gamma}_2|^{-1}). \end{cases} \tag{2.102}$$

Therefore, by substituting $\boldsymbol{\gamma}$, $g_1(\boldsymbol{\gamma}_1, \boldsymbol{\Gamma}_2)$, and $g_2(\boldsymbol{\gamma}_1, \boldsymbol{\Gamma}_2)$ in Eq. (2.102) into the general form of the conjugate distribution in Eq. (2.60), we can derive the function of mean $\boldsymbol{\mu}$ and $\boldsymbol{r}$ as follows:

$$\begin{aligned} &p(\boldsymbol{\gamma}_1, \boldsymbol{\Gamma}_2|\boldsymbol{\nu}_1, \mathbf{N}_2, \phi_1, \phi_2) \\ &\quad\propto \exp\left(\boldsymbol{\gamma}_1^\mathsf{T}\boldsymbol{\nu}_1 + \mathrm{tr}[\boldsymbol{\Gamma}_2^\mathsf{T}\mathbf{N}_2] - \phi_1 g_1(\boldsymbol{\gamma}_1, \boldsymbol{\Gamma}_2) - \phi_2 g_2(\boldsymbol{\gamma}_1, \boldsymbol{\Gamma}_2)\right) \\ &\quad\propto \exp\left(\boldsymbol{\mu}^\mathsf{T}\mathbf{R}\boldsymbol{\nu}_1 + \mathrm{tr}[\mathbf{R}^\mathsf{T}\mathbf{N}_2] - \frac{N\phi_1}{2}\boldsymbol{\gamma}_1^\mathsf{T}\boldsymbol{\Gamma}_2^{-1}\boldsymbol{\gamma}_1 - \frac{N\phi_2}{2}\log(2\pi|\boldsymbol{\Gamma}_2|^{-1})\right) \\ &\quad\propto \exp\left(\boldsymbol{\mu}^\mathsf{T}\mathbf{R}\boldsymbol{\nu}_1 + \mathrm{tr}[\mathbf{R}\mathbf{N}_2] - \frac{N\phi_1}{2}\boldsymbol{\mu}^\mathsf{T}\mathbf{R}\boldsymbol{\mu} - \frac{N\phi_2}{2}\log(|\mathbf{R}|^{-1})\right), \end{aligned} \tag{2.103}$$

where we omit the factor that does not depend on $\mathbf{R}$ and $\boldsymbol{\mu}$. Similarly to Example 2.6, we first use a complete square form of $\boldsymbol{\mu}$ to derive a Gaussian distribution from Eq. (2.103).

In Appendix B.4, we have the following formula for the complete square form of vectors:

$$\mathbf{x}^\mathsf{T}\mathbf{A}\mathbf{x} - 2\mathbf{x}^\mathsf{T}\mathbf{b} + c = (\mathbf{x} - \mathbf{u})^\mathsf{T}\,\mathbf{A}\,(\mathbf{x} - \mathbf{u}) + v, \tag{2.104}$$

where

$$\begin{aligned} \mathbf{u} &\triangleq \mathbf{A}^{-1}\mathbf{b} \\ v &\triangleq c - \mathbf{b}^\mathsf{T}\mathbf{A}^{-1}\mathbf{b}. \end{aligned} \tag{2.105}$$

Therefore, by $\mathbf{x} \rightarrow \boldsymbol{\mu}$, $\mathbf{A} \rightarrow N\phi_1\mathbf{R}$, and $\mathbf{b} \rightarrow \mathbf{R}\boldsymbol{\nu}_1$ in Eqs. (2.104) and (2.105), Eq. (2.103) is rewritten as follows:

$$p(\boldsymbol{\gamma}_1, \boldsymbol{\Gamma}_2|\boldsymbol{\nu}_1, \mathbf{N}_2, \phi_1, \phi_2) \propto |\mathbf{R}|^{\frac{N\phi_2}{2}} \exp\left(-\frac{N\phi_1}{2}\left(\boldsymbol{\mu} - \frac{\boldsymbol{\nu}_1}{N\phi_1}\right)^{\mathsf{T}} \mathbf{R}\left(\boldsymbol{\mu} - \frac{\boldsymbol{\nu}_1}{N\phi_1}\right) + \underbrace{\frac{\boldsymbol{\nu}_1^{\mathsf{T}}\mathbf{R}\boldsymbol{\nu}_1}{2N\phi_1} + \text{tr}[\mathbf{R}\mathbf{N}_2]}_{(*)}\right). \tag{2.106}$$

Now we focus on the $(*)$ term in Eq. (2.106). By using the matrix formula in Appendix B, $(*)$ is rewritten as

$$\begin{aligned}(*) &= \text{tr}\left[\frac{\boldsymbol{\nu}_1\boldsymbol{\nu}_1^{\mathsf{T}}\mathbf{R}}{2N\phi_1} + \mathbf{N}_2\mathbf{R}\right] \\ &= \text{tr}\left[\left(\frac{\boldsymbol{\nu}_1\boldsymbol{\nu}_1^{\mathsf{T}}}{2N\phi_1} + \mathbf{N}_2\right)\mathbf{R}\right].\end{aligned} \tag{2.107}$$

Thus, the conjugate prior distribution is rewritten as:

$$p(\boldsymbol{\gamma}_1, \boldsymbol{\Gamma}_2|\boldsymbol{\nu}_1, \mathbf{N}_2, \phi_1, \phi_2) \propto |\mathbf{R}|^{\frac{N\phi_2}{2}} \exp\left(-\frac{N\phi_1}{2}\left(\boldsymbol{\mu} - \frac{\boldsymbol{\nu}_1}{N\phi_1}\right)^{\mathsf{T}} \mathbf{R}\left(\boldsymbol{\mu} - \frac{\boldsymbol{\nu}_1}{N\phi_1}\right) + \text{tr}\left[\left(\frac{\boldsymbol{\nu}_1\boldsymbol{\nu}_1^{\mathsf{T}}}{2N\phi_1} + \mathbf{N}_2\right)\mathbf{R}\right]\right). \tag{2.108}$$

Therefore, Eq. (2.108) is represented as the following Gaussian–Wishart distribution in Appendix C.15:

$$\begin{aligned}&\mathcal{NW}(\boldsymbol{\mu}, \mathbf{R}|\boldsymbol{\mu}^0, \phi^{\mu}, \mathbf{R}^0, \phi^{\mathbf{R}}) \\ &\triangleq C_{\mathcal{NW}}(\phi^{\mu}, \mathbf{R}^0, \phi^{\mathbf{R}})|\mathbf{R}|^{\frac{\phi^{\mathbf{R}}-D}{2}} \\ &\quad \times \exp\left(-\frac{1}{2}\text{tr}\left[\mathbf{R}^0\mathbf{R}\right] - \frac{\phi^{\mu}}{2}(\boldsymbol{\mu} - \boldsymbol{\mu}^0)^{\mathsf{T}}\mathbf{R}(\boldsymbol{\mu} - \boldsymbol{\mu}^0)\right),\end{aligned} \tag{2.109}$$

where

$$\begin{cases} \phi^{\mu} = N\phi_1 \\ \boldsymbol{\mu}^0 = \dfrac{\boldsymbol{\nu}_1}{N\phi_1} \\ \phi^{\mathbf{R}} = N\phi_2 + D \\ \mathbf{R}^0 = -\dfrac{\boldsymbol{\nu}_1\boldsymbol{\nu}_1^{\mathsf{T}}}{N\phi_1} - 2\mathbf{N}_2. \end{cases} \tag{2.110}$$

Thus, we can derive the prior distribution as the Gaussian–Wishart distribution.

Now, we focus on the posterior distribution of $\boldsymbol{\mu}$ and $\mathbf{R}$. Similarly, the posterior distribution is represented as the same form of the Gaussian–Wishart distribution as the prior distribution (2.109), with hyperparameters $\hat{\phi}^{\mu}$, $\hat{\boldsymbol{\mu}}$, $\hat{\phi}^{\mathbf{R}}$, and $\hat{\mathbf{R}}$ as follows:

$$p(\boldsymbol{\mu}, \mathbf{R}|X) = \mathcal{NW}(\boldsymbol{\mu}, \mathbf{R}|\hat{\boldsymbol{\mu}}, \hat{\phi}^{\mu}, \hat{\mathbf{R}}, \hat{\phi}^{\mathbf{R}}). \tag{2.111}$$

Based on the conjugate distribution rule, Eq. (2.64), the hyperparameters of the posterior distribution are easily solved by just replacing $\boldsymbol{\nu}_1 \to \boldsymbol{\nu}_1 + \mathbf{t}(\mathbf{X})$, $\mathbf{N}_2 \to \mathbf{N}_2 + \mathbf{T}(\mathbf{X})$, and $\phi_m \to \phi_m + 1$ in Eq. (2.110) without complex calculations, as follows:

$$
\begin{aligned}
\hat{\phi}^{\mu} &= N(\phi_1 + 1) = \phi^{\mu} + N, \\
\hat{\boldsymbol{\mu}} &= \frac{\boldsymbol{\nu}_1 + \sum_{n=1}^{N} \mathbf{x}_n}{N(\phi_1 + 1)} = \frac{\phi^{\mu}\boldsymbol{\mu}^0 + \sum_{n=1}^{N} \mathbf{x}_n}{\phi^{\mu} + N}, \\
\hat{\phi}^{\mathbf{R}} &= N(\phi_2 + 1) + D = \phi^{\mathbf{R}} + N, \\
\hat{\mathbf{R}} &= -\frac{\left(\boldsymbol{\nu}_1 + \sum_{n=1}^{N} \mathbf{x}_n\right)\left(\boldsymbol{\nu}_1 + \sum_{n=1}^{N} \mathbf{x}_n\right)^{\mathsf{T}}}{N(\phi_1 + 1)} - 2\left(\mathbf{N}_2 - \frac{1}{2}\sum_{n=1}^{N} \mathbf{x}_n\mathbf{x}_n^{\mathsf{T}}\right) \\
&= -\hat{\phi}^{\mu}\hat{\boldsymbol{\mu}}\hat{\boldsymbol{\mu}}^{\mathsf{T}} + \frac{\boldsymbol{\nu}_1\boldsymbol{\nu}_1^{\mathsf{T}}}{N\phi_1} + \mathbf{R}^0 + \sum_{n=1}^{N} \mathbf{x}_n\mathbf{x}_n^{\mathsf{T}} \\
&= -\hat{\phi}^{\mu}\hat{\boldsymbol{\mu}}\hat{\boldsymbol{\mu}}^{\mathsf{T}} + \phi^{\mu}\boldsymbol{\mu}\boldsymbol{\mu}^{\mathsf{T}} + \mathbf{R}^0 + \sum_{n=1}^{N} \mathbf{x}_n\mathbf{x}_n^{\mathsf{T}}.
\end{aligned}
\tag{2.112}
$$

Thus, we derive the posterior distribution that is also represented as a Gaussian–Wishart distribution. Finally, the prior and posterior distributions of $\boldsymbol{\mu}$ and $\mathbf{R}$ in Eqs. (2.109) and (2.111) are also summarized as:

$$
\begin{cases}
p(\boldsymbol{\mu}, \mathbf{R}) = \mathcal{NW}(\boldsymbol{\mu}, \mathbf{R}|\boldsymbol{\mu}^0, \phi^{\mu}, \mathbf{R}^0, \phi^{\mathbf{R}}) \\
p(\boldsymbol{\mu}, \mathbf{R}|X) = \mathcal{NW}(\boldsymbol{\mu}, \mathbf{R}|\hat{\boldsymbol{\mu}}, \hat{\phi}^{\mu}, \hat{\mathbf{R}}, \hat{\phi}^{\mathbf{R}}),
\end{cases}
\tag{2.113}
$$

or

$$
\begin{cases}
p(\boldsymbol{\mu}, \mathbf{R}) = p(\boldsymbol{\mu}|\mathbf{R})p(\mathbf{R}) = \mathcal{N}(\boldsymbol{\mu}|\boldsymbol{\mu}^0, (\phi^{\mu}\mathbf{R})^{-1})\mathcal{W}(\mathbf{R}|\phi^{\mathbf{R}}, \mathbf{R}^0) \\
p(\boldsymbol{\mu}, \mathbf{R}|X) = p(\boldsymbol{\mu}|\mathbf{R}, X)p(\mathbf{R}|X) = \mathcal{N}(\boldsymbol{\mu}|\hat{\boldsymbol{\mu}}, (\hat{\phi}^{\mu}\mathbf{R})^{-1})\mathcal{W}(\mathbf{R}|\hat{\phi}^{\mathbf{R}}, \hat{\mathbf{R}}).
\end{cases}
\tag{2.114}
$$

Example 2.8 Conjugate distributions for multinomial distribution:
Based on the discussion of Eq. (2.50) in Example 2.4, the canonical form of the multivariate Gaussian distribution is represented as follows:

$$
\mathrm{Mult}(x_1, \cdots, x_J|\omega_1, \cdots, \omega_J) = h(x)\exp\left(\boldsymbol{\gamma}^{\mathsf{T}}\mathbf{t}(\mathbf{x})\right),
\tag{2.115}
$$

where

$$
\begin{cases}
\mathbf{t}(\mathbf{x}) = \mathbf{x} \\
h(x) = \dfrac{N!}{\prod_{j=1}^{J} x_j!} \\
\boldsymbol{\gamma} = \left[\log \frac{\omega_1}{1-\sum_{j=1}^{J-1}\omega_j}, \cdots, \log \frac{\omega_{J-1}}{1-\sum_{j=1}^{J-1}\omega_j}\right]^{\mathsf{T}} \\
g(\boldsymbol{\gamma}) = N\log\left(1 + \sum_{j=1}^{J-1}\exp(\gamma_j)\right).
\end{cases}
\tag{2.116}
$$

Note that we have the following constraints:

$$\sum_{j=1}^{J} x_j = N$$
$$\sum_{j=1}^{J} \omega_j = 1. \tag{2.117}$$

Therefore, by substituting $\boldsymbol{\gamma}$, $g_1(\boldsymbol{\gamma})$, and $g_2(\boldsymbol{\gamma})$ in Eq. (2.78) into the general form of the conjugate distribution in Eq. (2.56), we can derive the function of $\boldsymbol{\gamma}$ as follows:

$$\begin{aligned}
p(\boldsymbol{\gamma}|\boldsymbol{\nu}, \phi) &\propto \exp\left(\boldsymbol{\gamma}^{\mathsf{T}}\boldsymbol{\nu} - \phi g(\boldsymbol{\gamma})\right) \\
&= \exp\left(\left[\log \frac{\omega_1}{1-\sum_{j=1}^{J-1}\omega_1}, \cdots, \log \frac{\omega_{J-1}}{1-\sum_{j=1}^{J-1}\omega_j}\right]\boldsymbol{\nu} + N\phi \log\left(1 - \sum_{j=1}^{J-1}\omega_j\right)\right) \\
&= \exp\left(\left[\log \omega_1, \cdots, \log \omega_J\right]\left[\boldsymbol{\nu}^{\mathsf{T}}, N\phi - \textstyle\sum_{j=1}^{J-1}\nu_j\right]^{\mathsf{T}}\right) \\
&= \prod_{j=1}^{J} (\omega_j)^{\phi_j^{\omega}-1},
\end{aligned} \tag{2.118}$$

where hyperparameters $\{\phi_j^{\omega}\}_{j=1}^{J}$ are defined as follows:

$$\begin{aligned}
\phi_j^{\omega} &\triangleq \nu_j + 1 \text{ for } j = 1, \cdots, J-1 \\
\phi_J^{\omega} &\triangleq N\phi - \sum_{j=1}^{J-1}\nu_j + 1.
\end{aligned} \tag{2.119}$$

Thus, the conjugate prior distribution is represented as a Dirichlet distribution defined in Appendix C.4 as

$$\mathrm{Dir}(\{\omega_j\}_{j=1}^{J}|\{\phi_j^{\omega}\}_{j=1}^{J}) \triangleq \frac{\Gamma(\sum_{j=1}^{J}\phi_j^{\omega})}{\prod_{j=1}^{J}\Gamma(\phi_j^{\omega})}\prod_{j=1}^{J}(\omega_j)^{\phi_j^{\omega}-1}. \tag{2.120}$$

Based on the conjugate distribution rule, Eq. (2.63), the hyperparameters of the posterior distribution are easily solved by just replacing $\boldsymbol{\nu} \to \boldsymbol{\nu} + \mathbf{t}(\mathbf{x})$ and $\phi \to \phi + 1$ in Eq. (2.118) without complex calculations, as follows:

$$p(\boldsymbol{\gamma}|X) \to \mathrm{Dir}(\{\omega_j\}_{j=1}^{J}|\{\hat{\phi}_j^{\omega}\}_{j=1}^{J}), \tag{2.121}$$

where hyperparameters $\{\hat{\phi}_j^{\omega}\}_{j=1}^{J}$ are obtained as follows:

$$\begin{aligned}
\hat{\phi}_j^{\omega} &\triangleq \nu_j + x_j + 1 = \phi_j^{\omega} + x_j \\
\hat{\phi}_J^{\omega} &\triangleq N(\phi + 1) - \sum_{j=1}^{J-1}(\nu_j + x_j) + 1 \\
&= N\phi + x_J - \sum_{j=1}^{J-1}\nu_j + 1 \\
&= \phi_J^{\omega} + x_J.
\end{aligned} \tag{2.122}$$

Thus, we derive the posterior distribution that is represented as a Dirichlet distribution. Finally, the prior and posterior distributions of ω are given as:

$$\begin{cases} p(\{\omega_j\}_{j=1}^J) = \mathrm{Dir}\left(\{\omega_j\}_{j=1}^J \middle| \{\phi_j^\omega\}_{j=1}^J\right) \\ p(\{\omega_j\}_{j=1}^J | X) = \mathrm{Dir}\left(\{\omega_j\}_{j=1}^J \middle| \{\hat{\phi}_j^\omega\}_{j=1}^J\right). \end{cases} \tag{2.123}$$

Table 2.1 shows a recipe of the kind of distributions we use as a conjugate prior.

This section provides a solution of the posterior distribution for rather simple statistical models. However, in practical applications, we still face the problems of solving the equations, and often require the approximation to solve them efficiently. The next section explains a powerful approximation method, conditional independence, in Bayesian probabilities.

2.1.5 Conditional independence

Another important mathematical operation of the Bayesian approach, as well as the product and sum rules (Section 2.1.1), is called *conditional independence*. Let a, b, and c be probabilistic variables, the conditional independence of a and b on c is represented as follows:

$$p(a, b|c) = p(a|b, c)p(b|c) = p(a|c)p(b|a, c), \tag{2.124}$$

$$\approx p(a|c)p(b|c). \tag{2.125}$$

This is a useful assumption for the Bayesian approach when factorizing the joint probability distribution. For example, Eq. (2.124) based on the product rule needs to consider $p(b|a, c)$ or $p(a|b, c)$. Suppose a, b, and c are discrete elements of sets, i.e., $a \in \mathcal{A}$, $b \in \mathcal{B}$, and $c \in \mathcal{C}$, $p(b|a, c)$ or $p(a|b, c)$ considers the probability of all combinations of a, b, and c, which correspond to $|\mathcal{A}| \times |\mathcal{B}| \times |\mathcal{C}|$. The number of combinations is increased exponentially, if the number of valuables is increased. Therefore, it is computationally very expensive to obtain the conditional distribution, and almost impossible to consider

Table 2.1 Conjugate priors.

Likelihood function	Unknown variable	Conjugate prior
Gaussian	$\mu \in \mathbb{R}$	Gaussian C.5
Gaussian	$r \in \mathbb{R}_{>0}$	Gamma C.11
Gaussian	μ, r	Gaussian–gamma C.13
Multivariate Gaussian	$\boldsymbol{\mu} \in \mathbb{R}^D$	Multivariate Gaussian C.6
Multivariate Gaussian	$\mathbf{R} \in \mathbb{R}^{D \times D}$	Wishart C.14
Multivariate Gaussian	$\boldsymbol{\mu}, \mathbf{R}$	Gaussian–Wishart C.15
Multinomial	$\omega_i \in [0, 1], \sum_i \omega_i = 1$	Dirichlet C.4

large amounts of data as probabilistic valuables. Thus, the conditional independence approximation in Eq. (2.125) greatly reduces the computational complexity, and makes the Bayesian treatment of speech and language processing tractable.

By using the product rule, the conditional independence equation is rewritten as follows:

$$p(a|c) \approx \frac{p(a,b|c)}{p(b|c)} = \frac{p(a|b,c)p(b|c)}{p(b|c)} = p(a|b,c). \tag{2.126}$$

Thus,

$$p(a|b,c) \approx p(a|c) \tag{2.127}$$

is also equivalently used as the conditional independence assumption of Eq. (2.125).

The conditional independence is often used in the following sections to make the complicated relationship between probabilistic variables simple. For example, speech recognition has many probabilistic variables which come from acoustic and language models. It is very natural and effective to assume conditional independence between acoustic model and language model variables because these do not depend on each other explicitly.

Example 2.9 Naive Bayes classifier:
One of the simplest classifiers in the machine learning approach is the naive Bayes classifier. The approach is used for many applications including document classification (Lewis 1998, McCallum & Nigam 1998). For example, if we have N data $(x_1, x_2, \cdots x_N)$, and want to classify the data to a specific category $\hat{c}$, this can be performed by using the posterior distribution of category c as follows:

$$\hat{c} = \arg\max_{c} p(c|\{x_n\}_{n=1}^N). \tag{2.128}$$

The naive Bayes classifier approximates this posterior distribution with the product rule and conditional independence assumption as follows:

$$\begin{aligned} p(c|\{x_n\}_{n=1}^N) &\propto p(\{x_n\}_{n=1}^N|c)p(c) \\ &\approx \prod_{n-1}^{N} p(x_n|c)p(c). \end{aligned} \tag{2.129}$$

This approach approximates the posterior distribution $p(c|\{x_n\}_{n=1}^N)$ with the product of likelihood $p(x_n|c)$ for all samples and prior distribution $p(c)$. Since the naive Bayes classifier is very simple and easy to implement, it is often used as an initial attempt of the machine learning approach if we have training data with labels to obtain $p(x_n|c)$ for all c. For example, in document classification, a multinomial distribution is used to represent the likelihood function $p(x_n|c)$.

2.2 Graphical model representation

The previous sections (especially Sections 2.1.1 and 2.1.5) discuss how to provide the mathematical relationship between probabilistic variables in a Bayesian manner. This section briefly introduces a graphical model representation that visualizes the relationship between these probabilistic valuables to provide a more intuitive way of understanding the model. A graphical model framework is also widely used in Bayesian machine learning studies, and this book introduces basic graphical model descriptions, which are used in the following sections.

2.2.1 Directed graph

First, we simply consider the following joint distribution of a and b, which can be rewritten as the following two factorization forms based on the product rule:

$$p(a, b) = p(b|a)p(a), \tag{2.130}$$

$$= p(a|b)p(b). \tag{2.131}$$

Therefore, to obtain the joint distribution, we compute either Eq. (2.130) or (2.131) depending on the problem. The graphical model can separately represent these factorization forms intuitively. Figure 2.1 represents the graphical models of $p(b|a)p(a)$ and $p(a|b)p(b)$, respectively. The node represents a probabilistic variable, and the directed link represents the conditional dependency of two probabilistic variables. For example, Eq. (2.130) is composed of the conditional distribution $p(b|a)$ and then the corresponding graphical representation provides the directed link from node a to node b in Figure 2.1(a). Conversely, the conditional distribution $p(a|b)$ in Eq. (2.131) is represented by the directed link from node b to node a in Figure 2.1(b).

Thus, the graphical model specifies a unique factorization form of a joint distribution, intuitively. The graph composed of the directed link, which represents the conditional distribution, is called a *directed graph*. The graphical model can also deal with an *undirected graph*, which is a graphical representation of a Markov random field, but this book focuses on the directed graph representation, which is often used in the later applications.

2.2.2 Conditional independence in graphical model

As we discussed in Section 2.1.5, practical applications often need some approximations in the dependency of probabilistic variables to avoid a complicated dependency of the

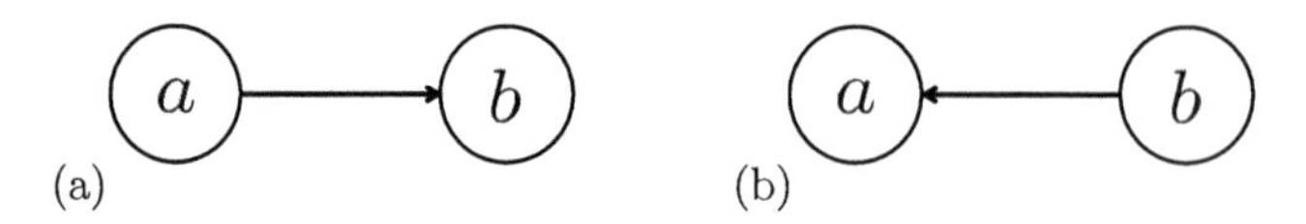

Figure 2.1 Graphical models of $p(b|a)p(a)$ and $p(a|b)p(b)$.

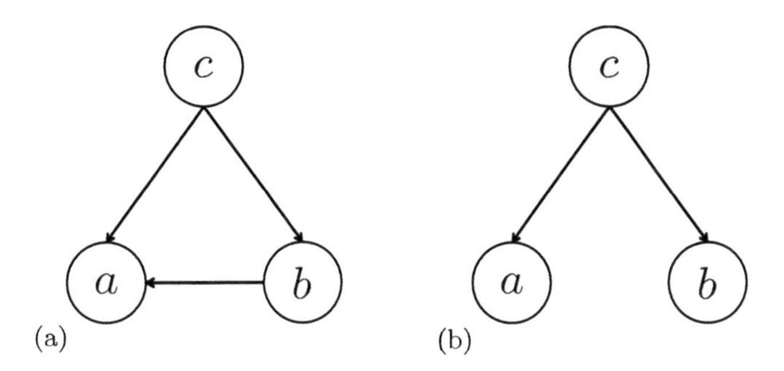

Figure 2.2 Graphical models of $p(a|b,c)p(b|c)p(c)$ and $p(a|c)p(b|c)p(c)$.

factorized distribution. We can represent this approximation in the graphical model representation. If we consider the joint distribution of a, b, and c, the joint distribution is, for example, represented as the following factorization form based on the product rule:

$$p(a,b,c) = p(a|b,c)p(b|c)p(c). \tag{2.132}$$

The graphical model of Eq. (2.132) is represented in Figure 2.2(a). Note that all nodes are connected to each other by directed links. This graph is called a *full connected graph.*

On the other hand, the joint distribution with the following conditional independence can also be represented as a graphical model in Figure 2.2(b):

$$p(a,b,c) = p(a,b|c)p(c) \approx p(a|c)p(b|c)p(c). \tag{2.133}$$

Note that the link between a and b has disappeared from Figure 2.2(b). Thus, the conditional independence in the graphical model is represented by pruning links in the graphs, which corresponds to reducing the dependencies in probabilistic variables, and leads to reduced computational cost.

In real applications, we need to consider large numbers of variables. For example, the naive Bayes classifier introduced in Example 2.9 has to consider $N+1$ probabilistic variables ($\{x_n\}_{n=1}^N$ and c):

$$p(x_1, \cdots, x_N|c)p(c) \approx p(x_1|c) \cdots p(x_N|c)p(c) = \prod_{n=1}^{N} p(x_n|c)p(c). \tag{2.134}$$

The graphical model of this case can be simplified from Figure 2.3(a) to 2.3(b) by using the plate. Based on the plate, we can represent a complicated relationship of probabilistic variables intuitively. In Section 8.2, we also consider the case when the number of probabilistic variables is dealt with as infinite in Bayesian nonparametrics. Then, the number of variables can be represented by using ∞ in a graphical model, as shown in Figure 2.4.

Thus, the graphical model can represent the dependencies of variables based on the product rule and conditional independence graphically. This dependency-network-based Bayesian method is also called a Bayesian network. In particular the Bayesian treatment that considers the dynamical relationship between probabilistic variables is also called a dynamic Bayesian network (Ghahramani 1998, Murphy 2002). A dynamic Bayesian

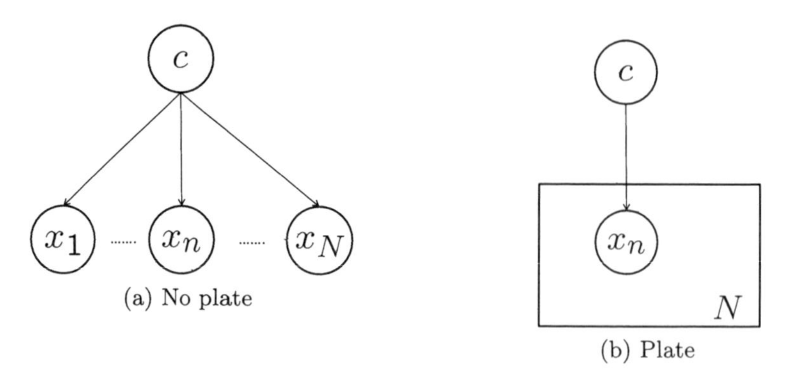

Figure 2.3 Graphical model of $p(x_1|c) \cdots p(x_N|c) = \prod_{n=1}^{N} p(x_n|c)$.

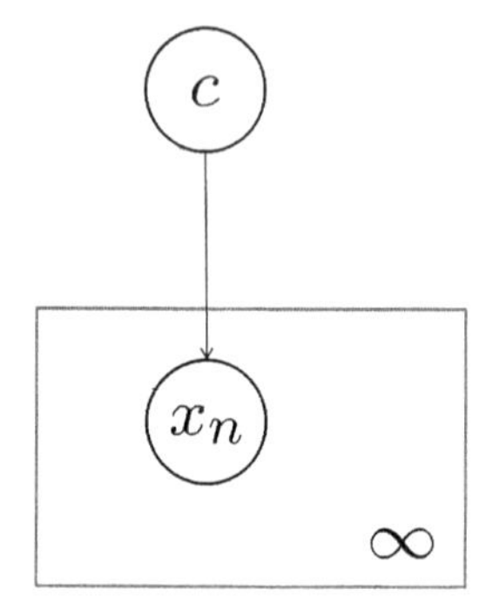

Figure 2.4 Graphical model of $p(x_1|c) \cdots p(x_\infty|c) = \prod_{n=1}^{\infty} p(x_n|c)$.

network provides efficient solutions to the time-series statistical models similarly to HMM and Kalman filters, which are also used in speech recognition (Zweig & Russell 1998, Nefian, Liang, Pi *et al.* 2002, Livescu, Glass & Bilmes 2003). It is helpful to understand probabilistic models, even when they are very complicated in the equation form.

2.2.3 Observation, latent variable, non-probabilistic variable

Previous sections deal with the graphical model of all probabilistic variables. However, our machine learning problems for speech and language processing have three types of variables: observation, latent variables, and non-probabilistic variables. For example, let x be an observation, z is a latent variable, and θ is a model parameter, which we don't deal with as a probabilistic variable in this section, unlike the full Bayesian approach. The probability distribution of x is represented as follows:

$$p(x|\theta) = \sum_{z} p(x, z|\theta) = \sum_{z} p(x|z, \theta)p(z|\theta). \tag{2.135}$$

The corresponding graphical model is represented in Figure 2.5(a). Note that three variables x, z, θ have different roles in this equation. For example, x is a final output of this

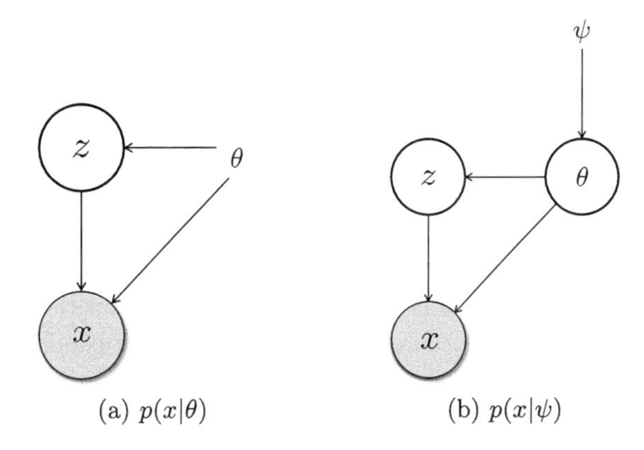

Figure 2.5 Graphical models that have observation x, latent variable z, model parameter θ, and hyperparameter ψ. Part (a) treats θ as a non-probabilistic variable, and (b) treats θ as a probabilistic variable to be marginalized.

equation as an observation, which is not marginalized, while z is a latent variable and should be marginalized. To distinguish the observation and latent variables, the node representing an observation is tinted. θ is not a probabilistic variable in this explanation, and so it is put in the graph *without a circle.*

Similarly, if we consider the same model, but treat θ as a probabilistic variable, θ is marginalized by the prior distribution of θ with hyperparameter ψ. The probability distribution of x is represented as follows:

$$
\begin{aligned}
p(x|\psi) &= \int \sum_z p(x, z, \theta|\psi) d\theta \\
&= \int \sum_z p(x|z, \theta) p(z|\theta) p(\theta|\psi) d\theta.
\end{aligned} \tag{2.136}
$$

Here we assume θ to be a continuous variable, and use the integral instead of the summation. We can regard θ as a latent variable in a broad sense, but the other sections distinguish the model parameters and latent variables. The corresponding graphical model is represented in Figure 2.5(b). Thus, by using the representations of observation, latent variables, and non-probabilistic variables, we can provide graphical models of various distributions other than joint distributions. These are basic rules of providing a directed graphical model from the corresponding probabilistic equation.

The directed graph basically describes how observation variables are generated conditioned on the other probabilistic variables. This statistical model of describing the generation of observation variables is called a *Generative model.* HMM, GMM, Kalman filter, n-gram, latent topic model, and deep belief network are typical examples of generative models that can generate speech feature vectors and word sequences. The next section also introduces another way of intuitively understanding our complicated statistical models by describing how observation variables are generated from the distributions in our models.

2.2.4 Generative process

This section also explains another representation of the Bayesian approach based on the generative process. This representation is used to generate the probabilistic variables in an algorithmic way. The generative process is used to express the joint distribution. The basic syntax of the generative process is as follows:

- Non-probabilistic variables: placed as "require";
- Latent variables: "drawn" from their probability distribution;
- Model parameters: "drawn" from their (prior) probability distribution;
- Observations: finally "drawn" from their probability distribution given sampled latent variables and model parameters.

If we also want to represent the marginalization of a probabilistic variable, we can use an additional syntax "Average" for the marginalization.

As an example of Eq. (2.136), Algorithm 1 represents the generative process of the joint distribution $p(x, z, \theta | \psi)$, which is represented as:

$$p(x, z, \theta | \psi) = p(x|z, \theta)p(z|\theta)p(\theta | \psi), \tag{2.137}$$

where x, z, θ, and ψ are observations, latent variables, model parameters, and hyperparameter (non-probabilistic variables), respectively.

Algorithm 1 Generative process of $p(x, z, \theta | \psi) = p(x|z, \theta)p(z|\theta)p(\theta | \psi)$

Require: ψ
1: Draw θ from $p(\theta | \psi)$
2: Draw z from $p(z|\theta)$
3: Draw x from $p(x|z, \theta)$

This generative process also helps us to understand models intuitively by understanding how probabilistic variables are generated algorithmically. Therefore, both the generative process and graphical model are often provided in the Bayesian approach to represent a complicated generative model. In some of the statistical models used in this book, we provide the generative process and graphical model to allow readers to understand the models intuitively.

2.2.5 Undirected graph

Another example of a graphical model is called an undirected graph (Figure 2.6), that represents the relationship of probabilistic variables but does not have explicit parent–child relationships compared with the directed graph. The network is called a Markov random field, and the probabilistic distribution is usually expressed by a potential function $\psi(a, b, c)$ (a positive, but otherwise arbitrary, real-valued function):

$$p(a, b, c) = \frac{1}{Z}\psi(a, b, c), \tag{2.138}$$

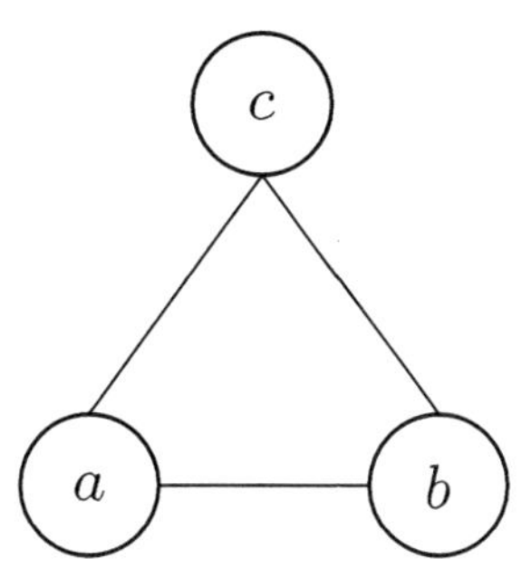

Figure 2.6 Undirected graph.

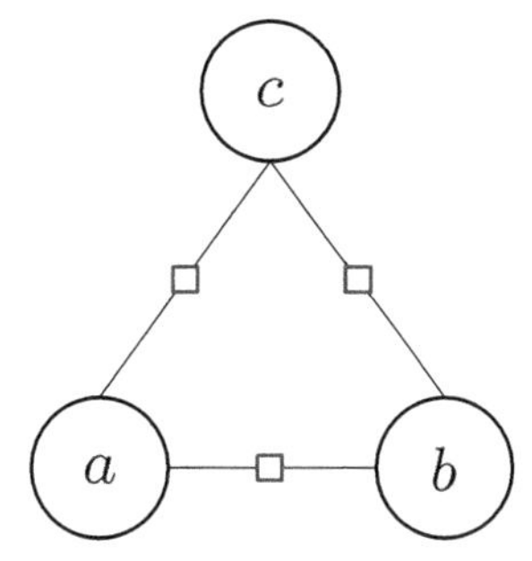

Figure 2.7 Factor graph.

where Z is a normalization constant of this distribution, and is called the *partition function* for this special case. This approach is often used as a context of a log linear discriminative model, where $\psi(a, b, c)$ is a linear function of a feature obtained by a, b, and c and the corresponding weight.

A *factor graph* is another class of graphical model representing the conditional independence relationship between variables. Actually, the factor graph can provide a more concrete representation of the joint distribution of variables than that of the undirected graph. The factor graph introduces additional square nodes to a graph, which can explicitly represent the dependency of several variables.

For example, the partition function can be represented by several cases, as shown in Figures 2.7 and 2.8. Both graphs are fully connected and can represent the joint distribution of a, b, and c. However, the partition function of Figure 2.7 is computed by using the three pairs of partition functions as follows:

$$p(a, b, c) = \frac{1}{Z}\psi(a, b)\psi(b, c)\psi(c, a). \tag{2.139}$$

The possible partition functions are $|\mathcal{A}| \times |\mathcal{B}| + |\mathcal{B}| \times |\mathcal{C}| + |\mathcal{A}| \times |\mathcal{C}|$. On the other hand, Figure 2.8 considers the potential function of the joint event for a, b, and c:

$$p(a, b, c) = \frac{1}{Z}\psi(a, b, c). \tag{2.140}$$

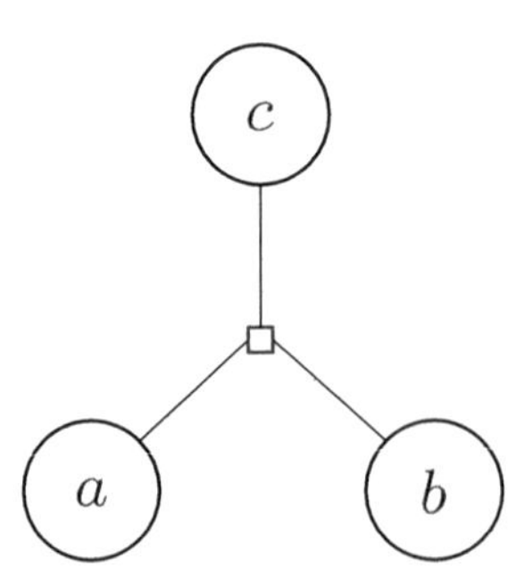

Figure 2.8 Factor graph.

The possible partition functions are $|\mathcal{A}| \times |\mathcal{B}| \times |\mathcal{C}|$. Therefore, if the number of possible variables ($|\mathcal{A}|$, $|\mathcal{B}|$, and $|\mathcal{C}|$) is very large, Figure 2.7 is a more compact representation since the number of possible functions would be smaller.

Thus, factor graphs are more specific about the precise form of the factorization of undirected graphs, and can be used to mainly represent some discriminative models (logistic regression, conditional random field (Lafferty, McCallum & Pereira 2001)). This book generally deals with generative models (HMM, GMM, n-gram, and latent topic model), and does not deal with these discriminative models. However, there are several important applications of discriminative models to speech and language processing (e.g., Gunawardana, Mahajan, Acero *et al.* 2005, Fosler & Morris 2008, Zweig & Nguyen 2009, Gales *et al.* 2012) in addition to the recent trend of deep neural networks (Hinton *et al.* 2012). We present an example of a Bayesian treatment of neural network acoustic models in Section 6.4. The fully Bayesian treatment of the other discriminative models in speech and language processing is an interesting future direction.

2.2.6 Inference on graphs

One of the powerful advantages of the graphical model representation is that once we fix a graphical model, we can infer all variables in the graph efficiently by using belief propagation if the graph does not have a loop.

For example, belief propagation provides a sum product algorithm that can efficiently compute the distribution $p(x_i)$ of the probabilistic variable in an arbitrary node by using message passing. In the HMM case, this sum product algorithm corresponds to the forward–backward algorithm, as discussed in Section 3.3.1. Similarly, belief propagation provides a max sum algorithm that can efficiently compute the arg max value ($\hat{x}_i = \arg\max_{x_i} p(x_i)$) in an arbitrary node by using message passing. Similarly to the sum product algorithm, the max sum algorithm corresponds to the Viterbi algorithm, as discussed in Section 3.3.2. A detailed discussion about the relationship between the forward–backward/Viterbi algorithms in the HMM and these algorithms can be found in Bishop (2006).

However, most of our applications have a loop in a graph, and we cannot use the exact inference based on the above algorithms. The following chapters introduce the approximations of the Bayesian inferences, and especially variational Bayes (VB), as discussed in Chapter 7, and Markov chain Monte Carlo (MCMC), as discussed in

Chapter 8, these being promising approaches to obtain approximate inferences in a graphical model. Actually, progress of the graphical model approach has been linked to the progress of these Bayesian inference techniques.

The other approximated approach to inference in a graphical model that contains cycles or loops is to use the sum-product algorithm for the graph even though there is no guarantee of convergence. This approach is called loopy belief propagation, and it is empirically known that it is convergent in some applications.

2.3 Difference between ML and Bayes

As discussed in previous sections, the Bayesian approach deals with all variables introduced for modeling as probabilistic variables. This is the unique difference between the Bayesian approach and the other standard statistical framework, the Maximum Likelihood (ML) approach. Actually this difference can yield various advantages over ML. This section overviews the advantage of the Bayesian approach over the ML approach in general. We discuss this, along with a general pattern recognition problem, as we consider practical speech and language processing issues in the following chapters.

Let $\mathbf{O}$, $\mathbf{Z}$, Θ, $\boldsymbol{M}$, and $\boldsymbol{W}$ be a set of observation features, latent variables, model parameters, model structure (hyperparameter) variables, and classification categories, respectively, details of which will be introduced in the following chapters. For comparison, we summarize the difference between the approaches in terms of model setting, training, and classification.

- **Model setting**
  - ML:
    Generative model distribution $p(\mathbf{O}, Z|\Theta, M)$.
  - Bayes:
    Generative model distribution $p(\mathbf{O}, Z|\Theta, M)$
    Prior distributions $p(\Theta|M)$ and $p(M)$.

  In addition to the generative model distribution, the Bayesian approach needs to set prior distributions.
- **Training**
  - ML: Point estimation
    $\hat{\Theta}$.
  - Bayes: Distribution estimation
    $p(\Theta|M, \mathbf{O})$ and $p(M|\mathbf{O})$.

  ML point-estimates are given by the optimal values $\hat{\Theta}$ by using the EM algorithm generally when the model has latent variables, while the Bayesian approach estimates posterior distributions. In addition, ML only focuses on model parameters Θ, but the Bayesian approach focuses on both model parameters Θ and model M.

- **Classification**
  - ML:

$$\arg\max_{W'} \sum_{Z'} p(\mathbf{O}', Z'|\hat{\Theta}, \hat{M}, W')p(W'). \quad (2.141)$$

  - Bayes:

$$\arg\max_{W'} \int \sum_{M,Z'} p(\mathbf{O}', Z'|\Theta, M, W')p(\Theta|M, \mathbf{O}, W)p(M|\mathbf{O}, W)p(W')d\Theta. \quad (2.142)$$

  Here, $\hat{\Theta}$ is obtained in the ML training step, and $\hat{M}$ is usually set in advance by an expert or optimized by evaluating the performance of model M using a development set. Equation (2.142) is obtained by the probabilistic sum and product rules and conditional independence, as discussed in the previous sections. Compared with ML, the Bayes approach marginalizes Θ and M through the expectations of the posterior distributions $p(\Theta|M, \mathbf{O}, W)$ and $p(M|\mathbf{O}, W)$, respectively.

Thus, the main differences between ML and Bayes are (i) use of prior distributions, (ii) use of distributions of model M, (iii) expectation with respect to probabilistic variables based on posterior distributions. These differences yield several advantages of the Bayesian approach over ML. The following sections describe the three main advantages.

2.3.1 Use of prior knowledge

First, we describe the most famous Bayesian advantage over ML based on the use of prior knowledge. Figure 2.9 depicts this advantage focusing on the estimation of model parameters (the mean and variance of a Gaussian). The dashed line shows the true distribution, and the solid line shows the estimated Gaussian distributions based on ML and Bayes. If data to be used to estimate parameters are not sufficient and biased to the

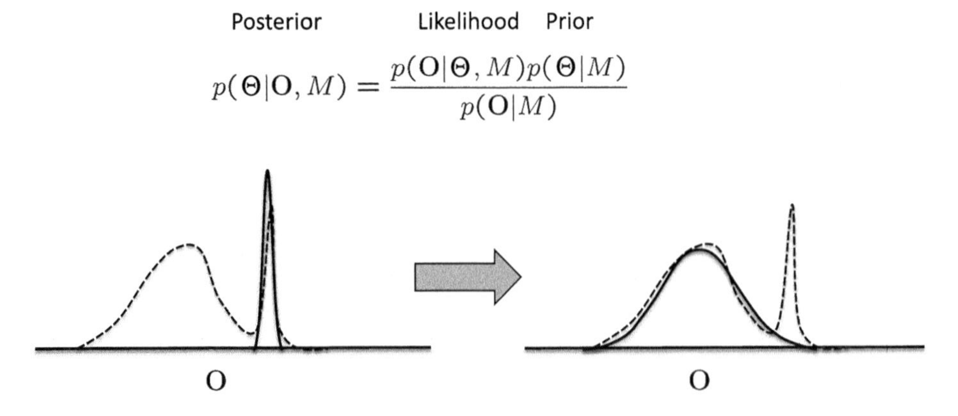

Figure 2.9 Use of prior knowledge.

small peak of the true distribution, ML tends to estimate the wrong parameter by using the biased data. This is because ML estimates only consider the likelihood function, which leads it to estimate the parameters that generate the observed data:

$$\Theta^{\text{ML}} = \arg\max_{\Theta} p(\mathbf{O}|\Theta, M). \tag{2.143}$$

Thus, ML can correctly estimate parameters only when the amount of data is sufficient.

On the other hand, the Bayesian approach also considers the prior distribution of model parameters $p(\Theta|M)$. Now, we consider point estimation using the maximum a-posteriori (MAP) value, instead of considering the Bayesian distribution estimation for simply comparing the prior effect with ML. For example, based on Eq. (3.345), the MAP estimate of Θ is represented as follows:

$$\begin{aligned}\Theta^{\text{MAP}} &= \arg\max_{\Theta} p(\Theta|\mathbf{O}, M) \\ &= \arg\max_{\Theta} p(\mathbf{O}|\Theta, M)p(\Theta|M).\end{aligned} \tag{2.144}$$

The result considers the prior distribution as a regularization term. So if we set a constraint on a distribution form by appropriate prior knowledge, we can recover a wrong estimation due to the sparse data problem in ML, and we can estimate the parameter correctly. Details are discussed in Chapter 4.

2.3.2 Model selection

The model selection is a unique function of the Bayesian approach, which determines a model structure from data automatically. For example, Figure 2.10 shows how many Gaussians we use to estimate the parameters. It is well known that likelihood values always increase as the number of parameters increases. Therefore, if we enforce use of the ML criterion for the model selection, ML tends to select too many Gaussians, which results in over-fitting. There are some extensions of ML to deal with model selection based on information criteria (e.g., Akaike 1974, Rissanen 1984). However, in most cases of speech and language processing, the ML framework usually optimizes model structure by evaluating the performance of the model using a development set. The development set is usually obtained from a part of training/test data. Although this

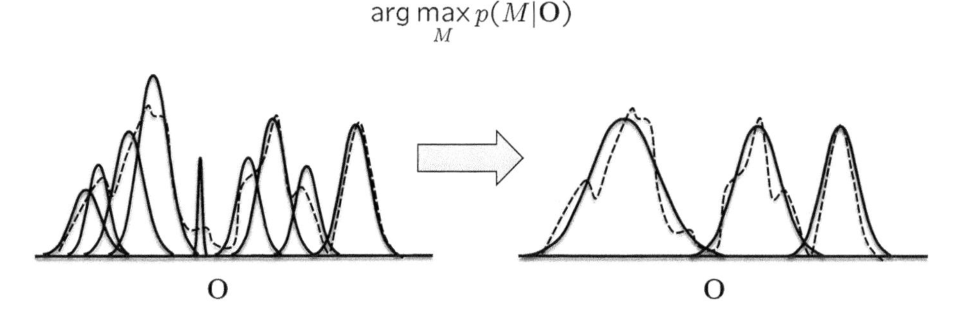

Figure 2.10 Model selection.

optimization is straightforward, it is very difficult to use in some of the applications when the performance evaluation is difficult (e.g., it has a large computational cost for evaluation or there is no objective performance measure).

The Bayesian approach can deal with model selection within the framework. For example, our Bayesian approach to acoustic modeling involves the posterior distribution of model $p(M|\mathbf{O})$, which will be described in Section 3.8.7. Once we obtain the posterior disquisition, we can select an appropriate model structure in terms of the MAP procedure introduced in Section 2.1.2, as follows:

$$M^{\text{MAP}} = \arg\max_{M} p(M|\mathbf{O}). \tag{2.145}$$

Thus, an appropriate model structure (e.g., the topology of HMMs (Stolcke & Omohundro 1993, Watanabe, Minami, Nakamura *et al.* 2004)) can be selected according to training data, without splitting them to create development data.

Instead of using the MAP procedure, we can use expectation based on $p(M|\mathbf{O})$. This is stricter in the Bayesian sense, and Eq. (2.142) actually includes the expectation over the posterior distribution of models. This approach corresponds to using multiple models with different model structures to classify unseen categories. However, in terms of the computational costs (needs large memory and computational time for the multiple model case), people usually carry out model selection by using the MAP procedure.

We also note that M involves other model variations than model structure as elements. For example, hyperparameters introduced in the model can be optimized by using the same MAP procedure or marginalized out by using the expectation. In particular, the optimization of hyperparameters through the posterior distributions of M is also a powerful example of the Bayesian advantage over ML.

2.3.3 Marginalization

The final Bayesian advantage over ML is the marginalization effect, which was discussed in the expectation effect of the Bayesian approach in the previous section. Since the stochastic fluctuation in the expectation absorbs estimation errors, the marginalization improves the robustness in estimation, classification, and regression over unknown data. In Figure 2.11, the left figure shows the maximum likelihood based distribution with $\hat{\Theta}$, while the right figure shows the example of marginalization over model

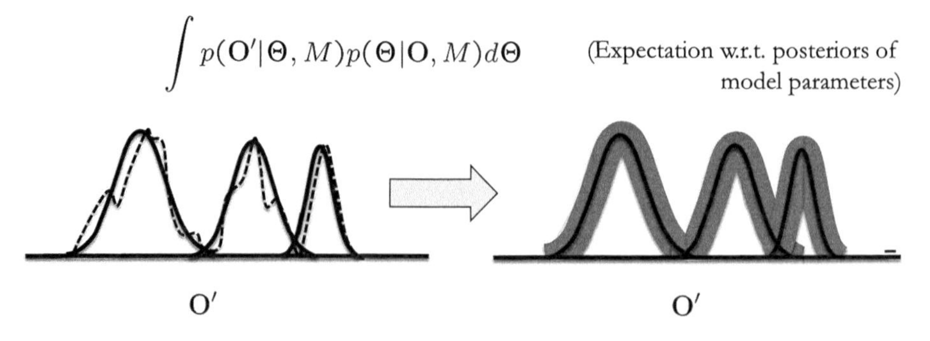

Figure 2.11 Marginalization over a set of model parameters Θ.

parameters Θ in the likelihood function $p(\mathbf{O}'|\Theta, M)$ given a model structure M. Since the right figure considers the probabilistic fluctuations of Θ by the posterior distribution $p(\Theta|\mathbf{O}, M)$, the marginalized function (expected with respect to Θ) can mitigate the error effects in estimating Θ with the variance, and make the likelihood function robust to unseen data. The marginalization can be performed to all probabilistic variables in a model including latent variables Z, model structure M (and hyperparameters), in addition to the model parameter Θ example in Figure 2.11, if we obtain the prior/posterior distributions of these probabilistic variables.

The marginalization is another unique advantage of the Bayesian approach over ML, whereby incorporating the uncertainty of variables introduced in a model based on probabilistic theory achieves robustness for unseen data. However, it requires an expectation with respect to variables that essentially needs to consider the integral or summations over the variables. Again, this is the main difficulty of the practical Bayesian approach, and it needs some approximations especially to utilize the Bayesian advantage of this marginalization effect.

Although marginalization is not usually performed for observation $\mathbf{O}$, observation features in speech and language processing often include noises, and marginalization over observation features is effective for some applications. For example, if we use a speech enhancement technique as a front-end denoising process of automatic speech recognition, the process inevitably includes noise estimation errors, and the errors can be propagated to speech recognition, which degrades the performance greatly. The approach called *uncertainty techniques* tries to mitigate the errors by using the following Bayesian marginalization of the conventional continuous-density HMM (CDHMM) likelihood function over observation features $\mathbf{O}$ (Droppo, Acero & Deng 2002, Delcroix, Nakatani & Watanabe 2009, Kolossa & Haeb-Umbach 2011):

$$p(\Theta, \Psi_{\mathbf{O}'}^{\text{uns}}, M) \approx \int p(\mathbf{O}'|\Theta, M) p(\mathbf{O}'|\Psi_{\mathbf{O}'}^{\text{uns}}) d\mathbf{O}'. \tag{2.146}$$

The main challenges of the uncertainty techniques are how to estimate feature uncertainties $\Psi_{\mathbf{O}'}^{\text{uns}}$ (the distribution of observation features $p(\mathbf{O}|\Psi_{\mathbf{O}'}^{\text{uns}})$ with hyperparameter $\Psi_{\mathbf{O}'}^{\text{uns}}$) and how to integrate the marginal likelihood function with the decoding algorithm of the HMM. The approaches have been successfully applied to noisy speech recognition tasks, and show improvements by mitigating the error effects in speech enhancement techniques (Barker, Vincent, Ma *et al.* 2013, Vincent, Barker, Watanabe *et al.* 2013).

Thus, we have explained the three main practical advantages of the Bayesian approaches. Note that all of the advantages are based on the posterior distributions, and obtaining the posterior distributions for our target applications is a main issue of the Bayesian approaches. Once we obtain the posterior distributions, Bayesian inference allows us to achieve robust performance for our applications.

2.4 Summary

This chapter introduces the selected Bayesian approaches used for speech and language processing by starting from the basic Bayesian probabilistic theory with graphical models, and concludes with a summarization of the Bayesian advantages over ML. The discussion is rather general, and to apply Bayesian approaches to our practical problems in speech and language processing, we still need to bridge a gap between the theoretical Bayesian approaches and these practical problems. This is the main goal of this book. The next chapter deals with basic statistical models used in speech and language processing based on ML, and it will be extended in the latter chapters toward this main goal.

3 Statistical models in speech and language processing

This chapter focuses on basic statistical models (Gaussian mixture models (GMM), hidden Markov models (HMM), n-gram models and latent topic models), which are widely used in speech and language processing. These are well-known generative models, and these probabilistic models can generate speech and language features based on their likelihood functions. We also provide parameter-learning schemes based on maximum likelihood (ML) estimation which is derived according to the expectation and maximization (EM) algorithm (Dempster *et al.* 1976). Basically, the following chapters extend these statistical models from ML schemes to Bayesian schemes. These models are fundamental for speech and language processing. We specifically build an automatic speech recognition (ASR) system based on these models and extend them to deal with different problems in speaker clustering, speech verification, speech separation and other natural language processing systems.

In this chapter, Section 3.1 first introduces the probabilistic approach to ASR, which aims to find the most likely word sequence W corresponding to the input speech feature vectors $\mathbf{O}$. Bayes decision theory provides a theoretical solution to build up a speech recognition system based on the posterior distribution of the word sequence $p(W|\mathbf{O})$ given speech feature vectors $\mathbf{O}$. Then the Bayes theorem decomposes the problem based on $p(W|\mathbf{O})$ into two problems based on two generative models of speech features $p(\mathbf{O}|W)$ (acoustic model) and language features $p(W)$ (language model), respectively. Therefore, the Bayes theorem changes the original problem to these two independent generative model problems.

Next, Section 3.2 introduces the HMM with the corresponding likelihood function as a generative model of speech features. The section first describes the discrete HMM, which has a multinomial distribution as a state observation distribution, and Section 3.2.4 introduces the GMM as a state observation distribution of the continuous density HMM for acoustic modeling. The GMM by itself is also used as a powerful statistical model for other speech processing approaches in the later chapters. Section 3.3 provides the basic algorithms of forward–backward and Viterbi algorithms. In Section 3.4, ML estimation of HMM parameters is derived according to the EM algorithm to deal with latent variables included in the HMM efficiently. Thus, we provide the conventional ML treatment of basic statistical models for acoustic models based on the HMM.

From Section 3.6, we go on to describe statistical language models as a generative model of language features. As a standard language model, we introduce n-gram models. Similarly to the HMM parameters, the n-gram parameters are also calculated by using the ML estimation. However, ML solutions to n-gram parameters are easily overestimated due to the intrinsic sparse data problems in natural languages. Therefore, the section also describes conventional (rather heuristic) smoothing techniques. Some of the smoothing techniques introduced here are revisited in later chapters to be interpreted as the Bayesian approach, where a Bayesian principle provides these smoothing techniques to regularize these models. In addition, it is well known that the number of n-gram parameters is exponentially increased with large n, which makes it impossible to model a whole document structure within the n-gram model.

Section 3.7 provides another generative model of language features, called the latent topic model, which deals with a statistical treatment of a document model. The section also discusses a way of combining such document models and n-gram models.

Finally, following the discussions of statistical acoustic and language models, Section 3.8 provides an example of applying the Bayesian approach to ASR, as a case study. The section provides an exact Bayesian manner of formulating the standard statistical model (HMM) in ASR, and introduces the posterior distributions of the variables used in acoustic models. The section points out the problem arising mainly due to the posterior distributions, which can be solved in the later chapters.

3.1 Bayes decision for speech recognition

This section describes a statistical speech recognition framework as an example of speech and language processing based on the Bayes decision theory. Before the Bayes decision discussion, we first introduce general speech recognition briefly.

Automatic speech recognition aims to extract helpful text information from speech signals, where both speech and text are represented by *sequential patterns*. These patterns correspond to the time-series signals which are observed in sequences of random variables. Speech recognition is processed in a temporal domain. There are some other technical data, e.g., music signal, video signal, text document, seismic signal, gene sequence, EEG signal, ECG signal, financial data, which are also collected in a time domain. The sequential pattern property is unique and different from data analysis in a spatial domain for general image processing and spectral signal processing. In general, speech and text data are driven under some specialized stochastic process and probabilistic model, e.g., HMMs (Rabiner & Juang 1986) are used to represent speech signals and n-gram models are used to characterize word sequences. In front-end processing, speech signals are first chunked into different time frames t with frame length 25 ms and then transformed to a sequence of speech feature vectors by using mel-frequency cepstral coefficients (MFCCs) (Davis & Mermelstein 1980) or perceptual linear prediction (PLP) coefficients (Hermansky 1990). Each feature vector $\mathbf{o}_t$ is regarded as a random vector consisting of entries with *continuous* value. However, the nth word w_n in a word sequence is a discrete value or label among all words in a dictionary $\mathcal{V}$ with vocabulary size $|\mathcal{V}|$. Thus, speech recognition, which involves acoustic and language models, handles the modeling of *continuous* data as well as *discrete* data.

Now we provide a mathematical notation for a speech recognition problem, which recognizes a speech utterance and outputs the corresponding word sequence in the utterance. Let $\mathbf{o}_t \in \mathbb{R}^D$ be a D dimensional feature vector at frame t, and $\mathbf{O} = \{\mathbf{o}_t | t = 1, \cdots, T\}$ be a speech feature sequence for T frames of one utterance. The number of dimensions (D) is usually 39, which consists of 12 dimensional MFCCs + log power, with delta and delta delta coefficients (Furui 1981). On the other hand, the corresponding word sequence is represented by $W = w_1^N = \{w_n | n = 1, \cdots, N\}$. Here, $w_n \in \mathcal{V}$ is the nth word in this word sequence with N words. The continuous-valued speech feature sequence $\mathbf{O}$ and the discrete-valued word sequence W are sequential patterns in an automatic speech recognition system. Based on the mathematical notations of the speech feature and word sequences, the Bayes decision theory is introduced to find a decision rule or *mapping function* $d(\cdot)$ which maps an input feature sequence $\mathbf{O}$ into an output word sequence W by

$$W = d(\mathbf{O}). \tag{3.1}$$

A popular decision rule is designed to find the most likely word sequence $\hat{W}$ corresponding to input feature sequence $\mathbf{O}$ based on the maximum a-posteriori (MAP) decision rule,

$$\hat{W} = d_{\text{MAP}}(\mathbf{O}) \triangleq \arg\max_{W} \; p(W|\mathbf{O}), \tag{3.2}$$

where $p(W|\mathbf{O})$ is the posterior distribution of W given $\mathbf{O}$. The posterior distribution is often rewritten as

$$\begin{aligned} \hat{W} = d_{\text{MAP}}(\mathbf{O}) &= \arg\max_{W} \; \frac{p(\mathbf{O}|W)p(W)}{p(\mathbf{O})} \\ &= \arg\max_{W} \; \underbrace{p(\mathbf{O}|W)}_{\text{acoustic model}} \times \underbrace{p(W)}_{\text{language model}} . \end{aligned} \tag{3.3}$$

The probabilistic product rule decomposes the posterior distribution into likelihood function $p(\mathbf{O}|W)$ and prior probability $p(W)$ based on acoustic model and language model, respectively. This is a well-known process, called the noisy channel model, that can deal with a speech recognition problem based on acoustic and language models. The same scheme of decomposition of this decision rule is widely used for other speech and language processing including machine translation (Brown, Cocke, Pietra *et al.* 1990, Brants, Popat, Xu *et al.* 2007), spell correction (Brill & Moore 2000), and voice conversion (Saito, Watanabe, Nakamura *et al.* 2012).

However, it is more general to follow a Bayesian perspective for pattern recognition and fulfil an optimal Bayes decision to estimate the decision rule $\hat{d}(\mathbf{O})$ of an input sentence $\mathbf{O}$ by minimizing the expected loss function or Bayes risk, which is defined by Lee & Huo (2000) as a functional of the decision rule:

$$\begin{aligned} r[d] &\triangleq \mathbb{E}_{(W,\mathbf{O})}[\ell(W, d(\mathbf{O}))] \\ &= \sum_{W} \int \ell(W, d(\mathbf{O}))p(W, \mathbf{O})d\mathbf{O} \end{aligned}$$

$$= \int p(\mathbf{O}) \left(\sum_W \ell(W, d(\mathbf{O})) p(W|\mathbf{O}) \right) d\mathbf{O}$$
$$= \sum_W p(W) \int \ell(W, d(\mathbf{O})) p(\mathbf{O}|W) d\mathbf{O}, \tag{3.4}$$

where $\mathbb{E}_{(W,\mathbf{O})}[\cdot]$ denotes the expectation function over joint distribution $p(W, \mathbf{O})$. For the later derivations, Eq. (3.4) provides the two equivalent equations in the third and fourth lines by using the product rule. The loss function satisfies this property:

$$0 \leq \ell(W, d(\mathbf{O}) = W) \leq \ell(W, d(\mathbf{O}) \neq W), \tag{3.5}$$

meaning that the loss due to misclassification $d(\mathbf{O}) \neq W$ is larger than or equal to the loss without misclassification $d(\mathbf{O}) = W$. Therefore, the optimal decision rule $\hat{d}(\mathbf{O})$ can be obtained by minimizing the Bayes risk, which corresponds to minimizing the expected loss function.

Bayes risk is expanded into two expressions, which are shown in the third and fourth equations in the right-hand-side of Eq. (3.4). Following the third equation in Eq. (3.4), we find that Bayes decision rule is equivalent to dealing with a minimization problem:

$$\min_{d \in \mathcal{D}} r[d] = \min_{d \in \mathcal{D}} \int p(\mathbf{O}) \left(\sum_W \ell(W, d(\mathbf{O})) p(W|\mathbf{O}) \right) d\mathbf{O}. \tag{3.6}$$

Here $\mathcal{D}$ denotes a set of all possible decision functions. This optimization can be solved by minimizing the expression in the brackets in the above equation, since the decision rule function does not depend on $\mathbf{O}$ in general. This minimization is satisfied by considering the following optimal decision rule given any $\mathbf{O}$:

$$\hat{d}(\mathbf{O}) = \arg \min_{d(\mathbf{O}) \in \mathcal{D}} \sum_W \ell(W, d(\mathbf{O})) p(W|\mathbf{O})$$
$$= \arg \min_{d(\mathbf{O}) \in \mathcal{D}} \mathbb{E}_{(W)}[\ell(W, d(\mathbf{O}))|\mathbf{O}]. \tag{3.7}$$

That is, finding the optimal decision rule $\hat{d}(\mathbf{O})$ in terms of minimizing the expected loss (in Eq. (3.4)) is equivalent to finding the optimal decision rule function in terms of the expected loss function given $\mathbf{O}$.

In Goel & Byrne (2000) and Chien, Huang, Shinoda *et al.* (2006), a minimum Bayes risk (MBR) classification was proposed to fulfil an optimal Bayes decision in Eq. (3.7) for automatic speech recognition by using a predefined loss function. In Goel & Byrne (2000), the word error rate (WER) loss function $\ell_{\text{WER}}(W, \mathbf{O})$ was calculated using Levenshtein distance between word sequence hypotheses. This function was used to build an MBR decision rule:

$$d_{\text{MBR}}(\cdot) = \arg \min_{d(\cdot)} \sum_W p(W)$$
$$\times \int \ell_{\text{WER}}(W, d(\mathbf{O})) p(\mathbf{O}|W) d\mathbf{O}, \tag{3.8}$$

which is derived according to the fourth equation in the right-hand-side of Eq. (3.4).

More popularly, a meaningful loss function for speech recognition is simply specified by a so-called zero-one loss function which treats misclassification of each observation sample $\mathbf{O}$ equally, namely by using

$$\ell_{01}(W, d(\mathbf{O})) = \begin{cases} 0 & \text{if correctly classified or } d(\mathbf{O}) = W \\ 1 & \text{if wrongly classified or } d(\mathbf{O}) \neq W. \end{cases} \tag{3.9}$$

Substituting Eq. (3.9) into the fourth line in Eq. (3.4) leads to zero-one Bayes risk:

$$\begin{aligned} r_{01}[d] &= \sum_W p(W) \int_{d(\mathbf{O}) \neq W} \ell_{01}(W, d(\mathbf{O})) p(\mathbf{O}|W) d\mathbf{O} \\ &+ \sum_W p(W) \int_{d(\mathbf{O}) = W} \ell_{01}(W, d(\mathbf{O})) p(\mathbf{O}|W) d\mathbf{O} \\ &= \sum_W p(W) \int_{d(\mathbf{O}) \neq W} p(\mathbf{O}|W) d\mathbf{O}. \end{aligned} \tag{3.10}$$

In the third line of Eq. (3.10), the expectation operation using zero-one loss function $r_{01}(d(\cdot))$ is calculated over all observations which are wrongly classified, i.e., $d(\mathbf{O}) \neq W$. This loss function corresponds to unconditional error probability, which is reasonable to act as a measure of goodness of the decision rule for speech recognition.

In addition, Eq. (3.10) is further rewritten as

$$\begin{aligned} r_{01}[d] &= \sum_W p(W) \left(1 - \int_{d(\mathbf{O}) = W} p(\mathbf{O}|W) d\mathbf{O}\right) \\ &= 1 - \sum_W \int_{d(\mathbf{O}) = W} p(W) p(\mathbf{O}|W) d\mathbf{O} \end{aligned} \tag{3.11}$$

by using the following properties:

$$\sum_W p(W) = 1, \tag{3.12}$$

$$\int p(\mathbf{O}|W) d\mathbf{O} = 1. \tag{3.13}$$

The resulting decision rule $d_{01}(\cdot)$ follows the minimum classification error criterion which leads to the MAP decision rule as addressed in Eq. (3.2), i.e.,

$$\hat{W} = d_{01}(\mathbf{O}) = d_{\text{MAP}}(\mathbf{O}) = \arg\max_W \; p(W|\mathbf{O}). \tag{3.14}$$

The most likely word sequence $\hat{W}$ is found so as to achieve the highest posterior probability for correctly classified observation sequence $d(\mathbf{O}) = W$.

Using an MAP decision rule, the probability measure $p(\mathbf{O}|W)$ calculates how likely the acoustic observation sequence $\mathbf{O}$ is, based on the word sequence hypothesis W. We also name $p(\mathbf{O}|W)$ as the acoustic likelihood function. There are many kinds of acoustic models which are assumed in calculation of the statistical model $p_\Theta(\mathbf{O}|W)$ based on a set of acoustic parameters Θ. In this chapter, the hidden Markov model (HMM) is considered for acoustic modeling and the HMM parameters Θ are plugged into the probability measure estimator $\hat{p}_\Theta(\mathbf{O}|W)$. On the other hand, the probability measure $p(W)$ is defined as the prior probability of word sequence W. This measure calculates the joint probability for a sequence of words based on a set of multinomial parameters or n-gram parameters Θ. The plug-in language model $\hat{p}_\Theta(W)$ is determined as a language model estimator. Here, acoustic parameters and linguistic parameters are both included in model parameters Θ. For an automatic speech recognition (ASR) system, we will estimate the acoustic parameters and the linguistic parameters from a set of training utterances $\mathbf{O}$ and their word transcriptions W according to the Maximum Likelihood (ML) estimation. We assume that these plug-in models $\{\hat{p}_\Theta(\mathbf{O}|W), \hat{p}_\Theta(W)\}$ given ML parameters Θ are true. The prediction of new test utterance $\mathbf{O}$ based on the estimated MAP decision rule $\hat{d}_{\text{MAP}}(\mathbf{O})$ is performed by

$$\begin{aligned} \hat{d}_{\text{MAP}}(\mathbf{O}) &= \arg\max_W \hat{p}(W|\mathbf{O}) \\ &= \arg\max_W \hat{p}_\Theta(\mathbf{O}|W)\hat{p}_\Theta(W). \end{aligned} \tag{3.15}$$

However, the point estimates of the ML-based acoustic model and language model Θ from the given observation space Ω_o using the collected training data may not generalize well for the unknown test data outside the training space Ω_o. The distributions $p(\mathbf{O}|W)$ and $p(W)$ may not be correctly assumed or may be over-trained or under-trained. From a Bayesian perspective, these issues could be tackled by treating acoustic parameters and linguistic parameters Θ as random variables. Consideration of these uncertainties is helpful for recognition of new test data. For this consideration, the expected loss function in Eq. (3.4) is calculated by additionally marginalizing over continuous parameters Θ:

$$r_{\text{BPC}}(d(\cdot)) = \mathbb{E}_{(W,\mathbf{O},\Theta)}[\ell(W, \mathbf{O}, \Theta)]. \tag{3.16}$$

The Bayesian predictive classification (BPC) rule,

$$d_{\text{BPC}}(\cdot) = \arg \min_{d(\cdot)\in\Omega_d} r_{\text{BPC}}(d(\cdot)), \tag{3.17}$$

(Jiang, Hirose & Huo 1999, Huo & Lee 2000, Lee & Huo 2000, Chien & Liao 2001) was proposed to establish a robust decision rule for unknown test speech. The zero-one loss function $\ell_{01}(\cdot)$ was applied in previous studies on the BPC rule. Details of BPC-based speech recognition will be addressed in Section 6.3.

In addition, Bayes decision theory was developed to estimate the *discriminative* acoustic model for speech recognition (Juang & Katagiri 1992). The idea is to minimize the expected loss function based on the logistic sigmoid function,

$$\ell(d_k(\mathbf{O}, \Theta)) = \frac{1}{1 + \exp^{-\alpha d_k(\mathbf{O},\Theta)}}, \tag{3.18}$$

which is a function of misclassification measure defined as

$$\begin{aligned} d_k(\mathbf{O}, \Theta) =& -g_k(\mathbf{O}; \Theta) \\ &+ \log\left(\frac{1}{K-1}\sum_{j,j\neq k} \exp(g_j(\mathbf{O}; \Theta))\right). \end{aligned} \tag{3.19}$$

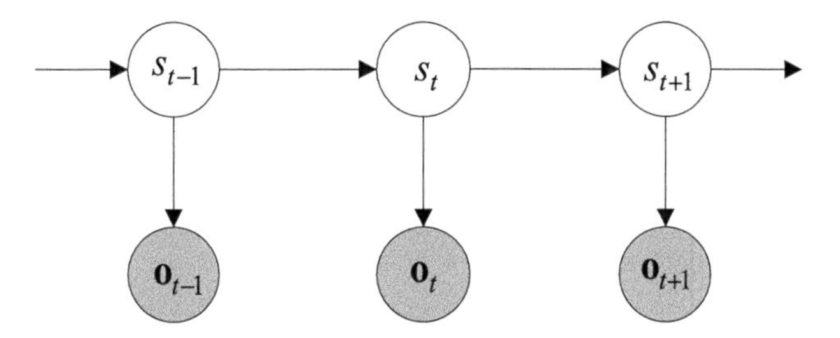

Figure 3.1 Graphical model of HMM without model parameters.

In Eq. (3.18), α is a tuning parameter and the misclassification measure of phone class C_k from training data $\mathbf{O} = \{\mathbf{o}_t\}$ is calculated by measuring the distance between the discriminant functions $g_k(\mathbf{O}, \Theta)$ of the target phone $\mathbf{o}_t \in C_k$ and its competing phones $\mathbf{o}_t \notin C_k$. The discriminant function of competing phones is averaged over all phones except the target phone. The discriminative acoustic model was trained according to the minimum classification error (MCE) criterion which is closely related to the minimum Bayes risk for optimal Bayes decision (Juang & Katagiri 1992).

3.2 Hidden Markov model

The previous section describes the Bayes decision theory and introduces acoustic model $p(\mathbf{O}|W)$ and language models $p(W)$. This section describes hidden Markov models (HMMs) (Rabiner & Juang 1986) (Figure 3.1) as a standard statistical acoustic model in detail. Before describing the HMM in detail, we first explain what HMM represents in speech recognition.

3.2.1 Lexical unit for HMM

The acoustic model $p(\mathbf{O}|W)$ means that we provide a likelihood function of the observations $\mathbf{O}$ given the word sequence W. However, since the number of all possible word sequences is an exponential order, we cannot prepare a likelihood function for each word sequence. Instead, we first introduce a lexical sequence $L = \{l_m \in \mathcal{L} | m = 1, \cdots, M\}$ that is composed of phonemes (e.g., /a/,/k/), context-dependent phonemes (e.g., /a/-/k/-/i/,/a/-/k/-/a/), or words (e.g., numbers, commands) as l_m. A discussion about the context-dependent phoneme (allophone) unit can be found in Section 6.5. We usually use a phoneme unit defined by linguistics, and the automatic discovery of the phoneme unit from speech data by using Bayesian nonparametrics can be found in Section 8.4. The definition of the lexical unit depends on application, but the standard acoustic model for LVCSR uses (context-dependent) phonemes as a lexical unit, and hereinafter in this chapter, we use phonemes to define a lexical unit. Then, $\mathcal{L}$ is a set of all distinct phonemes.

By using the lexical sequence L, we can revisit the MAP decision rule in Eq. (3.15) for ASR as follows:

$$\begin{aligned} d_{\mathrm{MAP}}(\mathbf{O}) &= \arg\max_W p(W|\mathbf{O}) \\ &= \arg\max_W p(\mathbf{O}|W)p(W) \\ &= \arg\max_W \sum_L p(\mathbf{O}, L|W)p(W). \end{aligned} \tag{3.20}$$

By using the product rule, and assuming that the likelihood function only depends on the lexical sequence L, it is rewritten as:

$$\begin{aligned} d_{\mathrm{MAP}}(\mathbf{O}) &= \arg\max_W \sum_L p(\mathbf{O}|L, W)p(L|W)p(W) \\ &\approx \arg\max_W \sum_L p(\mathbf{O}|L)p(L|W)p(W), \end{aligned} \tag{3.21}$$

where $p(L|W)$ is called a lexical model. Usually, the lexical model is not a probabilistic model, but is obtained deterministically by using a lexical dictionary, which provides a phoneme sequence (or multiple phoneme sequences) given a word. We further assume that the alignment of $\mathbf{O}$ for phoneme l_m is already given. This means that $\mathbf{O}$ is segmented to $\mathbf{O}_m = \{\mathbf{o}_{t_{m-1}+1}, \cdots, \mathbf{o}_{t_m}\}$ where $t_0 = 0, t_M = T$, and $\{t_m\}_{m=1}^{M-1}$ is given.[1]

By assuming that $\mathbf{O}_m$ is independent and identically distributed (iid) for phoneme l_m, the acoustic model is factorized by m as:

$$p(\mathbf{O}|L) = \prod_{m=1}^{M} p(\mathbf{O}_m|l_m). \tag{3.22}$$

This $p(\mathbf{O}_m|l_m)$ is an actual likelihood function that we deal with for ASR, and is represented by an HMM. Therefore, the acoustic model is composed of $|\mathcal{L}|$ HMMs where $|\mathcal{L}|$ denotes the number of distinct phonemes. The following section explains the HMM for one phoneme, and omits phoneme l_m in the explanation. In addition, since the alignment is already given, we omit the segmentation information m based on t_m, and use $\mathbf{O} = \{\mathbf{o}_t \in \mathbb{R}^D | t = 1, \cdots, T\}$ instead of $\mathbf{O}_m$ to be modeled by an HMM.

3.2.2 Likelihood function of HMM

This section describes a likelihood function of the HMM for a phoneme, where the HMM is a popular formalism for representation of sequential patterns. The likelihood function defined here is used to estimate HMM parameters.

In a set of D dimensional continuous-valued speech feature vectors $\mathbf{O}$, each observation $\mathbf{o}_t$ is represented under a Markov state s_t as illustrated in Figure 3.1. We assume

[1] In the actual search ($\arg\max_W$) process in Eq. (3.21), this alignment information is not fixed, and is searched at the same time. Many techniques have been developed to efficiently search this huge space considering the lexicon and word sequences (e.g., Ney, Haeb-Umbach, Tran *et al.* (1992) and weighted finite state transducer (WFST) based techniques (Mohri, Pereira & Riley 2002, Hori & Nakamura 2013)).

that each speech frame $\mathbf{o}_t$ is independent conditionally on its state label s_t. Starting from the state at the first time frame with an initial state probability π_{s_1}, which is a model parameter of HMM,

$$p(s_1) \triangleq \pi_{s_1}, \tag{3.23}$$

where each state at time t depends on the state at its previous time $t-1$. The state transition probability

$$p(s_t|s_{t-1}) \triangleq a_{s_{t-1}s_t} \tag{3.24}$$

is also introduced as HMM parameters to drive the change of HMM states or equivalently characterize the transition of acoustic events under a phone unit. The parameters of initial state probabilities and state transition probabilities should satisfy the following constraints:

$$\sum_{j=1}^{J} \pi_j = 1, \qquad \sum_{j=1}^{J} a_{ij} = 1, \qquad \forall i \tag{3.25}$$

respectively. A sequence of hidden states $S = \{s_t \in \{1, \cdots, J\} | t = 1, \cdots, T\}$ corresponding to a sequence of observations $\mathbf{O}$ is marginalized in calculation of likelihood function. HMMs basically involve a doubly stochastic process. One is for the observed speech frame sequence $\mathbf{O}$ and the other is for the corresponding HMM state sequence S, which is not observed and is regarded as a latent variable. The model which includes a latent variable is called the *latent model*.

Since S is not observed, the likelihood function of $\mathbf{O}$ is given by the summation of the joint distribution of $\mathbf{O}$ and S over all possible state sequences conditioned on a set of HMM parameters Θ:

$$p(\mathbf{O}|\Theta) = \sum_{S \in \mathcal{S}} p(\mathbf{O}, S|\Theta). \tag{3.26}$$

$\mathcal{S}$ denotes a set of all possible state sequences, which is often omitted when it is trivial. Note that the summation over all possible state sequences requires the exponential order of computations, which is not feasible in a practical use. The joint distribution $p(\mathbf{O}, S|\Theta)$ is also called *complete data likelihood*, where a set of observed data and latent variables $(\{\mathbf{O}, S\})$ is called *complete data*. The joint distribution (complete data likelihood) is a useful representation to provide approximate solutions for the latent model (Dempster *et al.* 1976, Huang, Ariki & Jack 1990). Therefore, we focus on the joint likelihood function of the HMM.

By using the product rule, the joint distribution in Eq. (3.26) is decomposed into the likelihood function $p(\mathbf{O}|S, \Theta)$ given the state sequence S, and the state sequence probability $P(S|\Theta)$ as follows:

$$p(\mathbf{O}, S|\Theta) = p(\mathbf{O}|S, \Theta)p(S|\Theta). \tag{3.27}$$

Since $\mathbf{o}_t$ is independent and identically distributed (iid) given the state sequence $S = \{s_1, \cdots, s_T\}$, the likelihood function $p(\mathbf{O}|S, \Theta)$ is represented as follows:

$$p(\mathbf{O}|S, \Theta) = \prod_{t=1}^{T} p(\mathbf{o}_t|\Theta_{s_t}). \tag{3.28}$$

Here, Θ_j is a set of state-dependent HMM parameters, where j is the index of the HMM state. Similarly, the state sequence probability $P(S|\Theta)$ given the state sequence $S = \{s_1, \cdots, s_T\}$ can be represented by the initial probability in Eq. (3.23) and transition probabilities in Eq. (3.24), as follows:

$$p(S|\Theta) = \pi_{s_1} \prod_{t=2}^{T} a_{s_{t-1}s_t}. \tag{3.29}$$

Thus, by substituting Eqs. (3.28), (3.29), and (3.27) into Eq. (3.26), we can obtain the following equation:

$$p(\mathbf{O}|\Theta) = \sum_{S=\{s_t\}\in\mathcal{S}} \left(\pi_{s_1} p(\mathbf{o}_1|\Theta_{s_1}) \prod_{t=2}^{T} a_{s_{t-1}s_t} p(\mathbf{o}_t|\Theta_{s_t}) \right). \tag{3.30}$$

This is the basic equation of the likelihood function of the HMM with the parameter $\Theta = \{\{\pi_j\}_{j=1}^{J}, \{\{a_{ij}\}_{j=1}^{J}\}_{i=1}^{J}, \{\Theta_j\}_{j=1}^{J}\}$. $p(\mathbf{o}_t|\Theta_j)$ is called the *emission probability*, and can be any distribution of $\mathbf{o}_t$.

If we use discrete values as observations for $\mathbf{o}_t$, the emission probability of the HMM is represented by a multinomial distribution. This HMM is called a *discrete HMM* (DHMM). In the DHMM, the HMM parameters are formed as $\Theta = \{\{\pi_j\}_{j=1}^{J}, \{\{a_{ij}\}_{j=1}^{J}\}_{i=1}^{J}, \{\{b_{jk}\}_{k=1}^{K}\}_{j=1}^{J}\}$ where the emission probabilities of the discrete observation values k given state j are represented as the multinomial distribution parameter b_{jk}. Assuming that the vector space of observations $\mathbf{O} = \{\mathbf{o}_t\}$ in state j is partitioned into K subspaces by clustering or vector quantization, we express the state observation probability b_{jk} using the probability $p(\mathbf{o}_t \in \mathcal{C}_k|\Theta_j)$, which is constrained by the property

$$\sum_{k=1}^{K} p(\mathbf{o}_t \in \mathcal{C}_k|\Theta_j) = \sum_{k=1}^{K} b_{jk} = 1. \tag{3.31}$$

Here $\mathcal{C}_k$ denotes a set of discrete feature vectors that belongs to a partitioned subspace k. This partition is undertaken by using Vector Quantization (VQ) techniques.

Historically, DHMM was first used for speech modeling with the VQ techniques (Matsui & Furui 1994, Jelinek 1997). However, since the speech feature vector is a continuous value, the Gaussian distribution or the mixture of Gaussian distributions is often used for an acoustic model. That is, an acoustic model, which dealt with discrete observation values obtained from VQ codes, was shifted to so-called *continuous density HMM* (CDHMM), where the HMM with Gaussian distribution based emission probabilities could model continuous observation values more appropriately. CDHMM is described in the next section.

3.2.3 Continuous density HMM

For the HMM framework addressed in Section 3.2, the state observation probability $p(\mathbf{o}_t|\Theta_j)$ is calculated according to the type of observation data. When using discrete observation symbols, each observation vector $\mathbf{o}_t \in C_k$ is simply represented by a codebook partition k which is determined through vector quantization over the vector space of all training samples $\mathbf{O}$. Nonetheless, the representation of a high dimensional feature vector $\mathbf{o}_t \in \mathbb{R}^D$ based on a set of discrete codebook labels is not adequate to fully reflect the randomness of the observation vector. More generally, we calculate the probability density function (pdf) of the observation vector $\mathbf{o}_t$ given an HMM state $s_t = j$ to determine the state observation probability $p(\mathbf{o}_t|\Theta_j)$. It is popular to represent the randomness of continuous-valued $\mathbf{o}_t$ using the multivariate Gaussian distribution $\mathcal{N}(\mathbf{o}_t|\boldsymbol{\mu}_j, \boldsymbol{\Sigma}_j)$ defined in Appendix C.6 as:

$$\begin{aligned} p(\mathbf{o}_t|\Theta_j) &= \mathcal{N}(\mathbf{o}_t|\boldsymbol{\mu}_j, \boldsymbol{\Sigma}_j) \\ &\triangleq \frac{1}{(2\pi)^{D/2}|\boldsymbol{\Sigma}_j|^{1/2}} \exp\left(-\frac{1}{2}(\mathbf{o}_t - \boldsymbol{\mu}_j)^\mathsf{T}\boldsymbol{\Sigma}_j^{-1}(\mathbf{o}_t - \boldsymbol{\mu}_j)\right), \end{aligned} \tag{3.32}$$

with state-dependent mean vector $\boldsymbol{\mu}_j \in \mathbb{R}^D$ and covariance matrix $\boldsymbol{\Sigma}_j \in \mathbb{R}^{D\times D}$.

However, a single Gaussian distribution is insufficient to represent the state-dependent observation space for an HMM state j because there are large amounts of training data collected from varying speakers with different genders, ages, accents, speaking rates, channel distortions and background noises, which are used to train the parameters of individual HMM states. Accordingly, a Gaussian mixture model (GMM) is adopted to represent the state-dependent observation space. This is based on a set of Gaussian distributions which reflect the variations of speech feature vectors within an HMM state due to various acoustic conditions. The state observation probability density function of a feature vector $\mathbf{o}_t$ at time t and in state j is expressed by GMM with K mixture components:

$$\begin{aligned} p(\mathbf{o}_t|s_t = j, \Theta_j) &= \sum_{k=1}^{K} p(\mathbf{o}_t, v_t = k|s_t = j, \Theta_j) \\ &= \sum_{k=1}^{K} p(v_t = k|s_t = j, \Theta_j)p(\mathbf{o}_t|s_t = j, v_t = k, \Theta_j) \\ &= \sum_{k=1}^{K} \omega_{jk}\mathcal{N}(\mathbf{o}_t|\boldsymbol{\mu}_{jk}, \boldsymbol{\Sigma}_{jk}). \end{aligned} \tag{3.33}$$

In Eq. (3.33), each mixture component of state j is expressed by a Gaussian distribution:

$$\begin{aligned} p(\mathbf{o}_t|s_t = j, v_t = k, \Theta_j) &= \mathcal{N}(\mathbf{o}_t|\boldsymbol{\mu}_{jk}, \boldsymbol{\Sigma}_{jk}) \\ &\triangleq \frac{1}{(2\pi)^{D/2}|\boldsymbol{\Sigma}_{jk}|^{1/2}} \exp\left(-\frac{1}{2}(\mathbf{o}_t - \boldsymbol{\mu}_{jk})^\mathsf{T}\boldsymbol{\Sigma}_{jk}^{-1}(\mathbf{o}_t - \boldsymbol{\mu}_{jk})\right), \end{aligned} \tag{3.34}$$

and the prior probability of a mixture component $v_t = k$ is defined as a mixture weight,

$$p(v_t = k|s_t = j, \Theta_j) \triangleq \omega_{jk}. \tag{3.35}$$

The state-dependent GMM parameters $\{\omega_{jk}, \boldsymbol{\mu}_{jk}, \boldsymbol{\Sigma}_{jk}\}_{k=1}^{K}$ consist of mixture weights ω_{jk}, mean vectors $\boldsymbol{\mu}_{jk}$, and covariance matrices $\boldsymbol{\Sigma}_{jk}$ for K Gaussian mixture components. The resulting realization of HMM using GMM as state observation probability is also called the *continuous density HMM* (CDHMM). The CDHMM parameters are formed as

$$\Theta \triangleq \{\{\pi_j\}_{j=1}^{J}, \{\{a_{ij}\}_{j=1}^{J}\}_{i=1}^{J}, \{\{\omega_{jk}, \boldsymbol{\mu}_{jk}, \boldsymbol{\Sigma}_{jk}\}_{k=1}^{K}\}_{j=1}^{J}\}, \tag{3.36}$$

with an additional constraint on the prior probability of mixture components or the mixture weights.

Typically, HMM covariance matrices are assumed to be diagonal in practical implementation, i.e., $\boldsymbol{\Sigma}_{jk} = \text{diag}(\Sigma_{jk1}, \cdots, \Sigma_{jkd}, \cdots, \Sigma_{jkD})$. Note that this book uses Σ to represent a diagonal component of covariance, or variance when $D = 1$. Then, the diagonal covariance version of Eq. (3.34) is factorized to the univariate Gaussian distribution (Appendix C.5) by the dimension index d as follows:

$$\begin{aligned} p(\mathbf{o}_t|s_t = j, v_t = k, \Theta_j) &= \prod_{d=1}^{D} \mathcal{N}(o_{td}|\mu_{jkd}, \Sigma_{jkd}) \\ &\triangleq \prod_{d=1}^{D} \frac{1}{(2\pi)^{1/2}(\Sigma_{jkd})^{1/2}} \exp\left(-\frac{1}{2\Sigma_{jkd}}(o_{td} - \mu_{jkd})^2\right). \end{aligned} \tag{3.37}$$

This diagonal covariance representation can represent the distribution with all scalar variables, which makes the calculation very simple compared with the vector and matrix representation in the multivariate Gaussian distribution in Eq. (3.34). In addition, the number of parameters for the covariance matrix is reduced from $D * (D + 1)/2$ to D, which can reduce the computational and memory costs. The full covariance Gaussian in Eq. (3.34) also requires computation of the inverse and the determinant of the covariance matrix ($\boldsymbol{\Sigma}_{jk}^{-1}$ and $|\boldsymbol{\Sigma}_{jk}|$), which makes the computation numerically unstable in addition to incurring the matrix computation cost. However, the speech feature vectors have correlations over dimensions, and the diagonal covariance assumption is not adequate. There are more sophisticated models such as the subspace mean and variance methods, aiming to bridge the gap between full and diagonal covariance modeling, which have been proposed (Gales 1999, Axelrod, Gopinath & Olsen 2002). In this book, we keep the vector and matrix representation for the multivariate Gaussian distribution for generality.

Based on the above likelihood discussion of the CDHMM for single observation vector $\mathbf{o}_t$ given HMM state $s_t = j$ at frame t, we consider the marginal likelihood for a whole observation sequence $\mathbf{O}$. Under CDHMM, we have an additional latent variable $v_t = k$ from the mixture component of GMM for each speech frame $\mathbf{o}_t$ at time t, compared with Eq. (3.30). Thus, the sequence of mixture components

$$V \triangleq \{v_t \in \{1, \cdots, K\}|t = 1, \cdots, T\}, \tag{3.38}$$

corresponding to the sequence of observed speech frames $\mathbf{O} = \{\mathbf{o}_t | t = 1, \cdots, T\}$, is introduced in estimation of HMM parameters in addition to $S = \{s_t \in \{1, \cdots, J\} | t = 1, \cdots, T\}$. Considering two sets of latent variables S and V, we consider the likelihood function of the CDHMM by following the formulation in Section 3.2.2. First, the marginal likelihood function of the CDHMM is represented as the joint distribution $p(\mathbf{O}, S, V|\Theta)$:

$$p(\mathbf{O}|\Theta) = \sum_{S,V} p(\mathbf{O}, S, V|\Theta), \tag{3.39}$$

where all possible state sequences S and mixture component sequences V are considered in calculation of the marginal distribution. The joint distribution $p(\mathbf{O}, S, V|\Theta)$ is also called the complete data likelihood, as we discussed in Eq. (3.26) with additional latent variable V. This equation is further decomposed by using the product rule as follows:

$$p(\mathbf{O}, S, V|\Theta) = p(\mathbf{O}|S, V, \Theta)p(V|S, \Theta)p(S|\Theta). \tag{3.40}$$

We provide an actual distribution for each probabilistic function.

Since the distributions $p(S|\Theta)$ are given in Eq. (3.29), we focus on $p(V|S, \Theta)$ and $p(\mathbf{O}|S, V, \Theta)$. Since v_t only depends on s_t, given the state sequence $S = \{s_1, \cdots, s_T\}$, $p(V|S, \Theta)$, it is represented by the factorized form with frame t as follows:

$$p(V|S, \Theta) = \prod_{t=1}^{T} p(v_t|s_t, \Theta) = \prod_{t=1}^{T} \omega_{s_t v_t}. \tag{3.41}$$

Similarly, from Eq. (3.34), $p(\mathbf{O}|S, V, \Theta)$ is also represented by the factorized form with frame t given S and V as:

$$\begin{aligned} p(\mathbf{O}|S, V, \Theta) &= \prod_{t=1}^{T} p(\mathbf{o}_t|s_t, v_t, \Theta) \\ &= \prod_{t=1}^{T} \mathcal{N}(\mathbf{o}_t|\boldsymbol{\mu}_{s_t v_t}, \boldsymbol{\Sigma}_{s_t v_t}). \end{aligned} \tag{3.42}$$

Thus, by substituting Eqs. (3.29), (3.41), and (3.42) into Eq. (3.40), the complete data likelihood function of the CDHMM is represented as

$$p(\mathbf{O}, S, V|\Theta) = \pi_{s_1} \omega_{s_1 v_1} \mathcal{N}(\mathbf{o}_1|\boldsymbol{\mu}_{s_1 v_1}, \boldsymbol{\Sigma}_{s_1 v_1}) \left(\prod_{t=2}^{T} a_{s_{t-1} s_t} \omega_{s_t v_t} \mathcal{N}(\mathbf{o}_t|\boldsymbol{\mu}_{s_t v_t}, \boldsymbol{\Sigma}_{s_t v_t}) \right). \tag{3.43}$$

Thus, the marginal likelihood of the CDHMM is extended from Eq. (3.30) as

$$p(\mathbf{O}|\Theta) = \sum_{S,V} \pi_{s_1} \omega_{s_1 v_1} \mathcal{N}(\mathbf{o}_1|\boldsymbol{\mu}_{s_1 v_1}, \boldsymbol{\Sigma}_{s_1 v_1}) \left(\prod_{t=2}^{T} a_{s_{t-1} s_t} \omega_{s_t v_t} \mathcal{N}(\mathbf{o}_t|\boldsymbol{\mu}_{s_t v_t}, \boldsymbol{\Sigma}_{s_t v_t}) \right). \tag{3.44}$$

This is the basic equation of the likelihood function of the CDHMM with the parameter $\Theta = \{\{\pi_j\}, \{a_{ij}\}, \{\omega_{jk}\}, \{\boldsymbol{\mu}_{jk}\}, \{\boldsymbol{\Sigma}_{jk}\}\}$. As we discussed in the previous section, the summation over all possible state and mixture component sequences requires an exponential order of computations, which is not tractable in practical use. Section 3.3 describes how to efficiently compute the marginal likelihood by using the forward and backward algorithms, and most probable state sequence based on the Viterbi algorithm. Section 3.4 also describes how to estimate the HMM parameters Θ efficiently based on these algorithms and the expectation and maximization algorithm.

Before moving to these explanations, as a subset model of the CDHMM, we introduce a simple GMM without an HMM, which is also widely used in speech and language processing.

3.2.4 Gaussian mixture model

The GMM is a simplified model of the CDHMM, without considering the state sequence S in the previous section. However, this simple GMM is still very powerful for modeling speech features. For example, the GMM is used in speech and noise models in speech enhancement, and in speaker models in speaker verification and clustering, which are discussed in Section 4.6. The GMM can also be widely used to model other signals than speech, e.g., image processing, and biosignals. Therefore, this section only introduces the marginal likelihood of a GMM to be used in later chapters, similar to that of the CDHMM in Eq. (3.44):

$$\begin{aligned} p(\mathbf{O}|\Theta) &= \sum_V p(\mathbf{O}, V|\Theta) = \sum_V p(\mathbf{O}|V, \Theta)p(V|\Theta) \\ &= \sum_V \prod_{t=1}^T p(\mathbf{o}_t|v_t, \Theta)p(v_t|\Theta) \\ &= \prod_{t=1}^T \sum_{v_t=1}^K \omega_{v_t} \mathcal{N}(\mathbf{o}_t|\boldsymbol{\mu}_{v_t}, \boldsymbol{\Sigma}_{v_t}), \end{aligned} \tag{3.45}$$

with the parameter $\Theta = \{\{\omega_k\}, \{\boldsymbol{\mu}_k\}, \{\boldsymbol{\Sigma}_k\}\}$. Since v_t is independent of $v_{t' \neq t}$ in a GMM, the sequential summation over V is independently applied to each $p(\mathbf{o}_t|v_t, \Theta)p(v_t|\Theta)$. So, we can factorize $p(\mathbf{O}|\Theta)$ into the following tth frame distribution given the kth mixture component :

$$\begin{aligned} p(\mathbf{o}_t|v_t = k, \Theta) &= \mathcal{N}(\mathbf{o}_t|\boldsymbol{\mu}_k, \boldsymbol{\Sigma}_k) \\ p(v_t = k|\Theta) &= \omega_k. \end{aligned} \tag{3.46}$$

Thus, unlike the HMM case that needs to consider the HMM state sequence S (computation based on an exponential order for the frame length), the GMM can compute the likelihood of all possible V by using the straightforward computation of Eq. (3.45) with the linear order computation of the frame length. As regards the parameter estimation, the EM algorithm is widely used, and this is discussed in Section 3.4.

Figure 3.2 illustrates a univariate GMM distribution $p(o) = \frac{1}{3}\mathcal{N}(o|\mu_1 = 2, \Sigma_1 = 36) + \frac{1}{3}\mathcal{N}(o|\mu_2 = 30, \Sigma_2 = 64) + \frac{1}{3}\mathcal{N}(o|\mu_3 = 45, \Sigma_3 = 16)$ which is plotted with

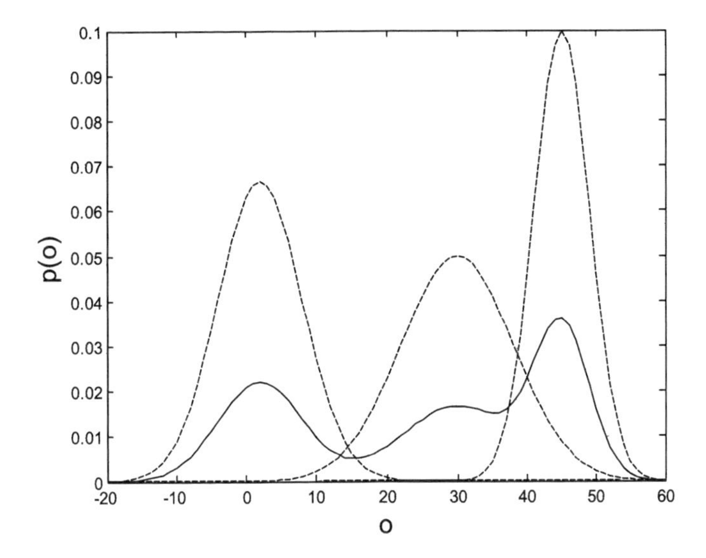

Figure 3.2 Distributions of three individual Gaussians (dashed line) and the corresponding GMM (solid line).

a dashed line and is formed as a mixture of three Gaussian distributions with different means $\{\mu_1, \mu_2, \mu_3\}$ and variances $\{\Sigma_1, \Sigma_2, \Sigma_3\}$, which are shown with a solid line. As shown in the figure, the GMM is a multi-modal distribution with multiple peaks, which can model multiple factors in speech features. Therefore, the GMM can accurately represent speech variations that cannot be represented by a single Gaussian distribution.

3.2.5 Graphical models and generative process of CDHMM

The previous sections explain the marginal likelihood functions of the GMM and CDHMM based on their joint likelihood distributions. As we discussed in the graphical model representation in Section 2.2, once we obtain the joint likelihood distribution of a model, we can obtain the corresponding graphical model and generative process of the model. This section provides the graphical model and generative process of the GMM and CDHMM, respectively, and these are used in the following chapters to deal with these models by using the Bayesian approach.

Graphical models and generative process of GMM

Based on the previous explanations, we can provide a generative process and graphical model of a K-component GMM discussed in Section 3.2.4 as an example. Given feature vectors $\mathbf{O}$, latent variables V, and model parameters Θ, the joint distribution of a complete data set $\{\mathbf{O}, V\}$ is represented from Eq. (3.45) as follows:

$$p(\mathbf{O}, V|\Theta) = \prod_{t=1}^{T} p(\mathbf{o}_t|v_t, \Theta)p(v_t|\Theta). \tag{3.47}$$

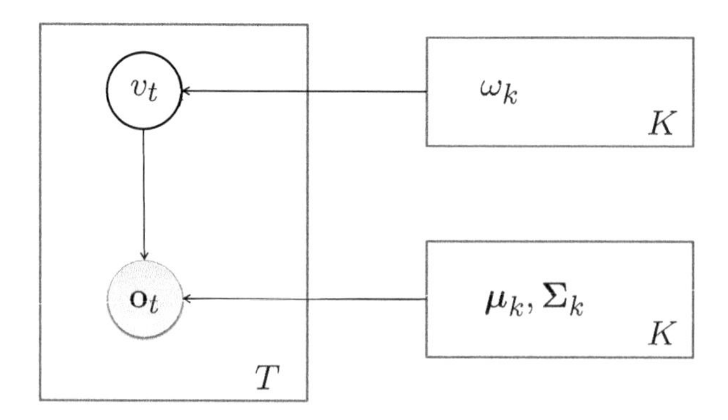

Figure 3.3 Graphical model of Gaussian mixture model.

Thus, the joint distribution of the GMM is parameterized as

$$p(\mathbf{O}, V|\Theta) = \prod_{t=1}^{T} \omega_{v_t} \mathcal{N}(\mathbf{o}_t|\boldsymbol{\mu}_{v_t}, \boldsymbol{\Sigma}_{v_t}). \tag{3.48}$$

Thus, we have the following three kinds of variables:

- **Observation**: $\mathbf{O} = \{\mathbf{o}_1, \cdots, \mathbf{o}_T\}$;
- **Latent**: $V = \{v_1, \cdots, v_T\}$;
- **Non-probabilistic**: $\Theta = \{\omega_k, \boldsymbol{\mu}_k, \boldsymbol{\Sigma}_k\}_{k=1}^K$.

The dependencies of the above variables are expressed in Eq. (3.48), and we provide the generative process of the GMM in Algorithm 2. Given the mixture weight ω_{v_t}, latent variable v_t is sampled from the multinomial distribution.

In addition, by using the plate for t and k, as discussed in Section 2.2.2, we can simply write the graphical model of the GMM, as shown in Figure 3.3.

Algorithm 2 Generative process of Gaussian mixture model

Require: T, $\{\omega_k, \boldsymbol{\mu}_k, \boldsymbol{\Sigma}_k\}_{k=1}^K$
1: **for** $t = 1, \cdots, T$ **do**
2: Draw v_t from $\text{Mult}(v_t|\{\omega_k\}_{k=1}^K)$
3: Draw $\mathbf{o}_t$ from $\mathcal{N}(\mathbf{o}_t|\boldsymbol{\mu}_{v_t}, \boldsymbol{\Sigma}_{v_t})$
4: **end for**

Graphical models and generative process of CDHMM

Similar to the GMM case, we can also provide a generative process and graphical model of a continuous density HMM discussed in Section 3.2 as another example. Given feature vectors $\mathbf{O}$, latent variables S and V, and model parameters Θ, the joint likelihood function of complete data $\mathbf{O}$, V, and S based on a continuous density HMM is represented as follows:

$$p(\mathbf{O}, S, V|\Theta) = \pi_{s_1} \omega_{s_1 v_1} \mathcal{N}(\mathbf{o}_1 | \boldsymbol{\mu}_{s_1 v_1}, \boldsymbol{\Sigma}_{s_1 v_1}) \times \prod_{t=2}^{T} a_{s_{t-1} s_t} \omega_{s_t v_t} \mathcal{N}(\mathbf{o}_t | \boldsymbol{\mu}_{s_t v_t}, \boldsymbol{\Sigma}_{s_t v_t}). \tag{3.49}$$

The CDHMM has the following three kinds of variables:

- **Observation**: $\mathbf{O} = \{\mathbf{o}_1, \cdots, \mathbf{o}_T\}$;
- **Latent**: $S = \{s_1, \cdots, s_T\}$ and $V = \{v_1, \cdots, v_T\}$;
- **Non-probabilistic**: $\Theta = \{\{\pi_j\}_{j=1}^J, \{\{a_{ij}\}_{j=1}^J\}_{i=1}^J, \{\{\omega_{jk}, \boldsymbol{\mu}_{jk}, \boldsymbol{\Sigma}_{jk}\}_{k=1}^K\}_{j=1}^J\}$.

Algorithm 3 Generative process of continuous density hidden Markov model

Require: T, Θ
1: Draw s_1 from $\mathrm{Mult}(s_1|\{\pi_j\}_{j=1}^J)$
2: Draw v_1 from $\mathrm{Mult}(v_1|\{\omega_{s_1 k}\}_{k=1}^K)$
3: Draw $\mathbf{o}_1$ from $\mathcal{N}(\mathbf{o}_1|\boldsymbol{\mu}_{s_1 v_1}, \boldsymbol{\Sigma}_{s_1 v_1})$
4: **for** $t = 2, \cdots, T$ **do**
5: Draw s_t from $\mathrm{Mult}(s_t|\{a_{s_{t-1} j}\}_{j=1}^J)$
6: Draw v_t from $\mathrm{Mult}(v_t|\{\omega_{s_t k}\}_{k=1}^K)$
7: Draw $\mathbf{o}_t$ from $\mathcal{N}(\mathbf{o}_t|\boldsymbol{\mu}_{s_t v_t}, \boldsymbol{\Sigma}_{s_t v_t})$
8: **end for**

Similarly to the GMM, dependencies of the above variables are expressed in Eq. (3.49), and we provide the generative process of the CDHMM in Algorithm 3. Figure 3.4 shows the graphical model of the CDHMM that uses the plates for the number of HMM states J, and the number of GMM components K. The graphical model of the CDHMM is rather complicated since the state transition makes it difficult to use the plate representation for t.

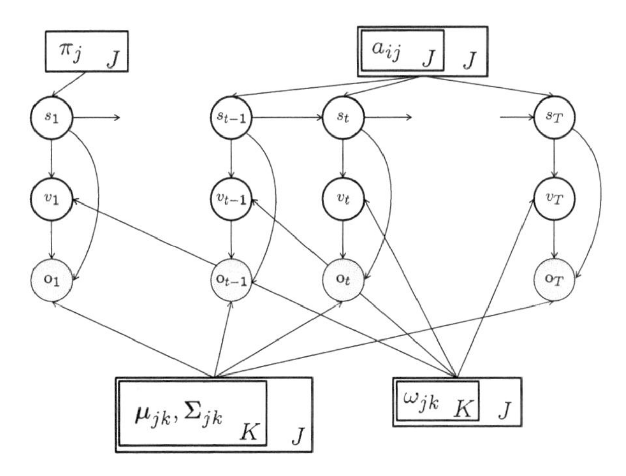

Figure 3.4 Graphical model of continuous density hidden Markov model given parameters Θ.

Thus, we provide the graphical model and generative process examples of our typical target model, CDHMM. Note that these are rather simple based on a non-Bayesian approach since we do not deal with the model parameters as probabilistic variables (do not use circle representations for them).

3.3 Forward–backward and Viterbi algorithms

This section describes how to compute the likelihood values of HMMs (given a latent variable) by using the famous forward–backward and Viterbi algorithms, which are used to solve the statistical speech recognition problem in Eq. (3.3). Although the direct computation of the likelihood values causes the combinatorial explosion problem, these algorithms can provide feasible computational costs. These algorithms are also used to estimate the model parameters, which is described in Section 3.4

3.3.1 Forward–backward algorithm

Basically, the direct computation of marginal likelihood $p(\mathbf{O}|\Theta) = \sum_S p(\mathbf{O}, S|\Theta)$ in Eq. (3.30) involves on the order of $2T \cdot J^T$ calculations, since at every $t = 1, 2, \cdots, T$, there are J possible states that can be reached (i.e., there are J^T possible state sequences), and for each such state sequence about $2T$ calculations are required for each term in the sum of Eq. (3.30). By considering the latent variable of mixture components of the CDHMM in Eq. (3.44), the direct computation of marginal likelihood $p(\mathbf{O}|\Theta) = \sum_{S,V} p(\mathbf{O}, S, V|\Theta)$ is further increased to the order of $2T \cdot J^T K^T$. However, this large computation problem could be tackled by applying the forward–backward algorithm (Rabiner & Juang 1986). According to this algorithm, the *forward variable* $\alpha_t(j)$ is defined by

$$\alpha_t(j) \triangleq p(\mathbf{o}_1, \cdots, \mathbf{o}_t, s_t = j|\Theta). \tag{3.50}$$

This forward variable is known as the probability of the partial observation sequence $\{\mathbf{o}_1, \mathbf{o}_2, \cdots, \mathbf{o}_t\}$ until time t and state j at time t given the current HMM parameters Θ. We also define an emission probability given HMM state j at frame t as

$$b_j(\mathbf{o}_t) \triangleq p(\mathbf{o}_t|s_t = j, \Theta_j). \tag{3.51}$$

In the CDHMM, this is represented by the likelihood function of the GMM, as described in Eq. (3.33). A forward procedure, which is a dynamic programming method, is inductively derived to speed up computation of $p(\mathbf{O}|\Theta)$ as follows:

Forward algorithm

- Initialization

$$\begin{aligned}\alpha_1(j) &= p(\mathbf{o}_1, s_1 = j|\Theta) \\ &= p(\mathbf{o}_1|s_1 = j, \Theta_j)p(s_1 = j|\Theta_j) \\ &= \pi_j b_j(\mathbf{o}_1), \quad 1 \le j \le J.\end{aligned} \tag{3.52}$$

- Induction

$$
\begin{aligned}
\alpha_t(j) &= p(\mathbf{o}_1, \cdots, \mathbf{o}_t, s_t = j|\Theta) \\
&= \sum_{i=1}^{J} p(\mathbf{o}_1, \cdots, \mathbf{o}_t, s_{t-1} = i, s_t = j|\Theta) \\
&= \sum_{i=1}^{J} p(\mathbf{o}_t|s_t = j, \mathbf{o}_1, \cdots, \mathbf{o}_{t-1}, s_{t-1} = i, \Theta) \\
&\qquad \times p(s_t = j|\mathbf{o}_1, \cdots, \mathbf{o}_{t-1}, s_{t-1} = i, \Theta) \\
&\qquad \times p(\mathbf{o}_1, \cdots, \mathbf{o}_{t-1}, s_{t-1} = i|\Theta) \\
&= p(\mathbf{o}_t|s_t = j, \Theta_j) \sum_{i=1}^{J} p(s_t = j|s_{t-1} = i, \Theta) p(\mathbf{o}_1, \cdots, \mathbf{o}_{t-1}, s_{t-1} = i|\Theta) \\
&= \left(\sum_{i=1}^{J} \alpha_{t-1}(i) a_{ij} \right) b_j(\mathbf{o}_t), \qquad \begin{matrix} 2 \le t \le T \\ 1 \le j \le J. \end{matrix}
\end{aligned}
\tag{3.53}
$$

- Termination

$$
\begin{aligned}
p(\mathbf{O}|\Theta) &= \sum_{j=1}^{J} p(\mathbf{o}_1, \cdots, \mathbf{o}_T, s_T = j|\Theta) \\
&= \sum_{j=1}^{J} \alpha_T(j).
\end{aligned}
\tag{3.54}
$$

Thus, we can compute the likelihood value $p(\mathbf{O}|\Theta)$ recursively by using the forward variable $\alpha_t(j)$. This iterative algorithm for computing the forward variable is called the *forward algorithm*. In Eq. (3.52) and Eq. (3.53), we see that the calculation of $\{\alpha_t(j)|1 \le t \le T, 1 \le j \le J\}$ requires on the order of J^2T calculations which is dramatically reduced from $2T \cdot J^T$ as required in the direct calculation. Figure 3.5 illustrates the sequence of operations for computation of the forward variable $\alpha_t(j)$.

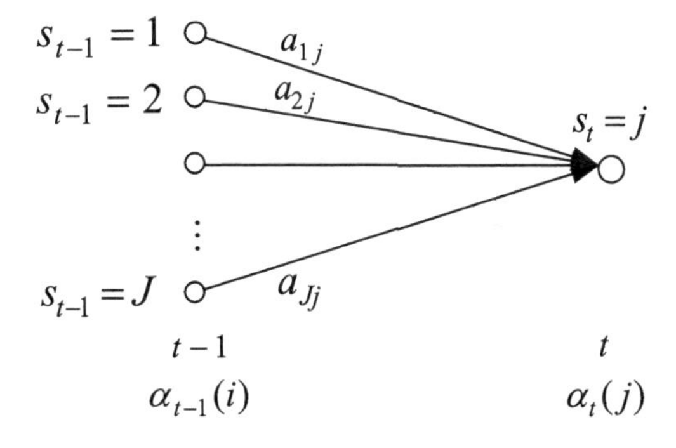

Figure 3.5 Propagation of forward variable from $\alpha_{t-1} = i$ to $\alpha_t = j$. All possible states at time $t-1$ are considered. Adapted from Rabiner & Juang (1993).

Similarly, we consider a backward variable which is defined as

$$\beta_t(j) \triangleq p(\mathbf{o}_{t+1}, \cdots, \mathbf{o}_T | s_t = j, \Theta). \tag{3.55}$$

This variable represents the probability of the partial observation sequence from $t + 1$ to the end, given state j at time t and model parameters Θ. Again, we can inductively derive the following backward procedure:

Backward algorithm

- Initialization

$$\beta_T(j) = 1, \quad 1 \leq j \leq J. \tag{3.56}$$

- Induction

  By using the sum and product rules, we can rewrite $\beta_t(i)$ as

$$\begin{aligned}
\beta_t(i) &= p(\mathbf{o}_{t+1}, \cdots, \mathbf{o}_T | s_t = i, \Theta) \\
&= \sum_{j=1}^{J} p(\mathbf{o}_{t+1}, \cdots, \mathbf{o}_T, s_{t+1} = j | s_t = i, \Theta) \\
&= \sum_{j=1}^{J} p(\mathbf{o}_{t+1}, \cdots, \mathbf{o}_T | s_{t+1} = j, s_t = i, \Theta) p(s_{t+1} = j | s_t = i, \Theta).
\end{aligned} \tag{3.57}$$

  By using the conditional independence property of the HMM,

$$\begin{aligned}
\beta_t(i) &= \sum_{j=1}^{J} p(\mathbf{o}_{t+1}, \cdots, \mathbf{o}_T | s_{t+1} = j, \Theta) p(s_{t+1} = j | s_t = i, \Theta) \\
&= \sum_{j=1}^{J} p(\mathbf{o}_{t+2}, \cdots, \mathbf{o}_T | s_{t+1} = j, \Theta) p(\mathbf{o}_{t+1} | s_{t+1} = j, \Theta_j) \\
&\qquad \times p(s_{t+1} = j | s_t = i, \Theta) \\
&= \sum_{j=1}^{J} a_{ij} b_j(\mathbf{o}_{t+1}) \beta_{t+1}(j), t = T-1, T-2, \cdots, 1, \quad 1 \leq i \leq J.
\end{aligned} \tag{3.58}$$

- Termination

$$\begin{aligned}
\beta_0 &\triangleq p(\mathbf{O} | \Theta) \\
&= \sum_{j=1}^{J} p(\mathbf{o}_1, \cdots, \mathbf{o}_T, s_1 = j | \Theta) \\
&= \sum_{j=1}^{J} p(\mathbf{o}_1, \cdots, \mathbf{o}_T | s_1 = j, \Theta) p(s_1 = j | \Theta) \\
&= \sum_{j=1}^{J} p(\mathbf{o}_1 | s_1 = j, \Theta) p(\mathbf{o}_2, \cdots, \mathbf{o}_t | s_1 = j, \Theta) p(s_1 = j | \Theta) \\
&= \sum_{j=1}^{J} \pi_j b_j(\mathbf{o}_1) \beta_1(j).
\end{aligned} \tag{3.59}$$

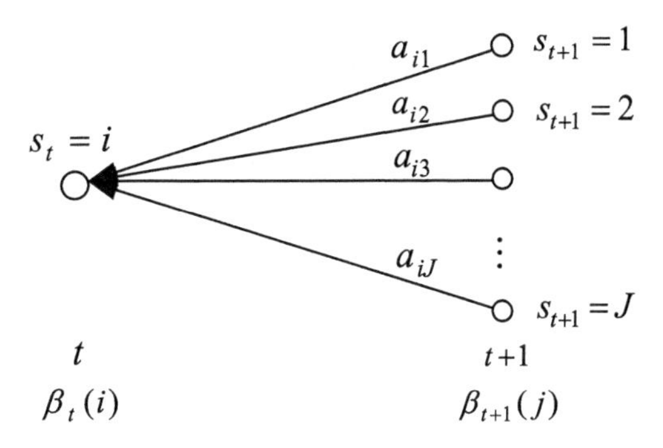

Figure 3.6 Propagation of backward variable from $\alpha_{t+1} = j$ to $\alpha_t = i$. All possible states at time $t+1$ are considered. Adapted from Rabiner & Juang (1993).

Thus, the backward algorithm can also compute the likelihood value $p(\mathbf{O}|\Theta)$ recursively by using the backward variable $\beta_t(j)$. In the initialization step of the backward procedure, $\beta_T(j)$ is arbitrarily assigned to be 1. In the induction step, the backward variable $\beta_t(i)$ is calculated in a backward fashion from $t = T-1$ to the beginning $t = 1$. Figure 3.6 shows the sequence of operations required for computation of the backward variable $\beta_t(i)$. This iterative algorithm for computing the backward variable is called the *backward algorithm.*

The forward variable $\alpha_t(j)$ and backward variable $\beta_t(j)$ are used to calculate the posterior probability of a specific case. For example, if we consider the posterior probability $p(s_t = j|\mathbf{O}, \Theta)$ when the HMM state is j at frame t, we first define

$$\gamma_t(j) \triangleq p(s_t = j|\mathbf{O}, \Theta), \tag{3.60}$$

which is useful in the later explanations. By using the sum and product rules, $\gamma_t(j)$ is represented by the likelihood ratio of

$$\begin{aligned} \gamma_t(j) &\triangleq \frac{p(s_t = j, \mathbf{O}|\Theta)}{p(\mathbf{O}|\Theta)} \\ &= \frac{p(\mathbf{O}, s_t = j|\Theta)}{\sum_{i=1}^{J} p(\mathbf{O}, s_t = i|\Theta)}. \end{aligned} \tag{3.61}$$

Now we focus on the joint distribution $p(s_t = j, \mathbf{O}|\Theta)$, which is rewritten by forward variable $\alpha_t(j)$ in Eq. (3.50) and backward variable $\beta_t(j)$ in Eq. (3.55), as follows:

$$\begin{aligned} p(s_t = j, \mathbf{O}|\Theta) &= p(\mathbf{O}|s_t = j, \Theta)p(s_t = j|\Theta) \\ &= p(\mathbf{o}_1, \cdots, \mathbf{o}_t|s_t = j, \Theta)p(\mathbf{o}_{t+1}, \cdots, \mathbf{o}_T|s_t = j, \Theta)p(s_t = j|\Theta) \\ &= \underbrace{p(\mathbf{o}_1, \cdots, \mathbf{o}_t, s_t = j|\Theta)}_{\alpha_t(j)} \underbrace{p(\mathbf{o}_{t+1}, \cdots, \mathbf{o}_T|s_t = j, \Theta)}_{\beta_t(j)}. \end{aligned} \tag{3.62}$$

From the first to second lines, we use the conditional independence assumption of the HMM. Thus, finally the posterior distribution $\gamma_t(j)$ is represented as:

$$\gamma_t(j) \triangleq p(s_t = j|\mathbf{O}, \Theta) = \frac{\alpha_t(j)\beta_t(j)}{\sum_{j'=1}^{J} \alpha_t(j')\beta_t(j')}. \tag{3.63}$$

This probability will be used to find the optimal state sequence based on the Viterbi algorithm, as mentioned in Section 3.3.2, as well as to estimate the HMM parameters based on the maximum likelihood method, as addressed in Section 3.4.

3.3.2 Viterbi algorithm

Finding the optimal state sequence $\hat{S} = \{\hat{s}_t | t = 1, \cdots, T\}$ of observation sequence $\mathbf{O}$, in terms of the maximum a-posteriori sense, is seen as a fundamental problem in the HMM framework. It is formulated as

$$\hat{S} = \arg\max_{S} p(S|\mathbf{O}, \Theta), \tag{3.64}$$

where $\arg\max_S$ find the most probable $\hat{S}$ from all possible state sequences, and this also involves on the order of J^T calculations, similar to the likelihood computation $p(\mathbf{O}|\Theta)$ in Section 3.3.1. In addition, since $p(\mathbf{O}|\Theta)$ does not depend on S, we can also rewrite Eq. (3.64) as

$$\begin{aligned} \hat{S} &= \arg\max_{S} p(S|\mathbf{O}, \Theta) = \arg\max_{S} p(S|\mathbf{O}, \Theta) p(\mathbf{O}|\Theta) \\ &= \arg\max_{S} p(S, \mathbf{O}|\Theta). \end{aligned} \tag{3.65}$$

That is, the optimal state sequences obtained in terms of the posterior distribution $p(S|\mathbf{O}, \Theta)$ and the joint distribution $p(S, \mathbf{O}|\Theta)$ are equivalent.

The optimal state sequence can be used to determine the segmentation of a speech sentence into phones or sub-phones, if we assume these as latent variables. In this case, the speech frames staying in the same state behave similarly, and the transition from one state to the other is treated as a segmentation boundary.

In addition, the optimal state sequence $\hat{S}$ is applied to approximate the calculation of the marginal likelihood function over S as follows:

$$p(\mathbf{O}|\Theta) = \sum_{S} p(S, \mathbf{O}|\Theta) \approx p(\hat{S}, \mathbf{O}|\Theta). \tag{3.66}$$

That is, the marginal likelihood function is approximately represented as the joint likelihood distribution of $\hat{S}$ and $\mathbf{O}$, which can be represented with Eq. (3.30) as

$$\begin{aligned} p(\hat{S}, \mathbf{O}|\Theta) &= \left(\pi_{\hat{s}_1} p(\mathbf{o}_1|\Theta_{\hat{s}_1}) \prod_{t=2}^{T} a_{\hat{s}_{t-1}\hat{s}_t} p(\mathbf{o}_t|\Theta_{\hat{s}_t}) \right) \\ &\triangleq \hat{p}. \end{aligned} \tag{3.67}$$

Therefore, the approximation of marginal likelihood using the optimal state sequence provides the solution to the *segmental* likelihood function $\hat{p}$ since likelihood calculation is based on the speech segmentation using $\hat{S}$. Considering all possible state sequences, $\{S\}$ is simplified to applying only the optimal state sequence $\hat{S}$. The number of calculations is further reduced to $2T$ as referred to Eq. (3.67).

We can individually determine the most likely state $\hat{s}_t$ for each time frame $1 \leq t \leq T$ according to the posterior probability as

$$\hat{s}_t = \arg \max_{1 \leq j \leq J} p(s_t = j | \mathbf{O}, \Theta). \tag{3.68}$$

The posterior probability $p(s_t = j | \mathbf{O}, \Theta)$ is calculated according to the forward-backward algorithm. As shown in Eq. (3.63), the joint distribution of observation sequence $\mathbf{O}$ and state $s_t = j$ at time t is calculated as a product of the forward variable, which accounts for partial observations $\{\mathbf{o}_1, \cdots, \mathbf{o}_t\}$ and state j at time t, and the backward variable, which accounts for the remainder of observations $\{\mathbf{o}_{t+1}, \cdots, \mathbf{o}_T\}$ given state j at time t with the normalization factor. However, Eq. (3.68) may not determine a *valid* optimal state sequence since the probability of occurrence of the *sequences* of states is not considered.

To cope with this problem, we need to find the single best state sequence by using Eq. (3.64) or an equivalent form of (3.65), directly. A dynamic programming method, called the Viterbi algorithm (Viterbi 1967), was proposed to efficiently find the single best state sequence. To do so, we define the highest probability along a single path, at time t, which accounts for the first t observations and ends in state j using a new notation:

$$\delta_t(j) \triangleq \max_{s_1, \cdots, s_{t-1}} p(s_1, \cdots, s_t = j, \mathbf{o}_1, \cdots, \mathbf{o}_t | \Theta). \tag{3.69}$$

By induction, a recursive formula of $\delta_{t+1}(j)$ from $\delta_t(j)$ is derived to calculate this probability. To derive the equation, we first focus on the joint distribution appearing in $\delta_{t+1}(j)$, which can be rewritten when $s_t = i$ and $s_{t+1} = j$ as:

$$\begin{aligned} & p(s_1, \cdots, i, j, \mathbf{o}_1, \cdots, \mathbf{o}_t, \mathbf{o}_{t+1} | \Theta) \\ & = p(s_1, \cdots, i, \mathbf{o}_1, \cdots, \mathbf{o}_t | \Theta) p(j, \mathbf{o}_{t+1} | s_1, \cdots, i, \mathbf{o}_1, \cdots, \mathbf{o}_t, \Theta) \\ & = p(s_1, \cdots, i, \mathbf{o}_1, \cdots, \mathbf{o}_t | \Theta) p(j | i, \Theta) p(\mathbf{o}_{t+1} | j, \Theta) \\ & = p(s_1, \cdots, i, \mathbf{o}_1, \cdots, \mathbf{o}_t | \Theta) a_{ij} b_j(\mathbf{o}_{t+1}). \end{aligned} \tag{3.70}$$

Here we use the conditional independence of the HMM from the second to the third lines. Thus, by using Eq. (3.69), $\delta_{t+1}(j)$ is computed recursively from $\delta_{t+1}(i)$ as:

$$\begin{aligned} \delta_{t+1}(j) & = \max_{s_1, \cdots, i} p(s_1, \cdots, i, \mathbf{o}_1, \cdots, \mathbf{o}_t | \Theta) a_{ij} b_j(\mathbf{o}_{t+1}) \\ & = \left(\max_i \delta_t(i) a_{ij} \right) b_j(\mathbf{o}_{t+1}). \end{aligned} \tag{3.71}$$

We need to keep track of the state that maximized Eq. (3.71) so as to backtrack to the single best state sequence in the following Viterbi algorithm:

- Initialization

$$\delta_1(i) = \pi_i b_i(\mathbf{o}_1), \quad \psi_1(i) = 0, \quad 1 \leq i \leq J. \tag{3.72}$$

- Recursion

$$\delta_t(j) = \left(\max_{1 \leq i \leq J} \delta_{t-1}(i) a_{ij} \right) \cdot b_j(\mathbf{o}_{t+1}),$$

$$\psi_t(j) = \left(\arg\max_{1\le i\le J} \delta_{t-1}(i)a_{ij}\right), \qquad \begin{array}{c} 2 \le t \le T \\ 1 \le j \le J. \end{array} \tag{3.73}$$

- Termination

$$\hat{p} = \max_{1\le j\le J} \delta_T(i),$$
$$\hat{s}_T = \arg\max_{1\le j\le J} \delta_T(i). \tag{3.74}$$

- State sequence backtracking

$$\hat{s}_t = \psi_{t+1}(\hat{s}_{t+1}), \qquad t = T-1, T-2, \cdots, 1. \tag{3.75}$$

In the termination step, the segmental likelihood function $\hat{p}$ is calculated and is equivalent to Eq. (3.67). It is noteworthy that this Viterbi algorithm is similar to the forward procedure in the forward–backward algorithm. The key difference is the maximization in Eq. (3.73) over previous states, which is used in place of a summation operation in Eq. (3.53). The variable $\delta_t(j)$ in the Viterbi algorithm is meaningfully related to the forward variable $\alpha_t(j)$ in the forward–backward algorithm.

Now, we summarize what we can compute from the HMM without taking on the combinatorial explosion problem. These values are used in the decoding step and the training step of estimating model parameters Θ, which is discussed in the next section.

- $p(\mathbf{O}|\Theta)$:
  The marginalized likelihood function from the forward or backward algorithm.
- $\gamma_t(j) \triangleq p(s_t = j|\mathbf{O}, \Theta)$:
  The posterior probability of $s_t = j$ from the forward–backward algorithm.
- $\hat{S} = \arg\max_S p(S|\mathbf{O}, \Theta) = \arg\max_S p(S, \mathbf{O}|\Theta)$:
  The optimal state sequence from the Viterbi algorithm.
- $p(\hat{S}, \mathbf{O}|\Theta)$:
  The segmental joint likelihood function from the Viterbi algorithm.

3.4 Maximum likelihood estimation and EM algorithm

Previous sections discuss the HMM-based speech modeling given model parameters Θ to compute the likelihood values and so on, efficiently based on the forward, backward, and Viterbi algorithms. One of the powerful properties of the HMM is that it also provides an efficient algorithm to obtain the model parameters based on the ML estimation from training data. The training algorithm is based on the EM algorithm, which can tackle the incomplete data problem in ML estimation. In what follows, we address how the EM algorithm is derived by applying Jensen's inequality and show the procedure of EM steps for estimation of HMM parameters, which involves latent variables in probabilistic functions (Nakagawa 1988, Huang *et al.* 1990, Rabiner & Juang 1993, Bilmes 1998).

3.4.1 Jensen's inequality

We start the discussion of HMM parameter estimation based on the Maximum Likelihood (ML) criterion, so that the optimal model parameters $\hat{\Theta}$ are computed by the likelihood function $p(\mathbf{O}|\Theta)$:[2]

$$\hat{\Theta} = \arg\max_{\Theta} p(\mathbf{O}|\Theta). \tag{3.76}$$

We assume that speech data $\mathbf{O}$ are observed and are generated by some distribution. However, as we discussed in Section 3.2, the HMM has additional hidden variables S (HMM state sequence) and V (GMM component sequence), which are not observed. In this situation where the model includes unobserved variables, $\mathbf{O}$ is called incomplete data (S and V). (Θ would also be included as unobserved variables in a broad sense, but this book considers latent variables as unobserved variables.). Conversely, the complete data $\mathbf{Y}$ are composed of the observed data $\mathbf{O}$ as well as the hidden variables $\{S, V\}$.

In general, the maximum likelihood estimation of the model parameter Θ for complete data $\mathbf{O}$ is difficult, since we need to consider the following equation:

$$\hat{\Theta} = \arg\max_{\Theta} \sum_{S,V} p(\mathbf{O}, S, V|\Theta). \tag{3.77}$$

As we discussed in Section 3.3, the summation over all possible S and V has to overcome the combinatorial explosion problem, and the direct optimization of Θ for this equation is not feasible. This problem is called the *incomplete data problem*.

According to the EM algorithm, the incomplete data problem in the ML estimation is resolved by iteratively and alternatively performing the expectation step (E-step) and the maximization step (M-step), which results in obtaining the local optimum solution of the model parameters. In the E-step, we calculate an auxiliary function $Q(\Theta'|\Theta)$, which is an expectation of the logarithm of the likelihood function using new HMM parameters Θ' given the current parameters Θ, i.e.

$$\begin{aligned} Q(\Theta'|\Theta) &= \mathbb{E}_{(S,V)}[\log p(\mathbf{O}, S, V|\Theta')|\mathbf{O}, \Theta] \\ &= \sum_{S}\sum_{V} p(S, V|\mathbf{O}, \Theta) \log p(\mathbf{O}, S, V|\Theta'). \end{aligned} \tag{3.78}$$

In the M-step, we maximize the auxiliary function instead of Eq. (3.77) with respect to the HMM parameters Θ' and estimate new parameters by

$$\hat{\Theta}' = \arg\max_{\Theta'} Q(\Theta'|\Theta). \tag{3.79}$$

The updated HMM parameters $\hat{\Theta}'$ are then treated as the current parameters for the next iteration of EM steps. This iterative estimation only obtains local optimum solutions, and not global optimum solutions. However, a careful setting of the initial model parameters would help the solution to reach appropriate parameter values, and moreover, the algorithm theoretically guarantees that the likelihood value is always increased as the number of iterations increases. This property is very useful in the implementation,

[2] The optimization with respect to the posterior distribution $p(\Theta|\mathbf{O})$ is more reasonable in the Bayesian sense, and it will be discussed in the following chapters.

since we can easily debug training source codes based on the EM algorithm by checking likelihood values.

Now, we prove how this indirect optimization for the auxiliary function $Q(\Theta'|\Theta)$ increases a likelihood value. For this proof, we first define the following logarithmic marginal function $L(\Theta')$:

$$L(\Theta') \triangleq \log p(\mathbf{O}|\Theta'). \tag{3.80}$$

Note that since the logarithmic function is a monotonically increasing function, we can check $L(\Theta')$ instead of $p(\mathbf{O}|\Theta')$ for the proof. We introduce the notation of complete data:

$$\mathbf{Y} \triangleq \{\mathbf{O}, S, V\}. \tag{3.81}$$

Then, the conditional distribution of $\mathbf{Y}$ given incomplete data $\mathbf{O}$ and parameters Θ has the following relationship from the product rule:

$$p(\mathbf{Y}|\mathbf{O}, \Theta') = \frac{p(\mathbf{Y}, \mathbf{O}|\Theta')}{p(\mathbf{O}|\Theta')} = \frac{p(\mathbf{Y}|\Theta')}{p(\mathbf{O}|\Theta')}, \tag{3.82}$$

where the likelihood function $p(\mathbf{Y}, \mathbf{O}|\Theta')$ is rewritten as $p(\mathbf{Y}|\Theta')$, since $\mathbf{Y}$ includes $\mathbf{O}$. Therefore, by taking the logarithmic operation for both sides in Eq. (3.82), we derive the equation for $L(\Theta')$ as

$$L(\Theta') = \log p(\mathbf{Y}|\Theta') - \log p(\mathbf{Y}|\mathbf{O}, \Theta'). \tag{3.83}$$

Then, by taking the expectation operation with respect to the posterior distribution of latent variables $p(S, V|\mathbf{O}, \Theta)$ for Eq. (3.83), Eq. (3.83) is represented as

$$\begin{aligned} L(\Theta') &= \mathbb{E}_{(S,V)}[\log p(\mathbf{Y}|\boldsymbol{\Theta}')|\mathbf{O}, \Theta] - \mathbb{E}_{(S,V)}[\log p(\mathbf{Y}|\mathbf{O}, \Theta')|\mathbf{O}, \Theta] \\ &= Q(\Theta'|\Theta) - H(\Theta'|\Theta), \end{aligned} \tag{3.84}$$

where $H(\Theta'|\Theta)$ is defined as follows:

$$H(\Theta'|\Theta) \triangleq \mathbb{E}_{(\mathbf{Y})}[\log p(\mathbf{Y}|\mathbf{O}, \Theta')|\mathbf{O}, \Theta]. \tag{3.85}$$

$Q(\Theta'|\Theta)$ is defined in Eq. (3.78). Then, we obtain this relation from Eq. (3.85):

$$Q(\Theta'|\Theta) = L(\Theta') + H(\Theta'|\Theta). \tag{3.86}$$

Here, Jensen's inequality is introduced to prove the inequality (Dempster *et al.* 1976),

$$H(\Theta'|\Theta) \le H(\Theta|\Theta). \tag{3.87}$$

- **Jensen's inequality**

  If x is a random variable with mean value $\mathbb{E}_{(x)}[x] = \mu$ and $f(x)$ is a *convex function*, then

  $$\mathbb{E}_{(x)}[f(x)] \ge f[\mathbb{E}_{(x)}(x)], \tag{3.88}$$

  with equality if and only if x is a degenerate distribution at μ.

The inequality in Eq. (3.87) is derived by

$$
\begin{aligned}
H(\Theta|\Theta) - H(\Theta'|\Theta) &= \mathbb{E}_{(\mathbf{Y})}\left[\log \frac{p(\mathbf{Y}|\mathbf{O}, \boldsymbol{\Theta})}{p(\mathbf{Y}|\mathbf{O}, \boldsymbol{\Theta}')}\middle| \mathbf{O}, \boldsymbol{\Theta}\right] \\
&= \int p(\mathbf{Y}|\mathbf{O}, \Theta)\left(-\log \frac{p(\mathbf{Y}|\mathbf{O}, \Theta')}{p(\mathbf{Y}|\mathbf{O}, \Theta)}\right) d\mathbf{Y} \\
&= \mathrm{KL}(p(\mathbf{Y}|\mathbf{O}, \Theta) \| p(\mathbf{Y}|\mathbf{O}, \Theta')) \\
&\geq -\log\left(\int p(\mathbf{Y}|\mathbf{O}, \Theta)\frac{p(\mathbf{Y}|\mathbf{O}, \Theta')}{p(\mathbf{Y}|\mathbf{O}, \Theta)} d\mathbf{Y}\right) = 0,
\end{aligned} \tag{3.89}
$$

where a convex function based on the negative logarithm $f(x) = -\log(x)$ is adopted. As shown in Eq. (3.89), the difference $H(\Theta|\Theta) - H(\Theta'|\Theta)$ is obtained as the *relative entropy* or *Kullback–Leibler (KL) divergence* (Kullback & Leibler 1951) between the distributions $p(\mathbf{Y}|\mathbf{O}, \Theta)$ and $p(\mathbf{Y}|\mathbf{O}, \Theta')$, that is

$$
\mathrm{KL}(p(\mathbf{Y}|\mathbf{O}, \Theta) \| p(\mathbf{Y}|\mathbf{O}, \Theta')) \triangleq \int p(\mathbf{Y}|\mathbf{O}, \Theta)\left(-\log \frac{p(\mathbf{Y}|\mathbf{O}, \Theta')}{p(\mathbf{Y}|\mathbf{O}, \Theta)}\right) d\mathbf{Y}. \tag{3.90}
$$

Given Eqs. (3.87) and (3.86), it is straightforward to see that if Θ' satisfies

$$
Q(\Theta'|\Theta) \geq Q(\Theta|\Theta), \tag{3.91}
$$

then we can prove that

$$
L(\Theta') = Q(\Theta'|\Theta) - H(\Theta'|\Theta) \geq Q(\Theta|\Theta) - H(\Theta|\Theta) = L(\Theta). \tag{3.92}
$$

Since $L(\Theta) = \log p(\mathbf{O}|\Theta)$ and the logarithmic function is a monotonic function,

$$
Q(\Theta'|\Theta) \geq Q(\Theta|\Theta) \Rightarrow p(\mathbf{O}|\Theta') \geq p(\mathbf{O}|\Theta). \tag{3.93}
$$

Since $\hat{\Theta}' = \arg\max_{\Theta'} Q(\Theta'|\Theta)$ always satisfies the inequality Eq. (3.91), we prove that the parameter estimated by the EM procedure, $\hat{\Theta}'$, always increases the likelihood value as:

$$
\hat{\Theta}' = \arg\max_{\Theta'} Q(\Theta'|\Theta) \Rightarrow p(\mathbf{O}|\Theta') \geq p(\mathbf{O}|\Theta). \tag{3.94}
$$

Such an EM procedure is bound to monotonically increase the auxiliary function $Q(\Theta'|\Theta)$ as well as the original likelihood function $p(\mathbf{O}|\Theta')$. Note that since it is not a direct optimization of the original likelihood function, the optimization of the auxiliary function leads to a local optimum solution to the ML parameter estimation.

3.4.2 Expectation step

To find ML estimates of HMM parameters, we expand the auxiliary function in Eq. (3.78) and rewrite it by substituting the joint distribution of complete data likelihood (Eq. (3.49)) into Eq. (3.78) as

$$Q(\Theta'|\Theta) = \mathbb{E}_{(S,V)}[\log p(\mathbf{O}, S, V|\Theta')|\mathbf{O}, \Theta]$$
$$= \sum_{S,V} p(S, V|\mathbf{O}, \Theta)\Bigg(\log \pi'_{s_1} + \left(\sum_{t=2}^{T} \log a'_{s_{t-1}s_t}\right)$$
$$+ \left(\sum_{t=1}^{T} \log \omega'_{s_t v_t} + \log \mathcal{N}(\mathbf{o}_t|\boldsymbol{\mu}'_{s_t v_t}, \boldsymbol{\Sigma}'_{s_t v_t})\right)\Bigg). \tag{3.95}$$

Note that four terms depend on the initial weight π_j, state transition probability a_{ij}, mixture weight w_{jk}, and Gaussian parameters $\{\boldsymbol{\mu}_{jk}, \boldsymbol{\Sigma}_{jk}\}$, respectively. We provide the solution for each term:

- $Q(\boldsymbol{\pi}'|\boldsymbol{\pi})$
  We first focus on the first term depending on the initial weight π_j, and define the following auxiliary function for π_j:

$$Q(\boldsymbol{\pi}'|\boldsymbol{\pi}) \triangleq \sum_{S,V} p(S, V|\mathbf{O}, \Theta) \log \pi'_{s_1}. \tag{3.96}$$

Since π'_{s_1} only depends on s_1, we obtain the following equation that marginalizes $p(S, V|\mathbf{O}, \Theta)$ over $S_{\backslash s_1} = \{s_2, \cdots, s_T\}$ and V as:

$$\sum_{S_{\backslash s_1}, V} p(S, V|\mathbf{O}, \Theta) = p(s_1|\mathbf{O}, \Theta). \tag{3.97}$$

Therefore, $Q(\boldsymbol{\pi}'|\boldsymbol{\pi})$ can be rewritten as

$$Q(\boldsymbol{\pi}'|\boldsymbol{\pi}) = \sum_{s_1} p(s_1|\mathbf{O}, \Theta) \log \pi'_{s_1}$$
$$= \sum_{j=1}^{J} p(s_1 = j|\mathbf{O}, \Theta) \log \pi'_j$$
$$= \sum_{j=1}^{J} \gamma_1(j) \log \pi'_j, \tag{3.98}$$

where $\gamma_1(j)$ is an occupation probability defined in Eq. (3.60) as:

$$\gamma_1(j) \triangleq p(s_1 = j|\mathbf{O}, \Theta). \tag{3.99}$$

This is computed from the forward–backward algorithm.

- $Q(\mathbf{A}'|\mathbf{A})$
  Next, we focus on the second term in Eq. (3.95), which depends on the state transition probability a_{ij}, and define the following auxiliary function for a_{ij}:

$$Q(\mathbf{A}'|\mathbf{A}) \triangleq \sum_{S,V} p(S, V|\mathbf{O}, \Theta) \sum_{t=1}^{T-1} \log a'_{s_t s_{t+1}}. \tag{3.100}$$

Here we replace the summation from $\sum_{t=2}^{T} \log a'_{s_{t-1}s_t}$ to $\sum_{t=1}^{T-1} \log a'_{s_t s_{t+1}}$ for notational convention. Similar to $Q(\boldsymbol{\pi}'|\boldsymbol{\pi})$, we obtain

$$\sum_{S_{\backslash s_t, s_{t+1}}, V} p(S, V|\mathbf{O}, \Theta) = p(s_t, s_{t+1}|\mathbf{O}, \Theta). \tag{3.101}$$

Therefore, by replacing the summation over t with the summation over S, V, we obtain

$$\begin{aligned} Q(\mathbf{A}'|\mathbf{A}) &= \sum_{t=1}^{T-1} \sum_{S,V} p(S, V|\mathbf{O}, \Theta) \log a'_{s_t s_{t+1}} \\ &= \sum_{t=1}^{T-1} \sum_{s_t, s_{t+1}} p(s_t, s_{t+1}|\mathbf{O}, \Theta) \log a'_{s_t s_{t+1}} \\ &= \sum_{t=1}^{T-1} \sum_{i=1}^{J} \sum_{j=1}^{J} p(s_t = i, s_{t+1} = j|\mathbf{O}, \Theta) \log a'_{ij} \\ &= \sum_{t=1}^{T-1} \sum_{i=1}^{J} \sum_{j=1}^{J} \xi_t(i,j) \log a'_{ij}, \end{aligned} \tag{3.102}$$

where $\xi_t(i,j)$ is an expected transition probability from $s_t = i$ to $s_{t+1} = j$, and is defined as:

$$\xi_t(i,j) \triangleq p(s_t = i, s_{t+1} = j|\mathbf{O}, \Theta). \tag{3.103}$$

Note that by using this technique, the summation over all possible sequences (that leads to a combinatorial explosion) can be replaced with a summation over the number of HMM states and the number of frames. Thus, this auxiliary function can be computed feasibly. For this computation, we need to obtain $\xi_t(i,j)$, which is discussed later as a variant of the forward–backward algorithm.

- $Q(\boldsymbol{\omega}'|\boldsymbol{\omega})$

  The third term in Eq. (3.95) depends on the mixture weight ω_{jk}, and so we define the following auxiliary function for ω_{jk}:

$$Q(\boldsymbol{\omega}'|\boldsymbol{\omega}) \triangleq \sum_{S,V} p(S, V|\mathbf{O}, \Theta) \sum_{t=1}^{T} \log \omega'_{s_t v_t}. \tag{3.104}$$

Similarly to the case of a_{ij}, we first obtain the following equation:

$$\sum_{S_{\backslash s_t}, V_{\backslash v_t}} p(S, V|\mathbf{O}, \Theta) = p(s_t, v_t|\mathbf{O}, \Theta). \tag{3.105}$$

Therefore,

$$\begin{aligned} Q(\boldsymbol{\omega}'|\boldsymbol{\omega}) &= \sum_{t=1}^{T} \sum_{S,V} p(S, V|\mathbf{O}, \Theta) \log \omega'_{s_t v_t} \\ &= \sum_{t=1}^{T} \sum_{s_t, v_t} p(s_t, v_t|\mathbf{O}, \Theta) \log \omega'_{s_t v_t} \end{aligned}$$

$$= \sum_{t=1}^{T} \sum_{j=1}^{J} \sum_{k=1}^{K} p(s_t = j, v_t = k | \mathbf{O}, \Theta) \log \omega'_{jk}$$

$$= \sum_{t=1}^{T} \sum_{j=1}^{J} \sum_{k=1}^{K} \gamma_t(j,k) \log \omega'_{jk}, \tag{3.106}$$

where $\gamma_t(j,k)$ is an expected occupation probability at $s_t = j$ and $v_t = k$, and is defined as:

$$\gamma_t(j,k) \triangleq p(s_t = j, v_t = k | \mathbf{O}, \Theta). \tag{3.107}$$

The computation of $\gamma_t(j,k)$ is also discussed later as a variant of the forward–backward algorithm.

- $Q(\boldsymbol{\mu}', \boldsymbol{\Sigma}' | \boldsymbol{\mu}, \boldsymbol{\Sigma})$
  Finally, the fourth term in Eq. (3.95) depends on the Gaussian parameters $\boldsymbol{\mu}_{jk}$ and $\boldsymbol{\Sigma}_{jk}$, and defines the following auxiliary function for $\boldsymbol{\mu}_{jk}$ and $\boldsymbol{\Sigma}_{jk}$:

$$Q(\boldsymbol{\mu}', \boldsymbol{\Sigma}' | \boldsymbol{\mu}, \boldsymbol{\Sigma}) \triangleq \sum_{S,V} p(S, V | \mathbf{O}, \Theta) \sum_{t=1}^{T} \log \mathcal{N}(\mathbf{o}_t | \boldsymbol{\mu}'_{s_t v_t}, \boldsymbol{\Sigma}'_{s_t v_t}). \tag{3.108}$$

Similarly to $Q(\boldsymbol{\omega}' | \boldsymbol{\omega})$, by using Eq. (3.105), $Q(\boldsymbol{\mu}', \boldsymbol{\Sigma}' | \boldsymbol{\mu}, \boldsymbol{\Sigma})$ can be rewritten with $\gamma_t(j,k)$ as

$$Q(\boldsymbol{\mu}', \boldsymbol{\Sigma}' | \boldsymbol{\mu}, \boldsymbol{\Sigma}) = \sum_{t=1}^{T} \sum_{j=1}^{J} \sum_{k=1}^{K} \gamma_t(j,k) \log \mathcal{N}(\mathbf{o}_t | \boldsymbol{\mu}'_{jk}, \boldsymbol{\Sigma}'_{jk}). \tag{3.109}$$

By using the definition of the multivariate Gaussian distribution in Appendix C.6:

$$\mathcal{N}(\mathbf{x} | \boldsymbol{\mu}, \boldsymbol{\Sigma}) \triangleq (2\pi)^{-\frac{D}{2}} |\boldsymbol{\Sigma}|^{-\frac{1}{2}} \exp\left(-\frac{1}{2} (\mathbf{x} - \boldsymbol{\mu})^{\mathsf{T}} \boldsymbol{\Sigma}^{-1} (\mathbf{x} - \boldsymbol{\mu}) \right). \tag{3.110}$$

Equation (3.109) can be rewritten as

$$\begin{aligned}
& Q(\boldsymbol{\mu}', \boldsymbol{\Sigma}' | \boldsymbol{\mu}, \boldsymbol{\Sigma}) \\
&= \sum_{t=1}^{T} \sum_{j=1}^{J} \sum_{k=1}^{K} \gamma_t(j,k) \log \left((2\pi)^{-\frac{D}{2}} |\boldsymbol{\Sigma}'_{jk}|^{-\frac{1}{2}} \right. \\
&\quad \left. \times \exp\left(-\frac{1}{2} (\mathbf{o}_t - \boldsymbol{\mu}'_{jk})^{\mathsf{T}} (\boldsymbol{\Sigma}'_{jk})^{-1} (\mathbf{o}_t - \boldsymbol{\mu}'_{jk}) \right) \right) \\
&\propto \sum_{t=1}^{T} \sum_{j=1}^{J} \sum_{k=1}^{K} -\frac{\gamma_t(j,k)}{2} \left(\log \left(|\boldsymbol{\Sigma}'_{jk}| \right) + (\mathbf{o}_t - \boldsymbol{\mu}'_{jk})^{\mathsf{T}} (\boldsymbol{\Sigma}'_{jk})^{-1} (\mathbf{o}_t - \boldsymbol{\mu}'_{jk}) \right),
\end{aligned} \tag{3.111}$$

where $\propto$ denotes the proportional relationship in the logarithmic domain. That is, the normalization factor in the linear domain is changed for the normalization constant in the logarithmic domain, and we shall continue to use $\propto$ in this book. Therefore, the normalization constant term that does not depend on $\boldsymbol{\mu}$, $\boldsymbol{\Sigma}$ is omitted from this expression.

Thus, we summarize the auxiliary function $Q(\Theta'|\Theta)$:

$$Q(\Theta'|\Theta) = Q(\pi'|\pi) + Q(\mathbf{A}'|\mathbf{A}) + Q(\omega'|\omega) + Q(\mu', \Sigma'|\mu, \Sigma), \tag{3.112}$$

where each term is defined as follows:

$$Q(\pi'|\pi) = \sum_{j=1}^{J} \gamma_1(j) \log \pi'_j, \tag{3.113}$$

$$Q(\mathbf{A}'|\mathbf{A}) = \sum_{t=1}^{T-1} \sum_{i=1}^{J} \sum_{j=1}^{J} \xi_t(i,j) \log a'_{ij}, \tag{3.114}$$

$$Q(\omega'|\omega) = \sum_{t=1}^{T} \sum_{j=1}^{J} \sum_{k=1}^{K} \gamma_t(j,k) \log \omega'_{jk}, \tag{3.115}$$

$$Q(\mu', \Sigma'|\mu, \Sigma) \propto \sum_{t=1}^{T} \sum_{j=1}^{J} \sum_{k=1}^{K} -\frac{\gamma_t(j,k)}{2} \Big(\log |\Sigma'_{jk}| + (\mathbf{o}_t - \mu'_{jk})^{\mathsf{T}} (\Sigma'_{jk})^{-1} (\mathbf{o}_t - \mu'_{jk}) \Big). \tag{3.116}$$

As an equivalent form of Eq. (3.116), we also write the following auxiliary function $Q(\mu', \mathbf{R}'|\mu, \mathbf{R})$, which replaces the covariance matrix Σ with the precision matrix $\mathbf{R} = \Sigma^{-1}$ as:

$$Q(\mu', \mathbf{R}'|\mu, \mathbf{R}) \propto \sum_{t=1}^{T} \sum_{j=1}^{J} \sum_{k=1}^{K} -\frac{\gamma_t(j,k)}{2} \times \Big(-\log |\mathbf{R}'_{jk}| + (\mathbf{o}_t - \mu'_{jk})^{\mathsf{T}} \mathbf{R}'_{jk} (\mathbf{o}_t - \mu'_{jk}) \Big). \tag{3.117}$$

This equivalent representation is used to make the computation simple in the following sections. Note that Eqs. (3.113)–(3.117) are represented by the following posterior distributions:

$$\xi_t(i,j) = p(s_t = i, s_{t+1} = j|\mathbf{O}, \Theta), \tag{3.118}$$

$$\gamma_t(j,k) = p(s_t = j, v_t = k|\mathbf{O}, \Theta). \tag{3.119}$$

Similarly to $\gamma_t(j) = p(s_t = j|\mathbf{O}, \Theta)$, as discussed in Section 3.3.1, these values are also computed efficiently by using the forward–backward algorithm.

First, $\gamma_t(j)$ has the following relationships with $\gamma_t(j,k)$ and $\xi_t(i,j)$:

$$\gamma_t(j) \triangleq p(s_t = j|\mathbf{O}, \Theta) = \sum_{k=1}^{K} \gamma_t(j,k) = \sum_{i=1}^{J} \xi_t(i,j). \tag{3.120}$$

The posterior probability $\gamma_t(j)$ can be calculated by using Eq. (3.63) based on the forward variables and backward variables $\{\alpha_t(j), \beta_t(j)\}$, but this can also be computed from $\xi_t(i,j)$ or $\gamma_t(j,k)$.

The posterior probability $\gamma_t(j,k)$ of occupying state j and Gaussian component k in Eq. (3.119) can be computed by the forward–backward algorithm. By using the sum and product rules, $\gamma_t(j,k)$ is represented by the posterior distribution $p(s_t = j|\mathbf{O}, \Theta)$ and joint likelihood function of $p(\mathbf{O}, v_t = k|s_t = j, \Theta)$ as:

$$\begin{aligned}\gamma_t(j,k) &= p(s_t = j|\mathbf{O}, \Theta)p(v_t = k|s_t = j, \mathbf{O}, \Theta) \\ &= p(s_t = j|\mathbf{O}, \Theta)\frac{p(\mathbf{O}, v_t = k|s_t = j, \Theta)}{\sum_{k'=1}^{K} p(\mathbf{O}, v_t = k'|s_t = j, \Theta)}.\end{aligned} \tag{3.121}$$

By using Eq. (3.33) for $p(\mathbf{O}, v_t = k|s_t = j, \Theta)$ and Eq. (3.63) for $p(s_t = j|\mathbf{O}, \Theta)$, Eq. (3.121) is represented as follows:

$$\gamma_t(j,k) = \frac{\alpha_t(j)\beta_t(j)}{\sum_{j'=1}^{J} \alpha_t(j')\beta_t(j')} \cdot \frac{\omega_{jk}\mathcal{N}(\mathbf{o}_t|\boldsymbol{\mu}_{jk}, \boldsymbol{\Sigma}_{jk})}{\sum_{k'=1}^{K} \omega_{jk'}\mathcal{N}(\mathbf{o}_t|\boldsymbol{\mu}_{jk'}, \boldsymbol{\Sigma}_{jk'})}. \tag{3.122}$$

In a similar manner, we express the posterior probability $\xi_t(i,j)$ in Eq. (3.118) by

$$\xi_t(i,j) = \frac{p(s_t = i, s_{t+1} = j, \mathbf{O}|\Theta)}{\sum_{i'=1}^{J} \sum_{j'=1}^{J} p(s_t = i', s_{t+1} = j', \mathbf{O}|\Theta)}. \tag{3.123}$$

Now we focus on the joint likelihood function $p(s_t = i, s_{t+1} = j, \mathbf{O}|\Theta)$, which is factorized by

$$\begin{aligned}&p(s_t = i, s_{t+1} = j, \mathbf{O}|\Theta) \\ &\quad = p(\mathbf{O}|s_t = i, s_{t+1} = j, \Theta)p(s_t = i, s_{t+1} = j|\Theta) \\ &\quad = p(\mathbf{O}|s_t = i, s_{t+1} = j, \Theta)p(s_{t+1} = j|s_t = i, \Theta)p(s_t = i|\Theta).\end{aligned} \tag{3.124}$$

By using the conditional independence assumption of the HMM, we can rewrite the equation as:

$$\begin{aligned}&p(s_t = i, s_{t+1} = j, \mathbf{O}|\Theta) \\ &\quad = p(\mathbf{o}_1, \cdots, \mathbf{o}_t|s_t = i, \Theta)p(\mathbf{o}_{t+1}, \cdots, \mathbf{o}_T|s_t = i, s_{t+1} = j, \Theta) \\ &\qquad \times p(s_{t+1} = j|s_t = i, \Theta)p(s_t = i|\Theta) \\ &\quad = p(\mathbf{o}_1, \cdots, \mathbf{o}_t|s_t = i, \Theta)p(\mathbf{o}_{t+1}|s_{t+1} = j, \Theta)p(\mathbf{o}_{t+2}, \cdots, \mathbf{o}_T|s_{t+1} = j, \Theta) \\ &\qquad \times p(s_{t+1} = j|s_t = i, \Theta)p(s_t = i|\Theta) \\ &\quad = \underbrace{p(\mathbf{o}_1, \cdots, \mathbf{o}_t, s_t = i, \Theta)}_{=\alpha_t(i)}\underbrace{p(\mathbf{o}_{t+1}|s_{t+1} = j, \Theta)}_{=b_j(\mathbf{o}_{t+1})}\underbrace{p(\mathbf{o}_{t+2}, \cdots, \mathbf{o}_T|s_{t+1} = j, \Theta)}_{=\beta_{t+1}(j)} \\ &\qquad \times \underbrace{p(s_{t+1} = j|s_t = i, \Theta)}_{=a_{ij}}.\end{aligned} \tag{3.125}$$

Thus, by using the forward variable $\alpha_t(i)$ in Eq. (3.50) and the backward variable $\beta_{t+1}(j)$ in Eq. (3.55), Eq. (3.123) is finally written as

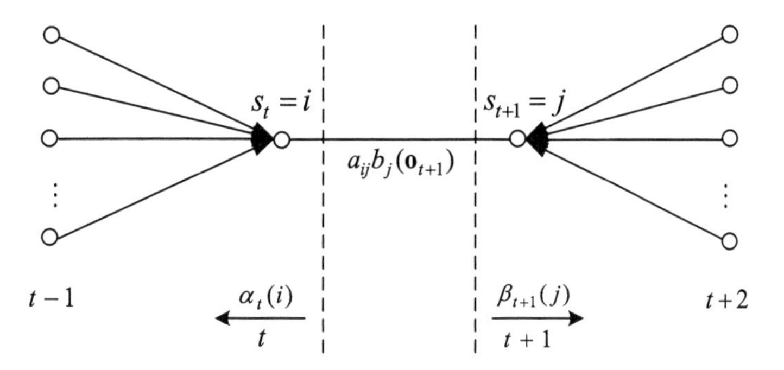

Figure 3.7 Calculation of posterior probability $\xi_t(i,j)$ based on forward variable $\alpha_t(i)$ and backward variable $\beta_{t+1}(j)$. Adapted from Rabiner & Juang (1993).

$$\begin{aligned}\xi_t(i,j) &= \frac{\alpha_t(i) a_{ij} b_j(\mathbf{o}_{t+1}) \beta_{t+1}(j)}{\sum_{i'=1}^{J} \sum_{j'=1}^{J} \alpha_t(i') a_{i'j'} b_{j'}(\mathbf{o}_{t+1}) \beta_{t+1}(j')} \\ &= \frac{\alpha_t(i) a_{ij} \left(\sum_{k=1}^{K} \omega_{jk} \mathcal{N}(\mathbf{o}_{t+1} | \boldsymbol{\mu}_{jk}, \boldsymbol{\Sigma}_{jk})\right) \beta_{t+1}(j)}{\sum_{i'=1}^{J} \sum_{j'=1}^{J} \alpha_t(i') a_{i'j'} \left(\sum_{k=1}^{K} \omega_{j'k} \mathcal{N}(\mathbf{o}_{t+1} | \boldsymbol{\mu}_{j'k}, \boldsymbol{\Sigma}_{j'k})\right) \beta_{t+1}(j')}. \end{aligned} \tag{3.126}$$

Figure 3.7 illustrates how the posterior probability $\xi_t(i,j)$ of visiting states i and j in consecutive time frames is calculated by applying the forward–backward algorithm.

In Eq. (3.63), Eq. (3.122) and Eq. (3.123), the posterior probabilities $\gamma_t(j)$, $\gamma_t(j,k)$, and $\xi_t(i,j)$ are calculated, respectively, through a kind of soft computation, i.e., the assignment information is represented by the probabilistic values for the elements i, j, and k. Instead of this soft computation, a simple and efficient approximation is to find the *segmental ML estimates* based on a hard computation where only the single best state sequence $\hat{S} = \{\hat{s}_t\}$ and mixture component sequence $\hat{V} = \{\hat{v}_t\}$ are considered with 0 (not assigned) or 1 (assigned) values for the elements i, j, and k. This computation complexity is significantly reduced. For this consideration, the calculation of posterior probabilities is simplified as

$$\gamma_t(j) = \delta(\hat{s}_t, j), \tag{3.127}$$

$$\gamma_t(j,k) = \delta(\hat{s}_t, j)\delta(\hat{v}_t, k), \tag{3.128}$$

$$\xi_t(i,j) = \delta(\hat{s}_t, i)\delta(\hat{s}_{t+1}, j), \tag{3.129}$$

where $\delta(a, a')$ denotes a Kronecker delta function that returns 1 when $a = a'$ and 0 otherwise. These probabilities are 1 for the cases of the best states i, j, and Gaussians k, and 0 for all of the other cases. Note that $\hat{s}_t$ is computed by using the Viterbi algorithm that maximizes the following segmental joint likelihood function as discussed in Section 3.3.2:

$$\hat{S} = \{\hat{s}_1, \cdots, \hat{s}_T\} = \arg\max_{S} p(S, \mathbf{O}|\Theta). \tag{3.130}$$

The value of $\hat{v}_t$, given HMM state j, is computed by

$$\hat{v}_t = \arg\max_{k} \omega_{jk} \mathcal{N}(\mathbf{o}_t | \boldsymbol{\mu}_{jk}, \boldsymbol{\Sigma}_{jk}). \tag{3.131}$$

The E-step in the EM algorithm is completed by calculating these posterior probabilities.

3.4.3 Maximization step

In the maximization step, we aim to maximize $Q(\boldsymbol{\pi}'|\boldsymbol{\pi})$, $Q(A'|A)$, $Q(\boldsymbol{\omega}'|\boldsymbol{\omega})$, and $Q(\boldsymbol{\mu}', \boldsymbol{\Sigma}'|\boldsymbol{\mu}, \boldsymbol{\Sigma})$ with respect to π', A', $\boldsymbol{\omega}'$, and $\{\boldsymbol{\mu}', \boldsymbol{\Sigma}'\}$ so as to estimate ML parameters π', A', $\boldsymbol{\omega}'$, and $\{\boldsymbol{\mu}', \boldsymbol{\Sigma}'\}$, respectively. However, the constraints of probability parameters

$$\sum_{j=1}^{J} \pi'_j = 1, \quad \sum_{j=1}^{J} a'_{ij} = 1, \forall i, \quad \sum_{k=1}^{K} \omega'_{jk} = 1, \forall j, \tag{3.132}$$

have to be imposed in the constrained optimization problem. For example, when considering the estimation of initial state probabilities $\boldsymbol{\pi}' = \{\pi'_j\}$, we construct a Lagrange function (or Lagrangian):

$$\widetilde{Q}(\boldsymbol{\pi}'|\boldsymbol{\pi}) = \sum_{j=1}^{J} \gamma_1(j) \log \pi'_j + \eta \left(\sum_{j=1}^{J} \pi'_j - 1 \right), \tag{3.133}$$

by combining the original auxiliary function in Eq. (3.113) and the constraint in Eq. (3.132) with an additional Lagrange multiplier η as a scaling factor. Then we differentiate this Lagrangian with respect to individual probability parameter π'_j and set it to zero to obtain

$$\begin{aligned} \frac{\partial \widetilde{Q}(\boldsymbol{\pi}'|\boldsymbol{\pi})}{\partial \pi'_j} &= \gamma_1(j) \frac{1}{\pi'_j} + \eta = 0 \\ \Rightarrow \quad \hat{\pi}'_j &= -\frac{1}{\eta} \gamma_1(j). \end{aligned} \tag{3.134}$$

By substituting Eq. (3.134) into the constrains in Eq. (3.132), we obtain

$$\begin{aligned} \sum_{j=1}^{J} \pi'_j &= \sum_{j=1}^{J} \left(-\frac{1}{\eta} \right) \gamma_1(j) = 1 \\ \Rightarrow \quad \eta &= -\sum_{j=1}^{J} \gamma_1(j). \end{aligned} \tag{3.135}$$

The ML estimate of new initial state probability is derived by substituting Eq. (3.135) into Eq. (3.134):

$$\hat{\pi}'_j = \frac{\gamma_1(j)}{\sum_{j'=1}^{J} \gamma_1(j')} = \gamma_1(j). \tag{3.136}$$

In the same manner, we derive the ML estimates of new state transition probability and new mixture component probability, which are provided by

$$\hat{a}'_{ij} = \frac{\sum_{t=1}^{T-1} \xi_t(i,j)}{\sum_{t=1}^{T-1} \sum_{i'=1}^{J} \xi_t(i',j)} = \frac{\sum_{t=1}^{T-1} \xi_t(i,j)}{\sum_{t=1}^{T-1} \gamma_t(j)}, \tag{3.137}$$

$$\hat{\omega}'_{jk} = \frac{\sum_{t=1}^{T} \gamma_t(j,k)}{\sum_{t=1}^{T} \sum_{k'=1}^{K} \gamma_t(j,k')} = \frac{\sum_{t=1}^{T} \gamma_t(j,k)}{\sum_{t=1}^{T} \gamma_t(j)}. \tag{3.138}$$

These results show that the initial state probability, state transition probability, and mixture weight can be computed using the ratio of the occupation statistics.

On the other hand, when estimating new HMM mean vectors $\boldsymbol{\mu}' = \{\boldsymbol{\mu}'_{jk}\}$, we individually differentiate $Q(\boldsymbol{\mu}', \boldsymbol{\Sigma}'|\boldsymbol{\mu}, \boldsymbol{\Sigma})$ in Eq. (3.117) with respect to $\boldsymbol{\mu}'_{jk}$ for each Gaussian component k in state j and set it to zero:

$$\begin{aligned}
&\frac{\partial Q(\boldsymbol{\mu}', \boldsymbol{\Sigma}'|\boldsymbol{\mu}, \boldsymbol{\Sigma})}{\partial \boldsymbol{\mu}'_{jk}} \\
&= \frac{\partial}{\partial \boldsymbol{\mu}'_{jk}} \sum_{t=1}^{T} \sum_{j'=1}^{J} \sum_{k'=1}^{K} -\frac{\gamma_t(j', k')}{2} \left(\log \left(|\boldsymbol{\Sigma}'_{j'k'}| \right) + (\mathbf{o}_t - \boldsymbol{\mu}'_{j'k'})^{\mathsf{T}} (\boldsymbol{\Sigma}'_{j'k'})^{-1} (\mathbf{o}_t - \boldsymbol{\mu}'_{j'k'}) \right) \\
&= \frac{\partial}{\partial \boldsymbol{\mu}'_{jk}} \sum_{t=1}^{T} -\frac{\gamma_t(j, k)}{2} (\mathbf{o}_t - \boldsymbol{\mu}'_{jk})^{\mathsf{T}} (\boldsymbol{\Sigma}'_{jk})^{-1} (\mathbf{o}_t - \boldsymbol{\mu}'_{jk}) \\
&= (\boldsymbol{\Sigma}'_{jk})^{-1} \sum_{t=1}^{T} \gamma_t(j, k)(\mathbf{o}_t - \boldsymbol{\mu}'_{jk}) = 0,
\end{aligned} \tag{3.139}$$

where we use the following vector derivative rule in Eq. (B.9):

$$\frac{\partial \mathbf{a}^{\mathsf{T}} \mathbf{b}}{\partial \mathbf{a}} = \frac{\partial \mathbf{b}^{\mathsf{T}} \mathbf{a}}{\partial \mathbf{a}} = \mathbf{b}. \tag{3.140}$$

Therefore, the ML estimate of the HMM mean vector is derived as shown by

$$\begin{aligned}
&\sum_{t=1}^{T} \gamma_t(j, k)(\mathbf{o}_t - \boldsymbol{\mu}'_{jk}) = 0 \\
&\Rightarrow \quad \hat{\boldsymbol{\mu}}'_{jk} = \frac{\sum_{t=1}^{T} \gamma_t(j, k) \mathbf{o}_t}{\sum_{t=1}^{T} \gamma_t(j, k)}.
\end{aligned} \tag{3.141}$$

Note that the mean vector is represented as the first-order expected value of $\mathbf{o}_t$ by using the occupation probability of $\gamma_t(j, k)$.

At the same time, the new HMM covariance matrices $\boldsymbol{\Sigma}' = \{\boldsymbol{\Sigma}'_{jk}\}$ or their inverse matrices $(\boldsymbol{\Sigma}')^{-1} = \{(\boldsymbol{\Sigma}'_{jk})^{-1} \triangleq \mathbf{R}'_{jk}\}$ (also called the precision matrices) are estimated by differentiation of $Q(\boldsymbol{\mu}', \mathbf{R}'|\boldsymbol{\mu}, \mathbf{R})$ in Eq. (3.117) with respect to $\mathbf{R}'_{jk}$ for each Gaussian k at each state j and setting it to zero:

$$\begin{aligned}
&\frac{\partial Q(\boldsymbol{\mu}', \mathbf{R}'|\boldsymbol{\mu}, \mathbf{R})}{\partial \mathbf{R}'_{jk}} \\
&= \frac{\partial}{\partial \mathbf{R}'_{jk}} \sum_{t=1}^{T} -\frac{\gamma_t(j, k)}{2} \left(-\log \left(|\mathbf{R}'_{jk}| \right) + (\mathbf{o}_t - \boldsymbol{\mu}'_{jk})^{\mathsf{T}} \mathbf{R}'_{jk} (\mathbf{o}_t - \boldsymbol{\mu}'_{jk}) \right) \\
&= \frac{1}{2} \sum_{t=1}^{T} \gamma_t(j, k)(\mathbf{R}'_{jk})^{-1} - \frac{1}{2} \sum_{t=1}^{T} \gamma_t(j, k)(\mathbf{o}_t - \boldsymbol{\mu}'_{jk})(\mathbf{o}_t - \boldsymbol{\mu}'_{jk})^{\mathsf{T}} = 0,
\end{aligned} \tag{3.142}$$

where we use the following matrix derivative rule in Eqs. (B.8) and (B.10):

$$\frac{\partial \log |\mathbf{A}|}{\partial \mathbf{A}} = \mathbf{A}^{-1}, \tag{3.143}$$

$$\frac{\partial \mathbf{a}^\mathsf{T}\mathbf{C}\mathbf{b}}{\partial \mathbf{C}} = \mathbf{a}\mathbf{b}^\mathsf{T}. \tag{3.144}$$

Thus, the new estimates of the HMM precision and covariance matrices are derived by using the ML estimate $\hat{\boldsymbol{\mu}}'_{jk}$ for $\boldsymbol{\mu}'_{jk}$ in Eq. (3.142) as

$$\begin{aligned}
\hat{\mathbf{R}}'_{jk} &= \left(\frac{\sum_{t=1}^{T} \gamma_t(j,k)(\mathbf{o}_t - \hat{\boldsymbol{\mu}}'_{jk})(\mathbf{o}_t - \hat{\boldsymbol{\mu}}'_{jk})^\mathsf{T}}{\sum_{t=1}^{T} \gamma_t(j,k)}\right)^{-1}, \\
\hat{\boldsymbol{\Sigma}}'_{jk} &= \frac{\sum_{t=1}^{T} \gamma_t(j,k)(\mathbf{o}_t - \hat{\boldsymbol{\mu}}'_{jk})(\mathbf{o}_t - \hat{\boldsymbol{\mu}}'_{jk})^\mathsf{T}}{\sum_{t=1}^{T} \gamma_t(j,k)}.
\end{aligned} \tag{3.145}$$

Interestingly, the calculation of the derived covariance matrices $\boldsymbol{\Sigma} = \{\boldsymbol{\Sigma}_{jk}\}$ in Eq. (3.145) is interpreted as the weighted ensemble expectation and covariance matrices, as well as the calculation of the ML mean vectors $\boldsymbol{\mu} = \{\boldsymbol{\mu}_{jk}\}$ in Eq. (3.141). The occupation probability $\gamma_t(j,k)$ of state j and mixture component k at time t is treated as the weighting factor in calculation of the weighted expectation function. Note that Eq. (3.145) is

$$\begin{aligned}
\hat{\boldsymbol{\Sigma}}'_{jk} &= \frac{\sum_{t=1}^{T} \gamma_t(j,k)\mathbf{o}_t\mathbf{o}_t^\mathsf{T}}{\sum_{t=1}^{T} \gamma_t(j,k)} - 2\frac{\sum_{t=1}^{T} \gamma_t(j,k)\mathbf{o}_t(\hat{\boldsymbol{\mu}}'_{jk})^\mathsf{T}}{\sum_{t=1}^{T} \gamma_t(j,k)} + \hat{\boldsymbol{\mu}}'_{jk}(\hat{\boldsymbol{\mu}}'_{jk})^\mathsf{T} \\
&= \frac{\sum_{t=1}^{T} \gamma_t(j,k)\mathbf{o}_t\mathbf{o}_t^\mathsf{T}}{\sum_{t=1}^{T} \gamma_t(j,k)} - 2\hat{\boldsymbol{\mu}}'_{jk}(\hat{\boldsymbol{\mu}}'_{jk})^\mathsf{T} + \hat{\boldsymbol{\mu}}'_{jk}(\hat{\boldsymbol{\mu}}'_{jk})^\mathsf{T} \\
&= \frac{\sum_{t=1}^{T} \gamma_t(j,k)\mathbf{o}_t\mathbf{o}_t^\mathsf{T}}{\sum_{t=1}^{T} \gamma_t(j,k)} - \hat{\boldsymbol{\mu}}'_{jk}(\hat{\boldsymbol{\mu}}'_{jk})^\mathsf{T}.
\end{aligned} \tag{3.146}$$

Thus, the ML estimate of the covariance matrix is computed from the second-order statistic and the ML estimate of the mean vector.

Now, we summarize the ML estimates of the CDHMM as follows:

$$\hat{\pi}'_j = \gamma_1(j), \tag{3.147}$$

$$\hat{a}'_{ij} = \frac{\sum_{t=1}^{T-1} \xi_t(i,j)}{\sum_{t=1}^{T-1} \sum_{j'=1}^{J} \xi_t(i,j')}, \tag{3.148}$$

$$\hat{\omega}'_{jk} = \frac{\sum_{t=1}^{T} \gamma_t(j,k)}{\sum_{t=1}^{T} \sum_{k'=1}^{K} \gamma_t(j,k')}, \tag{3.149}$$

$$\hat{\boldsymbol{\mu}}'_{jk} = \frac{\sum_{t=1}^{T} \gamma_t(j,k)\mathbf{o}_t}{\sum_{t=1}^{T} \gamma_t(j,k)}, \tag{3.150}$$

$$\hat{\boldsymbol{\Sigma}}'_{jk} = \frac{\sum_{t=1}^{T} \gamma_t(j,k)\mathbf{o}_t\mathbf{o}_t^\mathsf{T}}{\sum_{t=1}^{T} \gamma_t(j,k)} - \hat{\boldsymbol{\mu}}'_{jk}(\hat{\boldsymbol{\mu}}'_{jk})^\mathsf{T}. \tag{3.151}$$

If we consider the diagonal covariance matrix, the dth diagonal element of Eq. (3.151) is modified as follows:

$$\hat{\Sigma}'_{jkd} = \frac{\sum_{t=1}^{T} \gamma_t(j,k) o_{td}^2}{\sum_{t=1}^{T} \gamma_t(j,k)} - (\hat{\mu}'_{jkd})^2. \tag{3.152}$$

To compute these estimated values, we need to compute the following values:

$$\begin{aligned} \xi(i,j) &\triangleq \sum_{t=1}^{T-1} \xi_t(i,j) \\ \gamma(j,k) &\triangleq \sum_{t=1}^{T} \gamma_t(j,k) \\ \boldsymbol{\gamma}_{jk} &\triangleq \sum_{t=1}^{T} \gamma_t(j,k) \mathbf{o}_t \\ \boldsymbol{\Gamma}_{jk} &\triangleq \sum_{t=1}^{T} \gamma_t(j,k) \mathbf{o}_t \mathbf{o}_t^{\mathsf{T}}. \end{aligned} \tag{3.153}$$

These statistics are sufficient to compute the parameters, and are called *sufficient statistics* of the CDHMM. In particular, $\gamma(j,k)$, $\boldsymbol{\gamma}_{jk}$, and $\boldsymbol{\Gamma}_{jk}$ are called the 0th, 1st, and 2nd order sufficient statistics of the Gaussian at HMM state j and mixture component k, respectively. The sufficient statistic is first mentioned in Section 2.1.3, where the estimation problems are rather simple, and they do not include latent variables. In the latent models, the probabilistic assignment information of the occupation probabilities $\gamma_1(j)$, $\gamma_t(j,k)$, $\xi_t(i,j)$ is important to obtain the sufficient statistics. These statistics, composed of the occupation probabilities $\gamma_1(j)$, $\gamma_t(j,k)$, $\xi_t(i,j)$, are computed from the forward–backward algorithm, as discussed in the expectation step (Section 3.4.2).

Based on these sufficient statistics, we can rewrite the auxiliary function as follows:

$$Q(\boldsymbol{\pi}'|\boldsymbol{\pi}) = \sum_{j=1}^{J} \gamma_1(j) \log \pi'_j, \tag{3.154}$$

$$Q(\mathbf{A}'|\mathbf{A}) = \sum_{i=1}^{J} \sum_{j=1}^{J} \xi(i,j) \log a'_{ij}, \tag{3.155}$$

$$Q(\boldsymbol{\omega}'|\boldsymbol{\omega}) = \sum_{j=1}^{J} \sum_{k=1}^{K} \gamma(j,k) \log \omega'_{jk}, \tag{3.156}$$

$$\begin{aligned} Q(\boldsymbol{\mu}', \boldsymbol{\Sigma}'|\boldsymbol{\mu}, \boldsymbol{\Sigma}) = \sum_{j=1}^{J} \sum_{k=1}^{K} & -\frac{\gamma(j,k)}{2} \left(D \log(2\pi) + \log |\boldsymbol{\Sigma}'_{jk}| + \boldsymbol{\mu}'^{\mathsf{T}}_{jk} (\boldsymbol{\Sigma}'_{jk})^{-1} \boldsymbol{\mu}'_{jk} \right) \\ & + \boldsymbol{\mu}'^{\mathsf{T}}_{jk} (\boldsymbol{\Sigma}'_{jk})^{-1} \boldsymbol{\gamma}_{jk} - \frac{1}{2} \mathrm{tr} \left[(\boldsymbol{\Sigma}'_{jk})^{-1} \boldsymbol{\Gamma}_{jk} \right]. \end{aligned} \tag{3.157}$$

These forms are used in the analytical discussions in later chapters.

This total EM algorithm for iteratively estimating the HMM parameters is called the *Baum–Welch algorithm* (Baum, Petrie, Soules *et al.* 1970), and is based on the forward-backward algorithm, the accumulation of the sufficient statistics, and the update of the HMM parameters, as shown in Algorithm 4. The Baum–Welch algorithm can be extended based on the Bayesian approach (see Section 4.3 and Section 7.3).

Algorithm 4 Baum–Welch algorithm

Require: $\Theta \leftarrow \Theta^{\text{init}}$
1: **repeat**
2: Compute the forward variable $\alpha_t(j)$ from the forward algorithm
3: Compute the backward variable $\beta_t(j)$ from the backward algorithm
4: Compute the occupation probabilities $\gamma_1(j)$, $\gamma_t(j,k)$, and $\xi_t(i,j)$
5: Accumulate the sufficient statistics $\xi(i,j)$, $\gamma(j,k)$, $\boldsymbol{\gamma}_{jk}$, and $\boldsymbol{\Gamma}_{jk}$
6: Estimate the new HMM parameters $\hat{\Theta}'$
7: Update the HMM parameters $\Theta \leftarrow \hat{\Theta}'$
8: **until** Convergence

In the implementation of Viterbi or segmental ML estimation, we apply the current HMM parameter estimates $\Theta = \{\pi_j, a_{ij}, \omega_{jk}, \boldsymbol{\mu}_{jk}, \boldsymbol{\Sigma}_{jk}\}$ and the Viterbi algorithm to perform Viterbi decoding to find the state alignment. Given the best state sequence $\hat{S}$ and mixture component sequence $\hat{V}$, new ML estimates $\Theta' = \{\boldsymbol{\mu}'_{jk}, \boldsymbol{\Sigma}'_{jk}\}$ are computed as the ensemble mean vector and covariance matrix, where $\sum_{t=1}^{T} \gamma_t(j,k)$ is seen as the count N_{jk} of training samples $\mathbf{O} = \{\mathbf{o}_t\}$ which are aligned in state j and Gaussian component k (Juang & Rabiner 1990). This method of using the Viterbi training instead of the forward–backward algorithm to obtain (a part of) occupation probabilities in the training step is called *segmental K-means training* or *Viterbi training*.

We should note that to make the Baum–Welch algorithm work in real speech data, we need to consider some heuristics in the ML EM algorithm. For example, how to provide initial values is always an important question, and it is usually resolved by using a simple algorithm (e.g., K-means clustering or random initialization). The other important issue comes from the data sparseness, e.g., some mixture components or hidden states cannot have sufficient data assigned in the Viterbi or forward–backward algorithm, which makes the parameter estimation (especially covariance parameter estimation) unstable. This can be heuristically avoided by setting a threshold to update these parameters, or setting minimum threshold values for covariance parameters. These problems can be solved by the Bayesian approaches.

Despite the above problems, the Baum–Welch algorithm based on the EM algorithm is widely used in current CDHMM training. This Baum–Welch algorithm has several advantages. For example, the algorithm is unique in that the computational cost of the E step is much more than that of the M-step, since the E step computational cost depends on the training data size. However, the E-step can be parallelized with many computers by distributing a chunk of data to a computer and computing a sufficient statistic of the chunk independently of the other computers. Therefore, the Baum–Welch algorithm has a very nice data scalability, which enables it to use a large amount of training data. In addition, within the algorithm, the likelihood always increases by the EM theory. Therefore, by monitoring the likelihood values, we can easily detect errors and bugs during a training procedure.

3.5 Maximum likelihood linear regression for hidden Markov model

As we discussed in the previous section, CDHMM parameters can be estimated by statistical ML methods, the effectiveness of which depends on the quality and quantity of available data that should distribute according to the statistical features of the intended signal space or conditions. As there is no sure way of collecting sufficient data to cover all conditions, adaptive training of HMM parameters from a set of previously obtained parameters to a new set that befits a specific environment with a small amount of new data is an important research issue.

In speech recognition, one approach is to view the adaptation of model parameters to new data (e.g., speaker adaptation) as a transformation problem; that is, the new set of model parameters is a transformed version of the old set: $\Theta_{n+1} = f(\Theta_n, \{\mathbf{o}\}_n)$, where $\{\mathbf{o}\}_n$ denotes the new set of data available at moment n for the existing model parameters Θ_n to adapt to. Most frequently and practically, the function f is chosen to be of an affine transformation type (Digalakis, Ritischev & Neumeyer 1995, Leggetter & Woodland 1995):

$$\boldsymbol{\theta}_{n+1} = \mathbf{A}\boldsymbol{\theta}_n + \mathbf{b}, \tag{3.158}$$

where various parts of the model parameters, e.g., the mean vectors or the variances, are envisaged in a vector space. The adaptation algorithm therefore involves deriving the affine map components, $\mathbf{A}$ and $\mathbf{b}$, from the adaptation data $\{\mathbf{o}\}_n$. A number of algorithms have been proposed for this purpose (see Lee & Huo (2000), and Shinoda (2010) for details).

The linear regression method for HMM parameters estimates the affine transformation parameters from a set of adaptation data, usually limited in size. The transformation with the estimated parameters is then applied to the previously trained HMMs, resulting in the set of "adapted models." Note that for automatic speech recognition, the number of the Gaussian distributions or simply Gaussians, which are used as component distributions in forming state-dependent mixture distributions, is typically in the thousands or more. If each mean vector in the set of Gaussians is to be modified by a unique transformation matrix, the number of "adaptation parameters" can be quite large. The main problem of this method is thus how to improve "generalization capability" by avoiding the over-training problem when the amount of adaptation data is small. To solve the problem, we introduce a model selection approach.

The model selection approach was originally proposed within the estimation of linear transformation parameters by using the maximum likelihood EM algorithm, as discussed in Section 3.4. The technique is called maximum likelihood linear regression (MLLR). MLLR proposes to share one linear transformation in a cluster of many Gaussians in the HMM set, thereby effectively reducing the number of free parameters that can then be trained with a small amount of adaptation data. The Gaussian clusters

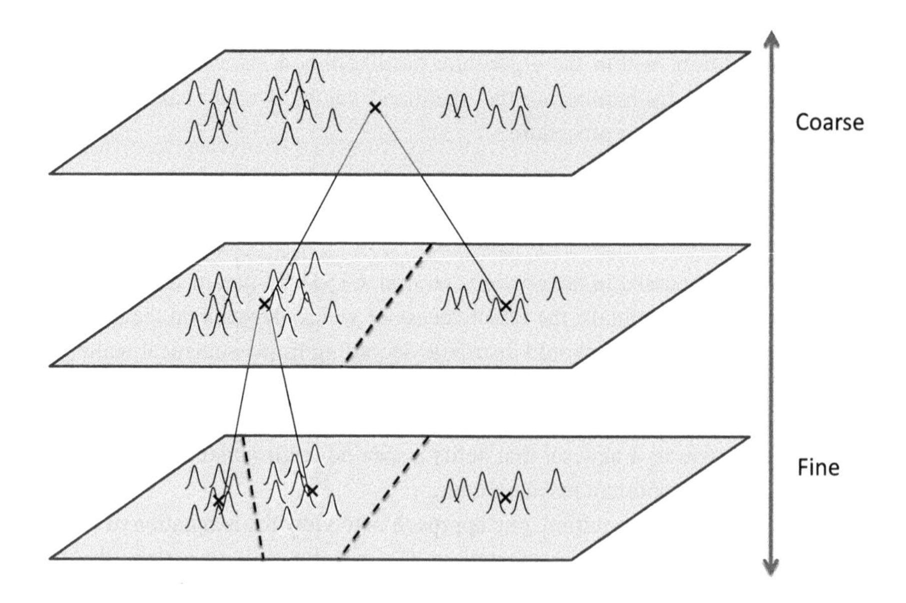

Figure 3.8 Gaussian tree representation of linear regression parameters.

are usually constructed as a tree structure, as shown in Figure 3.8, which is predetermined and fixed throughout adaptation. This tree (called a regression tree) is constructed based on a centroid splitting algorithm, described in Young, Evermann, Gales *et al.* (2006). This algorithm first makes two centroid vectors from a random perturbation of the global mean vector computed from Gaussians assigned to a target leaf node. Then it splits a set of these Gaussians according to the Euclidean distance between Gaussian mean vectors and two centroid vectors. The two sets of Gaussians obtained are assigned to child nodes, and this procedure is continued to finally build a tree.

The utility of the tree structure is commensurate with the amount of adaptation data; namely, if we have a small amount of data, it uses only coarse clusters (e.g., the root node of a tree in the top layer of Figure 3.8) where the number of free parameters in the linear transformation matrices is small. On the other hand, if we have a sufficiently large amount of data, it can use fine clusters where the number of free parameters in the linear transformation matrices is large, potentially improving the precision of the estimated parameters. This framework needs to select appropriate Gaussian clusters according to the amount of data, i.e., it needs an appropriate model selection function. Usually, model selection is performed by setting a threshold value manually (e.g., the total number of speech frames assigned to a set of Gaussians in a node).

3.5.1 Linear regression for hidden Markov models

This section briefly explains a solution for the linear regression parameters for HMMs within a maximum likelihood EM algorithm framework. It uses a solution based on a variance normalized representation of Gaussian mean vectors to simplify the solution.[3] In this section, we only focus on the transformation of Gaussian mean vectors in CDHMMs.

First, we review the basic EM algorithm of the conventional HMM parameter estimation, as discussed in Section 3.4, to set the notational convention and to align with the subsequent development of the MLLR approach. Let $\mathbf{O} \triangleq \{\mathbf{o}_t \in \mathbb{R}^D | t = 1, \cdots, T\}$ be a sequence of D dimensional feature vectors for T speech frames. The latent variables in a continuous density HMM are composed of HMM states and mixture components of GMMs. A sequence of HMM states is represented by $S \triangleq \{s_t | t = 1, \cdots, T\}$, where the value of s_t denotes an HMM state index at frame t. Similarly, a sequence of mixture components is represented by $V \triangleq \{v_t | t = 1, \cdots, T\}$, where the value of v_t denotes a mixture component index at frame t. As introduced in Eq. (3.78), the EM algorithm deals with the following auxiliary function as an optimization function instead of directly using the model likelihood:

$$\begin{aligned} Q(\Theta'|\Theta) &= \mathbb{E}_{(S,V)}[\log p(\mathbf{O}, S, V|\Theta')|\mathbf{O}, \Theta] \\ &= \sum_S \sum_V p(S, V|\mathbf{O}, \Theta) \log p(\mathbf{O}, S, V|\Theta'), \end{aligned} \tag{3.159}$$

where Θ is a set of HMM parameters and $p(\mathbf{O}, S, V|\Theta)$ is a complete data likelihood given Θ. $p(S, V|\mathbf{O}, \Theta)$ is the posterior distribution of the latent variables given the previously estimated HMM parameters Θ. Equation (3.78) is an expected value, and is efficiently computed by using the forward–backward algorithm as the E-step of the EM algorithm, as we discussed in Section 3.3.

The M-step of the EM algorithm estimates HMM parameters, as follows:

$$\Theta^{\mathrm{ML}} = \arg\max_{\Theta'} Q(\Theta'|\Theta). \tag{3.160}$$

The E-step and the M-step are performed iteratively until convergence, and finally we obtain the HMM parameters as a close approximation of the stationary point solution.

Now we focus on the linear transformation parameters within the EM algorithm. We prepare a transformation parameter matrix $\mathbf{W}_j \in \mathbb{R}^{D \times (D+1)}$ for each leaf node j in a Gaussian tree. Here, we assume that the Gaussian tree is pruned by a model selection approach as a model structure M, and the set of leaf nodes in the pruned tree is represented by $\mathcal{J}_M$. Hereinafter, we use Z to denote a joint event of S and V (i.e., $Z \triangleq \{S, V\}$). This will much simplify the following development pertaining to the adaptation of the mean and the covariance parameters. Similarly to Eq. (3.159), the auxiliary function with respect to a set of transformation parameters $\mathbf{\Lambda}_{\mathcal{J}_M} = \{\mathbf{W}_j | j = 1, \cdots, |\mathcal{J}_M|\}$ can be represented as follows:

$$\begin{aligned} Q(\mathbf{\Lambda}'_{\mathcal{J}_M}|\mathbf{\Lambda}_{\mathcal{J}_M}) &= \mathbb{E}_{(Z)}\left[\log p(\mathbf{O}, Z|\mathbf{\Lambda}'_{\mathcal{J}_M}, \Theta)\right] \\ &= \sum_{k=1}^{K} \sum_{t=1}^{T} \gamma_t(k) \log \mathcal{N}(\mathbf{o}_t|\boldsymbol{\mu}_k^{ad'}, \mathbf{\Sigma}_k). \end{aligned} \tag{3.161}$$

[3] This is first described in Gales & Woodland (1996) as normalized domain MLLR. The structural Bayes approach Shinoda & Lee (2001) for bias vector estimation in HMM adaptation also uses this normalized representation.

Here k denotes a unique mixture component index of all Gaussians in the target HMMs (for all phoneme HMMs in a speech recognition case), and K is the total number of Gaussians. $\gamma_t(k) \triangleq p(v_t = k|\mathbf{O}; \Theta, \mathbf{\Lambda}_{\mathcal{J}_M})$ is the posterior probability of mixture component k at t, derived from the previously estimated transformation parameters $\mathbf{\Lambda}_{\mathcal{J}_M}$.[4] Expression $\boldsymbol{\mu}_k^{ad}$ is a transformed mean vector with $\mathbf{\Lambda}_{\mathcal{J}_M}$, and the concrete form of this vector is discussed in the next paragraph. In the Q function, we disregard the parameters of the state transition probabilities and the mixture weights, since they do not depend on the optimization with respect to $\mathbf{\Lambda}_{\mathcal{J}_M}$. Expression $\mathcal{N}(\cdot|\boldsymbol{\mu}, \mathbf{\Sigma})$ denotes a Gaussian distribution with mean parameter $\boldsymbol{\mu}$ and covariance matrix parameter $\mathbf{\Sigma}$, and is defined in Appendix C.6 as follows:

$$\mathcal{N}(\mathbf{o}_t|\boldsymbol{\mu}_k^{ad}, \mathbf{\Sigma}_k) \triangleq C_{\mathcal{N}}(\mathbf{\Sigma}_k) \exp\left(-\frac{1}{2}\mathrm{tr}\left[(\mathbf{\Sigma}_k)^{-1}(\mathbf{o}_t - \boldsymbol{\mu}_k^{ad})(\mathbf{o}_t - \boldsymbol{\mu}_k^{ad})^{\intercal}\right]\right). \quad (3.162)$$

We use the trace based representation. Factor $C_{\mathcal{N}}(\mathbf{\Sigma}_k)$ is a normalization factor, and is defined as follows:

$$C_{\mathcal{N}}(\mathbf{\Sigma}_k) \triangleq (2\pi)^{-\frac{D}{2}}|\mathbf{\Sigma}_k|^{-\frac{1}{2}}. \quad (3.163)$$

In the following paragraphs, we derive Eq. (3.161) as a function of $\mathbf{\Lambda}_{\mathcal{J}_M}$ to optimize $\mathbf{\Lambda}_{\mathcal{J}_M}$.

We consider the concrete form of the transformed mean vector $\boldsymbol{\mu}_k^{\mathrm{ad}}$ based on the variance normalized representation. We first define the Cholesky decomposition matrix $\mathbf{C}_k$ as follows:

$$\mathbf{\Sigma}_k \triangleq \mathbf{C}_k(\mathbf{C}_k)^{\intercal}, \quad (3.164)$$

where $\mathbf{C}_k$ is a $D \times D$ triangular matrix. If the Gaussian k is included in a set of Gaussians $\mathcal{K}_j$ in leaf node j (i.e., $k \in \mathcal{K}_j$), the affine transformation of a Gaussian mean vector in a covariance normalized space $(\mathbf{C}_k)^{-1}\boldsymbol{\mu}_k^{\mathrm{ad}}$ is represented as follows:

$$\begin{aligned} (\mathbf{C}_k)^{-1}\boldsymbol{\mu}_k^{\mathrm{ad}} &= \mathbf{W}_j \begin{pmatrix} 1 \\ (\mathbf{C}_k)^{-1}\boldsymbol{\mu}_k^{\mathrm{ini}} \end{pmatrix} \\ \Rightarrow \boldsymbol{\mu}_k^{\mathrm{ad}} &= \mathbf{C}_k\mathbf{W}_j \begin{pmatrix} 1 \\ (\mathbf{C}_k)^{-1}\boldsymbol{\mu}_k^{\mathrm{ini}} \end{pmatrix} \triangleq \mathbf{C}_k\mathbf{W}_j\boldsymbol{\xi}_k, \end{aligned} \quad (3.165)$$

where $\boldsymbol{\xi}_k$ is an augmented normalized vector of an initial (non-adapted) Gaussian mean vector $\boldsymbol{\mu}_k^{\mathrm{ini}}$ and j is a leaf node index that holds a set of Gaussians. Thus, transformation parameter $\mathbf{W}_j$ is shared among a set of Gaussians $\mathcal{K}_j$. The clustered structure of the Gaussians is usually represented as a binary tree where a set of Gaussians belongs to each node.

[4] k denotes a combination of all possible HMM states and mixture components. In the common HMM representation (e.g., in this chapter), k can be represented by these two indexes in Eq. (3.109).

Now we focus on how to obtain the Q function of $\mathbf{\Lambda}_{\mathcal{J}_M}$. By following the equations in Example 2.3 with considering the occupation probability $\gamma_t(k)$ in Eq. (2.39), Eq. (3.161) is represented as follows:

$$
\begin{aligned}
&Q(\mathbf{\Lambda}'_{\mathcal{J}_M}|\mathbf{\Lambda}_{\mathcal{J}_M}) \\
&= \sum_{j\in\mathcal{J}_M}\sum_{k\in\mathcal{K}_j}\sum_{t=1}^{T}\gamma_t(k)\log\mathcal{N}(\mathbf{o}_t|\boldsymbol{\mu}_k^{\text{ad}'},\mathbf{\Sigma}_k) \\
&= \sum_{j\in\mathcal{J}_M}\sum_{k\in\mathcal{K}_j}\Bigg((\boldsymbol{\mu}_k^{\text{ad}'})^{\mathsf{T}}(\mathbf{\Sigma}_k)^{-1}\sum_{t=1}^{T}\gamma_t(k)\mathbf{o}_t - \frac{1}{2}\text{tr}\left[(\mathbf{\Sigma}_k)^{-1}\sum_{t=1}^{T}\gamma_t(k)\mathbf{o}_t\mathbf{o}_t^{\mathsf{T}}\right] \\
&\qquad + \sum_{t=1}^{T}\gamma_t(k)\left(\log C_{\mathcal{N}}(\mathbf{\Sigma}_k) - \frac{1}{2}(\boldsymbol{\mu}_k^{\text{ad}'})^{\mathsf{T}}(\mathbf{\Sigma}_k)^{-1}\boldsymbol{\mu}_k^{\text{ad}'}\right)\Bigg) \\
&= \sum_{j\in\mathcal{J}_M}\sum_{k\in\mathcal{K}_j}\Bigg((\boldsymbol{\mu}_k^{\text{ad}'})^{\mathsf{T}}(\mathbf{\Sigma}_k)^{-1}\boldsymbol{\gamma}_k - \frac{1}{2}\text{tr}\left[(\mathbf{\Sigma}_k)^{-1}\mathbf{\Gamma}_k\right] \\
&\qquad + \gamma_k\left(\log C_{\mathcal{N}}(\mathbf{\Sigma}_k) - \frac{1}{2}(\boldsymbol{\mu}_k^{\text{ad}'})^{\mathsf{T}}(\mathbf{\Sigma}_k)^{-1}\boldsymbol{\mu}_k^{\text{ad}'}\right)\Bigg),
\end{aligned}
\tag{3.166}
$$

where γ_k, $\boldsymbol{\gamma}_k$, and $\mathbf{\Gamma}_k$ are defined as follows:

$$
\begin{cases}
\gamma_k = \sum_{t=1}^{T}\gamma_t(k) \\
\boldsymbol{\gamma}_k = \sum_{t=1}^{T}\gamma_t(k)\mathbf{o}_t \\
\mathbf{\Gamma}_k = \sum_{t=1}^{T}\gamma_t(k)\mathbf{o}_t\mathbf{o}_t^{\mathsf{T}}.
\end{cases}
\tag{3.167}
$$

As introduced in Eq. (3.153), these are the 0th, 1st, and 2nd order sufficient statistics of Gaussians in HMMs, respectively. Then, the Q function of $\mathbf{\Lambda}_{\mathcal{J}_M}$ is represented by substituting Eq. (3.165) into Eq. (3.166) as follows:

$$
\begin{aligned}
&Q(\mathbf{\Lambda}'_{\mathcal{J}_M}|\mathbf{\Lambda}_{\mathcal{J}_M}) \\
&= \sum_{j\in\mathcal{J}_M}\sum_{k\in\mathcal{K}_j}\Bigg((\mathbf{C}_k\mathbf{W}'_j\boldsymbol{\xi}_k)^{\mathsf{T}}(\mathbf{\Sigma}_k)^{-1}\boldsymbol{\gamma}_k - \frac{1}{2}\text{tr}\left[(\mathbf{\Sigma}_k)^{-1}\mathbf{\Gamma}_k\right] \\
&\qquad + \gamma_k\left(\log C_{\mathcal{N}}(\mathbf{\Sigma}_k) - \frac{1}{2}(\mathbf{C}_k\mathbf{W}'_j\boldsymbol{\xi}_k)^{\mathsf{T}}(\mathbf{\Sigma}_k)^{-1}\mathbf{C}_k\mathbf{W}'_j\boldsymbol{\xi}_k\right)\Bigg) \\
&= \sum_{j\in\mathcal{J}_M}\sum_{k\in\mathcal{K}_j}\Bigg(\text{tr}\left[\mathbf{W}'^{\mathsf{T}}_j(\mathbf{C}_k)^{-1}\boldsymbol{\gamma}_k\boldsymbol{\xi}_k^{\mathsf{T}}\right] - \frac{1}{2}\text{tr}\left[(\mathbf{\Sigma}_k)^{-1}\mathbf{\Gamma}_k\right] \\
&\qquad + \gamma_k\left(\log C_{\mathcal{N}}(\mathbf{\Sigma}_k) - \frac{1}{2}\text{tr}\left[\mathbf{W}'^{\mathsf{T}}_j\mathbf{W}'_j\boldsymbol{\xi}_k\boldsymbol{\xi}_k^{\mathsf{T}}\gamma_k\right]\right)\Bigg)
\end{aligned}
$$

$$= \sum_{j \in \mathcal{J}_M} \left(\sum_{k \in \mathcal{K}_j} \gamma_k \log C_{\mathcal{N}}(\mathbf{\Sigma}_k) - \frac{1}{2} \mathrm{tr} \left[\mathbf{W}_j'^{\mathsf{T}} \mathbf{W}_j' \Xi_j - 2 \mathbf{W}_j'^{\mathsf{T}} \mathbf{Z}_j + \sum_{k \in \mathcal{K}_j} (\mathbf{\Sigma}_k)^{-1} \mathbf{\Gamma}_k \right] \right), \tag{3.168}$$

where Ξ_j and $\mathbf{Z}_j$ are 0th and 1st order sufficient statistics of linear regression parameters defined as:

$$\begin{cases} \Xi_j \triangleq \sum_{k \in \mathcal{K}_j} \boldsymbol{\xi}_k \boldsymbol{\xi}_k^{\mathsf{T}} \gamma_k \\ \mathbf{Z}_j \triangleq \sum_{k \in \mathcal{K}_j} (\mathbf{C}_k)^{-1} \boldsymbol{\gamma}_k \boldsymbol{\xi}_k^{\mathsf{T}}. \end{cases} \tag{3.169}$$

Here $\mathbf{Z}_j$ is a $D \times (D+1)$ matrix and Ξ_j is a $(D+1) \times (D+1)$ symmetric matrix. To derive Eq. (3.168), we use the fact that the trace of the scalar value is equal to the original scalar value, the cyclic property, and the distributive property of the trace as in Appendix B:

$$a = \mathrm{tr}[a], \tag{3.170}$$

$$\mathrm{tr}[\mathbf{ABC}] = \mathrm{tr}[\mathbf{BCA}], \tag{3.171}$$

$$\mathrm{tr}[\mathbf{A}(\mathbf{B} + \mathbf{C})] = \mathrm{tr}[\mathbf{AB} + \mathbf{AC}]. \tag{3.172}$$

We also use the definition of the Cholesky decomposition in Eq. (3.164).

Since Eq. (3.168) is represented as a quadratic form with respect to $\mathbf{W}_j$, we can obtain the optimal $\mathbf{W}_j^{\mathrm{ML}}$ in the sense of ML, similar to the discussion in Section 3.4.3. By differentiating the Q function with respect to $\mathbf{W}_j$, we can derive the following equation:

$$\frac{\partial}{\partial \mathbf{W}_j'} Q(\mathbf{\Lambda}'_{\mathcal{J}_M} | \mathbf{\Lambda}_{\mathcal{J}_M}) = 0. \Rightarrow \mathbf{Z}_j - \mathbf{W}_j^{\mathrm{ML}} \Xi_j = 0. \tag{3.173}$$

Here, we use the following matrix formulas for the derivation in Appendix B.3:

$$\begin{aligned} &\frac{\partial}{\partial \mathbf{X}} \mathrm{tr}[\mathbf{X}'\mathbf{A}] = \mathbf{A} \\ &\frac{\partial}{\partial \mathbf{X}} \mathrm{tr}[\mathbf{X}'\mathbf{X}\mathbf{A}] = 2\mathbf{X}\mathbf{A}. \quad (\mathbf{A} \text{ is a symmetric matrix}) \end{aligned} \tag{3.174}$$

Thus, we can obtain the following analytical solution:

$$\mathbf{W}_j^{\mathrm{ML}} = \mathbf{Z}_j \Xi_j^{-1}. \tag{3.175}$$

Therefore, the optimized mean vector parameter is represented as:

$$\boldsymbol{\mu}_k^{\mathrm{ad}^{\mathrm{ML}}} = \mathbf{C}_k \mathbf{Z}_j \Xi_j^{-1} \boldsymbol{\xi}_k. \tag{3.176}$$

Therefore, $\boldsymbol{\mu}_k^{\mathrm{ad}}$ is analytically obtained by using the statistics ($\mathbf{Z}_j$ and Ξ_j in Eq. (3.169)) and initial HMM parameters ($\mathbf{C}_k$ and $\boldsymbol{\xi}_k$). This solution corresponds to the M-step of the EM algorithm, and the E-step is performed by the forward–backward algorithm, similar to that of HMMs, to compute these statistics. The training procedure is shown in Algorithm 5.

Algorithm 5 Maximum likelihood linear regression

Require: Θ and $\mathbf{\Lambda}_{\mathcal{J}_M} \leftarrow \mathbf{\Lambda}_{\mathcal{J}_M}^{\text{init}}$
1: **repeat**
2: Compute the occupation probability $\gamma_t(k)$.
3: Accumulate the sufficient statistics γ_k, $\boldsymbol{\gamma}_k$, $\mathbf{\Gamma}_k$, $\mathbf{Z}_j$, and Ξ_j
4: Estimate the transformation parameters $\mathbf{\Lambda}_{\mathcal{J}_M}^{\text{ML}}$
5: Update the HMM parameters Θ
6: **until** Convergence

MLLR is one of the most popular techniques for acoustic modeling, and there are many variants of transformation types for HMMs, e.g., Sankar & Lee (1996), Chien, Lee & Wang (1997), Chen, Liau, Wang *et al.* (2000), Mak, Kwok & Ho (2005) and Delcroix *et al.* (2009). In addition to speech recognition, there are many other applications which are based on the adaptive training of HMMs (e.g., speech synthesis (Tamura, Masuko, Tokuda *et al.* 2001), speaker verification (Stolcke, Ferrer, Kajarekar *et al.* 2005), face recognition (Sanderson, Bengio & Gao 2006) and activity recognition (Maekawa & Watanabe 2011)).

3.6 *n*-gram with smoothing techniques

As we discussed in Section 3.1, a language model (LM) is known as crucial prior information for large vocabulary continuous speech recognition (LVCSR), according to the Bayes decision rule:

$$\hat{W} = d^{\text{MAP}}(\mathbf{O}) = \arg\max_{W} \underbrace{p(\mathbf{O}|W)}_{\text{acoustic model}} \times \underbrace{p(W)}_{\text{language model}}. \tag{3.177}$$

Many other applications include document classification, information retrieval, optical character recognition, machine translation, writing correction, and bio-informatics. An overview of language modeling in LVCSR systems has been given in Chen & Goodman (1999), Kita (1999), Rosenfeld (2000), Bellegarda (2004), and Saon & Chien (2012*b*).

A language model is a probability distribution $p(W)$ over a sequence of word strings $W = \{w_1, \cdots, w_i, \cdots, w_J\} \triangleq w_1^J$ that describes how likely it is that the word sequence W occurs as a sentence in some domain of interest.[5] Recall that word w is represented by a string, and it is an element of a set of distinct words $\mathcal{V}$, which is also called a vocabulary or dictionary. Here, w_i is a word at position i. The word string $w_i \in \mathcal{V}$ and the continuous speech vector $\mathbf{o}_t \in \mathbb{R}^D$ are both sequential patterns but in different data types and different time scales.

[5] Some languages do not have word boundaries explicitly marked by white-space in a text (e.g., Japanese and Chinese). Therefore, to process a text for language molding, these languages need an additional word segmentation step (Sproat, Gale, Shih & Chang 1996, Matsumoto, Kitauchi, Yamashita *et al.* 1999, Kudo 2005).

To get used to this notation, we provide several examples to represent the following word sequence:

my wife used my car.

This sentence has five words and one period at the end of the sentence. By regarding the period as one word,[6] this sentence is totally composed of a six-word sequence (i.e., $J = 6$), and can be represented by

$$w_1^6 = \{w_1 = \text{"my"}, w_2 = \text{"wife"}, w_3 = \text{"used"}, w_4 = \text{"my"}, w_5 = \text{"car"}, w_6 = \text{"."}\}. \tag{3.178}$$

Note that the vocabulary for this sentence is composed of distinct unique words represented as:

$$\mathcal{V} = \{\text{"my"}, \text{"wife"}, \text{"used"}, \text{"car"}, \text{"."}\} \tag{3.179}$$

and the vocabulary size in this example is

$$|\mathcal{V}| = 5. \tag{3.180}$$

We can also introduce the following summation over vocabulary $\mathcal{V}$, which is important in this section:

$$\sum_{w_i \in \mathcal{V}} f(w_i). \tag{3.181}$$

This summation is performed over each vocabulary in Eq. (3.179), and not over a position i in Eq. (3.178).

Basically, the prior word probability is employed to characterize the regularities in natural language. The probability of a word sequence $\{w_1, \cdots, w_J\}$ is represented based on the product rule as:

$$\begin{aligned} p_\Theta(w_1, \cdots, w_J) &= p(w_J|w_1, \cdots, w_{J-1})p(w_1, \cdots, w_{J-1}) \\ &= p(w_J|w_1, \cdots, w_{J-1})p(w_{J-1}|w_1, \cdots, w_{J-2})p(w_1, \cdots, w_{J-2}) \\ &\vdots \\ &= \prod_{i=1}^{J} p(w_i|w_1^{i-1}), \end{aligned} \tag{3.182}$$

where Θ denotes the n-gram parameters, namely the n-gram probabilities, which is explained later. Here, to describe the word sequence from ith word to nth word, we use the following notation:

$$\{w_i, \cdots, w_n\} \triangleq w_i^n. \tag{3.183}$$

We also define the following special cases:

$$\begin{aligned} w_i^i &= w_i \\ w_i^n &= \emptyset, \quad \text{when } i > n \end{aligned} \tag{3.184}$$

[6] In the implementation, we additionally define the start of a sentence with an auxiliary symbol for practical use, which makes the number of words seven in this example.

where $\emptyset$ denotes an empty set. For example, the word sequences $\{w_1, \cdots, w_{i-1}\}$ are represented as

$$w_1^{i-1} \triangleq \{w_1, \cdots, w_{i-1}\} \tag{3.185}$$

and so on. When $i = 1$ in Eq. (3.182), the conditional distribution is written as:

$$p(w_1|w_1^0) \triangleq p(w_1), \text{ where } w_1^0 = \emptyset. \tag{3.186}$$

However, it makes the model larger, as the number of words in a sequence is larger, and we need to model it with the fixed model size. Thus, Eq. (3.182) is approximated with the $(n-1)$th order Markov assumption by multiplying the probabilities of a predicted word w_i conditional on its preceding $n-1$ words $\{w_{i-n+1}, \cdots, w_{i-1}\}$:

$$p_\Theta(w_1^J) \approx \prod_{i=1}^{J} p(w_i|w_{i-n+1}^{i-1}). \tag{3.187}$$

This model is called an n-gram model. Usually n is taken to be from 2 to 5, which depends on the size of the training data and applications. When $n = 1$ in Eq. (3.187), the conditional distribution is written as:

$$p(w_i|w_i^{i-1}) \triangleq p(w_i), \quad w_i^{i-1} \triangleq \emptyset, \tag{3.188}$$

which is called a *unigram* model that does not depend on any history of words.

The n-gram parameter is defined as the weight given to a conditional word sequence. The probabilistic distribution of $p(w_i|w_{i-n+1}^{i-1})$ is parameterized as with a multinomial distribution as:

$$p(w_i|w_{i-n+1}^{i-1}) \triangleq \theta_{w_i|w_{i-n+1}^{i-1}}, \tag{3.189}$$

where

$$\begin{aligned} \sum_{w_i \in \mathcal{V}} \theta_{w_i|w_{i-n+1}^{i-1}} &= 1 \\ \theta_{w_i|w_{i-n+1}^{i-1}} &\geq 0 \quad \forall w_i \in \mathcal{V}. \end{aligned} \tag{3.190}$$

Note that the number of distinct n-gram parameters for $\theta_{w_i|w_{i-n+1}^{i-1}}$ would be the index to the power of the vocabulary size $|\mathcal{V}|$, i.e., $|\mathcal{V}|^n$. In this section, we use the following notation to present a set of n-gram parameters:

$$\begin{aligned} \Theta_n &\triangleq \{\theta_{w_i|w_{i-n+1}^{i-1}} | \forall w_i \in \mathcal{V}, \cdots, w_{i-n+1} \in \mathcal{V}\} \\ &\triangleq \{p(w_i|w_{i-n+1}^{i-1})\}. \end{aligned} \tag{3.191}$$

The number of parameters is a very large since the vocabulary size of LVCSR would be more than 50 000, and the main problem of language modeling is how to compactly represent these parameters.

The straightforward way to estimate the multinomial distribution of an n-gram $\theta_{w_i|w_{i-n+1}^{i-1}} = p(w_i|w_{i-n+1}^{i-1})$ from a text corpus $\mathcal{D}$ is to compute the ML estimate by

$$\begin{aligned}\theta^{\text{ML}}_{w_i|w_{i-n+1}^{i-1}} &= p^{\text{ML}}(w_i|w_{i-n+1}^{i-1}) \\ &= \arg\max_{\theta_{w_i|w_{i-n+1}^{i-1}}} p(\mathcal{D}|\theta_{w_i|w_{i-n+1}^{i-1}}),\end{aligned} \tag{3.192}$$

where the multinomial likelihood function is obtained from the definition in Appendix C.2 by

$$p(\mathcal{D}|\{\theta_{w_i|w_{i-n+1}^{i-1}}|w_i \in \mathcal{V}\}) = \prod_{w_i\in\mathcal{V}} (\theta_{w_i|w_{i-n+1}^{i-1}})^{c(w_{i-n+1}^{i})}, \tag{3.193}$$

where $c(w_{i-n+1}^{i-1})$ denotes the number of occurrences of word sequence w_{i-n+1}^{i-1} in training corpus $\mathcal{D}$. To estimate the parameter, similarly to the state transitions and mixture weights in the HMM in Section 3.4.3, we introduce a Lagrange multiplier η and solve the constrained optimization problem by maximizing

$$\sum_{w_i\in\mathcal{V}} c(w_{i-n+1}^{i}) \log \theta_{w_i|w_{i-n+1}^{i-1}} + \eta\left(\sum_{w_i\in\mathcal{V}} \theta_{w_i|w_{i-n+1}^{i-1}} - 1\right). \tag{3.194}$$

Setting the derivative of Eq. (3.194) with respect to $\theta_{w_i|w_{i-n+1}^{i-1}}$ to zero, we obtain

$$\theta_{w_i|w_{i-n+1}^{i-1}} = -\frac{1}{\eta} c(w_{i-n+1}^{i}). \tag{3.195}$$

By substituting this result into constraint Eq. (3.190), we find the value of the Lagrange multiplier

$$\begin{aligned}\sum_{w_i\in\mathcal{V}} \theta_{w_i|w_{i-n+1}^{i-1}} &= -\frac{1}{\eta}\sum_{w_i\in\mathcal{V}} c(w_{i-n+1}^{i}) = 1 \\ \Rightarrow \eta &= -\sum_{w_i\in\mathcal{V}} c(w_{i-n+1}^{i}),\end{aligned} \tag{3.196}$$

and the ML solution in the form of

$$\begin{aligned}\theta^{\text{ML}}_{w_i|w_{i-n+1}^{i-1}} &= \frac{c(w_{i-n+1}^{i})}{\sum_{w_i} c(w_{i-n+1}^{i})} \\ &= \frac{c(w_{i-n+1}^{i})}{c(w_{i-n+1}^{i-1})}.\end{aligned} \tag{3.197}$$

Without loss of generality, we neglect the notation Θ in the following expressions. The goal of the most popularly used language models, trigram models, is to determine the probability of a word given the previous two words $p(w_i|w_{i-2}, w_{i-1})$, which is estimated as the number of times the word sequence $\{w_{i-2}, w_{i-1}, w_i\}$ occurs in some corpus of training data divided by the number of times the word sequence $\{w_{i-2}, w_{i-1}\}$ occurs.

However, again the number of n-gram parameters depends on the number of word combinations in a word sequence $\{w_{i-n+1}, \cdots, w_{i-1}, w_i\}$, which is counted as the number of different words $w_n \in \mathcal{V}$ at different temporal positions from $i-n+1$ to i. This number is exponentially increased by involving large n. Although n-gram is effective at exploiting local lexical regularities, it suffers from the inadequacies of *training data* and *long-distance information* due to too many word combinations and too narrow an n-gram window size, respectively. These limitations substantially weaken the regularization of the trained n-gram models and the prediction for unseen words. The limitation of n-gram window size could be resolved by exploiting large-span latent semantic information (Hofmann 1999*b*, Bellegarda 2000), which is addressed in Section 3.7. In what follows, we address different *smoothing* solutions to the problem of insufficient training data in the n-gram model.

3.6.1 Class-based model smoothing

A simple and meaningful approach to tackle the data sparseness problem is to consider the transition probabilities between classes rather than words, namely to adopt the class-based n-gram language model (Brown, Desouza, Mercer *et al.* 1992, Chen 2009):

$$p(w_i|w_{i-n+1}^{i-1}) \approx p(w_i|c_i)p(c_i|c_{i-n+1}^{i-1}), \tag{3.198}$$

where $c_i \in \mathcal{C}$ is the class assignment of word w_i, $p(w_i|c_i)$ is the probability of word w_i, generated from class c_i, and $p(c_i|c_{i-n+1}^{i-1})$ is the class n-gram. The class assignments of different words are determined beforehand according to the word clustering using the metric of *mutual information*. An existing linguistic class (e.g., part of speech) is also used to provide the class assignments.

The word probability given a class $p(w_i|c_i)$ is usually estimated by using the ML estimation, similarly to Eq. (3.197), as:

$$p^{\mathsf{ML}}(w_i|c_i) = \theta_{w_i|c_i}^{\mathsf{ML}} = \frac{c(w_i, c_i)}{\sum_{w_i \in \mathcal{V}} c(w_i, c_i)}, \tag{3.199}$$

where $c(w_i, c_i)$ is the number of word counts labeled by both $\{w_i, c_i\}$ in the corpus $\mathcal{D}$. The class n-gram probability $p(c_i|c_{i-n+1}^{i-1})$ is estimated by using the ML estimation (replacing the word counts in Eq. (3.197) with class counts), or a smoothing technique, explained in the following sections.

The model parameters are composed of

$$\Theta = \{\{p(w_i|c_i)\}, \{p(c_i|c_{i-n+1}^{i-1})\}\}. \tag{3.200}$$

Both are represented by the multinomial distributions. Since the number of distinct classes $|\mathcal{C}|$ is much smaller than the vocabulary size $|\mathcal{V}|$, the model size is significantly reduced. Model parameters could be reliably estimated. For example, the number of parameters of $\{\{p(w_i|c_i)\}$ is $|\mathcal{V}||\mathcal{C}|$, and the number of parameters of $\{p(c_i|c_{i-n+1}^{i-1})\}$ is $|\mathcal{C}|^n$. Since $|\mathcal{V}| \gg |\mathcal{C}|$, the total number of parameters of a class n-gram model $|\mathcal{V}||\mathcal{C}| + |\mathcal{C}|^n$ is much smaller than an n-gram model $|\mathcal{V}|^n$. The class-based n-gram is also seen as a smoothed language model.

3.6.2 Jelinek–Mercer smoothing

As reported in Chen & Goodman (1999), it is usual to deal with the issue of data sparseness in an n-gram model by using a linear interpolation method where the nth order language model $p^{\text{interp}}(w_i|w_{i-n+1}^{i-1})$ is estimated by interpolating with the $(n-1)$th order language model in a form of

$$p^{\text{interp}}(w_i|w_{i-n+1}^{i-1}) = \lambda_{w_{i-n+1}^{i-1}} p^{\text{ML}}(w_i|w_{i-n+1}^{i-1}) + (1 - \lambda_{w_{i-n+1}^{i-1}}) p^{\text{interp}}(w_i|w_{i-n+2}^{i-1}), \tag{3.201}$$

where $\lambda_{w_{i-n+1}^{i-1}}$ denotes the interpolation weight which is estimated for each w_{i-n+1}^{i-1} in accordance with the ML method. The reason the interpolation weight $\lambda_{w_{i-n+1}^{i-1}}$ does not depend on w_i comes from the constraint of the sum-to-one property of an n-gram model. The nth order smoothed model is calculated recursively as a linear interpolation between the nth order ML model and the $(n-1)$th order smoothed model, that is

$$\begin{aligned} p^{\text{interp}}(w_i|w_{i-n+1}^{i-1}) &= \lambda_{w_{i-n+1}^{i-1}} p^{\text{ML}}(w_i|w_{i-n+1}^{i-1}) + (1 - \lambda_{w_{i-n+1}^{i-1}}) \\ &\quad \times \left(\lambda_{w_{i-n+2}^{i-1}} p^{\text{ML}}(w_i|w_{i-n+2}^{i-1}) + (1 - \lambda_{w_{i-n+2}^{i-1}}) p^{\text{interp}}(w_i|w_{i-n+3}^{i-1})\right) \\ &= \cdots . \end{aligned} \tag{3.202}$$

To carry through this recursive process to the end, we can take the smoothed unigram or first-order model to be the ML distribution in Eq. (3.192), i.e., when $n = 1$,

$$\begin{aligned} p^{\text{interp}}(w_i|w_i^{i-1}) &= p^{\text{interp}}(w_i) \\ &= p^{\text{ML}}(w_i) = \theta_{w_i}^{\text{ML}}, \end{aligned} \tag{3.203}$$

where $\theta_{w_i}^{\text{ML}}$ is obtained from Eq. (3.197) as:

$$\theta_{w_i}^{\text{ML}} = \frac{c(w_i)}{\sum_{w_i} c(w_i)}. \tag{3.204}$$

Or we can take the smoothed zeroth-order model to be the discrete uniform distribution (Appendix C.1),

$$p^{\text{interp}}(w_i) = \lambda p^{\text{ML}}(w_i) + (1 - \lambda)\text{Unif}(w_i), \tag{3.205}$$

where the uniform distribution is defined with the vocabulary size $|\mathcal{V}|$ as:

$$\text{Unif}(w_i) \triangleq \frac{1}{|\mathcal{V}|}. \tag{3.206}$$

Now, let us consider the original interpolation equation Eq. (3.201). The parameter $\lambda_{w_{i-n+1}^{i-1}}$ is estimated individually for the w_{i-n+1}^{i-1} that maximizes the probability of some of the data. Practically, the selection could be done for buckets of parameters. In general, this class of interpolated models is also known as the n-gram model with Jelinek–Mercer smoothing (Jelinek & Mercer 1980), which is a standard form of *interpolation smoothing*. The smoothing techniques in the following sections (Witten–Bell (WB) smoothing in Section 3.6.3, absolute discount in Section 3.6.4, Kneser–Ney (KN) smoothing in Eq. (3.232), and PLSA smoothing in Eq. (3.319)) provide specific interpolation weights in this standard form, as shown in Table 3.1.

An important constraint of the n-gram model is that the summation over w_i goes to 1, as shown in Eq. (3.190). The Jelinek–Mercer smoothing (and the following smoothing techniques) satisfies this condition, i.e.,

$$\sum_{w_i \in \mathcal{V}} p^{\text{interp}}(w_i | w_{i-n+1}^{i-1}) = 1. \tag{3.207}$$

We can prove this condition recursively from the unigram case to the n-gram case. First, it is obvious that for the unigram models for the Eqs. (3.203) and (3.205) cases:

$$\sum_{w_i \in \mathcal{V}} p^{\text{interp}}(w_i) = 1. \tag{3.208}$$

Next, the bi-gram case is also proved by:

$$\begin{aligned}
\sum_{w_i \in \mathcal{V}} p^{\text{interp}}(w_i | w_{i-1}) &= \sum_{w_i \in \mathcal{V}} \left(\lambda_{w_{i-1}} p^{\text{ML}}(w_i | w_{i-1}) + (1 - \lambda_{w_{i-1}}) p^{\text{interp}}(w_i) \right) \\
&= \lambda_{w_{i-1}} \left(\sum_{w_i \in \mathcal{V}} p^{\text{ML}}(w_i | w_{i-1}) \right) + (1 - \lambda_{w_{i-1}}) \left(\sum_{w_i \in \mathcal{V}} p^{\text{interp}}(w_i) \right) \\
&= \lambda_{w_{i-1}} + (1 - \lambda_{w_{i-1}}) = 1,
\end{aligned} \tag{3.209}$$

where we use Eq. (3.208) and $\sum_{w_i} p^{\text{ML}}(w_i|\cdot) = 1$. That is proven in the n-gram case, trivially. The important property of this proof of the sum-to-one condition is that the summation over w_i does not depend on the interpolation weight $\lambda_{w_{i-n+1}^{i-1}}$.

3.6.3 Witten–Bell smoothing

Witten–Bell smoothing (Witten & Bell 1991) is considered to be an instance of *interpolation smoothing* as addressed in Section 3.6.2 by setting a specific value for the interpolation parameter $\lambda_{w_{i-n+1}^{i-1}}$. The Witten–Bell smoothing first defines the following number based on the number of unique words that follow the history w_{i-n+1}^{i-1}:

$$N_{1+}(w_{i-n+1}^{i-1}, \bullet) \triangleq |\{w_i | c(w_{i-n+1}^{i-1}, w_i) > 0\}|. \tag{3.210}$$

The notation N_{1+} represents the number of distinct words that have one or more counts, and the $\bullet$ represents any possible words at i with this condition. By using $N_{1+}(w_{i-n+1}^{i-1}, \bullet)$, the Witten–Bell smoothing assigns the factor $1 - \lambda_{w_{i-n+1}^{i-1}}$ for the lower-order model (the second term in Eq. (3.201)) where

$$\begin{aligned}
1 - \lambda_{w_{i-n+1}^{i-1}} &\triangleq \frac{N_{1+}(w_{i-n+1}^{i-1}, \bullet)}{\sum_{w_i} c(w_{i-n+1}^{i}) + N_{1+}(w_{i-n+1}^{i-1}, \bullet)} \\
&= \frac{N_{1+}(w_{i-n+1}^{i-1}, \bullet)}{c(w_{i-n+1}^{i-1}) + N_{1+}(w_{i-n+1}^{i-1}, \bullet)}.
\end{aligned} \tag{3.211}$$

This factor is interpreted as the frequency with which we should use the lower-order model to predict the next word. It is meaningful that more unique words appearing after history words w_{i-n+1}^{i-1}, i.e., larger $N_{1+}(w_{i-n+1}^{i-1}, \bullet)$, implies that more reliable values of the $(n-1)$-grams are estimated, and the higher weight factor $1 - \lambda_{w_{i-n+1}^{i-1}}$ should be assigned for $(n-1)$-grams. Similarly, the rest of the weight $\lambda_{w_{i-n+1}^{i-1}}$ (the first term in Eq. (3.201)) is represented from Eq. (3.211) as

$$\lambda_{w_{i-n+1}^{i-1}} = 1 - \frac{N_{1+}(w_{i-n+1}^{i-1}, \bullet)}{\sum_{w_i} c(w_{i-n+1}^{i}) + N_{1+}(w_{i-n+1}^{i-1}, \bullet)} = \frac{\sum_{w_i} c(w_{i-n+1}^{i})}{\sum_{w_i} c(w_{i-n+1}^{i}) + N_{1+}(w_{i-n+1}^{i-1}, \bullet)}. \tag{3.212}$$

This is the modified ML estimate with the unique word count $N_{1+}(w_{i-n+1}^{i-1}, \bullet)$.

According to the interpolation smoothing in Eq. (3.201) and Eqs. (3.211) and (3.212), the Witten–Bell smoothing is expressed by

$$p^{\mathsf{WB}}(w_i|w_{i-n+1}^{i-1}) = \frac{c(w_{i-n+1}^{i}) p^{\mathsf{ML}}(w_i|w_{i-n+1}^{i-1}) + N_{1+}(w_{i-n+1}^{i-1}, \bullet) p^{\mathsf{WB}}(w_i|w_{i-n+2}^{i-1})}{\sum_{w_i} c(w_{i-n+1}^{i}) + N_{1+}(w_{i-n+1}^{i-1}, \bullet)}. \tag{3.213}$$

Note that the factor used in the Witten–Bell smoothing in Eq. (3.211) only depends on w_{i-n+1}^{i-1}, and has the same dependency as $\lambda_{w_{i-n+1}^{i-1}}$. Therefore, it is trivial that the Witten–Bell smoothing satisfies the sum-to-one condition, as it is a special solution of the interpolation smoothing (Eq. (3.201)) that satisfies the sum-to-one condition.

3.6.4 Absolute discounting

Absolute discounting is also considered to be an instance of *interpolation smoothing* as addressed in Section 3.6.2. However, the equation form is not represented as the Jelinek–Mercer form, which we set down again as follows for comparison:

$$p^{\mathsf{interp}}(w_i|w_{i-n+1}^{i-1}) = \lambda_{w_{i-n+1}^{i-1}} p^{\mathsf{ML}}(w_i|w_{i-n+1}^{i-1}) + (1 - \lambda_{w_{i-n+1}^{i-1}}) p^{\mathsf{interp}}(w_i|w_{i-n+2}^{i-1}). \tag{3.214}$$

Recall that Eq. (3.214) has the weight λ depending on the previous word sequence w_{i-n+1}^{i-1}, and is composed of the ML probability $p^{\mathsf{ML}}(w_i|w_{i-n+1}^{i-1})$ and lower-order probability $p^{\mathsf{interp}}(w_i|w_{i-n+2}^{i-1})$. However, in absolute discounting, instead of multiplying the higher-order ML model by a factor $\lambda_{w_{i-n+1}^{i-1}}$, the higher-order distribution is created by subtracting a fixed discount d for the case of non-zero count. The absolute discounting is defined by

$$p^{\mathsf{ABS}}(w_i|w_{i-n+1}^{i-1}) \triangleq \frac{\max\{c(w_{i-n+1}^{i}) - d, 0\}}{\sum_{w_i} c(w_{i-n+1}^{i})} + (1 - \lambda_{w_{i-n+1}^{i-1}}) p^{\mathsf{ABS}}(w_i|w_{i-n+2}^{i-1}), \tag{3.215}$$

where $0 \leq d \leq 1$ denotes a discounting parameter.

The interpolation weight $(1 - \lambda_{w_{i-n+1}^{i-1}})$ is formed with the unique word count $N_{1+}(w_{i-n+1}^{i-1}, \bullet)$ defined in Eq. (3.210) as:

$$1 - \lambda_{w_{i-n+1}^{i-1}} = \frac{dN_{1+}(w_{i-n+1}^{i-1}, \bullet)}{\sum_{w_i} c(w_{i-n+1}^{i})}. \tag{3.216}$$

Note that the weight does not depend on w_i, but w_{i-n+1}^{i-1}, similarly to the Jelinek–Mercer form in Eq. (3.214). The way to find weight parameter $1 - \lambda_{w_{i-n+1}^{i-1}}$ is again based on the sum-to-one condition. Consider the condition

$$\sum_{w_i} p^{\text{ABS}}(w_i|\cdot) = 1. \tag{3.217}$$

Then, by taking the summation over w_i for both sides in Eq. (3.215), Eq. (3.215) can be rewritten as:

$$\begin{aligned}
1 &= \sum_{w_i} p^{\text{ABS}}(w_i|w_{i-n+1}^{i-1}) \\
&= \sum_{w_i} \frac{\max\{c(w_{i-n+1}^{i}) - d, 0\}}{\sum_{w_i} c(w_{i-n+1}^{i})} + (1 - \lambda_{w_{i-n+1}^{i-1}}) \sum_{w_i} p^{\text{ABS}}(w_i|w_{i-n+2}^{i-1}) \\
&= \frac{\sum_{w_i} \max\{c(w_{i-n+1}^{i}) - d, 0\}}{\sum_{w_i} c(w_{i-n+1}^{i})} + (1 - \lambda_{w_{i-n+1}^{i-1}}).
\end{aligned} \tag{3.218}$$

Therefore,

$$\begin{aligned}
1 - \lambda_{w_{i-n+1}^{i-1}} &= 1 - \frac{\sum_{w_i} \max\{c(w_{i-n+1}^{i}) - d, 0\}}{\sum_{w_i} c(w_{i-n+1}^{i})} \\
&= \frac{\sum_{w_i} c(w_{i-n+1}^{i}) - \sum_{w_i} \max\{c(w_{i-n+1}^{i}) - d, 0\}}{\sum_{w_i} c(w_{i-n+1}^{i})} \\
&= \frac{\sum_{w_i} \left(c(w_{i-n+1}^{i}) - \max\{c(w_{i-n+1}^{i}) - d, 0\}\right)}{\sum_{w_i} c(w_{i-n+1}^{i})}.
\end{aligned} \tag{3.219}$$

Now we focus on the numerator in Eq. (3.219) that represents the total discount value from d. By considering the cases when $c(w_{i-n+1}^{i}) > 0$ and $c(w_{i-n+1}^{i}) = 0$, we can derive the following equation:

$$c(w_{i-n+1}^{i}) - \max\{c(w_{i-n+1}^{i}) - d, 0\} = \begin{cases} d & \text{if } c(w_{i-n+1}^{i}) > 0 \\ 0 & \text{if } c(w_{i-n+1}^{i}) = 0. \end{cases} \tag{3.220}$$

Therefore, by substituting Eq. (3.220) into Eq. (3.219), and by using the unique word count $N_{1+}(w_{i-n+1}^{i-1}, \bullet)$, Eq. (3.219) is finally represented as:

$$\begin{aligned}
1 - \lambda_{w_{i-n+1}^{i-1}} &= \frac{\sum_{w_i | c(w_{i-n+1}^{i}) > 0} d}{\sum_{w_i} c(w_{i-n+1}^{i})} \\
&= \frac{dN_{1+}(w_{i-n+1}^{i-1}, \bullet)}{\sum_{w_i} c(w_{i-n+1}^{i})},
\end{aligned} \tag{3.221}$$

where $\sum_{w_i | c(w_{i-n+1}^i) > 0}$ means that the summation is undertaken for a subset of distinct words in the vocabulary $\mathcal{V}$ that satisfies the condition $c(w_{i-n+1}^i) > 0$. Since d does not depend on w_i, the numerator can be represented with d times the number of distinct words w_i that satisfies the condition $c(w_{i-n+1}^i) > 0$, which corresponds to $N_{1+}(w_{i-n+1}^{i-1}, \bullet)$. Thus, we prove Eq. (3.216).

This weight means that the discount d in an observed n-gram event $c(w_{i-n+1}^{i-1} w_i) > 0$ at current word w_i is distributed to compensate for those unseen events $c(w_{i-n+1}^{i-1} w_i) = 0$ where a lower-order model $p^{\text{ABS}}(w_i | w_{i-n+2}^{i-1})$ is adopted. The discount d is shared for all n-grams and could be measured by using the total number of n-grams with exactly one and two counts, i.e., $c(w_{i-n+1}^{i-1}) = 1$ and $c(w_{i-n+1}^{i-1}) = 2$ in the training data, respectively (Ney, Essen & Kneser 1994).

In summary, the absolute discounting is represented by

$$p^{\text{ABS}}(w_i | w_{i-n+1}^{i-1}) \triangleq \frac{\max\{c(w_{i-n+1}^i) - d, 0\}}{\sum_{w_i} c(w_{i-n+1}^i)} + \frac{d N_{1+}(w_{i-n+1}^{i-1}, \bullet)}{\sum_{w_i} c(w_{i-n+1}^i)} p^{\text{ABS}}(w_i | w_{i-n+2}^{i-1}). \tag{3.222}$$

Note that the absolute discounting does not include the exact ML probabilities (although they can be included by re-arranging the first term). The weight parameter for the lower-order model is proportional to the value $N_{1+}(w_{i-n+1}^{i-1}, \bullet)$, which has a similarity to that in Witten–Bell smoothed n-grams, as seen in Eq. (3.211).

Thus, we have explained Witten–Bell smoothing in Section 3.6.3 and absolute discount in Section 3.6.4 as instances of the interpolation (Jelinek–Mercer) smoothing techniques. The next section introduces another type of well-known smoothing technique called *backoff smoothing*, with Katz smoothing as an example.

3.6.5 Katz smoothing

Katz smoothing (Katz 1987) was developed by intuitively combining higher-order models with lower-order models through scaling of the ML distribution. Taking a bi-gram as an example, the Katz smoothing is performed by calculating the probability by considering the cases where the co-occurrence count $c(w_{i-1}, w_i)$ of w_{i-1} and w_i is zero or positive as follows:

$$p^{\text{KZ}}(w_i | w_{i-1}) \triangleq \begin{cases} d_{c(w_{i-1}, w_i)} p^{\text{ML}}(w_i | w_{i-1}) & \text{if } c(w_{i-1}, w_i) > 0 \\ \alpha(w_{i-1}) p^{\text{KZ}}(w_i) & \text{if } c(w_{i-1}, w_i) = 0, \end{cases} \tag{3.223}$$

where $d_r < 1$ is a discount ratio that reduces the probability estimated from the ML. The discount ratio is usually is obtained by the Good–Turing estimate (Good 1953), and this book does not describe it in detail. The expression $\alpha(w_{i-1})$ is a scaling factor that only depends on the previous word w_{i-1}, while $p^{\text{KZ}}(w_i)$ is a unigram probability, and we usually use the ML unigram probability for the bi-gram case, i.e.,

$$p^{\text{KZ}}(w_i) = p^{\text{ML}}(w_i). \tag{3.224}$$

Now we focus on the scaling factor $\alpha(w_{i-1})$, which is obtained by the sum-to-one condition of $p^{\mathsf{KZ}}(w_i|\cdot)$ and $p^{\mathsf{ML}}(w_i|\cdot)$ that must be satisfied for any probabilistic distributions. By summing over w_i in both sides of Eq. (3.223), the equation can be rewritten as

$$\sum_{w_i} p^{\mathsf{KZ}}(w_i|w_{i-1}) = \begin{cases} \sum_{w_i} d_{c(w_{i-1},w_i)} p^{\mathsf{ML}}(w_i|w_{i-1}) & \text{if } c(w_{i-1},w_i) > 0 \\ \sum_{w_i} \alpha(w_{i-1}) p^{\mathsf{ML}}(w_i) & \text{if } c(w_{i-1},w_i) = 0. \end{cases} \tag{3.225}$$

This leads to the following equation:

$$1 = \sum_{w_i|c(w_{i-1},w_i)>0} d_{c(w_{i-1},w_i)} p^{\mathsf{ML}}(w_i|w_{i-1}) + \alpha(w_{i-1}) \sum_{w_i|c(w_{i-1},w_i)=0} p^{\mathsf{ML}}(w_i), \tag{3.226}$$

where $\sum_{w_i|c(w_{i-1},w_i)>0}$ or $\sum_{w_i|c(w_{i-1},w_i)=0}$ means that the summation is undertaken for a subset of distinct words in the vocabulary $\mathcal{V}$ that satisfies the condition $c(w_{i-1},w_i) > 0$ or $c(w_{i-1},w_i) = 0$. Thus, we obtain that

$$\begin{aligned} \alpha(w_{i-1}) &= \frac{1 - \sum_{w_i|c(w_{i-1},w_i)>0} d_{c(w_{i-1},w_i)} p^{\mathsf{ML}}(w_i|w_{i-1})}{\sum_{w_i|c(w_{i-1},w_i)=0} p^{\mathsf{ML}}(w_i)} \\ &= \frac{1 - \sum_{w_i|c(w_{i-1},w_i)>0} d_{c(w_{i-1},w_i)} p^{\mathsf{ML}}(w_i|w_{i-1})}{1 - \sum_{w_i|c(w_{i-1},w_i)>0} p^{\mathsf{ML}}(w_i)}. \end{aligned} \tag{3.227}$$

This smoothing technique can be generalized to the n-gram probability as

$$p^{\mathsf{KZ}}(w_i|w_{i-n+1}^{i-1}) \triangleq \begin{cases} d_{w_{i-n+1}^{i}} p^{\mathsf{ML}}(w_i|w_{i-n+1}^{i-1}) & \text{if } c(w_{i-n+1}^{i}) > 0 \\ \alpha(w_{i-n+1}^{i-1}) p^{\mathsf{KZ}}(w_i|w_{i-n+2}^{i-1}) & \text{if } c(w_{i-n+1}^{i}) = 0, \end{cases} \tag{3.228}$$

where

$$\alpha(w_{i-n+1}^{i-1}) = \frac{1 - \sum_{w_i|c(w_{i-n+1}^{i})>0} d_{c(w_{i-n+1}^{i})} p^{\mathsf{ML}}(w_i|w_{i-n+1}^{i-1})}{1 - \sum_{w_i|c(w_{i-n+1}^{i})>0} p^{\mathsf{KZ}}(w_i|w_{i-n+2}^{i-1})}. \tag{3.229}$$

This smoothing scheme is known as a realization of *backoff smoothing*. This smoothing is obtained by the weighted multiplication of the ML probability and is different from interpolation (Jelinek–Mercer) smoothing, as discussed in Section 3.6.2, which is obtained by the weighted summation of the ML probability. Note that both smoothing techniques include some free parameters (weight λ in interpolation smoothing and the discount factor d in backoff smoothing), but the other parameters are obtained by using the sum-to-one constraint of the probability distribution. Similarly to the meaning of λ, backoff smoothing relies more on the lower-order n-gram probability when d is small (close to 0), while backoff smoothing relies more on the ML n-gram probability when d is large (close to 1). The next section describes a famous (modified) Kneser–Ney smoothing, which can be realized with both interpolation and backoff smoothing methods.

3.6.6 Kneser–Ney smoothing

Kneser–Ney (KN) smoothing (Kneser & Ney 1995) can be interpreted as an extension of the absolute discount approach, as discussed in Section 3.6.4. When the highest-order probability case is considered, it is same as the absolute discount. However, the unique property of KN smoothing is that it provides a special lower-order probability based on the numbers of distinct words.

For example, in the highest-order probability case, the original Kneser–Ney method was developed as the following backoff smoothed model:

$$p^{\mathrm{KN}}(w_i|w_{i-n+1}^{i-1}) = \begin{cases} \frac{\max\{c(w_{i-n+1}^{i})-d,0\}}{\sum_{w_i} c(w_{i-n+1}^{i})} & \text{if } c(w_{i-n+1}^{i}) > 0 \\ \gamma(w_{i-n+1}^{i-1}) p^{\mathrm{KN}}(w_i|w_{i-n+2}^{i-1}) & \text{if } c(w_{i-n+1}^{i}) = 0, \end{cases} \tag{3.230}$$

where $\gamma(w_{i-n+1}^{i-1})$ is chosen to make the distribution sum to 1 and has the form

$$\gamma(w_{i-n+1}^{i-1}) = \frac{dN_{1+}(w_{i-n+1}^{i-1}, \bullet)}{\sum_{w_i} c(w_{i-n+1}^{i})}. \tag{3.231}$$

Alternatively, the Kneser–Ney model could be estimated according to the interpolation smoothing scheme in the highest-order probability case based on

$$p^{\mathrm{KN}}(w_i|w_{i-n+1}^{i-1}) = \frac{\max\{c(w_{i-n+1}^{i}) - d, 0\}}{\sum_{w_i} c(w_{i-n+1}^{i})} + \frac{dN_{1+}(w_{i-n+1}^{i-1}, \bullet)}{\sum_{w_i} c(w_{i-n+1}^{i})} p^{\mathrm{KN}}(w_i|w_{i-n+2}^{i-1}). \tag{3.232}$$

Note that the first term of Eq. (3.232) is calculated from a modification of the ML probability, and Eq. (3.232) is exactly the same as the absolute discounting in Eq. (3.222), except for the lower-order probability $p^{\mathrm{KN}}(w_i|w_{i-n+2}^{i-1})$. However, in the lower-order probability, instead of using Eq. (3.232) recursively by changing n to $n-1$, the Kneser–Ney smoothing provides more intuitive probability than the ML probability for the first term by considering the continuation of words.

For a simple explanation, we consider the bi-gram case (the highest-order probability is a bi-gram probability), i.e., Eq. (3.232) is represented as:

$$p^{\mathrm{KN}}(w_i|w_{i-1}) = \frac{\max\{c(w_{i-1}, w_i) - d, 0\}}{\sum_{w_i} c(w_{i-1}, w_i)} + \frac{dN_{1+}(w_{i-1}, \bullet)}{\sum_{w_i} c(w_{i-1}, w_i)} p^{\mathrm{KN}}(w_i). \tag{3.233}$$

The question here is whether we really use the unigram probability for $p^{\mathrm{KN}}(w_i)$. This is often illustrated with an example of the bi-gram "San Francisco" (Chen & Goodman 1999, Jurafsky 2014). In the training data of the *Wall Street Journal* (WSJ0) corpus (Paul & Baker 1992), the bi-gram "San Francisco" appears 3222 times, while the word "Francisco" appeared 3329 times. The other word "glasses" appears 185 times. That is

$$\begin{aligned} c(w_{i-1} = \text{“San”}, w_i = \text{“Francisco”}) &= 3222 \\ c(w_i = \text{“Francisco”}) &= 3329 \\ c(w_i = \text{“glasses”}) &= 185. \end{aligned} \tag{3.234}$$

From these statistics, we can say that:

- The word "Francisco" almost always (96.8%) follows "San," and does not follow the other words in most cases.
- However, the word "Francisco" is more common (18 times more) than "glasses."

Then, let us consider the following sentences:

1. **I can't see without my reading "glasses."**
2. **I can't see without my reading "Francisco."**

We consider the case when there are no "reading glasses" and "reading Francisco" in a training corpus, and the lower-order probability $p^{\mathsf{KN}}(w_i)$ (the second term in (3.233)) is used to compute these sentence probabilities. Intuitively, we want to make the probability of the first sentence with "glasses" higher than the other. However, if we use the count-oriented (ML-like) probability for $p^{\mathsf{KN}}(w_i)$, the sentence with "Francisco" is assigned higher probability because the simple word count of "Francisco" is much larger than that of "glasses." This problem can be avoided by considering the *continuity* of word sequences rather than word counts in the lower-order probability.[7]

The KN smoothing provides the following unigram probability that does not use the word count (ML-like) information:

$$p^{\mathsf{KN}}(w_i) \triangleq \frac{N_{1+}(\bullet, w_i)}{N_{1+}(\bullet, \bullet)}. \tag{3.235}$$

Here, from the definition of the unique count N_{1+} in Eq. (3.210), $N_{1+}(\bullet, w_i)$ and $N_{1+}(\bullet, \bullet)$ are represented as follows:

$$N_{1+}(\bullet, w_i) = |\{w_{i-1}|c(w_{i-1}, w_i) > 0\}|, \tag{3.236}$$

$$N_{1+}(\bullet, \bullet) = |\{\{w_i, w_{i-1}\}|c(w_{i-1}, w_i) > 0\}| = \sum_{w_i} N_{1+}(\bullet, w_i). \tag{3.237}$$

In the previous example,

$$\begin{aligned} N_{1+}(\bullet, w_i = \text{"Francisco"}) &= 58 \\ N_{1+}(\bullet, w_i = \text{"glasses"}) &= 88. \end{aligned} \tag{3.238}$$

The unique count of $N_{1+}(\bullet, w_i = \text{"glasses"})$ is larger than that of $N_{1+}(\bullet, w_i = \text{"Francisco"})$. Thus, the probability of the sentence with "glasses" becomes larger when we use the number of unique words in (3.235), which is a more intuitive result in this example.

The probability based on the unique counts can be generalized from the unigram case in Eq. (3.235) to higher-order n-gram probabilities, except for the highest-order n-gram that uses the absolute discounting from Eq. (3.232). For example, $p^{\mathsf{KN}}(w_i|w_{i-n+2}^{i-1})$, which is the second term in Eq. (3.232), is represented as follows:

$$\begin{aligned} p^{\mathsf{KN}}(w_i|w_{i-n+2}^{i-1}) = &\frac{\max\{N_{1+}(\bullet, w_{i-n+2}^{i-1}) - d, 0\}}{N_{1+}(\bullet, w_{i-n+2}^{i-1}, \bullet)} \\ &+ \frac{dN_{1+}(w_{i-n+2}^{i-1}, \bullet)}{N_{1+}(\bullet, w_{i-n+2}^{i-1}, \bullet)} p^{\mathsf{KN}}(w_i|w_{i-n+3}^{i-1}), \end{aligned} \tag{3.239}$$

[7] Actually "reading glasses" has appeared twice in the WSJ0 corpus, and the lower-order probability cannot be used so much in the WSJ0 language model when we set d very small. This is another solution to resolve this discontinuity problem by correcting a very large size of corpus.

where

$$N_{1+}(\bullet, w_{i-n+2}^{i-1}) \triangleq |\{w_{i-n+1}|c(w_{i-n+1}^{i-1}) > 0\}|$$
$$N_{1+}(\bullet, w_{i-n+2}^{i-1}, \bullet) \triangleq |\{w_{i-n+1}, w_i|c(w_{i-n+1}^{i}) > 0\}| = \sum_{w_i} N_{1+}(\bullet, w_{i-n+2}^{i-1}). \tag{3.240}$$

Thus, we can obtain the interpolation version of the Kneser–Ney smoothing for all cases, which is summarized as follows:

- The highest-order n-gram probability

$$p^{\mathsf{KN}}(w_i|w_{i-n+1}^{i-1}) = \frac{\max\{c(w_{i-n+1}^{i}) - d, 0\}}{\sum_{w_i} c(w_{i-n+1}^{i})} + \frac{dN_{1+}(w_{i-n+1}^{i-1}, \bullet)}{\sum_{w_i} c(w_{i-n+1}^{i})} p^{\mathsf{KN}}(w_i|w_{i-n+2}^{i-1}); \tag{3.241}$$

- Lower-order n-gram probability

$$p^{\mathsf{KN}}(w_i|w_{i-n+2}^{i-1}) = \frac{\max\{N_{1+}(\bullet, w_{i-n+2}^{i-1}) - d, 0\}}{N_{1+}(\bullet, w_{i-n+2}^{i-1}, \bullet)} + \frac{dN_{1+}(w_{i-n+2}^{i-1}, \bullet)}{N_{1+}(\bullet, w_{i-n+2}^{i-1}, \bullet)} p^{\mathsf{KN}}(w_i|w_{i-n+3}^{i-1}); \tag{3.242}$$

- Unigram probability

$$p^{\mathsf{KN}}(w_i) = \frac{N_{1+}(\bullet, w_i)}{N_{1+}(\bullet, \bullet)}. \tag{3.243}$$

In general, the interpolated n-gram model yields better performance than the backoff n-gram model (Chen & Goodman 1999).

Modified Kneser–Ney smoothing

In Kneser–Ney smoothing and discounting smoothing, a discount parameter d in Eq. (3.241) plays an important role in striking a balance between the target-order n-gram probability and the lower-order one. Then, it is a simple question whether the single discount parameter d is precise enough to handle the balance. Basically, the distribution of distinct words follows the *power-law* property where only a few frequent words form a high proportion of the total, but a lot of rare words occur only *one* or *two* times, i.e.,

$$|\{w_i|c(w_i) = 1\}| \gg |\{w_i|c(w_i) = 2\}| \gg |\{w_i|c(w_i) > 2\}|. \tag{3.244}$$

This power-law property comes from the "rich-get-richer behavior" in natural language, and this property also applies to n-grams, as well as words. Thus, Kneser–Ney smoothing is modified by separately using three discount parameters d_1, d_2, and d_{3+} for those

n-grams with one, two, and three or more counts, respectively, instead of using a single discount d for all non-zero counts, as follows:

$$d(c) \triangleq \begin{cases} 0 & \text{if } c = 0 \\ d_1 & \text{if } c = 1 \\ d_2 & \text{if } c = 2 \\ d_{3+} & \text{if } c \geq 3. \end{cases} \tag{3.245}$$

These parameters are empirically determined as (Chen & Goodman 1999):

$$d_{\text{base}} = \frac{m_1}{m_1 + 2m_2}, \tag{3.246}$$

$$d_1 = 1 - 2d_{\text{base}} \frac{m_2}{m_1}, \tag{3.247}$$

$$d_2 = 2 - 3d_{\text{base}} \frac{m_3}{m_2}, \tag{3.248}$$

$$d_{3+} = 3 - 4d_{\text{base}} \frac{m_4}{m_3}, \tag{3.249}$$

where m_j is the total number of n-grams appearing j times in a training corpus, i.e.,

$$m_j \triangleq \sum_{w_{i-n}^i : c(w_{i-n}^i) = j} c(w_{i-n}^i). \tag{3.250}$$

Based on the new discounting parameter $d(c)$ in Eq. (3.245), a modified Kneser–Ney (MKN) smoothing is conducted to estimate the smoothed n-grams using

$$p^{\mathsf{MKN}}(w_i | w_{i-n+1}^{i-1}) = \frac{c(w_{i-n+1}^i) - d(c(w_{i-n+1}^{i-1}))}{\sum_{w_i} c(w_{i-n+1}^i)} + \gamma(w_{i-n+1}^{i-1}) p^{\mathsf{MKN}}(w_i | w_{i-n+2}^{i-1}), \tag{3.251}$$

where $\gamma(w_{i-n+1}^{i-1})$ is derived by considering the sum-to-one condition. By taking the summation over w_i in both sides of Eq. (3.251), Eq. (3.251) is represented as follows:

$$\begin{aligned} \sum_{w_i} p^{\mathsf{MKN}}(w_i | w_{i-n+1}^{i-1}) = 1 \\ &= \sum_{w_i} \frac{c(w_{i-n+1}^i) - d(c(w_{i-n+1}^{i-1}))}{\sum_{w_i} c(w_{i-n+1}^i)} \\ &\quad + \gamma(w_{i-n+1}^{i-1}) \sum_{w_i} p^{\mathsf{MKN}}(w_i | w_{i-n+2}^{i-1}) \\ &= \sum_{w_i} \frac{c(w_{i-n+1}^i) - d(c(w_{i-n+1}^{i-1}))}{\sum_{w_i} c(w_{i-n+1}^i)} + \gamma(w_{i-n+1}^{i-1}). \end{aligned} \tag{3.252}$$

Now we focus on the first term of the 4th line in Eq. (3.252). By using Eq. (3.245), this term can be represented as:

$$\begin{aligned}
&\sum_{w_i} c(w_{i-n+1}^{i}) - d(c(w_{i-n+1}^{i-1})) \\
&\quad = \sum_{w_i} c(w_{i-n+1}^{i}) - d_1 \sum_{w_i | c(w_{i-n+1}^{i-1})=1} 1 - d_2 \sum_{w_i | c(w_{i-n+1}^{i-1})=2} 1 - d_{3+} \sum_{w_i | c(w_{i-n+1}^{i-1})\geq 3} 1 \\
&\quad = \sum_{w_i} c(w_{i-n+1}^{i}) - d_1 N_1(w_{i-n+1}^{i-1}, \bullet) - d_2 N_2(w_{i-n+1}^{i-1}, \bullet) - d_{3+} N_{3+}(w_{i-n+1}^{i-1}, \bullet).
\end{aligned} \tag{3.253}$$

Thus, by substituting Eq. (3.253) into Eq. (3.252), we obtain

$$\gamma(w_{i-n+1}^{i-1}) = \frac{d_1 N_1(w_{i-n+1}^{i-1}, \bullet) + d_2 N_2(w_{i-n+1}^{i-1}, \bullet) + d_{3+} N_{3+}(w_{i-n+1}^{i-1}, \bullet)}{\sum_{w_i} c(w_{i-n+1}^{i})}. \tag{3.254}$$

The MKN smoothed n-grams were shown to perform better than the previously described smoothed n-grams in terms of perplexity and word error rates for LVCSR tasks (Chen & Goodman 1999).

Table 3.1 summarizes different language model smoothing methods and their corresponding smoothing techniques including interpolation smoothing and backoff smoothing. (Modified) Kneser–Ney smoothing, addressed in Section 3.6.6, can be implemented in both interpolation smoothing and backoff smoothing. Bayesian approaches extend these standard n-gram language models to MAP estimation of an n-gram language model in Section 4.7, a hierarchical Dirichlet language model in Section 5.3, and a hierarchical Pitman–Yor language model in Section 8.5. Since n-gram language models always have to address the sparse data problem, Bayesian approaches provide an elegant solution to deal with the problem theoretically.

The next section considers another generative model of a text that can deal with document information rather than a simple word sequence modeled by an n-gram model.

Table 3.1 Summary of different language model smoothing methods in terms of interpolation smoothing and backoff smoothing.

	Interpolation smoothing	Backoff smoothing
Jelinek–Mercer smoothing	Section 3.6.2	
Witten–Bell smoothing	Section 3.6.3	
Absolute discount	Section 3.6.4	
Katz smoothing		Section 3.6.5
Kneser–Ney smoothing	Eq. (3.232)	Eq. (3.230)
PLSA smoothing	Eq. (3.319)	

3.7 Latent semantic information

One of the main limitations of the standard n-gram model is the inadequacy of long-distance information caused by n-gram window size. How to extract semantically meaningful information outside an n-gram window becomes crucial to achieve large-span language modeling. Traditionally, the cache-based language model (Kuhn & De Mori 1990) was proposed to exploit long-distance information where the short-term pattern in history words is continuously captured for word prediction. On the other hand, we may characterize long-distance information by finding long-term semantic dependencies which are seen as global constraints. The short-term statistics within an n-gram window serve as local constraints. Combining local and global constraints provides complete and structural information for word sequence probabilities.

In the literature, *latent semantic analysis* (LSA) (Berry, Dumais & O'Brien 1995) has been popular for many years (Manning & Schütze 1999) to construct latent topic space in information retrieval areas to evaluate the similarity between a query and a document in that space. LSA was extended to probabilistic latent semantic analysis (PLSA) (Hofmann 1999*b*, Hofmann 2001) by dealing with latent semantics as latent variables, an approach which is based on the maximum likelihood theory with the EM algorithm. Thus, PLSA provides an additional generative model to an n-gram for a text that considers long-term semantic information in the text. LSA and PLSA have been applied to develop large-span language models or topic-based language models in Bellegarda (2000) and in Gildea & Hofmann (1999), respectively, by combining these with n-gram language models, and these are addressed below.

3.7.1 Latent semantic analysis

Latent semantic analysis (LSA) introduces an additional longer-term index *document* d to word w. The document usually holds several to thousands of sentences, and the definition (e.g., a paragraph, article, journal paper, etc.) depends on target applications. Suppose there are M documents in a corpus $\mathcal{D}$, which include words with vocabulary size $|\mathcal{V}|$. LSA focuses on the $|\mathcal{V}| \times M$ word-document matrix $\mathbf{W}$:

$$\mathbf{W} = \begin{bmatrix} \omega_{1,1} & \cdots & \omega_{1,M} \\ \vdots & \vdots & \vdots \\ \omega_{|\mathcal{V}|,1} & \cdots & \omega_{|\mathcal{V}|,M} \end{bmatrix} = \begin{bmatrix} \omega_1 & \cdots & \omega_M \end{bmatrix}. \tag{3.255}$$

The element of a word-document matrix $\omega_{v,m}$ is often represented based on a co-occurrence based count $c(w_{(v)}, d_m)$, which is the number of occurrences of word $w_{(v)}$ in document d_m. The word $w_{(v)} \in \mathcal{V}$ is a word in vocabulary $\mathcal{V}$ pointed by an ordered index $v \in \{1, \cdots, |\mathcal{V}|\}$, and (v) in the subscript does not denote a position in a word sequence. The co-occurrence element is often weighted by considering the importance of words in documents. As we discuss later, tf–idf (term frequency–inverse document frequency) or information-theoretic measure is used as an instance of the co-occurrence element. Note that in this word-document matrix representation, we only consider the count information (or related information) for each distinct word $w_{(v)}$ and do not consider the sequential information of words (e.g., w_i) in a document. This feature representation of words is also called *bag of words* (Lewis 1998, Joachims 2002, Blei *et al.* 2003). The column vector $\boldsymbol{\omega}_m \in \mathbb{R}^{|\mathcal{V}|}$ can represent information about document m with a vector. This representation is called a vector space model of a text.

Document similarity

The problem of dealing with this matrix is that the column vector $\boldsymbol{\omega}_m$ is a sparse representation from natural language, and it is difficult to obtain the semantic information from this representation. For example, if we consider the similarity between documents d_m and $d_{m'}$, from the word-document matrix $\mathbf{W}$ with well-known cosine similarity, the cosine similarity is defined as

$$\cos(\mathbf{a}, \mathbf{b}) \triangleq \frac{\mathbf{a}^\mathsf{T}\mathbf{b}}{\|\mathbf{a}\| \|\mathbf{b}\|}. \tag{3.256}$$

The cosine similarity has the following property, and it is often used to measure the similarity in natural language processing:

$$-1 \leq \cos(\mathbf{a}, \mathbf{b}) \leq 1, \quad \cos(\mathbf{a}, \mathbf{a}) = 1. \tag{3.257}$$

The similarity between documents d_m and $d_{m'}$ based on the $|\mathcal{V}|$ dimensional space can be calculated as follows:

$$\mathrm{Sim}_{|\mathcal{V}|}(d_m, d_{m'}) = \cos(\boldsymbol{\omega}_m, \boldsymbol{\omega}_{m'}) = \frac{\sum_{v=1}^{|\mathcal{V}|} \omega_{vm}\omega_{vm'}}{\|\boldsymbol{\omega}_m\| \|\boldsymbol{\omega}_{m'}\|}. \tag{3.258}$$

Since ω_{vm} and $\omega_{vm'}$ are very sparse, most of the products are zero, and the cosine similarity cannot obtain meaningful similarity scores. In addition, the number of dimensions (vocabulary size $|\mathcal{V}|$, the number of documents M, or both) is too large to use. Therefore, LSA conducts a singular value decomposition (SVD) over the $|\mathcal{V}| \times M$ word-document matrix $\mathbf{W}$ and obtains

$$\mathbf{W} \approx \mathbf{U}\mathbf{S}\mathbf{V}^\mathsf{T}, \tag{3.259}$$

as shown in Figure 3.9. In Eq. (3.259), $\mathbf{S}$ is a $K \times K$ diagonal matrix with reduced dimension, $K < \min(|\mathcal{V}|, M)$, $\mathbf{U}$ is a $|\mathcal{V}| \times K$ matrix whose columns are the first K eigenvectors derived from word-by-word correlation matrix $\mathbf{W}\mathbf{W}^\mathsf{T}$, and $\mathbf{V}$ is an $M \times K$ matrix whose columns are the first K eigenvectors derived from the document-by-document correlation matrix $\mathbf{W}^\mathsf{T}\mathbf{W}$.

Document similarity in LSA

After the projection, instead of focusing on the original $|\mathcal{V}| \times M$ word-document matrix $\mathbf{W}$, LSA focuses on the factored $K \times M$ matrix $\mathbf{V}^\mathsf{T}$. Each column of $\mathbf{S}\mathbf{V}^\mathsf{T}$ characterizes the location of a particular document in the reduced K-dimensional semantic space. Therefore, we define the following document vector $\mathbf{v}_m \in \mathbb{R}^K$ as an mth column vector of $\mathbf{V}^\mathsf{T}$, which is a lower dimensional vector than $\boldsymbol{\omega}_m$:

$$\mathbf{V}^\mathsf{T} \triangleq \begin{bmatrix} \mathbf{v}_1 & \cdots & \mathbf{v}_m & \cdots & \mathbf{v}_M \end{bmatrix}. \tag{3.260}$$

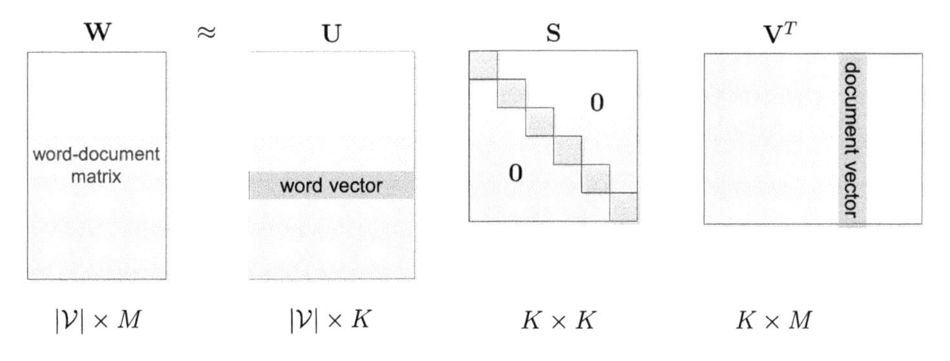

Figure 3.9 Singular value decomposition for latent semantic analysis.

This can be weighted by the diagonal matrix $\mathbf{S}$. Therefore, the similarity between documents d_m and $d_{m'}$ based on this lower K-dimensional representation is calculated as

$$\text{Sim}_K(d_m, d_{m'}) = \cos(\mathbf{S}\mathbf{v}_m, \mathbf{S}\mathbf{v}_{m'}) = \frac{\mathbf{v}_m^{\mathsf{T}}\mathbf{S}^2\mathbf{v}_{m'}}{\|\mathbf{S}\mathbf{v}_m\|\,\|\mathbf{S}\mathbf{v}_{m'}\|}. \tag{3.261}$$

Compared with Eq. (3.258), $\mathbf{S}\mathbf{v}_m$ is a denser feature, and can provide more (semantically) meaningful information (Manning & Schütze 1999).

Now, we consider the application based on information retrieval. We have a query q, which is composed of several words and can be regarded as an instance of a document. Then the problem of information retrieval is to search similar documents to query q, and computing the similarity between q and existing document d_m is very important. To evaluate the cosine similarity between the existing document d_m and this query q in the K-dimensional space, we first transform the corresponding occurrence vector $\boldsymbol{\omega}_q$ to the K-dimensional vector $\mathbf{v}_q$ as follows:

$$\mathbf{v}_q = \mathbf{S}^{-1}\mathbf{U}^{\mathsf{T}}\boldsymbol{\omega}_q. \tag{3.262}$$

Thus, the cosine similarity between q and d_m can be computed based on Eq. (3.261) as follows:

$$\begin{aligned} &\text{Sim}_K(d_m, q) = \cos(\mathbf{S}\mathbf{v}_m, \mathbf{S}\mathbf{v}_q) \propto \mathbf{v}_m^{\mathsf{T}}\mathbf{S}\mathbf{S}\mathbf{S}^{-1}\mathbf{U}^{\mathsf{T}}\boldsymbol{\omega}_q = (\mathbf{S}\mathbf{v}_m)^{\mathsf{T}}\mathbf{U}^{\mathsf{T}}\boldsymbol{\omega}_q \\ \Rightarrow\ &\text{Sim}_K(d_m, q) = \frac{(\mathbf{S}\mathbf{v}_m)^{\mathsf{T}}\mathbf{U}^{\mathsf{T}}\boldsymbol{\omega}_q}{\|\mathbf{S}\mathbf{v}_m\|\,\|\mathbf{U}^{\mathsf{T}}\boldsymbol{\omega}_q\|}. \end{aligned} \tag{3.263}$$

Thus, we can also obtain the similarity between the query q and document d_m with more (semantically) meaningful space, which can be used for information retrieval.

Word-document matrix

The discussion so far does not explicitly introduce a suitable value for the co-occurrence element (v, m) of the word-document matrix $\mathbf{W}$. Actually it is empirically determined, and the straightforward representation is based on the co-occurrence based count $c(w_{(v)}, d_m)$:

$$[\mathbf{W}]_{(v,m)} = \omega_{v,m} = c(w_{(v)}, d_m). \tag{3.264}$$

This is the most simple representation of **W**, and this representation is actually extended as a probabilistic version of LSA (PLSA) in the next section. Another popular representation is based on tf–idf (Salton & Buckley 1988, Manning & Schütze 1999), which represents the element as follows:

$$\begin{aligned}[\mathbf{W}]_{(v,m)} = \omega_{v,m} &= \text{tf}(w_{(v)}, d_m)\text{idf}(w_{(v)}) \\ &= \frac{c(w_{(v)}, d_m)}{\sum_{j=1}^{|\mathcal{V}|} c(w_{(j)}, d_m)} \log \frac{M}{|\{d_m | c(w_{(v)}, d_m) > 0\}|},\end{aligned} \tag{3.265}$$

where $\text{tf}(w_{(v)}, d_m)$ is called *term frequency*, defined as

$$\text{tf}(w_{(v)}, d_m) \triangleq \frac{c(w_{(v)}, d_m)}{\sum_{j=1}^{|\mathcal{V}|} c(w_{(j)}, d_m)}. \tag{3.266}$$

This is computed from the co-occurrence based count $c(w_{(v)}, d_m)$ in Eq. (3.264) with a normalization factor. The quantity $\text{idf}(w_{(v)})$ is called the *inverse document frequency*, and it is computed from the number of documents $|\{d_m | c(w_{(v)}, d_m) > 0\}|$ that include word $w_{(v)}$:

$$\text{idf}(w_{(v)}) \triangleq \log \frac{M}{|\{d_m | c(w_{(v)}, d_m) > 0\}|}. \tag{3.267}$$

$\text{idf}(w_{(v)})$ would score a lower weight for the co-occurrence element when the word $w_{(v)}$ has appeared in many documents, since such a word would be less important.

Bellegarda (2000) also proposes another weight based on an *information-theoretic measure* as follows:

$$[\mathbf{W}]_{(v,m)} = \omega_{v,m} = (1 - \varepsilon_{w_{(v)}}) \frac{c(w_{(v)}, d_m)}{\sum_{j=1}^{|\mathcal{V}|} c(w_{(v)}, d_m)}. \tag{3.268}$$

Compared with Eq. (3.265), the inverse document frequency in Eq. (3.265) is replaced with $(1 - \varepsilon_{w_{(v)}})$, where ε_v denotes the *normalized entropy* of word $w_{(v)}$ in a corpus, defined as

$$\varepsilon_{w_{(v)}} \triangleq -\frac{1}{\log M} \sum_{m=1}^{M} \frac{c(w_{(v)}, d_m)}{\sum_{j=1}^{M} c(w_{(v)}, d_j)} \log \frac{c(w_{(v)}, d_m)}{\sum_{j=1}^{M} c(w_{(v)}, d_j)}. \tag{3.269}$$

The entropy of $w_{(v)}$ would be increased when $w_{(v)}$ is distributed among many documents, which decreases the weight $(1 - \varepsilon_{w_{(v)}})$. Therefore, similarly to tf–idf, the word distributed in many documents would be less important, and have lower weight for the co-occurrence element.

3.7.2 LSA language model

The LSA language model was proposed to capture long-range word dependencies through discovery of latent topics from a text corpus. When incorporating LSA into an n-gram model, the prediction of word w_i making use of n-gram probability $p(w_i | w_{i-n+1}^{i-1})$ is calculated from two information sources in history:

1. The n-gram history words

$$h_{i-1}^{(n)} \triangleq w_{i-n+1}^{i-1}. \tag{3.270}$$

2. The long-term topic information of all history words w_1^{i-1}, which is represented by co-occurrence count (Eq. (3.264)), tf–idf (Eq. (3.265)), or information-theoretic measure (Eq. (3.268)), as discussed in the previous section:

$$h_{i-1}^{(l)} \triangleq \boldsymbol{\omega}_{i-1}. \tag{3.271}$$

The K-dimensional projected vector $\tilde{\mathbf{v}}_{i-1} \in \mathbb{R}^D$, from the history vector in the original space $\boldsymbol{\omega}_{i-1} \in \mathbb{R}^{|\mathcal{V}|}$ onto the LSA space, is obtained based on Eq. (3.262) as:

$$\tilde{\mathbf{v}}_{i-1} \triangleq \mathbf{S}\mathbf{v}_{i-1} = \mathbf{U}^{\mathsf{T}}\boldsymbol{\omega}_{i-1}. \tag{3.272}$$

By using these two types of history information, the LSA language model is represented by using the product and sum rules as:

$$\begin{aligned} p^{\mathsf{LSA}}(w_i|w_{i-n+1}^{i-1}) &\triangleq p(w_i|h_{i-1}^{(n)}, h_{i-1}^{(l)}) \\ &= \frac{p(w_i|h_{i-1}^{(n)})p(h_{i-1}^{(l)}|w_i, h_{i-1}^{(n)})}{p(h_{i-1}^{(l)}|h_{i-1}^{(n)})} \\ &= \frac{p(w_i|h_{i-1}^{(n)})p(h_{i-1}^{(l)}|w_i, h_{i-1}^{(n)})}{\sum_{w_j \in \mathcal{V}} p(w_j|h_{i-1}^{(n)})p(h_{i-1}^{(l)}|w_j, h_{i-1}^{(n)})}. \end{aligned} \tag{3.273}$$

From the definition of $h_{i-1}^{(n)}$ in Eq. (3.270), $p(w_i|h_{i-1}^{(n)})$ is represented as the following n-gram probability:

$$p(w_i|h_{i-1}^{(n)}) = p(w_i|w_{i-n+1}^{i-1}). \tag{3.274}$$

This can be calculated based on Section 3.6. Next, we focus on the distribution of the long-term topic information $p(h_{i-1}^{(l)}|w_i, h_{i-1}^{(n)})$ in Eq. (3.273). By using the definitions of $h_{i-1}^{(l)}$ and $h_{i-1}^{(n)}$ in Eqs. (3.270) and (3.271), the distribution can be rewritten as:

$$p(h_{i-1}^{(l)}|w_i, h_{i-1}^{(n)}) = p(\boldsymbol{\omega}_{i-1}|w_{i-n+1}^{i}) \approx p(\boldsymbol{\omega}_{i-1}|w_i). \tag{3.275}$$

The above approximation assumes that the document vector $\boldsymbol{\omega}_{i-1}$ only depends on w_i and does not depend on the word history w_{i-n+1}^{i-1}, in accordance with the discussion in Bellegarda (2000). This approximation would be effective when w_i is a content word. In addition, by using the Bayes theorem, $p(\boldsymbol{\omega}_{i-1}|w_i)$ can be further rewritten as follows:

$$p(\boldsymbol{\omega}_{i-1}|w_i) \propto \frac{p(w_i|\boldsymbol{\omega}_{i-1})}{p(w_i)}, \tag{3.276}$$

where $p(w_i)$ is easily calculated from a unigram probability. The $p(w_i|\boldsymbol{\omega}_{i-1})$ is a unigram probability given the document vector. Thus, the LSA language model is finally represented as follows:

$$p^{\mathsf{LSA}}(w_i|w_{i-n+1}^{i-1}) = \frac{p(w_i|w_{i-n+1}^{i-1})\frac{p(w_i|\boldsymbol{\omega}_{i-1})}{p(w_i)}}{\sum_{w_j} p(w_j|w_{i-n+1}^{i-1})\frac{p(w_j|\boldsymbol{\omega}_{i-1})}{p(w_j)}}. \tag{3.277}$$

Here, how to compute $p(w_i|\boldsymbol{\omega}_{i-1})$ is an important problem. This is performed in the projected LSA space, rather than the original high dimensional space.

The $p(w_i|\boldsymbol{\omega}_{i-1})$ is determined by the cosine measure between word w_i and pseudo document w_1^{i-1} in the LSA space, as we discussed in Eq. (3.261). First, w_i is converted with the corresponding co-occurrence vector in the $|\mathcal{V}|$ dimensional space as

$$\boldsymbol{\omega}_{w_i} = [0, \cdots, 0, \overset{\overset{v}{\vee}}{1}, 0, \cdots, 0]^{\mathsf{T}}. \tag{3.278}$$

Vector $\boldsymbol{\omega}_{w_i}$ is a one shot vector in the $|\mathcal{V}|$ dimensional space, where the element is 1 when $w_i = w_{(v)}$ and 0 otherwise. The document vector in the LSA space for $\boldsymbol{\omega}_{w_i}$ is obtained from Eq. (3.262) as

$$\tilde{\mathbf{v}}_{w_i} = \mathbf{S}\mathbf{v}_{w_i} = \mathbf{U}^{\mathsf{T}}\boldsymbol{\omega}_{w_i}. \tag{3.279}$$

Therefore, based on this equation and Eq. (3.272), the cosine similarity in the LSA K-dimensional space is obtained as:

$$\mathrm{Sim}_K(w_i, h_{i-1}^{(l)}) = \cos\left(\tilde{\mathbf{v}}_{w_i}, \tilde{\mathbf{v}}_{i-1}\right). \tag{3.280}$$

However, since a cosine value goes to negative, it cannot be used as a probability of $p(w_i|\boldsymbol{\omega}_{i-1})$ that must satisfy the non-negativity. Coccaro & Jurafsky (1998) propose the following value as a probabilistic distribution:

$$p(w_i|\boldsymbol{\omega}_{i-1}) = \frac{\left(\cos\left(\tilde{\mathbf{v}}_{w_i}, \tilde{\mathbf{v}}_{i-1}\right) - \min_{w_i' \in \mathcal{V}} \cos\left(\tilde{\mathbf{v}}_{w_i'}, \tilde{\mathbf{v}}_{i-1}\right)\right)^{\gamma}}{Z}, \tag{3.281}$$

where Z is a normalization constant. The minimum cosine similarity prevents a negative value, and it is also scaled by a tuning parameter γ.

In Eq. (3.280), the pseudo document vector $\boldsymbol{\omega}_{i-1}$ is recursively updated from w_{i-1} to w_i by using

$$\boldsymbol{\omega}_i = \frac{n_i - 1}{n_i}\boldsymbol{\omega}_{i-1} + \frac{1 - \varepsilon_{w_i}}{n_i}\boldsymbol{\omega}_{w_i}, \tag{3.282}$$

where n_i denotes the number of words appearing in word sequence w_1^i:

$$n_i \triangleq \sum_{v=1}^{|\mathcal{V}|} c(w_{(v)}, w_1^i), \tag{3.283}$$

and ε_{w_i} is a normalized entropy, as we defined in Eq. (3.269). Eq. (3.282) is obtained as a pseudo document vector in matrix $\mathbf{W}$, which is defined according to the entry value of matrix $\mathbf{W}$ as given in Eq. (3.268). This updating is equivalent to computing

$$\tilde{\mathbf{v}}_i = \mathbf{S}\mathbf{v}_i = \frac{1}{n_i}[(n_i - 1)\tilde{\mathbf{v}}_{i-1} + (1 - \varepsilon_{w_i})\tilde{\mathbf{v}}_{w_i}]. \tag{3.284}$$

This equation is obtained in accordance with Eqs. (3.272) and (3.279). Having the updating formula for $\boldsymbol{\omega}_{i-1}$ or $\tilde{\mathbf{v}}_{i-1}$ for pseudo document vector or history words w_1^{i-1} and the projection vector $\tilde{\mathbf{v}}_{w_i}$ of current word w_i in common topic space, we calculate the n-gram based on topic information in Eq. (3.280) and substitute it into an LSA language model according to Eq. (3.273).

Thus, we can consider a long-term effect in a language model by using LSA. However, to integrate it with a language model, we need to provide a probabilistic treatment for LSA, e.g., Eq. (3.281), which is a rather heuristic approach. The next section introduces a fully probabilistic formulation of LSA, PLSA with an elegant inference and statistical learning methods based on the EM algorithm.

3.7.3 Probabilistic latent semantic analysis

LSA was extended to the probabilistic latent semantic analysis (PLSA) and applied for document representation in Hofmann (1999*b*) and Hofmann (2001). We first introduce a J_m-length word sequence, given a document d_m, as

$$w_1^{J_m} = \{w_1, \cdots, w_i, \cdots, w_{J_m}\} = \{w_i \in \mathcal{V} | i = 1, \cdots, J_m\} \text{ for } m = 1, \cdots, M. \quad (3.285)$$

We also introduce a set of M documents as

$$d_1^M = \{d_1, \cdots, d_M\} = \{d_m | m = 1, \cdots, M\}. \quad (3.286)$$

The first step of the ML procedure within the statistical model framework is to provide a joint likelihood function of $\mathcal{D}$ of the word sequence and document, which can be represented as:

$$p(\mathcal{D}) = p(\{w_1^{J_m}\}_{m=1}^M, d_1^M), \quad (3.287)$$

where $\mathcal{D}$ is a set of the word sequences and documents in a corpus, which is defined as:

$$\mathcal{D} \triangleq \{\{w_1^{J_m}\}_{m=1}^M, d_1^M\}. \quad (3.288)$$

Equation (3.287) can be factorized by using the product rule and the conditional independence assumption of a word generative model given document d_m as:

$$\begin{aligned} p(\mathcal{D}) &= p(\{w_1^{J_m}\}_{m=1}^M | d_1^M) p(d_1^M) \\ &= \prod_{m=1}^M p(w_1^{J_m} | d_m) p(d_m). \end{aligned} \quad (3.289)$$

This can be modeled by the word-document representation (e.g., word-document matrix $\mathbf{W}$ in Eq. (3.255)), as we discussed in Section 3.7.1. However, similarly to the LSA case, this representation has too many sparse variables, and it is difficult to deal with this representation directly. Instead, we introduce a latent topic variable $k \in \{1, \cdots, K\}$, where K is the number of topics and each topic corresponds to a higher concept of words (e.g., politics and sports, if we deal with news articles). This model is called a *latent topic model*. Then we also introduce a latent topic sequence $z_1^{J_m}$ for a corresponding word sequence $w_1^{J_m}$ for a document d_m as

$$\begin{aligned} z_1^{J_m} &= \{z_i \in \{1, \cdots, K\} | i = 1, \cdots, J_m\} \\ Z &= \{z_1^{J_m}\}_{m=1}^M. \end{aligned} \quad (3.290)$$

Then, we model that the words $w_1^{J_m}$ are conditionally independent given the corresponding topic $z_1^{J_m}$, and the probability of w_i given d_m is represented by the following mixture model:

$$p(w_i|d_m) = \sum_{k=1}^{K} p(w_i|z_i = k)p(z_i = k|d_m). \tag{3.291}$$

This model factorizes the word probability into the topic-dependent unigram probability $p(w_i|z_i = k)$ and the topic proportion probability $p(z_i = k|d_m)$. This factorization corresponds to representing the huge size of a $|\mathcal{V}| \times M$ word-document matrix with the reduced subspace K used in an LSA, as discussed in Section 3.7.1.

Thus, we can construct a likelihood function of $\mathcal{D}$ as

$$p(\mathcal{D}|\Theta) = \prod_{m=1}^{M} \prod_{i=1}^{J_m} \sum_{k=1}^{K} p(w_i|z_i = k)p(z_i = k|d_m)p(d_m). \tag{3.292}$$

This model for representing a generative process of the word sequences and documents is called probabilistic latent semantic analysis (PLSA). Here, the PLSA model parameters $\Theta = \{p(w_{(v)}|k), p(k|d_m)\}$ consist of two sets of multinomial parameters. The complete data likelihood function of $\mathcal{D}$ and Z can be represented as:

$$p(\mathcal{D}, Z|\Theta) = \prod_{m=1}^{M} \prod_{i=1}^{J_m} p(w_i|z_i)p(z_i|d_m)p(d_m). \tag{3.293}$$

Figure 3.10 provides a graphical model of PLSA and Algorithm 6 provides a generative process of PLSA.

Algorithm 6 Generative process of probabilistic latent semantic analysis.

Require: $M, J_m, \omega_m^d, \omega_k^z, \omega_{kv}^w$
1: **for** $m = 1, \cdots, M$ **do**
2: Draw d_m from $\text{Mult}(\cdot|\{\omega_m^d\}_{m=1}^M)$
3: **for** $i = 1, \cdots, J_m$ **do**
4: Draw z_i from $\text{Mult}(\cdot|\{\omega_k^z\}_{k=1}^K)$
5: Draw w_i from $\text{Mult}(\cdot|\{\omega_{kv}^w\}_{v=1}^{|\mathcal{V}|})$
6: **end for**
7: **end for**

Parameter estimation based on EM algorithm

Once we obtain the likelihood function, maximum likelihood (ML) theory is applied to estimate PLSA parameters by

$$\Theta^{\text{ML}} = \arg\max_{\Theta} p(\mathcal{D}|\Theta). \tag{3.294}$$

Since PLSA has a latent variable z, it can be efficiently solved by using the EM algorithm, as we discussed in Section 3.4. The EM algorithm is introduced to find ML

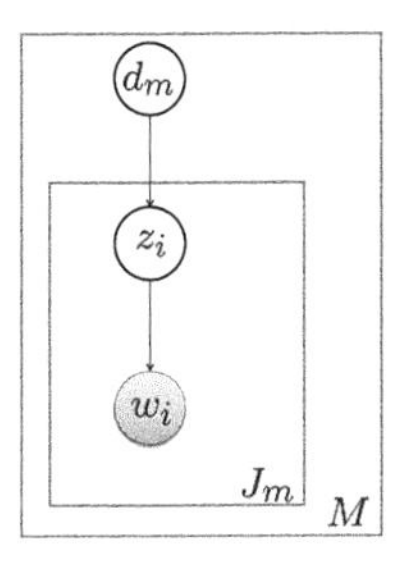

Figure 3.10 Graphical model of probabilistic latent semantic analysis.

estimates Θ^{ML}. In the E-step of the EM algorithm, we calculate an auxiliary function of new parameters Θ' given the current parameters Θ. The function is obtained by replacing HMM latent variables S and V, and speech feature observation $\mathbf{O}$ with PLSA latent variable Z and document observation $\mathcal{D}$, giving

$$
\begin{aligned}
Q^{\mathrm{ML}}(\Theta'|\Theta) &= \mathbb{E}_{(Z)}[\log p(\mathcal{D}, Z|\Theta')|\mathcal{D}, \Theta] \\
&= \sum_{Z} p(Z|\mathcal{D}, \Theta) \log p(\mathcal{D}, Z|\Theta').
\end{aligned} \tag{3.295}
$$

By substituting Eq. (3.293) into Eq. (3.295), the auxiliary function can be rewritten as

$$
\begin{aligned}
Q^{\mathrm{ML}}(\Theta'|\Theta) &= \sum_{Z} p(Z|\mathcal{D}, \Theta) \log \left(\prod_{m=1}^{M} \prod_{i=1}^{J_m} p'(w_i|z_i) p'(z_i|d_m) p'(d_m) \right) \\
&= \sum_{Z} p(Z|\mathcal{D}, \Theta) \sum_{m=1}^{M} \sum_{i=1}^{J_m} \log p'(w_i|z_i) p'(z_i|d_m) p'(d_m),
\end{aligned} \tag{3.296}
$$

where $p'(\cdot)$ means a multinomial distribution with parameter Θ'. By using a similar trick to that used in Eq. (3.102), we can change the order of the summations, and by executing the summation over $Z_{\backslash z_i}$, Eq. (3.296) can be further rewritten as

$$
\begin{aligned}
Q^{\mathrm{ML}}(\Theta'|\Theta) &= \sum_{m=1}^{M} \sum_{i=1}^{J_m} \sum_{Z} p(Z|\mathcal{D}, \Theta) \log \left(p'(w_i|z_i) p'(z_i|d_m) p'(d_m) \right) \\
&= \sum_{m=1}^{M} \sum_{i=1}^{J_m} \sum_{z_i} p(z_i|\mathcal{D}, \Theta) \log \left(p'(w_i|z_i) p'(z_i|d_m) p'(d_m) \right) \\
&= \sum_{m=1}^{M} J_m \log p'(d_m) + \sum_{m=1}^{M} \sum_{i=1}^{J_m} \sum_{z_i} p(z_i|\mathcal{D}, \Theta) \log \left(p'(w_i|z_i) p'(z_i|d_m) \right).
\end{aligned} \tag{3.297}
$$

PLSA assumes that the above probabilistic distributions $p'(w_i|z_i)$ and $p'(z_i|d_m)$ are represented by using the multinomial distribution, which is defined in Appendix C.2 as:

$$\mathrm{Mult}(x_j|\{\omega_j\}_{j=1}^{J}) \triangleq \omega_j. \tag{3.298}$$

Therefore, the topic-dependent unigram distribution $p'(w_i = w_{(v)}|z_i = k)$ and topic proportion distribution $p'(z_i = k|d_m)$ are parameterized as:

$$\begin{aligned} p'(w_i = w_{(v)}|z_i = k) &= \omega'_{vk} \\ p'(z_i = k|d_m) &= \omega'_{km}. \end{aligned} \tag{3.299}$$

Thus, the parameter Θ is represented as the following set of these multinomial distributions:

$$\Theta' \triangleq \{\omega'_{vk}, \omega'_{km}|v = 1, \cdots, |\mathcal{V}|, k = 1, \cdots, K, m = 1, \cdots, M\}. \tag{3.300}$$

Now, we focus on the posterior distribution of latent variable $p(z_i|\mathcal{D}, \Theta)$. From the graphical model, z_i has the following conditional independence property:

$$\begin{aligned} p(z_i = k|\mathcal{D}, \Theta) &= p(z_i = k|w_i = w_{(v)}, d_m, \Theta) \\ &= p(k|w_{(v)}, d_m, \Theta). \end{aligned} \tag{3.301}$$

The second probability means that the probability of topic k is given by the word $w_{(v)}$ with d_m and Θ, and it is iid for the position i in d_m. Then, we can rewrite Eq. (3.297) as the following equation:

$$\begin{aligned} &Q^{\mathrm{ML}}(\Theta'|\Theta) \\ &= \sum_{m=1}^{M} J_m \log p'(d_m) + \sum_{m=1}^{M}\sum_{v=1}^{|\mathcal{V}|} c(w_{(v)}, d_m) \sum_{k=1}^{K} p(k|w_{(v)}, d_m, \Theta) \log\left(\omega'_{vk}\omega'_{km}\right), \end{aligned} \tag{3.302}$$

where $c(w_{(v)}, d_m)$ is a co-occurrence count of word $w_{(v)}$ in document d_m. Note that the summation over words in a document $\sum_{i=1}^{J_m}$ is replaced with the summation over distinct words within a dictionary $\sum_{v=1}^{|\mathcal{V}|}$. Equation (3.302) is factorized into the topic-dependent unigram parameter ω'_{vk} and the topic proportion parameter ω'_{km} as follows:

$$\begin{aligned} Q^{\mathrm{ML}}(\Theta'|\Theta) &= \sum_{m=1}^{M} J_m \log p'(d_m) \\ &\quad + \underbrace{\sum_{m=1}^{M}\sum_{v=1}^{|\mathcal{V}|} c(w_{(v)}, d_m) \sum_{k=1}^{K} p(k|w_{(v)}, d_m, \Theta) \log\left(\omega'_{vk}\right)}_{\triangleq Q^{\mathrm{ML}}(\{\omega'_{vk}\}|\{\omega_{vk}\})} \\ &\quad + \underbrace{\sum_{m=1}^{M}\sum_{v=1}^{|\mathcal{V}|} c(w_{(v)}, d_m) \sum_{k=1}^{K} p(k|w_{(v)}, d_m, \Theta) \log\left(\omega'_{km}\right)}_{\triangleq Q^{\mathrm{ML}}(\{\omega'_{km}\}|\{\omega_{km}\})} \\ &\triangleq \sum_{m=1}^{M} J_m \log p'(d_m) + Q^{\mathrm{ML}}(\{\omega'_{vk}\}|\{\omega_{vk}\}) + Q^{\mathrm{ML}}(\{\omega'_{km}\}|\{\omega_{km}\}). \end{aligned} \tag{3.303}$$

Two multinomial parameters ω_{vk} and ω_{km} should meet the constraints for probability parameters,

$$\sum_{v=1}^{|\mathcal{V}|} \omega_{vk} = 1$$
$$\sum_{k=1}^{K} \omega_{km} = 1. \tag{3.304}$$

Thus, Eq. (3.302) provides the auxiliary function of a PLSA with the parameter constraint Eq. (3.304). Therefore, similarly to Section 3.4, we can iteratively estimate parameters based on the EM algorithm.

M-step

In the M-step, we maximize the extended auxiliary function with respect to new parameters Θ' under constraints of Eq. (3.304). Again, as we discussed in Section 3.4.3, this constrained optimization problem is solved through introducing a Lagrange multiplier η and establishing the extended auxiliary function of $Q^{\mathrm{ML}}(\{\omega'_{vk}\}|\{\omega_{vk}\}$ in Eq. (3.303) as

$$\begin{aligned}
&\bar{Q}(\{\omega'_{vk}\}|\{\omega_{vk}\}) \\
&= \sum_{m=1}^{M} \sum_{v=1}^{|\mathcal{V}|} c(w_{(v)}, d_m) \sum_{k=1}^{K} p(k|w_{(v)}, d_m, \Theta) \log\left(\omega'_{vk}\right) + \eta \left(\sum_{v=1}^{|\mathcal{V}|} \omega'_{vk} - 1 \right).
\end{aligned} \tag{3.305}$$

By differentiating Eq. (3.305) with respect to ω'_{vk} and setting it to zero, we obtain

$$\begin{aligned}
\frac{\partial \bar{Q}(\{\omega'_{vk}\}|\{\omega_{vk}\})}{\partial \omega'_{vk}} &= \sum_{m=1}^{M} c(w_{(v)}, d_m) p(k|w_{(v)}, d_m, \Theta) \frac{1}{\omega'_{vk}} + \eta = 0 \\
\Rightarrow \omega'_{vk} &= -\frac{1}{\eta} \sum_{m=1}^{M} c(w_{(v)}, d_m) p(k|w_{(v)}, d_m, \Theta).
\end{aligned} \tag{3.306}$$

Again, substituting this equation into constraint Eq. (3.304), we obtain the value of Lagrange multiplier η,

$$\begin{aligned}
\sum_{v=1}^{|\mathcal{V}|} \omega'_{vk} &= -\frac{1}{\eta} \sum_{v=1}^{|\mathcal{V}|} \sum_{m=1}^{M} c(w_{(v)}, d_m) p(k|w_{(v)}, d_m, \Theta) = 1 \\
\Rightarrow \eta &= -\sum_{v=1}^{|\mathcal{V}|} \sum_{m=1}^{M} c(w_{(v)}, d_m) p(k|w_{(v)}, d_m, \Theta).
\end{aligned} \tag{3.307}$$

Consequently, we find the following ML estimate of ω'_{vk} (i.e., the topic-dependent unigram proportion probability) for a PLSA topic model:

$$p^{\mathrm{ML}'}(w_i = w_{(v)}|z_i = k) = \omega'_{vk} = \frac{\sum_{m=1}^{M} c(w_{(v)}, d_m) p(k|w_{(v)}, d_m, \Theta)}{\sum_{v=1}^{|\mathcal{V}|} \sum_{m=1}^{M} c(w_{(v)}, d_m) p(k|w_{(v)}, d_m, \Theta)}. \tag{3.308}$$

Similarly, the parameter of topic proportion probability ω'_{km} is estimated based on the M-step as

$$p^{\mathrm{ML}'}(z_i = k|d_m) = \omega'_{km} = \frac{\sum_{v=1}^{|\mathcal{V}|} c(w_{(v)}, d_m)p(k|w_{(v)}, d_m, \Theta)}{\sum_{k'=1}^{K}\sum_{v=1}^{|\mathcal{V}|} c(w_{(v)}, d_m)p(k'|w_{(v)}, d_m, \Theta)}$$
$$= \frac{\sum_{v=1}^{|\mathcal{V}|} c(w_{(v)}, d_m)p(k|w_{(v)}, d_m, \Theta)}{\sum_{v=1}^{|\mathcal{V}|} c(w_{(v)}, d_m)}. \tag{3.309}$$

Thus, we obtain the ML solutions of the PLSA parameters based on the EM algorithm.

E-step

The E-step needs to consider the following posterior distribution of the latent topic k with the previously estimated PLSA parameter Θ:

$$p(k|w_{(v)}, d_m, \Theta). \tag{3.310}$$

By using the sum and product rules and the conditional independence (Eq. (3.291)), the posterior distribution is rewritten as

$$p(k|w_{(v)}, d_m, \Theta) = \frac{p(w_{(v)}, k|d_m, \Theta)}{\sum_{k'=1}^{K} p(w_{(v)}, k'|d_m, \Theta)}$$
$$= \frac{p(w_{(v)}|k, \Theta)p(k|d_m, \Theta)}{\sum_{k'=1}^{K} p(w_{(v)}|k', \Theta)p(k'|d_m, \Theta)}. \tag{3.311}$$

By using the multinomial parameters ω_{vk} and ω_{km}, this can be represented by

$$p(k|w_{(v)}, d_m, \Theta) = \frac{\omega_{vk}\omega_{km}}{\sum_{k'=1}^{K} \omega_{vk'}\omega_{k'm}}. \tag{3.312}$$

Thus, the posterior distribution can be computed by using the previously estimated PLSA parameter Θ. We summarize the parameter estimation algorithm of the PLSA in Algorithm 7. In a practical implementation, storing the above posterior distribution $p(k|w_{(v)}, d_m, \Theta)$ explicitly needs a huge size of memory (that corresponds to store $M \times |\mathcal{V}| \times K$ data), and it is impossible to do that for large-scale text data. Instead, the PLSA parameter update is performed by using a matrix multiplication based operation. Actually, this PLSA parameter update corresponds to the multiplicative matrix update of non-negative matrix factorization (NMF) (Lee & Seung 1999) based on the KL divergence cost function (Gaussier & Goutte 2005).

PLSA is another well-known generative model in addition to the n-gram model that generates words with exponential family distributions (multinomial distribution). Since the generative model is represented with exponential family distributions, PLSA can be extended by using Bayesian approaches, as we discussed in Section 2.1.3. This extension is called the latent Dirichlet allocation (Blei *et al.* 2003), where we use a Dirichlet distribution as a conjugate prior, which is discussed in Section 7.6. We also show that PLSA parameters can be estimated efficiently by using the EM algorithm. However, PLSA does not consider the short-span language property, unlike an n-gram model, and the next section discusses how to involve the PLSA and n-gram for the use of a language model.

Algorithm 7 EM algorithm for PLSA

Require: $\Theta \leftarrow \Theta^{\text{init}}$, $\{c(w_{(v)}, d_m)|v = 1, \cdots, |\mathcal{V}|, m = 1, \cdots, M\}$
1: **repeat**
2: **for** $m = 1, \cdots, M$ **do**
3: **for** $v = 1, \cdots, |\mathcal{V}|$ **do**
4: **for** $k = 1, \cdots, K$ **do**
5: Compute the posterior probability $p(k|w_{(v)}, d_m, \Theta)$
6: **end for**
7: **end for**
8: **end for**
9: **for** $v = 1, \cdots, |\mathcal{V}|$ **do**
10: **for** $k = 1, \cdots, K$ **do**
11: Compute the topic-dependent unigram probability
12: $p^{\text{ML}'}(w_i = w_{(v)}|z_i = k) = \omega'_{vk}$
13: **end for**
14: **end for**
15: **for** $m = 1, \cdots, M$ **do**
16: **for** $k = 1, \cdots, K$ **do**
17: Compute the topic proportion probability
18: $p^{\text{ML}'}(z_i = k|d_m) = \omega'_{km}$
19: **end for**
20: **end for**
21: $\Theta \leftarrow \Theta'$
22: **until** Convergence

3.7.4 PLSA language model

In Gildea & Hofmann (1999), Akita & Kawahara (2004) and Mrva & Woodland (2004), a PLSA framework was applied to estimate the n-gram model, which is seen as a combination of large-span and short-span language models. That is, we consider the short-span language model (n-gram)

$$p(w_i|w_{i-n+1}^{i-1}) \tag{3.313}$$

and long-span language model

$$p(w_i|\mathcal{D}) \approx \sum_{k=1}^{K} p(w_i|k)p(k|\mathcal{D}) \tag{3.314}$$

to obtain the probability of w_i. As an alternative, a long-span language model $p(w_i|\mathcal{D})$, a cache-based language model (Kuhn & De Mori 1990), or a trigger-based language model (Lau, Rosenfeld & Roukos 1993) are used as conventional approaches. As an example of the document $\mathcal{D}$, we consider the history of all words from the beginning w_1^{i-1}. Given word sequence w_1^{i-1}, latent topic probabilities of the history $p(k|w_1^{i-1})$ are inferred by using a PLSA model and incorporated into an n-gram language model.

To obtain the latent topic probabilities $p(k|w_1^{i-1})$, maximum likelihood PLSA parameters $\Theta = \{p(w_{(v)}|k), p(k|d_m)\} = \{\omega_{vk}, \omega_{km}|v = 1, \cdots, |\mathcal{V}|, k = 1, \cdots, K, m = 1, \cdots, M\}$ are estimated in advance from a large amount of training documents in the training step, as we discussed in the previous section. Then, given Θ as an initial parameter set of Algorithm 7, we estimate $p(k|w_1^{i-1})$ in the test step. However, to avoid over-training as we only use w_i^{i-1} and to avoid a high computational cost in the test step, we only update $p^{\mathrm{ML}'}(z_i = k|d_m) = \omega'_{km}$ with fewer iterations than the full estimation of the PLSA parameter Θ in the training step. In addition, we use the following *online EM* algorithm (Neal & Hinton 1998) to update topic posterior probabilities to make the estimation in an on-line manner for every word position i:

$$\begin{aligned} p(k|w_1^{i-1}) &= \frac{1}{i+1} \frac{p(w_{i-1} = v|k)p(k|w_1^{i-2})}{\sum_{k'=1}^{K} p(w_{i-1} = v|k')p(k'|w_1^{i-2})} + \frac{i}{i+1} p(k|w_1^{i-2}) \\ &= \frac{1}{i+1} \frac{\omega_{vk} p(k|w_1^{i-2})}{\sum_{k'=1}^{K} \omega_{vk'} p(k'|w_1^{i-2})} + \frac{i}{i+1} p(k|w_1^{i-2}). \end{aligned} \tag{3.315}$$

Here, $p(k|w_1^{i-2})$ in the first and second terms in the right-hand-side is obtained by the previous estimation, and the first term is computed on-the-fly based on a pre-computation of $p(w_{i-1}|k)$, which is obtained from ω_{vk} in Eq. (3.308). The factors $\frac{1}{i+1}$ and $\frac{i}{i+1}$ denote a linear interpolation ratio, and the contribution of the second term becomes larger when the length of the word sequence i is larger. These factors are derived as a specific solution of the on-line EM algorithm, and Mrva & Woodland (2004) suggest modifying these factors for practical use.

In an initialization stage (i.e., $w_1^{i=0} = \emptyset$ in Eq. (3.184)), the initial topic posterior probability could be obtained from a prior topic probability in the training stage as follows:

$$\begin{aligned} p(k|w_1^{i=0}) &= p(k) \\ &= \sum_{m=1}^{M} p(k|d_m)p(d_m) \\ &= \frac{\sum_{m=1}^{M} \sum_{v=1}^{|\mathcal{V}|} c(w_{(v)}, d_m) p(k|d_m)}{\sum_{m=1}^{M} \sum_{v=1}^{|\mathcal{V}|} c(w_{(v)}, d_m)} \\ &= \frac{\sum_{m=1}^{M} \sum_{v=1}^{|\mathcal{V}|} c(w_{(v)}, d_m) \omega_{km}}{\sum_{m=1}^{M} \sum_{v=1}^{|\mathcal{V}|} c(w_{(v)}, d_m)}, \end{aligned} \tag{3.316}$$

where $p(d_m)$ is estimated from the maximum likelihood equation based on the word co-occurrence counts $c(d_m) = \sum_{v=1}^{|\mathcal{V}|} c(w_{(v)}, d_m)$, i.e.,

$$p(d_m) = \frac{c(d_m)}{\sum_{m=1}^{M} c(d_m)}. \tag{3.317}$$

Once we obtain the topic proportion probability $p(k|w_1^{i-1})$, the PLSA based n-gram probability is finally calculated by

$$p^{\mathsf{PLSA}}(w_i = v|w_1^{i-1}) = \sum_{k=1}^{K} p(w_i = v|k)p(k|w_1^{i-1})$$
$$= \sum_{k=1}^{K} \omega_{vk} p(k|w_1^{i-1}). \tag{3.318}$$

Instead of hard-clustering computation in class-based n-gram in Eq. (3.198), PLSA n-gram in Eq. (3.318) performs a so-called *soft-clustering* computation by marginalizing the topic index k in a Bayesian sense. This model is known as a large-span language model because long-distance topics $k = 1, \cdots, K$ and history words w_1^{i-1} within the n-gram window as well as outside the n-gram window are all taken into account for word prediction. However, since the computation can be performed by considering the history words w_1^{i-1}, it is not pre-computed, unlike n-gram models, which makes the implementation of the PLSA language model harder than n-gram models in ASR.

Smoothing with n-grams

PLSA n-grams could be further improved by combining with ML n-grams based on additional linear interpolation with factor λ:

$$\hat{p}(w_i|w_1^{i-1}) = \lambda p^{\mathsf{ML}}(w_i|w_{i-n+1}^{i-1})$$
$$+ (1 - \lambda)p^{\mathsf{PLSA}}(w_i|w_1^{i-1}). \tag{3.319}$$

The interpolation parameter λ could be found by maximizing likelihood or tuned by maximizing the ASR performance of some validation data.

We can also use a unigram rescaling technique, which is approximated based on the conditional independence assumption of w_1^{i-1} and w_{i-n+1}^{i-1}:

$$\hat{p}(w_i|w_1^{i-1}) \approx p^{\mathsf{ML}}(w_i|w_{i-n+1}^{i-1}) \frac{p^{\mathsf{PLSA}}(w_i|w_1^{i-1})}{p^{\mathsf{ML}}(w_i)}. \tag{3.320}$$

Note that this representation does not hold the sum-to-one condition, and it must be normalized.

A more precise unigram rescaling technique can be performed by using the dynamic unigram marginal (also known as the minimum discrimination information (MDI) adaptation), which can consider backoff probabilities in an n-gram probability (Kneser, Peters & Klakow 1997, Niesler & Willett 2002, Tam & Schultz 2006). First, we define the following unigram rescaling factor π for word w_i:

$$\pi(w_i, w_1^{i-1}) \triangleq \left(\frac{p^{\mathsf{PLSA}}(w_i|w_1^{i-1})}{p^{\mathsf{ML}}(w_i)} \right)^{\rho}. \tag{3.321}$$

Then the new n-gram probability with a history of n-gram word sequence w_{i-n+1}^{i-1} is represented as follows:

$$p(w_i|w_{i-n+1}^{i-1}, w_1^{i-1}) = \begin{cases} \frac{\pi(w_i, w_1^{i-1})}{C_0(w_{i-n+1}^{i-1}, w_1^{i-1})} p(w_i|w_{i-n+1}^{i-1}) & \text{if} \quad c(w_{i-n+1}^{i}) > 0 \\ \frac{1}{C_1(w_{i-n+1}^{i-1}, w_1^{i-1})} p(w_i|w_{i-n+2}^{i-1}, w_1^{i-1}) & \text{otherwise,} \end{cases} \tag{3.322}$$

where $p(w_i|w_{i-n+1}^{i-1})$ is an n-gram probability obtained in advance, and C_0 and C_1 are normalization constants, which can be computed by:

$$C_0(w_{i-n+1}^{i-1}, w_1^{i-1}) = \frac{\sum_{\{w_i|c(w_{i-n+1}^{i})>0\}} \pi(w_i, w_1^{i-1}) p(w_i|w_{i-n+1}^{i-1})}{\sum_{\{w_i|c(w_{i-n+1}^{i})>0\}} p(w_i|w_{i-n+1}^{i-1})}, \tag{3.323}$$

and

$$C_1(w_{i-n+1}^{i-1}, w_1^{i-1}) = \frac{1 - \sum_{\{w_i|c(w_{i-n+1}^{i})>0\}} p(w_i|w_{i-n+2}^{i-1}, w_1^{i-1})}{1 - \sum_{\{w_i|c(w_{i-n+1}^{i})>0\}} p(w_i|w_{i-n+1}^{i-1})}. \tag{3.324}$$

Then, $p(w_i|w_{i-n+2}^{i-1})$ is also iteratively calculated by $\pi(w_i, w_1^{i-1})$, $n-1$-gram probability $p(w_i|w_{i-n+2}^{i-1})$, and $p(w_i|w_{i-n+3}^{i-1})$. Thus, the unigram rescaled language model is obtained by modifying the backoff coefficients.

In this chapter, we address several language model methods tackling the issues of small sample size and long-distance information. Different language model smoothing methods including interpolation smoothing, backoff smoothing and unigram smoothing are summarized. The Bayesian language modeling based on MAP estimation and VB learning is addressed in Section 4.7 and Section 7.7, respectively.

3.8 Revisit of automatic speech recognition with Bayesian manner

In Chapter 2, we introduce the basic mathematical tools in the Bayesian approach, including the sum and product rules, and the conditional independence, and provide simple ways to obtain the posterior distributions. In addition, throughout the discussions from Section 3.1 to Section 3.7, we also introduce statistical models of speech and language, which are mainly used for ASR. Based on these mathematical tools and statistical models, this section revisits methods to formulate the whole ASR framework in a Bayesian manner, as consistently as possible, by only using the sum rule, product rule, and conditional independence, as discussed in Section 2.1. In particular, we focus on how to train our ASR model based on the Bayesian manner. The aim of this section is to show the limitation of simply using basic Bayesian tools for ASR, which leads to the requirement of approximated Bayesian inference techniques in the following chapters.

3.8.1 Training and test (unseen) data for ASR

To deal with the ASR problem within a machine learning framework, we need to consider two sets of data: *training* data and *test* data. The training data usually consist of a sequence of speech feature vectors ($\mathbf{O}$) and the corresponding label sequence (W) for all utterances. The test data consist of only a sequence of speech feature vectors ($\mathbf{O}'$) for all utterances, and there is a candidate label sequence (W') among all possible word sequences. The ASR problem is to correctly estimate the corresponding label sequence ($\hat{W}'$). Therefore, we summarize these four variables as:

- $\mathbf{O}$: speech feature sequence for training (observed);
- W: word sequence for training (observed);
- $\mathbf{O}'$: speech feature sequence for test (observed);
- W': a candidate of word sequences for test (not observed).

Since we have various possible candidates for W', it is natural to consider the following conditional probabilistic distribution of W', given the other observed variables, as:

$$p(W'|\mathbf{O}', \mathbf{O}, W). \tag{3.325}$$

Therefore, this section starts with the formulation based on this conditional probability.

Note that if we want to estimate the correct word sequence, it is performed by using the well-known MAP decision rule, as we discussed in Eq. (3.2):

$$\hat{W}' = d^{\text{MAP}}(\mathbf{O}') = \arg\max_{W'} p(W'|\mathbf{O}', \mathbf{O}, W). \tag{3.326}$$

Therefore, the following section focuses on $p(W'|\mathbf{O}', \mathbf{O}, W)$ in more detail.

3.8.2 Bayesian manner

Recalling the discussion in Section 2.1, the Bayesian manner deals with these variables as the arguments of probability distributions as follows:

$$\begin{aligned}
\mathbf{O} &\rightarrow p(\mathbf{O}) \\
W &\rightarrow p(W) \\
\mathbf{O}' &\rightarrow p(\mathbf{O}') \\
W' &\rightarrow p(W').
\end{aligned} \tag{3.327}$$

We should also recall that these probabilistic variables (x: continuous variable and n: discrete variable) have the following mathematical properties:

$$p(n) \geq 0, \tag{3.328}$$

$$\sum_n p(n) = 1, \tag{3.329}$$

$$p(x) \geq 0, \tag{3.330}$$

$$\int p(x)dx = 1. \tag{3.331}$$

These properties yield various benefits for probabilistic pattern recognition problems. Based on this probabilistic variable treatment, we can apply the sum, product rules for all the variables, and we can also use the Bayes theorem for Eq. (3.325).

MAP decision rule

By replacing a with word sequence W' and b with speech feature vector $\mathbf{O}'$ in Bayes theorem Eq. (2.7) in Section 2.1.1, we can derive the following noisy channel model of speech recognition based on the well-known MAP decision rule (Eq. (3.2)):

$$\hat{W}' = d^{\mathrm{MAP}}(\mathbf{O}') = \arg\max_{W'} p(W'|\mathbf{O}'), \tag{3.332}$$

$$= \arg\max_{W'} \frac{p(\mathbf{O}'|W')p(W')}{p(\mathbf{O}')}, \tag{3.333}$$

$$= \arg\max_{W'} p(\mathbf{O}'|W')p(W'). \tag{3.334}$$

Here, the denominator of Eq. (3.333) is disregarded with the arg max operation, since $p(\mathbf{O}')$ does not depend on W'.

The product rule (Bayes theorem) based on the MAP decision theory decomposes the posterior distribution $p(W'|\mathbf{O}')$ into the two generative models of $\mathbf{O}'$ and W', i.e., acoustic model $p(\mathbf{O}'|W')$ and language model $p(W')$. Solving Eq. (3.332) by directly obtaining $p(W'|\mathbf{O}')$ is called the *discriminative approach*. The approach tries to infer what a speaker wants to say in his/her brain from the observation $\mathbf{O}'$, as shown in Figure 3.11.

On the other hand, the *generative approach* obtains $p(W'|\mathbf{O}')$ indirectly via the generative models $p(\mathbf{O}'|W')$ and $p(W')$ based on Eq. (3.333). Therefore, the approach tries to imitate the generation process of $\mathbf{O}'$ given W' in the acoustic model and W itself in the language model, as shown in Figure 3.12. Since the generation process comes from a physical phenomenon or linguistic phenomena, we can involve various knowledge of the phenomena (e.g., articulatory models in speech production and grammatical or semantic models in linguistics) in principle. The rest of this section further discusses the Bayesian formulation along with the generative approach.

$\hat{W}' = \underset{W'}{\mathrm{argmax}}\, p(W'|\mathrm{O}')$

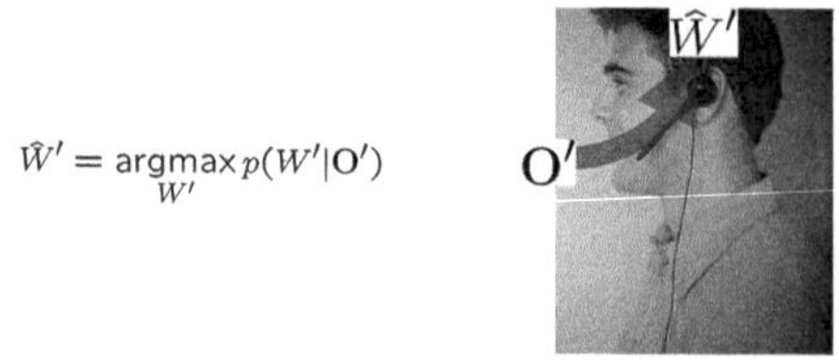

Figure 3.11 Probabilistic speech recognition: discriminative approach.

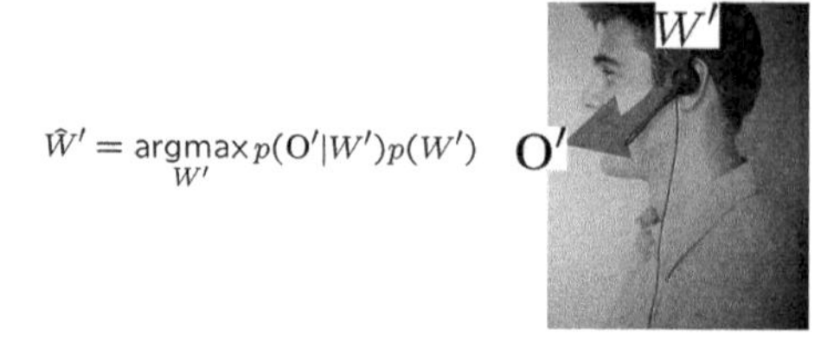

Figure 3.12 Probabilistic speech recognition: generative approach.

3.8.3 Learning generative models

Bayesian formulation reasonably extends the generative models to teach them from training data, by adding training data ($\mathbf{O}$ and W) to conditional variables in the target distribution, namely,

$$\begin{aligned} p(\mathbf{O}'|W') &\to p(\mathbf{O}'|W', \mathbf{O}, W) \\ p(W') &\to p(W'|\mathbf{O}, W). \end{aligned} \tag{3.335}$$

Usually, we also use the reasonable conditional independence assumption for the language model expressed by $p(W'|\mathbf{O}, W) \approx p(W'|W)$. Now, we focus on the acoustic model $p(\mathbf{O}'|W', \mathbf{O}, W)$, and make this abstract distribution more concrete.

3.8.4 Sum rule for model

To make the distribution more concrete, we usually provide a *model* with the distribution. For example, when we have some data to analyze, we first start to consider what kind of models we use to make the problems concrete by considering HMMs or GMMs for speech feature sequences, and n-gram for word sequences. However, we do not know whether the model provided is correct or not, and it should be tested by different model settings. Thus, the model settings can be regarded as probabilistic variables in the Bayesian formulation. In addition, the variation of setting model topologies, distribution forms of prior and posterior distributions, and hyperparameters of these distributions can also be treated as probabilistic variables, as shown in Table 3.2. We call this probabilistic variable to denote the model setting *model* variable M.

For example, once we decide to use an HMM for speech feature modeling, we have many arbitrary properties for the model topology: (i) whether we use a word unit, context-independent phoneme unit, or context-dependent phoneme unit; (ii) left to right HMM, including skip transitions, or fully connected ergodic HMM; (iii) how many shared-triphone HMM states and how many Gaussians in these states. Some of the model variables dealing with this model structure are called *model structure* variables, and so-called structure learning in the machine learning field tries to optimize the model structure using a more general framework than the Bayesian framework.

Table 3.2 Examples of model variables.

Variation	Examples
Model types	HMM, GMM, SVM, neural network, etc.
Model topologies	# HMM states, # Gaussians, # hidden layers, etc.
Priors/posteriors	Gaussian, Laplace, Gamma, Bernoulli, etc.
Hyperparameters	parameters of priors/posteriors, kernel parameters, etc.

The model variable M can be involved in our abstract distribution (Eq. (3.335)) of the acoustic model by using the sum rule (Eq. (2.3)) as follows:

$$p(\mathbf{O}'|W', \mathbf{O}, W) = \sum_{M} p(\mathbf{O}', M|W', \mathbf{O}, W). \tag{3.336}$$

As M would include continuous values (e.g., hyperparameters), the marginalization over M should involve both summation and integration. However, for simplicity, we use the summation in this formulation.

3.8.5 Sum rule for model parameters and latent variables

Once we set a model variable to a specific value, we can also provide the corresponding model parameters Θ and latent variables Z for training data and Z' for test data. For example, once we decide a model for the specific setting based on a standard HMM–GMM with a fixed model topology, these variables can also be involved in the distribution with M (Eq. (3.336)) by using the sum rule (Eq. (2.3)) as follows:

$$p(\mathbf{O}'|W', \mathbf{O}, W) = \int \sum_{M,Z,Z'} p(\mathbf{O}', M, \Theta, Z, Z'|W', \mathbf{O}, W) d\Theta. \tag{3.337}$$

However, it is really difficult to deal with this joint distribution, and we need to factorize the distribution.

3.8.6 Factorization by product rule and conditional independence

First, we factorize Eq. (3.337) by using the product rule (Eq. (2.4)), as follows:

$$\begin{aligned} p(\mathbf{O}'|W', \mathbf{O}, W) &= \int \sum_{M,Z,Z'} p(\mathbf{O}', M, \Theta, Z, Z'|W', \mathbf{O}, W) d\Theta \\ &= \int \sum_{M,Z,Z'} p(\mathbf{O}', Z'|Z, \Theta, M, W', \mathbf{O}, W) p(Z|\Theta, M, W', \mathbf{O}, W) \\ &\qquad \times p(\Theta|M, W', \mathbf{O}, W) p(M|W', \mathbf{O}, W) d\Theta. \end{aligned} \tag{3.338}$$

In this formulation, we keep the joint distribution of $\mathbf{O}'$ and Z' and do not factorize them. This is because the distribution corresponds to the complete data likelihood function, which is useful to handle latent models based on the EM algorithm, as discussed in Section 3.4.

Since the dependencies of the distributions in Eq. (3.338) are very complicated, we use the reasonable conditional independence assumptions for these distributions as follows:

$$p(\mathbf{O}', Z'|Z, \Theta, M, W', \mathbf{O}, W) \approx p(\mathbf{O}', Z'|\Theta, M, W'), \tag{3.339}$$

$$p(Z|\Theta, M, W', \mathbf{O}, W) \approx p(Z|\Theta, M, \mathbf{O}, W), \tag{3.340}$$

$$p(\Theta|M, W', \mathbf{O}, W) \approx p(\Theta|M, \mathbf{O}, W), \tag{3.341}$$

$$p(M|W', \mathbf{O}, W) \approx p(M|\mathbf{O}, W). \tag{3.342}$$

Equation (3.339) means that the test data are generated by the Θ and M and do not depend on the training data ($\mathbf{O}$ and W) and their latent variable Z, explicitly. The other assumptions just use the conditional independence of W'. By using the assumptions, we can approximate the original distribution of $p(\mathbf{O}'|W', \mathbf{O}, W)$ as follows:

$$\begin{aligned}\text{Eq. (3.338)} &\approx \int \sum_{M,Z,Z'} p(\mathbf{O}', Z'|M, \Theta, W')p(Z|\Theta, M, \mathbf{O}, W) \\ &\qquad \times p(\Theta|M, \mathbf{O}, W)p(M|\mathbf{O}, W)d\Theta \\ &\approx \int \sum_{M,Z'} p(\mathbf{O}', Z'|M, \Theta, W')p(\Theta|M, \mathbf{O}, W)p(M|\mathbf{O}, W)d\Theta, \end{aligned} \tag{3.343}$$

where we use the fact that $\sum_Z p(Z|\Theta, M, \mathbf{O}, W) = 1$ since Z does not depend on the other distributions. We also introduce lexical category c, and further factorize the equation as:

$$\text{Eq. (3.343)} \approx \int \sum_{M,Z'} \prod_{c'} p(\mathbf{O}', Z'|M, \Theta, c') \prod_{c} p(\Theta|M, \mathbf{O}, c)p(M|\mathbf{O}, c)d\Theta. \tag{3.344}$$

Thus, the acoustic model can be represented by the joint distribution of $\mathbf{O}'$ and Z' (complete data likelihood) and the posterior distributions of model parameters Θ and model M. Since $p(\mathbf{O}', Z'|M, \Theta, c')$ does not depend on training data ($\mathbf{O}$ and W), which can be obtained by setting the model and its parameters, we only focus on the two posterior distributions in Eq. (3.344).

3.8.7 Posterior distributions

The posterior distributions of models and model parameters can be rewritten by the following equations by using the sum and product rules (Eqs. (2.3) and (2.4)).

- **Model parameter Θ:**

$$p(\Theta|M, \mathbf{O}) = \frac{p(\mathbf{O}|\Theta, M)p(\Theta|M)}{p(\mathbf{O}|M)}, \tag{3.345}$$

$$\propto \sum_Z p(\mathbf{O}, Z|\Theta, M)p(\Theta|M). \tag{3.346}$$

- **Model M:**

$$p(M|\mathbf{O}) = \frac{p(\mathbf{O}|M)p(M)}{\sum_M p(\mathbf{O}|M)p(M)}, \tag{3.347}$$

$$\propto \sum_Z \int p(\mathbf{O}, Z|\Theta, M)p(\Theta|M)d\Theta p(M). \tag{3.348}$$

Note that the posterior distributions are represented by the following two types of distributions:

- Joint distribution of training data $\mathbf{O}$ and Z:
  $p(\mathbf{O}, Z|\Theta, M)$;
- Prior distributions of Θ and M :
  $p(\Theta|M)$ and $p(M)$.

Again, the joint distribution can be obtained by setting the model and its parameters. Once we set the prior distributions $p(\Theta|M)$ and $p(M)$, we can obtain the posterior distributions $p(\Theta|M, \mathbf{O})$ and $p(M|\mathbf{O})$ by solving (3.346) and (3.348). Then, the acoustic model likelihood can be computed by using Eq. (3.344).

3.8.8 Difficulties in speech and language applications

Although the Bayesian approach is powerful, it is very difficult to realize. One of the most critical aspects is that we cannot solve the above equations. The practical Bayesian approach rests on how to find appropriate approximations:

- *Chapter 4*: *Maximum a-posteriori approximation;*
- *Chapter 5: Evidence approximation;*
- *Chapter 6: Asymptotic approximation;*
- *Chapter 7: Variational Bayes;*
- *Chapter 8: Markov chain Monte Carlo.*

In the machine learning field, other approximations are also studied actively (e.g., loopy belief propagation (Murphy *et al.* 1999), expectation propagation (Minka 2001)). This book introduces the above approximations in speech and language applications.

Part II

Approximate inference

4 Maximum a-posteriori approximation

Maximum a-posteriori (MAP) approximation is a well-known and widely used approximation for Bayesian inference. The approximation covers all variables including model parameters Θ, latent variables Z, and classification categories C (word sequence W in the automatic speech recognition case). For example, the Viterbi algorithm ($\arg\max_Z p(Z|\mathbf{O})$) in the continuous density hidden Markov model (CDHMM), as discussed in Section 3.3.2, corresponds to the MAP approximation of latent variables, while the forward–backward algorithm, as discussed in Section 3.3.1, corresponds to an exact inference of these variables. As another example, the MAP decision rule ($\arg\max_C p(C|\mathbf{O})$) in Eq. (3.2) also corresponds to the MAP approximation of inferring the posterior distribution of classification categories. Since the final goal of automatic speech recognition is to output the word sequence, the MAP approximation of the word sequence matches the final goal.[1] Thus, the MAP approximation can be applied to all probabilistic variables in speech and language processing as an essential technique.

This chapter starts to discuss the MAP approximation of Bayesian inference in detail, but limits the discussion only to model parameters Θ in Section 4.1. In the MAP approximation for model parameters, the prior distributions work as a regularization of these parameters, which makes the estimation of the parameters more robust than that of the maximum likelihood (ML) approach. Another interesting property of the MAP approximation for model parameters is that we can easily involve the inference of latent variables by extending the EM algorithm from ML to MAP estimation. Section 4.2 describes the general EM algorithm with the MAP approximation by following the ML-based EM algorithm, as discussed in Section 3.4. Based on the general MAP–EM algorithm, Section 4.3 provides MAP–EM solutions for CDHMM parameters, and introduces the well-known applications based on speaker adaptation. Section 4.5 describes the parameter smoothing method in discriminative training of the CDHMM, which actually corresponds to the MAP solution for discriminative parameter estimation. Section 4.6 focuses on the MAP estimation of GMM parameters, which is a subset of the MAP estimation of CDHMM parameters. It is used to construct speaker GMMs that are used

[1] However, if we consider some other spoken language processing applications given automatic speech recognition inputs (e.g., dialog, machine translation, and information retrieval), we need to consider how to provide $p(W|\mathbf{O})$ rather than $\hat{W} = \arg\max_W p(W|\mathbf{O})$ to avoid propagating any speech recognition errors to the post-processing applications.

for speaker verification. Section 4.7 provides an MAP solution of n-gram parameters that leads to one instance of interpolation smoothing, as discussed in Section 3.6.2. Finally, Section 4.8 deals with the adaptive MAP estimation of latent topic model parameters.

4.1 MAP criterion for model parameters

This section begins with a general discussion of the MAP approximation for model parameters Θ. For simplicity, we first review the posterior distribution of model parameters given observations $\mathbf{O}$ without latent variables Z. Instead of estimating posterior distributions, the MAP estimation focuses on the following parameter estimation:

$$\Theta^{\mathrm{MAP}} = \arg\max_{\Theta} p(\Theta|\mathbf{O}). \tag{4.1}$$

This corresponds to estimating the model parameter Θ^{MAP} given training data $\mathbf{O}$. By using the product and sum rules, as discussed in Section 2.1.1, we can rewrite the above equation as follows:

$$\begin{aligned} \Theta^{\mathrm{MAP}} &= \arg\max_{\Theta} p(\Theta|\mathbf{O}) \\ &= \arg\max_{\Theta} \frac{p(\mathbf{O}|\Theta)p(\Theta)}{\int p(\mathbf{O}|\Theta)p(\Theta)d\Theta} \\ &= \arg\max_{\Theta} \underbrace{p(\mathbf{O}|\Theta)}_{\text{likelihood}} \times \underbrace{p(\Theta)}_{\text{prior}}. \end{aligned} \tag{4.2}$$

Since $p(\mathbf{O}) = \int p(\mathbf{O}|\Theta)p(\Theta)d\Theta$ does not depend on Θ, we can avoid computing this integral directly.[2] Furthermore, if we use an exponential family distribution for a likelihood function and a conjugate distribution for a prior distribution, as discussed in Section 2.1.3, the MAP estimate is represented as the mode of the corresponding conjugate posterior distribution, analytically. This is an advantage of using conjugate distributions.[3]

Equation (4.2) also tells us that the posterior distribution is composed of the likelihood function and the prior distribution, thus the estimation is based on the maximum likelihood function with the additional contribution of the prior distribution. That is, the prior distribution acts to regularize model parameters in the ML estimation, as we discussed in Section 2.3.1 as the best-known Bayesian advantage over ML. For example, let us consider the likelihood function $p(O|\Theta) = \prod_t \mathcal{N}(o_t|\mu, 1)$ as a one-dimensional Gaussian distribution with mean μ and precision as 1, and the prior distribution $p(\Theta)$ as a

[2] This term is called the evidence, which is neglected in the MAP approximation. However, the importance of the evidence is discussed in Chapter 5.

[3] In other words, it is not simple to obtain the mode of the posterior distribution, if we do not use the conjugate distribution, since we cannot obtain the posterior distribution analytically. For example, if we use the Laplace distribution for the prior distribution, the mode of posterior distributions cannot be obtained analytically, and we need some numerical computation.

one-dimensional Gaussian distribution of the mean vector μ^0 and the scale parameter r. Then, the MAP estimation can be represented as follows:

$$\begin{aligned}
\arg\max_{\mu} \log p(O|\Theta) + \log p(\Theta) \\
&= \arg\max_{\mu} \log \left(\prod_t \mathcal{N}(o_t|\mu, 1) \right) + \log \mathcal{N}(\mu|\mu^0, r^{-1}) \\
&= \arg\max_{\mu} \sum_t (o_t - \mu)^2 + r(\mu - \mu^0)^2,
\end{aligned} \tag{4.3}$$

where from the second to the third lines, we use the definition of a Gaussian distribution (Appendix C.5) as follows:

$$\mathcal{N}(x|\mu, r^{-1}) \triangleq (2\pi)^{-\frac{1}{2}}(r)^{\frac{1}{2}} \exp\left(-\frac{r(x-\mu)^2}{2} \right). \tag{4.4}$$

Thus, the optimization problem of the MAP solution corresponds to solving the minimum mean square error (MMSE) estimation with the l^2 regularization term around μ^0. The scale parameter r behaves as a tuning parameter to balance the MMSE estimation and the regularization term. These parameters are called regularization parameters, which can be hyperparameters of the prior distribution. Equation (4.3) can be analytically solved by using the conjugate distribution rule, as discussed in Section 2.1.4, or by using the following derivative method:

$$\begin{aligned}
\frac{\partial}{\partial \mu} \sum_{t=1}^{T} (o_t - \mu)^2 + r(\mu - \mu^0)^2 &= -2 \sum_{t=1}^{T} (o_t - \mu) + 2r(\mu - \mu^0) \\
&= -2 \left(\sum_{t=1}^{T} o_t + r\mu^0 \right) + 2(T + r)\mu = 0.
\end{aligned} \tag{4.5}$$

We obtain the MAP estimate of μ analytically as:

$$\begin{aligned}
\mu^{\text{MAP}} &= \frac{\sum_t^T o_t + r\mu^0}{T + r} \\
&= \frac{\mu^{\text{ML}} + \frac{r}{T}\mu^0}{1 + \frac{r}{T}}.
\end{aligned} \tag{4.6}$$

Thus, the regularization term sets a constraint for the ML estimate $\mu^{\text{ML}} = \frac{\sum_t^T o_t}{T}$ with a regularization constant r.

Similarly, if we use a Laplace distribution as a prior distribution, the prior distribution works as an l^1 regularization term. The Laplace distribution is defined as follows (Appendix C.10):

$$\text{Lap}(x|\mu, \beta) \triangleq \frac{1}{2\beta} \exp\left(-\frac{|x - \mu|}{\beta} \right). \tag{4.7}$$

Therefore, using $\mathrm{Lap}(\mu|\mu^0, \beta)$ instead of $\mathcal{N}(\mu|\mu^0, r^{-1})$, Eq. (4.3) is rewritten as follows:

$$\begin{aligned}
&\arg\max_{\mu} \log p(O|\Theta) + \log p(\Theta) \\
&\qquad = \arg\max_{\mu} \log \left(\prod_t \mathcal{N}(o_t|\mu, 1) \right) + \log \mathrm{Lap}(\mu|\mu^0, \beta) \\
&\qquad = \arg\max_{\mu} \sum_t (o_t - \mu)^2 + \frac{1}{\beta} |\mu - \mu^0|. \qquad (4.8)
\end{aligned}$$

Thus, the prior distribution effect in the MAP parameter estimation is often regarded as a regularization of parameters. Consequently, Eq. (4.3) can incorporate the prior knowledge of parameters via hyperparameters $\boldsymbol{\mu}^0$, and the MAP approximation retains the Bayesian advantage of use of prior knowledge, as discussed in Section 2.3.1. Note that Eq. (4.8) is not differentiable with respect to μ, and it does not have a well-defined conjugate distribution. Therefore, the MAP estimation with the Laplace prior (l^1 regularization) is often undertaken by a numerical method.

Now, we introduce a useful mathematical operation for the MAP approximation of model parameters. To compute the expected values of the posterior distribution with respect to model parameters Θ, the MAP approximation can use the following posterior distribution represented by a Dirac delta function:

$$p(\Theta|\mathbf{O}) = \delta(\Theta - \Theta^{\mathrm{MAP}}), \qquad (4.9)$$

where the Dirac delta function has the following property:

$$\int f(\mathbf{a})\delta(\mathbf{a} - \mathbf{a}^*) = f(\mathbf{a}^*). \qquad (4.10)$$

This posterior distribution intuitively corresponds to having a location parameter with the MAP estimate Θ^{MAP} and very small (0) variance. If the model parameters are represented by discrete variables, we can use the Kronecker delta function. Therefore, once we obtain the MAP estimation of model parameters Θ^{MAP}, we can compute the expected value of function $f(\Theta)$ based on Eq. (4.9) as follows:

$$\begin{aligned}
\mathbb{E}_{(\Theta)}[f(\Theta)|\mathbf{O}] &= \int f(\Theta)p(\Theta)d\Theta = \int f(\Theta)\delta(\Theta - \Theta^{\mathrm{MAP}})d\Theta \\
&= f(\Theta^{\mathrm{MAP}}). \qquad (4.11)
\end{aligned}$$

Here we use Eq. (4.10) to calculate the integral. Since this is equivalent to just plugging in the MAP estimates to the $f(\Theta)$, this procedure is called plug-in MAP (Lee & Huo 2000). For example, if we use the likelihood function of unseen data $\mathbf{O}'$ for $f(\Theta)$, Eq. (4.11) is rewritten as follows:

$$\mathbb{E}_{(\Theta)}[p(\mathbf{O}'|\Theta)|\mathbf{O}] = p(\mathbf{O}'|\Theta^{\mathrm{MAP}}). \qquad (4.12)$$

That is, the likelihood function of unseen data can be obtained by simply replacing Θ with the MAP estimate Θ^{MAP}. This can be used as a likelihood function to compute likelihood values in prediction and classification steps. The Dirac delta-function-based posterior representation is very useful, since the representation connects the analytical relationship between the MAP-based *point* estimation and the Bayesian *distribution* estimation.

Thus, the MAP approximation does not need to solve the marginalization explicitly in the training and prediction/classification steps. Although the approximation lacks the Bayesian advantages of the model selection and marginalization, as discussed in Section 2.3, it still has the most effective Bayesian advantage over ML, namely *use of prior knowledge*. Equation (4.3) also shows that the effect of the prior distribution in the MAP estimation works as a regularization. Therefore, the MAP approximation is widely used in practical Bayesian applications. In addition, the MAP approximation is simply extended to deal with latent variables based on the EM algorithm, which is a key technique in training statistical models in speech and language processing, as we discussed in Chapter 3. The next section discusses the MAP version of the EM algorithm.

4.2 MAP extension of EM algorithm

As we discussed in Section 3.4, most statistical models used in speech and language processing have to deal with latent variables Z, e.g., HMM states and mixture components of the CDHMM in acoustic modeling, and latent topics in language modeling. The maximum likelihood approach has an efficient solution based on the EM algorithm, which optimizes the auxiliary function $Q(\Theta'|\Theta)$ instead of a (log) likelihood function. This section describes the EM extension of MAP parameter estimation in general.

4.2.1 Auxiliary function

Following the discussion in Section 3.4, we prove that the EM steps ultimately lead to the local optimum value Θ^{MAP} in terms of the MAP criterion. First, since the logarithmic function is a monotonic function, the MAP criterion in Eq. (4.2) is represented as follows:

$$\begin{aligned}\Theta^{\text{MAP}} &= \arg\max_{\Theta} p(\mathbf{O}|\Theta)p(\Theta) \\ &= \arg\max_{\Theta} \log\left(p(\mathbf{O}|\Theta)p(\Theta)\right). \end{aligned} \tag{4.13}$$

By introducing latent variable Z, the above equation can be written as

$$\Theta^{\text{MAP}} = \arg\max_{\Theta} \log\left(\sum_{Z} p(\mathbf{O}, Z|\Theta)p(\Theta)\right). \tag{4.14}$$

As discussed in the ML–EM algorithm, the summation over latent variable $\sum_Z$ is computationally very difficult since the latent variable in speech and language processing is represented as a possible sequence, and the number of these variables is exponential. Therefore, we need to avoid having to compute $\sum_Z$ directly.

Similarly to the ML–EM algorithm in Section 3.4, in M-step, we maximize the MAP version of the auxiliary function with respect to the parameters Θ, and estimate new parameters by

$$\Theta^{\text{MAP}} = \arg\max_{\Theta'} Q^{\text{MAP}}(\Theta'|\Theta). \tag{4.15}$$

The updated parameters Θ' are then treated as the current parameters for the next iteration of EM steps. The $Q^{\text{MAP}}(\Theta'|\Theta)$ is defined as the expectation of the joint distribution $p(\mathbf{O}, Z, \Theta')$ with respect to $p(Z|\mathbf{O}, \Theta)$ as follows:

$$\begin{aligned} Q^{\text{MAP}}(\Theta'|\Theta) &\triangleq \mathbb{E}_{(Z)}[\log p(\mathbf{O}, Z, \Theta')|\mathbf{O}, \Theta] \\ &= \mathbb{E}_{(Z)}[\log p(\mathbf{O}, Z|\Theta')p(\Theta')|\mathbf{O}, \Theta] \\ &= \underbrace{\mathbb{E}_{(Z)}[\log p(\mathbf{O}, Z|\Theta')|\mathbf{O}, \Theta]}_{Q^{\text{ML}}(\Theta'|\Theta)} + \log p(\Theta'), \end{aligned} \tag{4.16}$$

where $\log p(\Theta')$ does not depend on Z, and can be separated from the expectation operation. Compared with the ML auxiliary function $Q^{\text{ML}}(\Theta'|\Theta)$ (Eq. (3.78)), we have an additional term $\log p(\Theta')$, which comes from a prior distribution of model parameters.

Now we explain how optimization of the auxiliary function Q^{MAP} leads to the local optimization of $p(\mathbf{O}|\Theta)p(\Theta)$ or $p(\Theta|\mathbf{O})$. For the explanation, we define the logarithmic function of the joint distribution $p(\mathbf{O}, \Theta') = p(\mathbf{O}|\Theta')p(\Theta')$ as follows:

$$L^{\text{MAP}}(\Theta') \triangleq \log\left(p(\mathbf{O}|\Theta)p(\Theta)\right). \tag{4.17}$$

This is similar to $L(\Theta')$ in Eq. (3.83), but $L^{\text{MAP}}(\Theta')$ has an additional factor from $p(\Theta)$. Now, we represent $p(\mathbf{O}|\Theta)$ in the above equation with the distributions of latent variable Z based on the product rule of probabilistic variables, as follows:

$$p(\mathbf{O}|\Theta') = \frac{p(\mathbf{O}, Z|\Theta')}{p(Z|\mathbf{O}, \Theta')}. \tag{4.18}$$

Therefore, by substituting Eq. (4.18) into Eq. (4.17), we obtain

$$L^{\text{MAP}}(\Theta') = \log p(\mathbf{O}, Z|\Theta') - \log p(Z|\mathbf{O}, \Theta') + \log p(\Theta'). \tag{4.19}$$

Now we perform the expectation operation with respect to $p(Z|\mathbf{O}, \Theta)$ for both sides of the equation, and obtain the following relationship:

$$\begin{aligned} L^{\text{MAP}}(\Theta') &= \mathbb{E}_{(Z)}[\log p(\mathbf{O}, Z|\Theta')|\mathbf{O}, \Theta] - \mathbb{E}_{(Z)}[\log p(Z|\mathbf{O}, \Theta')|\mathbf{O}, \Theta] + \log p(\Theta') \\ &= \underbrace{\mathbb{E}_{(Z)}[\log p(\mathbf{O}, Z|\Theta')|\mathbf{O}, \Theta] + \log p(\Theta')}_{Q^{\text{MAP}}(\Theta'|\Theta)} \underbrace{- \mathbb{E}_{(Z)}[\log p(Z|\mathbf{O}, \Theta')|\mathbf{O}, \Theta]}_{H(\Theta'|\Theta)}, \end{aligned} \tag{4.20}$$

where $L^{\text{MAP}}(\Theta')$ and $\log p(\Theta')$ are not changed since these do not depend on Z. Note that the third term of Eq. (4.20) is represented as $H(\Theta'|\Theta)$, which is exactly the same definition as Eq. (3.85). Thus, we derive a similar equation to that of the ML auxiliary function Eq. (3.86):

$$Q^{\text{MAP}}(\Theta'|\Theta) = L(\Theta')^{\text{MAP}} + H(\Theta'|\Theta). \tag{4.21}$$

Since $H(\Theta'|\Theta)$ is the same as in Eq. (3.86) and has a bound based on the Jensen's inequality, we can apply the same discussion to $Q^{\text{MAP}}(\Theta'|\Theta)$ to show that $Q^{\text{MAP}}(\Theta'|\Theta)$ is the auxiliary function of the MAP criterion.

Thus, we prove that the E-step performing the expectation and M-step maximizing the auxiliary function with respect to the model parameters Θ, always increase the joint likelihood value as:

$$\Theta^{\text{MAP}} = \arg\max_{\Theta'} Q^{\text{MAP}}(\Theta'|\Theta) \Rightarrow p(\mathbf{O}, \Theta^{\text{MAP}}) \geq p(\mathbf{O}, \Theta). \tag{4.22}$$

This leads to a local optimization of the joint likelihood function $p(\mathbf{O}, \Theta)$, which corresponds to the MAP criterion in Eq. (4.13). We call this procedure the MAP–EM algorithm.

4.2.2 A recipe

Based on the previous discussions, we summarize in the text box a general procedure to obtain the MAP estimation of model parameters. The following section describes the concrete form of MAP–EM solutions for CDHMMs similarly to Section 3.4.

1. Set a likelihood function for a statistical model (generative model) with model parameters.
2. Set appropriate prior distributions for model parameters (possibly conjugate distributions to obtain analytical results based on the conjugate distribution discussion in Section 2.1.4).
3. Solve the parameter estimation by maximizing the MAP objective function:
   i. Solve posterior distributions for model parameters when we can use conjugate priors;
   ii. Solve the parameter estimation by getting the modes of posterior distributions.

4.3 Continuous density hidden Markov model

This section describes the MAP estimation of HMM parameters based on the MAP–EM algorithm (Lee *et al.* 1991, Gauvain & Lee 1994). Following the general procedure for MAP estimation (as set out in Section 4.2.2), we first review a likelihood function of the CDHMM, as discussed in Section 3.2.3. Then we provide a concrete form of the prior distribution $p(\Theta)$ for full and diagonal covariance cases. Then, according to the derivation of the ML–EM algorithm in Section 3.4, we also derive the concrete form solutions of the MAP–EM algorithm of the CDHMM.

4.3.1 Likelihood function

We first provide the complete data likelihood function with speech feature sequence $\mathbf{O} = \{\mathbf{o}_t \in \mathbb{R}^D | t = 1, \cdots, T\}$, HMM state sequence $S = \{s_t \in \{1, \cdots, J\} | t = 1, \cdots, T\}$, and GMM component sequence $V = \{v_t \in \{1, \cdots, K\} | t = 1, \cdots, T\}$, which is introduced in Eq. (3.43) as follows:

$$p(\mathbf{O}, S, V|\Theta) = \pi_{s_1} \omega_{s_1 v_1} p(\mathbf{o}_1 | \Theta_{s_1 v_1}) \prod_{t=2}^{T} a_{s_{t-1} s_t} \omega_{s_t v_t} p(\mathbf{o}_t | \Theta_{s_t v_t}). \tag{4.23}$$

Recall that a set of HMM parameters Θ holds:

- Initial state probability π_j;
- State transition probability a_{ij};
- Mixture weight ω_{jk};
- Gaussian mean vector $\boldsymbol{\mu}_{jk}$;
- Gaussian covariance matrix $\boldsymbol{\Sigma}_{jk}$.

The next section provides prior distribution $p(\Theta)$.

4.3.2 Conjugate priors (full covariance case)

The prior distribution is considered to be the following joint distribution form:

$$p(\Theta) = p\left(\{\pi_j\}_{j=1}^J), \{\{a_{ij}\}_{i=1}^J\}_{j=1}^J, \{\{\omega_{jk}, \boldsymbol{\mu}_{jk}, \boldsymbol{\Sigma}_{jk}\}_{j=1}^J\}_{k=1}^K\right). \tag{4.24}$$

However, since it is difficult to handle this joint distribution, we usually factorize it by assuming conditional independence for each HMM state and mixture component. Then, the prior distribution is rewritten as follows:

$$\begin{aligned} p(\Theta) &= p(\boldsymbol{\pi})p(A)p(\boldsymbol{\omega})p(\boldsymbol{\mu}, \mathbf{R}) \\ &= p(\{\pi_j\}_{j=1}^J) \left(\prod_{i=1}^J p(\{a_{ij}\}_{j=1}^J) \right) \left(\prod_{j=1}^J p(\{\omega_{jk}\}_{k=1}^K) \right) \left(\prod_{j=1}^J \prod_{k=1}^K p(\boldsymbol{\mu}_{jk}, \boldsymbol{\Sigma}_{jk}) \right), \end{aligned} \tag{4.25}$$

where we also assume that π_j, a_{ij}, ω_{jk}, and $\{\boldsymbol{\mu}_{jk}, \boldsymbol{\Sigma}_{jk}\}$ are independent of each other, although we keep the dependency of $\boldsymbol{\mu}_{jk}$ and $\boldsymbol{\Sigma}_{jk}$.

Now we provide the concrete forms of the above prior distributions for the HMM parameters based on the conjugate distribution discussion in Section 2.1.4. We first focus on the prior distributions of the initial state probability π_j, state transition probability a_{ij}, and Gaussian mixture weight ω_{jk}. Note that these probabilistic variables have the same constraint that $\pi_j \geq 0, \sum_{j=1}^J \pi_j = 1$, $a_{ij} \geq 0, \sum_{j=1}^J a_{ij} = 1$, and $\omega_{jk} \geq 0, \sum_{k=1}^K \omega_{jk} = 1$. In addition, these are represented by a multinomial distribution in the complete data likelihood function. Therefore, according to Table 2.1, these are represented by a Dirichlet distribution with hyperparameters as follows:

$$p(\{\pi_j\}_{j=1}^J) = \mathrm{Dir}(\{\pi_j\}_{j=1}^J | \{\phi_j^\pi\}_{j=1}^J),$$
$$p(\{a_{ij}\}_{j=1}^J) = \mathrm{Dir}(\{a_{ij}\}_{j=1}^J | \{\phi_{ij}^a\}_{j=1}^J),$$
$$p(\{\omega_{jk}\}_{k=1}^K) = \mathrm{Dir}(\{\omega_{jk}\}_{k=1}^K | \{\phi_{jk}^\omega\}_{k=1}^K), \tag{4.26}$$

where $\phi_j^\pi \geq 0$, $\phi_{ij}^a \geq 0$, and $\phi_{jk}^\omega \geq 0$.

Next we consider the prior distribution of Gaussian mean vector $\boldsymbol{\mu}_{jk}$ and Gaussian precision matrix $\boldsymbol{\Sigma}_{jk}$. For simplicity, we focus on the precision matrix $\mathbf{R}$, which is the inverse matrix of the covariance matrix $\boldsymbol{\Sigma}$, i.e.,

$$\mathbf{R}_{jk} \triangleq (\boldsymbol{\Sigma}_{jk})^{-1}. \tag{4.27}$$

According to Table 2.1, the joint prior distribution of Gaussian mean vector $\boldsymbol{\mu}_{jk}$ and Gaussian precision matrix $\mathbf{R}_{jk}$ can be written as follows:

$$\begin{aligned} p(\boldsymbol{\mu}_{jk}, \mathbf{R}_{jk}) &= p(\boldsymbol{\mu}_{jk}|\mathbf{R}_{jk})p(\mathbf{R}_{jk}) \\ &= \mathcal{N}(\boldsymbol{\mu}_{jk}|\boldsymbol{\mu}_{jk}^0, (\phi_{jk}^{\boldsymbol{\mu}}\mathbf{R}_{jk})^{-1})\mathcal{W}(\mathbf{R}_{jk}|\mathbf{R}_{jk}^0, \phi_{jk}^{\mathbf{R}}), \end{aligned} \tag{4.28}$$

where $\mathcal{W}(\cdot)$ is a Wishart distribution, which is defined in Appendix C.14. Note that the prior distribution $p(\boldsymbol{\mu}_{jk}|\mathbf{R}_{jk})$ of the mean vector depends on covariance matrix $\mathbf{R}_{jk}$, and cannot be factorized independently. Instead, these parameters are represented by the joint prior distribution $p(\boldsymbol{\mu}_{jk}, \mathbf{R}_{jk})$ with the Gaussian–Wishart distribution (Appendix C.15) or the product form in Eq. (4.28), as we discussed in Section 2.1.4. We can also provide the prior distribution for the original covariance matrix $\boldsymbol{\Sigma}$ by using the inverse-Wishart distribution instead of the Wishart distribution in Eq. (4.28). Both representations yield the same result in the MAP estimation of HMM parameters.

Consequently, the conjugate prior distribution of a CDHMM is represented by the following factorization form with each parameter:

$$\begin{aligned} p(\Theta) = &\,\mathrm{Dir}(\{\pi_j\}_{j=1}^J | \{\phi_j^\pi\}_{j=1}^J) \\ &\times \left(\prod_{i=1}^J \mathrm{Dir}(\{a_{ij}\}_{j=1}^J | \{\phi_{ij}^a\}_{j=1}^J)\right)\left(\prod_{j=1}^J \mathrm{Dir}(\{\omega_{jk}\}_{k=1}^K | \{\phi_{jk}^\omega\}_{k=1}^K)\right) \\ &\times \left(\prod_{j=1}^J\prod_{k=1}^K \mathcal{N}(\boldsymbol{\mu}_{jk}|\boldsymbol{\mu}_{jk}^0, (\phi_{jk}^{\boldsymbol{\mu}}\mathbf{R}_{jk})^{-1})\mathcal{W}(\mathbf{R}_{jk}|\mathbf{R}_{jk}^0, \phi_{jk}^{\mathbf{R}})\right). \end{aligned} \tag{4.29}$$

Note that the prior distribution of a CDHMM is represented by three types of distributions, i.e., Dirichlet, Gaussian, and Wishart distributions. The prior distribution has five scalar hyperparameters $\phi^\pi, \phi^a, \phi^\omega, \phi^{\boldsymbol{\mu}}, \phi^{\mathbf{R}}$, one vector hyperparameter $\boldsymbol{\mu}^0$, and one matrix hyperparameter $\mathbf{R}^0$. A set of these hyperparameters is written as Ψ in this chapter, i.e.,

$$\Psi \triangleq \{\phi_j^\pi, \phi_{ij}^a, \phi_{jk}^\omega, \phi_{jk}^{\boldsymbol{\mu}}, \phi_{jk}^{\mathbf{R}}, \boldsymbol{\mu}_{jk}^0, \mathbf{R}_{jk}^0 | i = 1, \cdots, J, j = 1, \cdots, J, k = 1, \cdots, K\}. \tag{4.30}$$

In the following sections, we sometimes represent the prior distribution as $p(\Theta|\Psi)$ instead of $p(\Theta)$ to deal with the hyperparameter dependency on the prior distributions explicitly.

4.3.3 Conjugate priors (diagonal covariance case)

In practical use, the Gaussians in a CDHMM are often represented by a diagonal covariance matrix, as we discussed in Section 3.2.3. To deal with a conjugate distribution of a diagonal covariance matrix, we need to provide a specific distribution rather than the Wishart distribution since the off-diagonal elements are always zero, and it is not suitable to represent these random variables as the Wishart distribution. Instead, we use the gamma distribution for each diagonal component. We first define the $d-d$ element of the precision matrix as r_d:

$$r_d \triangleq [\mathbf{R}]_{dd}. \tag{4.31}$$

Then the joint prior distribution of Eq. (4.28) is factorized by a feature dimension, and it is replaced with the gamma distribution as follows:

$$\begin{aligned} p(\boldsymbol{\mu}_{jk}, \mathbf{R}_{jk}) &= \prod_{d=1}^{D} p(\mu_{jkd}|r_{jkd})p(r_{jkd}) \\ &= \prod_{d=1}^{D} \mathcal{N}(\mu_{jkd}|\mu^0_{jkd}, (\phi^{\boldsymbol{\mu}}_{jk} r_{jkd})^{-1}) \,\mathrm{Gam}(r_{jkd}|r^0_{jkd}, \phi^{\mathbf{R}}_{jk}), \end{aligned} \tag{4.32}$$

where a set of hyperparameters Ψ is represented as

$$\Psi \triangleq \{\phi^{\pi}_{j}, \phi^{a}_{ij}, \phi^{\omega}_{jk}, \phi^{\boldsymbol{\mu}}_{jk}, \phi^{\mathbf{R}}_{jk}, \boldsymbol{\mu}^0_{jk}, \mathbf{r}^0_{jk} | i = 1, \cdots, J, j = 1, \cdots, J, k = 1, \cdots, K\}, \tag{4.33}$$

where

$$\mathbf{r}^0 \triangleq [r^0_1, \cdots, r^0_D]^{\intercal}. \tag{4.34}$$

Similarly to the full covariance case, Eq. (4.32) can also be represented by a Gaussian-gamma distribution (Appendix C.13).

The dependency of the hyperparameter ϕ is not unique and can be arranged by considering applications due to the flexible parameterization of an exponential family distribution, as discussed in Section 2.1.3. For example, $\phi^{\boldsymbol{\mu}}_{jk}$ and $\phi^{\mathbf{R}}_{jk}$ can be changed depending on a dimension d (i.e., $\phi^{\boldsymbol{\mu}}_{jkd}$, $\phi^{\mathbf{R}}_{jkd}$). These make the model more precise, but need more effort to set $\phi^{\boldsymbol{\mu}}_{jkd}$, $\phi^{\mathbf{R}}_{jkd}$ for all dimensions manually or automatically. Actually, these values are often shared among all js and ks (i.e., $\phi^{\boldsymbol{\mu}}_{jk} \to \phi^{\boldsymbol{\mu}}$ etc.).

4.3.4 Expectation step

Once we set the prior distributions and likelihood function, by following the recipe in Section 4.2.2, we can perform the MAP–EM algorithm to estimate the model parameter Θ. This section considers the concrete form of the MAP expectation step. This procedure is very similar to Section 3.4 except for the additional consideration of the prior distribution $p(\Theta)$. The auxiliary function used in the MAP–EM algorithm is represented as

$$Q^{\mathrm{MAP}}(\Theta'|\Theta) = Q^{\mathrm{ML}}(\Theta'|\Theta) + \log p(\Theta'). \tag{4.35}$$

According to Section 3.4, the auxiliary function $Q^{\text{ML}}(\Theta'|\Theta)$ is factorized by a sum of four individual auxiliary functions as:

$$\begin{aligned}
Q^{\text{ML}}(\Theta'|\Theta) &= \sum_S \sum_V p(S, V|\mathbf{O}, \Theta) \Bigg[\log \pi'_{s_1} + \log \omega'_{s_1 v_1} + \log p(\mathbf{o}_1 | \boldsymbol{\mu}'_{s_1 v_1}, \boldsymbol{\Sigma}'_{s_1 v_1}) \\
&\quad + \sum_{t=2}^{T} \left(\log a'_{s_{t-1} s_t} + \log \omega'_{s_t v_t} + \log p(\mathbf{o}_t | \boldsymbol{\mu}'_{s_t v_t}, \boldsymbol{\Sigma}'_{s_t v_t}) \right) \Bigg] \\
&= Q^{\text{ML}}(\boldsymbol{\pi}'|\boldsymbol{\pi}) + Q^{\text{ML}}(A'|A) + Q^{\text{ML}}(\boldsymbol{\omega}'|\boldsymbol{\omega}) + Q^{\text{ML}}(\boldsymbol{\mu}', \mathbf{R}'|\boldsymbol{\mu}, \mathbf{R}).
\end{aligned} \tag{4.36}$$

Similarly, from Eq. (4.29), the prior distribution of all model parameters $p(\Theta)$ can be decomposed into the four individual prior distributions as

$$\begin{aligned}
\log p(\Theta) &= \underbrace{\log \left(\text{Dir}(\{\pi_j\}_{j=1}^J | \{\phi_j^\pi\}_{j=1}^J) \right)}_{\triangleq p(\boldsymbol{\pi})} + \underbrace{\log \prod_{i=1}^{J} \left(\text{Dir}(\{a_{ij}\}_{j=1}^J | \{\phi_{ij}^a\}_{j=1}^J) \right)}_{\triangleq p(A)} \\
&\quad + \underbrace{\log \prod_{i=1}^{J} \left(\text{Dir}(\{\omega_{jk}\}_{k=1}^K | \{\phi_{jk}^\omega\}_{k=1}^K) \right)}_{\triangleq p(\boldsymbol{\omega})} \\
&\quad + \underbrace{\log \prod_{j=1}^{J} \prod_{k=1}^{K} \left(\mathcal{N}(\boldsymbol{\mu}_{jk} | \boldsymbol{\mu}_{jk}^0, (\phi_{jk}^{\boldsymbol{\mu}} \mathbf{R}_{jk})^{-1}) \mathcal{W}(\mathbf{R}_{jk} | \mathbf{R}_{jk}^0, \phi_{jk}^{\mathbf{R}}) \right)}_{\triangleq p(\boldsymbol{\mu}, \mathbf{R})}.
\end{aligned} \tag{4.37}$$

Therefore, by using the factorization forms of Eqs. (4.36) and (4.37), similarly to the ML case, Eq. (4.36) is also represented as a sum of four individual auxiliary functions defined as follows:

$$\begin{aligned}
Q^{\text{MAP}}(\Theta'|\Theta) &= Q^{\text{MAP}}(\boldsymbol{\pi}'|\boldsymbol{\pi}) + Q^{\text{MAP}}(A'|A) \\
&\quad + Q^{\text{MAP}}(\boldsymbol{\omega}'|\boldsymbol{\omega}) + Q^{\text{MAP}}(\boldsymbol{\mu}', \mathbf{R}'|\boldsymbol{\mu}, \mathbf{R}),
\end{aligned} \tag{4.38}$$

where

$$\begin{aligned}
Q^{\text{MAP}}(\boldsymbol{\pi}'|\boldsymbol{\pi}) &= Q^{\text{ML}}(\boldsymbol{\pi}'|\boldsymbol{\pi}) + \log p(\boldsymbol{\pi}) \\
&= \sum_{j=1}^{J} \xi_1(j) \log \pi'_j + \log \left(\text{Dir}(\{\pi_j\}_{j=1}^J | \{\phi_j^\pi\}_{j=1}^J) \right),
\end{aligned} \tag{4.39}$$

$$\begin{aligned}
Q^{\text{MAP}}(A'|A) &= Q^{\text{ML}}(A'|A) + \log p(A) \\
&= \sum_{t=1}^{T} \sum_{i=1}^{J} \sum_{j=1}^{J} \xi_t(i,j) \log a'_{ij} + \sum_{i=1}^{J} \log \left(\text{Dir}(\{a'_{ij}\}_{j=1}^J | \{\phi_{ij}^a\}_{j=1}^J) \right),
\end{aligned} \tag{4.40}$$

$$
\begin{aligned}
Q^{\mathrm{MAP}}(\boldsymbol{\omega}'|\boldsymbol{\omega}) &= Q^{\mathrm{ML}}(\boldsymbol{\omega}'|\boldsymbol{\omega}) + \log p(\boldsymbol{\omega}) \\
&= \sum_{t=1}^{T}\sum_{j=1}^{J}\sum_{k=1}^{K}\gamma_t(j,k)\log\omega'_{j,k} + \sum_{j=1}^{J}\log\left(\mathrm{Dir}(\{\omega'_{jk}\}_{k=1}^{K}|\{\phi^{\omega}_{jk}\}_{k=1}^{K})\right),
\end{aligned} \tag{4.41}
$$

$$
\begin{aligned}
Q^{\mathrm{MAP}}(\boldsymbol{\mu}',\mathbf{R}'|\boldsymbol{\mu},\mathbf{R}) &= Q^{\mathrm{ML}}(\boldsymbol{\mu}',\mathbf{R}'|\boldsymbol{\mu},\mathbf{R}) + \log p(\boldsymbol{\mu},\mathbf{R}) \\
&\propto \sum_{t=1}^{T}\sum_{j=1}^{J}\sum_{k=1}^{K}\frac{\gamma_t(j,k)}{2}\left[\log|\mathbf{R}'_{jk}| - (\mathbf{o}_t-\boldsymbol{\mu}'_{jk})^{\intercal}\mathbf{R}'_{jk}(\mathbf{o}_t-\boldsymbol{\mu}'_{jk})\right] \\
&\quad + \sum_{j=1}^{J}\sum_{k=1}^{K}\log\left(\mathcal{N}(\boldsymbol{\mu}'_{jk}|\boldsymbol{\mu}^0_{jk},(\phi^{\boldsymbol{\mu}}_{jk}\mathbf{R}'_{jk})^{-1})\mathcal{W}(\mathbf{R}'_{jk}|\mathbf{R}^0_{jk},\phi^{\mathbf{R}}_{jk})\right),
\end{aligned} \tag{4.42}
$$

where $\xi_1(j)$, $\xi_t(i,j)$, and $\gamma_t(j,k)$ are the posterior probabilities, which are introduced in Eqs. (3.99), (3.118), (3.119) as follows:

$$
\begin{aligned}
\xi_1(j) &\triangleq p(s_1=j|\mathbf{O},\Theta) \\
\xi_t(i,j) &\triangleq p(s_t=i,s_{t+1}=j|\mathbf{O},\Theta) \\
\gamma_t(j,k) &\triangleq p(s_t=j,v_t=k|\mathbf{O},\Theta).
\end{aligned} \tag{4.43}
$$

Note that Θ is estimated by using the MAP estimation Θ^{MAP} for Θ instead of the ML estimation, which is discussed in Section 4.3.6.

We can also obtain the auxiliary function of diagonal covariance Gaussians instead of Eq. (4.42) by using a Gaussian-gamma distribution as a prior distribution, as discussed in Eq. (4.32) as follows:

$$
\begin{aligned}
&Q^{\mathrm{MAP}}(\boldsymbol{\mu}',\mathbf{R}'|\boldsymbol{\mu},\mathbf{R}) \\
&\propto \sum_{t=1}^{T}\sum_{j=1}^{J}\sum_{k=1}^{K}\sum_{d=1}^{D}\frac{\gamma_t(j,k)}{2}\left[\log r'_{jk} - (o_{td}-\mu'_{jkd})^2 r'_{jkd}\right] \\
&\quad + \sum_{j=1}^{J}\sum_{k=1}^{K}\sum_{d=1}^{D}\log\left(\mathcal{N}(\mu'_{jkd}|\mu^0_{jkd},(\phi^{\boldsymbol{\mu}}_{jk}r'_{jkd})^{-1})\mathrm{Gam}_2\left(r'_{jkd}\middle|\phi^{\mathbf{R}}_{jk},r_{jkd}\right)\right).
\end{aligned} \tag{4.44}
$$

Here we use the gamma distribution $\mathrm{Gam}_2(y|\phi,r^0)$ described in Eq. (C.81) instead of the original gamma distribution defined in Eq. (C.74), which provides a good relationship with the Wishart distribution, i.e., if $\mathbf{R}$ is a scalar value (the number of dimension $D=1$), the hyperparameters of the Wishart distribution become the same as ϕ^r_{jkd} and r^0_{jk},[4] as we discussed in Example 2.6. Note that the vector and matrix operations in Eq. (4.42) are represented as scalar operations with the summation over the dimension. This is a very good property for which to obtain the analytical solutions due to the simplicity of the scalar calculations. In addition, this representation avoids vector and matrix computations, which also makes implementation simple. This section provides

[4] We can set a hyperparameter ϕ which depends on each element d, i.e., $\phi^{\mathbf{R}}_{jk}\to\phi^{\mathbf{R}}_{jkd}$. Considering the compatibility with the Wishart distribution, this book uses $\phi^{\mathbf{R}}_{jk}$, which is independent of d.

both full and diagonal covariance solutions, but the diagonal covariance solution is used for most of our applications.

4.3.5 Maximization step

The maximization step in the ML–EM algorithm obtains the maximum values of parameters by using derivative techniques, as discussed in Section 3.4.3. In this section, we provide other solutions for this problem which:

1. Calculate the posterior distributions;
2. Obtain the mode values of the posterior distributions, which are used as the MAP estimates.

In general, it is difficult to analytically obtain the posterior distributions. However, since we use the conjugate prior distributions of a CDHMM, as discussed in Section 2.1.3, we can easily obtain the posterior distributions for these problems.

Initial weight

We first focus on $Q^{\mathrm{MAP}}(\boldsymbol{\pi}'|\boldsymbol{\pi})$ in Eq. (4.39):

$$Q^{\mathrm{MAP}}(\boldsymbol{\pi}'|\boldsymbol{\pi}) = \sum_{j=1}^{J} \xi_1(j) \log \pi'_j + \log \left(\mathrm{Dir}(\{\pi'_j\}_{j=1}^{J} | \{\phi_j^{\pi}\}_{j=1}^{J}) \right). \tag{4.45}$$

Recall that the Dirichlet distribution (Appendix C.4) is represented as follows:

$$\mathrm{Dir}(\{\pi_j\}_{j=1}^{J} | \{\phi_j^{\pi}\}_{j=1}^{J}) = C_{\mathrm{Dir}}(\{\phi_j^{\pi}\}_{j=1}^{J}) \prod_{j=1}^{J} (\pi_j)^{\phi_j^{\pi} - 1}. \tag{4.46}$$

Then, by substituting Eq. (4.46) into Eq. (4.45), Eq. (4.45) is re-written as follows:

$$\begin{aligned} Q^{\mathrm{MAP}}(\boldsymbol{\pi}'|\boldsymbol{\pi}) &= \sum_{j=1}^{J} \xi_1(j) \log \pi'_j + (\phi_j^{\pi} - 1) \log \pi'_j + \log C_{\mathrm{Dir}}(\{\phi_j^{\pi}\}_{j=1}^{J}) \\ &= \sum_{j=1}^{J} (\xi_1(j) + \phi_j^{\pi} - 1) \log \pi'_j + \log C_{\mathrm{Dir}}(\{\phi_j^{\pi}\}_{j=1}^{J}) \\ &= \log \prod_{j=1}^{J} (\pi'_j)^{\xi_1(j) + \phi_j^{\pi} - 1} + \log C_{\mathrm{Dir}}(\{\phi_j^{\pi}\}_{j=1}^{J}). \end{aligned} \tag{4.47}$$

Comparing the result with Eq. (4.46), it is the same function form with different hyperparameters. Therefore, the auxiliary function $Q^{\mathrm{MAP}}(\boldsymbol{\pi}'|\boldsymbol{\pi})$ is represented by the following Dirichlet distribution:

$$\begin{aligned} Q^{\mathrm{MAP}}(\boldsymbol{\pi}'|\boldsymbol{\pi}) = \log \left(\mathrm{Dir}(\{\pi'_j\}_{j=1}^{J} | \{\hat{\phi}_j^{\pi}\}_{j=1}^{J}) \right) - \log C_{\mathrm{Dir}}(\{\hat{\phi}_j^{\pi}\}_{j=1}^{J}) \\ + \log C_{\mathrm{Dir}}(\{\phi_j^{\pi}\}_{j=1}^{J}) \end{aligned}$$

$$
\begin{aligned}
&= \log\left(\mathrm{Dir}(\{\pi'_j\}_{j=1}^J | \{\hat{\phi}^\pi_j\}_{j=1}^J)\right) + \log \frac{C_{\mathrm{Dir}}(\{\phi^\pi_j\}_{j=1}^J)}{C_{\mathrm{Dir}}(\{\hat{\phi}^\pi_j\}_{j=1}^J)} \\
&\propto \log\left(\mathrm{Dir}(\{\pi'_j\}_{j=1}^J | \{\hat{\phi}^\pi_j\}_{j=1}^J)\right),
\end{aligned} \tag{4.48}
$$

where

$$
\hat{\phi}^\pi_j \triangleq \phi^\pi_j + \xi_1(j). \tag{4.49}
$$

We finally omit the ratio of the normalization factor of the prior and posterior distributions, which do not depend on π'_j. Actually, this Dirichlet distribution corresponds to the posterior distribution of π with new hyperparameter $\hat{\phi}$. This result is similar to that of the conjugate prior and posterior distributions for multinomial likelihood function, as we discussed in Example 2.8.

Once we obtain the analytical form of the posterior distribution, the MAP estimate can be obtained as the mode of the Dirichlet distribution (Appendix C.4):

$$
\pi_j^{\mathrm{MAP}} = \frac{\phi^\pi_j + \xi_1(j) - 1}{\sum_{j'=1}^J (\phi^\pi_{j'} + \xi_1(j') - 1)}. \tag{4.50}
$$

Thus, we obtain the MAP estimate of the initial weight π^{MAP}, which is proportional to the hyperparameter $\hat{\phi}^\pi$. We discuss the meaning of this solution in Section 4.3.7.

State transition

Similarly, the auxiliary function of state transition parameters A is obtained as follows:

$$
\begin{aligned}
Q^{\mathrm{MAP}}(A'|A) &= \log\left(\prod_{i=1}^J \mathrm{Dir}(\{a'_{ij}\}_{j=1}^J | \{\hat{\phi}^a_{ij}\}_{j=1}^J)\right) + \log\left(\prod_{i=1}^J \frac{C_{\mathrm{Dir}}(\{\phi^a_{ij}\}_{j=1}^J)}{C_{\mathrm{Dir}}(\{\hat{\phi}^a_{ij}\}_{j=1}^J)}\right) \\
&\propto \log\left(\prod_{i=1}^J \mathrm{Dir}(\{a'_{ij}\}_{j=1}^J | \{\hat{\phi}^a_{ij}\}_{j=1}^J)\right),
\end{aligned} \tag{4.51}
$$

where

$$
\hat{\phi}^a_{ij} \triangleq \phi^a_{ij} + \sum_{t=1}^{T-1} \xi_t(i,j). \tag{4.52}
$$

Therefore, the mode of the Dirichlet distribution is obtained as:

$$
a_{ij}^{\mathrm{MAP}} = \frac{\phi^a_{ij} + \sum_{t=1}^{T-1} \xi_t(i,j) - 1}{\sum_{j'=1}^J (\phi^a_{ij'} + \sum_{t=1}^{T-1} \xi_t(i,j') - 1)}. \tag{4.53}
$$

The solution is similar to the initial weight in Eq. (4.50), and it is computed from the statistics of the accumulated posterior values of the state transition $\sum_{t=1}^{T-1} \xi_t(i,j)$ and prior parameter ϕ^a_{ij}.

Again, the result indicates that the auxiliary function of the state transition is represented by the same Dirichlet distribution as that used in the prior distribution with different hyperparameters. This result corresponds to the conjugate distribution analysis for multinomial distribution, as we discussed in Section 2.8. Therefore, although we need to handle the latent variables within a MAP–EM framework based on the iterative calculation, the conjugate prior provides an analytic solution in the M-step, and it makes the estimation process efficient.

Mixture weight

Similar to the state transition, the auxiliary function of the mixture weight parameters $\boldsymbol{\omega}$ is as follows:

$$
\begin{aligned}
Q^{\mathrm{MAP}}(\boldsymbol{\omega}'|\boldsymbol{\omega}) &= \sum_{t=1}^{T}\sum_{j=1}^{J}\sum_{k=1}^{K}\gamma_t(j,k)\log\omega'_{jk} + \log\left(\prod_{j=1}^{J}\mathrm{Dir}(\{\omega'_{jk}\}_{k=1}^{K}|\{\phi^{\omega}_{jk}\}_{k=1}^{K})\right) \\
&= \log\left(\prod_{j=1}^{J}\mathrm{Dir}(\{\omega'_{jk}\}_{k=1}^{K}|\{\hat{\phi}^{\omega}_{jk}\}_{k=1}^{K})\right) + \log\left(\prod_{j=1}^{J}\frac{C_{\mathrm{Dir}}(\{\phi^{\omega}_{jk}\}_{k=1}^{K})}{C_{\mathrm{Dir}}(\{\hat{\phi}^{\omega}_{jk}\}_{k=1}^{K})}\right) \\
&\propto \log\left(\prod_{j=1}^{J}\mathrm{Dir}(\{\omega'_{jk}\}_{k=1}^{K}|\{\hat{\phi}^{\omega}_{jk}\}_{k=1}^{K})\right),
\end{aligned}
\tag{4.54}
$$

where

$$
\hat{\phi}^{\omega}_{jk} \triangleq \phi^{\omega}_{jk} + \sum_{t=1}^{T}\gamma_t(j,k). \tag{4.55}
$$

Therefore, the mode of the Dirichlet distribution is obtained as:

$$
\omega_{jk}^{\mathrm{MAP}} = \frac{\phi^{\omega}_{jk} + \sum_{t=1}^{T}\gamma_t(j,k) - 1}{\sum_{k'=1}^{K}(\phi^{\omega}_{jk'} + \sum_{t=1}^{T}\gamma_t(j,k') - 1)}. \tag{4.56}
$$

Again, it is computed from the statistics of the accumulated posterior values of the state occupancy $\sum_{t=1}^{T}\gamma_t(j,k)$ and prior parameter ϕ^{ω}_{jk}.

Mean vector and covariance matrix

Finally, we focus on the auxiliary function of the mean vector $\boldsymbol{\mu}$ and precision matrix $\mathbf{R}$. Recall that the multivariate Gaussian distribution (Appendix C.6) is represented as follows:

$$
\mathcal{N}(\mathbf{x}|\boldsymbol{\mu},\mathbf{R}^{-1}) = C_{\mathcal{N}}(\mathbf{R}^{-1})\exp\left(-\frac{1}{2}(\mathbf{x}-\boldsymbol{\mu})^{\mathsf{T}}\mathbf{R}(\mathbf{x}-\boldsymbol{\mu})\right), \tag{4.57}
$$

and the Wishart distribution (Appendix C.14) is represented as follows:

$$
\mathcal{W}(\mathbf{Y}|\mathbf{R}^0,\phi) = C_{\mathcal{W}}(\mathbf{R}^0,\phi)|\mathbf{Y}|^{\frac{\phi-D-1}{2}}\exp\left(-\frac{1}{2}\mathrm{tr}\left[\mathbf{R}^0\mathbf{Y}\right]\right). \tag{4.58}
$$

Then, $Q^{\mathrm{MAP}}(\boldsymbol{\mu}',\mathbf{R}'|\boldsymbol{\mu},\mathbf{R})$ is represented by using Eqs. (4.42) with the normalization constant, (4.57), and (4.58), as follows:

$$
\begin{aligned}
Q^{\mathrm{MAP}}(\boldsymbol{\mu}',\mathbf{R}'|\boldsymbol{\mu},\mathbf{R}) &= \sum_{t=1}^{T}\sum_{j=1}^{J}\sum_{k=1}^{K}\frac{\gamma_t(j,k)}{2}\left(\log|\mathbf{R}'_{jk}| - (\mathbf{o}_t-\boldsymbol{\mu}'_{jk})^{\mathsf{T}}\mathbf{R}'_{jk}(\mathbf{o}_t-\boldsymbol{\mu}'_{jk})\right) \\
&+ \sum_{j=1}^{J}\sum_{k=1}^{K}\log\left(\mathcal{N}(\boldsymbol{\mu}'_{jk}|\boldsymbol{\mu}^0_{jk},(\phi^{\mu}_{jk}\mathbf{R}'_{jk})^{-1})\mathcal{W}(\mathbf{R}'_{jk}|\mathbf{R}^0_{jk},\phi^{\mathbf{R}}_{jk})\right) \\
&- \sum_{t=1}^{T}\sum_{j=1}^{J}\sum_{k=1}^{K}\frac{\gamma_t(j,k)D}{2}\log(2\pi)
\end{aligned}
$$

$$
\begin{aligned}
&= \sum_{t=1}^{T}\sum_{j=1}^{J}\sum_{k=1}^{K}\frac{\gamma_t(j,k)}{2}\left(\log|\mathbf{R}'_{jk}| - (\mathbf{o}_t - \boldsymbol{\mu}'_{jk})^\mathsf{T}\mathbf{R}'_{jk}(\mathbf{o}_t - \boldsymbol{\mu}'_{jk})\right) \\
&\quad + \frac{1}{2}\sum_{j=1}^{J}\sum_{k=1}^{K}\Bigg(\log\left|\mathbf{R}'_{jk}\right| - (\boldsymbol{\mu}'_{jk} - \boldsymbol{\mu}^0_{jk})^\mathsf{T}\phi^{\boldsymbol{\mu}}_{jk}\mathbf{R}'_{jk}(\boldsymbol{\mu}'_{jk} - \boldsymbol{\mu}^0_{jk}) \\
&\qquad\qquad + (\phi^{\mathbf{R}}_{jk} - D - 1)\log|\mathbf{R}'_{jk}| - \mathrm{tr}\left[\mathbf{R}^0_{jk}\mathbf{R}'_{jk}\right]\Bigg) \\
&\quad + \sum_{j=1}^{J}\sum_{k=1}^{K}\Bigg(-\sum_{t=1}^{T}\frac{\gamma_t(j,k)D}{2}\log(2\pi) - \frac{D}{2}\log(2\pi) + \frac{D}{2}\log\phi^{\boldsymbol{\mu}}_{jk} \\
&\qquad\qquad + \log C_{\mathcal{W}}(\mathbf{R}^0_{jk}, \phi^{\mathbf{R}}_{jk})\Bigg),
\end{aligned} \tag{4.59}
$$

where the final line includes the terms that do not depend on $\boldsymbol{\mu}$ and $\mathbf{R}$. Then we rearrange Eq. (4.59) so that we can write it in a probabilistic form. First, we omit j, k, and $'$ in Eq. (4.59) for simplicity, and consider the following function:

$$
\begin{aligned}
g(\boldsymbol{\mu}, \mathbf{R}) &\triangleq \frac{1}{2}\Bigg(\sum_{t=1}^{T}\gamma_t\left(\log|\mathbf{R}| - (\mathbf{o}_t - \boldsymbol{\mu})^\mathsf{T}\mathbf{R}(\mathbf{o}_t - \boldsymbol{\mu})\right) \\
&\qquad + \log|\mathbf{R}| - (\boldsymbol{\mu} - \boldsymbol{\mu}^0)^\mathsf{T}\phi^{\boldsymbol{\mu}}\mathbf{R}(\boldsymbol{\mu} - \boldsymbol{\mu}^0) \\
&\qquad + (\phi^{\mathbf{R}} - D - 1)\log|\mathbf{R}| - \mathrm{tr}\left[\mathbf{R}^0\mathbf{R}\right]\Bigg) \\
&= \frac{1}{2}\left(\sum_{t=1}^{T}\gamma_t\log|\mathbf{R}| + \log|\mathbf{R}| + (\phi^{\mathbf{R}} - D - 1)\log|\mathbf{R}| - \mathrm{tr}\left[\mathbf{R}^0\mathbf{R}\right]\right) \\
&\quad \underbrace{-\frac{1}{2}\left((\mathbf{o}_t - \boldsymbol{\mu})^\mathsf{T}\mathbf{R}(\mathbf{o}_t - \boldsymbol{\mu}) - (\boldsymbol{\mu} - \boldsymbol{\mu}^0)^\mathsf{T}\phi^{\boldsymbol{\mu}}\mathbf{R}(\boldsymbol{\mu} - \boldsymbol{\mu}^0)\right)}_{\triangleq f(\boldsymbol{\mu}, \mathbf{R})}.
\end{aligned} \tag{4.60}
$$

Then, we focus on $f(\boldsymbol{\mu}, \mathbf{R})$ that has the terms in $g(\boldsymbol{\mu}, \mathbf{R})$ that depend on $\boldsymbol{\mu}$. $f(\boldsymbol{\mu}, \mathbf{R})$ is re-written as follows:

$$
\begin{aligned}
-2f(\boldsymbol{\mu}, \mathbf{R}) &= \boldsymbol{\mu}^\mathsf{T}\left(\left(\sum_t \gamma_t + \phi^{\boldsymbol{\mu}}\right)\mathbf{R}\right)\boldsymbol{\mu} - 2\boldsymbol{\mu}^\mathsf{T}\left(\mathbf{R}\sum_t \gamma_t\mathbf{o}_t + \phi^{\boldsymbol{\mu}}\mathbf{R}\boldsymbol{\mu}^0\right) \\
&\quad + \sum_t \gamma_t\mathbf{o}_t^\mathsf{T}\mathbf{R}\mathbf{o}_t + (\boldsymbol{\mu}^0)^\mathsf{T}\phi^{\boldsymbol{\mu}}\mathbf{R}\boldsymbol{\mu}^0.
\end{aligned} \tag{4.61}
$$

Since this is a quadratic form of $\boldsymbol{\mu}$, it can be represented by a Gaussian distribution by arranging Eq. (4.61) into the complete square form. Although it is complicated to deal with the complete square form for vectors, we can use the following complete square rules found in Eqs. (B.16) and (B.17):

$$
\mathbf{x}^\mathsf{T}\mathbf{A}\mathbf{x} - 2\mathbf{x}^\mathsf{T}\mathbf{b} + c = (\mathbf{x} - \mathbf{u})^\mathsf{T}\mathbf{A}(\mathbf{x} - \mathbf{u}) + v, \tag{4.62}
$$

where

$$\mathbf{u} \triangleq \mathbf{A}^{-1}\mathbf{b}$$
$$v \triangleq c - \mathbf{b}^{\mathsf{T}}\mathbf{A}^{-1}\mathbf{b}. \tag{4.63}$$

Therefore, by using this rule (e.g., $\mathbf{x} \to \boldsymbol{\mu}$, $\mathbf{A} \to \left(\sum_t \gamma_t + \phi^{\mu}\right)\mathbf{R}$, $\mathbf{b} \to \mathbf{R}\sum_t \gamma_t \mathbf{o}_t + \phi^{\mu}\mathbf{R}\boldsymbol{\mu}^0$, and $c \to \sum_t \gamma_t \mathbf{o}_t^{\mathsf{T}}\mathbf{R}\mathbf{o}_t + (\boldsymbol{\mu}^0)^{\mathsf{T}}\phi^{\mu}\mathbf{R}\boldsymbol{\mu}^0$), Eq. (4.61) is rewritten with the complete square form as follows:

$$\begin{aligned}
-2f(\boldsymbol{\mu}, \mathbf{R}) &= \boldsymbol{\mu}^{\mathsf{T}}\left(\left(\sum_t \gamma_t + \phi^{\mu}\right)\mathbf{R}\right)\boldsymbol{\mu} - 2\boldsymbol{\mu}^{\mathsf{T}}\left(\mathbf{R}\sum_t \gamma_t \mathbf{o}_t + \phi^{\mu}\mathbf{R}\boldsymbol{\mu}^0\right) \\
&\quad + \sum_t \gamma_t \mathbf{o}_t^{\mathsf{T}}\mathbf{R}\mathbf{o}_t + (\boldsymbol{\mu}^0)^{\mathsf{T}}\phi^{\mu}\mathbf{R}\boldsymbol{\mu}^0 \\
&= \left(\boldsymbol{\mu} - \hat{\boldsymbol{\mu}}\right)^{\mathsf{T}}\left(\hat{\phi}^{\mu}\mathbf{R}\right)\left(\boldsymbol{\mu} - \hat{\boldsymbol{\mu}}\right) + v(\mathbf{R}),
\end{aligned} \tag{4.64}$$

where $\hat{\phi}^{\mu}$, $\hat{\boldsymbol{\mu}}$, and $v(\mathbf{R})$ are defined as follows:

$$\begin{aligned}
\hat{\phi}^{\mu} &\triangleq \phi^{\mu} + \sum_t \gamma_t, \\
\hat{\boldsymbol{\mu}} &\triangleq \left(\left(\sum_t \gamma_t + \phi^{\mu}\right)\mathbf{R}\right)^{-1}\left(\mathbf{R}\sum_t \gamma_t \mathbf{o}_t + \phi^{\mu}\mathbf{R}\boldsymbol{\mu}^0\right) \\
&= \frac{\phi^{\mu}\boldsymbol{\mu}^0 + \sum_t \gamma_t \mathbf{o}_t}{\phi^{\mu} + \sum_t \gamma_t}, \\
v(\mathbf{R}) &\triangleq \sum_t \gamma_t \mathbf{o}_t^{\mathsf{T}}\mathbf{R}\mathbf{o}_t + (\boldsymbol{\mu}^0)^{\mathsf{T}}\phi^{\mu}\mathbf{R}\boldsymbol{\mu}^0 - \left(\mathbf{R}\sum_t \gamma_t \mathbf{o}_t + \phi^{\mu}\mathbf{R}\boldsymbol{\mu}^0\right)^{\mathsf{T}}\hat{\boldsymbol{\mu}} \\
&= \sum_t \gamma_t \mathbf{o}_t^{\mathsf{T}}\mathbf{R}\mathbf{o}_t + (\boldsymbol{\mu}^0)^{\mathsf{T}}\phi^{\mu}\mathbf{R}\boldsymbol{\mu}^0 - \hat{\boldsymbol{\mu}}^{\mathsf{T}}\hat{\phi}^{\mu}\mathbf{R}\hat{\boldsymbol{\mu}}.
\end{aligned} \tag{4.65}$$

Note that $\hat{\phi}^{\mu}$ and $\hat{\boldsymbol{\mu}}$ correspond to the mean and covariance hyperparameters of the conjugate Gaussian distribution of $\boldsymbol{\mu}$. Thus, $f(\boldsymbol{\mu}, \mathbf{R})$ is rewritten with the definition of the Gaussian distribution in Eq. (4.57) as follows:

$$\begin{aligned}
f(\boldsymbol{\mu}, \mathbf{R}) &= -\frac{1}{2}\left(\boldsymbol{\mu} - \hat{\boldsymbol{\mu}}\right)^{\mathsf{T}}\left(\hat{\phi}^{\mu}\mathbf{R}\right)\left(\boldsymbol{\mu} - \hat{\boldsymbol{\mu}}\right) - \frac{1}{2}v(\mathbf{R}) \\
&= \log \mathcal{N}(\boldsymbol{\mu}|\hat{\boldsymbol{\mu}}, (\hat{\phi}^{\mu}\mathbf{R})^{-1}) - \log C_{\mathcal{N}}(\mathbf{R}^{-1}) - \frac{1}{2}v(\mathbf{R}) \\
&= \log \mathcal{N}(\boldsymbol{\mu}|\hat{\boldsymbol{\mu}}, (\hat{\phi}^{\mu}\mathbf{R})^{-1}) + \frac{D}{2}\log(2\pi) - \frac{D}{2}\log \hat{\phi}^{\mu} - \frac{1}{2}\log|\mathbf{R}| - \frac{1}{2}v(\mathbf{R}),
\end{aligned} \tag{4.66}$$

where $v(\mathbf{R})$ is used to obtain the analytic form of $\mathbf{R}$ with the rest of the $\mathbf{R}$-dependent terms in Eq. (4.60). That is, the auxiliary function (Eq. (4.60)) with Eq. (4.66) is represented as follows:

$$g(\boldsymbol{\mu}, \mathbf{R}) = \log \mathcal{N}(\boldsymbol{\mu}|\hat{\boldsymbol{\mu}}, (\hat{\phi}^{\mu}\mathbf{R})^{-1}) + \frac{D}{2}\log(2\pi) - \frac{D}{2}\log\hat{\phi}^{\mu}$$
$$\times \underbrace{\frac{1}{2}\left(-v(\mathbf{R}) + \left(\sum_t \gamma_t + \phi^{\mathbf{R}} - D - 1\right)\log|\mathbf{R}| - \mathrm{tr}\left[\mathbf{R}^0\mathbf{R}\right]\right)}_{\triangleq h(\mathbf{R})}. \quad (4.67)$$

Now we focus on $h(\mathbf{R})$ that has the terms in the second line in Eq. (4.67). By using the definition of $v(\mathbf{R})$ in Eq. (4.65), we can rewrite $h(\mathbf{R})$ as follows:

$$\begin{aligned} h(\mathbf{R}) &= \frac{1}{2}\left(-v(\mathbf{R}) + \left(\sum_t \gamma_t + \phi^{\mathbf{R}} - D - 1\right)\log|\mathbf{R}| - \mathrm{tr}\left[\mathbf{R}^0\mathbf{R}\right]\right) \\ &= \frac{1}{2}\left(-\sum_t \gamma_t \mathbf{o}_t^{\mathsf{T}}\mathbf{R}\mathbf{o}_t - (\boldsymbol{\mu}^0)^{\mathsf{T}}\phi^{\mu}\mathbf{R}\boldsymbol{\mu}^0 + \hat{\boldsymbol{\mu}}^{\mathsf{T}}\hat{\phi}^{\mu}\mathbf{R}\hat{\boldsymbol{\mu}} - \mathrm{tr}\left[\mathbf{R}^0\mathbf{R}\right]\right) \\ &\quad + \frac{1}{2}\left(\sum_t \gamma_t + \phi^{\mathbf{R}} - D - 1\right)\log|\mathbf{R}|. \end{aligned} \quad (4.68)$$

Since Eq. (4.68) includes the trace operation, it is difficult to re-arrange this equation. Therefore, by using the trace rule of $a = \mathrm{tr}[a]$ (Eq. (B.1)), we represent all terms except for the $\log|\mathbf{R}|$ term in Eq. (4.68) as follows:

$$\begin{aligned} h(\mathbf{R}) = \frac{1}{2}\Bigg(&-\mathrm{tr}\left[\sum_t \gamma_t \mathbf{o}_t\mathbf{o}_t^{\mathsf{T}}\mathbf{R}\right] - \mathrm{tr}\left[\phi^{\mu}\boldsymbol{\mu}^0(\boldsymbol{\mu}^0)^{\mathsf{T}}\mathbf{R}\right] \\ &+ \mathrm{tr}\left[\hat{\phi}^{\mu}\hat{\boldsymbol{\mu}}\hat{\boldsymbol{\mu}}^{\mathsf{T}}\mathbf{R}\right] - \mathrm{tr}\left[\mathbf{R}^0\mathbf{R}\right]\Bigg) \\ &+ \frac{1}{2}\left(\sum_t \gamma_t + \phi^{\mathbf{R}} - D - 1\right)\log|\mathbf{R}|. \end{aligned} \quad (4.69)$$

In addition, by using the trace rules of $\mathrm{tr}[\mathbf{ABC}] = \mathrm{tr}[\mathbf{BCA}]$ and $\mathrm{tr}[\mathbf{A}+\mathbf{B}] = \mathrm{tr}[\mathbf{A}] + \mathrm{tr}[\mathbf{B}]$ (Eqs. (B.2) and (B.3)), Eq. (4.69) is finally represented by comparing $h(\mathbf{R})$ with the definition of the Wishart distribution (Eq. (4.58)), as follows:

$$\begin{aligned} h(\mathbf{R}) &= -\frac{1}{2}\mathrm{tr}\Bigg[\underbrace{\left(\sum_t \gamma_t \mathbf{o}_t\mathbf{o}_t^{\mathsf{T}} + \phi^{\mu}\boldsymbol{\mu}^0(\boldsymbol{\mu}^0)^{\mathsf{T}} - \hat{\phi}^{\mu}\hat{\boldsymbol{\mu}}\hat{\boldsymbol{\mu}}^{\mathsf{T}} + \mathbf{R}^0\right)}_{\triangleq\hat{\mathbf{R}}}\mathbf{R}\Bigg] \\ &\quad + \frac{1}{2}\Bigg(\underbrace{\left(\sum_t \gamma_t + \phi^{\mathbf{R}}\right)}_{\triangleq\hat{\phi}^{\mathbf{R}}} - D - 1\Bigg)\log|\mathbf{R}| \\ &= \log \mathcal{W}(\mathbf{R}|\hat{\mathbf{R}}, \hat{\phi}^{\mathbf{R}}) - \log C_{\mathcal{W}}(\hat{\mathbf{R}}, \hat{\phi}^{\mathbf{R}}). \end{aligned} \quad (4.70)$$

Thus, Eq. (4.70) can be represented as the Wishart distribution with the following hyperparameters:

$$\hat{\phi}^{\mathbf{R}} \triangleq \phi^{\mathbf{R}} + \sum_t \gamma_t,$$
$$\hat{\mathbf{R}} \triangleq \sum_t \gamma_t \mathbf{o}_t \mathbf{o}_t^{\mathsf{T}} + \phi^{\boldsymbol{\mu}} \boldsymbol{\mu}^0 (\boldsymbol{\mu}^0)^{\mathsf{T}} - \hat{\phi}^{\boldsymbol{\mu}} \hat{\boldsymbol{\mu}} \hat{\boldsymbol{\mu}}^{\mathsf{T}} + \mathbf{R}^0. \tag{4.71}$$

Note that these hyperparameters are almost equivalent to those of the conjugate distribution analysis discussed in Eq. (2.112). The only difference is that the statistics in Eq. (4.71) are computed from the expectation value of the posterior distribution of a latent variable γ_t.

Here, $g(\boldsymbol{\mu}, \mathbf{R})$ in the original Q function, Eq. (4.59), is represented by Eqs. (4.67) and (4.70) as follows:

$$g(\boldsymbol{\mu}, \mathbf{R}) = \log \mathcal{N}(\boldsymbol{\mu}|\hat{\boldsymbol{\mu}}, (\hat{\phi}^{\boldsymbol{\mu}}\mathbf{R})^{-1}) + \frac{D}{2}\log(2\pi) - \frac{D}{2}\log \hat{\phi}^{\boldsymbol{\mu}}$$
$$+ \log \mathcal{W}(\mathbf{R}|\hat{\mathbf{R}}, \hat{\phi}^{\mathbf{R}}) - \log C_{\mathcal{W}}(\hat{\mathbf{R}}, \hat{\phi}^{\mathbf{R}}). \tag{4.72}$$

Thus, we have found that $Q^{\mathrm{MAP}}(\boldsymbol{\mu}', \mathbf{R}'|\boldsymbol{\mu}, \mathbf{R})$ can be represented with the same distribution form as the prior distributions, which is represented by Gaussian and Wishart distributions as follows:

$$Q^{\mathrm{MAP}}(\boldsymbol{\mu}', \mathbf{R}'|\boldsymbol{\mu}, \mathbf{R})$$
$$= \sum_{j=1}^{J}\sum_{k=1}^{K} \log \left(\mathcal{N}(\boldsymbol{\mu}'_{jk}|\hat{\boldsymbol{\mu}}_{jk}, (\hat{\phi}^{\boldsymbol{\mu}}_{jk}\mathbf{R}'_{jk})^{-1}) \mathcal{W}(\mathbf{R}'_{jk}|\hat{\mathbf{R}}_{jk}, \hat{\phi}^{\mathbf{R}}_{jk}) \right)$$
$$+ \sum_{j=1}^{J}\sum_{k=1}^{K} \left(-\sum_{t=1}^{T} \frac{\gamma_t(j,k)D}{2}\log(2\pi) + \frac{D}{2}\log\frac{\phi^{\boldsymbol{\mu}}_{jk}}{\hat{\phi}^{\boldsymbol{\mu}}_{jk}} + \log\frac{C_{\mathcal{W}}(\mathbf{R}^0_{jk}, \phi^{\mathbf{R}}_{jk})}{C_{\mathcal{W}}(\hat{\mathbf{R}}_{jk}, \hat{\phi}^{\mathbf{R}}_{jk})} \right)$$
$$\propto \sum_{j=1}^{J}\sum_{k=1}^{K} \log \left(\mathcal{N}(\boldsymbol{\mu}'_{jk}|\hat{\boldsymbol{\mu}}_{jk}, (\hat{\phi}^{\boldsymbol{\mu}}_{jk}\mathbf{R}'_{jk})^{-1}) \mathcal{W}(\mathbf{R}'_{jk}|\hat{\mathbf{R}}_{jk}, \hat{\phi}^{\mathbf{R}}_{jk}) \right), \tag{4.73}$$

where we recover the indexes j, k, and $'$. By using the definition of the Gaussian–Wishart distribution in Appendix C.15, $Q^{\mathrm{MAP}}(\boldsymbol{\mu}', \mathbf{R}'|\boldsymbol{\mu}, \mathbf{R})$ can also be represented as:

$$Q^{\mathrm{MAP}}(\boldsymbol{\mu}', \mathbf{R}'|\boldsymbol{\mu}, \mathbf{R}) \propto \sum_{j=1}^{J}\sum_{k=1}^{K} \log \left(\mathcal{NW}(\boldsymbol{\mu}'_{jk}, \mathbf{R}'_{jk}|\hat{\boldsymbol{\mu}}_{jk}, \hat{\phi}^{\boldsymbol{\mu}}_{jk}, \hat{\mathbf{R}}_{jk}, \hat{\phi}^{\mathbf{R}}_{jk}) \right). \tag{4.74}$$

By summarizing the result of Eqs. (4.65) and (4.71), the hyperparameters of these distributions are defined as:

$$\hat{\phi}^{\boldsymbol{\mu}}_{jk} \triangleq \phi^{\boldsymbol{\mu}}_{jk} + \sum_t \gamma_t(j,k),$$

$$\hat{\boldsymbol{\mu}}_{jk} \triangleq \frac{\phi^{\boldsymbol{\mu}}_{jk}\boldsymbol{\mu}^0_{jk} + \sum_t \gamma_t(j,k)\mathbf{o}_t}{\phi^{\boldsymbol{\mu}}_{jk} + \sum_t \gamma_t(j,k)},$$

$$\hat{\phi}^{\mathbf{R}}_{jk} \triangleq \phi^{\mathbf{R}}_{jk} + \sum_t \gamma_t(j,k),$$

$$\hat{\mathbf{R}}_{jk} \triangleq \sum_t \gamma_t(j,k)\mathbf{o}_t\mathbf{o}_t^\mathsf{T} + \phi^{\boldsymbol{\mu}}_{jk}\boldsymbol{\mu}^0_{jk}(\boldsymbol{\mu}^0_{jk})^\mathsf{T} - \hat{\phi}^{\boldsymbol{\mu}}_{jk}\hat{\boldsymbol{\mu}}_{jk}\hat{\boldsymbol{\mu}}^\mathsf{T}_{jk} + \mathbf{R}^0_{jk}. \tag{4.75}$$

The MAP estimates of these values are obtained by considering the modes of the Gaussian–Wishart distribution in Appendix C.15. The modes of $\boldsymbol{\mu}'_{jk}$ and $\boldsymbol{\Sigma}'_{jk}$ are represented as:

$$\boldsymbol{\mu}^{\mathrm{MAP}}_{jk} = \hat{\boldsymbol{\mu}}_{jk},$$

$$\boldsymbol{\Sigma}^{\mathrm{MAP}}_{jk} = \left(\mathbf{R}^{\mathrm{MAP}}_{jk}\right)^{-1} = (\hat{\phi}^{\mathbf{R}}_{jk} - D - 1)^{-1}\hat{\mathbf{R}}_{jk}. \tag{4.76}$$

Note that $\boldsymbol{\Sigma}^{\mathrm{MAP}}_{jk}$ cannot be obtained when $\hat{\phi}^{\mathbf{R}}_{jk} - D - 1 \leq 0$. Thus, we can analytically obtain the M-step solutions of CDHMM parameters (i.e., initial weight, state transition, mixture weight, mean vector and covariance matrix) in the MAP sense, thanks to the conjugate prior distributions. We summarize the solutions below. The hyperparameters of the posterior distributions are represented with Gaussian sufficient statistics and the prior hyperparameters as:

$$\begin{cases} \hat{\phi}^{\pi}_j \triangleq \phi^{\pi}_j + \xi_1(j), \\ \hat{\phi}^{a}_{ij} \triangleq \phi^{a}_{ij} + \sum_{t=1}^{T-1} \xi_t(i,j), \\ \hat{\phi}^{\omega}_{jk} \triangleq \phi^{\omega}_{jk} + \sum_{t=1}^{T} \gamma_t(j,k), \\ \hat{\phi}^{\boldsymbol{\mu}}_{jk} \triangleq \phi^{\boldsymbol{\mu}}_{jk} + \sum_{t=1}^{T} \gamma_t(j,k), \\ \hat{\boldsymbol{\mu}}_{jk} \triangleq \dfrac{\phi^{\boldsymbol{\mu}}_{jk}\boldsymbol{\mu}^0_{jk} + \sum_{t=1}^{T}\gamma_t(j,k)\mathbf{o}_t}{\phi^{\boldsymbol{\mu}}_{jk} + \sum_{t=1}^{T}\gamma_t(j,k)}, \\ \hat{\phi}^{\mathbf{R}}_{jk} \triangleq \phi^{\mathbf{R}}_{jk} + \sum_{t=1}^{T} \gamma_t(j,k), \\ \hat{\mathbf{R}}_{jk} \triangleq \sum_{t=1}^{T} \gamma_t(j,k)\mathbf{o}_t\mathbf{o}_t^\mathsf{T} + \phi^{\boldsymbol{\mu}}_{jk}\boldsymbol{\mu}^0_{jk}(\boldsymbol{\mu}^0_{jk})^\mathsf{T} - \hat{\phi}^{\boldsymbol{\mu}}_{jk}\hat{\boldsymbol{\mu}}_{jk}\hat{\boldsymbol{\mu}}^\mathsf{T}_{jk} + \mathbf{R}^0_{jk}. \end{cases} \tag{4.77}$$

A set of hyperparameters $\hat{\Psi}$ is defined as follows:

$$\hat{\Psi} \triangleq \{\hat{\phi}^{\pi}_j, \hat{\phi}^{a}_{ij}, \hat{\phi}^{\omega}_{jk}, \hat{\phi}^{\boldsymbol{\mu}}_{jk}, \hat{\phi}^{\mathbf{R}}_{jk}, \hat{\boldsymbol{\mu}}_{jk}, \hat{\mathbf{R}}_{jk} | i = 1, \cdots, J, j = 1, \cdots, J, k = 1, \cdots, K\}. \tag{4.78}$$

Then, the MAP solutions of HMM parameters are represented with the posterior hyperparameters as follows:

$$\begin{cases} \pi_j^{\text{MAP}} &= \dfrac{\hat{\phi}_j^{\pi} - 1}{\sum_{j'=1}^{J}(\hat{\phi}_{j'}^{\pi} - 1)}, \\ a_{ij}^{\text{MAP}} &= \dfrac{\hat{\phi}_{ij}^{a} - 1}{\sum_{j'=1}^{J}(\hat{\phi}_{ij'}^{a} - 1)}, \\ \omega_{jk}^{\text{MAP}} &= \dfrac{\hat{\phi}_{jk}^{\omega} - 1}{\sum_{k'=1}^{K}(\hat{\phi}_{jk'}^{\omega} - 1)}, \\ \boldsymbol{\mu}_{jk}^{\text{MAP}} &= \hat{\boldsymbol{\mu}}_{jk}, \\ \boldsymbol{\Sigma}_{jk}^{\text{MAP}} &= (\hat{\phi}_{jk}^{\mathbf{R}} - D - 1)^{-1}\hat{\mathbf{R}}_{jk}. \end{cases} \tag{4.79}$$

Compared with the ML M-step of CDHMM parameters in Eq. (3.151), these solutions are more complicated, and actually incur more computational cost than that of the ML solution. However, the computational cost of the M-step is much smaller than that of the E-step, and the additional computational cost of the MAP estimate can be disregarded in practical use.

Mean vector and diagonal covariance matrix

In practise, we often use the diagonal covariance matrix for a multivariate Gaussian distribution for HMMs, as we discussed in Section 3.2.3. This section provides the MAP solution for the diagonal covariance case (Gauvain & Lee 1991). Since the one-dimensional solution of the full-covariance Gaussian posterior distribution in the previous discussion corresponds to that of the diagonal covariance Gaussian posterior distribution of a diagonal element, we can obtain the hyperparameters of the posterior distribution of the CDHMM parameters by using $D \to 1$ for each diagonal component. We also summarize the solution of the MAP estimates of HMM parameters of the diagonal covariance Gaussian case below. The hyperparameters of the posterior distributions are represented with Gaussian sufficient statistics and the prior hyperparameters as:

$$\begin{cases} \hat{\phi}_j^{\pi} &\triangleq \phi_j^{\pi} + \xi_1(j), \\ \hat{\phi}_{ij}^{a} &\triangleq \phi_{ij}^{a} + \sum_{t=1}^{T-1} \xi_t(i,j), \\ \hat{\phi}_{jk}^{\omega} &\triangleq \phi_{jk}^{\omega} + \sum_{t=1}^{T} \gamma_t(j,k), \\ \hat{\phi}_{jk}^{\boldsymbol{\mu}} &\triangleq \phi_{jk}^{\boldsymbol{\mu}} + \sum_{t=1}^{T} \gamma_t(j,k), \\ \hat{\boldsymbol{\mu}}_{jk} &\triangleq \frac{\phi_{jk}^{\boldsymbol{\mu}}\boldsymbol{\mu}_{jk}^{0} + \sum_{t=1}^{T}\gamma_t(j,k)\mathbf{o}_t}{\phi_{jk}^{\boldsymbol{\mu}} + \sum_{t=1}^{T}\gamma_t(j,k)}, \\ \hat{\phi}_{jk}^{\mathbf{R}} &\triangleq \phi_{jk}^{\mathbf{R}} + \sum_{t=1}^{T} \gamma_t(j,k), \\ \hat{r}_{jkd} &\triangleq \sum_{t=1}^{T} \gamma_t(j,k) o_{td}^2 + \phi_{jk}^{\boldsymbol{\mu}}(\mu_{jkd}^{0})^2 - \hat{\phi}_{jk}^{\boldsymbol{\mu}}(\hat{\mu}_{jkd})^2 + r_{jkd}^{0}. \end{cases} \tag{4.80}$$

In this case, a set of hyperparameters $\hat{\Psi}$ is defined as follows:

$$\hat{\Psi} \triangleq \{\hat{\phi}_j^{\pi}, \hat{\phi}_{ij}^{a}, \hat{\phi}_{jk}^{\omega}, \hat{\phi}_{jk}^{\boldsymbol{\mu}}, \hat{\phi}_{jk}^{\mathbf{R}}, \hat{\boldsymbol{\mu}}_{jk}, \hat{\mathbf{r}}_{jk} | i = 1, \cdots, J, j = 1, \cdots, J, k = 1, \cdots, K\}, \quad (4.81)$$

where

$$\hat{\mathbf{r}} \triangleq [\hat{r}_1, \cdots, \hat{r}_D]^{\mathsf{T}}. \quad (4.82)$$

Then, the MAP solutions of HMM parameters can be represented with the posterior hyperparameters as follows:

$$\begin{cases} \pi_j^{\text{MAP}} = \dfrac{\hat{\phi}_j^{\pi} - 1}{\sum_{j'=1}^{J}(\hat{\phi}_{j'}^{\pi} - 1)}, \\ a_{ij}^{\text{MAP}} = \dfrac{\hat{\phi}_{ij}^{a} - 1}{\sum_{j'=1}^{J}(\hat{\phi}_{ij'}^{a} - 1)}, \\ \omega_{jk}^{\text{MAP}} = \dfrac{\hat{\phi}_{jk}^{\omega} - 1}{\sum_{k'=1}^{K}(\hat{\phi}_{jk'}^{\omega} - 1)}, \\ \boldsymbol{\mu}_{jk}^{\text{MAP}} = \hat{\boldsymbol{\mu}}_{jk}, \\ \Sigma_{jkd}^{\text{MAP}} = \dfrac{\hat{r}_{jkd}}{\hat{\phi}_{jk}^{\mathbf{R}} - 2}. \end{cases} \quad (4.83)$$

Thus, we obtain the MAP estimates of HMM parameters in both diagonal and full covariance matrix cases.

4.3.6 Sufficient statistics

As we discussed in Eq. (4.43), the posterior probabilities of the state transition $\xi_t(i,j)$, and mixture occupation $\gamma_t(j,k)$ can be computed by plugging the MAP estimates Θ^{MAP} obtained by using Eqs. (4.79) or (4.83) into the variables in Sections 3.3 and 3.4.2. Note that the analytical results here are exactly the same as those in the ML–EM algorithm (except for the MAP estimates Θ^{MAP}): we list these for convenience.

First, the MAP forward variable $\alpha_t(j)$ is computed by using the following equation:

- Initialization

$$\begin{aligned} \alpha_1(j) &= p(\mathbf{o}_1, s_1 = j | \Theta^{\text{MAP}}) \\ &= \pi_j^{\text{MAP}} b_j^{\text{MAP}}(\mathbf{o}_1), \quad 1 \leq j \leq J. \end{aligned} \quad (4.84)$$

- Induction

$$\begin{aligned} \alpha_t(j) &= p(\mathbf{o}_1, \cdots, \mathbf{o}_t, s_t = j | \Theta^{\text{MAP}}) \\ &= \left(\sum_{i=1}^{J} \alpha_{t-1}(i) a_{ij}^{\text{MAP}} \right) b_j^{\text{MAP}}(\mathbf{o}_t), \quad \begin{matrix} 2 \leq t \leq T \\ 1 \leq j \leq J. \end{matrix} \end{aligned} \quad (4.85)$$

- Termination

$$p(\mathbf{O}|\Theta^{\text{MAP}}) = \sum_{j=1}^{J} \alpha_T(j). \tag{4.86}$$

Here, $b_j^{\text{MAP}}(\mathbf{o}_t)$ is a GMM emission probability distribution with the MAP estimate parameters defined as:

$$b_j^{\text{MAP}}(\mathbf{o}_t) \triangleq \sum_{k=1}^{K} \omega_{jk}^{\text{MAP}} \mathcal{N}(\mathbf{o}_t|\boldsymbol{\mu}_{jk}^{\text{MAP}}, \boldsymbol{\Sigma}_{jk}^{\text{MAP}}). \tag{4.87}$$

The MAP backward variable $\beta_t(j)$ is computed by using the following equations:

- Initialization

$$\beta_T(j) = 1, \quad 1 \leq j \leq J. \tag{4.88}$$

- Induction

$$\begin{aligned}\beta_t(i) &= p(\mathbf{o}_{t+1}, \cdots, \mathbf{o}_T|s_t = i, \Theta^{\text{MAP}}) \\ &= \sum_{j=1}^{J} a_{ij}^{\text{MAP}} b_j^{\text{MAP}}(\mathbf{o}_{t+1})\beta_{t+1}(j), \\ & t = T-1, T-2, \cdots, 1, \quad 1 \leq i \leq J.\end{aligned} \tag{4.89}$$

- Termination

$$\begin{aligned}\beta_0 &\triangleq p(\mathbf{O}|\Theta^{\text{MAP}}) \\ &= \sum_{j=1}^{J} \pi_j^{\text{MAP}} b_j^{\text{MAP}}(\mathbf{o}_1)\beta_1(j).\end{aligned} \tag{4.90}$$

Therefore, based on the MAP forward and backward variables, we can compute the posterior probabilities as follows:

$$\xi_t(i,j) = \frac{\alpha_t(i)a_{ij}^{\text{MAP}}\left(\sum_{k=1}^{K} \omega_{jk}^{\text{MAP}} \mathcal{N}(\mathbf{o}_t|\boldsymbol{\mu}_{jk}^{\text{MAP}}, \boldsymbol{\Sigma}_{jk}^{\text{MAP}})\right)\beta_{t+1}(j)}{\sum_{i'=1}^{J}\sum_{j'=1}^{J} \alpha_t(i')a_{i'j'}^{\text{MAP}}\left(\sum_{k=1}^{K} \omega_{j'k}^{\text{MAP}} \mathcal{N}(\mathbf{o}_t|\boldsymbol{\mu}_{j'k}^{\text{MAP}}, \boldsymbol{\Sigma}_{j'k}^{\text{MAP}})\right)\beta_{t+1}(j')}, \tag{4.91}$$

$$\gamma_t(j,k) = \frac{\alpha_t(j)\beta_t(j)}{\sum_{j'=1}^{J} \alpha_t(j')\beta_t(j')} \cdot \frac{\omega_{jk}^{\text{MAP}} \mathcal{N}(\mathbf{o}_t|\boldsymbol{\mu}_{jk}^{\text{MAP}}, \boldsymbol{\Sigma}_{jk}^{\text{MAP}})}{\sum_{k'=1}^{K} \omega_{jk'}^{\text{MAP}} \mathcal{N}(\mathbf{o}_t|\boldsymbol{\mu}_{jk'}^{\text{MAP}}, \boldsymbol{\Sigma}_{jk'}^{\text{MAP}})}. \tag{4.92}$$

Once we have computed the posterior probabilities, we can compute the following sufficient statistics:

$$\begin{cases} \sum_{t=1}^{T-1} \xi_t(i,j) & \triangleq \xi_{ij}, \\ \sum_{t=1}^{T} \gamma_t(j,k) & \triangleq \gamma_{jk}, \\ \sum_{t=1}^{T} \gamma_t(j,k)\mathbf{o}_t & \triangleq \boldsymbol{\gamma}_{jk}^{(1)}, \\ \sum_{t=1}^{T} \gamma_t(j,k)\mathbf{o}_t\mathbf{o}_t^{\mathsf{T}} & \triangleq \boldsymbol{\Gamma}_{jk}^{(2)}, \\ \sum_{t=1}^{T} \gamma_t(j,k)o_{td}^2 & \triangleq \gamma_{jkd}^{(2)}. \end{cases} \tag{4.93}$$

The superscripts $^{(1)}$ and $^{(2)}$ denote the 1st and 2nd order statistics.

This total EM algorithm for iteratively estimating the HMM parameters based on the MAP estimate is set out in Algorithm 8. Compared with the ML Baum–Welch algorithm, Algorithm 4, it requires the hyperparameters Ψ of the CDHMMs. Section 7.3 also introduces a variant of the Baum–Welch algorithm based on variational Bayes. Note

Algorithm 8 MAP Baum–Welch algorithm

Require: Ψ and $\Theta^{\mathrm{MAP}} \leftarrow \Theta^{\mathrm{init}}$
1: **repeat**
2: Compute the forward variable $\alpha_t(j)$ from the forward algorithm
3: Compute the backward variable $\beta_t(j)$ from the backward algorithm
4: Compute the occupation probabilities $\gamma_1(j)$, $\gamma_t(j,k)$, and $\xi_t(i,j)$
5: Accumulate the sufficient statistics $\xi(i,j)$, $\gamma(j,k)$, $\boldsymbol{\gamma}_{jk}^{(1)}$, and $\boldsymbol{\Gamma}_{jk}^{(2)}$ (or $\gamma_{jkd}^{(2)}$)
6: Estimate the new hyperparameters $\hat{\Psi}$
7: Estimate the new HMM parameters $(\Theta^{\mathrm{MAP}})'$
8: Update the HMM parameters $\Theta^{\mathrm{MAP}} \leftarrow (\Theta^{\mathrm{MAP}})'$
9: **until** Convergence

again that the MAP E-step (computing forward variables, occupation probabilities, and accumulation) is exactly the same as that of the ML E-step, and retains the nice property of the parallelization and data scalability. In addition, since the E-step computation is dominant in the algorithm, the computational costs of the ML and MAP Baum–Welch algorithms are almost same.

4.3.7 Meaning of the MAP solution

This section discusses the meaning of the MAP solution obtained by Eqs. (4.79) and (4.83). We consider the two extreme case of the MAP solution, where the amount of data is small and large. That is, we consider the small data limit as $\xi_{ij}, \gamma_{jk} \rightarrow 0$. On the other hand, the large data limit corresponds to $\xi_{ij}, \gamma_{jk} \rightarrow \infty$.

- Mixture weight

  We first focus on the MAP estimate of the mixture weight ω, but the discussion can be applied to the state transition a.

  Large sample:

  The MAP estimate of the mixture weight is represented as follows:

$$\omega_{jk}^{\mathrm{MAP}} = \frac{\phi_{jk}^{\omega} + \gamma_{jk} - 1}{\sum_{k'=1}^{K}(\phi_{jk'}^{\omega} + \gamma_{jk'} - 1)} = \frac{\gamma_{jk}\left(\frac{\phi_{jk}^{\omega}-1}{\gamma_{jk}} + 1\right) - 1}{\sum_{k'=1}^{K}\gamma_{jk'}\left(\frac{\phi_{jk'}^{\omega}-1}{\gamma_{jk'}} + 1\right) - 1}. \tag{4.94}$$

Since $\frac{\phi^{\omega}_{jk}-1}{\gamma_{jk}} \rightarrow 0$ in the large sample case, the MAP estimate $\omega^{\mathrm{MAP}}_{jk}$ approaches the ML estimate $\omega^{\mathrm{ML}}_{jk}$ when γ_{jk} is sufficiently larger than $\phi^{\omega}_{jk} - 1$:

$$\omega^{\mathrm{MAP}}_{jk} \approx \frac{\gamma_{jk}}{\sum_{k'=1}^{K} \gamma_{jk'}} = \omega^{\mathrm{ML}}_{jk} \quad (\gamma_{jk} \gg \phi^{\omega}_{jk} - 1). \tag{4.95}$$

Small sample:
Similarly, the MAP estimate $\omega^{\mathrm{MAP}}_{jk}$ approaches the following value when γ_{jk} is sufficiently smaller than $\phi^{\omega}_{jk} - 1$:

$$\omega^{\mathrm{MAP}}_{jk} \approx \frac{\phi^{\omega}_{jk} - 1}{\sum_{k'=1}^{K} (\phi^{\omega}_{jk'} - 1)} \quad (\gamma_{jk} \ll \phi^{\omega}_{jk} - 1). \tag{4.96}$$

The weight is only computed from the prior hyperparameter ϕ^{ω}_{jk}. Thus, the mixture weight approaches the ML estimate when the amount of data is large, while it approaches the weight obtained only from the prior hyperparameters when the amount is small. Hyperparameter ϕ^{ω}_{jk} can be regarded as a scale.

- Mean
  By using Eqs. (4.93) and (4.79), the MAP estimate of the mean vector can be rewritten as follows:

$$\begin{aligned} \boldsymbol{\mu}^{\mathrm{MAP}}_{jk} &= \frac{\phi^{\mu}_{jk}\boldsymbol{\mu}^{0}_{jk} + \sum_t \gamma_t(j,k)\mathbf{o}_t}{\phi^{\mu}_{jk} + \sum_t \gamma_t(j,k)} \\ &= \frac{\phi^{\mu}_{jk}\boldsymbol{\mu}^{0}_{jk} + \boldsymbol{\gamma}^{(1)}_{jk}}{\phi^{\mu}_{jk} + \gamma_{jk}} \\ &= \frac{\frac{\phi^{\mu}_{jk}}{\gamma_{jk}}\boldsymbol{\mu}^{0}_{jk} + \boldsymbol{\mu}^{\mathrm{ML}}_{jk}}{\frac{\phi^{\mu}_{jk}}{\gamma_{jk}} + 1}. \end{aligned} \tag{4.97}$$

This equation means that the MAP estimate $\boldsymbol{\mu}^{\mathrm{MAP}}_{jk}$ is linearly interpolated by the ML estimate $\boldsymbol{\mu}^{\mathrm{ML}}_{jk}$ and the hyperparameter $\boldsymbol{\mu}^{0}_{jk}$, as shown in Figure 4.1. $\frac{\phi^{\mu}_{jk}}{\gamma_{jk}}$ is an interpolation ratio, and it has a specific meaning when the amount of data is sufficiently large ($\gamma_{jk} \gg \phi^{\omega}_{jk}$) or small ($\gamma_{jk} \ll \phi^{\omega}_{jk}$).

Large sample:

$$\boldsymbol{\mu}^{\mathrm{MAP}}_{jk} \approx \boldsymbol{\mu}^{\mathrm{ML}}_{jk}. \tag{4.98}$$

Similarly to the discussion of the mixture weight, the MAP estimate of the mean vector theoretically converges to the ML estimate.

Small sample:

$$\boldsymbol{\mu}^{\mathrm{MAP}}_{jk} \approx \boldsymbol{\mu}^{0}_{jk}. \tag{4.99}$$

This is a good property of the MAP estimate. Although the ML estimate with a small amount of data incorrectly estimates the mean vector, which degrades the

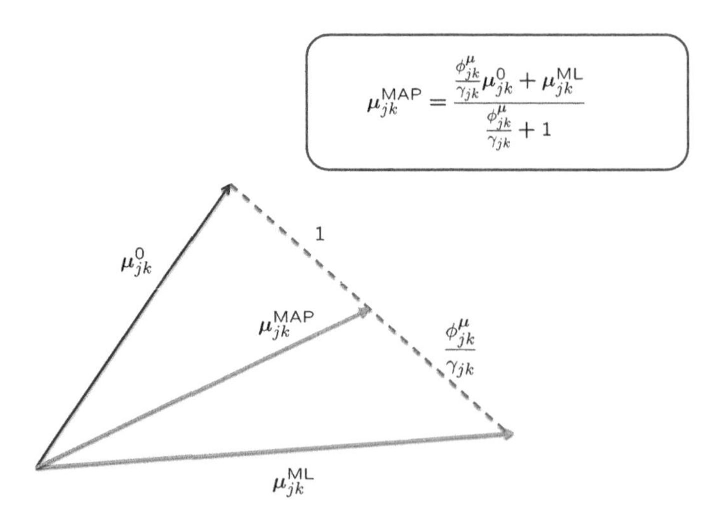

Figure 4.1 Geometric meaning of the MAP estimate of the Gaussian mean vector. It is represented as the linear interpolation of the prior mean vector $\boldsymbol{\mu}_{jk}^{0}$ and the ML estimate of the mean vector $\boldsymbol{\mu}_{jk}^{\mathrm{ML}}$. The interpolation ratio depends on the hyperparameter $\phi_{jk}^{\boldsymbol{\mu}}$ and the amount of data γ_{jk} assigned to the Gaussian.

performance drastically, the MAP estimate can smooth the incorrect estimation based on the hyperparameter $\boldsymbol{\mu}_{jk}^{0}$. In the practical situation, we also often have a zero count problem (i.e., $\gamma_{jk} = 0$), which makes the ML estimate singular due to the zero divide. However, the MAP solution of the mean vector avoids this problem and provides a reasonable estimate obtained from the hyperparameter $\boldsymbol{\mu}_{jk}^{0}$.

- Covariance matrix

By using Eqs. (4.93) and (4.79), the MAP estimate of the covariance matrix can be rewritten as follows:

$$\boldsymbol{\Sigma}_{jk}^{\mathrm{MAP}} = (\phi_{jk}^{\mathbf{R}} + \gamma_{jk} - D - 1)^{-1} \times \left(\boldsymbol{\Gamma}_{jk}^{(2)} + \phi_{jk}^{\boldsymbol{\mu}}\boldsymbol{\mu}_{jk}^{0}(\boldsymbol{\mu}_{jk}^{0})^{\mathsf{T}} - \hat{\phi}_{jk}^{\boldsymbol{\mu}}\hat{\boldsymbol{\mu}}_{jk}\hat{\boldsymbol{\mu}}_{jk}^{\mathsf{T}} + \mathbf{R}_{jk}^{0}\right). \tag{4.100}$$

Large sample:

$$\boldsymbol{\Sigma}_{jk}^{\mathrm{MAP}} \approx (\gamma_{jk})^{-1}\left(\boldsymbol{\Gamma}_{jk}^{(2)} - \gamma_{jk}\boldsymbol{\mu}_{jk}^{\mathrm{ML}}(\boldsymbol{\mu}_{jk}^{\mathrm{ML}})^{\mathsf{T}}\right) = \boldsymbol{\Sigma}_{jk}^{\mathrm{ML}}. \tag{4.101}$$

The result is the same when we use the diagonal covariance.

Small sample (full covariance):

$$\boldsymbol{\Sigma}_{jk}^{\mathrm{MAP}} \approx (\phi_{jk}^{\mathbf{R}} - D - 1)^{-1}\left(\phi_{jk}^{\boldsymbol{\mu}}\boldsymbol{\mu}_{jk}^{0}(\boldsymbol{\mu}_{jk}^{0})^{\mathsf{T}} - \phi_{jk}^{\boldsymbol{\mu}}\boldsymbol{\mu}_{jk}^{0}(\boldsymbol{\mu}_{jk}^{0})^{\mathsf{T}} + \mathbf{R}_{jk}^{0}\right) = (\phi_{jk}^{\mathbf{R}} - D - 1)^{-1}\mathbf{R}_{jk}^{0}. \tag{4.102}$$

Unlike the mean vector case, the covariance matrix of the small sample limit is represented by the two hyperparameters $\mathbf{R}_{jk}^{0}$ and $\phi_{jk}^{\mathbf{R}}$. To make the solution meaningful, we need to set $\phi_{jk}^{\mathbf{R}} > D + 1$ to avoid a negative or zero value of the variance.

- *Small sample* (diagonal covariance):
  By using Eq. (4.83), we can also obtain the variance parameter for dimension d as follows:

$$\Sigma_{jkd}^{\text{MAP}} \approx \frac{r_{jkd}^{0}}{\phi_{jk}^{\mathbf{R}} - 2}. \tag{4.103}$$

  Similarly to the diagonal case, we need to set $\phi_{jk}^{\mathbf{R}} > 2$ to avoid a negative or zero value of the variance.

In summary, the MAP solutions can smooth the estimation values with the hyperparameters when the amount of data is small, and the solutions approach the ML estimates when the amount of data is large. A similar discussion has already been presented in Section 2.1.4, as a general property of the Bayesian approach.

The MAP estimation of the CDHMM parameters can be applied to general CDHMM training. However, since CDHMM is usually trained with a sufficient amount of training data, we do not have to use MAP estimation, and ML estimation is enough in most cases, which corresponds to the case of the large sample limitation in the above discussion. However, we often face the case when the amount of data is small at an adaptation scenario. The following section introduces one of the most successful applications of the MAP estimation for *speaker adaptation*.

4.4 Speaker adaptation

Speaker adaptation is one of the most important techniques in speech recognition, mainly to deal with speaker variations in speech (Lee & Huo 2000, Shinoda 2010). The speech features of a speaker are different from those of another speaker, which degrades the performance of speech recognition. A straightforward solution for this problem is to build a speaker-dependent acoustic model for a specific person. However, it is difficult to collect sufficient training data with labels.

Speaker adaptation aims to solve the problem by first building a speaker-independent (SI) acoustic model Θ^{SI} by using many speakers' data, and updates the model as a speaker-dependent (SD) acoustic model Θ^{SD} with a small amount of data of the target speaker, as shown in Figure 4.2. The speaker-independent acoustic model is usually made by the conventional maximum likelihood procedure, as we discussed in Chapter 3, or discriminative training. It is also obtained by using so-called speaker adaptive training (Anastasakos, McDonough, Schwartz *et al.* 1996) or cluster adaptive training (Gales, Center & Heights 2000), which normalizes the speaker characteristics (or some other characteristics (e.g., noises, speaking styles) obtained from clustering of speech utterances) by using a variant of the maximum likelihood linear regression (MLLR) technique, which is discussed in Section 3.5.1.

4.4.1 Speaker adaptation by a transformation of CDHMM

Once we have SI model parameters Θ^{SI}, the problem is how to estimate the SD model parameters Θ^{SD} without over-training. Basically, the number of SI model parameters

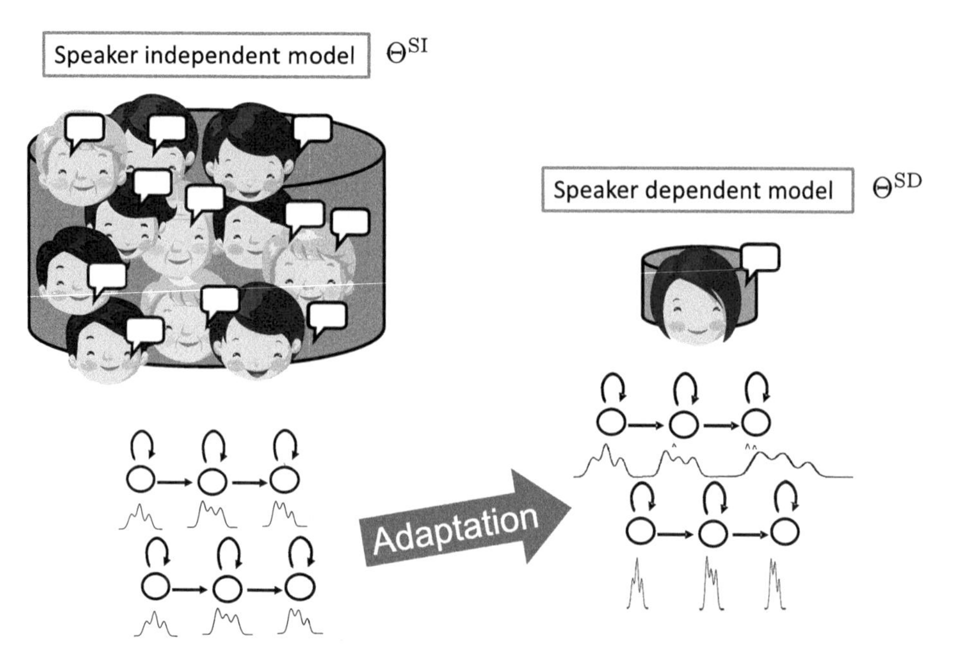

Figure 4.2 Speaker adaptation of HMM parameters. The initial HMMs are trained with many speakers, and then the HMMs are adapted to the target speaker's model with a small amount of adaptation data.

is very large. For example, in the famous speech recognition task using read speech of *Wall Street Journal* (*WSJ*) sentences, the number of CDHMM parameters amounts to several millions or more. On the other hand, the amount of speech data for the target speaker with text labels would be a few minutes at most, and the number of frames corresponds to the order of tens of thousands, and is even smaller than the number of standard CDHMM parameters. The following ML estimate with the EM algorithm, as we discussed in Section 3.4, causes serious over-training:

$$\Theta^{\mathrm{SD,\,ML}} = \arg\max_{\Theta^{\mathrm{SD}'}} Q^{\mathrm{ML}}(\Theta^{\mathrm{SD}'}|\Theta^{\mathrm{SD}}). \tag{4.104}$$

There are several approaches to overcoming the problem by using the maximum likelihood linear regression (MLLR) (Digalakis *et al.* 1995, Leggetter & Woodland 1995, Gales & Woodland 1996), as discussed in Section 3.5, eigenvoice approaches (Kuhn, Junqua, Ngyuen *et al.* 2000), and so on. These approaches set a parametric constraint of fewer CDHMM parameters, and estimate these parameters (Λ) instead of CDHMM parameters with ML indirectly, that is:

$$\Lambda^{\mathrm{ML}} = \arg\max_{\Lambda'} Q^{\mathrm{ML}}(\Lambda'|\Lambda;\Theta^{\mathrm{SI}}). \tag{4.105}$$

Detailed discussions of these adaptation techniques can be found in Lee & Huo (2000) and Shinoda (2010). We can also consider the Bayesian treatment of this indirect estimation of transformation parameters Λ, which is discussed in Section 7.4.

4.4.2 MAP-based speaker adaptation

In speaker adaptation using the MAP estimation (MAP adaptation), we directly estimate the SD CDHMM parameters Θ^{SD}, unlike the MLLR and eigenvoice techniques.[5] The MAP estimation can avoid the over-training problem. Then, we use SI CDHMM parameters Θ^{SI} as hyperparameters of the prior distributions, i.e.,

$$\Theta^{\text{SD, MAP}} = \arg\max_{\Theta^{\text{SD}'}} Q^{\text{MAP}}(\Theta^{\text{SD}'}|\Theta^{\text{SD}}; \Psi(\Theta^{\text{SI}})), \tag{4.106}$$

$$= \arg\max_{\Theta^{\text{SD}'}} Q^{\text{ML}}(\Theta^{\text{SD}'}|^{\text{SD}}) + \log p(\Theta^{\text{SD}'}|\Psi(\Theta^{\text{SI}})), \tag{4.107}$$

where $p(\Theta^{\text{SD}'}|\Psi(\Theta^{\text{SI}}))$ is a prior distribution with hyperparameters of the prior distribution, and it is set as a conjugate distribution of CDHMM, as discussed in Section 4.3.3. We discuss below how to set Θ^{SI} to hyperparameters Ψ in detail.

Let π_j^{SI}, a_{ij}^{SI}, ω_{jk}^{SI}, $\boldsymbol{\mu}_{jk}^{\text{SI}}$, and Σ_{jkd}^{SI} be the SI CDHMM parameters with diagonal covariance. Although there are several ways to determine hyperparameters from the speaker-independent HMM parameters, we can set the following relationship between hyperparameters and SI parameters by using Eqs. (4.96), (4.99), and (4.103):

$$\begin{cases} \dfrac{\phi_j^{\pi}-1}{\sum_{j'=1}^{J}(\phi_{j'}^{\pi}-1)} & = \pi_j^{\text{SI}}, \\ \dfrac{\phi_{ij}^{a}-1}{\sum_{j'=1}^{J}(\phi_{ij'}^{a}-1)} & = a_{ij}^{\text{SI}}, \\ \dfrac{\phi_{jk}^{\omega}-1}{\sum_{k'=1}^{K}(\phi_{jk'}^{\omega}-1)} & = \omega_{jk}^{\text{SI}}, \\ \boldsymbol{\mu}_{jk}^{0} & = \boldsymbol{\mu}_{jk}^{\text{SI}}, \\ \dfrac{r_{jkd}^{0}}{\phi_{jk}^{\mathbf{R}}-2} & = \Sigma_{jk}^{\text{SI}}. \end{cases} \tag{4.108}$$

This equation is obtained based on the constraint that we can obtain the SI performance when the amount of adaptation data for the target speaker is zero. To satisfy the above equations, we can use the following hyperparameter setting:

$$\begin{cases} \phi_j^{\pi} & = \lambda\pi_j^{\text{SI}} + 1, \\ \phi_{ij}^{a} & = \lambda a_{ij}^{\text{SI}} + 1, \\ \phi_{jk}^{\omega} & = \lambda\omega_{jk}^{\text{SI}} + 1, \\ \phi_{jk}^{\boldsymbol{\mu}} & = \phi, \\ \boldsymbol{\mu}_{jk}^{0} & = \boldsymbol{\mu}_{jk}^{\text{SI}}, \\ r_{jkd}^{0} & = \Sigma_{jkd}^{\text{SI}}(\phi - 2). \end{cases} \tag{4.109}$$

[5] There are several approaches combining indirect adaptation via the estimation of transformation parameters and MAP-based direct estimation of CDHMM parameters (Digalakis & Neumeyer 1996, Takahashi & Sagayama 1997).

Note that the above hyperparameter setting has two additional parameters ϕ and λ. These are often set with fixed values (e.g., $\phi = 10$, $\lambda = 1$). Thus, by substituting Eq. (4.109) into Eq. (4.80) and (4.83), the MAP estimates of SD HMM parameters are obtained as:

$$\begin{cases} \pi_j^{\text{SD, MAP}} &= \frac{\hat{\phi}_j^{\pi} - 1}{\sum_{j'=1}^{J} (\hat{\phi}_{j'}^{\pi} - 1)} \\ &= \frac{\lambda \pi_j^{\text{SI}} + \xi_1(j)}{\sum_{j'=1}^{J} (\lambda \pi_{j'}^{\text{SI}} + \xi_1(j'))}, \\ a_{ij}^{\text{SD, MAP}} &= \frac{\hat{\phi}_{ij}^{a} - 1}{\sum_{j'=1}^{J} (\hat{\phi}_{ij'}^{a} - 1)} \\ &= \frac{\lambda a_{ij}^{\text{SI}} + \sum_{t=1}^{T} \xi_t(i,j)}{\sum_{j'=1}^{J} (\lambda a_{ij'}^{\text{SI}} + \sum_{t=1}^{T} \xi_t(i,j'))}, \\ \omega_{jk}^{\text{SD, MAP}} &= \frac{\hat{\phi}_{jk}^{\omega} - 1}{\sum_{k'=1}^{K} (\hat{\phi}_{jk'}^{\omega} - 1)} \\ &= \frac{\lambda \omega_{jk}^{\text{SI}} + \sum_{t=1}^{T} \gamma_t(j,k)}{\sum_{k'=1}^{K} (\lambda \omega_{jk'}^{\text{SI}} + \sum_{t=1}^{T} \gamma_t(j,k'))}, \\ \boldsymbol{\mu}_{jk}^{\text{SD, MAP}} &= \hat{\boldsymbol{\mu}}_{jk} \\ &= \frac{\phi \boldsymbol{\mu}_{jk}^{\text{SI}} + \sum_{t=1}^{T} \gamma_t(j,k) \mathbf{o}_t}{\phi + \sum_{t=1}^{T} \gamma_t(j,k)}, \\ \Sigma_{jkd}^{\text{SD, MAP}} &= \frac{\hat{r}_{jkd}}{\hat{\phi}_{jk}^{\mathbf{R}} - 2} \\ &= \frac{\sum_{t=1}^{T} \gamma_t(j,k) o_{td}^2 + \phi (\mu_{jkd}^{\text{SI}})^2 - (\phi + \sum_{t=1}^{T} \gamma_t(j,k)) (\mu_{jkd}^{\text{SD, MAP}})^2 + \Sigma_{jkd}^{\text{SI}} (\phi - 2)}{\phi + \sum_{t=1}^{T} \gamma_t(j,k) - 2}. \end{cases} \tag{4.110}$$

Gauvain & Lee (1994) compare speaker adaptation performance by employing ML and MAP estimations of acoustic model parameters using the DARPA Naval Resources Management (RM) task (Price, Fisher, Bernstein *et al.* 1988). With 2 minutes of adaptation data, the ML word error rate was 31.5 % and was worse than the speaker independent word error rate (13.9 %) due to the over-training effect. However, the MAP word error rate was 8.7 %, clearly showing the effectiveness of the MAP approach. MAP estimation has also been used in speaker verification based on universal background models (Reynolds, Quatieri & Dunn 2000), which is described in Section 4.6, and in the discriminative training of acoustic models in speech recognition as a parameter smoothing technique (Povey 2003), which is described in the next section.

4.5 Regularization in discriminative parameter estimation

This section describes another well-known application of MAP estimation in discriminative training of CDHMM parameters. Discriminative training is based on discriminative criteria, which minimizes the ASR errors directly rather than maximizing likelihood values (Juang & Katagiri 1992), and improves the performance further from the ML-based CDHMM. However, discriminative training of CDHMM parameters always has a problem of over-estimation, and the regularization effect of the MAP estimation helps to avoid this problem. Discriminative training has been studied by many researchers, and there are many approaches to realize it for ASR based on different discriminative criteria and optimization techniques (e.g., maximum mutual information (MMI) criterion (Bahl, Brown, de Souza *et al.* 1986), MMI with extended Baum–Welch algorithm (Normandin 1992), minimum classification error (MCE) criterion (Juang & Katagiri 1992), MCE with various gradient methods (McDermott, Hazen, Le Roux *et al.* 2007), minimum phone error (MPE) criterion (Povey & Woodland 2002), and the unified interpretation of these techniques (Schlüter, Macherey, Müller *et al.* 2001, Nakamura, McDermott, Watanabe *et al.* 2009)).

This section explains the regularization effect of the MMI estimation of HMM parameters with the extended Baum–Welch algorithm (Povey & Woodland 2002). In this section, we limit the discussion of discriminative training to focus on introducing the application of MAP estimation.

4.5.1 Extended Baum–Welch algorithm

The MMI estimation of HMM parameters can be performed by the extended Baum–Welch algorithm or variants of gradient based methods. The MMI estimation starts from the following objective function based on the posterior distribution of the word sequence:

$$\begin{aligned}\mathcal{F}^{\text{MMI}}(\Theta) &= \sum_{r=1}^{R} \log p(W_r|\mathbf{O}_r;\Theta) \\ &= \sum_{r=1}^{R} \log \frac{\sum_{S_{W_r}} \left(p\left(\mathbf{O}_r, S_{W_r}|\Theta\right)\right)^{\kappa} p_L(W_r)}{\sum_{W}\sum_{S_W} \left(p\left(\mathbf{O}_r, S_W|\Theta\right)\right)^{\kappa} p_L(W)},\end{aligned} \tag{4.111}$$

where $\mathbf{O}_r = \{\mathbf{o}_t|t = 1, \cdots, T_r\}$ is the rth utterance's acoustic feature sequence whose length is T_r. The total number of the utterances is R. W_r is a correct word sequence of the utterance r, and S_{W_r} is a set of all possible state sequences given W_r.[6] Similarly, W is a word sequence hypothesis, and the summation over W is performed among all possible word sequences. κ is the acoustic score scale, and $p\left(\mathbf{O}_r, S_{W_r}|\Theta\right)$ is an acoustic likelihood, and p_L is the language model probability. Θ is a set of all acoustic model (CDHMM) parameters for all context-dependent phonemes, unlike the definition of the CDHMM parameters for single context-dependent phonemes in Section 4.2. The MMI estimate of Θ can be obtained by optimizing this objective function as follows:

[6] The summation over state sequences S_{W_r} in the numerator in Eq. (4.111) is often approximated by the Viterbi sequence without the summation obtained by the Viterbi algorithm in Section 3.3.2, i.e.,

$$\sum_{S_{W_r}} p\left(\mathbf{O}_r, S_{W_r}|\Theta\right) \approx \max_{S_{W_r}} p\left(\mathbf{O}_r, S_{W_r}|\Theta\right). \tag{4.112}$$

Similarly, the exact summation over W in the denominator is almost impossible in the large-scale ASR, and it is also approximated by the summation over pruned word sequences in a lattice, which is obtained after the ASR decoding process.

$$\Theta^{\mathrm{DT}} = \arg\max_{\Theta} \mathcal{F}^{\mathrm{MMI}}(\Theta). \tag{4.113}$$

When we only consider the numerator of Eq. (4.111) in this optimization, that corresponds to the maximum likelihood estimation of the CDHMM parameters.

By using the extended Baum–Welch algorithm, a new mean vector and variance at dimension d are iteratively updated from the previously estimated $\mu_{jkd}^{\mathrm{DT}}[\tau]$ and $\Sigma_{jkd}^{\mathrm{DT}}[\tau]$[7] at the τ iteration step, as follows:

$$\begin{aligned}
\boldsymbol{\mu}_{jk}^{\mathrm{DT}}[\tau+1] &= \frac{\boldsymbol{\gamma}_{jk}^{(1),\mathrm{num}} - \boldsymbol{\gamma}_{jk}^{(1),\mathrm{den}} + D\boldsymbol{\mu}_{jk}^{\mathrm{DT}}[\tau]}{\gamma_{jk}^{\mathrm{num}} - \gamma_{jk}^{\mathrm{den}} + D}, \\
\Sigma_{jkd}^{\mathrm{DT}}[\tau+1] &= \frac{\gamma_{jkd}^{(2),\mathrm{num}} - \gamma_{jkd}^{(2),\mathrm{den}} + D\left(\Sigma_{jkd}^{\mathrm{DT}}[\tau] + \left(\mu_{jkd}^{\mathrm{DT}}[\tau]\right)^2\right)}{\gamma_{jk}^{\mathrm{num}} - \gamma_{jk}^{\mathrm{den}} + D} - \left(\mu_{jkd}^{\mathrm{DT}}[\tau+1]\right)^2.
\end{aligned} \tag{4.114}$$

The derivation of the extended Baum–Welch algorithm can also be found in Section 5.2.8. Here, $\gamma_{jk}^{\mathrm{num}}$, $\boldsymbol{\gamma}_{jk}^{(1),\mathrm{num}}$, and $\gamma_{jkd}^{(2),\mathrm{num}}$ are the Gaussian sufficient statistics defined in Eq. (4.93), but these are obtained from the numerator of the lattice. Similarly, $\gamma_{jk}^{\mathrm{den}}$, $\boldsymbol{\gamma}_{jk}^{(1),\mathrm{den}}$, and $\gamma_{jkd}^{(2),\mathrm{den}}$ are obtained from the denominator of the lattice. D is a smoothing parameter used with the previous estimated parameters.

By comparison with the ML estimates of $\boldsymbol{\mu}$ and Σ in Eq. (3.151), which is only computed from the Gaussian sufficient statistics, the MMI estimates are computed from $\boldsymbol{\mu}_{jk}^{\mathrm{DT}}[\tau]$ and $\Sigma_{jkd}^{\mathrm{DT}}[\tau]$, and the numerator and denominator statistics. This is the main difference between ML and MMI estimation methods. However, by setting D, $\gamma_{jk}^{\mathrm{den}}$, $\boldsymbol{\gamma}_{jk}^{(1),\mathrm{den}}$, and $\gamma_{jkd}^{(2),\mathrm{den}}$ to 0, Eq. (4.114) is close to the ML estimates if we consider that the numerator statistics can be regarded as the statistics used in the ML–EM, i.e.,

$$\begin{aligned}
\lim_{D,\gamma^{\mathrm{den}}\to 0} \boldsymbol{\mu}_{jk}^{\mathrm{DT}}[\tau+1] &= \frac{\boldsymbol{\gamma}_{jk}^{(1),\mathrm{num}}}{\gamma_{jk}^{\mathrm{num}}} \approx \boldsymbol{\mu}_{jk}^{\mathrm{ML}}, \\
\lim_{D,\gamma^{\mathrm{den}}\to 0} \Sigma_{jkd}^{\mathrm{DT}}[\tau+1] &= \frac{\gamma_{jkd}^{(2),\mathrm{num}}}{\gamma_{jk}^{\mathrm{num}}} - \left(\frac{\gamma_{jkd}^{(1),\mathrm{num}}}{\gamma_{jk}^{\mathrm{num}}}\right)^2 \approx \Sigma_{jkd}^{\mathrm{ML}}.
\end{aligned} \tag{4.115}$$

Therefore, the MMI estimate can also involve the ML-like solution in a specific limitation.

We can further provide an interesting interpretation of the MMI estimate. First we focus on the following difference statistics between the numerator and denominator statistics:

$$\begin{aligned}
\delta_{jk} &\triangleq \gamma_{jk}^{\mathrm{num}} - \gamma_{jk}^{\mathrm{den}}, \\
\boldsymbol{\delta}_{jk}^{(1)} &\triangleq \boldsymbol{\gamma}_{jk}^{(1),\mathrm{num}} - \boldsymbol{\gamma}_{jk}^{(1),\mathrm{den}}, \\
\delta_{jkd}^{(2)} &\triangleq \gamma_{jkd}^{(2),\mathrm{num}} - \gamma_{jkd}^{(2),\mathrm{den}}.
\end{aligned} \tag{4.116}$$

[7] Note again that Σ means the diagonal component of the covariance matrix, and does not mean the standard deviation σ.

Then, we can rewrite Eq. (4.114) with these difference statistics as follows:

$$\boldsymbol{\mu}_{jk}^{\mathrm{DT}}[\tau+1] = \frac{\boldsymbol{\delta}_{jk}^{(1)} + D\boldsymbol{\mu}_{jk}^{\mathrm{DT}}[\tau]}{\delta_{jk}+D},$$

$$\Sigma_{jkd}^{\mathrm{DT}}[\tau+1] = \frac{\delta_{jkd}^{(2)} + D\left(\Sigma_{jkd}^{\mathrm{DT}}[\tau] + \left(\mu_{jkd}^{\mathrm{DT}}[\tau]\right)^2\right)}{\delta_{jk}+D} - \left(\mu_{jkd}^{\mathrm{DT}}[\tau+1]\right)^2. \tag{4.117}$$

Therefore, Eq. (4.117) means that the MMI estimates are represented by the linear interpolation between the difference-based Gaussian statistics and the previously estimated parameters. D plays a role of tuning the linear interpolation ratio.

Note that the MMI estimates are based on the difference statistics, and if the denominator statistics are large, the difference statistics become small, and the MMI estimates would meet an over-training problem. The smoothing based on the D with the previous estimated parameters could mitigate the over-training problem, and the combination of the MMI estimation with the MAP estimation can further mitigate it.

4.5.2 MAP interpretation of i-smoothing

In MMI and MPE training (Povey & Woodland 2002), the following smoothing terms are introduced for the numerator statistics in Eq. (4.114):

$$\begin{aligned}
\gamma_{jk}^{'\mathrm{num}} &= \gamma_{jk}^{\mathrm{num}} + \eta, \\
\boldsymbol{\gamma}_{jk}^{'(1),\mathrm{num}} &= \boldsymbol{\gamma}_{jk}^{(1),\mathrm{num}} + \eta\boldsymbol{\mu}_{jk}^{0}, \\
\gamma_{jkd}^{'(2),\mathrm{num}} &= \gamma_{jkd}^{(2),\mathrm{num}} + \eta\left((\mu_{jkd}^{0})^2 + \Sigma_{jkd}^{0}\right),
\end{aligned} \tag{4.118}$$

where η is called the i-smoothing factor. This section reviews this statistics update, which can be interpreted as the MAP estimation where η behaves as a hyperparameter in the MAP estimation. μ_{jkd}^{0} and Σ_{jkd}^{0} are obtained from the maximum likelihood estimation (i.e., $\mu_{jkd}^{0} = \mu_{jkd}^{\mathrm{ML}}$ and $\Sigma_{jkd}^{0} = \Sigma_{jkd}^{\mathrm{ML}}$), or estimation based on discriminative training.

To derive the above smoothing factor, we first consider the conjugate distribution of the diagonal-covariance Gaussian distribution, which is based on the Gaussian–gamma distribution, as shown in Table 2.1. The Gaussian–gamma distribution is defined in Appendix C.13 as follows:

$$\begin{aligned}
&\mathcal{N}\mathrm{Gam}(\mu, r|\mu^0, \phi^\mu, r^0, \phi^r) \\
&= C_{\mathcal{N}\mathrm{Gam}}(\phi^\mu, r^0, \phi^r) r^{\frac{\phi^r-1}{2}} \exp\left(-\frac{r^0 r}{2} - \frac{\phi^\mu r(\mu-\mu^0)^2}{2}\right),
\end{aligned} \tag{4.119}$$

where we omit state index j, mixture component k, and dimension index d for simplicity. By setting hyperparameters μ^0, ϕ^μ, r^0, and ϕ^r with the following variables:

$$\begin{cases} \phi^{\mu} = \eta, \\ r^0 = \Sigma^0 \eta, \\ \phi^r = \eta + 2, \end{cases} \tag{4.120}$$

the prior distribution is represented as follows:

$$\begin{aligned} p(\mu, \Sigma) &= \mathcal{N}\mathrm{Gam}(\mu, r|\mu^0, \eta, \Sigma^0\eta, \eta + 2) \\ &\propto r^{\frac{\eta+\frac{1}{2}}{2}} \exp\left(-\frac{\Sigma^0 r\eta}{2} - \frac{\eta r(\mu - \mu^0)^2}{2}\right). \end{aligned} \tag{4.121}$$

By using this prior distribution similarly to the MAP auxiliary function in Section 4.2, the MMI objective function with the prior distribution can be obtained as follows:

$$\mathcal{F}^{\mathrm{MMI}}(\mu, \Sigma) + \log p(\mu, \Sigma). \tag{4.122}$$

Based on the extended Baum–Welch calculation with the additional prior distribution, we can obtain the following update equation (Povey & Woodland 2002):

$$\begin{aligned} \mu^{\mathrm{DT}}[\tau + 1] &= \frac{\gamma^{(1),\mathrm{num}} + \eta\mu^0 - \gamma^{(1),\mathrm{den}} + D\mu^{\mathrm{DT}}[\tau]}{\gamma^{\mathrm{num}} + \eta - \gamma^{\mathrm{den}} + D}, \\ \Sigma^{\mathrm{DT}}[\tau + 1] &= \frac{\gamma^{(2),\mathrm{num}} + \eta\left((\mu^0)^2 + \Sigma^0\right) - \gamma^{(2),\mathrm{den}} + D\left(\Sigma^{\mathrm{DT}}[\tau] + \left(\mu^{\mathrm{DT}}[\tau]\right)^2\right)}{\gamma^{\mathrm{num}} + \eta - \gamma^{\mathrm{den}} + D} \\ &\quad - \left(\mu^{\mathrm{DT}}[\tau + 1]\right)^2. \end{aligned} \tag{4.123}$$

This equation is based on Eq. (4.117), with the effect of prior distribution $p(\mu, \Sigma)$ through Eq. (4.118).

Below, we discuss this update equation with the MAP solution by following the similar discussion in the previous section based on the ML–EM conversion. By setting D, γ^{den}, $\gamma^{(1),\mathrm{den}}$, and $\gamma^{(2),\mathrm{den}}$ to 0, Eq. (4.123) is represented as follows:

$$\lim_{D,\gamma^{\mathrm{den}} \to 0} \mu^{\mathrm{DT}}[\tau + 1] = \frac{\gamma^{(1),\mathrm{num}} + \eta\mu^0}{\gamma^{\mathrm{num}} + \eta}, \tag{4.124}$$

$$\begin{aligned} \lim_{D,\gamma^{\mathrm{den}} \to 0} \Sigma^{\mathrm{DT}}[\tau + 1] &= \frac{\gamma^{(2),\mathrm{num}} + \eta\left((\mu^0)^2 + \Sigma^0\right)}{\gamma^{\mathrm{num}} + \eta} \\ &\quad - \left(\frac{\gamma^{(1),\mathrm{num}} + \eta\mu^0}{\gamma^{\mathrm{num}} + \eta}\right)^2. \end{aligned} \tag{4.125}$$

By comparing the MAP solutions for μ and Σ in Eq. (4.83), we find that Eqs. (4.124) and (4.125) correspond to the MAP solutions. Thus, we have found that the i-smoothing in the MMI and MPE solutions can be interpreted as MAP. From Eq. (4.123), the smoothing terms that come from D can also be similarly interpreted as MAP, when we consider the following Gaussian–gamma prior distribution:

$$p(\mu, \Sigma) = \mathcal{N}\mathrm{Gam}(\mu, r|\mu^{\mathrm{DT}}[\tau], D, \Sigma^{\mathrm{DT}}[\tau]D, D + 2). \tag{4.126}$$

Therefore, we could also provide the MAP interpretation of the D-related terms in the extended Baum–Welch algorithm. However, how to provide this prior distribution in the objective function is not trivial, and the theoretical analysis of this interpretation in the discriminative training framework is an interesting open question.

Povey & Woodland (2002) use this i-smoothing technique with MMI and MPE estimation methods, and report 1% absolute WER improvement from the MMI estimation method without the i-smoothing technique. In addition, using the speaker independent HMM parameters as prior hyperparameters (Povey, Gales, Kim *et al.* 2003, Povey, Woodland & Gales 2003) also realizes discriminative acoustic model adaptation based on the MAP estimation. There are several studies of using Bayesian approaches to discriminative training of acoustic models (e.g., based on minimum relative entropy discrimination (Kubo, Watanabe, Nakamura *et al.* 2010)), and Section 5.2 also introduces the Bayesian sensing HMM with discriminative training based on the evidence framework.

4.6 Speaker recognition/verification

In this section we focus on text-independent speaker recognition or speaker verification systems, and show how MAP estimation is used. Speaker recognition is a similar problem to automatic speech recognition. Let $\mathbf{O} \in \mathbb{R}^D$ be a sequence of D dimensional speech feature vectors. Usually $\mathbf{O}$ is one utterance by a specific speaker. Similarly to speech recognition, MFCC is usually used as a feature. The speaker recognition task is to estimate speaker label $\hat{c}$ among a speaker set $\mathcal{C}$ by using the maximum a-posteriori estimation for the posterior distribution:

$$\hat{c} = \arg\max_{c \in \mathcal{C}} p(c|\mathbf{O}), \tag{4.127}$$

where $p(c|\mathbf{O})$ is obtained from a statistical speaker model, and is discussed later. This is similar to a speech recognition problem, as shown in Eq. (3.2) by replacing the estimation target from the word sequence W with the speaker index c. Since the output is not structured compared with W, speaker recognition can be realized by relatively simple models compared with ASR.

Speaker verification is to determine whether $\mathbf{O}$ is spoken by a target speaker s, which is regarded as single-speaker detection. This is reformulated as a basic test between two hypotheses:

- H_0 : $\mathbf{O}$ is from the hypothesized speaker s;
- H_1 : $\mathbf{O}$ is *not* from the hypothesized speaker s.

And the verification is performed by comparing the posterior distributions of H_0 and H_1 as follows:

$$\frac{p(H_0|\mathbf{O})}{p(H_1|\mathbf{O})} \begin{cases} \geq \epsilon & \text{accept} \quad H_0 \\ < \epsilon & \text{reject} \quad H_0 \end{cases}, \tag{4.128}$$

where ϵ is a decision threshold. By using the Bayes rule, we can rewrite Eq. (4.128) with a likelihood ratio as follows:

$$\frac{p(H_0|\mathbf{O})}{p(H_1|\mathbf{O})} = \frac{p(\mathbf{O}|H_0)p(H_0)}{p(\mathbf{O}|H_1)p(H_1)} \approx \frac{p(\mathbf{O}|H_0)}{p(\mathbf{O}|H_1)}, \tag{4.129}$$

where we disregard the contribution of the prior distributions of each hypothesis. Thus, by using the generative model of the H_0 and H_1, we can compare the two hypotheses. We use GMM as these generative models. Hence, $p(\mathbf{O}|H_0)$ and $p(\mathbf{O}|H_1)$ are represented from $p(\mathbf{O}|\Theta_{H_0})$ and $p(\mathbf{O}|\Theta_{H_1})$, where Θ_{H_0} and Θ_{H_1} are sets of GMM parameters.

4.6.1 Universal background model

The generative model of H_1 must consider the characteristics of many speakers. That can be achieved by training the GMM parameters Θ_{H_1} from many speakers. The GMM of H_1 is called the *universal background model* (UBM), and Θ^{UBM} denotes its GMM parameters. Therefore, the likelihood for test data $\mathbf{O}$ can be computed by:

$$p(\mathbf{O}|H_1) = p(\mathbf{O}|\Theta^{\mathrm{UBM}}) = \prod_{t=1}^{T}\sum_{k=1}^{K} \omega_k^{\mathrm{UBM}} \mathcal{N}(\mathbf{o}_t|\boldsymbol{\mu}_k^{\mathrm{UBM}}, \boldsymbol{\Sigma}_k^{\mathrm{UBM}}), \tag{4.130}$$

where Θ^{UBM} are computed from many training data $\mathcal{O}$ in advance, uttered by various speakers to train Θ^{UBM} with the ML training as follows:

$$\Theta^{\mathrm{UBM}} = \arg\max_{\Theta} p(\mathcal{O}|\Theta), \tag{4.131}$$

which can be performed efficiently by using the EM algorithm, as we discussed in Section 3.4.

While we use many data $\mathcal{O}$ to train Θ^{UBM}, the hypothesis speaker model H_0 with model parameters Θ^{HYP} can be trained by using only a small amount of data $\mathbf{O}$ (e.g., one utterance). Although ML has an over-training problem in this setting, MAP estimation can avoid the problem, and estimate Θ^{HYP} as follows:

$$\begin{aligned}\Theta^{\mathrm{HYP}} &= \arg\max_{\Theta} p(\Theta|\mathbf{O}) \\ &= \arg\max_{\Theta} p(\mathbf{O}|\Theta)p(\Theta|\Psi(\Theta^{\mathrm{UBM}})).\end{aligned} \tag{4.132}$$

where $p(\Theta|\Psi(\Theta^{\mathrm{UBM}}))$ is a prior distribution and $\Psi(\Theta^{\mathrm{UBM}})$ are hyperparameters of GMM parameters. Note that some of the hyperparameters are obtained from the UBM parameters Θ^{UBM}. This is a very similar technique to MAP adaptation of CDHMM parameters, as we discussed in Section 4.4.2, where the target model is based on a speaker-dependent CDHMM while the prior model is based on a speaker-independent CDHMM. Equation (4.132) is a subset solution of CDHMM, as we discussed in Section 4.3, and the Θ^{HYP} is obtained as follows:

$$\begin{cases} \omega_k^{\mathrm{HYP}} &= \frac{\hat{\phi}_k^{\omega}-1}{\sum_{k'=1}^{K}(\hat{\phi}_{k'}^{\omega}-1)}, \\ \boldsymbol{\mu}_k^{\mathrm{HYP}} &= \hat{\boldsymbol{\mu}}_k, \\ \Sigma_{kd}^{\mathrm{HYP}} &= \frac{\hat{r}_{kd}}{\hat{\phi}_k^{\mathbf{R}}-2}, \end{cases} \tag{4.133}$$

where

$$\begin{cases} \hat{\phi}_k^{\omega} & \triangleq \phi_k^{\omega} + \sum_{t=1}^{T} \gamma_t(k), \\ \hat{\phi}_k^{\mu} & \triangleq \phi_k^{\mu} + \sum_{t=1}^{T} \gamma_t(k), \\ \hat{\mu}_k & \triangleq \frac{\phi_k^{\mu}\mu_k^0 + \sum_{t=1}^{T} \gamma_t(k)\mathbf{o}_t}{\phi_k^{\mu} + \sum_{t=1}^{T} \gamma_t(k)}, \\ \hat{\phi}_k^{\mathbf{R}} & \triangleq \phi_k^{\mathbf{R}} + \sum_{t=1}^{T} \gamma_t(k), \\ \hat{r}_{kd} & \triangleq \sum_{t=1}^{T} \gamma_t(k) o_{td}^2 + \phi_k^{\mu}(\mu_{kd}^0)^2 - \hat{\phi}_k^{\mu}(\hat{\mu}_{kd})^2 + r_{kd}^0. \end{cases} \tag{4.134}$$

Reynolds *et al.* (2000) suggest using specific hyperparameter settings for ϕ_k^{ω}, ϕ_k^{μ}, μ_k^0, r_{kd}^0 to obtain the following forms:

$$\begin{cases} \omega_k^{\text{HYP}} = \frac{\alpha_k^w \gamma_k / T + (1 - \alpha_k^w)\omega_k^{\text{UBM}}}{\sum_{k'=1}^{K} \alpha_{k'}^w \gamma_{k'} / T + (1 - \alpha_{k'}^w)\omega_{k'}^{\text{UBM}}}, \\ \mu_k^{\text{HYP}} = \alpha_k^m \gamma_k^{(1)} + (1 - \alpha_k^m)\mu_k^{\text{UBM}}, \\ \Sigma_{kd}^{\text{HYP}} = \alpha_k^v \gamma_{kd}^{(2)} + (1 - \alpha_k^v)((\mu_{kd}^{\text{UBM}})^2 + \Sigma_{kd}^{\text{UBM}}) - (\mu_{kd}^{\text{HYP}})^2, \end{cases} \tag{4.135}$$

where α_k^w, α_k^m, and α_k^v are hyperparameters, and can be controlled by a tuning parameter β, as follows:

$$\alpha_k^{w,m,v} = \frac{\gamma_k}{\gamma_k + \beta}. \tag{4.136}$$

Note that this solution also has the MAP property of avoiding sparse data problems.

The hypothesis test in Eq. (4.128) can be performed by considering the likelihood ratio test of UBM and HYP GMMs as

$$\frac{p(\mathbf{O}|\Theta^{\text{HYP}})}{p(\mathbf{O}|\Theta^{\text{UBM}})} \begin{cases} \geq \epsilon & \text{accept} \quad H_0 \\ < \epsilon & \text{reject} \quad H_0. \end{cases} \tag{4.137}$$

Thus, we have shown that MAP estimation plays an important role in speaker verification based on UBM, especially in estimating the hypothesis speaker model.

4.6.2 Gaussian super vector

The MAP estimation of speaker models is further developed by using the Gaussian super vector technique (Campbell, Sturim & Reynolds 2006). The idea of this approach is to consider the MAP estimated GMM parameters as a feature of speaker verification or speaker recognition. The verification/recognition is performed by using a multi-class Support Vector Machine (SVM) (Vapnik 1995), cosine similarity scoring, or other simple classifier. Suppose we have $\mathbf{O}_n$ features, the GMM–UBM process can create the following super vector by concatenating the Gaussian mean vector $\{\mu_{k,n}^{\text{HYP}}|k = 1, \cdots, K\}$, estimated from $\mathbf{O}_n$:

$$\boldsymbol{\mu}_n \triangleq \begin{bmatrix} \boldsymbol{\mu}_{1,n}^{\text{HYP}} \\ \boldsymbol{\mu}_{2,n}^{\text{HYP}} \\ \vdots \\ \boldsymbol{\mu}_{K,n}^{\text{HYP}} \end{bmatrix}. \tag{4.138}$$

The super vector is also obtained by using the vectorized form of the transformation matrix estimated by using the MLLR algorithm (Stolcke *et al.* 2005). The technique is widely used for speaker and language recognition tasks (Kinnunen & Li 2010), and it can usually be used with factor analysis techniques (Kenny 2010, Dehak, Kenny, Dehak *et al.* 2011) by representing the super vector with the speaker-specific and other (channel) factors:

$$\boldsymbol{\mu}_n = \boldsymbol{\mu} + \underbrace{\mathbf{U}_1\mathbf{x}_1}_{\text{speaker}} + \underbrace{\mathbf{U}_2\mathbf{x}_{2n}}_{\text{channel}} + \boldsymbol{\epsilon}_n, \tag{4.139}$$

where the speaker and channel specific factors are also represented by the linear model with the transformation matrices $\mathbf{U}_1$ and $\mathbf{U}_2$. $\boldsymbol{\mu}$ is a bias vector, and $\boldsymbol{\epsilon}_n$ is a noise vector. The approach is also applied to video processing (Shinoda & Inoue 2013). Thus, MAP estimation is still used as an important component of speaker verification tasks, but the techniques have been developed further based on the above factor analysis. Section 7.5 describes a VB solution of this factor analysis.

4.7 *n*-gram adaptation

MAP estimation is also used to obtain the n-gram language model (Federico 1996, Masataki, Sagisaka, Hisaki *et al.* 1997). In the n-gram language model, the generative probability of the word sequence $w_1^N = \{w_i \in \mathcal{V} | i = 1, \cdots, N\}$ with vocabulary $\mathcal{V}$ can be basically represented as a product of multinomial distributions, as discussed in Section 3.6:

$$\begin{aligned} p_{\Theta}(w_1^N) &= \prod_{i=1}^{N} p(w_i|w_1^{i-1}) \approx \prod_{i=1}^{N} p(w_i|w_{i-n+1}^{i-1}) \\ &\approx \prod_{i=1}^{N} \text{Mult}(w_i|\theta_{w_i|w_{i-n+1}^{i-1}}), \end{aligned} \tag{4.140}$$

where $\Theta = \{\theta_{w_i|w_{i-n+1}^{i-1}}\}$ denotes the n-gram parameters. As discussed in Eq. (3.197), the ML estimate of Θ is obtained by using the number of occurrences $c(w_{i-n+1}^{i})$ of word sequence w_{i-n+1}^{i} in training corpus $\mathcal{D}$:

$$\begin{aligned} \theta^{\text{ML}}_{w_i|w_{i-n+1}^{i-1}} &= \arg \max_{\theta_{w_i|w_{i-n+1}^{i-1}}} p(\mathcal{D}|\theta_{w_i|w_{i-n+1}^{i-1}}) \\ &= \frac{c(w_{i-n+1}^{i})}{\sum_{w_i} c(w_{i-n+1}^{i})}. \end{aligned} \tag{4.141}$$

Note that we do not consider the smoothing techniques in this section to make the discussion simple.

4.7.1 MAP estimation of *n*-gram parameters

The MAP extension from the above ML framework can be performed by considering the posterior distribution and introducing the prior distribution as follows:

$$\begin{aligned}\theta^{\mathsf{MAP}}_{w_i|w^{i-1}_{i-n+1}} &= \arg\max_{\theta_{w_i|w^{i-1}_{i-n+1}}} p(\theta_{w_i|w^{i-1}_{i-n+1}}|\mathcal{D}) \\ &= \arg\max_{\theta_{w_i|w^{i-1}_{i-n+1}}} p(\mathcal{D}|\theta_{w_i|w^{i-1}_{i-n+1}})p(\theta_{w_i|w^{i-1}_{i-n+1}}).\end{aligned} \tag{4.142}$$

Since $p(\mathcal{D}|\theta_{w_i|w^{i-1}_{i-n+1}})$ is a multinomial distribution, as discussed in Section 2.1.4, we use the following Dirichlet distribution for $p(\theta_{w_i|w^{i-1}_{i-n+1}})$ in Appendix C.4:

$$p(\theta_{w_i|w^{i-1}_{i-n+1}}) = \mathrm{Dir}(\{\theta_{w_i|w^{i-1}_{i-n+1}}\}_{w_i}|\{\phi_{w_i|w^{i-1}_{i-n+1}}\}_{w_i}). \tag{4.143}$$

Thus, we can analytically solve Eq. (4.142) as follows:

$$\theta^{\mathsf{MAP}}_{w_i|w^{i-1}_{i-n+1}} = \frac{\phi_{w_i|w^{i-1}_{i-n+1}} - 1 + c(w^i_{i-n+1})}{\sum_{w_i}\phi_{w_i|w^{i-1}_{i-n+1}} - 1 + c(w^i_{i-n+1})}. \tag{4.144}$$

This is a similar result to the MAP solutions of mixture weights or transition probabilities in Section 4.3.5. Since the *n*-gram parameter estimation in this setting does not include the latent variables, we can obtain the solution without using the EM algorithm. Therefore, the difference between Eq. (4.144) and those in Section 4.3.5 is between using discrete counts $c(w^i_{i-n+1})$ or EM-based expected counts γ and ξ, which are continuous values. Note that the parameters represented by a Dirichlet distribution always satisfy the sum-to-one condition required for *n*-gram language modeling.

4.7.2 Adaptation method

Similarly to MAP estimation based speaker adaptation for HMM parameters, MAP estimation of *n*-gram parameters can be used for speaker/task adaptations (Federico 1996, Masataki *et al.* 1997). Let $\mathcal{D}^{\mathrm{SI}}$ be the speaker (or task) independent corpus, and $\mathcal{D}^{\mathrm{SD}}$ be the speaker (or task) dependent corpus. The following hyperparameter setting is often used:

$$\phi_{w_i|w^{i-1}_{i-n+1}} = \alpha\sum_{w_i} c^{\mathrm{SI}}(w^i_{i-n+1})\theta^{\mathrm{SI,\,ML}}_{w_i|w^{i-1}_{i-n+1}} + 1. \tag{4.145}$$

Here $\theta^{\mathrm{SI,\,ML}}_{w_i|w^{i-1}_{i-n+1}}$ is obtained from the ML estimation in Eq. (4.141) by using $\mathcal{D}^{\mathrm{SI}}$. Similarly, $c^{\mathrm{SI}}(w^i_{i-n+1})$ is a word count obtained from $\mathcal{D}^{\mathrm{SI}}$. Then, the numerator of the MAP estimation in Eq. (4.144) is rewritten as:

$$
\begin{aligned}
\theta^{\mathsf{MAP}}_{w_i|w_{i-n+1}^{i-1}} &\propto \alpha \sum_{w_i} c^{\mathrm{SI}}(w_{i-n+1}^{i})\theta^{\mathrm{SI,\,ML}}_{w_i|w_{i-n+1}^{i-1}} + c(w_{i-n+1}^{i}) \\
&\propto \alpha \sum_{w_i} c^{\mathrm{SI}}(w_{i-n+1}^{i})\theta^{\mathrm{SI,\,ML}}_{w_i|w_{i-n+1}^{i-1}} + \sum_{w_i} c^{\mathrm{SD}}(w_{i-n+1}^{i})\theta^{\mathrm{SD,\,ML}}_{w_i|w_{i-n+1}^{i-1}} \\
&\propto \frac{\alpha \sum_{w_i} c^{\mathrm{SI}}(w_{i-n+1}^{i})}{\alpha \sum_{w_i} c^{\mathrm{SI}}(w_{i-n+1}^{i}) + \sum_{w_i} c^{\mathrm{SD}}(w_{i-n+1}^{i})}\theta^{\mathrm{SI,\,ML}}_{w_i|w_{i-n+1}^{i-1}} \\
&\quad + \frac{\sum_{w_i} c^{\mathrm{SD}}(w_{i-n+1}^{i})}{\alpha \sum_{w_i} c^{\mathrm{SI}}(w_{i-n+1}^{i}) + \sum_{w_i} c^{\mathrm{SD}}(w_{i-n+1}^{i})}\theta^{\mathrm{SD,\,ML}}_{w_i|w_{i-n+1}^{i-1}}.
\end{aligned} \tag{4.146}
$$

Note that

$$
\begin{aligned}
&\frac{\sum_{w_i} c^{\mathrm{SD}}(w_{i-n+1}^{i})}{\alpha \sum_{w_i} c^{\mathrm{SI}}(w_{i-n+1}^{i}) + \sum_{w_i} c^{\mathrm{SD}}(w_{i-n+1}^{i})} \\
&\quad + \frac{\alpha \sum_{w_i} c^{\mathrm{SI}}(w_{i-n+1}^{i})}{\alpha \sum_{w_i} c^{\mathrm{SI}}(w_{i-n+1}^{i}) + \sum_{w_i} c^{\mathrm{SD}}(w_{i-n+1}^{i})} = 1.
\end{aligned} \tag{4.147}
$$

Therefore, Eq. (4.146) can be regarded as a well-known linear interpolation of two n-gram language model parameters $\theta^{\mathrm{SI,\,ML}}$ and $\theta^{\mathrm{SD,\,ML}}$, i.e.,

$$
\begin{aligned}
\theta^{\mathsf{MAP}}_{w_i|w_{i-n+1}^{i-1}} &= \alpha(w_{i-n+1}^{i})\theta^{\mathrm{SI,\,ML}}_{w_i|w_{i-n+1}^{i-1}} \\
&\quad + \left(1 - \alpha(w_{i-n+1}^{i})\right)\theta^{\mathrm{SD,\,ML}}_{w_i|w_{i-n+1}^{i-1}},
\end{aligned} \tag{4.148}
$$

where $\alpha(w_{i-n+1}^{i})$ is a linear interpolation ratio defined as:

$$
\alpha(w_{i-n+1}^{i}) \triangleq \frac{\alpha \sum_{w_i} c^{\mathrm{SI}}(w_{i-n+1}^{i})}{\alpha \sum_{w_i} c^{\mathrm{SI}}(w_{i-n+1}^{i}) + \sum_{w_i} c^{\mathrm{SD}}(w_{i-n+1}^{i})}. \tag{4.149}
$$

The linear interpolation ratio depends on the count of each corpus and hyperparameter α.

This linear interpolation based MAP solution can be regarded as an instance of well-known interpolation smoothing techniques in n-gram language modeling (Chen & Goodman 1999, Rosenfeld 2000), as discussed in Section 3.6.2. The analytical result shows that the linear interpolation technique can be viewed as the MAP estimation of n-gram parameters in a Bayesian sense.

4.8 Adaptive topic model

We are facing the era of big data. The volume of data collections grows vastly. Statistical document modeling becomes increasingly important in language processing areas. As addressed in Section 3.7.3, probabilistic latent semantic analysis (PLSA) has been developed to represent a set of documents according to the maximum likelihood (ML) principle. The semantics and statistics can be effectively captured for document representation. However, PLSA is highly sensitive to the target domain, which is continuously changing in real-world applications. Similarly to the adaptation of hidden Markov models to a new speaker in Section 4.3 and the adaptation of n-gram models to a new recognition task in Section 4.7, we are interested in adapting the topic-based document model using PLSA to a new application domain from a set of application-specific documents.

A Bayesian PLSA framework (Chien & Wu 2008) is presented to establish an adaptive topic model to improve document representation by incrementally extracting the up-to-date latent semantic information to match the changing domains at run time. The Dirichlet distribution is introduced to serve as the *conjugate priors* for PLSA parameters, which are multinomial distributed. The reproducible prior/posterior distributions facilitate two kinds of adaptation applications. One is *corrective training* while the other is *incremental learning*. An incremental PLSA is constructed to accomplish the parameter estimation as well as the hyperparameter updating. Differently from standard PLSA using an ML estimate, the Bayesian PLDA is capable of performing dynamic document indexing and modeling. The mechanism of adapting a topic model based on Bayesian PLSA is similar to the mechanisms of folding-in (Berry *et al.* 1995) and SVD updating (Bellegarda 2002) based on latent semantic analysis (LSA)(Berry *et al.* 1995, Bellegarda 2000), which is known as a nonparametric approach. The updating and downdating in an SVD-based LSA framework could not be directly applied for an ML-based PLSA framework. To add up-to-date or remove out-of-date knowledge, the adaptive PLSA is developed for document modeling. The goal of adaptive PLSA aims to use the newly collected documents, called adaptation documents, to adapt an existing PLSA model to match the domains of new queries or documents in information retrieval systems. In Chien & Wu (2008), adaptive PLSA is shown to be superior to adaptive LSA in information retrieval tasks. In what follows, we address the methods of maximum a-posteriori estimation and quasi-Bayes estimation designed for corrective training and incremental learning, respectively.

4.8.1 MAP estimation for corrective training

Corrective training is intended to use batch collection data to correct the ML-based PLSA parameters Θ^{ML} to fit new domain knowledge via the MAP estimation. In a topic model based on PLSA, two sets of multinomial parameters $\Theta = \{p(w_{(v)}|k), p(k|d_m)\}$ have been estimated in the training phase subject to the constraints of multinomial distributions as given in Eq. (3.304). The first one is the topic-dependent unigram probability $p(w_{(v)}|k)$ of a vocabulary word $w_{(v)}$, and the second one is the posterior probability $p(k|d_m)$ of topic k given an observed document d_m. According to MAP estimation, we adapt PLSA parameters $\Theta = \{p(w_{(v)}|k), p(k|d_m)\}$ by maximizing the a-posteriori probability or the sum of logarithms of likelihood function $p(\mathcal{D}|\Theta)$ of adaptation words and documents $\mathcal{D} = \{w_{(v)}, d_m | v = 1, \cdots, |\mathcal{V}|, m = 1, \cdots, M\}$ and prior distribution $p(\Theta)$:

$$\begin{aligned}\Theta^{\text{MAP}} &= \arg\max_{\Theta} p(\Theta|\mathcal{D}) \\ &= \arg\max_{\Theta} \log p(\mathcal{D}|\Theta) + \log p(\Theta).\end{aligned} \tag{4.150}$$

Here, prior distribution $p(\Theta)$ represents the randomness of multinomial parameters $\{p(w_{(v)}|k)\}$ and $\{p(k|d_m)\}$. Again, it is mathematically attractive to select the *conjugate prior* as the candidate for Bayesian inference. The Dirichlet distribution is known as the conjugate prior for multinomial parameters. Owing to the selection of conjugate prior, two properties of Bayesian learning could be obtained: 1) a closed-form solution for *rapid adaptation*; and 2) a reproducible prior/posterior distribution pair for *incremental learning*. Assuming parameters $\{p(w_{(v)}|k)\}$ and $\{p(k|d_m)\}$ are independent, the prior distribution of the entire parameter set based on Dirichlet density is expressed by

$$p(\Theta|\Psi) \propto \prod_{k=1}^{K} \left[\prod_{v=1}^{|\mathcal{V}|} p(w_{(v)}|k)^{\alpha_{vk}-1} \prod_{m=1}^{M} p(k|d_m)^{\beta_{km}-1} \right], \tag{4.151}$$

where $\Psi = \{\alpha_{vk}, \beta_{km}\}$ denote the hyperparameters of Dirichlet densities. Following the EM algorithm, we implement Eq. (4.151) by calculating the posterior auxiliary function

$$Q^{\text{MAP}}(\Theta'|\Theta) = \mathbb{E}_{(Z)}[\log p(\mathcal{D}, Z|\Theta')|\mathcal{D}, \Theta] + \log p(\Theta'|\Psi). \tag{4.152}$$

By imposing the constraints of multinomial parameters in Eq. (3.304) into the constrained optimization, we form the extended auxiliary function as

$$\begin{aligned} &\tilde{Q}^{\text{MAP}}(\Theta'|\Theta) \\ &\propto \sum_{k=1}^{K} \sum_{v=1}^{|\mathcal{V}|} \Bigg[\left(\sum_{m=1}^{M} c(w_{(v)}, d_m) p(z_{w_{(v)}} = k|w_{(v)}, d_m) + (\alpha_{vk} - 1) \right) \\ &\qquad \times \log p'(w_{(v)}|k) \Bigg] + \eta_w \left(1 - \sum_{v=1}^{|\mathcal{V}|} p'(w_{(v)}|k) \right) \\ &+ \sum_{k=1}^{K} \sum_{m=1}^{M} \Bigg[\left(\sum_{v=1}^{|\mathcal{V}|} c(w_{(v)}, d_m) p(z_{w_{(v)}} = k|w_{(v)}, d_m) + (\beta_{km} - 1) \right) \\ &\qquad \times \log p'(k|d_m) \Bigg] + \eta_d \left(1 - \sum_{k=1}^{K} p'(k|d_m) \right), \end{aligned} \tag{4.153}$$

which is manipulated and extended from the ML auxiliary function $Q^{\text{ML}}(\Theta'|\Theta)$ given in Eq. (3.295). In Eq. (4.153), η_w and η_d denote the Lagrange multipliers for two constraints of multinomial parameters. Then we differentiate Eq. (4.153) with respect to individual multinomial parameters $p'(w_{(v)}|k)$ and set it to zero:

$$\frac{\partial \tilde{Q}^{\text{MAP}}(\Theta'|\Theta)}{\partial p'(w_{(v)}|k)} = \frac{\sum_{m=1}^{M} c(w_{(v)}, d_m) p(z_{w_{(v)}} = k|w_{(v)}, d_m) + (\alpha_{vk} - 1)}{p'(w_{(v)}|k)} - \eta_w = 0, \tag{4.154}$$

and obtain

$$p'(w_{(v)}|k) = \frac{1}{\eta_w} \left[\sum_{m=1}^{M} c(w_{(v)}, d_m) p(z_{w_{(v)}} = k|w_{(v)}, d_m) + (\alpha_{vk} - 1) \right]. \tag{4.155}$$

By substituting this result into the constraint $\sum_{v=1}^{|\mathcal{V}|} p'(w_{(v)}|k) = 1$, we find the Lagrange parameter

$$\eta_w = \sum_{v=1}^{|\mathcal{V}|}\left[\sum_{m=1}^{M} c(w_{(v)}, d_m)p(z_{w_{(v)}} = k|w_{(v)}, d_m) + (\alpha_{vk} - 1)\right]. \tag{4.156}$$

Accordingly, we derive the MAP estimates of two PLSA parameters in closed form:

$$p^{\text{MAP}}(w_{(v)}|k) = \frac{\sum_{m=1}^{M} c(w_{(v)}, d_m)p(z_{w_{(v)}} = k|w_{(v)}, d_m) + (\alpha_{vk} - 1)}{\sum_{j=1}^{|\mathcal{V}|}\left[\sum_{m=1}^{M} c(w_{(j)}, d_m)p(z_{w_j} = k|w_{(j)}, d_m) + (\alpha_{jk} - 1)\right]}. \tag{4.157}$$

$$\begin{aligned} p^{\text{MAP}}(k|d_m) &= \frac{\sum_{v=1}^{|\mathcal{V}|} c(w_{(v)}, d_m)p(z_{w_{(v)}} = k|w_{(v)}, d_m) + (\beta_{km} - 1)}{\sum_{j=1}^{K}\left[\sum_{v=1}^{|\mathcal{V}|} c(w_{(v)}, d_m)p(z_{w_{(v)}} = j|w_{(v)}, d_m) + (\beta_{jm} - 1)\right]} \\ &= \frac{\sum_{v=1}^{|\mathcal{V}|} c(w_{(v)}, d_m)p(z_{w_{(v)}} = k|w_{(v)}, d_m) + (\beta_{km} - 1)}{c(d_m) + \sum_{j=1}^{K}(\beta_{jm} - 1)}, \end{aligned} \tag{4.158}$$

where the posterior probability $p(z_{w_{(v)}} = k|w_{(v)}, d_m)$ is calculated according to Eq. (3.311) by using adaptation documents $\mathcal{D}$ based on the current estimates $\Theta = \{p(w_{(v)}|k), p(k|d_m)\}$. MAP estimates in Eq. (4.158) are seen as an extension of ML estimates of Eq. (3.308) and Eq. (3.309) by interpolating with the prior statistics $\{\alpha_{vk}\}$ and $\{\beta_{km}\}$, respectively. If prior density is non-informative or adaptation data $\mathcal{D}$ are abundant, MAP estimates are reduced to ML estimates. The MAP PLSA algorithm is developed for corrective training or batch adaptation, and adapts the existing parameters to Θ^{MAP} in a single epoch. In MAP PLSA, the Dirichlet priors and their hyperparameters $\Psi = \{\alpha_{vk}, \beta_{km}\}$ are adopted to characterize the variations of topic-dependent document and word probabilities. These priors are used to express the environmental variations. MAP PLSA involves both word-level $p(w_{(v)}|k)$ and document-level $p(k|d_m)$ parameters. In general, MAP parameters Θ^{MAP} perform better than ML parameters Θ^{ML} when classifying future documents with new terms, topics, and domains.

4.8.2 Quasi-Bayes estimation for incremental learning

Using MAP estimation, only a single learning epoch is performed to correct PLSA parameters. Batch learning is performed. However, batch learning cannot catch the continuously changing domain knowledge or deal with the non-stationary documents collected from real-world applications. An adaptive information system should continuously update system parameters with new words and topics. Out-of-date words or documents should fade away from the system as time moves on. Accordingly, we tackle the updating and downdating problems simultaneously for latent semantic indexing. MAP PLSA could not incrementally accumulate statistics for adaptive topic modeling. It is more interesting to develop an incremental learning algorithm to track the changing topics and domains in test documents. A learning procedure is executed repeatedly in different epochs. Incremental learning is also known as sequential learning or online learning, which is important for speaker adaptation in automatic speech recognition systems where speaker characteristics gradually change with time (Huo & Lee 1997, Chien 1999).

To implement incremental learning for adaptive topic modeling, we continuously estimate PLSA parameters in different learning epochs using the incrementally observed adaptation documents. At the n learning epoch, we estimate PLSA parameters by maximizing the posterior distribution using a sequence of adaptation documents $\mathfrak{D}^n = \{\mathcal{D}_1, \cdots, \mathcal{D}_n\}$:

$$\begin{aligned}(\Theta^{(n)})^{\text{QB}} &= \arg\max_{\Theta} p(\Theta|\mathfrak{D}^n) \\ &= \arg\max_{\Theta} p(\mathcal{D}_n|\Theta)p(\Theta|\mathfrak{D}^{n-1}) \\ &\approx \arg\max_{\Theta} p(\mathcal{D}_n|\Theta)p(\Theta|\Psi^{(n-1)}),\end{aligned} \tag{4.159}$$

where the posterior distribution $p(\Theta|\mathfrak{D}^{n-1})$ is approximated by the closest tractable prior distribution $p(\Theta|\Psi^{(n-1)})$ with *sufficient statistics* or hyperparameters $\Psi^{(n-1)}$ which are evolved from history documents $\mathfrak{D}^{n-1}$. This estimation method is also called the quasi-Bayes (QB) estimation (Huo & Lee 1997, Chien 1999). QB estimation provides recursive learning of PLSA parameters,

$$\Theta^{(1)} \to \Theta^{(2)} \to \cdots \to \Theta^{(n)}, \tag{4.160}$$

from incrementally observed documents,

$$\mathcal{D}_1 \to \mathcal{D}_2 \to \cdots \to \mathcal{D}_n. \tag{4.161}$$

At each epoch, we only use the current block of documents $\mathcal{D}_n = \{w_{(v)}^{(n)}, d_m^{(n)} | v = 1, \cdots, |\mathcal{V}|, m = 1, \cdots, M_n\}$ and the accumulated statistics $\Psi^{(n-1)}$ to update PLSA parameters from $\Theta^{(n-1)}$ to $\Theta^{(n)}$. Current block data $\mathcal{D}_n$ are released after accumulating statistics from $\Psi^{(n-1)}$ to $\Psi^{(n)}$. Memory and computation requirements are reduced at each epoch. The key technique in QB PLSA comes from the introduction of *incremental hyperparameters*. If we substitute the hyperparameters $\Psi^{(n-1)} = \{\alpha_{vk}^{(n-1)}, \beta_{km}^{(n-1)}\}$ into Eq. (4.157) and Eq. (4.158), the QB estimates $(\Theta^{(n)})^{\text{QB}} = \{p^{\text{QB}}(w_{(v)}^{(n)}|k), p^{\text{QB}}(k|d_m^{(n)})\}$ are obtained. This QB PLSA method is geared with the updating mechanism of hyperparameters, which is derived by the E-step for QB estimation in Eq. (4.159). By referring to the auxiliary function of MAP PLSA in Eq. (4.152) and Eq. (4.153), the QB auxiliary function of new estimates $(\Theta^{(n)})' = \{p'(w_{(v)}^{(n)}|k), p'(k|d_m^{(n)})\}$ given current estimates $\Theta^{(n)} = \{p(w_{(v)}^{(n)}|k), p(k|d_m^{(n)})\}$ is defined by

$$
\begin{aligned}
Q^{\mathrm{QB}}&((\Theta^{(n)})'|\Theta^{(n)}) \\
&\propto \sum_{k=1}^{K}\sum_{v=1}^{|\mathcal{V}|}\Bigg[\Bigg(\sum_{m=1}^{M_n} c(w_{(v)}^{(n)}, d_m^{(n)})p(k|w_{(v)}^{(n)}, d_m^{(n)}) \\
&\qquad + (\alpha_{vk}^{(n-1)} - 1)\Bigg)\log p'(w_{(v)}^{(n)}|k)\Bigg] \\
&+ \sum_{k=1}^{K}\sum_{m=1}^{M_n}\Bigg[\Bigg(\sum_{v=1}^{|\mathcal{V}|} c(w_{(v)}^{(n)}, d_m^{(n)})p(k|w_{(v)}^{(n)}, d_m^{(n)}) \\
&\qquad + (\beta_{km}^{(n-1)} - 1)\Bigg)\log p'(k|d_m^{(n)})\Bigg].
\end{aligned}
\tag{4.162}
$$

It is important that the exponential of the QB auxiliary function in Eq. (4.162) can be arranged as a new Dirichlet distribution:

$$
\begin{aligned}
\exp&\left\{Q^{\mathrm{QB}}((\Theta^{(n)})'|\Theta^{(n)})\right\} \\
&\propto \prod_{k=1}^{K}\left[\prod_{v=1}^{\mathcal{V}} p'(w_{(v)}^{(n)}|k)^{\alpha_{vk}^{(n)}-1}\prod_{m=1}^{M_n} p'(k|d_m^{(n)})^{\beta_{km}^{(n)}-1}\right],
\end{aligned}
\tag{4.163}
$$

with the updated hyperparameters $\Psi^{(n)} = \{\alpha_{vk}^{(n)}, \beta_{km}^{(n)}\}$ derived thus:

$$
\alpha_{vk}^{(n)} = \sum_{m=1}^{M_n} c(w_{(v)}^{(n)}, d_m^{(n)})p(k|w_{(v)}^{(n)}, d_m^{(n)}) + \alpha_{vk}^{(n-1)}, \tag{4.164}
$$

$$
\beta_{km}^{(n)} = \sum_{v=1}^{|\mathcal{V}|} c(w_{(v)}^{(n)}, d_m^{(n)})p(k|w_{(v)}^{(n)}, d_m^{(n)}) + \beta_{km}^{(n-1)}, \tag{4.165}
$$

where the posterior probability

$$
p(k|w_{(v)}^{(n)}, d_m^{(n)}) = \frac{p(w_{(v)}^{(n)}|k)p(k|d_m^{(n)})}{\sum_{j=1}^{K} p(w_{(v)}^{(n)}|j)p(j|d_m^{(n)})} \tag{4.166}
$$

is obtained by using the current block of adaptation data $\mathcal{D}_n = \{w_{(v)}^{(n)}, d_m^{(n)}\}$ based on current QB estimates $\Theta^{(n)} = \{p(w_{(v)}^{(n)}|k), p(k|d_m^{(n)})\}$. Importantly, a *reproducible distribution pair* of a prior in Eq. (4.151) and a posterior in Eq. (4.163) is established. This property is crucial to activate the updating mechanism of hyperparameters for incremental learning. New hyperparameters $\Psi^{(n)} = \{\alpha_{vk}^{(n)}, \beta_{km}^{(n)}\}$ are estimated by combining the previous hyperparameters $\Psi^{(n-1)} = \{\alpha_{vk}^{(n-1)}, \beta_{km}^{(n-1)}\}$ with the accumulated statistics from adaptation documents $\mathcal{D}_n = \{w_{(v)}^{(n)}, d_m^{(n)}\}$ at learning epoch n.

Table 4.1 Numbers of training, adaptation, and test documents for five populous classes in the Reuters-21578 dataset.

	Acquisitions	Crude	Earn	Money-fx	Trade
Number of training documents	825	196	1447	284	189
Number of added documents per epoch	275	65	475	85	60
Number of test documents	719	189	1087	180	117

Basically, QB estimation finds the point estimate for adaptive topic modeling. This estimation is seen as an extended realization of MAP estimation by activating the mechanism of hyperparameter updating so that incremental learning is established to compensate the non-stationary variations in observation data which may be speech signals, word sequences, or text documents. Incremental learning based on QB estimation is helpful for speech and language applications including speech recognition, information retrieval, and others.

4.8.3 System performance

The performance of the adaptive topic model was evaluated through the tasks of corrective training and incremental learning. The evaluation was performed for the application of document categorization. Table 4.1 shows the set-up of experimental data of the Reuters-21578 dataset. We collected training, adaptation, and test documents from the five most populous categories in the Reuters-21578 dataset for system evaluation. Preprocessing stages of stemming and stop word removal were done. In the task of incremental learning, we used one-third of the adaptation documents at each epoch and investigated the effect of incremental learning in three learning epochs. The performance of corrective training and incremental learning was compared. Training samples of each category were roughly partitioned into half for training and the other half for adaptation. A fivefold cross validation over training and adaptation sets was performed. In the implementation, we determined PLSA probability for each test document. The cosine similarity of feature vectors between a test document and a given class model was calculated for pattern classification. The class feature vector consisted of PLSA probabilities averaging over all documents corresponding to a class. The classification error rate was computed over all test documents in five populous classes. We obtained the classification error rates for PLSA (Hofmann 1999*b*, 2001) (3.47%), SVD updating (Bellegarda 2002) (3.39%), MAP PLSA (Chien & Wu 2008) (3.04%), QB PLSA (Chien & Wu 2008) at 1st learning epoch (3.13%), QB PLSA at 2nd learning epoch (3.04%), and QB PLSA at 3rd learning epoch (3%). Corrective training using SVD updating and MAP PLSA and incremental learning using QB PLSA at different epochs decrease the classification error rates. SVD updating is worse than MAP PLSA and QB PLSA. Incremental learning using QB PLSA performs slightly better than batch learning using MAP PLSA.

4.9 Summary

This chapter introduced various applications of MAP approximation for model parameter posteriors. Since the approach can be easily realized from the existing ML based approaches by simply considering the regularization term based on model parameter priors, it is widely used for speech and language processing including acoustic and language modeling in ASR, speaker verification, and document processing. Although the MAP approximation can utilize the most famous Bayesian advantage of "use of prior knowledge," it does not deal with probabilistic variables explicitly with the marginalization, and it does not fully utilize the Bayesian concept. The following chapters consider more strict Bayesian approaches for speech and language processing.

5 Evidence approximation

In a maximum a-posteriori (MAP) approximation as addressed in Chapter 4, we treat model parameters Θ of a target model M as unknown but deterministic variables which are estimated by maximizing the posterior probability $p(\Theta|\mathbf{O})$ using some observation data $\mathbf{O}$. Prior distribution $p(\Theta|\Psi)$ with heuristically determined hyperparameters Ψ is introduced. This is known as the *point estimation* of a target model. However, from a full Bayesian perspective, we treat model parameters as random variables where the randomness is represented by a prior distribution $p(\Theta|\Psi)$. In contrast with point estimation in a MAP approximation based on heuristic hyperparameters, the *distribution estimation* is implemented for full Bayesian learning. According to the distribution estimation, we find the whole prior distribution $p(\Theta|\Psi)$, or equivalently estimate the corresponding hyperparameters Ψ from observations $\mathbf{O}$ in an empirical fashion. In this implementation, the *marginalization* of likelihood function $p(\mathbf{O}|\Theta)$ with respect to model parameters Θ should be calculated for model construction, as follows:[1]

$$p(\mathbf{O}|\Psi) = \int p(\mathbf{O}|\Theta)p(\Theta|\Psi)d\Theta. \tag{5.1}$$

Rather than trusting the point estimate in MAP approximation, the resulting *evidence function* $p(\mathbf{O}|\Psi)$ in *evidence framework* considers all possible values of model parameters when making a prediction of $\mathbf{O}$ as new observation data or training data depending on the task. In cases of complicated model structure and coupled latent variables, this evidence function is prone to be intractable and should be approximated to estimate the optimal hyperparameters Ψ. For the applications of speech and language processing, we focus on acoustic modeling and language modeling in accordance with the distribution estimation based on evidence approximation.

In what follows, Section 5.1 first presents the evidence framework and addresses the type-2 maximum likelihood estimation for general pattern recognition. The optimal hyperparameters are estimated based on this framework. The optimization of evidence function or marginal likelihood is then extended to *sparse Bayesian learning* for acoustic modeling based on Bayesian sensing hidden Markov models (Saon & Chien 2012*a*) in Section 5.2. In this section, the scheme of automatic relevance determination is introduced and illustrated. In addition, evidence approximation to a hierarchical Dirichlet language model (MacKay & Peto 1995) is detailed in Section 5.3. The optimal

[1] This chapter regards model structure M and parameters of prior distribution Ψ as the hyperparameters in a broad sense. Equation (5.1) formulates the likelihood function given Ψ, but the discussion can be applied to the case of using M.

hyperparameters are obtained for the acoustic model as well as the language model. These Bayesian models are beneficial for noise-robust speech recognition and large vocabulary continuous speech recognition.

5.1 Evidence framework

This section begins with a general discussion of the evidence framework. We first review the well-known Bayes theorem, as discussed in Section 2.1. The evidence function $p(\mathbf{O}|\Psi)$, given prior parameters Ψ, is introduced in the Bayes theorem, which relates posterior distribution $p(\Theta|\mathbf{O},\Psi)$ to the likelihood function $p(\mathbf{O}|\Theta)$, prior distribution $p(\Theta|\Psi)$, and the evidence function $p(\mathbf{O}|\Psi)$ by

$$\begin{aligned} p(\Theta|\mathbf{O},\Psi) &= \frac{p(\mathbf{O}|\Theta)p(\Theta|\Psi)}{p(\mathbf{O}|\Psi)} \\ &= \frac{p(\mathbf{O}|\Theta)p(\Theta|\Psi)}{\int p(\mathbf{O}|\Theta)p(\Theta|\Psi)d\Theta}. \end{aligned} \tag{5.2}$$

In words:

$$\text{Posterior} = \frac{\text{Likelihood} \times \text{Prior}}{\text{Evidence}}. \tag{5.3}$$

The evidence function is a marginal likelihood function which takes all values of model parameters Θ into account. It is precisely the normalization term that appears in the denominator in Bayes' theorem as shown in Eq. (5.2).

Although the evidence function has appeared in the previous sections (e.g., the MAP approximation in Eq. (4.2)), it has not been explicitly considered so far. However, the evidence function $p(\mathbf{O}|\Psi)$ can directly link the hyperparameters Ψ and observations $\mathbf{O}$ by marginalizing the model parameters Θ, and can be used to infer the hyperparameters Ψ in the Bayesian framework.

5.1.1 Bayesian model comparison

This section also considers the model structure M in the evidence framework. In MacKay (1992*a*), the evidence framework was proposed to conduct a Bayesian model comparison, where the best model or model structure M is selected according to the posterior distribution of M as

$$\hat{M} = \arg\max_{M} p(M|\mathbf{O}) = \arg\max_{M} p(\mathbf{O}|M)p(M). \tag{5.4}$$

Here $p(M)$ is a prior distribution of the model structure M. In the case that each model is equally probable or has uniform probability, i.e., $p(M) = \text{constant}$, different models M are ranked according to the evidence function $p(\mathbf{O}|M)$, which is obtained by marginalizing the model parameters Θ based on the product and sum rules as

$$\begin{aligned} p(\mathbf{O}|M) &= \int p(\mathbf{O},\Theta|M)d\Theta \\ &= \int p(\mathbf{O}|\Theta,M)p(\Theta|M)d\Theta. \end{aligned} \tag{5.5}$$

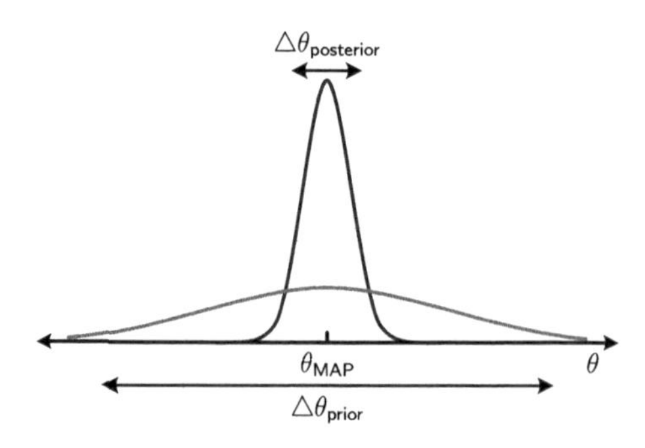

Figure 5.1 Evidence approximation. Adapted from Bishop (2006).

The marginalization of likelihood function over model parameters Θ or model structure is calculated to come out with a meaningful objective for model selection. However, we may obtain some insight into the model evidence by making a simple approximation to the integral over parameters Θ (MacKay 1992*a*, Bishop 2006). Consider the case of a model with a posterior distribution which is sharply peaked around the most probable value Θ_{MAP}, with the width $\Delta\Theta_{\text{posterior}}$. Then the posterior distribution is represented as $p(\Theta|\mathbf{O}) = \frac{1}{\Delta\Theta_{\text{posterior}}}$. Similarly, we also assume that the prior is flat with width $\Delta\Theta_{\text{prior}}$, and we have $p(\Theta) = \frac{1}{\Delta\Theta_{\text{prior}}}$. Therefore, by using the Bayes theorem, the evidence function in Eq. (5.5) can be approximated without solving the integral as:

$$\begin{aligned} p(\mathbf{O}) &= \int p(\mathbf{O}|\Theta)p(\Theta)d\Theta = \frac{p(\mathbf{O}|\Theta)p(\Theta)}{p(\Theta|\mathbf{O})} \\ &\approx p(\mathbf{O}|\Theta_{\text{MAP}})\frac{\Delta\Theta_{\text{posterior}}}{\Delta\Theta_{\text{prior}}}, \end{aligned} \tag{5.6}$$

where we omit the model structure index M for simplicity. The approximation to model evidence is illustrated by Figure 5.1. Without loss of generality, the notation Θ in Eq. (5.6) is treated as a single parameter. Taking the logarithm, we obtain

$$\log p(\mathbf{O}) \approx \log p(\mathbf{O}|\Theta^{\text{MAP}}) + \underbrace{\log\left(\frac{\Delta\Theta_{\text{posterior}}}{\Delta\Theta_{\text{prior}}}\right)}_{\text{Occam factor}}. \tag{5.7}$$

In this approximation, the first term represents the goodness-of-fit to the data $\mathbf{O}$ given the most probable parameter value Θ^{MAP}. The second term is known as the *Occam factor* (MacKay 1992*a*) which penalizes the model according to its complexity. Theoretically, we have the property $\Delta\Theta_{\text{posterior}} < \Delta\Theta_{\text{prior}}$. The Occam factor is negative and it increases in magnitude as the ratio $\frac{\Delta\Theta_{\text{posterior}}}{\Delta\Theta_{\text{prior}}}$ gets smaller. Thus, if parameters are finely tuned to the data in posterior distribution, then the penalty term is large. In practice, the model complexity or the Occam factor is multiplied by the number of adaptive parameters N in Θ. The optimal model complexity, as determined by the maximum evidence, is given by a trade-off between these two competing terms. A refined version of this evidence approximation could be further derived as:

$$\log p(\mathbf{O}) \approx \log p(\mathbf{O}|\Theta^{\mathsf{MAP}}) + \underbrace{\log p(\Theta^{\mathsf{MAP}}) + \frac{N}{2}\log(2\pi) - \frac{1}{2}\log|\mathbf{H}|}_{\text{Occam factor}}, \tag{5.8}$$

by using the *Laplace approximation*, which is detailed in Section 6.1. In Eq. (5.8), $\mathbf{H}$ is the Hessian matrix:

$$\mathbf{H} = -\nabla\nabla \log p(\mathbf{O}|\Theta^{\mathsf{MAP}})p(\Theta^{\mathsf{MAP}}) = -\nabla\nabla \log p(\Theta^{\mathsf{MAP}}|\mathbf{O}), \tag{5.9}$$

which is the second derivative of the negative log posterior. The determinant of this matrix plays an important role in the Occam factor, which penalizes the model complexity.

We can attain further insight into Bayesian model comparison and understand how the marginal likelihood is favorable to the models with intermediate complexity by considering Figure 5.2 (Bishop 2006). Here, the horizontal axis is a one-dimensional representation of the data space $\mathbf{O}$. We consider three models M_1, M_2, and M_3 in which M_1 is the simplest and M_3 is the most complex. When generating a particular data set from a specific model, we first choose the values of the parameters from their prior distribution $p(\Theta)$. Then, given these parameter values, we sample the data from $p(\mathbf{O}|\Theta)$. A simple or *shallow* model has little variability and so will generate data sets using $p(\mathbf{O})$, which is confined to a relatively small region in space $\mathbf{O}$. In contrast, a complex or *deep* model can generate a variety of data sets, and so its distribution $p(\mathbf{O})$ is spread over a large region of the data space $\mathbf{O}$. In this example, for the particular observed data set $\mathbf{O}_M$, the model M_2 with intermediate complexity has the largest evidence.

5.1.2 Type-2 maximum likelihood estimation

The complexity of a model M is generally defined by the scope of data set $\mathbf{O}$ that model M could predict. This scope is not only determined by the model size (e.g., model structure, model order, or number of parameters N), but is also affected by the hyperparameters Ψ of the parameters Θ which are used to generate the observations $\mathbf{O}$. Instead of point estimation of model parameters Θ in MAP approximation based on heuristically determined hyperparameters, the evidence approximation conducts distribution estimation which determines the optimal prior distribution $\hat{p}(\Theta|\Psi)$ as a whole, or equivalently

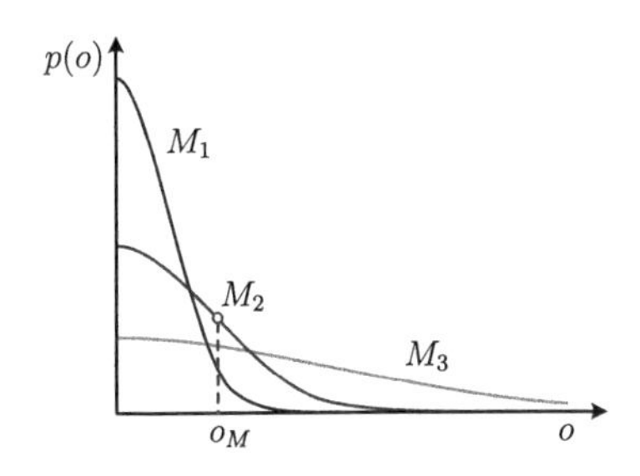

Figure 5.2 Model comparison based on the maximum evidence. Adapted from Bishop (2006).

infers the optimal hyperparameters $\hat{\Psi}$ corresponding to the prior distribution according to

$$\begin{aligned}\hat{\Psi} = \Psi^{\text{ML2}} &= \arg\max_{\Psi} p(\mathbf{O}|\Psi) \\ &= \arg\max_{\Psi} \int p(\mathbf{O}|\Theta, \Psi)p(\Theta|\Psi)d\Theta. \end{aligned} \tag{5.10}$$

Differently from conventional ML estimation for model parameters Θ, the type-2 maximum likelihood (ML2) estimation aims to search the optimal hyperparameters Ψ^{ML2} from the observation data $\mathbf{O}$. In the literature, a point estimation based on ML or MAP conducts the so-called *level-1 inference*, while the distribution estimation based on ML2 undertakes the *level-2 inference* (MacKay 1995, Kwok 2000). The level-3 inference is performed to rank different models M according to the posterior probability $p(M|\mathbf{O})$ or the evidence function $p(\mathbf{O}|M)$ with equally probable model M. Three levels of inference can be iterated.

The evidence framework or the ML2 estimation has been developed for different learning machines including linear regression/classification networks (MacKay 1992*b*, Bishop 2006), feed-forward neural network (NN) (MacKay 1992*c*), support vector machine (SVM) (Kwok 2000), and hidden Markov model (Zhang, Liu, Chien *et al.* 2009). For the cases of linear regression and neural network regression models, the optimal hyperparameters of weight parameters and modeling errors are estimated by maximizing the evidence function of training data which is marginalized with respect to the weight parameters. In Zhang *et al.* (2009), the optimal hyperparameters were derived for the mixture model of exponential family distributions and then realized to build the robust HMMs for noisy speech recognition. Practically, these hyperparameters Ψ are interpreted as the *regularization parameter* λ, which plays a crucial role in the regularized regression models. The regularized models are developed to prevent the *over-fitting problem* in conventional models based on ML estimation or least-squares regression. In what follows, the evidence framework is illustrated to be closely connected to the *regularization theory*, which has been developed to regularize model structure and deal with the over-fitting problem when building generative models for speech recognition and other information systems.

5.1.3 Regularization in regression model

Model regularization is a fundamental issue in pattern recognition. It aims to estimate the smoothed parameters or construct a generalized model which has good prediction capability for future unseen data. The gap or mismatch between training data and test data can be compensated by tackling this issue. The over-fitting problem or the ill-posed problem is also resolved by following the regularization theory. In the regularized least-squares (RLS) model $f(\cdot)$, the over-fitting problem is avoided by incorporating a regularization term $E_w(\mathbf{w})$ into the training objective, and this penalizes too complex a model. Here, model parameters Θ are rewritten by an N-dimensional weight vector $\mathbf{w}$. The regularization term is determined by the weight parameters $\mathbf{w}$ and the corresponding model structure M. The simplest form of this regularizer is given by a sum-of-squares of the parameter elements:

$$E_w(\mathbf{w}) = \frac{1}{2}\mathbf{w}^\mathsf{T}\mathbf{w} = \frac{1}{2}\|\mathbf{w}\|^2 = \frac{1}{2}\sum_{j=0}^{N-1} w_j^2. \tag{5.11}$$

The sum-of-squares error function is calculated by introducing training samples $\mathbf{O} = \{\mathbf{o}_t | t = 1, \cdots, T\}$ given by model parameters $\mathbf{w}$, i.e.

$$E_o(\mathbf{w}) = \frac{1}{2}\sum_{t=1}^{T}(f(\mathbf{o}_t, \mathbf{w}) - y_t)^2, \tag{5.12}$$

where y_t is the target value of the observation $\mathbf{o}_t$ at time t. Accordingly, the training objective for RLS parameters $\mathbf{w}^{\mathsf{RLS}}$ is yielded as a regularized least-squares function which is formed by

$$\mathbf{w}^{\mathsf{RLS}} = \arg\min_{\mathbf{w}} \{E_o(\mathbf{w}) + \lambda E_w(\mathbf{w})\}. \tag{5.13}$$

Notably, a regularization parameter λ is introduced in Eq. (5.13) to balance the trade-off between the sum-of-squares error function $E_o(\mathbf{w})$ and the model complexity penalty function $E_w(\mathbf{w})$. Minimization of the training objective in Eq. (5.13) eventually obtains a set of parameters $\mathbf{w}$ which works simultaneously towards the goals of fitting the data and reducing the norm of the solution. Regularization theory is beneficial for model selection. Regularization parameter λ is generally selected by applying the *cross validation* method using a small set of validation data which is outside the training data $\mathbf{O}$.

Nevertheless, it is more attractive to pursue Bayesian interpretation of model regularization. Considering the same regression problem, but now under a probabilistic framework, we assume that the modeling error $f(\mathbf{o}_t, \mathbf{w}) - y_t$ has a Gaussian distribution:

$$p(y_t|\mathbf{o}_t, \mathbf{w}, \beta) = \mathcal{N}(f(\mathbf{o}_t, w) - y_t)|0, \beta^{-1}), \tag{5.14}$$

and the parameter vector $\mathbf{w}$ comes from a Gaussian distribution

$$p(\mathbf{w}|\alpha) = \mathcal{N}(\mathbf{w}|\mathbf{0}, \alpha^{-1}\mathbf{I}), \tag{5.15}$$

where $\mathbf{I}$ is the $N \times N$ identity matrix and β and α are the precision parameters for modeling error and parameter vector, respectively. Here, the hyperparameters $\Psi = \{\alpha, \beta\}$ consist of α and β. The MAP estimate of model parameters $\mathbf{w}^{\mathsf{MAP}}$ is obtained by maximizing the posterior distribution

$$p(\mathbf{w}|\mathbf{o}_t, y_t, \alpha, \beta) \propto p(y_t|\mathbf{o}_t, \mathbf{w}, \beta)p(\mathbf{w}|\alpha), \tag{5.16}$$

or equivalently minimizing the negative log posterior distribution,

$$\frac{\beta}{2}\sum_{t=1}^{T}\{f(\mathbf{o}_t, \mathbf{w}) - y_t\}^2 + \frac{\alpha}{2}\mathbf{w}^\mathsf{T}\mathbf{w}, \tag{5.17}$$

by using training samples $\mathbf{O} = \{\mathbf{o}_t | t = 1, \cdots, T\}$. It is interesting to see that maximizing the posterior distribution in Eq. (5.17) is equivalent to minimizing the regularized least-squares error function in Eq. (5.13) with a regularization parameter $\lambda = \alpha/\beta$.

Readers may refer to MacKay (1992*c*) and Bishop (2006) for detailed solution to optimal hyperparameters $\Psi^{\text{ML2}} = \{\alpha^{\text{ML2}}, \beta^{\text{ML2}}\}$ of a linear regression model with regression function

$$f(\mathbf{o}_t, \mathbf{w}) = w_0 + \sum_{j=1}^{N-1} w_j \phi_j(\mathbf{o}_t) = \mathbf{w}^{\mathsf{T}} \boldsymbol{\phi}(\mathbf{o}_t). \tag{5.18}$$

A linear combination of fixed non-linear functions of the input variable $\mathbf{o}_t$ is considered. Here, $\boldsymbol{\phi} = [\phi_0, \cdots, \phi_{N-1}]^{\mathsf{T}}$ denotes the basis functions and $\phi_0(\mathbf{o}_t) = 1$ is assigned. For the case of a neural network regression model, ML2 estimation of hyperparameters is addressed in Section 6.4.

The comparison among RLS estimation, MAP estimation, and ML2 estimation is further investigated below. According to RLS estimation, level-1 inference is performed to find model parameters $\mathbf{w}^{\text{RLS}}$ by using training data $\mathbf{O}$ while the hyperparameter λ is estimated in level-2 inference via a cross validation scheme by using additional *validation data*. However, Bayesian inference is implemented to calculate model parameters $\mathbf{w}^{\text{MAP}}$ in level-1 inference based on MAP estimation. In level-2 inference, the hyperparameters $\Psi^{\text{ML2}} = \{\alpha^{\text{ML2}}, \beta^{\text{ML2}}\}$ are inferred using ML2 estimation. By applying MAP and ML2 methods, the same training data $\mathbf{O}$ are used to estimate parameters $\mathbf{w}^{\text{MAP}}$ and hyperparameters Ψ^{ML2}, respectively. It is desirable that *no validation data are required* by using the Bayesian approach. Besides, RLS and MAP methods fulfil the point estimation and assume that the estimates $\hat{\mathbf{w}} = \mathbf{w}_{\text{RLS}}$ and $\hat{\mathbf{w}} = \mathbf{w}^{\text{MAP}}$ are true values for prediction of new data $\mathbf{O}$ in a test phase according to likelihood function $p(\mathbf{O}|\hat{\mathbf{w}})$. Instead of relying on single parameter values $\hat{\mathbf{w}}$ in the RLS or MAP method, the ML2 method implements the distribution estimation and directly infers the optimal hyperparameters $\hat{\Psi} = \Psi^{\text{ML2}}$ by maximizing the predictive distribution or the marginal likelihood of training data $p(\mathbf{O}|\Psi) = \int p(\mathbf{O}|\mathbf{w}, \Psi)p(\mathbf{w}|\Psi)d\mathbf{w}$, where all possible values of parameters $\mathbf{w}$ are considered. In a test phase, the same marginal distribution $p(\mathbf{O}|\hat{\Psi})$, given the estimated hyperparameters $\hat{\Psi}$, is applied for prediction of new data $\mathbf{O}$.

5.1.4 Evidence framework for HMM and SVM

Although the evidence framework is only addressed for a linear regression model $f(\cdot)$, extensions to the classification models including HMM and SVM have been proposed in Zhang *et al.* (2009) and Kwok (2000), respectively. When ML2 estimation is applied for the HMM framework, we estimate the hyperparameters of continuous-density HMM parameters including initial state probabilities $\{\pi_j\}$, state transition probabilities $\{a_{ij}\}$, mixture weights $\{\omega_{jk}\}$, mean vectors $\{\boldsymbol{\mu}_{jk}\}$, and covariance matrices $\{\boldsymbol{\Sigma}_{jk}\}$. For the probability parameters $\{\pi_j\}$, $\{a_{ij}\}$ and $\{\omega_{jk}\}$ in multinomial distributions, ML2 estimation is performed to find the corresponding hyperparameters Ψ which are the parameters of *Dirichlet priors* for multinomial or discrete variables of states j, state pairs (i, j), and mixture components k, respectively. For the remaining Gaussian parameters $\{\boldsymbol{\mu}_{jk}\}$ and $\{\boldsymbol{\Sigma}_{jk}\}$ of continuous feature vectors $\{\mathbf{o}_t\}$, ML2 estimation aims to calculate the corresponding hyperparameters Ψ which are the parameters of *Gaussian–Wishart priors* for Gaussian mean vectors $\{\boldsymbol{\mu}_{jk}\}$ and precision (or inverse covariance) matrices $\{\boldsymbol{\Sigma}_{jk}^{-1}\}$. In general,

Dirichlet distribution is known as the *conjugate prior* for multinomial variables while Gaussian–Wishart distribution is seen as the conjugate prior for Gaussian variables. By following this guideline, the closed-form solution to the integral in marginal likelihood does exist, so that the optimization of marginal distribution with respect to individual hyperparameters has an analytical solution. These hyperparameters characterize the uncertainties of HMM parameters which could be applied for robust speech recognition according to the Bayesian predictive classification as addressed in Eq. (3.16) and Section 6.3. This approach is different from conventional BPC based on the hyperparameters which are heuristically determined or calculated in an ensemble way (Huo & Lee 2000, Chien & Liao 2001).

The support vector machine (SVM) approach is based on the idea of structural risk minimization, which shows that the generalization error is bounded by the sum of the training set error and the Vapnik-Chervonenkis (VC) dimension of the learning machine (Abu-Mostafa 1989, Vapnik 1995). By minimizing this upper bound, generalization to future data is improved. Generalization error is related not to the number of inputs, but to the margin with which it separates the data. SVM has been successfully applied in many classification problems including speech recognition (Ganapathiraju, Hamaker & Picone 2004). Although SVM is a nonparametric method, the probabilistic framework and Bayesian perspective have been introduced to deal with the selection of two tuning parameters or hyperparameters, including:

- a regularization parameter λ, which determines the trade-off between minimizing the training errors and minimizing the model complexity;
- a kernel parameter, which implicitly defines the high dimensional feature space to be used.

Conventionally, these hyperparameters are empirically selected by hand or via cross validation. The evidence framework has been applied to find the optimal regularization parameter for SVM (Kwok 2000). Next, we address the detailed estimation of hyperparameters for two practical solutions to speech recognition. One is developed for sparse Bayesian acoustic modeling while the other is proposed for hierarchical Dirichlet language modeling.

5.2 Bayesian sensing HMMs

Speech recognition systems are usually constructed by collecting large amounts of training data and estimating a large number of model parameters to achieve the desired recognition accuracy on test data. A large set of context-dependent Gaussian components (several hundred thousand components is usually the norm) is trained to build context-dependent phone models. GMMs with Gaussian mean vectors and diagonal covariance matrices may not be an accurate representation of high dimensional acoustic features. The Gaussian components may be overdetermined. The mismatch between training data and test conditions may not be carefully compensated. The uncertainty of estimated HMM parameters may not be properly characterized. A Bayesian learning approach is introduced to tackle these issues based on the *basis representation*. ML2 estimation is conducted to estimate the automatic relevance determination (ARD) parameter (MacKay 1995, Tipping 2001) which is the state-dependent hyperparameter of weight parameter in basis representation. Sparse Bayesian learning is performed by using the ARD parameter.

5.2.1 Basis representation

An acoustic feature vector $\mathbf{o}$ can be viewed as lying in a vector space spanned by a set of basis vectors. Such a basis representation has been popular for regression problems in machine learning and for signal recovery in the signal processing literature. This approach is now increasingly important for acoustic feature representation. Compressive sensing and sparse representation are popular topics in the signal processing community. The basic idea of compressive sensing is to encode a feature vector $\mathbf{o} \in \mathbb{R}^D$ based on a set of over-determined dictionary or basis vectors $\mathbf{\Phi} = [\boldsymbol{\phi}_1, \cdots, \boldsymbol{\phi}_N]$ via

$$\mathbf{o} = w_1\boldsymbol{\phi}_1 + \cdots + w_N\boldsymbol{\phi}_N = \mathbf{\Phi}\mathbf{w}, \tag{5.19}$$

where the sensing weights $\mathbf{w} = [w_1, \cdots, w_N]^\top$ are sparse and the basis vectors $\mathbf{\Phi}$ are formed by training samples. A relatively small set of relevant basis vectors is used for sparse representation based on this exemplar-based method. The sparse solution to $\mathbf{w}$ can be derived by optimizing the ℓ_1-regularized objective function (Sainath, Ramabhadran, Picheny *et al.* 2011). However, the exemplar-based method is a memory-based method, which is time-consuming with high memory cost. It is also important to integrate HMMs into sparse representation of continuous speech frames $\mathbf{O} = \{\mathbf{o}_t | t = 1, \cdots, T\}$.

5.2.2 Model construction

Bayesian sensing HMMs (BS-HMMs) (Saon & Chien 2012*a*) are developed by incorporating Markov chains into the basis representation of continuous speech. A Bayesian sensing framework is presented for speech recognition. The underlying aspect of BS-HMMs is to measure an observed feature vector $\mathbf{o}_t$ of a speech sentence $\mathbf{O}$ based on a compact set of state-dependent dictionary $\mathbf{\Phi}_j = \left[\boldsymbol{\phi}_{j1}, \cdots, \boldsymbol{\phi}_{jN}\right]$ at state j. For each frame, the reconstruction error between measurement $\mathbf{o}_t$ and its representation $\mathbf{\Phi}_j\mathbf{w}_t$, where $\mathbf{w}_t = [w_{t1}, \cdots, w_{tN}]^\top$, is assumed to be Gaussian distributed with zero mean and a state-dependent covariance matrix or inverse precision matrix $\mathbf{\Sigma}_j = \mathbf{R}_j^{-1}$. The state likelihood function with time-dependent sensing weights $\mathbf{w}_t$ is defined by

$$p(\mathbf{o}_t|\Theta_j) \triangleq \mathcal{N}(\mathbf{o}_t|\mathbf{\Phi}_j\mathbf{w}_t, \mathbf{R}_j^{-1}). \tag{5.20}$$

The Bayesian perspective in BS-HMMs has its origin from the relevance vector machine (RVM) (Tipping 2001). Figure 5.3 illustrates the graphical model based on BS-HMMs. The RVM is known as a sparse Bayesian learning approach for regression and classification problems. We would like to apply RVM to conduct sparse basis representation and combine it with HMMs to characterize the dynamics in the time domain. Therefore,

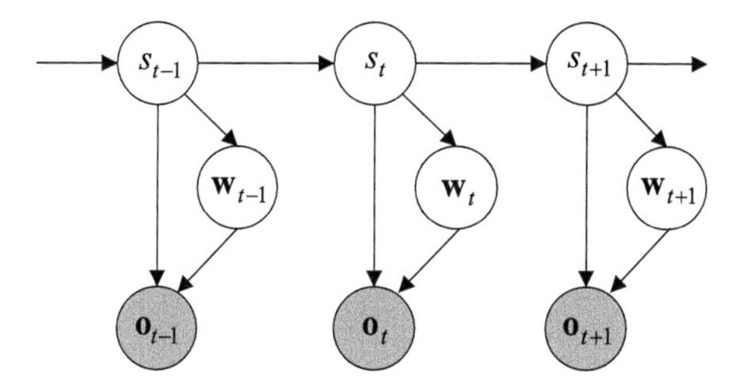

Figure 5.3 Graphical model of BS-HMM.

BS-HMM parameters are obtained by $\Theta = \{\pi_j, a_{ij}, \mathbf{\Phi}_j, \mathbf{R}_j\}$. Obviously, similarly to conventional GMM-based HMMs, we can extend BS-HMM to deal with a *mixture model* for basis representation where the mixture weight ω_{jk}, the basis vectors $\mathbf{\Phi}_{jk}$, and the precision matrix $\mathbf{R}_{jk}$ of individual mixture component k are incorporated. In what follows, we neglect the extension to a mixture model and exclude the time-dependent weight parameters $\mathbf{w}_t$ from the parameter set Θ.

5.2.3 Automatic relevance determination

However, the sensing weights $\mathbf{w}_t$ play a crucial role in basis representation, and so we introduce Bayesian compressive sensing (Ji, Xue & Carin 2008) for acoustic modeling. The idea of Bayesian learning in BS-HMMs is to yield the *distribution estimates* of the speech feature vectors $\mathbf{o}_t$ due to the variations of sensing weights $\mathbf{w}_t$ in basis representation. A Gaussian prior with zero mean and state-dependent diagonal precision matrix $\mathcal{A}_j = \text{diag}\{\alpha_{jn}\}$ is introduced to characterize the weight vector, i.e.

$$\begin{aligned} p\left(\mathbf{w}_t|\mathcal{A}_j\right) &= \mathcal{N}(\mathbf{w}_t|\mathbf{0}, \text{diag}\{\alpha_{jn}^{-1}\}) \\ &= \prod_{n=1}^{N} \mathcal{N}(w_{tn}|0, \alpha_{jn}^{-1}). \end{aligned} \tag{5.21}$$

Considering a *hierarchical prior* model where precision parameter α_{jn} is represented by a gamma prior with parameters a and b in Appendix C.11:

$$p(\alpha_{jn}) = \text{Gam}(\alpha_{jn}|a, b) = \frac{1}{\Gamma(a)} b^a \alpha_{jn}^{a-1} \exp(-b\alpha_{jn}). \tag{5.22}$$

The marginal prior distribution is derived as a Student's t-distribution as defined in Appendix C.16:

$$\begin{aligned} p(w_{tn}|a, b) &= \int_0^\infty \mathcal{N}(w_{tn}|0, \alpha_{jn}^{-1}) \text{Gam}(\alpha_{jn}|a, b) d\alpha_{jn} \\ &= \frac{b^a}{\Gamma(a)} \left(\frac{1}{2\pi}\right)^{1/2} \left(b + \frac{w_{tn}^2}{2}\right)^{-a-1/2} \Gamma(a + 1/2) \\ &= \text{St}\left(w_{tn} \middle| 0, \frac{b}{a}, 2a\right), \end{aligned} \tag{5.23}$$

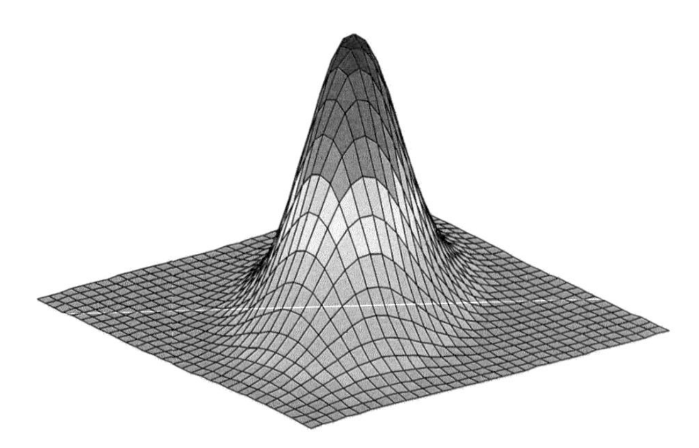

Figure 5.4 An example of two-dimensional Gaussian distribution.

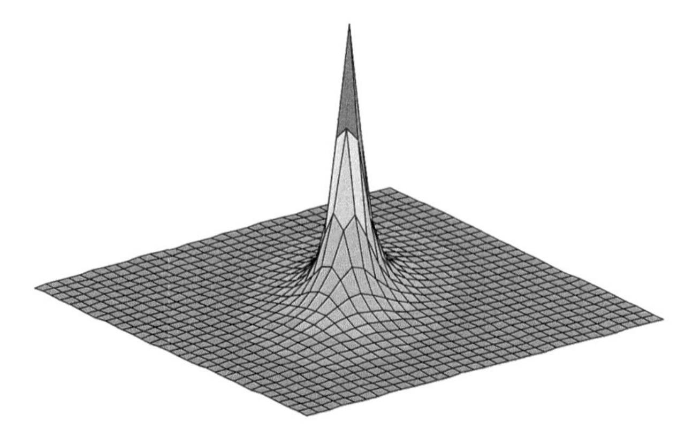

Figure 5.5 An example of two-dimensional Student's t-distribution.

which is known as a *sparse prior* distribution, since this distribution has a heavy tail and steep peak. Student's t-distribution, illustrated in Figure 5.5, is a heavy tailed distribution and is more robust to outliers than a Gaussian distribution (Figure 5.4) (Bishop 2006).

Correspondingly, if the precision parameter α_{jn} in $\mathcal{N}(w_{tn}|0, \alpha_{jn}^{-1})$ is large, the weight parameter w_{tn} is likely to be zero, $w_{tn} \to 0$, which implies that the associated basis vector $\boldsymbol{\phi}_{jn}$ is irrelevant to Bayesian basis representation of a target observation $\mathbf{o}_t$. The physical meaning of automatic relevance determination (ARD) is then reflected by the precision parameter α_{jn} in state-dependent hyperparameters, $\Psi = \{\mathcal{A}_j = \text{diag}\{\alpha_{jn}\}\}$. We simply call α_{jn} the ARD parameter. According to the ARD scheme, only relevant basis vectors are selected to represent sequence data $\mathbf{O}$. Sparse Bayesian learning (SBL) can be realized by using an ARD scheme (Tipping 2001). In the implementation, the values of state-dependent ARD parameters $\{\alpha_{j1}, \cdots, \alpha_{jN}\}$ can be used to rank or select salient basis vectors $\{\boldsymbol{\phi}_{j1}, \cdots, \boldsymbol{\phi}_{jN}\}$ which are relevant to a target HMM state j. The larger the estimated value α_{jn}, the more likely it is that the basis vector $\boldsymbol{\phi}_{jn}$ should be pruned from the parameter set. The compressed model can be achieved by applying this property (Saon & Chien 2011). The ARD parameter serves as a *compression factor* for model complexity control. One can initially train a large model and then prune it to a smaller size by removing basis elements which correspond to the larger ARD values. Although we appear to be utilizing a non-sparse Gaussian prior over the weights, in truth the hierarchical formulation implies that the real weight prior is clearly recognized as encouraging sparsity. Considering this sparse Bayesian basis representation, BS-HMM parameters and hyperparameters are formed by $\{\Theta, \Psi\} = \{\pi_j, a_{ij}, \mathbf{\Phi}_j, \mathbf{R}_j, \mathcal{A}_j | j = 1, \cdots, J\}$ consisting of initial state probability π_j, state transition probability a_{ij}, basis vectors $\mathbf{\Phi}_j$, and precision matrices of reconstruction errors $\mathbf{R}_j$ and sensing weights $\mathcal{A}_j$. Level-1 inference and level-2 inference are done simultaneously in BS-HMMs.

5.2.4 Model inference

We estimate BS-HMM parameters and hyperparameters from the observed speech data $\mathbf{O}$ according to the type-2 ML (ML2) estimation:

$$\{\pi_j^{\mathsf{ML2}}, a_{ij}^{\mathsf{ML2}}, \mathbf{\Phi}_j^{\mathsf{ML2}}, \mathbf{R}_j^{\mathsf{ML2}}, \mathcal{A}_j^{\mathsf{ML2}}\} = \arg \max_{\{\pi_j, a_{ij}, \mathbf{\Phi}_j, \mathbf{R}_j, \mathcal{A}_j\}} p(\mathbf{O} | \{\pi_j, a_{ij}, \mathbf{\Phi}_j, \mathbf{R}_j, \mathcal{A}_j\}). \tag{5.24}$$

Without loss of generality, we view Eq. (5.24) as the ML2 estimation because the marginal likelihood with respect to sensing weights $\mathbf{w}_t$ is calculated at each frame t in likelihood function $p(\mathbf{O} | \{\pi_j, a_{ij}, \mathbf{\Phi}_j, \mathbf{R}_j, \mathcal{A}_j\})$. However, the marginalization over π_j, a_{ij}, $\mathbf{\Phi}_j$, and $\mathbf{R}_j$ is not considered. Since the optimization procedure is affected by an incomplete data problem, the EM algorithm (Dempster *et al.* 1976) is applied to find the optimal solution to $\{\pi_j, a_{ij}, \mathbf{\Phi}_j, \mathbf{R}_j, \mathcal{A}_j\}$. In E-step, an auxiliary function is calculated by averaging the log likelihood function of the new estimates $\{\Theta', \Psi'\}$, given the old estimates $\{\Theta, \Psi\}$ over all latent variables $\{S, V\}$:

$$\begin{aligned} Q(\Theta', \Psi' | \Theta, \Psi) &= \mathbb{E}_{(S,V)}[\log p(\mathbf{O}, S, V | \Theta', \Psi') | \mathbf{O}, \Theta, \Psi] \\ &= \sum_S \sum_V p(S, V | \mathbf{O}, \Theta, \Psi) \log p(\mathbf{O}, S, V | \Theta', \Psi'). \end{aligned} \tag{5.25}$$

In the M-step, we maximize the auxiliary function with respect to new parameters and hyperparameters $\{\Theta', \Psi'\}$,

$$\{\Theta', \Psi'\} = \arg \max_{\{\Theta', \Psi'\}} Q(\Theta', \Psi' | \Theta, \Psi), \tag{5.26}$$

to find optimal parameters and hyperparameters. The auxiliary function is expanded by

$$\begin{aligned} &\sum_S \sum_V p(S, V | \mathbf{O}, \Theta, \Psi) \left[\sum_{t=1}^{T} (\log a'_{s_{t-1} s_t} + \log p(\mathbf{o}_t | \Theta'_{s_t}, \Psi'_{s_t})) \right] \\ &= \sum_j \sum_{t=1}^{T} \gamma_t(j) \left[\log a'_{s_{t-1} j} + \log \int p(\mathbf{o}_t | \mathbf{w}_t, \mathbf{\Phi}'_j, \mathbf{R}'_j) p(\mathbf{w}_t | \mathcal{A}'_j) d\mathbf{w}_t \right], \end{aligned} \tag{5.27}$$

where $\gamma_t(j) = p(s_t = j | \mathbf{O}, \Theta, \Psi)$ is the posterior probability of being in state j at time t given the current parameters and hyperparameters $\{\Theta, \Psi\}$ generating measurements $\mathbf{O}$.

We tacitly use the convention $a_{s_0 s_1} = \pi_{s_1}$. Since the estimation of initial state probability π_j and state transition probability a_{ij} is the same as that in HMMs, we neglect the estimation of $\{\pi_j, a_{ij}\}$ hereafter.

5.2.5 Evidence function or marginal likelihood

The key issue in the E-step is to calculate the frame-based evidence function or marginal likelihood $p(\mathbf{o}_t|\Theta'_{s_t}, \Psi'_{s_t}) = p(\mathbf{o}_t|\mathbf{\Phi}'_j, \mathbf{R}'_j, \mathcal{A}'_j)$, which is marginalized over sensing weights $\mathbf{w}_t$ at state $s_t = j$ and is proportional to

$$
\begin{aligned}
&\int |\mathbf{R}'_j|^{1/2} \exp\left[-\frac{1}{2}(\mathbf{o}_t - \mathbf{\Phi}'_j\mathbf{w}_t)^{\mathsf{T}}\mathbf{R}'_j(\mathbf{o}_t - \mathbf{\Phi}'_j\mathbf{w}_t)\right] \\
&\quad \times |\mathcal{A}'_j|^{1/2} \exp\left[-\frac{1}{2}\mathbf{w}_t^{\mathsf{T}}\mathcal{A}'_j\mathbf{w}_t\right] d\mathbf{w}_t \\
&= |\mathbf{R}'_j|^{1/2}|\mathcal{A}'_j|^{1/2} \int \exp\left[-\frac{1}{2}\left(\mathbf{o}_t^{\mathsf{T}}\mathbf{R}'_j\mathbf{o}_t\right.\right. \\
&\quad \left.\left. - 2\mathbf{o}_t^{\mathsf{T}}\mathbf{R}'_j\mathbf{\Phi}'_j\mathbf{w}_t + \mathbf{w}_t^{\mathsf{T}}\left((\mathbf{\Phi}'_j)^{\mathsf{T}}\mathbf{R}'_j\mathbf{\Phi}'_j + \mathcal{A}'_j\right)\mathbf{w}_t\right)\right] d\mathbf{w}_t \\
&= |\mathbf{R}'_j|^{1/2}|\mathcal{A}'_j|^{1/2} \exp\left[-\frac{1}{2}(\mathbf{o}_t^{\mathsf{T}}\mathbf{R}'_j\mathbf{o}_t)\right] \\
&\quad \times \int \exp\left[-\frac{1}{2}\begin{pmatrix} \mathbf{w}_t^{\mathsf{T}}(\mathbf{\Sigma}'_j)^{-1}\mathbf{w}_t - 2(\mathbf{o}_t^{\mathsf{T}}\mathbf{R}'_j\mathbf{\Phi}'_j\mathbf{\Sigma}'_j)(\mathbf{\Sigma}'_j)^{-1}\mathbf{w}_t \\ +(\mathbf{o}_t^{\mathsf{T}}\mathbf{R}'_j\mathbf{\Phi}'_j\mathbf{\Sigma}'_j)(\mathbf{\Sigma}'_j)^{-1}(\mathbf{\Sigma}'_j(\mathbf{\Phi}'_j)^{\mathsf{T}}\mathbf{R}'_j\mathbf{o}_t) \\ -(\mathbf{o}_t^{\mathsf{T}}\mathbf{R}'_j\mathbf{\Phi}'_j\mathbf{\Sigma}'_j)(\mathbf{\Sigma}'_j)^{-1}(\mathbf{\Sigma}'_j(\mathbf{\Phi}'_j)^{\mathsf{T}}\mathbf{R}'_j\mathbf{o}_t) \end{pmatrix}\right] d\mathbf{w}_t \\
&\propto |\mathbf{R}'_j|^{1/2}|\mathcal{A}'_j|^{1/2}|\mathbf{\Sigma}'_j|^{1/2} \exp\left[-\frac{1}{2}\mathbf{o}_t^{\mathsf{T}}(\mathbf{R}'_j - \mathbf{R}'_j\mathbf{\Phi}'_j\mathbf{\Sigma}'_j(\mathbf{\Phi}'_j)^{\mathsf{T}}\mathbf{R}'_j)\mathbf{o}_t\right] \\
&= |\mathbf{R}'_j|^{1/2}|\mathcal{A}'_j|^{1/2}|\mathbf{\Sigma}'_j|^{1/2} \exp\left[-\frac{1}{2}(\mathbf{o}_t^{\mathsf{T}}\mathbf{R}'_j\mathbf{o}_t - (\mathbf{m}'_{tj})^{\mathsf{T}}(\mathbf{\Sigma}'_j)^{-1}\mathbf{m}'_{tj})\right].
\end{aligned} \tag{5.28}
$$

In Eq. (5.28), the notations

$$(\mathbf{\Sigma}'_j)^{-1} \triangleq (\mathbf{\Phi}'_j)^{\mathsf{T}}\mathbf{R}'_j\mathbf{\Phi}'_j + \mathcal{A}'_j, \tag{5.29}$$

$$\mathbf{m}'_{tj} \triangleq \mathbf{\Sigma}'_j(\mathbf{\Phi}'_j)^{\mathsf{T}}\mathbf{R}'_j\mathbf{o}_t, \tag{5.30}$$

are introduced, and the integral of a Gaussian distribution

$$\mathcal{N}(\mathbf{w}_t|\mathbf{\Sigma}'_j(\mathbf{\Phi}'_j)^{\mathsf{T}}\mathbf{R}'_j\mathbf{o}_t, \mathbf{\Sigma}'_j) \tag{5.31}$$

is manipulated. By applying the Woodbury matrix inversion (Eq. (B.20))

$$(\mathbf{A} + \mathbf{UCV})^{-1} = \mathbf{A}^{-1} - \mathbf{A}^{-1}\mathbf{U}(\mathbf{C}^{-1} + \mathbf{VA}^{-1}\mathbf{U})^{-1}\mathbf{VA}^{-1}, \tag{5.32}$$

given the dimensionally compatible matrices $\mathbf{A}$, $\mathbf{U}$, $\mathbf{C}$, and $\mathbf{V}$, the marginal likelihood $p(\mathbf{o}_t|\mathbf{\Phi}'_j, \mathbf{R}'_j, \mathcal{A}'_j)$ is derived as

$$
\begin{aligned}
&\mathcal{N}(\mathbf{o}_t|\mathbf{0}, (\mathbf{R}'_j - \mathbf{R}'_j\mathbf{\Phi}'_j\left((\mathbf{\Phi}'_j)^{\mathsf{T}}\mathbf{R}'_j\mathbf{\Phi}'_j + \mathcal{A}'_j\right)^{-1}(\mathbf{\Phi}'_j)^{\mathsf{T}}\mathbf{R}'_j)^{-1}) \\
&= \mathcal{N}(\mathbf{o}_t|\mathbf{0}, (\mathbf{R}'_j)^{-1} + \mathbf{\Phi}'_j(\mathcal{A}'_j)^{-1}(\mathbf{\Phi}'_i)^{\mathsf{T}}),
\end{aligned} \tag{5.33}
$$

which is a Gaussian likelihood function with zero mean. The equality of the determinant

$$|\mathbf{R}'_j - \mathbf{R}'_j\boldsymbol{\Phi}'_j\boldsymbol{\Sigma}'_j(\boldsymbol{\Phi}'_j)^\mathsf{T}\mathbf{R}'_j| = |\mathbf{R}'_j||\mathcal{A}'_j||\boldsymbol{\Sigma}'_j| \tag{5.34}$$

is held. This implies that frame discrimination among different states is done solely on the basis of the covariance matrix. Apparently, the covariance matrix $(\mathbf{R}'_j)^{-1} + \boldsymbol{\Phi}'_j(\mathcal{A}'_j)^{-1}(\boldsymbol{\Phi}'_j)^\mathsf{T}$ is positive definite, so that $p(\mathbf{o}_t|\boldsymbol{\Phi}'_j, \mathbf{R}'_j, \mathcal{A}'_j)$ is a valid probability density function. For diagonal $\mathbf{R}'_j$, the marginal likelihood is seen as a new Gaussian distribution with a factor analyzed covariance matrix $(\mathbf{R}'_j)^{-1}+\boldsymbol{\Phi}'_j(\mathcal{A}'_j)^{-1}(\boldsymbol{\Phi}'_j)^\mathsf{T}$, where the factor loading matrix $\boldsymbol{\Phi}'_j(\mathcal{A}'_j)^{-1/2}$ is seen as a rank-N correction to $(\mathbf{R}'_j)^{-1}$ (Saon & Chien 2011).

5.2.6 Maximum a-posteriori sensing weights

In BS-HMMs, we can determine the maximum a-posteriori (MAP) estimate of Bayesian sensing weights $\mathbf{w}_t^{\mathsf{MAP}}$ for each observation $\mathbf{o}_t$ from

$$\begin{aligned}
\mathbf{w}_t^{\mathsf{MAP}} &= \arg\max_{\mathbf{w}_t} p(\mathbf{w}_t|\mathbf{o}_t, \Theta', \Psi') \\
&= \arg\max_{\mathbf{w}_t} p(\mathbf{o}_t|\mathbf{w}_t, \boldsymbol{\Phi}'_j, \mathbf{R}'_j)p(\mathbf{w}_t|\mathcal{A}'_j) \\
&= \boldsymbol{\Sigma}'_j(\boldsymbol{\Phi}'_j)^\mathsf{T}\mathbf{R}'_j\mathbf{o}_t \triangleq \mathbf{m}_{tj},
\end{aligned} \tag{5.35}$$

which is seen as a weighted product in vector space of observation $\mathbf{o}_t$ and transposed basis vectors $(\boldsymbol{\Phi}'_j)^\mathsf{T}$. The notations $\mathbf{m}'_{tj}$ (or equivalently $\mathbf{w}_t^{\mathsf{MAP}}$) and $\boldsymbol{\Sigma}'_j$ are the mean vector and the covariance matrix of the posterior distribution $p(\mathbf{w}_t|\mathbf{o}_t, \Theta', \Psi')$, respectively. The precision matrix for $\mathbf{w}_t$ is modified from $\mathcal{A}'_j$ of a-priori density $p(\mathbf{w}_t)$ to $(\boldsymbol{\Phi}'_j)^\mathsf{T}\mathbf{R}'_j\boldsymbol{\Phi}'_j + \mathcal{A}'_j$ of the a-posteriori distribution $p(\mathbf{w}_t|\mathbf{o}_t)$. The difference term $(\boldsymbol{\Phi}'_j)^\mathsf{T}\mathbf{R}'_j\boldsymbol{\Phi}'_j$ comes from the likelihood function $p(\mathbf{o}_t|\mathbf{w}_t)$, and is caused by the measurement $\boldsymbol{\Phi}'_j\mathbf{w}_t$ for observation $\mathbf{o}_t$ at frame t represented by new basis vectors $\boldsymbol{\Phi}'_j$ of state j. This is meaningful because Bayesian learning performs subjective inference, naturally increasing the model precision.

5.2.7 Optimal parameters and hyperparameters

By substituting Eq. (5.28) into Eq. (5.27), the optimal BS-HMM parameters and hyperparameters are estimated by maximizing the expanded auxiliary function with respect to individual parameters and hyperparameters $\{\boldsymbol{\Phi}'_j, \mathbf{R}'_j, \mathcal{A}'_j\}$. The auxiliary function in a BS-HMM state is simplified to

$$\begin{aligned}
&Q(\boldsymbol{\Phi}'_j, \mathbf{R}'_j, \mathcal{A}'_j|\boldsymbol{\Phi}_j, \mathbf{R}_j, \mathcal{A}_j) \\
&\quad = \sum_{t=1}^{T} \gamma_t(j) \log \int p(\mathbf{o}_t|\mathbf{w}_t, \boldsymbol{\Phi}'_j, \mathbf{R}'_j)p(\mathbf{w}_t|\mathcal{A}'_j)d\mathbf{w}_t \\
&\quad \propto \sum_{t=1}^{\mathsf{T}} \gamma_t(j)\Big[\log|\mathbf{R}'_j| + \log|\mathcal{A}'_j| \\
&\qquad\qquad + \log|\boldsymbol{\Sigma}_j| - \mathbf{o}_t^\mathsf{T}\mathbf{R}'_j\mathbf{o}_t + (\mathbf{m}'_{tj})^\mathsf{T}(\boldsymbol{\Sigma}'_j)^{-1}\mathbf{m}'_{tj}\Big].
\end{aligned} \tag{5.36}$$

Let us first consider the maximization of Eq. (5.28) with respect to $N \times N$ hyperparameter matrix $\mathcal{A}'_j$. We take the gradient of Eq. (5.28) with respect to $\mathcal{A}'_j$ and set it to zero to obtain

$$\sum_{t=1}^{T} \gamma_t(j) \left[(\mathcal{A}'_j)^{-1} - \boldsymbol{\Sigma}'_j - \underbrace{\boldsymbol{\Sigma}'_j(\boldsymbol{\Phi}'_j)^\intercal \mathbf{R}'_j \mathbf{o}_t}_{\mathbf{m}'_{tj}} \cdot \underbrace{\mathbf{o}_t^\intercal \mathbf{R}'_j \boldsymbol{\Phi}'_j \boldsymbol{\Sigma}'_j}_{(\mathbf{m}'_{tj})^\intercal} \right] = 0. \tag{5.37}$$

The value of $\mathcal{A}'_j$ that maximizes the auxiliary function satisfies

$$\begin{aligned}(\mathcal{A}_j^{\mathsf{ML2}})^{-1} &= \boldsymbol{\Sigma}'_j + \frac{\sum_{t=1}^{T} \gamma_t(j) \mathbf{m}'_{tj} (\mathbf{m}'_{tj})^\intercal}{\sum_{t=1}^{T} \gamma_t(j)} \\ &\triangleq \mathbf{F}^{\mathsf{ML2}}(\mathcal{A}'_j).\end{aligned} \tag{5.38}$$

Notably, Eq. (5.38) is an implicit solution to $\mathcal{A}'_j$ because $\mathbf{F}^a$ is a function of $\mathcal{A}'_j$. The hyperparameter $(\mathcal{A}_j^{\mathsf{ML2}})^{-1}$ of sensing weights is obtained by adding the covariance matrix $\boldsymbol{\Sigma}'_j$ of posterior $p(\mathbf{w}_t|\mathbf{o}_t, \Theta', \Psi')$ and the weighted autocorrelation of MAP sensing weights $\{\mathbf{m}'_{tj} \triangleq \mathbf{w}_t^{\mathsf{MAP}}\}$.

To find an ML2 estimate of basis vectors $\boldsymbol{\Phi}'_j$, we maximize Eq. (5.36) by taking the gradient of the terms related to $\boldsymbol{\Phi}'_j$ and setting it to zero, which leads to

$$\begin{aligned}\frac{\partial}{\partial \boldsymbol{\Phi}'_j} \Bigg[\sum_{t=1}^{T} \gamma_t(j) \Bigg[&- \log |(\boldsymbol{\Phi}'_j)^\intercal \mathbf{R}'_j \boldsymbol{\Phi}'_j + \mathcal{A}'_j| \\ &+ \mathbf{o}_t^\intercal \mathbf{R}'_j \boldsymbol{\Phi}'_j \left((\boldsymbol{\Phi}'_j)^\intercal \mathbf{R}'_j \boldsymbol{\Phi}'_j + \mathcal{A}'_j \right)^{-1} (\boldsymbol{\Phi}'_j)^\intercal \mathbf{R}'_j \mathbf{o}_t \Bigg] \Bigg] = 0,\end{aligned} \tag{5.39}$$

where the gradients of the two terms are derived as the $D \times N$ matrices given by

$$\begin{aligned}\frac{\partial}{\partial \boldsymbol{\Phi}'_j} &\log |(\boldsymbol{\Phi}'_j)^\intercal \mathbf{R}'_j \boldsymbol{\Phi}'_j + \mathcal{A}'_j| \\ &= 2\mathbf{R}'_j \boldsymbol{\Phi}'_j \left((\boldsymbol{\Phi}'_j)^\intercal \mathbf{R}'_j \boldsymbol{\Phi}'_j + \mathcal{A}'_j \right)^{-1} = 2\mathbf{R}'_j \boldsymbol{\Phi}'_j \boldsymbol{\Sigma}'_j,\end{aligned} \tag{5.40}$$

$$\begin{aligned}\frac{\partial}{\partial \boldsymbol{\Phi}'_j} &\left[\mathbf{o}_t^\intercal \mathbf{R}'_j \boldsymbol{\Phi}'_j \left((\boldsymbol{\Phi}'_j)^\intercal \mathbf{R}'_j \boldsymbol{\Phi}'_j + \mathcal{A}'_j \right)^{-1} (\boldsymbol{\Phi}'_j)^\intercal \mathbf{R}'_j \mathbf{o}_t \right] \\ &= \frac{\partial}{\partial \boldsymbol{\Phi}'_j} \mathrm{tr}\{\boldsymbol{\Sigma}'_j (\boldsymbol{\Phi}'_j)^\intercal \mathbf{R}'_j \mathbf{o}_t \mathbf{o}_t^\intercal \mathbf{R}'_j \boldsymbol{\Phi}'_j\} \\ &= -2\mathbf{R}'_j \boldsymbol{\Phi}'_j \boldsymbol{\Sigma}'_j (\boldsymbol{\Phi}'_j)^\intercal \mathbf{R}'_j \mathbf{o}_t \mathbf{o}_t^\intercal \mathbf{R}'_j \boldsymbol{\Phi}'_j \boldsymbol{\Sigma}'_j + 2\mathbf{R}'_j \mathbf{o}_t \mathbf{o}_t^\intercal \mathbf{R}'_j \boldsymbol{\Phi}'_j \boldsymbol{\Sigma}'_j.\end{aligned} \tag{5.41}$$

The optimal solution $\boldsymbol{\Phi}_j^{\mathsf{ML2}}$, which is a $D \times N$ matrix, satisfies

$$\begin{aligned}\sum_{t=1}^{T} \gamma_t(j) &\left[-\mathbf{R}'_j \boldsymbol{\Phi}'_j \boldsymbol{\Sigma}'_j - \mathbf{R}'_j \boldsymbol{\Phi}'_j \boldsymbol{\Sigma}'_j (\boldsymbol{\Phi}'_j)^\intercal \mathbf{R}'_j \mathbf{o}_t \mathbf{o}_t^\intercal \mathbf{R}'_j \boldsymbol{\Phi}'_j \boldsymbol{\Sigma}'_j + \mathbf{R}'_j \mathbf{o}_t \mathbf{o}_t^\intercal \mathbf{R}'_j \boldsymbol{\Phi}'_j \boldsymbol{\Sigma}'_j \right] \\ &= \sum_{t=1}^{T} \gamma_t(j) \mathbf{R}'_j \left[-\boldsymbol{\Phi}'_j \boldsymbol{\Sigma}'_j - \boldsymbol{\Phi}'_j \mathbf{m}'_{tj} (\mathbf{m}'_{tj})^\intercal + \mathbf{o}_t (\mathbf{m}'_{tj})^\intercal \right] = 0.\end{aligned} \tag{5.42}$$

Similarly to the solution to $\mathcal{A}'_j$, the ML2 estimate can be expressed in an implicit form written by a function $\mathbf{F}^{\phi}(\mathbf{\Phi}'_j)$, namely

$$\begin{aligned}\mathbf{\Phi}_j^{\mathsf{ML2}} &= \left[\sum_{t=1}^{T} \gamma_t(j)\mathbf{o}_t(\mathbf{m}'_{tj})^{\mathsf{T}}\right]\left[\sum_{t=1}^{T} \gamma_t(j)(\mathbf{\Sigma}'_j + \mathbf{m}'_{tj}(\mathbf{m}'_{tj})^{\mathsf{T}})\right]^{-1} \\ &= \frac{\sum_{t=1}^{T} \gamma_t(j)\mathbf{o}_t(\mathbf{m}'_{tj})^{\mathsf{T}}}{\sum_{t=1}^{T} \gamma_t(j)} \cdot \mathcal{A}_j^{\mathsf{ML2}} \\ &\triangleq \mathbf{F}^{\mathsf{ML2}}(\mathbf{\Phi}'_j). \end{aligned} \tag{5.43}$$

This solution is viewed as a weighted operation in the outer product space of observations $\{\mathbf{o}_t\}$ and MAP sensing weights $\{\mathbf{m}'_{tj} \triangleq \mathbf{w}_t^{\mathsf{MAP}}\}$. The posterior probabilities $\{\gamma_t(j)\}$ and the ML2 hyperparameters $\mathcal{A}_j^{\mathsf{ML2}}$ serve as the weights and the rotation operator of the weighted average, respectively.

To find the ML2 estimate of precision matrix $\mathbf{R}'_j$, we maximize Eq. (5.36) with respect to $\mathbf{R}'_j$ and obtain

$$\begin{aligned}\sum_{t=1}^{T} \gamma_t(j)\Big[&(\mathbf{R}'_j)^{-1} - \mathbf{\Phi}'_j\mathbf{\Sigma}'_j(\mathbf{\Phi}'_j)^{\mathsf{T}} - \mathbf{o}_t\mathbf{o}_t^{\mathsf{T}} \\ &+ \frac{\partial}{\partial \mathbf{R}'_j}\left(\mathbf{o}_t^{\mathsf{T}}\mathbf{R}'_j\mathbf{\Phi}'_j\mathbf{\Sigma}'_j(\mathbf{\Phi}'_j)^{\mathsf{T}}\mathbf{R}'_j\mathbf{o}_t\right)\Big] = 0,\end{aligned} \tag{5.44}$$

where

$$\begin{aligned}\frac{\partial}{\partial \mathbf{R}'_j}\left(\mathbf{o}_t^{\mathsf{T}}\mathbf{R}'_j\mathbf{\Phi}'_j\mathbf{\Sigma}'_j(\mathbf{\Phi}'_j)^{\mathsf{T}}\mathbf{R}'_j\mathbf{o}_t\right) &= \frac{\partial}{\partial \mathbf{R}'_j}\mathrm{tr}\{\mathbf{\Sigma}'_j(\mathbf{\Phi}'_j)^{\mathsf{T}}\mathbf{R}'_j\mathbf{o}_t\mathbf{o}_t^{\mathsf{T}}\mathbf{R}'_j\mathbf{\Phi}'_j\} \\ &= -\mathbf{\Phi}'_j\mathbf{\Sigma}'_j(\mathbf{\Phi}'_j)^{\mathsf{T}}\mathbf{R}'_j\mathbf{o}_t\mathbf{o}_t^{\mathsf{T}}\mathbf{R}'_j\mathbf{\Phi}'_j\mathbf{\Sigma}'_j(\mathbf{\Phi}'_j)^{\mathsf{T}} \\ &\quad + \mathbf{\Phi}'_j\mathbf{\Sigma}'_j(\mathbf{\Phi}'_j)^{\mathsf{T}}\mathbf{R}'_j\mathbf{o}_t\mathbf{o}_t^{\mathsf{T}} + \mathbf{o}_t\mathbf{o}_t^{\mathsf{T}}\mathbf{R}'_j\mathbf{\Phi}'_j\mathbf{\Sigma}'_j(\mathbf{\Phi}'_j)^{\mathsf{T}}.\end{aligned} \tag{5.45}$$

We derive the ML2 solution as

$$\begin{aligned}(\mathbf{R}_j^{\mathsf{ML2}})^{-1} &= \frac{\sum_{t=1}^{T}\gamma_t(j)\left[\begin{array}{c}\mathbf{\Phi}'_j\mathbf{\Sigma}'_j(\mathbf{\Phi}'_j)^{\mathsf{T}} + \mathbf{o}_t\mathbf{o}_t^{\mathsf{T}} \\ +\mathbf{\Phi}'_j\mathbf{\Sigma}'_j(\mathbf{\Phi}'_j)^{\mathsf{T}}\mathbf{R}'_j\mathbf{o}_t\mathbf{o}_t^{\mathsf{T}}\mathbf{R}'_j\mathbf{\Phi}'_j\mathbf{\Sigma}'_j(\mathbf{\Phi}'_j)^{\mathsf{T}} \\ -\mathbf{\Phi}'_j\mathbf{\Sigma}'_j(\mathbf{\Phi}'_j)^{\mathsf{T}}\mathbf{R}'_j\mathbf{o}_t\mathbf{o}_t^{\mathsf{T}} - \mathbf{o}_t\mathbf{o}_t^{\mathsf{T}}\mathbf{R}'_j\mathbf{\Phi}'_j\mathbf{\Sigma}'_j(\mathbf{\Phi}'_j)^{\mathsf{T}}\end{array}\right]}{\sum_{t=1}^{T}\gamma_t(j)} \\ &= \mathbf{\Phi}'_j\mathbf{\Sigma}'_j(\mathbf{\Phi}'_j)^{\mathsf{T}} + \frac{\sum_{t=1}^{T}\gamma_t(j)(\mathbf{o}_t - \mathbf{\Phi}'_j\mathbf{m}'_{tj})(\mathbf{o}_t - \mathbf{\Phi}'_j\mathbf{m}'_{tj})^{\mathsf{T}}}{\sum_{t=1}^{T}\gamma_t(j)} \\ &\triangleq \mathbf{F}^{\mathsf{ML2}}(\mathbf{R}'_j),\end{aligned} \tag{5.46}$$

which is also an implicit solution to $\mathbf{R}'_j$ since the right-hand-side (RHS) of Eq. (5.46) depends on $\mathbf{R}'_j$. Note that the RHS of Eq. (5.46) is symmetric positive definite. The first term is a scaled covariance matrix $\mathbf{\Sigma}'_j$ of the posterior distribution $p(\mathbf{w}_t|\mathbf{o}_t, \Theta', \Psi')$, which is doubly transformed by $\mathbf{\Phi}'_j$. The second term is interpreted as a covariance matrix weighted by posterior probabilities $\gamma_t(j)$, and is calculated using observations $\{\mathbf{o}_t\}$ and "mean" vectors $\boldsymbol{\Phi}'_j \mathbf{m}'_{tj}$. This corresponds to performing *Bayesian sensing*, again using the MAP estimates $\{\mathbf{m}'_{tj} \triangleq \mathbf{w}_t^{\mathsf{MAP}}\}$.

We can see that type-2 ML (ML2) estimates of BS-HMM parameters and hyperparameters are consistently formulated as the *implicit solutions*, which are beneficial for efficient implementation and good convergence in parameter estimation. Differently from conventional basis representation, where basis vectors and sensing weights are found separately, BS-HMMs provide a multivariate Bayesian approach to hybrid estimation of the compact basis vectors and the precision matrices of sensing weights under a consistent objective function. No training examples are stored for memory-based implementation.

To improve LVCSR performance based on BS-HMMs, there have been several extensions developed for acoustic modeling. As mentioned in Section 5.2.2, the mixture model of BS-HMMs can be extended by considering multiple sets of basis vectors per state. A mixture model of basis vectors is included for acoustic modeling. Using this mixture model, each observation $\mathbf{o}_t$ at state $s_t = j$ is expressed by

$$p(\mathbf{o}_t|\Theta_j) \triangleq \sum_{k=1}^{K} \omega_{jk} \mathcal{N}(\mathbf{o}_t|\boldsymbol{\Phi}_{jk}\mathbf{w}_t, \mathbf{R}_{jk}^{-1}), \tag{5.47}$$

where ω_{jk} is the mixture weight of jth component with the constraint

$$\sum_{k=1}^{K} \omega_{jk} = 1. \tag{5.48}$$

Here, the reconstruction error of an observation vector $\mathbf{o}_t$ due to the jth component with basis vectors $\boldsymbol{\Phi}_{jk} = [\boldsymbol{\phi}_{jk1}, \cdots, \boldsymbol{\phi}_{jkN}]$ is assumed to be Gaussian distributed with zero mean and precision matrix $\mathbf{R}_{jk}$. In addition, BS-HMMs can be constructed by incorporating a non-zero mean vector $\boldsymbol{\mu}_j^w$ in the prior density of sensing weights, i.e.,

$$p(\mathbf{w}_t|\mathbf{0}, \mathcal{A}_j) \rightarrow p(\mathbf{w}_t|\boldsymbol{\mu}_j^w, \mathcal{A}_j). \tag{5.49}$$

Similarly to the Maximum Likelihood Linear Regression (MLLR) adaptation for HMMs, BS-HMMs are developed for speaker adaptation where the nth BS-HMM basis vector is transformed by

$$\hat{\boldsymbol{\phi}}_{jn} = \mathbf{M}\tilde{\boldsymbol{\phi}}_{jn}, \tag{5.50}$$

where $\mathbf{M}$ is a $D \times (D + 1)$ regression matrix and $\tilde{\boldsymbol{\phi}}_{jn} = [\boldsymbol{\phi}_{jn}^{\mathsf{T}} \; 1]^{\mathsf{T}}$ is the extended basis vector. The type-2 ML estimation can be applied to calculate the optimal solutions to non-zero mean vector $\boldsymbol{\mu}_j^w$ and regression matrix $\mathbf{M}$.

5.2.8 Discriminative training

Finally, BS-HMMs are sophisticated, incorporating both model-space and feature-space discriminative training, which is crucial to improve classification of confusing patterns in pattern recognition systems. Developing discriminative training for BS-HMMs is important for LVCSR. Instead of the goodness-of-fit criterion using marginal likelihood function, the objective function for discriminative training is established according to the mutual information between observation data $\mathbf{O}$ and the sequence of reference words W^r (Bahl *et al.* 1986, Povey & Woodland 2002, Povey, Kanevsky, Kingsbury *et al.* 2008):

$$\begin{aligned}\mathcal{F}(\Theta) &\triangleq I_\Theta(\mathbf{O}, W^r) = \log \frac{p_\Theta(\mathbf{O}, W^r)}{p_\Theta(\mathbf{O})p(W^r)} \\ &= \log p_\Theta(\mathbf{O}|W^r) - \log \sum_W p_\Theta(\mathbf{O}|W)p(W) \\ &\triangleq \mathcal{F}^{\mathsf{num}}(\Theta) - \mathcal{F}^{\mathsf{den}}(\Theta),\end{aligned} \tag{5.51}$$

which consists of a numerator term $\mathcal{F}^{\mathsf{num}}(\Theta)$ and a denominator term $\mathcal{F}^{\mathsf{den}}(\Theta)$. The Maximum Mutual Information (MMI) estimation of BS-HMMs Θ^{MMI} is performed for discriminative training. To solve the optimization problem, we calculate the *weak-sense auxiliary function* (Povey & Woodland 2002, Povey 2003), where the HMM state sequence S is incorporated as follows:

$$\begin{aligned}Q(\Theta'|\Theta) &= Q^{\mathsf{num}}(\Theta'|\Theta) - Q^{\mathsf{den}}(\Theta'|\Theta) + Q^{\mathsf{sm}}(\Theta'|\Theta) \\ &= \sum_S p(S|\mathbf{O}, W^r, \Theta) \log p(S, \mathbf{O}|\Theta') \\ &\quad - \sum_S \sum_W p(S, W|\mathbf{O}, \Theta) \log p(S, \mathbf{O}|\Theta') \\ &\quad + Q^{\mathsf{sm}}(\Theta'|\Theta).\end{aligned} \tag{5.52}$$

The property of weak-sense auxiliary function turns out to meet the condition

$$\left.\frac{\partial Q(\Theta'|\Theta)}{\partial \Theta'}\right|_{\Theta'=\Theta} = \left.\frac{\partial \mathcal{F}(\Theta')}{\partial \Theta'}\right|_{\Theta'=\Theta}, \tag{5.53}$$

where the mode of MMI auxiliary function and its weak-sense auxiliary function have the same value. The smoothing function $Q^{\mathsf{sm}}(\Theta'|\Theta)$ in Eq. (5.52) is added to ensure that the objective function $Q(\Theta'|\Theta)$ is improved by this extended EM algorithm. For this, the smoothing function should satisfy

$$\left.\frac{\partial Q^{\mathsf{sm}}(\Theta'|\Theta)}{\partial \Theta'}\right|_{\Theta'=\Theta} = 0. \tag{5.54}$$

In what follows, we address the discriminative training of basis vectors $\mathbf{\Phi}_j$ of state j. The same procedure can be applied to estimate the discriminative precision matrix of reconstruction errors $\mathbf{R}_j$.

One possible choice of smoothing function meeting Eq. (5.54) is formed by the Kullback–Leibler divergence $\mathrm{KL}(\cdot\|\cdot)$ between marginal likelihoods of the current estimate $p(\mathbf{o}_t|\Theta)$ and the new estimate $p(\mathbf{o}_t|\Theta')$ given by

$$\begin{aligned}Q^{\mathsf{sm}}(\{\mathbf{\Phi}'_j\}|\{\mathbf{\Phi}_j\}) &\triangleq -\sum_{j=1}^{J} D_j \mathrm{KL}(p(\mathbf{o}|\mathbf{\Phi}_j)\|p(\mathbf{o}|\mathbf{\Phi}'_j)) \\ &\propto \sum_{j=1}^{J} D_j \int p(\mathbf{o}|\mathbf{\Phi}_j) \log p(\mathbf{o}|\mathbf{\Phi}'_j) d\mathbf{o}\end{aligned}$$

$$
\begin{aligned}
&\propto \sum_{j=1}^{J} D_j \bigg[\log |\mathbf{R}'_j| - \int p(\mathbf{o}|\boldsymbol{\Phi}_j) \\
&\quad \times (\mathbf{o} - \boldsymbol{\Phi}_j \mathbf{w} + \boldsymbol{\Phi}_j \mathbf{w} - \boldsymbol{\Phi}'_j \mathbf{w})^{\mathsf{T}} \\
&\quad \times \mathbf{R}'_j (\mathbf{o} - \boldsymbol{\Phi}_j \mathbf{w} + \boldsymbol{\Phi}_j \mathbf{w} - \boldsymbol{\Phi}'_j \mathbf{w}) d\mathbf{o} \bigg] \\
&= \sum_{j=1}^{J} D_j \bigg[\log |\mathbf{R}'_j| - \int p(\mathbf{o}|\boldsymbol{\Phi}_j) \\
&\quad \times (\mathbf{o} - \boldsymbol{\Phi}_j \mathbf{w})^{\mathsf{T}} \mathbf{R}'_j (\mathbf{o} - \boldsymbol{\Phi}_j \mathbf{w}) d\mathbf{o} \\
&\quad - (\boldsymbol{\Phi}_j \mathbf{w} - \boldsymbol{\Phi}'_j \mathbf{w})^{\mathsf{T}} \mathbf{R}'_j (\boldsymbol{\Phi}_j \mathbf{w} - \boldsymbol{\Phi}'_j \mathbf{w}) \bigg].
\end{aligned} \tag{5.55}
$$

Here, D_j is a state-dependent smoothing constant. Typically, the smoothing function is mathematically intractable when applying marginal likelihoods. Noting this, we can approximate the marginal likelihood by using an average plug-in MAP estimate:

$$
\mathbf{w} \approx \frac{1}{T} \sum_{t=1}^{T} \mathbf{w}_t^{\text{MAP}}, \tag{5.56}
$$

obtained by taking an ensemble average of the MAP estimates in Eq. (5.35) using all observation frames $\{\mathbf{o}_1, \cdots, \mathbf{o}_T\}$. Ignoring the terms independent of $\boldsymbol{\Phi}_j$, the smoothing function is obtained by substituting the approximate MAP estimate of Eq. (5.56) into

$$
Q^{\mathsf{sm}}(\{\boldsymbol{\Phi}'_j\}|\{\boldsymbol{\Phi}_j\}) = -\sum_{j=1}^{J} D_j \mathbf{w}^{\mathsf{T}} (\boldsymbol{\Phi}_j - \boldsymbol{\Phi}'_j)^{\mathsf{T}} \mathbf{R}'_j (\boldsymbol{\Phi}_j - \boldsymbol{\Phi}'_j) \mathbf{w}. \tag{5.57}
$$

As a result, the weak-sense auxiliary function is expressed in terms of state occupation posteriors as follows:

$$
\begin{aligned}
Q(\{\boldsymbol{\Phi}'_j\}|\{\boldsymbol{\Phi}_j\}) &= \sum_{t=1}^{T} \sum_{j=1}^{J} (\gamma_t^{\mathsf{num}}(j) - \gamma_t^{\mathsf{den}}(j)) \\
&\quad \times \bigg[\log |\mathcal{A}'_j| + \log |\mathbf{R}'_j| + \log |\boldsymbol{\Sigma}'_j| \\
&\qquad - \mathbf{o}_t^{\mathsf{T}} \mathbf{R}'_j \mathbf{o}_t + (\mathbf{m}'_{tj})^{\mathsf{T}} (\boldsymbol{\Sigma}'_j)^{-1} \mathbf{m}'_{tj} \bigg] \\
&\quad - \sum_{j=1}^{J} D_j \mathbf{w}^{\mathsf{T}} (\boldsymbol{\Phi}_j - \boldsymbol{\Phi}'_j)^{\mathsf{T}} \mathbf{R}'_j (\boldsymbol{\Phi}_j - \boldsymbol{\Phi}'_j) \mathbf{w},
\end{aligned} \tag{5.58}
$$

where the state occupation posteriors of staying in state $s_t = j$ at time t in the numerator and denominator terms are calculated by

$$
\gamma_t^{\mathsf{num}}(j) \triangleq p(s_t = j|\mathbf{O}, W^r, \Theta), \tag{5.59}
$$

$$
\gamma_t^{\mathsf{den}}(j) \triangleq \sum_{W} p(s_t = j|\mathbf{O}, W, \Theta), \tag{5.60}
$$

given the reference word sequence W^r and all possible word sequences $\{W\}$, respectively. The current estimates $\{\mathbf{\Phi}_j\}$ are used in this calculation. To find an MMI estimate for basis parameters, we differentiate Eq. (5.58) with respect to $\mathbf{\Phi}'_j$ and set it to zero:

$$\begin{aligned}\frac{\partial}{\partial \mathbf{\Phi}'_j} Q(\mathbf{\Phi}'_j|\mathbf{\Phi}_j) \propto \mathbf{R}'_j \sum_{t=1}^{T} (\gamma_t^{\text{num}}(j) - \gamma_t^{\text{den}}(j)) \\ \times (-\mathbf{\Phi}'_j \mathbf{\Sigma}'_j - \mathbf{\Phi}'_j \mathbf{m}'_{tj}(\mathbf{m}'_{tj})^{\mathsf{T}} + \mathbf{o}_t(\mathbf{m}'_{tj})^{\mathsf{T}}) \\ - \mathbf{R}'_j D_j (\mathbf{\Phi}'_j - \mathbf{\Phi}_j)\mathbf{w}\mathbf{w}^{\mathsf{T}} = 0,\end{aligned} \tag{5.61}$$

which is derived by considering the definition of variables $(\mathbf{\Sigma}'_j)^{-1}$ and $\mathbf{m}'_j$ in Eq. (5.29) and Eq. (5.30), respectively. Again, the implicit solution to an MMI estimate of $\mathbf{\Phi}'_j$ should satisfy

$$\begin{aligned}\mathbf{\Phi}_j^{\text{MMI2}} &= \left[\sum_{t=1}^{T} (\gamma_t^{\text{num}}(j) - \gamma_t^{\text{den}}(j)) \mathbf{o}_t (\mathbf{m}'_{tj})^{\mathsf{T}} + D_j \mathbf{\Phi}_j \mathbf{w}\mathbf{w}^{\mathsf{T}} \right] \\ &\quad \times \left[\sum_{t=1}^{T} (\gamma_t^{\text{num}}(j) - \gamma_t^{\text{den}}(j)) \right. \\ &\quad \left. \times (\mathbf{\Sigma}'_j + \mathbf{m}'_{tj}(\mathbf{m}'_{tj})^{\mathsf{T}}) + D_j \mathbf{w}\mathbf{w}^{\mathsf{T}} \right]^{-1} \\ &\triangleq \mathbf{F}^{\text{MMI2}}(\mathbf{\Phi}'_j).\end{aligned} \tag{5.62}$$

This solution is expressed as a recursive function $\mathbf{F}^{\text{MMI2}}$ of new basis parameter $\mathbf{\Phi}'_j$. Strictly speaking, an MMI estimation based on marginal likelihood is seen as a type 2 MMI (MMI2) estimation, which is different from conventional MMI training (Bahl *et al.* 1986, Povey & Woodland 2002, Povey *et al.* 2008) based on likelihood function without marginalization. In Eq. (5.62), the second terms in the numerator and the denominator come from smoothing function $Q^{\text{sm}}(\mathbf{\Phi}'_j|\mathbf{\Phi}_j)$ and serve to prevent instability in the MMI2 optimization procedure. The solution is highly affected by the difference between statistics for the reference hypothesis and statistics for the competing hypotheses $\gamma_t^{\text{num}}(j) - \gamma_t^{\text{den}}(j)$.

It is clear that *discriminative training* and *Bayesian learning* are simultaneously performed in a type 2 MMI estimation. By doing this, the performance of LVCSR can be significantly improved (Saon & Chien 2012*b*). The robustness to uncertainty of sensing weights in a basis representation can be assured. Some experimental results are described below.

5.2.9 System performance

Evaluation of BS-HMMs was performed by using the LVCSR task in the domain of Arabic broadcast news transcription which was part of the DARPA GALE program. In total, 1800 hours of manually transcribed Arabic broadcast news and conversations were used in this evaluation (Saon & Chien 2011). The results on several test sets were

Table 5.1 Comparison of the number of free parameters and word error rates for baseline acoustic models after ML training and BS-HMMs after ML2 training.

System	Nb. parameters	WER		
		DEV'07	DEV'08	DEV'09
Baseline 800K	64.8M	13.8%	16.4%	19.6%
Baseline 2.8M	226.8M	14.1%	16.2%	19.3%
BS-HMM 417K	148.5M	13.6%	16.0%	18.9%

reported: DEV'07 (2.5 hours), DEV'08 (3 hours) and DEV'09 (3 hours). The front-end processing was performed as mentioned in Saon & Chien (2012*b*). The vocal-tract-length-warped PLP (Hermansky 1990) cepstrum features were extracted with a context window of nine frames. The features were mean and variance normalized on a per speaker basis. Linear discriminant analysis was used to reduce the feature dimension to 40. The maximum likelihood training of the acoustic model was interleaved with the estimation of a global semi-tied covariance transform (Gales 1998). All models in this evaluation were estimated based on pentaphones and speaker adaptively trained with feature-space MLLR (fMLLR). Each pentaphone was modeled by a 3-state left-to-right HMM without state skipping. At test time, speaker adaptation was performed with vocal-tract-length normalization (Wegmann, McAllaster, Orloff *et al.* 1996), fMLLR and multiple regression MLLR transforms. The vocabulary contained 795K words. The decoding was done with 4-gram language models which were estimated with modified Kneser–Ney smoothing. The acoustic models were discriminatively trained in both feature space and model space according to the boosted MMI criterion (Povey, Kingsbury, Mangu *et al.* 2005). The baseline acoustic models had 5000 context-dependent HMM states and 800K 40-dimensional diagonal covariance Gaussians.

In the implementation, BS-HMM parameters were initialized by training a large HMM model with 2.8M diagonal covariance Gaussians by maximum likelihood method. The means of GMM were clustered and then treated as the initial basis $\mathbf{\Phi}_{jk}$ for state j and mixture component k. The resulting number of mixture components in BS-HMMs after the clustering step was 417K. The precision matrices $\mathbf{R}_{jk}$ and $\mathcal{A}_{jk}$ were assumed to be diagonal and were initialized to the identity matrix. Table 5.1 compares the performance of the baseline 800K Gaussians model and the 2.8M Gaussians model used to train baseline acoustic models after ML training and to seed BS-HMMs after ML2 training. The number of free parameters is included in this comparison. As we can see, BS-HMMs outperform both baseline systems in terms of word error rates (%). The possible reasons are twofold. The first one is that the covariance modeling in BS-HMMs is more accurate than that in HMMs. The second one is due to the Bayesian parameter updates which provide an effective smoothing.

In contrast, we conducted a model comparison for BS-HMMs according to the estimated hyperparameters of sensing weights. Model compression was performed by discarding 50% of the basis vectors Φ_{jkn} corresponding to the largest hyperparameters α_{jkn}. This results in a compressed model with approximately 91M free parameters (about

Table 5.2 Comparison of word error rates for original and compressed BS-HMMs before and after model-space discriminative training with MMI2 training.

Model	Training	DEV07	DEV08	DEV09
Original	ML2	12.0%	13.9%	17.4%
Compressed	ML2	12.4%	14.2%	17.6%
Original	MMI2	10.7%	11.9%	15.0%
Compressed	MMI2	10.4%	11.7%	14.8%

30% larger than the 800K baseline HMM). For both original and compressed models, we performed model-space discriminative training using the MMI2 criterion. Table 5.2 reports the recognition performance before and after model-space discriminative training. We find that the compressed models outperform the originals after discriminative training even though they start from a higher word error rate after ML2 estimation. However, discriminative training is significantly more expensive than ML estimation which makes it difficult to find the optimal model size.

5.3 Hierarchical Dirichlet language model

In what follows, we revisit the interpolation smoothing methods presented in Section 3.6 and used to deal with the small sample size problem in ML estimation of n-gram parameters $\Theta^{\mathsf{ML}} = \{p_{\mathsf{ML}}(w_i|w_{i-n+1}^{i-1})\}$. Different from the heuristic solutions to language model smoothing in Section 3.6, a *full Bayesian* language model is proposed to realize interpolation smoothing for n-grams. The theory of evidence approximation is developed and applied to construct the hierarchical Dirichlet language model (MacKay & Peto 1995, Kawabata & Tamoto 1996).

5.3.1 n-gram smoothing revisited

In general, the frequency estimator in a higher-order language model has large variance, because there are so many possible word combinations in an n-gram event $\{w_{i-n+1}^{i-1}, w_i\}$ that only a small fraction of them have been observed in the data. A simple linear interpolation scheme for an n-gram language model is performed by interpolating an ML model of an n-gram $p_{\mathsf{ML}}(w_i|w_{i-n+1}^{i-1})$ with that of an $(n-1)$-gram $p_{\mathsf{ML}}(w_i|w_{i-n+2}^{i-1})$ using

$$\hat{p}(w_i|w_{i-n+1}^{i-1}) = \lambda p_{\mathsf{ML}}(w_i|w_{i-n+1}^{i-1}) + (1-\lambda) p_{\mathsf{ML}}(w_i|w_{i-n+2}^{i-1}), \tag{5.63}$$

where λ denotes the interpolation weight, which can be determined empirically from validation data. Basically, it is not possible to make language models without making a-priori assumptions. The smoothed n-gram $\hat{p}(w_i|w_{i-n+1}^{i-1})$ in Eq. (5.63) can be seen as an integrated n-gram model which is determined from a hierarchical model based on the smoothing parameters a-posteriori from the data. In what follows, we would like to reverse-engineer the underlying model, which gives a probabilistic meaning to language model smoothing. We will explain the construction of a hierarchical prior model and illustrate how this model is used to justify the smoothed language model from a Bayesian perspective. A type-2 ML estimation is conducted to find optimal hyperparameters from training data. No cross validation procedure is required in a Bayesian language model.

5.3.2 Dirichlet prior and posterior

n-gram model parameters $\Theta = \{p(w_i|w_{i-n+1}^{i-1})\}$ are known as multinomial parameters which are used to predict n-gram events of word w_i appearing after observing history words w_{i-n+1}^{i-1}. The ML estimation of an n-gram $\theta_{w_i|w_{i-n+1}^{i-1}} = p(w_i|w_{i-n+1}^{i-1})$ from a text corpus $\mathcal{D}$ has been shown in Eq. (3.192). Here, we are interested in a Bayesian language model where the prior density of multinomial parameters is introduced. A parameter vector consists of N multinomial parameters:

$$\boldsymbol{\theta} = \text{vec}(\Theta) = [\theta_1, \cdots, \theta_N]^\intercal \text{ subject to}$$
$$0 \leq \theta_i \leq 1 \text{ and } \sum_{i=1}^{N} \theta_i = 1. \tag{5.64}$$

The *conjugate prior* over multinomial parameters $\boldsymbol{\theta}$ is specified by a Dirichlet prior with hyperparameters $\boldsymbol{\alpha} = [\alpha_1, \cdots, \alpha_N]^\intercal$, which is a multivariate distribution in the form of

$$p(\boldsymbol{\theta}|\boldsymbol{\alpha}) = \text{Dir}(\boldsymbol{\theta}|\boldsymbol{\alpha}) = \frac{1}{Z(\boldsymbol{\alpha})} \prod_{i=1}^{N} \theta_i^{\alpha_i - 1}. \tag{5.65}$$

We express the normalization constant of the Dirichlet distribution by

$$Z(\boldsymbol{\alpha}) = \frac{\prod_{i=1}^{N} \Gamma(\alpha_i)}{\Gamma(\sum_{i=1}^{N} \alpha_i)}. \tag{5.66}$$

The mean vector of Dirichlet distribution is given by

$$\int \boldsymbol{\theta} \text{Dir}(\boldsymbol{\theta}|\boldsymbol{\alpha}) d\boldsymbol{\theta} = \frac{\boldsymbol{\alpha}}{\sum_{i=1}^{N} \alpha_i}. \tag{5.67}$$

When we observe the training samples $\mathcal{D}$, the posterior distribution is derived as another Dirichlet distribution:

$$\begin{aligned} p(\boldsymbol{\theta}|\mathcal{D}, \boldsymbol{\alpha}) &= \frac{p(\mathcal{D}|\boldsymbol{\theta}) p(\boldsymbol{\theta}|\boldsymbol{\alpha})}{p(\mathcal{D}|\boldsymbol{\alpha})} \\ &= \frac{\prod_{i=1}^{N} \theta_i^{c(\theta_i)} \prod_{i=1}^{N} \theta_i^{\alpha_i - 1}}{p(\mathcal{D}|\boldsymbol{\alpha}) Z(\boldsymbol{\alpha})} \\ &= \frac{\prod_{i=1}^{N} \theta_i^{c(\theta_i) + \alpha_i - 1}}{Z(\mathbf{c} + \boldsymbol{\alpha})} \\ &= \text{Dir}(\boldsymbol{\theta}|\mathbf{c} + \boldsymbol{\alpha}), \end{aligned} \tag{5.68}$$

with the updated hyperparameters $\mathbf{c} + \boldsymbol{\alpha}$. In Eq. (5.68), each entry $c(\theta_i)$ of

$$\mathbf{c} = [c(\theta_1), \cdots, c(\theta_N)]^\intercal \tag{5.69}$$

denotes the number of occurrences of the ith n-gram event in θ_i in training data $\mathcal{D}$. This shows the property of a conjugate prior by using the Dirichlet distribution.

5.3.3 Evidence function

To obtain the predictive probability of a word w_i given history words w_{i-n+1}^{i-1} and training data $\mathcal{D}$, we apply the sum rule to calculate the evidence function or the marginal likelihood:

$$\begin{aligned} p(w_i|w_{i-n+1}^{i-1}, \mathcal{D}, \boldsymbol{\alpha}) &= \int p(w_i|w_{i-n+1}^{i-1}, \mathcal{D}, \boldsymbol{\alpha}) p(\boldsymbol{\theta}|\mathcal{D}, \boldsymbol{\alpha}) d\boldsymbol{\theta} \\ &= \int \theta_{w_i|w_{i-n+1}^{i-1}} \text{Dir}(\boldsymbol{\theta}|\mathbf{c} + \boldsymbol{\alpha}) d\boldsymbol{\theta} \\ &= \frac{c(\theta_{i|j}) + \alpha_i}{\sum_{k=1}^{N} [c(\theta_{k|j}) + \alpha_k]}, \end{aligned} \tag{5.70}$$

which is marginalized over all values of parameter $\boldsymbol{\theta}$. This predictive distribution is equivalent to calculating the *mean* of an n-gram parameter:

$$p(w_i|w_{i-n+1}^{i-1}) \triangleq \theta_{i|j}, \tag{5.71}$$

based on the *posterior distribution* $p(\boldsymbol{\theta}|\mathcal{D}, \boldsymbol{\alpha})$, which is a Dirichlet distribution with hyperparameters $\mathbf{c} + \boldsymbol{\alpha}$. Here, the n-gram probability $p(w_i|w_{i-n+1}^{i-1})$ is simply expressed by $\theta_{i|j}$ where j denotes the back-off smoothing information from the lower-order model $p(w_i|w_{i-n+1}^{i-2})$, which is addressed in Section 5.3.4.

We may further conduct the next level of inference by inferring the hyperparameters given the data. The posterior distribution of $\boldsymbol{\alpha}$ is expressed by

$$p(\boldsymbol{\alpha}|\mathcal{D}) = \frac{p(\mathcal{D}|\boldsymbol{\alpha}) p(\boldsymbol{\alpha})}{p(\mathcal{D})}. \tag{5.72}$$

The hierarchical prior/posterior model of parameters $\boldsymbol{\theta}$ and hyperparameters $\boldsymbol{\alpha}$ is constructed accordingly. The marginal likelihood over hyperparameters $\boldsymbol{\alpha}$ is yielded by

$$p(w_i|w_{i-n+1}^{i-1}, \mathcal{D}) = \int p(w_i|w_{i-n+1}^{i-1}, \mathcal{D}, \boldsymbol{\alpha}) p(\boldsymbol{\alpha}|\mathcal{D}) d\boldsymbol{\alpha}. \tag{5.73}$$

We may find the most probable MAP estimate $\boldsymbol{\alpha}^{\mathsf{MAP}}$ by

$$\boldsymbol{\alpha}^{\mathsf{MAP}} = \arg\max_{\boldsymbol{\alpha}} p(\boldsymbol{\alpha}|\mathcal{D}). \tag{5.74}$$

Then the marginal distribution is approximated as

$$p(w_i|w_{i-n+1}^{i-1}, \mathcal{D}) \approx p(w_i|w_{i-n+1}^{i-1}, \mathcal{D}, \boldsymbol{\alpha}^{\mathsf{MAP}}). \tag{5.75}$$

In addition, we would like to calculate the optimal hyperparameters via ML2 estimation from training data $\mathcal{D}$ based on the evidence framework. To do so, we need to determine the *evidence function* given hyperparameters $p(\mathcal{D}|\boldsymbol{\alpha})$. By referring to Eq. (5.68), this function is derived as

$$p(\mathcal{D}|\boldsymbol{\alpha}) = \frac{Z(\mathbf{c} + \boldsymbol{\alpha})}{Z(\boldsymbol{\alpha})} = \prod_j \left(\frac{\prod_{i=1}^{N} \Gamma(c(\theta_{i|j}) + \alpha_i)}{\Gamma(\sum_{i=1}^{N} c(\theta_{i|j}) + \alpha_i)} \cdot \frac{\Gamma(\sum_{i=1}^{N} \alpha_i)}{\prod_{i=1}^{N} \Gamma(\alpha_i)} \right), \tag{5.76}$$

which is viewed as a ratio of normalization constants of posterior probability $p(\boldsymbol{\theta}|\mathcal{D},\boldsymbol{\alpha})$ over prior probability $p(\boldsymbol{\theta}|\boldsymbol{\alpha})$.

5.3.4 Bayesian smoothed language model

It is important to illustrate the physical meaning of predictive distribution in Eq. (5.70). The hyperparameter α_i appears as an *effective initial count* for an n-gram event w_{i-n+1}^i. This marginal likelihood is integrated by the information sources from prior statistics $\boldsymbol{\alpha}$ as well as training data $\mathcal{D}$ or their counts of occurrences $\mathbf{c}$. On the other hand, the predictive distribution in Eq. (5.70) can be rewritten as

$$p(w_i|w_{i-n+1}^{i-1},\mathcal{D},\boldsymbol{\alpha}) = \frac{c(w_{i-n+1}^i)+\alpha_{w_i|w_{i-n+2}^{i-1}}}{\sum_{w_i}\left[c(w_{i-n+1}^i)+\alpha_{w_i|w_{i-n+2}^{i-1}}\right]}$$
$$= \lambda_{w_{i-n+1}^{i-1}} p_{\mathsf{ML}}(w_i|w_{i-n+1}^{i-1}) + (1-\lambda_{w_{i-n+1}^{i-1}})\frac{\alpha_{w_i|w_{i-n+2}^{i-1}}}{\sum_{w_i}\alpha_{w_i|w_{i-n+2}^{i-1}}}, \quad (5.77)$$

where $p_{\mathsf{ML}}(w_i|w_{i-n+1}^{i-1})$ denotes the ML model introduced in Eq. (3.192) and $1-\lambda_{w_{i-n+1}^{i-1}}$ implies the interpolation weight for prior statistics and is herein obtained as

$$1-\lambda_{w_{i-n+1}^{i-1}} = \frac{\sum_{w_i}\alpha_{w_i|w_{i-n+2}^{i-1}}}{\sum_{w_i}\left[c(w_{i-n+1}^i)+\alpha_{w_i|w_{i-n+2}^{i-1}}\right]}. \quad (5.78)$$

It is interesting to see that the predictive distribution in Eq. (5.77) is interpreted as the smoothed n-gram based on the *interpolation smoothing*. We build the tight connection between the Bayesian language model and a linearly smoothed language model as addressed in Section 3.6. The prior statistics or hyperparameters $\alpha_{w_i|w_{i-n+2}^{i-1}}$ should sufficiently reflect the backoff information from the low-order model $p(w_i|w_{i-n+2}^{i-1})$ when calculating the predictive n-gram probability $p(w_i|w_{i-n+1}^{i-1},\mathcal{D},\boldsymbol{\alpha})$. Comparing Eq. (5.78) and Eq. (3.211), the Bayesian language model is shown to be equivalent to the Witten–Bell smoothed language model in the case that the hyperparameters $\alpha_{w_i|w_{i-n+2}^{i-1}}$ are selected to meet the condition

$$\sum_{w_i}\alpha_{w_i|w_{i-n+2}^{i-1}} = N_{1+}(w_{i-n+1}^{i-1},\bullet). \quad (5.79)$$

(As before, the $\bullet$ represents any possible words at i that are summed over.)

Nevertheless, the advantage of the *Bayesian smoothed language model* is to automatically determine the optimal hyperparameters $\alpha_{w_i|w_{i-n+2}^{i-1}}$ from training data $\mathcal{D}$.

5.3.5 Optimal hyperparameters

According to the evidence framework, a type-2 ML estimation is carried out to find optimal hyperparameters $\boldsymbol{\alpha} = [\alpha_1,\cdots,\alpha_N]^\mathsf{T}$ by maximizing the evidence function

$$\boldsymbol{\alpha}^{\mathsf{ML2}} = \arg\max_{\boldsymbol{\alpha}} p(\mathcal{D}|\boldsymbol{\alpha}). \quad (5.80)$$

More specifically, we find individual parameter α_i^{ML2} through calculating the differentiation

$$\frac{\partial}{\partial \alpha_i} \log p(\mathcal{D}|\boldsymbol{\alpha}) = \sum_j \Bigg[\Psi(c(\theta_{i|j}) + \alpha_i) - \Psi\left(\sum_{i=1}^N c(\theta_{i|j}) + \alpha_i\right) + \Psi\left(\sum_{i=1}^N \alpha_i\right) - \Psi(\alpha_i) \Bigg], \tag{5.81}$$

where the di-gamma function

$$\Psi(x) \triangleq \frac{\partial}{\partial x} \log \Gamma(x) \tag{5.82}$$

is incorporated. We may use the conjugate gradient algorithm to find α_i^{ML2} or apply some approximation to derive an explicit optimization algorithm.

In general, it is reasonable that $\sum_{i=1}^N \alpha_i > 1$ and $\alpha_i < 1$. We can use the recursive formula of the di-gamma function (MacKay & Peto 1995),

$$\Psi(x+1) = \Psi(x) + \frac{1}{x}, \tag{5.83}$$

to combine the first and fourth terms in the brackets of Eq. (5.81) to obtain

$$\Psi(c(\theta_{i|j}) + \alpha_i) - \Psi(\alpha_i) = \frac{1}{c(\theta_{i|j}) - 1 + \alpha_i} + \frac{1}{c(\theta_{i|j}) - 2 + \alpha_i} + \cdots + \frac{1}{2 + \alpha_i} + \frac{1}{1 + \alpha_i} + \frac{1}{\alpha_i}. \tag{5.84}$$

The number of terms in the right-hand-side of Eq. (5.84) is $c(\theta_{i|j})$. Assuming α_i is smaller than 1, we can approximate Eq. (5.84), for $c(\theta_{i|j}) \geq 1$, by

$$\begin{aligned}
\Psi(c(\theta_{i|j}) + \alpha_i) - \Psi(\alpha_i) &= \frac{1}{\alpha_i} + \sum_{c=2}^{c(\theta_{i|j})} \left[\frac{1}{c - 1 + \alpha_i} \right] \\
&\approx \frac{1}{\alpha_i} + \sum_{c=2}^{c(\theta_{i|j})} \left[\frac{1}{c-1} - \frac{\alpha_i}{(c-1)^2} + O(\alpha_i^2) \right] \\
&= \frac{1}{\alpha_i} + \sum_{c=2}^{c(\theta_{i|j})} \frac{1}{c-1} - \alpha_i \sum_{c=2}^{c(\theta_{i|j})} \frac{1}{(c-1)^2} + O(\alpha_i^2).
\end{aligned} \tag{5.85}$$

Here, the function inside the brackets,

$$f(\alpha_i) = \frac{1}{c - 1 + \alpha_i}, \tag{5.86}$$

is approximated by a Taylor series at the point $\alpha_i = 0$. Further, we apply the approximation of a di-gamma function,

$$\Psi(x) \approx \log(x) - \frac{1}{2x} + O\left(\frac{1}{x^2}\right), \tag{5.87}$$

to approximate the second and third terms in Eq. (5.81) (MacKay & Peto 1995):

$$\begin{aligned}K(\boldsymbol{\alpha}) &= \sum_j \left\{ \Psi\left(\sum_{i=1}^N \alpha_i\right) - \Psi\left(\sum_{i=1}^N c(\theta_{i|j}) + \alpha_i\right)\right\} \\ &\approx \sum_j \log\left[\frac{\sum_{i=1}^N c(\theta_{i|j}) + \alpha_i}{\sum_{i=1}^N \alpha_i}\right] \\ &+ \frac{1}{2}\sum_j \left[\frac{\sum_{i=1}^N c(\theta_{i|j})}{\left(\sum_{i=1}^N \alpha_j\right)\left(\sum_{i=1}^N c(\theta_{i|j}) + \alpha_i\right)}\right].\end{aligned} \tag{5.88}$$

For each count c and word i, let N_{ci} be the number of back-off contexts j such that $c(\theta_{i|j}) \geq c$, and let c_i^{max} denote the largest c such that $N_{ci} > 0$. Denote the number of entries in row i of $c(\theta_{i|j})$ that are non-zero N_{1i} by V_i. We compute the quantities:

$$G_i = \sum_{c=2}^{c_i^{\mathsf{max}}} \frac{N_{ci}}{c-1}, \tag{5.89}$$

$$H_i = \sum_{c=2}^{c_i^{\mathsf{max}}} \frac{N_{ci}}{(c-1)^2}. \tag{5.90}$$

Finally, the solution to optimal hyperparameters $\boldsymbol{\alpha}^{\mathsf{ML2}}$ is obtained. The hyperparameter α_i^{ML2} corresponding to each word i should satisfy

$$\begin{aligned}\alpha_i^{\mathsf{ML2}} &= \frac{2V_i}{K(\boldsymbol{\alpha}) - G_i + \sqrt{(K(\boldsymbol{\alpha}) - G_i)^2 + 4H_iV_i}} \\ &\triangleq f_i(\boldsymbol{\alpha}).\end{aligned} \tag{5.91}$$

Again, this solution to ML2 estimation is expressed as an *implicit solution* because the right-hand-side of Eq. (5.91) is a function of hyperparameters $\boldsymbol{\alpha}$. Starting from the initial hyperparameters $\boldsymbol{\alpha}$ and the resulting function $K(\boldsymbol{\alpha})$, the individual hyperparameter is then updated to $\alpha_j^{(1)}$. The function $K(\boldsymbol{\alpha}^{(1)})$ is updated as well. The estimation procedure $\boldsymbol{\alpha} \rightarrow \boldsymbol{\alpha}^{(1)} \rightarrow \cdots$ converges very rapidly.

6 Asymptotic approximation

Asymptotic approximation is also well known in practical Bayesian approaches (De Bruijn 1970) for approximately obtaining the posterior distributions. For example, as we discussed in Chapter 2, the posterior distributions of a model parameter $p(\Theta|\mathbf{O})$ and a model $p(M|\mathbf{O})$ given an observation $\mathbf{O} = \{\mathbf{o}_t \in \mathbb{R}^D | t = 1, \cdots, T\})$ are usually difficult to solve. The approach assumes that we have enough data (i.e., T is sufficiently large), which also makes Bayesian inference mathematically tractable. As a particular example of asymptotic approximations, we introduce the Laplace approximation and Bayesian information criterion, which are widely used for speech and language processing.

The Laplace approximation is used to approximate a complex distribution as a Gaussian distribution (Kass & Raftery 1995, Bernardo & Smith 2009). It assumes that the posterior distribution is highly peaked at about its maximum value, which corresponds to the mode of the posterior distribution. Then the posterior distribution is modeled as a Gaussian distribution with the mode as a mean parameter. By using the approximation, we can obtain the posterior distributions analytically to some extent. Section 6.1 first explains the Laplace approximation in general. In Sections 6.3 and 6.4 we also discuss use of the Laplace approximation for analytically obtaining Bayesian predictive distributions for acoustic modeling and Bayesian extension of successful neural-network-based acoustic modeling, respectively.

Another example of this asymptotic approximation is the Bayesian information criterion (or Schwarz criterion (Schwarz 1978)). The Bayesian information criterion also assumes the large sample case, and approximates the posterior distribution of a model $p(M|\mathbf{O})$ with a simple equation. Since the Bayesian information criterion assumes the large sample case, it is also described as an instance of asymptotic approximations. Section 6.2 explains the Bayesian information criterion in general; it is used for model selection problems in speech processing. For example, Section 6.5 discusses the optimization of an appropriate model structure of hidden Markov models, and Section 6.6 discusses estimation of the number of speakers and detecting speech segments in conversations by regarding these approaches as model selection problems.

6.1 Laplace approximation

This section first describes the basic theory of the Laplace approximation. We first consider a simple case where a model does not have a latent variable. We focus on the posterior distributions of model parameters, but this approximation can be applied to the other continuous probabilistic variables. Let $\boldsymbol{\theta} \in \mathbb{R}^J$ be a J dimensional vector form of continuous model parameters and $\mathbf{O} = \{\mathbf{o}_t \in \mathbb{R}^D | t = 1, \cdots, T\}$ be a T-frame sequence of D dimensional feature vectors. We consider the posterior distribution of $\boldsymbol{\theta}$, which has the following distribution $p(\boldsymbol{\theta}|\mathbf{O})$:

$$p(\boldsymbol{\theta}|\mathbf{O}) \triangleq \exp(-f(\boldsymbol{\theta})), \tag{6.1}$$

where $f(\boldsymbol{\theta})$ is a continuous function over $\boldsymbol{\theta}$. Note that most parametric distributions can be represented by this equation. The Laplace approximation approximates $p(\boldsymbol{\theta}|\mathbf{O})$ as a Gaussian distribution with the mode of $p(\boldsymbol{\theta}|\mathbf{O})$ as a mean parameter, which can be obtained numerically or analytically for some specific distributions. Let $\boldsymbol{\theta}^{\mathrm{MAP}}$ be the mode (MAP value, as discussed in Chapter 4) of $p(\boldsymbol{\theta}|\mathbf{O})$, that is:

$$\begin{aligned} \boldsymbol{\theta}^{\mathrm{MAP}} &= \arg\max_{\boldsymbol{\theta}} p(\boldsymbol{\theta}|\mathbf{O}) \\ &= \arg\min_{\boldsymbol{\theta}} f(\boldsymbol{\theta}). \end{aligned} \tag{6.2}$$

Here we use Eq. (6.1). Since the mode $\boldsymbol{\theta}^{\mathrm{MAP}}$ is the minimum value in $f(\boldsymbol{\theta})$, $\boldsymbol{\theta}^{\mathrm{MAP}}$ also satisfies the following equation:

$$\nabla_{\boldsymbol{\theta}} f(\boldsymbol{\theta})|_{\boldsymbol{\theta}=\boldsymbol{\theta}^{\mathrm{MAP}}} = 0, \tag{6.3}$$

where $\nabla_{\boldsymbol{\theta}}$ is the gradient operator with respect to $\boldsymbol{\theta}$. Based on the property, a Taylor expansion around $\boldsymbol{\theta}^{\mathrm{MAP}}$ approximates $f(\boldsymbol{\theta})$ as follows:

$$\begin{aligned} f(\boldsymbol{\theta}) &\approx f(\boldsymbol{\theta}^{\mathrm{MAP}}) + (\boldsymbol{\theta} - \boldsymbol{\theta}^{\mathrm{MAP}})^{\mathsf{T}} \nabla_{\boldsymbol{\theta}} f(\boldsymbol{\theta})|_{\boldsymbol{\theta}=\boldsymbol{\theta}^{\mathrm{MAP}}} + \frac{1}{2}(\boldsymbol{\theta} - \boldsymbol{\theta}^{\mathrm{MAP}})^{\mathsf{T}} \mathbf{H} (\boldsymbol{\theta} - \boldsymbol{\theta}^{\mathrm{MAP}}) \\ &= f(\boldsymbol{\theta}^{\mathrm{MAP}}) + \frac{1}{2}(\boldsymbol{\theta} - \boldsymbol{\theta}^{\mathrm{MAP}})^{\mathsf{T}} \mathbf{H} (\boldsymbol{\theta} - \boldsymbol{\theta}^{\mathrm{MAP}}), \end{aligned} \tag{6.4}$$

where the second term in the first line is canceled by using Eq. (6.3). Equation (6.4) is a basic equation in this chapter. $\mathbf{H}$ is a $J \times J$ Hessian matrix defined as

$$\mathbf{H} \triangleq \nabla_{\boldsymbol{\theta}} \nabla_{\boldsymbol{\theta}} f(\boldsymbol{\theta})|_{\boldsymbol{\theta}=\boldsymbol{\theta}^{\mathrm{MAP}}}. \tag{6.5}$$

By using Eq. (6.1), it can also be represented as

$$\mathbf{H} \triangleq -\nabla_{\boldsymbol{\theta}} \nabla_{\boldsymbol{\theta}} \log p(\boldsymbol{\theta}|\mathbf{O})|_{\boldsymbol{\theta}=\boldsymbol{\theta}^{\mathrm{MAP}}}. \tag{6.6}$$

The Hessian matrix also appeared in Eq. (5.9), as the second derivative of the negative log posterior. The determinant of this matrix plays an important role in the Occam factor, which penalizes the model complexity. Note that the Hessian matrix is a symmetric matrix based on the following property:

$$\begin{aligned} [\mathbf{H}]_{ij} &= \frac{\partial}{\partial \theta_i} \frac{\partial}{\partial \theta_j} f(\boldsymbol{\theta})|_{\boldsymbol{\theta}=\boldsymbol{\theta}^{\mathrm{MAP}}} \\ &= \frac{\partial}{\partial \theta_j} \frac{\partial}{\partial \theta_i} f(\boldsymbol{\theta})|_{\boldsymbol{\theta}=\boldsymbol{\theta}^{\mathrm{MAP}}} = [\mathbf{H}]_{ji}. \end{aligned} \tag{6.7}$$

Thus, by substituting the Taylor expansion of $f(\boldsymbol{\theta})$ (Eq. (6.4)) into Eq. (6.1), we obtain the approximated form of $p(\boldsymbol{\theta}|\mathbf{O})$, as follows:

$$\begin{aligned} p(\boldsymbol{\theta}|\mathbf{O}) &= \exp(-f(\boldsymbol{\theta})) \\ &\approx \exp\left(-f(\boldsymbol{\theta}^{\text{MAP}}) - \frac{1}{2}(\boldsymbol{\theta} - \boldsymbol{\theta}^{\text{MAP}})^{\mathsf{T}}\mathbf{H}(\boldsymbol{\theta} - \boldsymbol{\theta}^{\text{MAP}})\right) \\ &\propto \exp\left(-\frac{1}{2}(\boldsymbol{\theta} - \boldsymbol{\theta}^{\text{MAP}})^{\mathsf{T}}\mathbf{H}(\boldsymbol{\theta} - \boldsymbol{\theta}^{\text{MAP}})\right). \end{aligned} \tag{6.8}$$

Here, $\exp\left(-f(\boldsymbol{\theta}^{\text{MAP}})\right)$ does not depend on θ, and we can neglect it. Thus, $p(\boldsymbol{\theta}|\mathbf{O})$ is approximated by the following Gaussian distribution with mean vector $\boldsymbol{\theta}^{\text{MAP}}$ and covariance matrix $\mathbf{H}^{-1}$:

$$p(\boldsymbol{\theta}|\mathbf{O}) \approx \mathcal{N}(\boldsymbol{\theta}|\boldsymbol{\theta}^{\text{MAP}}, \mathbf{H}^{-1}). \tag{6.9}$$

Note that since the dimensionality of the Hessian matrix $\mathbf{H}$ is the number of parameters J, and it is often large in our practical problems, we would have a numerical issue as to how to obtain $\mathbf{H}^{-1}$. To summarize the Laplace approximation, if we have an arbitrary distribution $p(\boldsymbol{\theta}|\mathbf{O})$ with given continuous distribution $f(\boldsymbol{\theta})$, the Laplace approximation provides an analytical Gaussian distribution by using the mode $\boldsymbol{\theta}^{\text{MAP}}$ and the Hessian matrix $\mathbf{H}$. $\boldsymbol{\theta}^{\text{MAP}}$ is usually obtained by some numerical computation or the MAP estimation.

Now, let us consider the relationship between MAP (Chapter 4) and Laplace approximations. As we discussed in Eq. (4.9), the MAP approximation of the posterior distribution is represented by a Dirac delta function as follows:

$$p^{\text{MAP}}(\boldsymbol{\theta}|\mathbf{O}) = \delta(\boldsymbol{\theta} - \boldsymbol{\theta}^{\text{MAP}}). \tag{6.10}$$

Therefore, Eq. (6.9) approaches the MAP solution, when the variance in the Gaussian distribution in Eq. (6.9) becomes very small. This often arises when the amount of training data is very large. In this sense, the Laplace approximation is a more precise approximation for the true posterior distribution than MAP, in terms of the asymptotic approximation. This precision of the Laplace approximation comes from consideration of the covariance matrix (Hessian matrix $\mathbf{H}$) effect in Eq. (6.9).

The Laplace approximation is widely used as an approximated Bayesian inference method for various topics (e.g., Bayesian logistic regression (Spiegelhalter & Lauritzen 1990, Genkin, Lewis & Madigan 2007), Gaussian processes (Rasmussen & Williams 2006), and Bayesian neural networks (MacKay 1992*c*)). Sections 6.3 and 6.4 discuss the applications of the Laplace approximation to Bayesian predictive classification and neural networks in acoustic modeling, respectively.

Note that the Laplace approximation is usually used to obtain the posterior distribution for non-exponential family distributions. Although the Laplace approximation can be used to deal with latent variable problems in the posterior distribution (e.g., Gaussian mixture model), the assumption of approximating a multiple peak distribution with a single peak Gaussian is not adequate in many cases. Therefore, the Laplace approximation is often used with the other approximations based on the EM algorithm (MAP–EM, VB–EM) or sampling algorithm (MCMC) that handle latent variable problems.

6.2 Bayesian information criterion

This section describes the Bayesian information criterion in general. We focus on the posterior distribution of model $p(M|\mathbf{O})$, which is introduced in Section 3.8.7. We first think about a simple case when there is no latent variable. Then, $p(M|\mathbf{O})$ is represented by using the Bayes rule and the sum rule as follows:

$$
\begin{aligned}
p(M|\mathbf{O}) &= \frac{p(\mathbf{O}|M)p(M)}{p(\mathbf{O})} \\
&= \frac{\int p(\mathbf{O}|\boldsymbol{\theta}, M)p(\boldsymbol{\theta}|M)d\boldsymbol{\theta}p(M)}{p(\mathbf{O})},
\end{aligned} \tag{6.11}
$$

where $p(M)$ denotes a prior distribution for a model M. Similarly to the formalization for Laplace approximation, we use a vector form of model parameters, that is $\boldsymbol{\theta} \in \mathbb{R}^J$, and J is the number of parameters. Then, $p(\boldsymbol{\theta}|M)$ denotes a prior distribution for a model parameter $\boldsymbol{\theta}$ given M, and $p(\mathbf{O}|\boldsymbol{\theta}, M)$ denotes a likelihood function given $\boldsymbol{\theta}$ and M.

Suppose we want to compare two types of models (M_1 and M_2) in terms of the posterior values of these models. Figure 6.1 compares fitting the same data with a single Gaussian (M_1) or two Gaussians (M_2). For this comparison, we can compute the following ratio of $p(M_1|\mathbf{O})$ and $p(M_2|\mathbf{O})$ and check whether the ratio is larger/less than 1:

$$
\begin{aligned}
\frac{p(M_1|\mathbf{O})}{p(M_2|\mathbf{O})} &= \frac{\frac{\int p(\mathbf{O}|\boldsymbol{\theta}_1, M_1)p(\boldsymbol{\theta}_1|M_1)d\boldsymbol{\theta}_1 p(M_1)}{p(\mathbf{O})}}{\frac{\int p(\mathbf{O}|\boldsymbol{\theta}_2, M_2)p(\boldsymbol{\theta}_2|M_2)d\boldsymbol{\theta}_2 p(M_2)}{p(\mathbf{O})}} \\
&= \frac{\int p(\mathbf{O}|\boldsymbol{\theta}_1, M_1)p(\boldsymbol{\theta}_1|M_1)d\boldsymbol{\theta}_1 p(M_1)}{\int p(\mathbf{O}|\boldsymbol{\theta}_2, M_2)p(\boldsymbol{\theta}_2|M_2)d\boldsymbol{\theta}_2 p(M_2)}.
\end{aligned} \tag{6.12}
$$

This factor is called the *Bayes factor* (Kass & Raftery 1993, 1995). To obtain the Bayes factor, we need to compute the marginal likelihood term:

$$
\int p(\mathbf{O}|\boldsymbol{\theta}, M)p(\boldsymbol{\theta}|M)d\boldsymbol{\theta}. \tag{6.13}
$$

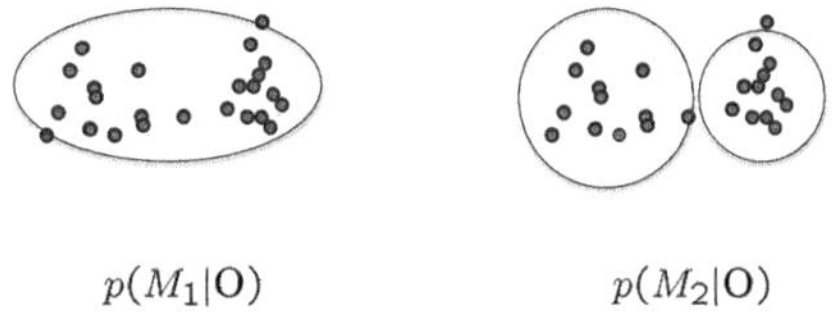

$p(M_1|\mathrm{O})$ $p(M_2|\mathrm{O})$

Figure 6.1 An example of model comparison, fitting the same data with a single Gaussian (M_1) or two Gaussians (M_2). By comparing the posterior probabilities of $p(M_1|\mathbf{O})$ and $p(M_2|\mathbf{O})$, we can select a more probable model.

Since it is difficult to solve this integral analytically, the following section approximates this calculation based on the asymptotic approximation, similarly to the Laplace approximation, as discussed in Section 6.1.

We first define the following continuous function g over $\boldsymbol{\theta}$:

$$p(\mathbf{O}|\boldsymbol{\theta}, M)p(\boldsymbol{\theta}|M) \triangleq \exp\left(-g(\boldsymbol{\theta})\right). \tag{6.14}$$

This is slightly different from the definition of function f in Eq. (6.1), as Eq. (6.1) focuses on the posterior distribution of the parameters. Then, similarly to the previous section, we define the following MAP value:

$$\begin{aligned}\boldsymbol{\theta}^* &= \arg\max_{\boldsymbol{\theta}} p(\mathbf{O}|\boldsymbol{\theta}, M)p(\boldsymbol{\theta}|M) \\ &= \arg\min_{\boldsymbol{\theta}} g(\boldsymbol{\theta}).\end{aligned} \tag{6.15}$$

By using the Taylor expansion result in Eq. (6.4) with $\boldsymbol{\theta}^*$, Eq. (6.13) can be represented as follows:

$$\begin{aligned}&\int p(\mathbf{O}|\boldsymbol{\theta}, M)p(\boldsymbol{\theta}|M)d\boldsymbol{\theta} \\ &\approx \exp\left(-g(\boldsymbol{\theta}^*)\right)\int \exp\left(-\frac{1}{2}(\boldsymbol{\theta}-\boldsymbol{\theta}^*)^{\mathsf{T}}\mathbf{H}(\boldsymbol{\theta}-\boldsymbol{\theta}^*)\right)d\boldsymbol{\theta} \\ &= \exp\left(-g(\boldsymbol{\theta}^*)\right)\frac{(2\pi)^{\frac{J}{2}}}{|\mathbf{H}|^{\frac{1}{2}}},\end{aligned} \tag{6.16}$$

where J is the number of dimensions of $\boldsymbol{\theta}$, that is the number of parameters. The Hessian matrix $\mathbf{H}$ is defined with $g(\boldsymbol{\theta})$ as

$$\mathbf{H} \triangleq \nabla_{\boldsymbol{\theta}}\nabla_{\boldsymbol{\theta}} g(\boldsymbol{\theta})|_{\boldsymbol{\theta}=\boldsymbol{\theta}^*}. \tag{6.17}$$

Or it is defined with $p(\mathbf{O}|\boldsymbol{\theta}, M)$ and $p(\boldsymbol{\theta}|M)$ as

$$\mathbf{H} \triangleq -\nabla_{\boldsymbol{\theta}}\nabla_{\boldsymbol{\theta}} \log\left(p(\mathbf{O}|\boldsymbol{\theta}, M)p(\boldsymbol{\theta}|M)\right)|_{\boldsymbol{\theta}=\boldsymbol{\theta}^*}. \tag{6.18}$$

The integral in Eq. (6.16) can be solved by using the following normalization property of a Gaussian distribution:

$$\begin{aligned}&\int \mathcal{N}(\boldsymbol{\theta}|\boldsymbol{\theta}^*, \mathbf{H}^{-1})d\boldsymbol{\theta} = \frac{|\mathbf{H}|^{\frac{1}{2}}}{(2\pi)^{\frac{J}{2}}}\int \exp\left(-\frac{1}{2}(\boldsymbol{\theta}-\boldsymbol{\theta}^*)^{\mathsf{T}}\mathbf{H}(\boldsymbol{\theta}-\boldsymbol{\theta}^*)\right)d\boldsymbol{\theta} = 1 \\ &\Longrightarrow \int \exp\left(-\frac{1}{2}(\boldsymbol{\theta}-\boldsymbol{\theta}^*)^{\mathsf{T}}\mathbf{H}(\boldsymbol{\theta}-\boldsymbol{\theta}^*)\right)d\boldsymbol{\theta} = \frac{(2\pi)^{\frac{J}{2}}}{|\mathbf{H}|^{\frac{1}{2}}}.\end{aligned} \tag{6.19}$$

Thus, from Eq. (6.16), the marginal likelihood is computed by using the determinant of the Hessian matrix $|\mathbf{H}|^{\frac{1}{2}}$.

Here, we focus on the determinant of the Hessian matrix $|\mathbf{H}|$. From the definition in Eq. (6.18), the Hessian matrix can be represented as:

$$\begin{aligned}\mathbf{H} &= -\nabla_{\boldsymbol{\theta}}\nabla_{\boldsymbol{\theta}} \log p(\mathbf{O}|\boldsymbol{\theta}, M)|_{\boldsymbol{\theta}=\boldsymbol{\theta}^*} - \nabla_{\boldsymbol{\theta}}\nabla_{\boldsymbol{\theta}} \log p(\boldsymbol{\theta}|M)|_{\boldsymbol{\theta}=\boldsymbol{\theta}^*} \\ &\approx -\nabla_{\boldsymbol{\theta}}\nabla_{\boldsymbol{\theta}} \log p(\mathbf{O}|\boldsymbol{\theta}, M)|_{\boldsymbol{\theta}=\boldsymbol{\theta}^*}.\end{aligned} \tag{6.20}$$

Thus, the Hessian matrix is decomposed to the logarithmic likelihood and prior term, and we neglect the prior term that does not depend on the amount of data. If we assume that the observed data are independently and identically distributed (iid), the likelihood term of T frame feature vectors is approximated when T is very large as[1]

$$\nabla_{\boldsymbol{\theta}} \nabla_{\boldsymbol{\theta}} \log p(\mathbf{O}|\boldsymbol{\theta}, M)|_{\boldsymbol{\theta}=\boldsymbol{\theta}^*} \approx -T\mathbf{I}, \tag{6.21}$$

where $\mathbf{I}$ is a Fisher information matrix obtained from a single observation, which does not depend on the amount of data. Therefore, we can approximate the determinant of the Hessian matrix as:

$$|\mathbf{H}|^{-\frac{1}{2}} \approx |T\mathbf{I}|^{-\frac{1}{2}} = T^{-\frac{J}{2}}|\mathbf{I}|. \tag{6.22}$$

This approximation implies that the Hessian matrix approaches zero when the amount of data is large. That is

$$\mathcal{N}(\boldsymbol{\theta}|\boldsymbol{\theta}^*, \mathbf{H}^{-1}) \to \delta(\boldsymbol{\theta} - \boldsymbol{\theta}^*) \text{ when } T \to \infty. \tag{6.23}$$

Therefore, the Laplace approximation can be equivalent to the MAP approximation when the amount of data is large, which is a reasonable outcome.

In the BIC approximation, by substituting Eq. (6.22) into Eq. (6.16), we obtain

$$\log \left(\int p(\mathbf{O}|\boldsymbol{\theta}, M) p(\boldsymbol{\theta}|M) \right) \approx \log p(\mathbf{O}|\boldsymbol{\theta}^*, M) + \log p(\boldsymbol{\theta}^*|M) + \frac{J}{2}\log(2\pi) - \frac{J}{2}\log T - \frac{J}{2}\log|\mathbf{I}|. \tag{6.24}$$

Further, by neglecting the terms that do not depend on T, we finally obtain the following simple equation:

$$\log p(M|\mathbf{O}) \propto \log \left(\int p(\mathbf{O}|\boldsymbol{\theta}, M) p(\boldsymbol{\theta}|M) \right) \approx \underbrace{\log p(\mathbf{O}|\boldsymbol{\theta}^*, M)}_{\text{log likelihood}} - \underbrace{\frac{J}{2}\log T}_{\text{penalty term}} . \tag{6.25}$$

Here, $\boldsymbol{\theta}^*$ is often approximated by the ML estimate $\boldsymbol{\theta}^{\mathrm{ML}}$ instead of $\boldsymbol{\theta}^{\mathrm{MAP}}$, since $\boldsymbol{\theta}^{\mathrm{MAP}}$ approaches $\boldsymbol{\theta}^{\mathrm{ML}}$ when $T \to \infty$, as discussed in Section 4.3.7. Model selection based on this equation is said to be based on the *Bayesian information criterion* (BIC), or Schwartz information criterion, which approximates the logarithmic function of the posterior distribution of a model $p(M|\mathbf{O})$. The first term on the right-hand-side is a log-likelihood term with the MAP estimate (it is often approximated by the ML estimate) of model parameter $\boldsymbol{\theta}$. It is well known that this log-likelihood value is always increased when the number of parameters increases. Therefore, the log-likelihood value from the maximum likelihood criterion cannot be used for model selection since it will always prefer the model that has the largest number of parameters. On the other hand, Eq. (6.25), which is based on the Bayesian criterion, has an additional term, which

[1] This proof is not obvious. See Ghosh, Delampady & Samanta (2007) for more detailed derivations.

is proportional to the number of model parameters, denoted by J, and the logarithmic number of observation frames, denoted by $\log T$. This term provides the penalty so that the total value in Eq. (6.25) is penalized not to select a model with too many parameters.

Comparing Eq. (6.25) with the regularized form of the objective function in Eq. (4.3) in the MAP estimation, Eq. (6.25) can be viewed as having an l^0 regularization term:

$$\begin{aligned}\log p(M|\mathbf{O}) &\approx \log p(\mathbf{O}|\boldsymbol{\theta}^*, M) - \frac{J}{2}\log T \\ &= \underbrace{\log p(\mathbf{O}|\boldsymbol{\theta}^*, M)}_{\text{log likelihood}} - \underbrace{\frac{\|\boldsymbol{\theta}\|_0}{2}\log T}_{l^0 \text{ regularization term}} \quad .\end{aligned} \tag{6.26}$$

Therefore, the BIC objective function can also be discussed within a regularization perspective. Another well-known model selection criterion, the Akaike information criterion (AIC) (Akaike 1974), is similarly represented by

$$\begin{aligned}-\frac{1}{2}\text{AIC} &= \log p(\mathbf{O}|\boldsymbol{\theta}^*, M) - J \\ &= \log p(\mathbf{O}|\boldsymbol{\theta}^*, M) - \|\boldsymbol{\theta}\|_0.\end{aligned} \tag{6.27}$$

It also has the l^0 regularization term, but it does not depend on the amount of data T.

Using the BIC, we can simplify the model comparison based on the Bayes factor. That is, by substituting Eq. (6.25) into the logarithm of the Bayes factor (Eq. (6.12)) we can also obtain the following equation:

$$\log\left(\frac{p(M_1|\mathbf{O})}{p(M_2|\mathbf{O})}\right) \approx \log p(\mathbf{O}|\boldsymbol{\theta}_1^*, M_1) - \log p(\mathbf{O}|\boldsymbol{\theta}_2^*, M_2) - \frac{J_1 - J2}{2}\log T. \tag{6.28}$$

Therefore, we can perform the model comparison by checking the sign of Eq. (6.28). If the sign is positive, M_1 is a more appropriate model in the BIC sense.

However, since we usually cannot assume that a single Gaussian distribution applies to a complicated model and/or that there are enough data to satisfy the large sample approximation, the BIC assumption is not valid for our practical problems. Therefore, in practical use, we introduce a tuning parameter λ which heuristically controls the balance of the log-likelihood and penalty terms, as follows:

$$\log\left(\int p(\mathbf{O}|\boldsymbol{\theta}, M)p(\boldsymbol{\theta}|M)\right) \approx \log p(\mathbf{O}|\boldsymbol{\theta}^*, M) - \lambda\frac{J}{2}\log T. \tag{6.29}$$

This actually works effectively for the model selection problems of HMMs and GMMs, including speech processing (Chou & Reichl 1999, Shinoda & Watanabe 2000) and image processing (Stenger, Ramesh, Paragios *et al.* 2001). We discuss the applications of BIC to speech processing in Sections 6.5 and 6.6.

6.3 Bayesian predictive classification

This section addresses how the Laplace approximation is developed to build the Bayesian predictive classification (BPC) approach which has been successfully applied for robust speech recognition in noisy environments (Jiang *et al.* 1999, Huo & Lee 2000, Lee & Huo 2000, Chien & Liao 2001). As mentioned in Section 3.1, the Bayes decision theory for speech recognition can be realized as the BPC rule where the expected loss function in Eq. (3.16) is calculated by marginalizing over the continuous density HMM parameters $\Theta = \{\boldsymbol{\pi} = \{\pi_j\}, A = \{a_{ij}\}, B = \{\omega_{jk}, \boldsymbol{\mu}_{jk}, \boldsymbol{\Sigma}_{jk}\}\}$, as discussed in Section 3.2.3, as well as the n-gram parameters $\Theta = \{p(w_i|w_{i-n+1}^{i-1})\}$, as discussed in Section 3.6. Here, we only focus on the acoustic models based on HMMs. The marginalization over all possible parameter values aims to construct a decision rule which considers the *prior uncertainty* of HMM parameters. The resulting speech recognition is robust to mismatch conditions between training and test data. In general, such marginalization could compensate some other ill-posed conditions due to the uncertainties and variations from the observed data and the assumed models. In this section, we first describe how a Bayesian decision rule is established for robust speech recognition and how the uncertainties of model parameters are characterized. Then, the approximate solution to the HMM-based decision rule is formulated by using the Laplace approximation. We also present some other extensions of the BPC decision rule.

6.3.1 Robust decision rule

In automatic speech recognition, we usually apply the plug-in maximum a-posteriori (MAP) decision rule, as given in Eq. (3.2), and combine the approximate acoustic model $\hat{p}_\Theta(\mathbf{O}|W)$ and language model $\hat{p}_\Theta(W)$ for finding the most likely word sequence W corresponding to an input speech sentence $\mathbf{O} = \{\mathbf{o}_t\}$. The maximum likelihood (ML) parameters Θ^{ML}, as described in Section 3.4, and the MAP parameters Θ^{MAP}, as formulated in Section 4.3, are treated as the point estimates and plugged into the MAP decision for speech recognition. However, in practical situations, speech recognition systems are occasionally applied in noisy environments. The training data may be collected in ill-posed conditions, where the HMM parameters may be over-trained with limited training data conditions, and can lose the generalization capability for future data conditions. Thus, the assumed model may not properly represent the real-world speech data. For this situation, the plug-in MAP decision, relying only on the single best HMM parameter values Θ^{ML} or Θ^{MAP}, is risky for speech recognition. Because of these considerations, we are motivated to establish a robust decision rule which accommodates the uncertainties of the assumed model and the collected data so that the error risks in speech recognition can be reduced.

In the HMM framework (Section 3.2), the prior uncertainty of GMM parameters $\Theta_j = \{\omega_{jk}, \boldsymbol{\mu}_{jk}, \boldsymbol{\Sigma}_{jk}|k = 1, \cdots, K\}$ in an HMM state j should be characterized from training data and then applied for a BPC decision of future data. In this section, we use the following notation to clearly distinguish training and future data:

$$\mathbf{O} : \text{future data},$$
$$\mathcal{O} : \text{training data.} \tag{6.30}$$

Let the distribution $p(\mathbf{o}_t|\Theta_j)$ of a future observation frame $\mathbf{o}_t$ under an HMM state j be distorted in an admissible set $\mathcal{P}_j(\epsilon_j)$ with a distortion level $\epsilon_j \leq 0$ which is denoted by

$$\mathcal{P}_j(\epsilon_j) = \{p(\mathbf{o}_t|\Theta_j)|\Theta_j \in \Omega(\epsilon_j)\}, \tag{6.31}$$

where $\Omega(\epsilon_j)$ denotes the admission region of the GMM parameter space. In the special case of no distortion ($\epsilon_j = 0$), $\mathcal{P}_j(0) = \{p(\mathbf{o}_t|\Theta_j^{(0)})\}$ is a singleton set which consists of the ideal and non-distorted distribution of an HMM state j with the parameters $\Theta_j^{(0)}$ estimated from a training set $\mathcal{O}$. However, in real-world applications, there exist many kinds of distortions ($\epsilon_j > 0$) between the trained models and the test observations. These distortions and variations should be modeled and incorporated in the Bayes decision to achieve robustness of speech recognition in adverse environments.

The Bayesian inference approach provides a good way to formalize this parameter uncertainty problem and formulate the solution to the robust decision rule. To do so, we intend to consider the uncertainty of the HMM parameters Θ to be random. An a-priori distribution $p(\Theta|\Psi)$ could serve as the prior knowledge about Θ, where $\Theta \in \Omega$ are located in a region of interest of the HMM parameter space Ω, and Ψ denotes the corresponding hyperparameters. Such prior information may come from subject considerations and/or from previous experience. Therefore, the BPC decision rule is established as

$$\begin{aligned} \hat{W} = d_{\mathrm{BPC}}(\mathbf{O}) &= \arg\max_W \tilde{p}(W|\mathbf{O}) \\ &= \arg\max_W \tilde{p}_\Theta(\mathbf{O}|W)p_\Theta(W), \end{aligned} \tag{6.32}$$

where

$$\tilde{p}(\mathbf{O}|W) = \int_\Omega p(\mathbf{O}|\Theta, W)p(\Theta|\mathcal{O}, W)d\Theta. \tag{6.33}$$

Here, $p(\Theta|\mathcal{O}, W)$ denotes the posterior distribution of HMM parameters Θ given the training utterances $\mathcal{O}$, which is written as

$$p(\Theta|\mathcal{O}, W) = \frac{p(\mathcal{O}|\Theta, W)p(\Theta)}{\int_\Omega p(\mathcal{O}|\Theta, W)p(\Theta)d\Theta}. \tag{6.34}$$

In Eq. (6.33), the marginalization is performed over all possible HMM parameter values Θ. This marginalized distribution is also called the *predictive distribution*. Model uncertainties are considered in the BPC decision rule. The optimum BPC decision rule was illustrated in Nadas (1985).

The crucial difference between the plug-in MAP decision and the BPC decision is that the former acts as if the estimated HMM parameters were the true ones, whereas the latter averages over the uncertainty of parameters. In an extreme case, if $p(\Theta|\mathcal{O}, W) = \delta(\Theta - \Theta^{\mathrm{ML/MAP}})$ with $\delta(\cdot)$ denoting the Dirac delta function with the ML or MAP estimate, the BPC decision rule coincides with the plug-in MAP decision rule based on ML HMM parameters $\Theta^{\mathrm{ML/MAP}}$, i.e., the predictive distribution can be approximated as:

$$\begin{aligned}\tilde{p}(\mathbf{O}|W) &= \int_\Omega p(\mathbf{O}|\Theta, W)p(\Theta|\mathcal{O}, W)d\Theta \\ &\approx \int_\Omega p(\mathbf{O}|\Theta, W)\delta(\Theta - \Theta^{\mathrm{ML/MAP}})d\Theta \\ &= p(\mathbf{O}|\Theta^{\mathrm{ML/MAP}}, W).\end{aligned} \tag{6.35}$$

Therefore, the plug-in MAP decision rule with the ML/MAP estimate corresponds to an approximation of the BPC decision rule.

Using the continuous density HMMs, the missing data problem happens so that the predictive distribution is calculated by considering all sequences of HMM states $S = \{s_t\}$ and mixture components $V = \{v_t\}$:

$$\begin{aligned}\tilde{p}(\mathbf{O}|W) &= \int p(\mathbf{O}|\Theta, W)p(\Theta|\mathcal{O}, W)d\Theta \\ &= \int p(\mathbf{O}|\Theta, W)p(\Theta|\Psi, W)d\Theta \\ &= \sum_{S,V} \int p(\mathbf{O}, S, V|\Theta, W)p(\Theta|\Psi, W)d\Theta \\ &= \sum_{S,V} \int p(\mathbf{O}|S, V, \Theta, W)p(S, V, |\Theta, W)p(\Theta|\Psi, W)d\Theta \\ &\triangleq \mathbb{E}_{(S,V,\Theta)}[p(\mathbf{O}|S, V, \Theta, W)],\end{aligned} \tag{6.36}$$

where the posterior distribution $p(\Theta|\mathcal{O}, W)$ given training data $\mathcal{O}$ is replaced by the prior distribution $p(\Theta|\Psi, W)$ given hyperparameters Ψ in accordance with an empirical Bayes theory. This predictive distribution is seen as an integration of likelihood function $p(\mathbf{O}|S, V, \Theta, W)$ over all possible values of $\{S, V, \Theta\}$. However, the computation of predictive distribution over all state sequences S and mixture component sequences V is complicated and expensive. A popular way to deal with this computation is to employ the Viterbi approximation, as we discussed in Section 3.3.2, to compute the approximate predictive distribution (Jiang *et al.* 1999):

$$\tilde{p}(\mathbf{O}|W) \approx \max_{S,V} \int p(\mathbf{O}, S, V|\Theta, W)p(\Theta|\Psi, W)d\Theta. \tag{6.37}$$

Therefore, the approximated predictive distribution only considers marginalization over the model parameter Θ. The following sections describe how to deal with model parameter marginalization.

6.3.2 Laplace approximation for BPC decision

In this section, Laplace approximation is adopted to approximate the integral in calculation of the predictive distribution $\tilde{p}(\mathbf{O}, S, V|W)$ in Eq. (6.36) or in Eq. (6.37), given state and mixture component sequences S and V. Let us define the continuous function $g(\Theta)$, similarly to Eqs. (6.1) and (6.14):

$$p(\mathbf{O}, S, V|\Theta, W)p(\Theta|\Psi, W) = \exp(-g(\Theta)). \tag{6.38}$$

The value of Θ that minimizes $g(\Theta)$ is obtained as a mode or an MAP estimate of HMM parameters,

$$\begin{aligned}\Theta^{\mathrm{MAP}} &= \arg\min_{\Theta} g(\Theta) \\ &= \arg\max_{\Theta} p(\mathbf{O}, S, V|\Theta, W)p(\Theta|\Psi, W).\end{aligned} \tag{6.39}$$

Note that we have a solution to the above optimization based on the MAP estimation, as we discussed in Section 4.3. Then since we have the equation $\nabla_\Theta g(\Theta)|_{\Theta=\Theta^{\mathrm{MAP}}} = 0$ from Eq. (6.39), a second-order Taylor expansion around Θ^{MAP} can be used to approximate $g(\Theta)$ as follows:

$$g(\Theta) \approx g(\Theta^{\mathrm{MAP}}) + \frac{1}{2}(\Theta - \Theta^{\mathrm{MAP}})^\intercal \mathbf{H}(\Theta - \Theta^{\mathrm{MAP}}), \tag{6.40}$$

where $\mathbf{H}$ denotes a Hessian matrix defined as

$$\mathbf{H} \triangleq \nabla_\Theta \nabla_\Theta g(\Theta)|_{\Theta=\Theta^{\mathrm{MAP}}}. \tag{6.41}$$

We should note that we use a vector representation of Θ (i.e., $\Theta \in \mathbb{R}^M$ and M is the number of all CDHMM parameters) by arranging all CDHMM parameters ($\{a_{ij}, \omega_{jk}, \boldsymbol{\mu}_{jk}, \boldsymbol{\Sigma}_{jk}\}$) to form a huge vector, to make the above gradient well defined.[2]

Therefore, we obtain the approximate predictive distribution, which is derived from

$$\begin{aligned}\tilde{p}(\mathbf{O}, S, V|W) &= \int \exp(-g(\Theta))d\Theta \\ &\approx \exp(-g(\Theta^{\mathrm{MAP}})) \int \exp\left(-\frac{1}{2}(\Theta - \Theta^{\mathrm{MAP}})^\intercal \mathbf{H}(\Theta - \Theta^{\mathrm{MAP}})\right)d\Theta.\end{aligned} \tag{6.42}$$

Similarly to Eq. (6.18), the predictive distribution in Eq. (6.42) is approximately obtained through arranging a multivariate Gaussian distribution in the integrand of Eq. (6.42) as follows:

$$\begin{aligned}\tilde{p}(\mathbf{O}, S, V|W) &\approx p(\mathbf{O}, S, V|\Theta^{\mathrm{MAP}}, W) \\ &\quad \times p(\Theta^{\mathrm{MAP}}|\Psi, W)(2\pi)^{M/2}|\mathbf{H}|^{-1/2}.\end{aligned} \tag{6.43}$$

As a result, the BPC decision based on Laplace approximation is implemented by Eq. (6.43). The logarithm of predictive distribution, $\log \tilde{p}(\mathbf{O}, S, V|W)$, is seen as follows:

$$\begin{aligned}&\log \tilde{p}(\mathbf{O}, S, V|W) \\ &\approx \log\left(p(\mathbf{O}, S, V|\Theta^{\mathrm{MAP}}, W)p(\Theta^{\mathrm{MAP}}|\Psi, W)\right) + \log\left((2\pi)^{M/2}|\mathbf{H}|^{-1/2}\right).\end{aligned} \tag{6.44}$$

[2] We should also note that most CDHMM parameters have some constraint (e.g., state transitions and mixture weights have positivity and sum-to-one constraint) and the gradient of these parameters should consider these constraints. However, the following example only considers mean vector parameters, and we do not have this constraint problem.

The likelihood of the first term can be obtained by using the MAP–EM (or Viterbi approximation), as we discussed in Section 4.3. The second term is based on the logarithm of the determinant of the Hessian matrix. From Eq. (6.9), the Hessian matrix can be interpreted as the precision (inverse covariance) matrix of the model parameters ($p(\Theta|\mathbf{O}) \approx \mathcal{N}(\Theta|\Theta^{\mathrm{MAP}}, \mathbf{H}^{-1})$). Therefore, if the model parameters are uncertain, the determinant of the covariance matrix of Θ becomes large (i.e., the determinant of the Hessian matrix $\mathbf{H}$ becomes small), the log likelihood of the predictive distribution is decreased. Thus, considering model uncertainty in the BPC decision is meaningfully reflected in Eq. (6.43).

6.3.3 BPC decision considering uncertainty of HMM means

To simplify the discussion, we only consider the uncertainty of the mean vectors in HMM parameters for BPC decoding, as the mean vectors are the most dominant parameters for the ASR performance. In addition, by only focusing on the mean vectors, we can avoid the difficulty of applying the Laplace approximation to the covariance and weight parameters, which have constraints, assuming the HMM means of state j[3] and mixture component k in dimension d are independently generated from Gaussian prior distributions with the Hessian matrix as the precision matrix.

The joint prior distribution $p(\Theta|\Psi, W)$ is expressed by

$$p(\{\mu_{jkd}\}|\{\mu^0_{jkd}, \Sigma^0_{jkd}\}, W) = \prod_{j=1}^{J}\prod_{k=1}^{K}\prod_{d=1}^{D} \frac{1}{\sqrt{2\pi\,\Sigma^0_{jkd}}} \exp\left[-\frac{(\mu_{jkd} - \mu^0_{jkd})^2}{2\Sigma^0_{jkd}}\right], \tag{6.45}$$

with a collection of hyperparameters including Gaussian mean vectors $\boldsymbol{\mu}^0_{jk} = \{\mu^0_{jkd}\}$ and diagonal covariance matrices $\boldsymbol{\Sigma}^0_{jk} = \mathrm{diag}\,\{\Sigma^0_{jkd}|d = 1, \cdots, D\}$ for different states and mixture components. The GMM mixture weights $\{\omega_{jk}\}$ and diagonal covariance matrices $\{\boldsymbol{\Sigma}_{jk} = \mathrm{diag}\,\{\Sigma_{jkd}|d = 1, \cdots, D\}\}$ of the original CDHMM are assumed to be deterministic without uncertainty. Note that the above prior distribution is conditional on the word sequence W, as it is required in the BPC decision rule in Eq. (6.36). This condition means that we only deal with the prior distributions of HMM parameters appearing in the state sequences obtained by hypothesized word sequence W. But this condition can be disregarded since this condition for the prior distribution does not matter in the following analysis.

Given an unknown test utterance $\mathbf{O} = \{\mathbf{o}_t|t = 1, \cdots, T\}$ and the unobserved state sequence $S = \{s_t|t = 1, \cdots, T\}$ and mixture component sequence $V = \{v_t|t = 1, \cdots, T\}$ obtained by hypothesized word sequence W, we combine the Gaussian likelihood function $p(\mathbf{O}|S, V, \{\mu'_{jkd}\}, W)$ and the Gaussian prior distribution $p(\{\mu'_{jkd}\}|\{\mu^0_{jkd}, \Sigma^0_{jkd}\}, W)$ based on the MAP–EM approach (Section 4.2) to give:

[3] Note that index j denotes all HMM states for all phonemes, unlike index j for a phoneme in Section 3.2. Therefore, the number of HMM states J appearing later in this section also means the number of all HMM states used in the acoustic models, which could be several thousand when we use a standard LVCSR setup.

$$\log p(\{\mu'_{jkd}\}|\mathbf{O}, W) \approx Q^{\text{MAP}}(\{\mu'_{jkd}\}|\{\mu_{jkd}\})$$
$$= \mathbb{E}_{(S,V)}[\log p(\mathbf{O}, S, V|\{\mu'_{jkd}\}, W)|\mathbf{O}, \{\mu_{jkd}\}]$$
$$+ \log p(\{\mu'_{jkd}\}|\{\mu^0_{jkd}, \Sigma^0_{jkd}\}, W). \tag{6.46}$$

Equation (6.46) is further rewritten to come up with an approximate posterior distribution which is arranged as a Gaussian distribution:

$$\log p(\{\mu'_{jkd}\}|\mathbf{O}, W) \approx \sum_{t=1}^{T}\sum_{j=1}^{J}\sum_{k=1}^{K} \gamma_t(j,k)\Big[\log p(\mathbf{o}_t|s_t = j, v_t = k, \mu'_{jkd}, W)$$
$$+ \log p(\mu'_{jkd}|\mu^0_{jkd}, \Sigma^0_{jkd}, W)\Big]$$
$$= \sum_{t=1}^{T}\sum_{j=1}^{J}\sum_{k=1}^{K} \log \mathcal{N}(\mu'_{jkd}|\hat{\mu}_{jkd}, \hat{\Sigma}_{jkd}), \tag{6.47}$$

with the hyperparameters derived as

$$\hat{\mu}_{jkd} = \frac{\Sigma_{jkd}}{c_{jk}\Sigma^0_{jkd} + \Sigma_{jkd}}\mu^0_{jkd} + \frac{c_{jk}\Sigma^0_{jkd}}{c_{jk}\Sigma^0_{jkd} + \Sigma_{jkd}}\bar{o}_{jkd}, \tag{6.48}$$
$$\frac{1}{\hat{\Sigma}_{jkd}} = \frac{1}{\Sigma^0_{jkd}} + \frac{c_{jk}}{\Sigma_{jkd}}, \tag{6.49}$$

where

$$\gamma_t(j,k) = p(s_t = j, v_t = k|\mathbf{O}, \{\mu_{jkd}\}, W), \tag{6.50}$$
$$c_{jk} = \sum_{t=1}^{T} \gamma_t(j,k), \tag{6.51}$$
$$\bar{\mathbf{o}}_{jk} = \frac{\sum_{t=1}^{T} \gamma_t(j,k)\mathbf{o}_t}{c_{jk}}. \tag{6.52}$$

The posterior distribution in Eq. (6.46) is obtained because we adopt the conjugate prior to model the uncertainty of HMM means. Recall the MAP estimation result of the mean vector in Eq. (6.53):

$$\hat{\mu}_{jkd} \triangleq \frac{\phi^{\mu}_{jkd}\mu^0_{jkd} + \sum_{t=1}^{T}\gamma_t(j,k)o_{td}}{\phi^{\mu}_{jk} + \sum_{t=1}^{T}\gamma_t(j,k)}. \tag{6.53}$$

Comparing Eqs. (6.53) and (6.48), it is apparent that Eq. (6.48) based on the BPC solution considers the effect of Σ^0_{jkd}.

An EM algorithm is iteratively applied to approximate the log posterior distribution of new estimate $\{\mu'_{jkd}\}$, $\log p(\{\mu'_{jkd}\}|\mathbf{O}, W)$, by using the posterior auxiliary function of new estimate $\{\mu'_{jkd}\}$ given the current estimate $\{\mu_{jkd}\}$, $Q^{\text{MAP}}(\{\mu'_{jkd}\}|\{\mu_{jkd}\})$. The calculation converges after several EM iterations. Given the updated occupation probability, the Gaussian posterior distribution of $\{\mu'_{jkd}\} = [\boldsymbol{\mu}'^{\mathsf{T}}_{11}, \boldsymbol{\mu}'^{\mathsf{T}}_{12}, \cdots, \boldsymbol{\mu}'^{\mathsf{T}}_{JK}]^{\mathsf{T}}$, which concatenates all Gaussian mean vectors, is calculated and used to find the approximation in Eq. (6.9):

$$p(\{\mu'_{jkd}\}|\mathbf{O}, W) \approx \mathcal{N}(\{\mu'_{jkd}\}|\{\hat{\mu}_{jkd}\}, \{\hat{\Sigma}_{jkd}\}) \\ \approx \mathcal{N}(\{\mu'_{jkd}\}|\{\mu^{\text{MAP}}_{jkd}\}, \{[(\mathbf{H}^{\mu})^{-1}]_{jkd}\}). \tag{6.54}$$

Therefore, we substitute

$$\mu^{\text{MAP}}_{jkd} = \hat{\mu}_{jkd}, \tag{6.55}$$

$$[(\mathbf{H}^{\mu})^{-1}]_{jkd} = \hat{\Sigma}_{jkd}, \tag{6.56}$$

into Eq. (6.43). This approximation avoids a need to directly compute the Hessian matrix, where the number of dimensions of the Hessian matrix is very large ($J \times K \times D$). The BPC-based recognition is accordingly implemented for robust speech recognition.

In Chien & Liao (2001), the BPC decision was extended to a transformation-based BPC decision where the uncertainties of transformation parameters were taken into account. The transformation parameters of HMM means and variances were characterized by the Gaussian–Wishart distribution. Using this decision, HMM parameters were treated as deterministic values. In Chien (2003), the transformation-based BPC was further combined with the maximum likelihood linear regression adaptation (Leggetter & Woodland 1995). The linear regression based BPC was proposed for noisy speech recognition. The transformation regression parameters were represented by the multivariate Gaussian distribution. The predictive distributions in these two methods were approximated by calculating the predictive distributions for individual frames $\mathbf{o}_t$ without involving the Laplace approximation.

6.4 Neural network acoustic modeling

Acoustic modeling based on the *deep neural network* (DNN) is now a new trend towards achieving high performance in automatic speech recognition (Dahl, Yu, Deng *et al.* 2012, Hinton *et al.* 2012). The context-dependent DNNs were combined with HMMs and have recently shown significant improvements over discriminatively trained HMMs with state-dependent GMMs (Seide, Li, Chen *et al.* 2011). In particular, the DNN was proposed to conduct a greedy and layer-wise pretraining of the weight parameters with either a supervised or unsupervised criterion (Hinton, Osindero & Teh 2006, Salakhutdinov 2009). This pretraining step prevents the supervised training of the network from being trapped in a poor local optimum. In general, there are five to seven hidden layers with thousands of sigmoid non-linear neurons in a DNN model. The output layer consists of softmax non-linear neurons. In addition to the pretraining scheme, the success of the DNN acoustic model also comes from the tandem processing, frame randomization (Seide *et al.* 2011) and (Hessian-free) sequence training (Kingsbury, Sainath & Soltau 2012, Veselỳ, Ghoshal, Burget *et al.* 2013). Without loss of generality, we address the artificial neural network (NN) acoustic model for ASR, based on the feed-forward multilayer perceptron with a single hidden layer, as

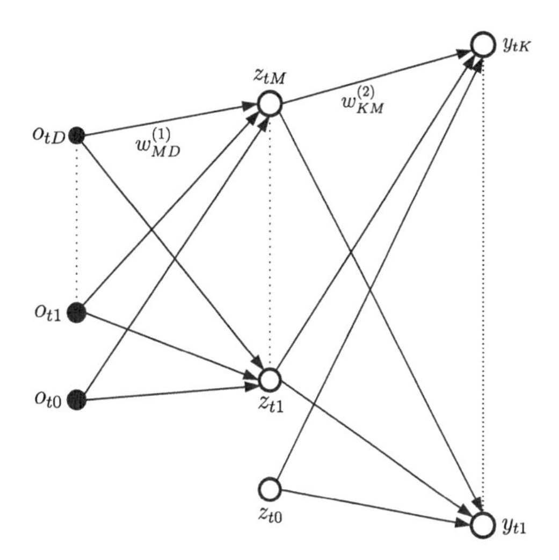

Figure 6.2 A neural network based on the multilayer perceptron with a single hidden layer. $\mathbf{o}_t \in \mathbb{R}^D$, $\mathbf{z}_t \in \mathbb{R}^M$, and $\mathbf{y}_t \in \mathbb{R}^K$ denote the input vector, hidden unit vector, and output vector at frame t, respectively. o_{t0} and z_{t0} are introduced to consider the bias terms in the transformation, and these are usually set with some fixed values (e.g., $o_{t0} = z_{t0} = 1$). $\mathbf{w}^{(1)} \in \mathbb{R}^{M(D+1)}$ and $\mathbf{w}^{(2)} \in \mathbb{R}^{K(M+1)}$ denote the weight vectors (to use a Laplace approximation, we represent these values as the vector representation rather than the matrix representation) from the input to hidden layers and from the hidden to output layers, respectively.

illustrated in Figure 6.2. We bring in the issue of model regularization in construction of NNs and present the *Bayesian neural networks* (MacKay 1992*c*, Bishop 2006) for robust speech recognition. This framework can obviously be extended to DNN-based speech recognition with many hidden layers. Importantly, we introduce a prior distribution to express the uncertainty of synaptic weights in Bayesian NNs. This uncertainty serves as a penalty factor to avoid too-complicated or over-trained models. The variations of acoustic models due to the heterogeneous training data could be compensated through Bayesian treatment. The Laplace approximation is applied to derive the predictive distribution for robust classification of a speech signal. In what follows, we first describe the combination of Bayesian NNs and HMMs for the application of speech recognition. MAP estimation is applied to find a solution to adaptive NN training. A Laplace approximation is developed to construct Bayesian NNs for HMM-based speech recognition.

6.4.1 Neural network modeling and learning

A standard neural network for a K-class classification problem is depicted in Figure 6.2, which is established as a feed-forward network function with a layer structure. This function maps the D input neurons $\mathbf{o}_t = \{o_{ti}\} \in \mathcal{R}^D$ to K output neurons $\mathbf{y}_t(\mathbf{o}_t, \mathbf{w}) = \{y_{tk}(\mathbf{o}_t, \mathbf{w})\} \in \mathcal{R}^K$ where

$$y_{tk}(\mathbf{o}_t, \mathbf{w}) = \frac{\exp(a_k(\mathbf{o}_t, \mathbf{w}))}{\sum_j \exp\left(a_j(\mathbf{o}_t, \mathbf{w})\right)} \tag{6.57}$$

is given by a softmax non-linear function, or the normalized exponential function which satisfies $0 \leq y_{tk} \leq 1$ and $\sum_{k=1}^{K} y_{tk} = 1$. In Eq. (6.57), $a_k(\mathbf{o}_t, \mathbf{w})$ is calculated as the output unit activation given by

$$a_k(\mathbf{o}_t, \mathbf{w}) = \sum_{j=0}^{M} w_{kj}^{(2)} \text{Sigmoid}\left(\sum_{i=0}^{D} w_{ji}^{(1)} o_{ti}\right), \tag{6.58}$$

where

$$\text{Sigmoid}(a) = \frac{1}{1 + \exp(-a)} \tag{6.59}$$

denotes the logistic sigmoid function and

$$\mathbf{w} = \{w_{ji}^{(1)}, w_{kj}^{(2)} | 1 \leq i \leq D, 1 \leq j \leq M, 1 \leq k \leq K\} \tag{6.60}$$

denotes the synaptic weights of the first and the second layers, respectively. For model training, the target values of output neurons are assigned as Bernoulli variables where value 1 in the kth neuron means that $\mathbf{o}_t$ belongs to class k and the other neurons have target value 0. In this figure, $\phi_{t0} = z_{t0} = 1$ and the corresponding weights $\{w_{j0}^{(1)}, w_{k0}^{(2)}\}$ denote the bias parameters in neurons. This model is also known as the *multilayer perceptron*. Such a neural network model performs well for pattern recognition for two reasons. One is the nested non-linear function, which can learn the complicated deep structure from real-world data through feed-forwarding the input signals to the output classes layer by layer. The second reason is that it adopts the softmax function or the logistic sigmoid function Sigmoid($\cdot$) as the neuron processing unit, which is beneficial for building high-performance discriminative models. In general, deep neural networks are affordable for learning the invariance information and extracting the hierarchy of concepts or features from data. Higher-level concepts are defined from lower-level ones. The logistic sigmoid function and softmax function play an important role for binary classification and multi-class classification, respectively. These functions are essential for calculating posterior probability and establishing the discriminative model. Basically, we collect a set of training samples and their target values $\{\mathbf{o}_t, \mathbf{t}_t\}$. The synaptic weights are estimated by calculating an error function $E(\mathbf{w})$ and minimizing it through the gradient descent algorithm,

$$\mathbf{w}^{(\tau+1)} = \mathbf{w}^{(\tau)} - \eta \nabla E(\mathbf{w}^{(\tau)}), \tag{6.61}$$

where η denotes the learning rate and τ denotes the iteration index. The sum-of-squares error function is calculated for the regression problem while the *cross entropy error function* is applied for the classification problem. The *error back-propagation* algorithm has been proposed to find gradient vector $\nabla E(\mathbf{w})$ and widely applied for training of individual synaptic weights $\mathbf{w} = \{w_{ji}^{(1)}, w_{kj}^{(2)}\}$ in different layers.

6.4.2 Bayesian neural networks and hidden Markov models

For the application of speech recognition, the NNs should be combined with the HMMs (denoted by NN–HMMs) to deal with *sequential training* of synaptic weights $\mathbf{w}$ from a sequence of speech feature vectors $\mathbf{O} = \{\mathbf{o}_t \in \mathbb{R}^D | t = 1, \cdots, T\}$. As discussed before, this feature can be obtained by stacking the neighboring MFCC or filter bank features. For example, if the original features are represented by $\mathbf{O}^{\text{org}} = \{\mathbf{o}_t^{\text{org}} \in \mathbb{R}^{D^{\text{org}}} | t = 1, \cdots, T\}$, the stacked feature for the NN input is represented by

$$\mathbf{o}_t = [(\mathbf{o}_{t-L}^{\text{org}})^\mathsf{T}, \cdots, (\mathbf{o}_t^{\text{org}})^\mathsf{T}, \cdots, (\mathbf{o}_{t-L}^{\text{org}})^\mathsf{T}]^\mathsf{T}. \tag{6.62}$$

Then, the number of feature dimensions becomes $D = (2L + 1)D^{\text{org}}$. L is set between five and seven depending on the task. This expanded feature can model the short-range dynamics of speech features, as we model in the conventional GMM framework by using the delta cepstrum or linear discriminant analysis techniques (Furui 1986, Haeb-Umbach & Ney 1992).

However, the stacking technique cannot fully model the long-range speech dynamics. The Markov states are used to characterize the temporal correlation in sequential patterns. We estimate the NN–HMM parameters $\Theta = \{\boldsymbol{\pi}, A, \mathbf{w}\}$, consisting of initial state probabilities $\boldsymbol{\pi}$, state transition probabilities A and the N-dimensional synaptic weights $\mathbf{w}$, by maximizing the likelihood function $p(\mathbf{O}|\Theta)$. Each output neuron calculates the posterior probability of a context dependent HMM state k given a feature vector $\mathbf{o}_t$ using synaptic weights $\mathbf{w}$, i.e.,

$$y_k(\mathbf{o}_t, \mathbf{w}) \triangleq p(s_t = k|\mathbf{o}_t, \mathbf{w}). \tag{6.63}$$

ML estimation of Θ has an incomplete data problem and should be solved according to the EM algorithm. The EM algorithm is an iterative procedure where the E-step is to calculate an auxiliary function of training utterances $\mathbf{O}$ using new NN–HMM parameters Θ' given parameters Θ at the current iteration:

$$\begin{aligned} Q^{\text{ML}}(\Theta'|\Theta) &= \mathbb{E}_{(S)}\{\log p(\mathbf{O}, S|\Theta')|\mathbf{O}, \Theta\} \\ &= \sum_S p(S|\mathbf{O}, \Theta) \left[\sum_{t=1}^T (\log a'_{s_{t-1}s_t} + \log p(\mathbf{o}_t|s_t, \mathbf{w}')) \right] \\ &\propto \sum_{t=1}^T \sum_{k=1}^K \gamma_t(k) \left[\log a'_{s_{t-1}s_t} + \log p(s_t = k|\mathbf{o}_t, \mathbf{w}') - \log p(s_t = k) \right], \end{aligned} \tag{6.64}$$

where $\gamma_t(k) = p(s_t = k|\mathbf{O}, \Theta)$ denotes the state occupation probability. NN–HMM parameters Θ' at a new iteration are then estimated by M-step:

$$\mathbf{w}' = \arg\min_{\mathbf{w}'} \{-Q^{\text{ML}}(\mathbf{w}'|\mathbf{w})\}. \tag{6.65}$$

Here, we only consider the estimation of the synaptic weights $\mathbf{w}$ by minimizing the corresponding negative auxiliary function:

$$\begin{aligned} -Q^{\text{ML}}(\mathbf{w}'|\mathbf{w}) &\propto -\sum_{t=1}^T \sum_{k=1}^K \gamma_t(k) \log p(s_t = k|\mathbf{o}_t, \mathbf{w}) \\ &= -\sum_{t=1}^T \sum_{k=1}^K \gamma_t(k) \log y_k(\mathbf{o}_t, \mathbf{w}). \end{aligned} \tag{6.66}$$

To implement the ML supervised training of NN–HMM parameters $\mathbf{w}$, we first perform Viterbi decoding of training speech $\mathbf{O} = \{\mathbf{o}_t | t = 1, \cdots, T\}$ by using true transcription and NN–HMM parameters Θ or $\mathbf{w}$ at the current iteration. Each frame $\mathbf{o}_t$ is assigned an associated HMM state or output neuron $s_t = k$. This Viterbi alignment is equivalent to assigning a label or target value $t_{tk} \triangleq \gamma_t(k)$ of a speech frame $\mathbf{o}_t$ at an output neuron k of either 0 or 1. That is, $\gamma_t(k) \in \{0, 1\}$ is treated as a Bernoulli target variable. The negative auxiliary function in ML estimation is accordingly viewed as the *cross entropy error function* between the Bernoulli target values based on current estimate $\mathbf{w}$ and the posterior distributions from the NN outputs using new estimate $\mathbf{w}'$. By minimizing the cross entropy between the desired outputs $\{\gamma_t(k)\}$ and the actual outputs $\{y_k(\mathbf{o}_t, \mathbf{w})\}$ over all time frames $1 \leq t \leq T$ and all context-dependent HMM states $1 \leq k \leq K$, we establish the discriminative acoustic models (NN–HMMs) for speech recognition. The ML NN–HMM parameters $\Theta^{\text{ML}} = \{\mathbf{w}^{\text{ML}}\}$ are estimated. The error back-propagation algorithm can be implemented to train the synaptic weights $\mathbf{w}^{\text{ML}}$ by minimizing $-Q(\mathbf{w}'|\mathbf{w})$, as given in Eq. (6.66).

More importantly, we address the maximum a-posteriori (MAP) estimation of NN–HMM parameters where the uncertainty of synaptic weights is compensated by using a prior distribution. For simplicity, we assume that the continuous values of weights in different layers come from a multivariate Gaussian distribution,

$$p(\mathbf{w}) = \mathcal{N}(\mathbf{w}|\boldsymbol{\mu}_w, \boldsymbol{\Sigma}_w), \tag{6.67}$$

where $\boldsymbol{\mu}_w$ denotes the mean vector and $\boldsymbol{\Sigma}_w$ denotes the covariance matrix. This prior information could be empirically inferred from training data. There are two Bayesian inference stages in the so-called Bayesian NN–HMMs. The first stage is to find the *point estimate* of synaptic weights according to the MAP estimation. This idea is similar to MAP estimation of HMM parameters (Gauvain & Lee 1994). MAP estimates of NN–HMM parameters are calculated for adaptive training or speaker adaptation. The second stage is to conduct the *distribution estimation*, and this stage tries to fulfil a full Bayesian analysis by calculating the marginal likelihood with respect to all values of synaptic weights. Robustness of speech recognition to variations of synaptic weights due to heterogeneous data is assured.

In the first inference stage, MAP estimates of model parameters $\Theta^{\text{MAP}} = \{\mathbf{w}^{\text{MAP}}\}$ are calculated by maximizing a-posteriori probability or the product of likelihood function and prior distribution. An EM algorithm is applied to find MAP estimates by maximizing the posterior auxiliary function $Q^{\text{MAP}}(\Theta'|\Theta)$. A MAP estimate of synaptic weights $\mathbf{w}^{\text{MAP}}$ is derived by minimizing the corresponding auxiliary function:

$$\begin{aligned} -Q^{\text{MAP}}(\mathbf{w}'|\mathbf{w}) \propto &- \sum_{t=1}^{T} \sum_{k=1}^{K} \gamma_t(k) \log p(s_t = k | \mathbf{o}_t, \mathbf{w}) \\ &+ \frac{1}{2}(\mathbf{w}' - \boldsymbol{\mu}_w)^{\intercal} \boldsymbol{\Sigma}_w^{-1} (\mathbf{w}' - \boldsymbol{\mu}_w), \end{aligned} \tag{6.68}$$

which is viewed as a kind of penalized cross entropy error function. The prior distribution provides subjective information for Bayesian learning, or equivalently, serves as a penalty function for model training. $\boldsymbol{\Sigma}_w$ plays the role of a metric matrix of the $\mathbf{w}' - \boldsymbol{\mu}_w$ distance. Again, the error back-propagation algorithm is applied to estimate the synaptic weights $\mathbf{w}'$ at a new iteration. Using this algorithm, the derivative of penalty function with respect to $\mathbf{w}'$ contributes the estimation of $\mathbf{w}^{\text{MAP}}$. The estimated $\mathbf{w}^{\text{MAP}}$ is treated as a point estimate or deterministic value for prediction or classification of future data.

6.4.3 Laplace approximation for Bayesian neural networks

In the second Bayesian inference stage, the synaptic weights $\mathbf{w}$ are treated as a latent random variable. The *hyperparameters* $\Psi = \{\boldsymbol{\mu}_w, \boldsymbol{\Sigma}_w\}$ are used to represent the uncertainty of random parameters $\mathbf{w}$. Such uncertainty information plays an important role in subjective inference, adaptive learning, and robust classification. Similarly to the BPC based on the standard HMMs as addressed in Section 6.3, we would like to present the BPC decision rule based on the combined NN–HMMs. To do so, we need to calculate the predictive distribution of test utterance $\mathbf{O}$ which is marginalized by considering all values of weight parameters $\mathbf{w}$ in the likelihood function $p(\mathbf{O}|\mathbf{w})$:

$$\begin{aligned}\tilde{p}(\mathbf{O}|\boldsymbol{\mu}_w, \boldsymbol{\Sigma}_w) &= \int p(\mathbf{O}, \mathbf{w}|\boldsymbol{\mu}_w, \boldsymbol{\Sigma}_w) d\mathbf{w} \\ &= \int p(\mathbf{O}|\mathbf{w}) p(\mathbf{w}|\boldsymbol{\mu}_w, \boldsymbol{\Sigma}_w) d\mathbf{w} \\ &\triangleq \mathbb{E}_{(\mathbf{w})}[p(\mathbf{O}|\mathbf{w})]. \end{aligned} \tag{6.69}$$

The integral is performed over the parameter space of $\mathbf{w}$. However, due to the non-linear feed-forward network function $y_k(\mathbf{o}_t, \mathbf{w}) = p(s_t = k|\mathbf{o}_t, \mathbf{w})$, the closed-form solution to the integral calculation does not exist and should be carried out by using the Laplace approximation. By applying the Laplace approximation for integral operation in BIC and BPC decision, as given in Eqs. (6.16) and (6.43), we develop the BPC decision based on the NN–HMMs according to the approximate predictive distribution:

$$\begin{aligned}\tilde{p}(\mathbf{O}|\boldsymbol{\mu}_w, \boldsymbol{\Sigma}_w) &\approx p(\mathbf{O}|\mathbf{w}^{\text{MAP}}) \cdot p(\mathbf{w}^{\text{MAP}}|\boldsymbol{\mu}_w, \boldsymbol{\Sigma}_w) \cdot |\mathbf{H}|^{-1/2} \\ &= p(\mathbf{w}^{\text{MAP}}|\mathbf{O}, \boldsymbol{\mu}_w, \boldsymbol{\Sigma}_w) \cdot |\mathbf{H}|^{-1/2} \\ &\approx \exp\left(Q^{\text{MAP}}(\mathbf{w}'|\mathbf{w})\right)|_{\mathbf{w}'=\mathbf{w}^{\text{MAP}}} \cdot |\mathbf{H}|^{-1/2}, \end{aligned} \tag{6.70}$$

where the mode $\mathbf{w}^{\text{MAP}}$ of posterior distribution $p(\mathbf{w}|\mathbf{O}, \boldsymbol{\mu}_w, \boldsymbol{\Sigma}_w)$ is obtained through the EM iteration procedure,

$$\mathbf{w}^{\text{MAP}} = \arg\max_{\mathbf{w}'} \{Q^{\text{MAP}}(\mathbf{w}'|\mathbf{w})\}, \tag{6.71}$$

and the Hessian matrix $\mathbf{H}$ is also calculated according to the EM algorithm through

$$\begin{aligned}\mathbf{H} &= \nabla_{\mathbf{w}'} \nabla_{\mathbf{w}'} \log\{p(\mathbf{O}|\mathbf{w}') p(\mathbf{w}'|\boldsymbol{\mu}_w, \boldsymbol{\Sigma}_w)\}|_{\mathbf{w}'=\mathbf{w}^{\text{MAP}}} \\ &\approx \nabla_{\mathbf{w}'} \nabla_{\mathbf{w}'} Q^{\text{MAP}}(\mathbf{w}'|\mathbf{w})|_{\mathbf{w}'=\mathbf{w}^{\text{MAP}}} \\ &= \nabla_{\mathbf{w}'} \nabla_{\mathbf{w}'} Q^{\text{ML}}(\mathbf{w}'|\mathbf{w})|_{\mathbf{w}'=\mathbf{w}^{\text{MAP}}} - \boldsymbol{\Sigma}_w^{-1}, \end{aligned} \tag{6.72}$$

which is the second-order differentiation of the log posterior distribution with respect to NN–HMM parameters at the mode $\mathbf{w}^{\text{MAP}}$. Here, the auxiliary function $Q^{\text{MAP}}(\mathbf{w}'|\mathbf{w})$ is introduced because the NN–HMM framework involves the missing data problem. As explained in Section 6.3.3, the EM algorithm is applied to iteratively approximate the posterior distribution at its mode by

$$p(\mathbf{w}'|\mathbf{O}, \boldsymbol{\mu}_w, \boldsymbol{\Sigma}_w)|_{\mathbf{w}'=\mathbf{w}^{\text{MAP}}} \approx \exp\left(Q^{\text{MAP}}(\mathbf{w}'|\mathbf{w})\right)|_{\mathbf{w}'=\mathbf{w}^{\text{MAP}}}. \tag{6.73}$$

Accordingly, we construct the BPC decision based on NN–HMMs, which can achieve robustness in automatic speech recognition.

6.5 Decision tree clustering

This section introduces the application of BIC-based model selection to decision tree clustering of context-dependent HMMs, as performed in Shinoda & Watanabe (1997) and Chou & Reichl (1999), without having to set a heuristic stopping criterion, as is done in Young, Odell & Woodland (1994).[4] This section first describes decision tree clustering based on the ML criterion, which determines the tying structure of context-dependent HMMs efficiently. Then, the approach is extended by using BIC.

6.5.1 Decision tree clustering using ML criterion

The decision tree method has been widely used to construct a tied state HMM effectively by utilizing the phonetic knowledge-based constraint and the binary tree search (Odell 1995) approaches. Here, we introduce a conventional decision tree method using the ML criterion.

Let $\Omega(n)$ denote a set of states that tree node n holds. We start with only a root node ($n = 0$) which holds a set of all context-dependent phone (e.g., triphone hereinafter, as shown in Figure 6.5) HMM states $\Omega(0)$ for an identical center phoneme. The set of triphone states is then split into two sets depending on question Q, $\Omega(n_Y^Q)$ and $\Omega(n_N^Q)$, which are held by two new nodes, n_Y^Q and n_N^Q, respectively, as shown in Figure 6.3. The partition is determined by an answer to a phonemic question Q, such as "is the preceding phoneme a vowel" and "is the following phoneme a nasal." A particular question is chosen so that the partition is the optimal of all the possibilities, based on the likelihood value. We continue this splitting successively for every new set of states to obtain a binary tree, as shown in Figure 6.4, where each leaf node holds a shared set of triphone states. The states belonging to the same cluster are merged into a single HMM state. A set of triphones is thus represented by a set of tied-state HMMs. An HMM in an acoustic model usually has a left-to-right topology with three or four temporal states. A decision tree is usually produced for each state in the sequence, and

[4] Acoustic model selections based on BIC and minimum description length (MDL) criteria have been independently proposed, but they are practically the same if we deal with Gaussian distributions. Therefore, they are identified in this book and referred to as BIC.

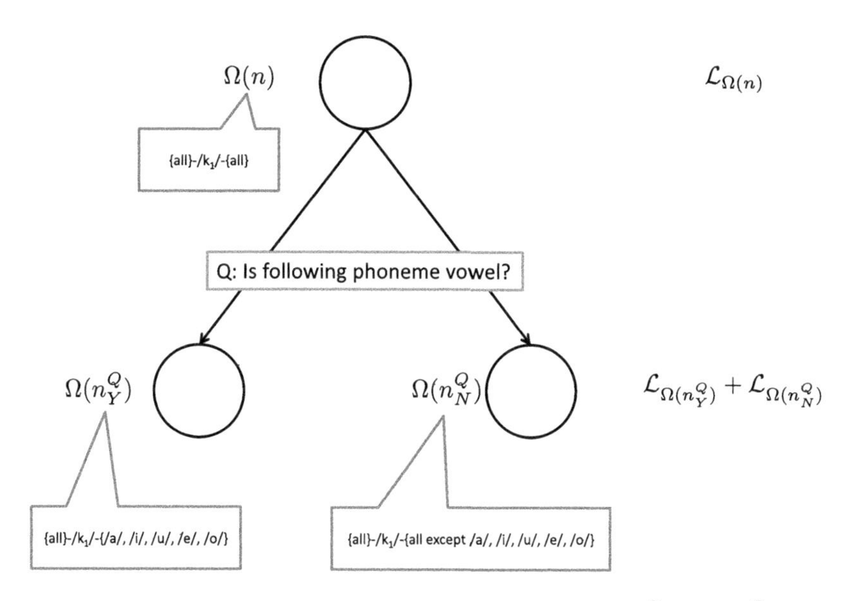

Figure 6.3 Splitting a set of HMM states $\Omega(n)$ in node n into two sets of states $\Omega(n_Y^Q)$ and $\Omega(n_N^Q)$ according to question Q. The objective function is changed from $\mathcal{L}_{\Omega(n)}$ to $\mathcal{L}_{\Omega(n_Y^Q)} + \mathcal{L}_{\Omega(n_N^Q)}$ after the split.

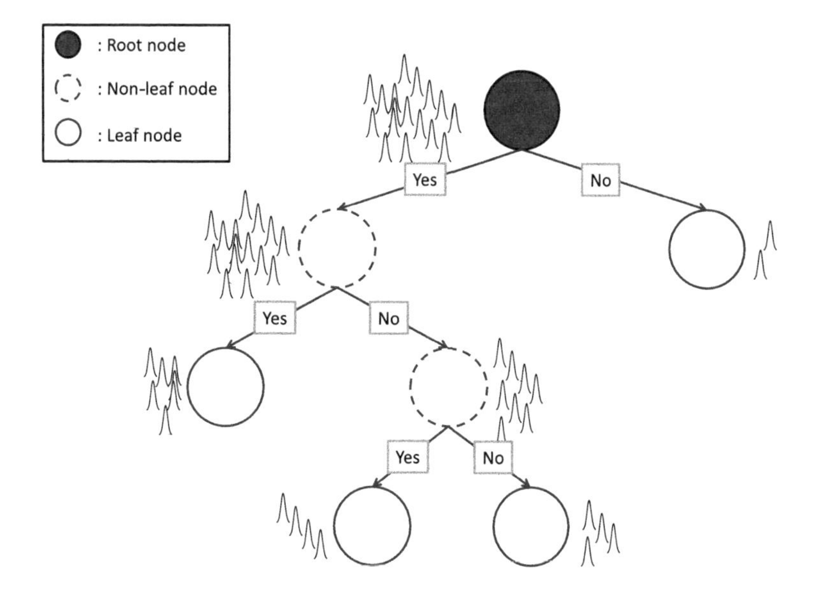

Figure 6.4 Binary decision tree. A context-dependent HMM state is represented by a single Gaussian distribution, and a set of states is assigned to each node. Two child nodes are obtained based on a yes/no answer to a (phonemic) question.

the trees are independent of each other, or a single decision tree is produced for all states.

The phonetic question concerns the preceding and following phoneme context, and is obtained through knowledge of the phonetics (Odell 1995) or it is also decided

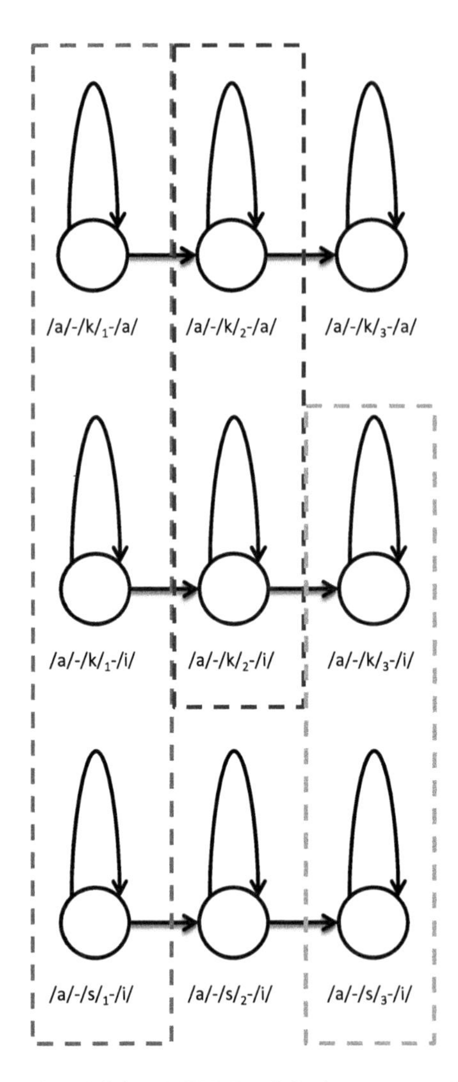

Figure 6.5 Examples of context-dependent (triphone) HMM and sharing structure. We prepare a three-state left-to-right HMM for each triphone, where /a/-/k/$_j$-/a/ denotes a jth state in a triphone HMM with preceding phoneme /a/, central phoneme /k/, and following phoneme /a/. In this example, triphone HMM states /a/-/k/$_1$-/a/, /a/-/k/$_1$-/i/, and /a/-/s/$_1$-/i/ are shared with the same Gaussian distribution.

in a data-driven manner (Povey, Ghoshal, Boulianne *et al.* 2011). Table 6.1 shows examples of the question. In a conventional ML-based approach, an appropriate question is obtained by maximizing a likelihood value as follows:

$$Q = \arg\max_{Q'} \Delta\mathcal{L}_{(Q')}, \tag{6.74}$$

where $\Delta\mathcal{L}_{(Q)}$ denotes the gain of log-likelihood when a state set in a node is split by a question Q. To calculate $\Delta\mathcal{L}_{(Q)}$, we assume the following constraints:

Table 6.1 Examples of phonetic questions for a $j(=1)$th HMM state of a phoneme /a/.

Question	Yes	No
Preceding phoneme is vowel?	{/a/, /i/, /u/, /e/, /o/} - $/a/_{j=1}$ - { all }	otherwise
Following phoneme is vowel?	{ all } - $/a/_{j=1}$ - {/a/, /i/, /u/, /e/, /o/}	otherwise
Following phoneme is media ?	{ all } - $/a/_{j=1}$ - {/b/, /d/, /g/}	otherwise
Preceding phoneme is back vowel?	{/u/, /o/} - $/a/_{j=1}$ - { all }	otherwise
⋮	⋮	⋮

- Data alignments $\gamma_t(j,k)$ and $\xi_t(i,j)$ for each state are fixed while splitting.
- Emission probability distribution in a state is represented by a single Gaussian distribution (i.e., $K = 1$).
- Covariance matrices have only diagonal elements.
- A contribution of state transitions a_{ij} and initial weights π_j for likelihood is disregarded.

These constraints simplify the likelihood calculation without using an iterative calculation, which greatly reduces the computational time.

Equation (6.74) corresponds to an approximation of the Bayes factor, as discussed in Eq. (6.12) to compare two models: M_1 models a set of triphones in the node n with a single Gaussian; and M_2 models two sets of triphones in the nodes n_Y^Q and n_N^Q separated by a question Q:

$$\Delta\mathcal{L}_{(Q')} = \log\left(\frac{p(M_2|\mathbf{O})}{p(M_1|\mathbf{O})}\right) \approx \log\left(\frac{p(\mathbf{O}|\Theta^{\mathrm{ML}}, M_2)}{p(\mathbf{O}|\Theta^{\mathrm{ML}}, M_1)}\right). \tag{6.75}$$

We obtain the gain of log-likelihood $\Delta\mathcal{L}_{(Q)}$ in Eq. (6.74) under the above constraints. Let $\mathbf{O}(i) = \{\mathbf{o}_t(i) \in \mathbb{R}^D : t = 1, \ldots, T(i)\}$ be a set of feature vectors that are assigned to HMM state i by the Viterbi algorithm. $T(i)$ denotes the frame number of training data assigned to state i, and D denotes the number of feature vector dimensions. From the constraints, log-likelihood $\mathcal{L}_\Omega$ for a training data set, assigned to state set Ω, is expressed by the following D dimensional Gaussian distribution:

$$\begin{aligned}\mathcal{L}_\Omega &= \log p(\{\mathbf{O}(i)\}_{i\in\Omega}|\boldsymbol{\mu}_\Omega, \boldsymbol{\Sigma}_\Omega) \\ &= \log \prod_{i\in\Omega}\prod_{t=1}^{T(i)} \mathcal{N}(\mathbf{O}_t(i)|\boldsymbol{\mu}_\Omega, \boldsymbol{\Sigma}_\Omega) \\ &= \log \prod_{i\in\Omega}\prod_{t=1}^{T(i)} (2\pi)^{-\frac{D}{2}}\, |\boldsymbol{\Sigma}_\Omega|^{-\frac{1}{2}}\, e^{-\frac{1}{2}(\mathbf{o}_t(i)-\boldsymbol{\mu}_\Omega)^\intercal \boldsymbol{\Sigma}_\Omega^{-1}(\mathbf{o}_t(i)-\boldsymbol{\mu}_\Omega)},\end{aligned} \tag{6.76}$$

where $\boldsymbol{\mu}_\Omega$ and $\boldsymbol{\Sigma}_\Omega$ denote a D dimensional mean vector and a $D \times D$ diagonal covariance matrix for a data set in Ω, respectively. ML estimates $\boldsymbol{\mu}_\Omega^{\mathrm{ML}}$ and $\boldsymbol{\Sigma}_\Omega^{\mathrm{ML}}$ are obtained by using the derivative technique. First, Eq. (6.76) is rewritten as

$$\mathcal{L}_\Omega = -\frac{D}{2}\log(2\pi)\sum_{i\in\Omega} T(i) - \frac{1}{2}\sum_{i\in\Omega}\sum_{t=1}^{T(i)}\sum_{d=1}^{D}\left(\log \Sigma_{\Omega d} + \frac{(o_{td}(i) - \mu_{\Omega d})^2}{\Sigma_{\Omega d}}\right), \tag{6.77}$$

where $\Sigma_{\Omega d}$ is a d-d element of the diagonal covariance matrix $\boldsymbol{\Sigma}_\Omega$. Then, by using the following derivative with respect to $\mu_{\Omega d}$,

$$\frac{\partial}{\partial \mu_{\Omega d}}\mathcal{L}_\Omega = -\frac{1}{2}\sum_{i\in\Omega}\sum_{t=1}^{T(i)} 2\frac{o_{td}(i) - \mu_{\Omega d}}{\Sigma_{\Omega d}} = 0, \tag{6.78}$$

we can obtain the ML estimate of the mean vector $\boldsymbol{\mu}_\Omega^{\mathrm{ML}}$ as:

$$\boldsymbol{\mu}_\Omega^{\mathrm{ML}} = \frac{\sum_{i\in\Omega}\sum_{t=1}^{T(i)} \mathbf{o}_t(i)}{T_\Omega}. \tag{6.79}$$

Similarly, by using the following derivative with respect to $\Sigma_{\Omega d}$,

$$\frac{\partial}{\partial \Sigma_{\Omega d}}\mathcal{L}_\Omega = -\frac{1}{2}\sum_{i\in\Omega}\sum_{t=1}^{T(i)}\left(\frac{1}{\Sigma_{\Omega d}} - \frac{(o_{td}(i) - \mu_{\Omega d})^2}{\Sigma_{\Omega d}^2}\right) = 0, \tag{6.80}$$

we can obtain the ML estimate of the diagonal component of the covariance matrix $\Sigma_{\Omega d}^{\mathrm{ML}}$ as:

$$\Sigma_{\Omega d}^{\mathrm{ML}} = \frac{\sum_{i\in\Omega}\sum_{t=1}^{T(i)}(o_{td}(i) - \mu_{\Omega d}^{\mathrm{ML}})^2}{T_\Omega}. \tag{6.81}$$

We summarize the above ML estimates as:

$$\begin{aligned}
\boldsymbol{\mu}_\Omega^{\mathrm{ML}} &= \frac{\sum_{i\in\Omega}\sum_{t=1}^{T(i)} \mathbf{o}_t(i)}{T_\Omega}, \\
[\boldsymbol{\Sigma}_\Omega^{\mathrm{ML}}]_{dd} = \Sigma_{\Omega d}^{\mathrm{ML}} &= \frac{\sum_{i\in\Omega}\sum_{t=1}^{T(i)}(o_{td}(i) - \mu_{\Omega d}^{\mathrm{ML}})^2}{T_\Omega}.
\end{aligned} \tag{6.82}$$

$T_\Omega \triangleq \sum_{i\in\Omega} T(i)$ denotes the frame number of training data assigned to HMM states belonging to Ω. $\Sigma_{\Omega d}^{\mathrm{ML}}$ denotes a d-d component for matrix $\boldsymbol{\Sigma}_\Omega^{\mathrm{ML}}$. By substituting Eq. (6.82) into Eq. (6.76), we can derive the following log-likelihood $\mathcal{L}_\Omega$ with the ML estimates of $\boldsymbol{\Sigma}_\Omega^{\mathrm{ML}}$:

$$\begin{aligned}
\mathcal{L}_\Omega &= -\frac{D}{2}\log(2\pi)\sum_{i\in\Omega} T(i) - \frac{1}{2}\sum_{i\in\Omega}\sum_{t=1}^{T(i)}\sum_{d=1}^{D}\left(\log \Sigma_{\Omega d} + \frac{(o_{td}(i) - \mu_{\Omega d})^2}{\Sigma_{\Omega d}}\right)\Bigg|_{\substack{\mu_{\Omega d}\to\mu_{\Omega d}^{\mathrm{ML}} \\ \Sigma_{\Omega d}\to\Sigma_{\Omega d}^{\mathrm{ML}}}} \\
&= -\frac{D}{2}\log(2\pi)T_\Omega - \frac{1}{2}T_\Omega\sum_{d=1}^{D}\log \Sigma_{\Omega d}^{\mathrm{ML}} - \frac{1}{2}T_\Omega\sum_{d=1}^{D}\frac{\Sigma_{\Omega d}^{\mathrm{ML}}}{\Sigma_{\Omega d}^{\mathrm{ML}}} \\
&= -\frac{T_\Omega}{2}\left(D(1 + \log(2\pi)) + \log\left|\boldsymbol{\Sigma}_\Omega^{\mathrm{ML}}\right|\right).
\end{aligned} \tag{6.83}$$

Note that the likelihood value is represented by the determinant of the ML estimate of the covariance matrix, which can easily be calculated as we use the diagonal covariance.

Therefore, a gain of log-likelihood $\Delta\mathcal{L}_{(Q)}$ can be solved as follows (Odell 1995):

$$\begin{aligned}\Delta\mathcal{L}_{(Q)} &= \mathcal{L}_{\Omega(n_Y^Q)} + \mathcal{L}_{\Omega(n_N^Q)} - \mathcal{L}_{\Omega(n)} \\ &= l(\Omega(n_Y^Q)) + l(\Omega(n_N^Q)) - l(\Omega(n)).\end{aligned} \tag{6.84}$$

Here l in Eq. (6.84) is defined as:

$$\begin{aligned} l(\Omega) &\triangleq -\frac{1}{2}\left(T_\Omega \log\left(\left|\boldsymbol{\Sigma}_\Omega^{\mathrm{ML}}\right|\right)\right) \\ &\text{for } \Omega = \{\Omega(n_Y^Q), \Omega(n_N^Q), \Omega(n)\},\end{aligned}$$

where we use the following relation:

$$T_{\Omega(n)} = T_{\Omega(n_Y^Q)} + T_{\Omega(n_N^Q)}. \tag{6.85}$$

Equations (6.84) and (6.85) show that $\Delta\mathcal{L}_{(Q)}$ can be calculated using the ML estimate $\boldsymbol{\Sigma}_\Omega^{\mathrm{ML}}$ and frame number T_Ω. Finally, the appropriate question in the sense of the ML criterion can be computed by substituting Eq. (6.84) into Eq. (6.74).

However, we cannot use this likelihood criterion for stopping the split of nodes in a tree. That is, the positivity of $\Delta\mathcal{L}_{(Q)}$ for any split causes the ML criterion to always select the model structure in which the number of states is the largest. That is, no states are shared at all. To avoid this, the ML criterion requires the following threshold to be set to stop splitting manually:

$$\Delta\mathcal{L} \leq \mathrm{Thd}. \tag{6.86}$$

There exist other approaches to stopping splitting manually by setting the number of total states, or the maximum depth of the tree, as well as a hybrid approach combining those approaches. However, the effectiveness of the thresholds in all of these manual approaches has to be judged by the performance of the development set.

6.5.2 Decision tree clustering using BIC

We consider automatic model selection based on the BIC criterion, which is widely used in model selection for various aspects of statistical modeling. Recall the following definition of a BIC-based objective function for $\mathbf{O} = \{\mathbf{o}_t | t = 1, \cdots T\}$ in Eq. (6.29):

$$\mathcal{L}^{\mathrm{BIC}} = \log p(\mathbf{O}|\boldsymbol{\theta}^*, M) - \lambda\frac{J}{2}\log T, \tag{6.87}$$

where $\boldsymbol{\theta}^*$ is the ML or MAP estimate of model parameters for model M, J is the number of model parameters, and λ is a tuning parameter. Similarly to Section 6.5.1, we use a single Gaussian for each node. Then, Eq. (6.87) for node Ω can be rewritten as follows:

$$\begin{aligned}\mathcal{L}_\Omega^{\mathrm{BIC}} &= \mathcal{L}_\Omega - \lambda D \log T_\Omega \\ &= -\frac{T_\Omega}{2}\left(D(1 + \log(2\pi)) + \log\left|\boldsymbol{\Sigma}_\Omega^{\mathrm{ML}}\right|\right) - \lambda D \log T_\Omega,\end{aligned} \tag{6.88}$$

where $J = 2D$ when we use the diagonal covariance matrix. Therefore, the gain of objective function $\Delta\mathcal{L}^{BIC}_{(Q)}$ using the BIC criterion (the logarithmic Bayes factor) is obtained while splitting a state set by question Q, as follows:

$$\Delta\mathcal{L}^{\mathrm{BIC}}_{(Q)} = \Delta\mathcal{L}_{(Q)} - \lambda D \log T_{\Omega(0)}, \tag{6.89}$$

where λ is a tuning parameter in the BIC criterion, and $T_{\Omega(0)}$ denotes the frame number of data assigned to a root node.[5] Equation (6.89) suggests that the BIC objective function penalizes the gain in log-likelihood on the basis of the balance between the number of free parameters and the amount of training data, and the penalty can be controlled by varying λ. Model structure selection is achieved according to the amount of training data by using $\Delta\mathcal{L}^{\mathrm{BIC}}_{(Q)}$ instead of using $\Delta\mathcal{L}_{(Q)}$ in Eq. (6.74), that is

$$Q = \arg\max_{Q'} \Delta\mathcal{L}^{\mathrm{BIC}}_{(Q')}. \tag{6.91}$$

We can also use $\Delta\mathcal{L}^{\mathrm{BIC}}_{(Q)}$ for stopping splitting when this value is negative, without using a threshold in Eq. (6.86), that is

$$\Delta\mathcal{L}^{\mathrm{BIC}}_{(Q)} \leq 0. \tag{6.92}$$

Thus, we can obtain the tied-state HMM by using the BIC model selection criterion.[6]

Note that the BIC criterion is an asymptotic criterion that is theoretically effective only when the amount of training data is sufficiently large. Therefore, in the case of a small amount of training data, model selection does not perform well because of the uncertain ML estimates $\boldsymbol{\mu}^{\mathrm{ML}}$ and $\boldsymbol{\Sigma}^{\mathrm{ML}}$. Shinoda & Watanabe (2000) show the effectiveness of the BIC criterion for decision tree clustering in a 5000 Japanese word recognition task by comparing the performance of acoustic models based on BIC with models based on heuristic stopping criteria (namely, the state occupancy count and the likelihood threshold). BIC selected 2069 triphone HMM states automatically with an 80.4 % recognition rate, while heuristic stopping criteria selected 1248 and 591 states with recognition rates of 77.9 % and 66.6 % in the best and worst cases, respectively. This result clearly shows the effectiveness of model selection using BIC. An extension of the BIC objective function by considering a tree structure is also discussed in Hu & Zhao (2007), and an extension based on variational Bayes is discussed in Section 7.3.6. In addition, BIC is used for Gaussian pruning in acoustic models Shinoda & Iso (2001), and speaker segmentation (Chen & Gopinath 1999), which is discussed in the next section.

[5] The following objective function

$$\Delta\mathcal{L}^{\mathrm{BIC}}_{(Q)} = \Delta\mathcal{L}_{(Q)} - \lambda D \log T_{\Omega(0)} \tag{6.90}$$

is derived in Shinoda & Watanabe (2000). There are several ways to set the penalty term from Eq. (6.89).

[6] The BIC criterion also has a tuning parameter λ in Eq. (6.87). However, the λ setting is more robust than the threshold setting in Eq. (6.86) (Shinoda & Watanabe 2000).

6.6 Speaker clustering/segmentation

BIC-based speaker segmentation is a particularly important technique for speaker diarization, which has been widely studied recently (Chen & Gopinath 1999, Anguera Miro, Bozonnet, Evans *et al.* 2012). The approach is usually performed in two steps; the first step segments a long sequence of speech to several chunks by using a model selection approach. Then the second step clusters these chunks to form speaker segments, which can also be performed by a model selection approach. Both steps use BIC as model selection. This section considers the first step of speaker segmentation based on the BIC criterion.

6.6.1 Speaker segmentation

First, we consider the simple segmentation problems of segmenting a speech sequence to two chunks where these chunks are uttered by two speakers, respectively, as shown in Figure 6.6. That is:

$$\{\mathbf{o}_1, \cdots, \mathbf{o}_T\} \rightarrow \{\mathbf{o}_1, \cdots, \mathbf{o}_t\} \text{ and } \{\mathbf{o}_{t+1}, \cdots, \mathbf{o}_T\}, \tag{6.93}$$

where t is a change point from one speaker to the other speakers or noises. Assuming that each speaker utterance is modeled by a single Gaussian, which is a very simple assumption but works well in a practical use, this problem can be regarded as a model selection problem where we have a set of candidates represented by:

- $M_0 : \mathbf{o}_1, \cdots, \mathbf{o}_T \sim \mathcal{N}(\boldsymbol{\mu}, \boldsymbol{\Sigma})$;
- $M_t : \mathbf{o}_1, \cdots, \mathbf{o}_t \sim \mathcal{N}(\boldsymbol{\mu}_{1,t}, \boldsymbol{\Sigma}_{1,t}), \mathbf{o}_{t+1}, \cdots, \mathbf{o}_T \sim \mathcal{N}(\boldsymbol{\mu}_{2,t}, \boldsymbol{\Sigma}_{2,t})$ for $t = [2, T-1]$.

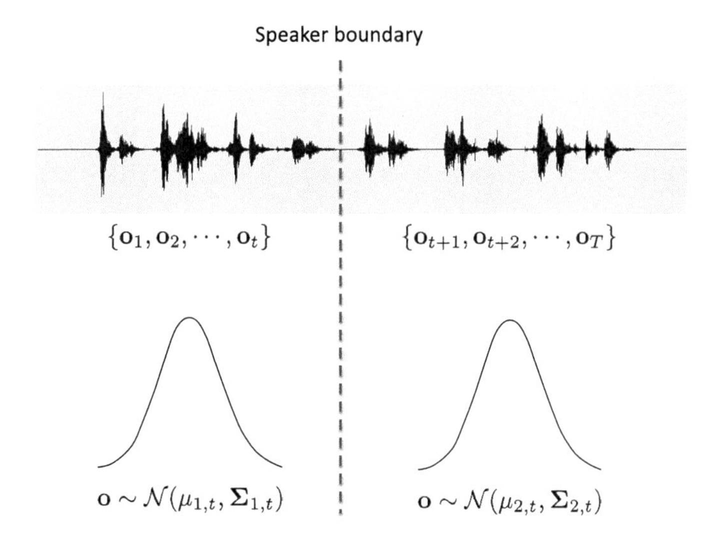

Figure 6.6 Speaker segmentation for audio data, which find a speaker boundary t given speech features $\{\mathbf{o}_1, \cdots, \mathbf{o}_T\}$. The BIC segmentation method assumes that each segmentation is represented by a Gaussian distribution.

The model M_0 means that the sequence can be represented by one Gaussian with mean vector $\boldsymbol{\mu}$ and the diagonal covariance matrix $\boldsymbol{\Sigma}$, that is, the sequence only has one speaker utterance. Similarly, the model M_t for $t = [2, T-1]$ means that the sequence has two speakers with a change point t with two Gaussians that have mean vectors $\boldsymbol{\mu}_{1,t}$ and $\boldsymbol{\mu}_{2,t}$, and diagonal covariance matrices $\boldsymbol{\Sigma}_{1,t}$ and $\boldsymbol{\Sigma}_{2,t}$, depending on the change point t.

Based on these hypothesized models, we can also consider the Bayes factor in Eq. (6.12) to compare two models of M_0 and M_t:

$$\log \left(\frac{p(M_t|\mathbf{O})}{p(M_0|\mathbf{O})} \right) \quad \text{for } t = [2, T-1]. \tag{6.94}$$

Similarly to the decision tree clustering, we can select an appropriate model from M_0 and $\{M_t\}_{t=2}^{T-1}$ by using BIC. Recall the following definition of a BIC-based objective function of model M for $\mathbf{O} = \{\mathbf{o}_t | t = 1, \cdots N\}$ in Eq. (6.29):

$$\mathcal{L}^{\text{BIC}} = \log p(\mathbf{O}|\boldsymbol{\theta}^*, M) - \lambda \frac{J}{2} \log N, \tag{6.95}$$

where $\boldsymbol{\theta}^*$ is the ML or MAP estimate of model parameters for model M, J is the number of model parameters, and λ is a tuning parameter. Since we use a Gaussian (with diagonal covariance matrix) for the likelihood function in Eq. (6.95), the BIC function for M_0 is written as follows:

$$\begin{aligned}
\mathcal{L}^{\text{BIC}}(M_0) &= \log \prod_{t=1}^{T} \mathcal{N}(\mathbf{o}_t|\boldsymbol{\mu}^{\text{ML}}, \boldsymbol{\Sigma}^{\text{ML}}) - \lambda \frac{2D}{2} \log T \\
&= -\frac{T}{2} \left(D(1 + \log(2\pi)) + \log \left| \boldsymbol{\Sigma}^{\text{ML}} \right| \right) - \lambda D \log T,
\end{aligned} \tag{6.96}$$

where $\boldsymbol{\Sigma}^{\text{ML}}$ is the ML estimate of the diagonal covariance matrix:

$$\begin{aligned}
\boldsymbol{\mu}^{\text{ML}} &= \frac{\sum_{t=1}^{T} \mathbf{o}_t}{T}, \\
\boldsymbol{\Sigma}_d^{\text{ML}} &= \frac{\sum_{t=1}^{T} (o_{td} - \mu_d^{\text{ML}})^2}{T}.
\end{aligned} \tag{6.97}$$

This result is obtained by assuming that we only have one state in Eq. (6.83).

Similarly, M_t is written as follows:

$$\begin{aligned}
\mathcal{L}^{\text{BIC}}(M_t) &= \log \prod_{t'=1}^{t} \mathcal{N}(\mathbf{o}_{t'}|\boldsymbol{\mu}_{1,t}^{\text{ML}}, \boldsymbol{\Sigma}_{1,t}^{\text{ML}}) + \log \prod_{t'=t+1}^{T} \mathcal{N}(\mathbf{o}_{t'}|\boldsymbol{\mu}_{2,t}^{\text{ML}}, \boldsymbol{\Sigma}_{2,t}^{\text{ML}}) - \lambda \frac{4D}{2} \log T \\
&= -\frac{t}{2} \left(D(1 + \log(2\pi)) + \log \left| \boldsymbol{\Sigma}_{1,t}^{\text{ML}} \right| \right) \\
&\quad - \frac{T-t}{2} \left(D(1 + \log(2\pi)) + \log \left| \boldsymbol{\Sigma}_{2,t}^{\text{ML}} \right| \right) - 2\lambda D \log T \\
&= -\frac{t}{2} \log \left| \boldsymbol{\Sigma}_{1,t}^{\text{ML}} \right| - \frac{T-t}{2} \log \left| \boldsymbol{\Sigma}_{2,t}^{\text{ML}} \right| - \frac{T}{2} \left(D(1 + \log(2\pi)) \right) - 2\lambda D \log T,
\end{aligned} \tag{6.98}$$

where $\mathbf{\Sigma}_{1,t}^{\mathrm{ML}}$ and $\mathbf{\Sigma}_{2,t}^{\mathrm{ML}}$ are the ML estimates of the diagonal covariance matrices, and are obtained as follows:

$$
\begin{aligned}
\boldsymbol{\mu}_{1,t}^{\mathrm{ML}} &= \frac{\sum_{t'=1}^{t} \mathbf{o}_{t'}}{t}, \\
\mathbf{\Sigma}_{1,td}^{\mathrm{ML}} &= \frac{\sum_{t'=1}^{t} (o_{t'd} - \mu_{1,td}^{\mathrm{ML}})^2}{t}, \\
\boldsymbol{\mu}_{2,t}^{\mathrm{ML}} &= \frac{\sum_{t'=t+1}^{T} \mathbf{o}_{t'}}{t}, \\
\mathbf{\Sigma}_{2,td}^{\mathrm{ML}} &= \frac{\sum_{t'=t+1}^{T} (o_{t'd} - \mu_{2,td}^{\mathrm{ML}})^2}{t}.
\end{aligned} \tag{6.99}
$$

Again, this result is obtained by assuming that we only have two separated states (M_0 and M_t) in Eq. (6.83).

Therefore, the difference of BIC values between M_0 and M_t is represented as follows:

$$
\begin{aligned}
\Delta \mathcal{L}^{\mathrm{BIC}}(t) &\triangleq \mathcal{L}^{\mathrm{BIC}}(M_t) - \mathcal{L}^{\mathrm{BIC}}(M_0) \\
&= -\frac{T}{2} \log \left| \mathbf{\Sigma}^{\mathrm{ML}} \right| + \frac{t}{2} \log \left| \mathbf{\Sigma}_{1,t}^{\mathrm{ML}} \right| + \frac{T-t}{2} \log \left| \mathbf{\Sigma}_{2,t}^{\mathrm{ML}} \right| - \lambda D \log T.
\end{aligned} \tag{6.100}
$$

This is the objective function of finding the segmentation boundary based on the BIC criterion. If the $\Delta \mathcal{L}^{\mathrm{BIC}}(t)$ is positive, t is a segmentation boundary, and the optimal boundary $\hat{t}$ is obtained by maximizing $\Delta \mathcal{L}^{\mathrm{BIC}}(t)$ as follows:

$$
\hat{t} = \arg\max_{t} \Delta \mathcal{L}^{\mathrm{BIC}}(t) \text{ if } \Delta \mathcal{L}^{\mathrm{BIC}}(t) > 0. \tag{6.101}
$$

Thus, we can find the optimal boundary of two speech segments by using BIC. If we consider multiple changing points based on BIC, we can use Algorithm 9. The approach has also been extended to deal with prior distributions (Watanabe & Nakamura 2009).

Algorithm 9 Detecting multiple changing points

Require: interval $[t_a, t_b] : t_a = 1; t_b = 2$
1: **while** $t_a < T$ **do**
2: Detect if there is one changing point in $[t_a, t_b]$ via BIC
3: **if** no change in $[t_a, t_b]$ **then**
4: $t_b = t_b + 1$
5: **else**
6: $t_a = \hat{t} + 1; t_b = t_a + 1$
7: **end if**
8: **end while**

6.6.2 Speaker clustering

Once we have the speech segments obtained by the BIC-based speech segmentation, we can again use the BIC criterion to cluster the speech segments, where the cluster obtained can be interpreted as a specific speaker. This can be obtained by similar techniques to the decision tree clustering of context-dependent HMMs in Section 6.5.2. The main difference between them is that the decision tree clustering is performed by *splitting* the nodes from the root node by using the (phonetic) question, while the speaker clustering starts to *merge* the leaf nodes, which are represented by a speech segment, successively to build a tree.

Let i be a speech segment index obtained by speech segmentation techniques, and we have in total J segments. Then, similarly to the previous sections, we can compute the approximated Bayes factor based on the following difference of the BIC value when merging the speech segments i and i' as

$$\Delta\mathcal{L}_{i,i'}^{\text{BIC}} \triangleq \mathcal{L}_{i,i'}^{\text{BIC}} - \mathcal{L}_{i}^{\text{BIC}} + \mathcal{L}_{i'}^{\text{BIC}}, \tag{6.102}$$

where the BIC values of $\mathcal{L}_{i}^{\text{BIC}}$ and $\mathcal{L}_{i'}^{\text{BIC}}$ are represented as:

$$\mathcal{L}_{i}^{\text{BIC}} = -\frac{T_i}{2}\left(D(1+\log(2\pi)) + \log\left|\boldsymbol{\Sigma}_i^{\text{ML}}\right|\right) - \lambda D \log T_i, \tag{6.103}$$

where T_i and $\boldsymbol{\Sigma}_i^{\text{ML}}$ are the number of frames assigned to the speech segment i and the ML estimate of the covariance matrix of the speech segment i, respectively. Similarly, the BIC value of $\mathcal{L}_{i,i'}^{\text{BIC}}$, which merges two nodes i and i', is represented as

$$\mathcal{L}_{i,i'}^{\text{BIC}} = -\frac{T_{i,i'}}{2}\left(D(1+\log(2\pi)) + \log\left|\boldsymbol{\Sigma}_{i,i'}^{\text{ML}}\right|\right) - \lambda D \log T_{i,i'}, \tag{6.104}$$

where $T_{i,i'} = T_i + T_{i'}$, and $\boldsymbol{\Sigma}_{i,i'}^{\text{ML}}$ is the ML estimate of the covariance matrix when we represent the speech segments i and i' as a single Gaussian.

The most appropriate combination of merging two nodes can be obtained by considering all possible i and i' combinations and selecting the combination

$$\hat{i}, \hat{i}' = \arg \max_{1 \leq i \leq J, i < i' < J} \Delta\mathcal{L}_{i,i'}^{\text{BIC}}. \tag{6.105}$$

And the selected $\hat{i}$ and $\hat{i}'$ can be merged to a new node. This process is iteratively performed while it satisfies $\Delta\mathcal{L}_{i,i'}^{\text{BIC}} > 0$. The final leaf nodes can represent a speaker cluster. This approach is also called *agglomerative clustering*, and BIC based agglomerative clustering is actually used to build state-of-the-art speaker diarization systems (Wooters & Huijbregts 2008), which provide the "who spoke when" information in speech data.

6.7 Summary

This chapter describes the asymptotic approximation of the Bayesian approach, and introduces the Laplace approximation and the BIC criterion. Both approaches are very powerful for approximating the Bayesian inference of acoustic model parameters in speech recognition and Bayesian inference of model selection problems in speech clustering and segmentation. However, the asymptotic approximation makes the single Gaussian assumption for the prior and posterior distributions and large sample assumptions for target data, which limits the application for speech and language processing. The next chapter provides another approximation technique, variational Bayes, which can handle the problem in the asymptotic approximation.

7 Variational Bayes

Variational Bayes (VB) was developed in the machine learning community in the 1990s (Attias 1999, Jordan, Ghahramani, Jaakkola *et al.* 1999) and has now become a standard technique to approximated Bayesian inference for latent models, based on the EM-like algorithm. In Chapter 4, we have also dealt with latent models based on the maximum a-posteriori (MAP) EM algorithm. However, the MAP approximation uses the point estimation of model parameters instead of the distribution estimation, which is far from a true Bayesian manner of regarding all the variables introduced in our problem as probabilistic random variables. Another approximation based on the asymptotic approximation in Chapter 6 assumes a complex posterior distribution as a single Gaussian distribution without latent variables, which is not a true assumption for many of our applications. The evidence approximation in Chapter 5 also does not explicitly deal with latent models (can be obtained by combining MAP, VB, or MCMC). Instead of considering the MAP, evidence, and asymptotic approximations, VB can efficiently approximate complicated integrals and expectations over model parameters, based on variational method within a specific family of distribution types (exponential family, as discussed in Section 2.1.3). The key idea of the variational technique is to find the lower bound of the marginal log likelihood, similar to the EM algorithm in Section 3.4, and obtain the posterior distributions directly based on the variational method.

This chapter starts to explain the general framework of VB in Section 7.1, and more specific pattern recognition problems in Section 7.2. Then this chapter goes on to provide a VB version of the EM algorithm for statistical models and model selection in speech and language processing, including speech recognition in Sections 7.3 and 7.4 and speaker verification in Section 7.5. Sections 7.6 and 7.7 also deal with latent topic models and their extensions; these try to capture long-range topic information from (spoken) documents, based on VB solutions.

7.1 Variational inference in general

This section starts by describing a general latent model with observation data $\mathbf{X} = \{\mathbf{x}_n | n = 1, \cdots, N\}$, and the set of all variables introduced in our model including latent variables, parameters, hyperparameters, and model structure Z. The latter sections specify Z with more specific variables. The goal of Bayesian inference is to obtain posterior distributions of any variables introduced in the problem, that is:

$$p(Z|\mathbf{X}). \tag{7.1}$$

As discussed in Section 2.1.2, once we obtain $p(Z|\mathbf{X})$, we can estimate various information by the MAP or expectation procedure. In VB, we consider an arbitrary posterior distribution $q(Z|\mathbf{X})$. We use the approximated posterior distribution denoted by $q(\cdot)$ to distinguish it from the true posterior distribution $p(\cdot)$. Then the problem is how to obtain a $q(Z|\mathbf{X})$ that is close to $p(Z|\mathbf{X})$, so as to obtain a well-approximated posterior distribution.

7.1.1 Joint posterior distribution

As a measure of evaluating the difference between two distributions, we use the Kullback–Leibler divergence (Kullback & Leibler 1951), as introduced in the ML–EM algorithm (Section 3.4). The Kullback–Leibler divergence between $q(Z|\mathbf{X})$ and $p(Z|\mathbf{X})$ is defined as follows:

$$\mathrm{KL}(q(Z|\mathbf{X})\|p(Z|\mathbf{X})) \triangleq \int q(Z|\mathbf{X}) \log \frac{q(Z|\mathbf{X})}{p(Z|\mathbf{X})} dZ. \tag{7.2}$$

Z can be a set of discrete variables or a set of both continuous and discrete variables, and, strictly speaking, we should use the summation sum_Z for discrete variables and $\int dZ$ for continuous variables in such a case. However, for simplicity we use $\int dZ$ instead of mixing integrals and summations in the following formulation.

The KL divergence (Eq. (7.2)) is represented as

$$\begin{aligned}\mathrm{KL}(q(Z|\mathbf{X})\|p(Z|\mathbf{X})) &= \int q(Z|\mathbf{X}) \log \frac{q(Z|\mathbf{X})}{\frac{p(\mathbf{X},Z)}{p(\mathbf{X})}} dZ \\ &= \log p(\mathbf{X}) - \underbrace{\int q(Z|\mathbf{X}) \log \frac{p(\mathbf{X},Z)}{q(Z|\mathbf{X})} dZ}_{\triangleq \mathcal{F}[q(Z|\mathbf{X})]},\end{aligned} \tag{7.3}$$

where

$$\mathcal{F}[q(Z|\mathbf{X})] \triangleq \int q(Z|\mathbf{X}) \log \frac{p(\mathbf{X},Z)}{q(Z|\mathbf{X})} dZ \tag{7.4}$$

is called the *variational lower bound.* The reason we call it the lower bound is that $\mathcal{F}[q(Z|\mathbf{X})]$ is a lower bound of the evidence (marginal log likelihood) $\log p(\mathbf{X})$ because of the non-negativity of the KL divergence, i.e.,

$$\mathrm{KL}(q(Z|\mathbf{X})\|p(Z|\mathbf{X})) \geq 0 \iff \log p(\mathbf{X}) \geq \mathcal{F}[q(Z|\mathbf{X})]. \tag{7.5}$$

Equation (7.3) means that we can obtain the optimal $q(Z|\mathbf{X})$ by maximizing the variational lower bound $\mathcal{F}[q(Z|\mathbf{X})]$ that corresponds to minimizing the KL divergence since the log evidence $\log p(\mathbf{X})$ does not depend on Z. That is

$$\tilde{q}(Z|\mathbf{X}) = \arg\max_{q(Z|\mathbf{X})} \mathcal{F}[q(Z|\mathbf{X})] \iff \arg\max_{q(Z|\mathbf{X})} \mathrm{KL}(q(Z|\mathbf{X})\|p(Z|\mathbf{X})). \tag{7.6}$$

To obtain the optimal $\tilde{q}(Z|\mathbf{X})$, we use a variational method for a *functional* $\mathcal{F}[q(Z|\mathbf{X})]$, which achieves mapping a function to a real (or complex) value, i.e., $f \mapsto a \in \mathbb{R}$. Thus, the approach is called *variational Bayes.* The variational method is discussed in Section 7.1.3.

7.1.2 Factorized posterior distribution

In practical applications, we need to consider the factorized form of the joint distribution $q(Z|\mathbf{X})$ to make the calculation simple. To do this, we consider the jth element of Z and make the following conditional independence assumption:

$$q(Z|\mathbf{X}) = \prod_{j=1}^{J} q(Z_j|\mathbf{X}). \tag{7.7}$$

J is the total number of elements. This is an essential approximation requirement of VB to make the problem practical. Z_j would be a subset of model parameters (e.g., Gaussian mean vector $\boldsymbol{\mu}$ and covariance matrix $\boldsymbol{\Sigma}$ in a particular component of GMM/CDHMM), or an HMM state indicator s_t at frame t, for instance. Note that we do not assume the factorization form for the true posterior $p(Z|\mathbf{X})$. Instead, the true posterior $p(Z_i|\mathbf{X})$ can be represented as the following marginalized distribution of $p(Z|\mathbf{X})$ over all Z_j except Z_i:

$$p(Z_i|\mathbf{X}) = \int \cdots \int p(Z|\mathbf{X}) \prod_{j \neq i}^{J} dZ_j \triangleq \int p(Z|\mathbf{X}) dZ_{\backslash i}, \tag{7.8}$$

where $Z_{\backslash i}$ denotes the complementary set of Z_i.

By using Eq. (7.8), the KL divergence between $q(Z_i|\mathbf{X})$ and $p(Z_i|\mathbf{X})$ (not the KL divergence between the joint distributions in Eq. (7.2)) is represented as follows:

$$\begin{aligned} \mathrm{KL}(q(Z_i|\mathbf{X})\|p(Z_i|\mathbf{X})) &= \int q(Z_i|\mathbf{X}) \log \frac{q(Z_i|\mathbf{X})}{\int p(Z|\mathbf{X}) dZ_{\backslash i}} dZ_i \\ &= \int q(Z_i|\mathbf{X}) \log \frac{q(Z_i|\mathbf{X})}{\int \frac{p(\mathbf{X},Z)}{p(\mathbf{X})} dZ_{\backslash i}} dZ_i \\ &= \log p(\mathbf{X}) - \int q(Z_i|\mathbf{X}) \log \frac{\int p(\mathbf{X}, Z) dZ_{\backslash i}}{q(Z_i|\mathbf{X})} dZ_i. \end{aligned} \tag{7.9}$$

Since Eq. (7.9) has integrals in the logarithmic function, it is difficult to deal with. Therefore, we use the following Jensen's inequality for a concave function f, distribution function $p(Y)$ ($\int p(Y)dY = 1$), and an arbitrary function $g(Y)$ introduced in Section 3.4.1:

$$f\left(\int p(Y)g(Y)dY\right) \geq \int p(Y)f(g(Y))dY. \tag{7.10}$$

In the special case of $f(\cdot) = \log(\cdot)$, $Y = Z_{\backslash i}$, $p(Y) = q(Z_{\backslash i}|\mathbf{X})$, and $g(Y) = \frac{p(\mathbf{X},Z)}{q(Z|\mathbf{X})}$, Eq. (7.10) can be rewritten as follows:

$$\begin{aligned} &\log\left(\int q(Z_{\backslash i}|\mathbf{X}) \frac{p(\mathbf{X}, Z)}{q(Z|\mathbf{X})} dZ_{\backslash i}\right) \\ &= \log\left(\frac{\int p(\mathbf{X}, Z) dZ_{\backslash i}}{q(Z_i|\mathbf{X})}\right) \geq \int q(Z_{\backslash i}|\mathbf{X}) \log\left(\frac{p(\mathbf{X}, Z)}{q(Z|\mathbf{X})}\right) dZ_{\backslash i}, \end{aligned} \tag{7.11}$$

where we use Eq. (7.7) to cancel $q(Z_{\backslash i}|\mathbf{X})$ in the fraction of the first line. We also use the following relationship, which is true when $p(Y) \geq 0$:

$$a(Y) \geq b(Y) \; \forall Y \Rightarrow \int p(Y)a(Y)dY \geq \int p(Y)b(Y)dY. \tag{7.12}$$

In the special case of

$$\begin{aligned} f(\cdot) &= \log(\cdot), \\ Y &= Z_i, \\ p(Y) &= q(Z_i|\mathbf{X}), \\ a(Y) &= \log\left(\frac{\int p(\mathbf{X}, Z)dZ_{\backslash i}}{q(Z_i|\mathbf{X})}\right), \\ b(Y) &= \int q(Z_{\backslash i}|\mathbf{X}) \log\left(\frac{p(\mathbf{X}, Z)}{q(Z|\mathbf{X})}\right) dZ_{\backslash i}, \end{aligned} \tag{7.13}$$

Eq. (7.12) is represented as

$$\begin{aligned} &\int q(Z_i|\mathbf{X}) \log\left(\frac{\int p(\mathbf{X}, Z)dZ_{\backslash i}}{q(Z_i|\mathbf{X})}\right) dZ_i \\ &\geq \int q(Z_i|\mathbf{X}) \int q(Z_{\backslash i}|\mathbf{X}) \log\left(\frac{p(\mathbf{X}, Z)}{q(Z|\mathbf{X})}\right) dZ_{\backslash i} dZ_i. \\ &= \int q(Z|\mathbf{X}) \log\left(\frac{p(\mathbf{X}, Z)}{q(Z|\mathbf{X})}\right) dZ = \mathcal{F}[q(Z|\mathbf{X})], \end{aligned} \tag{7.14}$$

where we use Eq. (7.7) to obtain $q(Z|\mathbf{X})$, and use the definition of the variational lower bound in Eq. (7.4). Therefore, by substituting Eq. (7.14) into the KL divergence Eq. (7.9), Eq. (7.9) is finally represented as follows:

$$\mathrm{KL}(q(Z_i|\mathbf{X}) \| p(Z_i|\mathbf{X})) \leq \log p(\mathbf{X}) - \mathcal{F}[q(Z|\mathbf{X})]. \tag{7.15}$$

Compared with Eq. (7.3) that is the equality relationship, Eq. (7.15) is the inequality relationship. Equation (7.15) still has the nice property that the maximization of the variational lower bound corresponds to reducing the KL divergence that then causes the approximated posterior $q(Z_i|\mathbf{X})$ to approach the true posterior $p(Z_i|\mathbf{X})$. That is, if we obtain the following posterior distribution:

$$\tilde{q}(Z_i|\mathbf{X}) = \arg \max_{q(Z_i|\mathbf{X})} \mathcal{F}[q(Z|\mathbf{X})], \tag{7.16}$$

$\tilde{q}(Z_i|\mathbf{X})$ could be a well-approximated posterior distribution in terms of reducing the KL divergence between the true posterior $p(Z_i|\mathbf{X})$ and approximated posterior $\tilde{q}(Z_i|\mathbf{X})$. In this section, a tilde $\tilde{}$ is added to indicate variationally optimized values or functions. However, the maximization of the posterior distribution in terms of the lower bound $\mathcal{F}[q(Z|\mathbf{X})]$ does not directly correspond to minimization of the KL divergence, and we cannot globally optimize $q(Z_i|\mathbf{X})$ in terms of the KL divergence when we use the lower bound as the objective functional. This is a shortcoming of the factorization approximation in Eq. (7.7), but it enables us to obtain the posterior distribution of each variable $q(Z_i|\mathbf{X})$, unlike the joint distribution $q(Z|\mathbf{X})$, which is more practical. The next section discusses how to optimize the approximated posterior by using the variational method.

7.1.3 Variational method

The variational method is based on functional differentiation, which is a technique for obtaining an optimal function based on a variational calculation, and is defined as follows:

Continuous function case

$$\frac{\delta}{\delta g(y)}\mathcal{H}[g(x)] = \lim_{\epsilon\to 0}\frac{\mathcal{H}[g(x)+\epsilon\delta(x-y)]-\mathcal{H}[g(x)]}{\varepsilon}, \tag{7.17}$$

where $g(x)$ is a continuous function to be optimized, $\mathcal{H}[g(x)]$ is a functional of $g(x)$ and $\delta(x-y)$ is a Dirac delta function.

Discrete function case

$$\frac{\delta}{\delta g_l}\mathcal{H}[g_n] = \lim_{\epsilon\to 0}\frac{\mathcal{H}[g_n+\epsilon\delta(n,l)]-\mathcal{H}[g_n]}{\varepsilon}. \tag{7.18}$$

Similarly, g_n is a discrete function to be optimized, and $\delta(n,l)$ is a Kronecker delta function. This section aims to obtain the following optimized posterior distribution based on the above variational method:

$$\begin{aligned}\tilde{q}(Z_i|\mathbf{X}) &= \arg\max_{q(Z_i|\mathbf{X})} \mathcal{F}[q(Z|\mathbf{X})] \\ &= \arg\max_{q(Z_i|\mathbf{X})} \int q(Z|\mathbf{X})\log\left(\frac{p(\mathbf{X},Z)}{q(Z|\mathbf{X})}\right)dZ. \end{aligned}\tag{7.19}$$

For simplicity of the calculation, we simplify $q(Z_i|\mathbf{X})$ to $q(Z_i)$ in this section.

If we consider $\int q(Z_i)dZ_i = 1$ constraint, the functional differentiation is represented by substituting $\mathcal{F}[q(Z)]$ and $q(Z_i)$ into $\mathcal{H}$ and $g(y)$ in Eq. (7.17), respectively, as follows:

$$\begin{aligned}&\frac{\delta}{\delta q(Z_i')}\left(\mathcal{F}[q(Z)] + K\left(\int q(Z_i)dZ_i - 1\right)\right) \\ &= \lim_{\epsilon\to 0}\frac{1}{\epsilon}\Bigg(\int\left(q(Z_i)+\epsilon\delta(Z_i-Z_i')\right)\mathbb{E}_{(Z_{\backslash i})}\left[\log\frac{p(\mathbf{X},Z)}{\left(q(Z_i)+\epsilon\delta(Z_i-Z_i')\right)q(Z_{\backslash i})}\right]dZ_i \\ &\quad - \mathcal{F}[q(Z)] + K\left(\int\left(q(Z_i)+\epsilon\delta(Z_i-Z_i')\right)dZ_i - 1\right) \\ &\quad - K\left(\int q(Z_i)dZ_i - 1\right)\Bigg), \end{aligned}\tag{7.20}$$

where K is a Lagrange multiplier, as introduced in Section 3.4.3 for the function derivative. We focus on the first term in the brackets in the second line of Eq. (7.20). The first term is rewritten as follows:

$$\int \left(q(Z_i) + \epsilon\delta(Z_i - Z_i')\right) \mathbb{E}_{(Z_{\backslash i})}\left[\log \frac{p(\mathbf{X}, Z)}{\left(q(Z_i) + \epsilon\delta(Z_i - Z_i')\right) q(Z_{\backslash i})}\right] dZ_i$$
$$= \int \left(q(Z_i) + \epsilon\delta(Z_i - Z_i')\right) \mathbb{E}_{(Z_{\backslash i})}\left[\log \frac{p(\mathbf{X}, Z)}{q(Z) + \frac{q(Z)}{q(Z_i)}\epsilon\delta(Z_i - Z_i')}\right] dZ_i$$
$$= \int \left(q(Z_i) + \epsilon\delta(Z_i - Z_i')\right) \mathbb{E}_{(Z_{\backslash i})}\left[\log \frac{p(\mathbf{X}, Z)}{q(Z)} - \log\left(1 + \epsilon\frac{\delta(Z_i - Z_i')}{q(Z_i)}\right)\right] dZ_i. \tag{7.21}$$

By expanding the logarithmic term in Eq. (7.21) with respect to ϵ, Eq. (7.21) can be represented thus:

Equation (7.21)

$$= \int \left(q(Z_i) + \epsilon\delta(Z_i - Z_i')\right) \left(\mathbb{E}_{(Z_{\backslash i})}\left[\log \frac{p(\mathbf{X}, Z)}{q(Z)}\right] - \epsilon\frac{\delta(Z_i - Z_i')}{q(Z_i)}\right) dZ_i + \mathbf{o}(\epsilon^2)$$
$$= \int q(Z_i)\mathbb{E}_{(Z_{\backslash i})}\left[\log \frac{p(\mathbf{X}, Z)}{q(Z)}\right] dZ_i - \epsilon \int \delta(Z_i - Z_i') dZ_i$$
$$+ \epsilon \int \delta(Z_i - Z_i')\mathbb{E}_{(Z_{\backslash i})}\left[\log \frac{p(\mathbf{X}, Z)}{q(Z)}\right] dZ_i + \mathbf{o}(\epsilon^2)$$
$$= \mathcal{F}[q(Z)] - \epsilon + \epsilon\mathbb{E}_{(Z_{\backslash i})}\left[\log \frac{p(\mathbf{X}, Z')}{q(Z')}\right] + \mathbf{o}(\epsilon^2)$$
$$= \mathcal{F}[q(Z)] + \epsilon\left(-1 + \mathbb{E}_{(Z_{\backslash i})}\left[\log \frac{p(\mathbf{X}, Z')}{q(Z')}\right]\right) + \mathbf{o}(\epsilon^2), \tag{7.22}$$

where $\mathbf{o}(\epsilon^2)$ denotes a set of terms of no less than the second power of ϵ. $Z' \triangleq \{Z_i', Z_{\backslash i}\}$, but in the following equations, we simply use Z instead of Z', as we do not have to distinguish them. Therefore, by substituting Eq. (7.22) into Eq. (7.20), it can be represented as:

Equation (7.20)

$$= \lim_{\epsilon \to 0} \frac{1}{\epsilon}\left(\epsilon\left(-1 + \mathbb{E}_{(Z_{\backslash i})}\left[\log \frac{p(\mathbf{X}, Z)}{q(Z)}\right] + K\right) + \mathbf{o}(\epsilon^2)\right)$$
$$= -1 + \mathbb{E}_{(Z_{\backslash i})}\left[\log \frac{p(\mathbf{X}, Z)}{q(Z)}\right] + K$$
$$= -1 + \mathbb{E}_{(Z_{\backslash i})}\left[\log p(\mathbf{X}, Z)\right] - \mathbb{E}_{(Z_{\backslash i})}\left[\log q(Z)\right] + K$$
$$= -1 + \mathbb{E}_{(Z_{\backslash i})}\left[\log p(\mathbf{X}, Z)\right] - \log q(Z_i) - \mathbb{E}_{(Z_{\backslash i})}\left[\log q(Z_{\backslash i})\right] + K. \tag{7.23}$$

We use Eq. (7.7) to factorize $q(Z_i)$ and $q(Z_{\backslash i})$. Therefore, the optimal posterior (VB posterior) $\widetilde{q}(Z_i)$ satisfies the relation whereby Eq. (7.23) $= 0$, and is obtained as:

$$\log \widetilde{q}(Z_i) = -1 + \mathbb{E}_{(Z_{\backslash i})}\left[\log p(\mathbf{X}, Z)\right] - \mathbb{E}_{(Z_{\backslash i})}\left[\log q(Z_{\backslash i})\right] + K. \tag{7.24}$$

Since only the second term in the right-hand-side depends on Z_i, the optimal VB posterior is finally derived as:

$$\widetilde{q}(Z_i|\mathbf{X}) \propto \exp\left(\mathbb{E}_{(Z_{\backslash i}|\mathbf{X})}\left[\log p(\mathbf{X}, Z)\right]\right), \tag{7.25}$$

or by considering the normalization constant, it is derived as

$$\widetilde{q}(Z_i|\mathbf{X}) = \frac{\exp\left(\mathbb{E}_{(Z_{\backslash i}|\mathbf{X})}\left[\log p(\mathbf{X}, Z)\right]\right)}{\int \exp\left(\mathbb{E}_{(Z_{\backslash i}|\mathbf{X})}\left[\log p(\mathbf{X}, Z)\right]\right) dZ_i}, \tag{7.26}$$

where the omitted notations are recovered ($q(Z_i) \to q(Z_i|\mathbf{X})$). Thus, we obtain the general form of the VB posterior distribution $\widetilde{q}(Z_i|\mathbf{X})$ by using the variational method. Equation (7.25) tells us that if we want to infer some probabilistic variables, we first need to prepare the joint distribution of the observation $\mathbf{X}$ and target variables. Note that $\widetilde{q}(Z_i|\mathbf{X})$ and the other posterior distributions $\widetilde{q}(Z_{\backslash i}|\mathbf{X}) = \prod_{j \neq i}^{J} q(Z_j|\mathbf{X})$ depend on each other due to the expectation in Eq. (7.25). Therefore, this optimization can be performed iteratively from the initial posterior distributions for all $\widetilde{q}(Z_i|\mathbf{X})$. The following sections provide more practical forms of VB posteriors.

7.2 Variational inference for classification problems

This section provides more specific formulations for our speech and language processing issues which focus more on pattern classification problems. Let $\mathbf{O}$ be a training data set of feature vectors, and Z be a set of discrete latent variables. Then, with a fixed model structure M, posterior distributions for model parameters $p(\Theta^{(c)}|\mathbf{O}, M)$ and $p(Z^{(c)}|\mathbf{O}, M)$ given category c are expressed as follows:[1]

$$p(\Theta^{(c)}|\mathbf{O}, M) = \sum_{Z} \int \frac{p(\mathbf{O}, Z|\Theta, M)p(\Theta|M)}{p(\mathbf{O}|M)} d\Theta^{(\backslash c)} \tag{7.27}$$

and

$$p(Z^{(c)}|\mathbf{O}, M) = \sum_{Z^{(\backslash c)}} \int \frac{p(\mathbf{O}, Z|\Theta, M)p(\Theta|M)}{p(\mathbf{O}|M)} d\Theta, \tag{7.28}$$

where $p(\Theta|M)$ is a prior distribution for Θ. Here, $\backslash c$ represents the set of all categories without c. In this section, we can also regard the prior hyperparameter setting as the model structure setting, and include its variations in index M. The posterior distributions for the model structure $p(M|\mathbf{O})$ are expressed as follows:

$$p(M|\mathbf{O}) = \sum_{Z} \int \frac{p(\mathbf{O}, Z|\Theta, M)p(\Theta|M)p(M)}{p(\mathbf{O})} d\Theta, \tag{7.29}$$

where $p(M)$ denotes a prior distribution for model structure M.

These equations cannot be solved analytically, because the acoustic model for speech recognition includes latent variables in HMMs and GMMs, as discussed in Section 3.2, and the total number of model parameters amounts to more than *one million*. In addition, these parameters depend on each other hierarchically. Solving all integrals and expectations numerically requires huge amounts of computation time. Therefore, when applying

[1] It is reasonable to deal with the prior distribution $p(\Theta|M)$ of model parameters given model M instead of $p(\Theta)$, since the actual functional form of model parameters is determined by model M. Conversely, it is very difficult to consider the prior distribution of model parameters $p(\Theta)$ without the model setting.

the Bayesian approach to acoustic modeling for speech recognition, an effective approximation technique is necessary. Therefore, this section focuses on the VB approach and derives general solutions for VB posterior distributions $q(\Theta|\mathbf{O},M)$, $q(Z|\mathbf{O},M)$, and $q(M|\mathbf{O})$ to approximate the corresponding true posteriors. To begin with, by following the general VB formulation in Section 7.1.2, we assume that

$$\begin{aligned} q(\Theta,Z|\mathbf{O},M) &= \prod_c q(\Theta^{(c)}|\mathbf{O}^{(c)},M)q(Z^{(c)}|\mathbf{O}^{(c)},M), \\ p(\Theta,Z|\mathbf{O},M) &= \prod_c p(\Theta^{(c)}|\mathbf{O}^{(c)},M)p(Z^{(c)}|\mathbf{O}^{(c)},M). \end{aligned} \tag{7.30}$$

This assumption means that probabilistic variables associated with each category are statistically independent from other categories. In addition, these posterior distributions depend on the model variable M, which is not marginalized. The speech data used are assumed to be well transcribed and the label information is assumed to be reliable. In addition, the frequently used feature extraction (e.g., MFCC) from the speech is good enough for the statistical independence assumption of the observation data to be guaranteed. Therefore, the assumption of class independence is reasonable.

7.2.1 VB posterior distributions for model parameters

This subsection discusses VB posterior distributions for model parameters with fixed model structure M. Initially, arbitrary posterior distribution $q(\Theta^{(c)}|\mathbf{O},M)$ is introduced, and the Kullback–Leibler (KL) divergence (Kullback & Leibler 1951) between $q(\Theta^{(c)}|\mathbf{O},M)$ and true posterior distribution $p(\Theta^{(c)}|\mathbf{O},M)$ is considered:

$$\mathrm{KL}(q(\Theta^{(c)}|\mathbf{O},M)\|p(\Theta^{(c)}|\mathbf{O},M)) = \int q(\Theta^{(c)}|\mathbf{O},M)\log\frac{q(\Theta^{(c)}|\mathbf{O},M)}{p(\Theta^{(c)}|\mathbf{O},M)}d\Theta^{(c)}. \tag{7.31}$$

Substituting Eq. (7.27) into Eq. (7.31), Eq. (7.31) is rewritten as follows:

$$\begin{aligned} &\mathrm{KL}(q(\Theta^{(c)}|\mathbf{O},M)\|p(\Theta^{(c)}|\mathbf{O},M)) \\ &= \int q(\Theta^{(c)}|\mathbf{O},M)\log\frac{q(\Theta^{(c)}|\mathbf{O},M)}{\sum_Z\int\frac{p(\mathbf{O},Z|\Theta,M)p(\Theta|M)}{p(\mathbf{O}|M)}d\Theta^{(\backslash c)}}d\Theta^{(c)} \\ &= \log p(\mathbf{O}|M) - \int q(\Theta^{(c)}|\mathbf{O},M) \\ &\qquad \times \log\frac{\sum_Z\int p(\mathbf{O},Z|\Theta,M)p(\Theta|M)d\Theta^{(\backslash c)}}{q(\Theta^{(c)}|\mathbf{O},M)}d\Theta^{(c)}. \end{aligned} \tag{7.32}$$

Then applying the continuous Jensen's inequality Eq. (7.10) to Eq. (7.32), the following inequality is obtained:

$$\begin{aligned} &\mathrm{KL}(q(\Theta^{(c)}|\mathbf{O},M)\|p(\Theta^{(c)}|\mathbf{O},M)) \\ &\leq \log p(\mathbf{O}|M) - \sum_Z\int q(\Theta^{(c)}|\mathbf{O},M)q(\Theta^{(\backslash c)}|\mathbf{O},M)q(Z|\mathbf{O},M) \\ &\qquad \times \log\frac{p(\mathbf{O},Z|\Theta,M)p(\Theta|M)}{q(\Theta^{(c)}|\mathbf{O},M)q(\Theta^{(\backslash c)}|\mathbf{O},M)q(Z)}d\Theta^{(c)}d\Theta^{(\backslash c)} \end{aligned}$$

$$= \log p(\mathbf{O}|M) - \sum_Z \int q(\Theta|\mathbf{O},M)q(Z|\mathbf{O},M)$$
$$\times \log \frac{p(\mathbf{O},Z|\Theta,M)p(\Theta|M)}{q(\Theta|\mathbf{O},M)q(Z|\mathbf{O},M)p(\mathbf{O}|M)} d\Theta. \tag{7.33}$$

From the third to the fourth line, we use the definition $d\Theta^{(c)}d\Theta^{(\backslash c)} \equiv d\Theta$ and the relation $q(\Theta^{(c)}|\mathbf{O},M)q(\Theta^{(\backslash c)}|\mathbf{O},M) = q(\Theta|\mathbf{O},M)$, which is derived from Eq. (7.30). Thus, we finally obtain the following inequality:

$$\mathrm{KL}(q(\Theta^{(c)}|\mathbf{O},M)\|p(\Theta^{(c)}|\mathbf{O},M))$$
$$\leq \log p(\mathbf{O}|M) - \mathcal{F}^M[q(\Theta|\mathbf{O},M), q(Z|\mathbf{O},M)], \tag{7.34}$$

where

$$\mathcal{F}^M[q(\Theta|\mathbf{O},M), q(Z|\mathbf{O},M)]$$
$$\triangleq \sum_Z \int q(\Theta|\mathbf{O},M)q(Z|\mathbf{O},M) \log \frac{p(\mathbf{O},Z|\Theta,M)p(\Theta|M)}{q(\Theta|\mathbf{O},M)q(Z|\mathbf{O},M)} d\Theta$$
$$= \mathbb{E}_{(\Theta,Z)}\left[\log \frac{p(\mathbf{O},Z|\Theta,M)p(\Theta|M)}{q(\Theta|\mathbf{O},M)q(Z|\mathbf{O},M)}\right]. \tag{7.35}$$

This corresponds to the variational lower bound, as discussed in Eq. (7.3). The inequality (7.34) is strict unless $q(\Theta|\mathbf{O},M) = p(\Theta|\mathbf{O},M)$ and $q(Z|\mathbf{O},M) = p(Z|\mathbf{O},M)$ (i.e., the arbitrary posterior distribution q is equivalent to the true posterior distribution p). From the assumption Eq. (7.30), $\mathcal{F}^M$ is decomposed into each category as follows:

$$\mathcal{F}^M[q(\Theta|\mathbf{O},M), q(Z|\mathbf{O},M)]$$
$$= \sum_c \mathbb{E}_{(\Theta^{(c)},Z^{(c)})}\left[\log \frac{p(\mathbf{O}^{(c)},Z^{(c)}|\Theta^{(c)},M)p(\Theta^{(c)}|M)}{q(\Theta^{(c)}|\mathbf{O}^{(c)},M)q(Z^{(c)}|\mathbf{O}^{(c)},M)}\right]$$
$$= \sum_c \mathcal{F}^{M,(c)}[q(\Theta^{(c)}|\mathbf{O}^{(c)},M), q(Z^{(c)}|\mathbf{O}^{(c)},M)]. \tag{7.36}$$

This indicates that the total objective function is calculated by summing all objective functions for each category.

From inequality Eq. (7.34), $q(\Theta^{(c)}|\mathbf{O},M)$ approaches $p(\Theta^{(c)}|\mathbf{O},M)$ as the right-hand-side decreases. Therefore, the optimal posterior distribution can be obtained by a variational method. Since the term $\log p(\mathbf{O}|M)$ can be disregarded, minimization is changed to maximization of $\mathcal{F}^M$ with respect to $q(\Theta^{(c)}|\mathbf{O},M)$, and is given by the following variational equation:

$$\frac{\delta}{\delta q(\Theta^{(c)}|\mathbf{O},M)} \mathcal{F}^M[q(\Theta|\mathbf{O},M), q(Z|\mathbf{O},M)]$$
$$= \frac{\delta}{\delta q(\Theta^{(c)}|\mathbf{O},M)} \mathcal{F}^{M,(c)}[q(\Theta^{(c)}|\mathbf{O}^{(c)},M), q(Z^{(c)}|\mathbf{O}^{(c)},M)] = 0. \tag{7.37}$$

From this equation, the optimal VB posterior distribution $\tilde{q}(\Theta^{(c)}|\mathbf{O},M)$ is obtained by using the variational method as follows:

$$\tilde{q}(\Theta^{(c)}|\mathbf{O},M) \propto p(\Theta^{(c)}|M) \exp\left(\mathbb{E}_{(Z^{(c)})}\left[\log p(\mathbf{O}^{(c)},Z^{(c)}|\Theta^{(c)},M)\right]\right). \tag{7.38}$$

This result can also be obtained by using the general formula of the variational posterior in Eq. (7.25). By using the replacement $Z_i \to \Theta^{(c)}$, Eq. (7.25) can be rewritten as follows:

$$\begin{aligned}
\tilde{q}(\Theta^{(c)}|\mathbf{O}^{(c)}, M) &\propto \exp\left(\mathbb{E}_{(\Theta^{(\backslash c)}, Z)}\left[\log p(\mathbf{O}, \Theta, Z|M)\right]\right) \\
&= \exp\left(\mathbb{E}_{(\Theta^{(\backslash c)}, Z)}\left[\log p(\mathbf{O}, Z|\Theta, M) p(\Theta|M)\right]\right) \\
&= \exp\left(\sum_{c' \neq c} \mathbb{E}_{(\Theta^{(c')}, Z^{(c')})}\left[\log p(\mathbf{O}^{(c')}, Z^{(c')}|\Theta^{(c')}, M) p(\Theta^{(c')}|M)\right]\right) \\
&\quad \times \exp\left(\mathbb{E}_{(Z^{(c)})}\left[\log p(\mathbf{O}^{(c)}, Z^{(c)}|\Theta^{(c)}, M) p(\Theta^{(c)}|M)\right]\right) \\
&\propto p(\Theta^{(c)}|M) \exp\left(\mathbb{E}_{(Z^{(c)})}\left[\log p(\mathbf{O}^{(c)}, Z^{(c)}|\Theta^{(c)}, M)\right]\right).
\end{aligned} \tag{7.39}$$

Here, we use the factorization property of the posterior distributions in Eq. (7.30). This result means that the optimal posterior distribution of model parameters $\tilde{q}(\Theta^{(c)}|\mathbf{O}^{(c)}, M)$ is obtained by its prior distribution $p(\Theta^{(c)}|M)$ and the expected complete data likelihood $p(\mathbf{O}^{(c)}, Z^{(c)}|\Theta^{(c)}, M)$.

7.2.2 VB posterior distributions for latent variables

A similar method is used for the optimal VB posterior distribution $\tilde{q}(Z^{(c)}|\mathbf{O}, M)$. An inequality similar to Eq. (7.37) is obtained by considering the KL divergence between the arbitrary posterior distribution $q(Z^{(c)}|\mathbf{O}, M)$ and the true posterior distribution $p(Z^{(c)}|\mathbf{O}, M)$ as follows:

$$\begin{aligned}
&\mathrm{KL}(q(Z^{(c)}|\mathbf{O}, M)\|p(Z^{(c)}|\mathbf{O}, M)) \\
&\quad \leq \log p(\mathbf{O}|M) - \mathcal{F}^M[q(\Theta|\mathbf{O}, M), q(Z|\mathbf{O}, M)].
\end{aligned} \tag{7.40}$$

The optimal VB posterior distribution $\tilde{q}(Z^{(c)}|\mathbf{O}, M)$ is also obtained by maximizing $\mathcal{F}^M$ with respect to $q(Z^{(c)}|\mathbf{O}, M)$ with the variational method as follows:

$$\tilde{q}(Z^{(c)}|\mathbf{O}, M) \propto \exp\left(\mathbb{E}_{(\Theta^{(c)})}\left[\log p(\mathbf{O}^{(c)}, Z^{(c)}|\Theta^{(c)}, M)\right]\right). \tag{7.41}$$

This result is also obtained by using the general formula of the variational posterior in Eq. (7.25). Compared with the result for $\tilde{q}(\Theta^{(c)}|\mathbf{O}^{(c)}, M)$ in Eq. (7.38), Eq. (7.41) does not need to prepare the prior distribution for Z.

7.2.3 VB–EM algorithm

Equations (7.38) and (7.41) are closed-form expressions, and these optimizations can be effectively performed by iterative calculations analogous to the expectation and maximization (EM) algorithm (Dempster *et al.* 1976), as discussed in Sections 3.4 and 4.2, which increases $\mathcal{F}^M$ at every iteration up to a converged value. Then, Eqs. (7.38) and (7.41), respectively, correspond to the maximization step (M-step) and the expectation

Table 7.1 Training specifications for ML and VB.

	Training	Min-max optimization	Objective function
ML	ML–EM	differential method	Q function
VB	VB–EM	variational method	$\mathcal{F}^M$ functional

step (E-step) in the VB approach. We call this algorithm the variational Bayes expectation and maximization (VB–EM) algorithm. Therefore, by substituting q into $\widetilde{q}$, these equations can be represented as follows:

$$\begin{cases} \widetilde{q}(\Theta^{(c)}|\mathbf{O}, M) \propto p(\Theta^{(c)}|M) \exp\left(\mathbb{E}_{(Z^{(c)})}\left[\log p(\mathbf{O}^{(c)}, Z^{(c)}|\Theta^{(c)}, M)\right]\right), \\ \widetilde{q}(Z^{(c)}|\mathbf{O}, M) \propto \exp\left(\mathbb{E}_{(\Theta^{(c)})}\left[\log p(\mathbf{O}^{(c)}, Z^{(c)}|\Theta^{(c)}, M)\right]\right). \end{cases} \tag{7.42}$$

Note that optimal posterior distributions for a particular category can be obtained simply by using the category's variables, i.e., we are not concerned with the other categories in the calculation, since Eq. (7.42) only depends on category c, which is based on the assumption given by Eq. (7.30).

Finally, to compare the VB approach with the conventional ML approach for training latent variable models, the training specifications for ML and VB are summarized in Table 7.1.

7.2.4 VB posterior distribution for model structure

The VB posterior distributions for a model structure are derived in the same way as in Section 7.2.1, and model selection is carried out employing the posterior distribution. Arbitrary posterior distribution $q(M|\mathbf{O})$ is introduced and the KL divergence between $q(M|\mathbf{O})$ and the true posterior distribution $p(M|\mathbf{O})$ is considered:

$$\mathrm{KL}(q(M|\mathbf{O})\|p(M|\mathbf{O})) = \sum_M q(M|\mathbf{O}) \log \frac{q(M|\mathbf{O})}{p(M|\mathbf{O})}. \tag{7.43}$$

Substituting Eq. (7.29) into Eq. (7.43) and using Jensen's inequality, the inequality of Eq. (7.43) can be obtained as follows:

$$\begin{aligned} &\mathrm{KL}\left(q(M|\mathbf{O})\|p(M|\mathbf{O})\right) \\ &\leq \log p(\mathbf{O}) + \mathbb{E}_{(M)}\left[\log \frac{q(M|\mathbf{O})}{p(M)} - \mathcal{F}^M[q(\Theta|\mathbf{O}, M), q(Z|\mathbf{O}, M)]\right]. \end{aligned} \tag{7.44}$$

Similarly to the discussion in Section 7.2.1, from the inequality Eq. (7.44), $q(M|\mathbf{O})$ approaches $p(M|\mathbf{O})$ as the right-hand-side decreases.

Compared with the posterior distributions of model parameters and latent variables, we cannot use the formula Eq. (7.25), since it is not practical to marginalize all possible model structures M. Therefore, the optimal posterior distribution for a model structure can again be obtained by a variational method, as explained in Section 7.1.3. If we consider the constraint $\sum_M q(M|\mathbf{O}) = 1$, the functional differentiation is represented by substituting respectively $\mathcal{F}^M$ and $q(M|\mathbf{O})$ into $\mathcal{H}$ and g_n in Eq. (7.18) as follows:

$$
\begin{aligned}
&\frac{\delta}{\delta q(M'|\mathbf{O})}\left(\mathbb{E}_{(M)}\left[\log\frac{q(M|\mathbf{O})}{p(M)}-\mathcal{F}^M\right]+K\left(\sum_M q(M|\mathbf{O})-1\right)\right)\\
&=\lim_{\epsilon\to 0}\frac{1}{\epsilon}\left(\sum_M (q(M|\mathbf{O})+\epsilon\delta_{MM'})\left(\log\frac{q(M|\mathbf{O})+\epsilon\delta_{MM'}}{p(M)}-\mathcal{F}^M\right)\right.\\
&\quad-\mathbb{E}_{(M)}\left[\log\frac{q(M|\mathbf{O})}{p(M)}-\mathcal{F}^M\right]\\
&\quad\left.+K\left(\sum_M q(M|\mathbf{O})+\epsilon\delta_{MM'}-1\right)-K\left(\sum_M q(M|\mathbf{O})-1\right)\right),
\end{aligned}
\tag{7.45}
$$

where K is a Lagrange multiplier. We focus on the first term in the brackets in the 2nd line of Eq. (7.45). This term can be rewritten as follows:

$$
\begin{aligned}
&\sum_M (q(M|\mathbf{O})+\epsilon\delta_{MM'})\left(\log\frac{q(M|\mathbf{O})+\epsilon\delta_{MM'}}{p(M)}-\mathcal{F}^M\right)\\
&=\sum_M (q(M|\mathbf{O})+\epsilon\delta_{MM'})\left(\log\frac{q(M|\mathbf{O})}{p(M)}+\log\left(1+\epsilon\frac{\delta_{MM'}}{q(M|\mathbf{O})}\right)-\mathcal{F}^M\right).
\end{aligned}
\tag{7.46}
$$

By expanding the logarithmic term in Eq. (7.46) with respect to ϵ, Eq. (7.46) is represented as:

$$
\begin{aligned}
&\text{Equation (7.46)}\\
&=\mathbb{E}_{(M)}\left[\log\frac{q(M|\mathbf{O})}{p(M)}-\mathcal{F}^M\right]+\epsilon\left(\log\frac{q(M'|\mathbf{O})}{p(M')}-\mathcal{F}^{M'}+1\right)+\mathbf{o}(\epsilon^2).
\end{aligned}
\tag{7.47}
$$

Therefore, by substituting Eq. (7.47) into Eq. (7.45), Eq. (7.45) is represented as:

$$
\text{Equation (7.45)}=\log\frac{q(M'|\mathbf{O})}{p(M')}-\mathcal{F}^{M'}+1+K.
\tag{7.48}
$$

Therefore, the optimal posterior (VB posterior) $\tilde{q}(M|\mathbf{O})$ satisfies the relation whereby Eq. (7.48) $=0$, and is obtained as:

$$
\log\frac{\tilde{q}(M|\mathbf{O})}{p(M)}-\mathcal{F}^M+1+K=0.
\tag{7.49}
$$

By disregarding the normalization constant, the optimal VB posterior is finally derived as:

$$
\tilde{q}(M|\mathbf{O})\propto p(M)\exp\left(\mathcal{F}^M[q(\Theta|\mathbf{O},M),q(Z|\mathbf{O},M)]\right).
\tag{7.50}
$$

Compared with Eqs. (7.38) and (7.41), the posterior obtained is represented by the total variational lower bound.

Assuming that $p(M)$ is a uniform distribution,[2] the proportional relation between $\tilde{q}(M|\mathbf{O})$ and $\mathcal{F}^M$ is obtained as follows, based on the convexity of the logarithmic function:

$$
\mathcal{F}^{M'}\geq\mathcal{F}^M\Leftrightarrow\tilde{q}(M'|\mathbf{O})\geq\tilde{q}(M|\mathbf{O}).
\tag{7.51}
$$

[2] We can set an informative prior distribution for $p(M)$ instead of the uniform distribution. Several prior distributions for model structure are considered in Chapter 8.

Therefore, an optimal model structure in the sense of maximum posterior probability estimation can be selected as follows:

$$\widetilde{M} = \arg\max_{M} \widetilde{q}(M|\mathbf{O}) = \arg\max_{M} \mathcal{F}^{M}. \tag{7.52}$$

This indicates that by maximizing total $\mathcal{F}^M$ with respect to both $q(\Theta|\mathbf{O}, M)$, $q(Z|\mathbf{O}, M)$, and M, we can obtain the optimal parameter distributions and select the optimal model structure simultaneously (Attias 1999, Ueda & Ghahramani 2002).

Thus, we analytically derive the variational posterior distributions of general latent models. The next section applies these solutions to the continuous density hidden Markov model (CDHMM), as we apply ML–EM and MAP–EM to CDHMMs in Sections 3.4 and 4.3, respectively.

7.3 Continuous density hidden Markov model

This section reformulates the CDHMM training for speech processing based on the VB framework (Valente & Wellekens 2003, Somervuo 2004, Watanabe *et al.* 2004). The four formulations are obtained by using the VB framework to perform acoustic model construction (model setting, training, and selection) and speech classification consistently, based on the Bayesian approach. Consequently, the conventional formulations based on the ML and MAP approaches in Sections 3.4 and 4.3 are replaced by formulations based on the Bayesian approach as follows:

- Set generative model distributions
  → *Set generative model distributions and prior distributions* (Section 7.3.1 and 7.3.2);

- ML/MAP Baum–Welch algorithm
  → *VB Baum–Welch algorithm* (Section 7.3.3);

- Log likelihood
  → *VB objective function* (Section 7.3.4);

- ML/MAP-based classification
  → *VB–BPC* (Section 7.3.5).

These four formulations are explained in the following four subsections, by applying the acoustic model for speech recognition to the general solution in Section 7.2.

7.3.1 Generative model

Similarly to the MAP–EM algorithm in Section 4.3, setting of the emission and prior distributions is required when calculating the VB posterior distributions. This section provides these distributions for CDHMM again, to provide the VB-based analytical solutions.

Let $\mathbf{O} = \{\mathbf{o}_t \in \mathbb{R}^D | t = 1, \ldots, T\}$ be a sequential speech data set for a speech segment of a phoneme category. Since the formulations for the posterior distributions are common to all phoneme categories, the phoneme category index c is omitted from this section to simplify the equation forms. D is used to denote the dimension number of the feature vector and T to denote the frame number. The complete data likelihood function with speech, HMM state, and GMM component sequences ($\mathbf{O}$, S, and V), which is introduced in Eqs. (3.44) and (4.23), is expressed by

$$p(\mathbf{O}, S, V|\Theta, M) = \prod_{t=1}^{T} a_{s_{t-1}s_t} \omega_{s_t v_t} p(\mathbf{o}_t|\Theta_{s_t v_t}, M), \tag{7.53}$$

where $a_{s_0 s_1} = \pi_{s_1}$.[3] Although we have many segments for each phoneme category and the generative model distribution must consider the product of each segment in Eq. (7.53), this is also omitted in this book. Here, S and V are sets of discrete latent variables, which are the concrete forms of Z in Section 7.2. The parameter a_{ij} denotes the state transition probability from state i to state j, and ω_{jk} is the kth weight factor of the Gaussian mixture for state j. In addition, $p(\mathbf{o}_t|\Theta_{jk})(= \mathcal{N}(\mathbf{o}_t|\boldsymbol{\mu}_{jk}, \boldsymbol{\Sigma}_{jk}))$ denotes the Gaussian with mean vector $\boldsymbol{\mu}_{jk}$ and covariance matrix $\boldsymbol{\Sigma}_{jk}$ defined as:

$$\mathcal{N}(\mathbf{o}_t|\boldsymbol{\mu}_{jk}, \boldsymbol{\Sigma}_{jk}) \triangleq (2\pi)^{-\frac{D}{2}} |\boldsymbol{\Sigma}_{jk}|^{-\frac{1}{2}} \exp\left(-\frac{1}{2}(\mathbf{o}_t - \boldsymbol{\mu}_{jk})^{\mathsf{T}} \boldsymbol{\Sigma}_{jk}^{-1} (\mathbf{o}_t - \boldsymbol{\mu}_{jk})\right). \tag{7.54}$$

$\Theta = \{a_{ij}, \omega_{jk}, \boldsymbol{\mu}_{jk}, \boldsymbol{\Sigma}_{jk}^{-1} | i, j = 1, \ldots, J, k = 1, \ldots, K\}$ is a set of model parameters. Here, J denotes the number of states in an HMM sequence and K denotes the number of Gaussian components in a state. This section only considers the diagonal covariance matrix case.

7.3.2 Prior distribution

Conjugate distributions, which are based on the exponential function, are as easy to use as prior distributions since the function forms of prior and posterior distributions become the same (Berger 1985, Gauvain & Lee 1994, Bernardo & Smith 2009), as discussed in Sections 2.1.3 and 4.3.3. Then a distribution is selected where the probabilistic variable constraint is the same as that of the model parameter. The state transition probability a_{ij} and the mixture weight factor ω_{jk} have the constraint that $\sum_j a_{ij} = 1$ and $\sum_k \omega_{jk} = 1$. Therefore, the Dirichlet distributions for π_j, a_{ij}, and ω_{jk} are used, where the variables of the Dirichlet distribution satisfy the above constraint. Similarly, the diagonal elements of the inverse covariance matrix $\boldsymbol{\Sigma}_{jk}^{-1}$ are always positive, and the gamma distribution is used. The range of the mean vector $\boldsymbol{\mu}_{jk}$ is from $-\infty$ to ∞, and the multivariate Gaussian distribution is used. Thus, as introduced in Eqs. (4.29) and (4.32), the prior distribution for a CDHMM with a diagonal covariance matrix is expressed as follows:

[3] This section does not explicitly provide the posterior solution of the initial weight, as it is trivial.

$$
\begin{aligned}
p(\Theta|M) &\triangleq \prod_{i=1}^{J} p(\{a_{ij'}\}_{j'=1}^{J}|M) \prod_{j=1}^{J} p(\{\omega_{jk'}\}_{k'=1}^{K}|M) \prod_{k=1}^{K} p(\boldsymbol{\mu}_{jk}, \boldsymbol{\Sigma}_{jk}|M) \\
&\triangleq \prod_{i=1}^{J} \mathrm{Dir}(\{a_{ij'}\}_{j'=1}^{J}|\{\phi_{ij'}^{a}\}_{j'=1}^{J}) \prod_{j=1}^{J} \mathrm{Dir}(\{\omega_{jk'}\}_{k'=1}^{K}|\{\phi_{jk'}^{\omega}\}_{k'=1}^{K}) \\
&\times \prod_{k=1}^{K} \prod_{d=1}^{D} \mathcal{N}(\mu_{jkd}|\mu_{jkd}^{0}, (\phi_{jk}^{\mu} r_{jkd})^{-1}) \mathrm{Gam}_2(r_{jkd}|\phi_{jk}^{r}, r_{jkd}^{0}).
\end{aligned} \tag{7.55}
$$

Here, $\Phi^0 \triangleq \{\phi_{ij}^{a}, \phi_{jk}^{\omega}, \phi_{jk}^{\mu}, \mu_{jkd}^{0}, \phi_{jk}^{r}, r_{jkd}^{0}|i,j = 1,\ldots,J, k = 1,\ldots,K, d = 1,\cdots,D\}$ is a set of prior parameters. In Eq. (7.55), $\mathrm{Dir}(\cdot)$ denotes a Dirichlet distribution and $\mathrm{Gam}_2(\cdot)$ denotes a gamma distribution. (It is different from the conventional definition of the gamma distribution, see Appendix C.11.) If the covariance matrix elements are off the diagonal, a Gaussian–Wishart distribution is used as the prior distribution of $\boldsymbol{\mu}_{jk}$ and $\boldsymbol{\Sigma}_{jk}$. The explicit forms of the distributions are defined as follows (Appendixes C.4, C.5, and C.11):

$$
\begin{cases}
\mathrm{Dir}(\{a_{ij}\}_{j=1}^{J}|\{\phi_{ij}^{a}\}_{j=1}^{J}) & \triangleq C_{\mathrm{Dir}}(\{\phi_{ij}^{a}\}_{j=1}^{J}) \prod_{j=1}^{J} (a_{ij})^{\phi_{ij}^{a}-1}, \\
\mathrm{Dir}(\{\omega_{jk}\}_{k=1}^{K}|\{\phi_{jk}^{\omega}\}_{k=1}^{K}) & \triangleq C_{\mathrm{Dir}}(\{\phi_{jk}^{\omega}\}_{k=1}^{K}) \prod_{k=1}^{K} (\omega_{jk})^{\phi_{jk}^{\omega}-1}, \\
\mathcal{N}(\mu_{jkd}|\mu_{jkd}^{0}, (\phi_{jk}^{\mu} r_{jkd})^{-1}) & \triangleq C_{\mathcal{N}}(\phi_{jk}^{\mu})(r_{jkd})^{\frac{1}{2}} \exp\left(-\frac{\phi_{jk}^{\mu} r_{jkd}(\mu_{jkd}-\mu_{jkd}^{0})^2}{2}\right), \\
\mathrm{Gam}_2(r_{jkd}|\phi_{jk}^{r}, r_{jkd}^{0}) & \triangleq C_{\mathrm{Gam}_2}(\phi_{jk}^{r}, r_{jkd}^{0})(r_{jkd})^{\frac{\phi_{jk}^{r}}{2}-1} \exp\left(-\frac{r_{jkd}^{0} r_{jkd}}{2}\right),
\end{cases} \tag{7.56}
$$

where the normalization constants are defined as follows:

$$
\begin{cases}
C_{\mathrm{Dir}}(\{\phi_{ij}^{a}\}_{j=1}^{J}) & \triangleq \frac{\Gamma(\sum_{j=1}^{J} \phi_{ij}^{a})}{\prod_{j=1}^{J} \Gamma(\phi_{ij}^{a})}, \\
C_{\mathrm{Dir}}(\{\phi_{jk}^{\omega}\}_{k=1}^{K}) & \triangleq \frac{\Gamma(\sum_{k=1}^{K} \phi_{jk}^{\omega})}{\prod_{k=1}^{K} \Gamma(\phi_{jk}^{\omega})}, \\
C_{\mathcal{N}}(\phi_{jk}^{\mu}) & \triangleq \left(\frac{\phi_{jk}^{\mu}}{2\pi}\right)^{\frac{1}{2}}, \\
C_{\mathrm{Gam}_2}(\phi_{jk}^{r}, r_{jkd}^{0}) & \triangleq \frac{\left(\frac{r_{jkd}^{0}}{2}2\right)^{\frac{\phi_{jk}^{r}}{2}}}{\Gamma\left(\frac{\phi_{jk}^{r}}{2}\right)}.
\end{cases} \tag{7.57}
$$

In the Bayesian approach, an important problem is how to set the prior parameters. Here, two kinds of prior parameters of μ^0 and r^0 are set using sufficient amounts of data from:

- Statistics of higher hierarchy acoustic models for the acoustic model construction task;
- Statistics of speaker independent models for the speaker adaptation task.

The other parameters ($\phi^a, \phi^\omega, \phi^\mu$, and ϕ^r) have a meaning as regarding tuning the balance between the values obtained from training data and the above statistics. These parameters are set appropriately based on experiments, as discussed in speaker adaptation (Section 4.4).

Finally, Algorithm 10 provides a generative process for a CDHMM with prior distribution. For simplicity, the initial weight, the hyperparameters, and the model structure are given in this generative process, but it can also be sampled from some distributions. Compared with Algorithm 3, CDHMM parameters are also sampled from the prior distributions.

Algorithm 10 Generative process for continuous density hidden Markov model with prior distributions

Require: Ψ, M, and $\{\pi_j\}_{j=1}^J$

1: **for** $i, j = 1, \cdots, J$ **do**
2: Draw a_{ij} from $\mathrm{Dir}(\{a_{ij}\}_{j=1}^J | \{\phi_{ij}^a\}_{j=1}^J)$
3: **end for**
4: **for** $j = 1, \cdots, J$ **do**
5: **for** $k = 1, \cdots, K$ **do**
6: Draw ω_{jk} from $\mathrm{Dir}(\{\omega_{jk}\}_{k=1}^K | \{\phi_{jk}^\omega\}_{k=1}^K)$
7: **for** $d = 1, \cdots, D$ **do**
8: Draw r_{jkd} from $\mathrm{Gam}_2(r_{jkd} | \phi_{jk}^r, r_{jkd}^0)$
9: Draw μ_{jkd} from $\mathcal{N}(\mu_{jkd} | \mu_{jkd}^0, (\phi_{jk}^\mu r_{jkd})^{-1})$
10: **end for**
11: **end for**
12: **end for**
13: Draw s_1 from $\mathrm{Mult}(s_1 | \{\pi_j\}_{j=1}^J)$
14: Draw v_1 from $\mathrm{Mult}(v_1 | \{\omega_{s_1 k}\}_{k=1}^K)$
15: Draw $\mathbf{o}_1$ from $\mathcal{N}(\mathbf{o}_1 | \boldsymbol{\mu}_{s_1 v_1}, \boldsymbol{\Sigma}_{s_1 v_1})$
16: **for** $t = 2, \cdots, T$ **do**
17: Draw s_t from $\mathrm{Mult}(s_t | \{a_{s_{t-1} j}\}_{j=1}^J)$
18: Draw v_t from $\mathrm{Mult}(v_t | \{\omega_{s_t k}\}_{k=1}^K)$
19: Draw $\mathbf{o}_t$ from $\mathcal{N}(\mathbf{o}_t | \boldsymbol{\mu}_{s_t v_t}, \boldsymbol{\Sigma}_{s_t v_t})$
20: **end for**

7.3.3 VB Baum–Welch algorithm

This subsection introduces concrete forms of the VB posterior distributions for model parameters $q(\Theta|\mathbf{O}, M)$ and for latent variables $q(Z|\mathbf{O}, M)$ in acoustic modeling, which are efficiently computed by VB iterative calculations within the VB framework. This calculation is effectively carried out by the VB Baum–Welch algorithm (MacKay 1997).

VB M-step

First, the VB M-step for acoustic model training is explained. This is solved by substituting the acoustic model setting in Section 7.3.1 into the general solution for the VB M-step in Section 7.2. From Eq. (7.42), the VB posterior distributions for the model parameters are represented as follows:

$$\widetilde{q}(\Theta|\mathbf{O}, M) \propto p(\Theta|M) \exp\left(\mathbb{E}_{(S,V)}\left[\log p(\mathbf{O}, S, V|\Theta, M)\right]\right). \tag{7.58}$$

Taking the logarithmic operation, Eq. (7.58) is represented as:

$$\begin{aligned}\log \widetilde{q}(\Theta|\mathbf{O}, M) &\propto \log p(\Theta|M) + \mathbb{E}_{(S,V)}\left[\log p(\mathbf{O}, S, V|\Theta, M)\right] \\ &\triangleq \tilde{Q}(\Theta).\end{aligned} \tag{7.59}$$

Here, $\tilde{Q}(\Theta)$ is a VB auxiliary function defined as follows:

$$\begin{aligned}\tilde{Q}(\Theta) &\triangleq \mathbb{E}_{(S,V)}\left[\log p(\mathbf{O}, S, V|\Theta, M)\right] + \log p(\Theta|M) \\ &= \sum_{S,V} \tilde{q}(S, V|\mathbf{O}, M) \log p(\mathbf{O}, S, V|\Theta, M) + \log p(\Theta|M).\end{aligned} \tag{7.60}$$

On the other hand, the MAP auxiliary function in Eq. (4.16) is defined as follows:

$$Q^{\mathrm{MAP}}(\Theta'|\Theta) \triangleq \sum_{S,V} p(S, V|\mathbf{O}, \Theta, M) \log p(\mathbf{O}, S, V|\Theta', M) + \log p(\Theta'|M). \tag{7.61}$$

Comparing the VB and MAP auxiliary functions, these are almost equivalent except the posterior distributions of latent variables, i.e., $\tilde{q}(S, V|\mathbf{O}, M)$ vs. $p(S, V|\mathbf{O}, \Theta, M)$. Since $\tilde{q}(S, V|\mathbf{O}, M)$ is obtained by marginalizing Θ, $\tilde{Q}(\Theta)$ is more appropriate in terms of the Bayesian treatment.

Therefore, this VB-M step is solved by using the result of the MAP-M step solution, except that $\tilde{q}(S, V|\mathbf{O}, M)$ is obtained by using the VB-E step. The calculated results for the optimal VB posterior distributions for the model parameters are summarized as follows:

$$\begin{aligned}\widetilde{q}(\Theta|M) &\triangleq \prod_{i=1}^{J} \widetilde{q}(\{a_{ij'}\}_{j'=1}^{J}|M) \prod_{j=1}^{J} \widetilde{q}(\{\omega_{jk'}\}_{k'=1}^{K}|M) \prod_{k=1}^{K} \widetilde{q}(\boldsymbol{\mu}_{jk}, \boldsymbol{\Sigma}_{jk}|M) \\ &\triangleq \prod_{i=1}^{J} \mathrm{Dir}(\{a_{ij'}\}_{j'=1}^{J}|\{\widetilde{\phi}^{a}_{ij'}\}_{j'=1}^{J}) \prod_{j=1}^{J} \mathrm{Dir}(\{\omega_{jk'}\}_{k'=1}^{K}|\{\widetilde{\phi}^{\omega}_{jk'}\}_{k'=1}^{K}) \\ &\quad \times \prod_{k=1}^{K} \prod_{d=1}^{D} \mathcal{N}(\mu_{jkd}|\widetilde{\mu}_{jkd}, (\widetilde{\phi}^{\mu}_{jk} r_{jkd})^{-1}) \mathrm{Gam}_2(r_{jkd}|\widetilde{\phi}^{r}_{jk}, \widetilde{r}_{jkd}).\end{aligned} \tag{7.62}$$

The concrete forms of the distributions are defined as follows:

$$\begin{cases}\mathrm{Dir}(\{a_{ij}\}_{j=1}^{J}|\{\widetilde{\phi}^{a}_{ij}\}_{j=1}^{J}) & \triangleq C_{\mathrm{Dir}}(\{\widetilde{\phi}^{a}_{ij}\}_{j=1}^{J}) \prod_{j=1}^{J} (a_{ij})^{\widetilde{\phi}^{a}_{ij}-1}, \\ \mathrm{Dir}(\{\omega_{jk}\}_{k=1}^{K}|\{\widetilde{\phi}^{\omega}_{jk}\}_{k=1}^{K}) & \triangleq C_{\mathrm{Dir}}(\{\widetilde{\phi}^{\omega}_{jk}\}_{k=1}^{K}) \prod_{k=1}^{K} (\omega_{jk})^{\widetilde{\phi}^{\omega}_{jk}-1}, \\ \mathcal{N}(\mu_{jkd}|\widetilde{\mu}_{jkd}, (\widetilde{\phi}^{\mu}_{jk} r_{jkd})^{-1}) & \triangleq C_{\mathcal{N}}(\widetilde{\phi}^{\mu}_{jk})(r_{jkd})^{\frac{1}{2}} \exp\left(-\frac{\widetilde{\phi}^{\mu}_{jk} r_{jkd}(\mu_{jkd}-\widetilde{\mu}_{jkd})^2}{2}\right), \\ \mathrm{Gam}_2(r_{jkd}|\widetilde{\phi}^{r}_{jk}, \widetilde{r}_{jkd}) & \triangleq C_{\mathrm{Gam}_2}(\widetilde{\phi}^{r}_{jk}, \widetilde{r}_{jkd})(r_{jkd})^{\frac{\widetilde{\phi}^{r}_{jk}}{2}-1} \exp\left(-\frac{\widetilde{r}_{jkd} r_{jkd}}{2}\right),\end{cases} \tag{7.63}$$

where the normalization constants are:

$$\begin{cases} C_{\mathrm{Dir}}(\{\widetilde{\phi}^a_{ij}\}_{j=1}^J) & \triangleq \dfrac{\Gamma(\sum_{j=1}^J \widetilde{\phi}^a_{ij})}{\prod_{j=1}^J \Gamma(\widetilde{\phi}^a_{ij})}, \\ C_{\mathrm{Dir}}(\{\widetilde{\phi}^\omega_{jk}\}_{k=1}^K) & \triangleq \dfrac{\Gamma(\sum_{k=1}^K \widetilde{\phi}^\omega_{jk})}{\prod_{k=1}^K \Gamma(\widetilde{\phi}^\omega_{jk})}, \\ C_{\mathcal{N}}(\widetilde{\phi}^\mu_{jk}) & \triangleq \left(\dfrac{\widetilde{\phi}^\mu_{jk}}{2\pi}\right)^{\frac{1}{2}}, \\ C_{\mathrm{Gam}_2}(\widetilde{\phi}^r_{jk}, \widetilde{r}_{jkd}) & \triangleq \dfrac{\left(\frac{\widetilde{r}_{jkd}}{2}\right)^{\frac{\widetilde{\phi}^r_{jk}}{2}}}{\Gamma\left(\frac{\widetilde{\phi}^r_{jk}}{2}\right)}. \end{cases} \tag{7.64}$$

Note that Eqs. (7.55) and (7.62) are members of the same function family, and the only difference is that the set of prior parameters Φ^0 in Eq. (7.55) is replaced with a set of posterior distribution parameters $\widetilde{\Phi}$ in Eq. (7.62), where $\widetilde{\Phi}$ is defined as:

$$\begin{aligned} \widetilde{\Phi} \triangleq & \{\widetilde{\phi}^a_{ij}, \widetilde{\phi}^\omega_{jk}, \widetilde{\phi}^\mu_{jk}, \widetilde{\mu}_{jkd}, \widetilde{\phi}^r_{jk}, \widetilde{r}_{jkd} \\ & |i,j = 1,\ldots,J, k = 1,\ldots,K, d = 1,\cdots,D\}. \end{aligned} \tag{7.65}$$

The conjugate prior distribution is adopted because the posterior distribution is theoretically a member of the same function family as the prior distribution, and is obtained analytically, which is a characteristic of the exponential distribution family, as discussed in Section 2.1.4. Here, $\widetilde{\Phi}$ values are calculated from:

$$\begin{aligned} \widetilde{\phi}^a_{ij} &= \phi^a_{ij} + \widetilde{\xi}_{ij}, \\ \widetilde{\phi}^\omega_{jk} &= \phi^\omega_{jk} + \widetilde{\gamma}_{jk}, \\ \widetilde{\phi}^\mu_{jk} &= \phi^\mu_{jk} + \widetilde{\gamma}_{jk}, \\ \widetilde{\mu}_{jkd} &= \frac{\phi^\mu_{jk}\mu^0_{jkd} + \gamma^{(1)}_{jkd}}{\phi^\mu_{jk} + \widetilde{\gamma}_{jk}}, \\ \widetilde{\phi}^r_{jk} &= \phi^r_{jk} + \widetilde{\gamma}_{jk}, \\ \widetilde{r}_{jkd} &= \widetilde{\gamma}^{(2)}_{jkd} + \phi^\mu_{jk}(\mu^0_{jkd})^2 - \widetilde{\phi}^\mu_{jk}(\widetilde{\mu}_{jkd})^2 + r^0_{jkd}. \end{aligned} \tag{7.66}$$

$\widetilde{\xi}_{ij}$ denotes the sufficient statistics of the transition matrix, and $\widetilde{\gamma}_{jk}$, $\widetilde{\gamma}^{(1)}_{jkd}$, and $\widetilde{\gamma}^{(2)}_{jkd}$ denote zeroth-, first-, and second-order sufficient statistics of the GMM, respectively, and are defined as follows:

$$\begin{cases} \widetilde{\xi}_{ij} & \triangleq \sum_{t=1}^{T-1} \widetilde{\xi}_t(i,j), \\ \widetilde{\gamma}_{jk} & \triangleq \sum_{t=1}^{T} \widetilde{\gamma}_t(j,k), \\ \widetilde{\gamma}^{(1)}_{jkd} & \triangleq \sum_{t=1}^{T} \widetilde{\gamma}_t(j,k)o_{td}, \\ \widetilde{\gamma}^{(2)}_{jkd} & \triangleq \sum_{t=1}^{T} \widetilde{\gamma}_t(j,k)(o_{td})^2. \end{cases} \tag{7.67}$$

These sufficient statistics $\widetilde{\Xi} \triangleq \{\widetilde{\xi}_{ij}, \widetilde{\gamma}_{jk}, \widetilde{\gamma}^{(1)}_{jkd}, \widetilde{\gamma}^{(2)}_{jkd} | i,j = 1, \ldots, J, k = 1, \ldots, K, d = 1, \cdots, D\}$ are computed by using $\widetilde{\xi}_t(i,j)$ and $\widetilde{\gamma}_t(j,k)$, defined as follows:

$$\begin{cases} \widetilde{\xi}_t(i,j) & \triangleq & \widetilde{q}(s_t = i, s_{t+1} = j | \mathbf{O}, M), \\ \widetilde{\gamma}_t(j,k) & \triangleq & \widetilde{q}(s_t = j, v_t = k | \mathbf{O}, M). \end{cases} \tag{7.68}$$

Here, $\widetilde{\xi}_t(i,j)$ is a VB transition posterior distribution, which denotes the transition probability from a state i to a state j at a frame t, and $\widetilde{\gamma}_t(j,k)$ is a VB occupation posterior distribution, which denotes the occupation probability of a mixture component k in a state j at a frame t, in the VB approach. These are similar to those defined in Eq. (3.119) by using the ML–EM algorithm of a CDHMM. Therefore, $\widetilde{\Phi}$ can be calculated from Φ^0, $\widetilde{\xi}_t(i,j)$, and $\widetilde{\gamma}_t(j,k)$, enabling $\widetilde{q}(\Theta|\mathbf{O}, M)$ to be obtained.

VB E-step

Before we focus on the calculation of $\widetilde{\xi}_t(i,j)$ and $\widetilde{\gamma}_t(j,k)$, we first focus on the posterior distribution of the joint distribution of the HMM state and mixture component sequences $\widetilde{q}(S, V|\Theta, M)$. From Eq. (7.25), $\widetilde{q}(S, V|\Theta, M)$ is represented as follows:

$$\widetilde{q}(S, V|\mathbf{O}, M) = \frac{\widetilde{q}(\mathbf{O}, S, V|M)}{\widetilde{q}(\mathbf{O}|M)} \propto \exp\left(\mathbb{E}_{(\Theta)}\left[\log p(\mathbf{O}, S, V|\Theta, M)\right]\right). \tag{7.69}$$

Since $\widetilde{q}(\mathbf{O}|M)$ does not depend on S and V, we find that the complete data likelihood marginalized by the model parameter Θ can also be obtained by considering the same expectation with Eq. (7.69):

$$\widetilde{q}(\mathbf{O}, S, V|M) \propto \exp\left(\mathbb{E}_{(\Theta)}\left[\log p(\mathbf{O}, S, V|\Theta, M)\right]\right). \tag{7.70}$$

Once we obtain the complete data likelihood $\widetilde{q}(\mathbf{O}, S, V|M)$, we can compute the posterior probabilities $\widetilde{\xi}_t(i,j)$ and $\widetilde{\gamma}_t(j,k)$ by using the forward–backward algorithm, similarly to Section 3.3. Therefore, we focus on how to compute the following expectation of the complete data log likelihood:

$$\mathbb{E}_{(\Theta)}\left[\log p(\mathbf{O}, S, V|\Theta, M)\right]. \tag{7.71}$$

By substituting Eq. (7.53) into Eq. (7.71), Eq. (7.71) can be represented as follows:

$$\begin{aligned} &\mathbb{E}_{(\Theta)}\left[\log p(\mathbf{O}, S, V|\Theta, M)\right] \\ &= \mathbb{E}_{(\Theta)}\left[\log \prod_{t=1}^{T} a_{s_{t-1}s_t} \omega_{s_t v_t} \mathcal{N}(\mathbf{o}_t|\boldsymbol{\mu}_{s_t v_t}, \boldsymbol{\Sigma}_{s_t v_t})\right] \\ &= \sum_{t=1}^{T} \mathbb{E}_{(\Theta)}\left[\log(a_{s_{t-1}s_t}) + \log(\omega_{s_t v_t}) + \log(\mathcal{N}(\mathbf{o}_t|\boldsymbol{\mu}_{s_t v_t}, \boldsymbol{\Sigma}_{s_t v_t}))\right] \\ &= \sum_{t=1}^{T} \mathbb{E}_{(\Theta)}\left[\log(a_{s_{t-1}s_t})\right] + \mathbb{E}_{(\Theta)}\left[\log(\omega_{s_t v_t})\right] + \mathbb{E}_{(\Theta)}\left[\log(\mathcal{N}(\mathbf{o}_t|\boldsymbol{\mu}_{s_t v_t}, \boldsymbol{\Sigma}_{s_t v_t}))\right]. \end{aligned} \tag{7.72}$$

Now we focus on the case when $s_{t-1} = i$, $s_t = j$, and $v_t = k$, where we need to compute the following equations:

$$\begin{aligned}
\log \widetilde{a}_{ij} &\triangleq \mathbb{E}_{(a_{ij})}\left[\log(a_{ij})\right], \\
\log \widetilde{\omega}_{jk} &\triangleq \mathbb{E}_{(\omega_{jk})}\left[\log(\omega_{jk})\right], \\
\log \widetilde{b}_{jk}(\mathbf{o}_t) &\triangleq \mathbb{E}_{(\boldsymbol{\mu}_{jk}, \boldsymbol{\Sigma}_{jk})}\left[\log(\mathcal{N}(\mathbf{o}_t|\boldsymbol{\mu}_{jk}, \boldsymbol{\Sigma}_{jk}))\right].
\end{aligned} \tag{7.73}$$

We can also define the following function based on Eq. (7.73):

$$\tilde{u}(\mathbf{O}, S, V|M) \triangleq \prod_{t=1}^{T} \widetilde{a}_{s_{t-1}s_t} \widetilde{\omega}_{s_t v_t} \widetilde{b}_{s_t v_t}(\mathbf{o}_t). \tag{7.74}$$

This equation behaves similarly to the likelihood function of $p(\mathbf{O}, S, V|\Theta, M)$ in Eq. (7.53). Note that $\tilde{u}(\mathbf{O}, S, V|M)$ is not properly normalized, and cannot be dealt with as a probabilistic distribution. However, from Eq. (7.70), $\tilde{u}(\mathbf{O}, S, V|M)$ is proportional to $\tilde{q}(\mathbf{O}, S, V|M)$, and this function has the following relationship from Eq. (7.69):

$$\begin{aligned}
\tilde{q}(\mathbf{O}, S, V|M) &= \tilde{q}(\mathbf{O}|M)\tilde{q}(S, V|\mathbf{O}, M) \\
&= \frac{\tilde{q}(\mathbf{O}|M)}{\sum_{S',V'} \tilde{u}(\mathbf{O}, S', V'|M)} \tilde{u}(\mathbf{O}, S, V|M) \\
&\triangleq \prod_{t=1}^{T} C_a \widetilde{a}_{s_{t-1}s_t} C_{\omega b} \widetilde{\omega}_{s_t v_t} \widetilde{b}_{s_t v_t}(\mathbf{o}_t),
\end{aligned} \tag{7.75}$$

where C_a and $C_{\omega b}$ are normalization constants of $\widetilde{a}_{ij}$ and $\widetilde{\omega}_{jk}\widetilde{b}_{jk}(\mathbf{o}_t)$ respectively for each frame, and satisfy the following condition:

$$\frac{\tilde{q}(\mathbf{O}|M)}{\sum_{S',V'} \tilde{u}(\mathbf{O}, S', V'|M)} = (C_a C_{\omega b})^T. \tag{7.76}$$

Note that it is not easy to obtain the normalization factors C_a and $C_{\omega b}$ explicitly, since it requires $\tilde{q}(\mathbf{O}|M)$ and $\sum_{S',V'} \tilde{u}(\mathbf{O}, S', V'|M)$. However, it will be shown later that the calculation of the occupation probabilities does not require computation of the normalization factors explicitly, but only requires $\widetilde{a}_{ij}$, $\widetilde{\omega}_{jk}$, and $\widetilde{b}_{jk}$. Therefore, we can compute various values from Eq. (7.74) (e.g., the forward and backward variables and the occupation probabilities), as discussed in Section 3.3. Thus, the following paragraphs provide the analytical solutions of $\widetilde{a}_{ij}$, $\widetilde{\omega}_{jk}$, and $\widetilde{b}_{jk}$ in detail.

State transition $\widetilde{a}_{ij}$

First, the integral over a_{ij} is solved from Eq. (7.63) by using a partial integral technique and a normalization constant:

$$\begin{aligned}
\log \widetilde{a}_{ij} &= \int \widetilde{q}(\{a_{ij'}\}_{j'=1}^{J}|M) \log a_{ij} \prod_{j'=1}^{J} da_{ij'} \\
&= C_{\mathrm{Dir}}(\{\widetilde{\phi}^a_{ij'}\}_{j'=1}^{J}) \int \log a_{ij} \prod_{j'=1}^{J} (a_{ij'})^{\widetilde{\phi}^a_{ij'}-1} da_{ij'}.
\end{aligned} \tag{7.77}$$

Then we use the following derivative formula:

$$\frac{\partial}{\partial \widetilde{\phi}^a_{ij}}(a_{ij})^{\widetilde{\phi}^a_{ij}-1} = (\log a_{ij})(a_{ij})^{\widetilde{\phi}^a_{ij}-1}. \tag{7.78}$$

By substituting Eq. (7.78) into Eq. (7.77), Eq. (7.77) can be rewritten as:

$$\begin{aligned}
\log \widetilde{a}_{ij} &= C_{\mathrm{Dir}}(\{\widetilde{\phi}^a_{ij'}\}^J_{j'=1}) \int \frac{\partial}{\partial \widetilde{\phi}^a_{ij}}(a_{ij})^{\widetilde{\phi}^a_{ij}-1} da_{ij} \int \prod^J_{j' \neq j}(a_{ij'})^{\widetilde{\phi}^a_{ij'}-1} da_{ij'} \\
&= C_{\mathrm{Dir}}(\{\widetilde{\phi}^a_{ij'}\}^J_{j'=1}) \frac{\partial}{\partial \widetilde{\phi}^a_{ij}} \int \prod^J_{j'=1}(a_{ij'})^{\widetilde{\phi}^a_{ij'}-1} da_{ij'} \\
&= C_{\mathrm{Dir}}(\{\widetilde{\phi}^a_{ij'}\}^J_{j'=1}) \frac{\partial}{\partial \widetilde{\phi}^a_{ij}} \frac{1}{C_{\mathrm{Dir}}(\{\widetilde{\phi}^a_{ij'}\}^J_{j'=1})},
\end{aligned} \tag{7.79}$$

where we replace the derivative and integral, and the integral can be performed to derive the inverse of the normalization constant of the Dirichlet distribution.

From Eq. (7.64), this derivative can be calculated as follows:

$$\begin{aligned}
&\frac{\partial}{\partial \widetilde{\phi}^a_{ij}} \frac{1}{C_{\mathrm{Dir}}(\{\widetilde{\phi}^a_{ij'}\}^J_{j'=1})} \\
&= \frac{\partial}{\partial \widetilde{\phi}^a_{ij}} \frac{\prod^J_{j'=1} \Gamma(\widetilde{\phi}^a_{ij'})}{\Gamma(\sum^J_{j'=1} \widetilde{\phi}^a_{ij'})} \\
&= \frac{\left(\frac{\partial}{\partial \widetilde{\phi}^a_{ij}} \Gamma(\widetilde{\phi}^a_{ij})\right) \prod^J_{j' \neq j} \Gamma(\widetilde{\phi}^a_{ij'}) \Gamma(\sum^J_{j'=1} \widetilde{\phi}^a_{ij'}) - \prod^J_{j'=1} \Gamma(\widetilde{\phi}^a_{ij'}) \left(\frac{\partial}{\partial \widetilde{\phi}^a_{ij}} \Gamma(\sum^J_{j'=1} \widetilde{\phi}^a_{ij'})\right)}{\left(\Gamma(\sum^J_{j'=1} \widetilde{\phi}^a_{ij'})\right)^2} \\
&= \frac{\Psi(\widetilde{\phi}^a_{ij}) \prod^J_{j'=1} \Gamma(\widetilde{\phi}^a_{ij'}) \Gamma(\sum^J_{j'=1} \widetilde{\phi}^a_{ij'}) - \prod^J_{j'=1} \Gamma(\widetilde{\phi}^a_{ij'}) \Psi(\sum^J_{j'=1} \widetilde{\phi}^a_{ij'}) \Gamma(\sum^J_{j'=1} \widetilde{\phi}^a_{ij'})}{\left(\Gamma(\sum^J_{j'=1} \widetilde{\phi}^a_{ij'})\right)^2} \\
&= \frac{\prod^J_{j'=1} \Gamma(\widetilde{\phi}^a_{ij'}) \left(\Psi(\widetilde{\phi}^a_{ij}) - \Psi(\sum^J_{j'=1} \widetilde{\phi}^a_{ij'})\right)}{\Gamma(\sum^J_{j'=1} \widetilde{\phi}^a_{ij'})} \\
&= \frac{1}{C_{\mathrm{Dir}}(\{\widetilde{\phi}^a_{ij'}\}^J_{j'=1})} \left(\Psi(\widetilde{\phi}^a_{ij}) - \Psi(\sum^J_{j'=1} \widetilde{\phi}^a_{ij'})\right),
\end{aligned} \tag{7.80}$$

where $\Psi(y)$ is a di-gamma function, which first appeared in Eq. (5.82), and is defined as

$$\Psi(y) \triangleq \frac{\partial}{\partial y} \log \Gamma(y) = \frac{\frac{\partial}{\partial y}\Gamma(y)}{\Gamma(y)}. \tag{7.81}$$

Thus, $\widetilde{a}_{ij}$ is finally obtained as follows:

$$\log \widetilde{a}_{ij} = \Psi(\widetilde{\phi}^a_{ij}) - \Psi(\sum^J_{j'=1} \widetilde{\phi}^a_{ij'}). \tag{7.82}$$

Mixture weight $\widetilde{\omega}_{jk}$

In a way similar to that used for $\widetilde{a}_{ij}$, the integral over ω_{jk} is solved from Eq. (7.63), and $\widetilde{\omega}_{jk}$ is obtained as follows:

$$\log \widetilde{\omega}_{jk} = \Psi(\widetilde{\phi}^{\omega}_{jk}) - \Psi(\sum_{k'=1}^{K} \widetilde{\phi}^{\omega}_{jk'}). \tag{7.83}$$

Gaussian distribution $\widetilde{b}_{jk}(\mathbf{o}_t)$

First, $\log \widetilde{b}_{jk}(\mathbf{o}_t)$ can be factorized for each dimension:

$$\begin{aligned}
\log \widetilde{b}_{jk}(\mathbf{o}_t) &= \mathbb{E}_{(\boldsymbol{\mu}_{jk}, \boldsymbol{\Sigma}_{jk})}\left[\log(\mathcal{N}(\mathbf{o}_t|\boldsymbol{\mu}_{jk}, \boldsymbol{\Sigma}_{jk}))\right] \\
&= \mathbb{E}_{(\boldsymbol{\mu}_{jk}, \boldsymbol{\Sigma}_{jk})}\left[\log\left(\prod_{d=1}^{D} \mathcal{N}(o_{td}|\mu_{jkd}, \Sigma_{jkd})\right)\right] \\
&= \sum_{d=1}^{D} \mathbb{E}_{(\mu_{jkd}, \Sigma_{jkd})}\left[\log\left(\mathcal{N}(o_{td}|\mu_{jkd}, \Sigma_{jkd})\right)\right].
\end{aligned} \tag{7.84}$$

Therefore, we focus on calculation of the d element. Since the calculation is more complicated than the two previous calculations, the indexes j, k, t, and d are removed to simplify the derivation. By using (7.63), $\log \widetilde{b}(o)$ can be rewritten as follows:

$$\begin{aligned}
&\log \widetilde{b}(o) \\
&= \int \mathcal{N}(\mu|\widetilde{\mu}, (\widetilde{\phi}^{\mu} r)^{-1}) \text{Gam}_2(r|\widetilde{\phi}^{r}, \widetilde{r}) \\
&\quad \times \left(-\frac{1}{2}\left(\log(2\pi) - \log(r) + r(o-\mu)^2\right)\right) d\mu dr.
\end{aligned} \tag{7.85}$$

Now we focus on the integral over mean parameter μ. To calculate the integral, we first rewrite the part that is related to μ as follows:

$$\begin{aligned}
&\int \mathcal{N}(\mu|\widetilde{\mu}, (\widetilde{\phi}^{\mu} r)^{-1}) r(o-\mu)^2 d\mu \\
&= r \int \mathcal{N}(\mu|\widetilde{\mu}, (\widetilde{\phi}^{\mu} r)^{-1})(o - \mu + \widetilde{\mu} - \widetilde{\mu})^2 d\mu \\
&= r \int \mathcal{N}(\mu|\widetilde{\mu}, (\widetilde{\phi}^{\mu} r)^{-1}) \left((\mu - \widetilde{\mu})^2 + (o - \widetilde{\mu})^2 - 2(\mu - \widetilde{\mu})(o - \widetilde{\mu})\right) d\mu.
\end{aligned} \tag{7.86}$$

The integral of the above terms can be analytically solved. We first consider the following partial derivative:

$$\frac{\partial}{\partial \widetilde{\phi}^{\mu} r} \exp\left(-\frac{1}{2}(\mu - \widetilde{\mu})^2 \widetilde{\phi}^{\mu} r\right) = -\frac{1}{2}(\mu - \widetilde{\mu})^2 \exp\left(-\frac{1}{2}(\mu - \widetilde{\mu})^2 \widetilde{\phi}^{\mu} r\right). \tag{7.87}$$

Therefore, the first integral can be represented as:

$$\begin{aligned}
&\int \mathcal{N}(\mu|\widetilde{\mu},(\widetilde{\phi}^\mu r)^{-1})(\mu-\widetilde{\mu})^2 d\mu \\
&= (2\pi)^{-\frac{1}{2}}(\widetilde{\phi}^\mu r)^{\frac{1}{2}} \int \exp\left(-\frac{1}{2}(\mu-\widetilde{\mu})^2\widetilde{\phi}^\mu r\right)(\mu-\widetilde{\mu})^2 d\mu \\
&= (2\pi)^{-\frac{1}{2}}(\widetilde{\phi}^\mu r)^{\frac{1}{2}}(-2) \int \frac{\partial}{\partial \widetilde{\phi}^\mu r} \exp\left(-\frac{1}{2}(\mu-\widetilde{\mu})^2\widetilde{\phi}^\mu r\right) d\mu. \qquad (7.88)
\end{aligned}$$

By replacing the integral with the partial derivative, we can solve the integral as:

$$\begin{aligned}
&\int \mathcal{N}(\mu|\widetilde{\mu},(\widetilde{\phi}^\mu r)^{-1})(\mu-\widetilde{\mu})^2 d\mu \\
&\qquad = (-2)(2\pi)^{-\frac{1}{2}}(\widetilde{\phi}^\mu r)^{\frac{1}{2}} \frac{\partial}{\partial \widetilde{\phi}^\mu r} \int \exp\left(-\frac{1}{2}(\mu-\widetilde{\mu})^2\widetilde{\phi}^\mu r\right) d\mu \\
&\qquad = (-2)(2\pi)^{-\frac{1}{2}}(\widetilde{\phi}^\mu r)^{\frac{1}{2}} \frac{\partial}{\partial \widetilde{\phi}^\mu r}(2\pi)^{\frac{1}{2}}(\widetilde{\phi}^\mu r)^{-\frac{1}{2}} \\
&\qquad = (-2)(\widetilde{\phi}^\mu r)^{\frac{1}{2}}\left(-\frac{1}{2}\right)(\widetilde{\phi}^\mu r)^{-\frac{3}{2}} = (\widetilde{\phi}^\mu r)^{-1}. \qquad (7.89)
\end{aligned}$$

The other two integrals are trivially solved as follows:

$$\begin{aligned}
\int \mathcal{N}(\mu|\widetilde{\mu},(\widetilde{\phi}^\mu r)^{-1})(o-\widetilde{\mu})^2 d\mu &= (o-\widetilde{\mu})^2, \\
\int \mathcal{N}(\mu|\widetilde{\mu},(\widetilde{\phi}^\mu r)^{-1})(\mu-\widetilde{\mu})(o-\widetilde{\mu}) d\mu &= 0. \qquad (7.90)
\end{aligned}$$

Therefore, Eq. (7.86) is solved as:

$$\begin{aligned}
&\int \mathcal{N}(\mu|\widetilde{\mu},(\widetilde{\phi}^\mu r)^{-1}) r(o-\mu)^2 d\mu \\
&\qquad = r\left((o-\widetilde{\mu})^2 + (\widetilde{\phi}^\mu r)^{-1}\right) = r(o-\widetilde{\mu})^2 + \frac{1}{\widetilde{\phi}^\mu}. \qquad (7.91)
\end{aligned}$$

Now, we focus on the integral over r, because the integral without $\log(r)$ can be easily computed by the result of the mean value of the gamma distribution in Appendix C.11 as:

$$\begin{aligned}
&\log \widetilde{b}(o) \\
&= \int \mathrm{Gam}_2(r|\widetilde{\phi}^r,\widetilde{r})\left(-\frac{1}{2}\left(\log(2\pi) - \log(r) + r(o-\widetilde{\mu})^2 + \frac{1}{\widetilde{\phi}^\mu}\right)\right) dr \\
&= -\frac{1}{2}\left(\log(2\pi) + \frac{\widetilde{\phi}^r}{\widetilde{r}}(o-\widetilde{\mu})^2 + \frac{1}{\widetilde{\phi}^\mu}\right) + \frac{1}{2}\int \mathrm{Gam}_2(r|\widetilde{\phi}^r,\widetilde{r})\log(r) dr. \qquad (7.92)
\end{aligned}$$

Therefore, we focus on the final term. From Eqs. (7.63) and (7.64), the concrete form of the gamma distribution, $\mathrm{Gam}_2(\cdot)$, is defined as follows:

$$\mathrm{Gam}_2(r|\widetilde{\phi}^r,\widetilde{r}) = C_{\mathrm{Gam}_2}(\widetilde{\phi}^r,\widetilde{r})(r)^{\frac{\widetilde{\phi}^r}{2}-1}\exp\left(-\frac{\widetilde{r}r}{2}\right), \qquad (7.93)$$

where

$$C_{\mathrm{Gam}_2}(\widetilde{\phi}^r,\widetilde{r}) = \frac{\left(\frac{\widetilde{r}}{2}\right)^{\frac{\widetilde{\phi}^r}{2}}}{\Gamma\left(\frac{\widetilde{\phi}^r}{2}\right)}. \qquad (7.94)$$

Similarly to the Dirichlet and Gaussian distributions, we consider the following derivative:

$$\frac{\partial}{\partial \widetilde{\phi}^r}(r)^{\frac{\widetilde{\phi}^r}{2}-1} = \frac{1}{2}(r)^{\frac{\widetilde{\phi}^r}{2}-1}\log(r). \tag{7.95}$$

Therefore, the integral is solved by using this relationship as:

$$\begin{aligned}
&\int \mathrm{Gam}_2(r|\widetilde{\phi}^r, \widetilde{r})\log(r)dr \\
&\quad = C_{\mathrm{Gam}_2}(\widetilde{\phi}^r, \widetilde{r})\int (r)^{\frac{\widetilde{\phi}^r}{2}-1}\exp\left(-\frac{\widetilde{r}r}{2}\right)\log(r)dr \\
&\quad = C_{\mathrm{Gam}_2}(\widetilde{\phi}^r, \widetilde{r})\int 2\frac{\partial}{\partial \widetilde{\phi}^r}(r)^{\frac{\widetilde{\phi}^r}{2}-1}\exp\left(-\frac{\widetilde{r}r}{2}\right)dr \\
&\quad = 2C_{\mathrm{Gam}_2}(\widetilde{\phi}^r, \widetilde{r})\frac{\partial}{\partial \widetilde{\phi}^r}\frac{1}{C_{\mathrm{Gam}_2}(\widetilde{\phi}^r, \widetilde{r})}.
\end{aligned} \tag{7.96}$$

The derivative with respect to $\widetilde{\phi}^r$ is calculated as follows:

$$\begin{aligned}
\frac{\partial}{\partial \widetilde{\phi}^r}\frac{\Gamma\left(\frac{\widetilde{\phi}^r}{2}\right)}{\left(\frac{\widetilde{r}}{2}\right)^{\frac{\widetilde{\phi}^r}{2}}} &= \frac{\frac{1}{2}\frac{\partial}{\partial \frac{\widetilde{\phi}^r}{2}}\Gamma\left(\frac{\widetilde{\phi}^r}{2}\right)\left(\frac{\widetilde{r}}{2}\right)^{\frac{\widetilde{\phi}^r}{2}} + \frac{1}{2}\log\left(\frac{\widetilde{r}}{2}\right)\left(\frac{\widetilde{r}}{2}\right)^{\frac{\widetilde{\phi}^r}{2}}\Gamma\left(\frac{\widetilde{\phi}^r}{2}\right)}{\left(\frac{\widetilde{r}}{2}\right)^{\widetilde{\phi}^r}} \\
&= \frac{\frac{1}{2}\Psi\left(\frac{\widetilde{\phi}^r}{2}\right)\Gamma\left(\frac{\widetilde{\phi}^r}{2}\right) + \frac{1}{2}\log\left(\frac{\widetilde{r}}{2}\right)\Gamma\left(\frac{\widetilde{\phi}^r}{2}\right)}{\left(\frac{\widetilde{r}}{2}\right)^{\frac{\widetilde{\phi}^r}{2}}},
\end{aligned} \tag{7.97}$$

where $\Psi(\cdot)$ is a di-gamma function defined in Eq. (7.81). Therefore,

$$2C_{\mathrm{Gam}_2}(\widetilde{\phi}^r, \widetilde{r})\frac{\partial}{\partial \widetilde{\phi}^r}\frac{1}{C_{\mathrm{Gam}_2}(\widetilde{\phi}^r, \widetilde{r})} = \Psi\left(\frac{\widetilde{\phi}^r}{2}\right) + \log\left(\frac{\widetilde{r}}{2}\right). \tag{7.98}$$

Thus, finally $\log \widetilde{b}(o)$ is obtained analytically as follows:

$$\begin{aligned}
&\log \widetilde{b}(o) \\
&= -\frac{1}{2}\int \mathrm{Gam}_2(r|\widetilde{\phi}^r, \widetilde{r})\left(\log(2\pi) + \frac{1}{\widetilde{\phi}^\mu} - \log(r) + r(o - \widetilde{\mu})^2\right)dr \\
&= -\frac{1}{2}\left(\log(2\pi) + \frac{1}{\widetilde{\phi}^\mu} - \Psi\left(\frac{\widetilde{\phi}^r}{2}\right)\right) - \frac{1}{2}\left(\log\left(\frac{\widetilde{r}}{2}\right) + (o - \widetilde{\mu})^2\frac{\widetilde{\phi}^r}{\widetilde{r}}\right).
\end{aligned} \tag{7.99}$$

Reverting to the indexes k, j, t, and d, $\log \widetilde{b}_{jk}(\mathbf{o}_t)$ is represented as

$$\begin{aligned}
\log \widetilde{b}_{jk}(\mathbf{o}_t) = &-\frac{D}{2}\left(\log(2\pi) + \frac{1}{\widetilde{\phi}^\mu_{jk}} - \Psi\left(\frac{\widetilde{\phi}^r_{jk}}{2}\right)\right) \\
&-\frac{1}{2}\sum_{d=1}^{D}\left(\log\left(\frac{\widetilde{r}_{jk}}{2}\right) + \frac{\widetilde{\phi}^r_{jk}(o_{td} - \widetilde{\mu}_{jkd})^2}{\widetilde{r}_{jkd}}\right).
\end{aligned} \tag{7.100}$$

Thus, we obtain $\widetilde{a}_{ij}, \widetilde{\omega}_{jk}$ and $\widetilde{b}_{jk}(\mathbf{o}_t)$, which are summarized as follows:

$$\begin{cases} \widetilde{a}_{ij} & \triangleq \exp\left(\Psi(\widetilde{\phi}^a_{ij}) - \Psi(\sum_{j'} \widetilde{\phi}^a_{ij'})\right), \\ \widetilde{\omega}_{jk} & \triangleq \exp\left(\Psi(\widetilde{\phi}^\omega_{jk}) - \Psi(\sum_{k'} \widetilde{\phi}^\omega_{jk'})\right), \\ \widetilde{b}_{jk}(\mathbf{o}_t) & \triangleq \exp\left(-\frac{D}{2}\left(\log(2\pi) + \frac{1}{\widetilde{\phi}^\mu_{jk}} - \Psi\left(\frac{\widetilde{\phi}^r_{jk}}{2}\right)\right)\right. \\ & \left. -\frac{1}{2}\sum_{d=1}^D \left(\log\left(\frac{\widetilde{r}_{jk}}{2}\right) + \frac{\widetilde{\phi}^r_{jk}(o_{td} - \widetilde{\mu}_{jkd})^2}{\widetilde{r}_{jkd}}\right)\right). \end{cases} \tag{7.101}$$

These variables are used to compute the VB transition probability $\widetilde{\xi}_t(i,j)$ and VB occupation probability $\widetilde{\gamma}_t(j,k)$.

VB transition probability $\widetilde{\xi}_t(i,j)$ and occupation probability $\widetilde{\gamma}_t(j,k)$ (VB E-step)

From Eq. (7.25), VB transition probability $\widetilde{\xi}_t(i,j)$ is represented as:

$$\widetilde{\xi}_t(i,j) \triangleq \widetilde{q}(s_t = i, s_{t+1} = j|\mathbf{O}, M). \tag{7.102}$$

Section 3.4.2 shows an efficient computation of the transition probability based on the complete data likelihood $p(\mathbf{O}, S, V|\Theta)$. Here we also consider how to obtain it based on the VB version of the complete data likelihood $\tilde{q}(\mathbf{O}, S, V|M)$, as introduced in Eq. (7.69). However, from Eq. (7.74), $\widetilde{a}_{ij}, \widetilde{\omega}_{jk}$, and $\widetilde{b}_{jk}(\mathbf{o}_t)$ can only compute the unnormalized likelihood function $\tilde{u}(\mathbf{O}, S, V|M)$, that is

$$\tilde{u}(\mathbf{O}, S, V|M) = \prod_{t=1}^T \widetilde{a}_{s_{t-1}s_t} \widetilde{\omega}_{s_t v_t} \widetilde{b}_{s_t v_t}(\mathbf{o}_t). \tag{7.103}$$

Therefore, as we discussed in Section 3.4.2, from the dependency of the HMM, we can represent $\widetilde{q}(\mathbf{O}, s_t = i, s_{t+1} = j|M)$ as follows:

$$\begin{aligned} &\widetilde{q}(s_t = i, s_{t+1} = j, \mathbf{O}|M) \\ &= \underbrace{\widetilde{q}(\mathbf{o}_1, \cdots, \mathbf{o}_t, s_t = i|M)}_{=\widetilde{\alpha}_t(i)} \underbrace{\widetilde{q}(\mathbf{o}_{t+1}|s_{t+1} = j, M)}_{=C_{\omega b}\sum_{k=1}^K \widetilde{\omega}_{jk}\widetilde{b}_{jk}(\mathbf{o}_{t+1})} \underbrace{\widetilde{q}(\mathbf{o}_{t+2}, \cdots, \mathbf{o}_T|s_{t+1} = j, M)}_{=\widetilde{\beta}_{t+1}(j)} \\ &\quad \times \underbrace{\widetilde{q}(s_{t+1} = j|s_t = i, M)}_{=C_a \widetilde{a}_{ij}}. \end{aligned} \tag{7.104}$$

Here, $\widetilde{\alpha}_t(i)$ is a forward variable at frame t in state i, as introduced in Eq. (3.50). Similarly, $\widetilde{\beta}_{t+1}(j)$ is a backward variable at frame $t+1$ in state j, as introduced in Eq. (3.55). The forward and backward variables based on the VB formulation are represented as described below.

First, the VB forward variable $\widetilde{\alpha}_t(j)$ is computed by using the following equation:

- Initialization

$$\begin{aligned} \widetilde{\alpha}_1(j) &= \widetilde{q}(\mathbf{o}_1, s_1 = j|M) \\ &= \widetilde{q}(\mathbf{o}_1|s_1 = j, M)\widetilde{q}(s_1 = j|M) \\ &= C_a \widetilde{a}_j C_{\omega b} \sum_{k=1}^K \widetilde{\omega}_{jk} \widetilde{b}_{jk}(\mathbf{o}_1), \quad 1 \le j \le J. \end{aligned} \tag{7.105}$$

Then, unnormalized forward variable $\tilde{\tilde{\alpha}}_1(j)$ is defined as:

$$\begin{aligned}\tilde{\tilde{\alpha}}_1(j) &\triangleq \tilde{a}_j \sum_{k=1}^{K} \tilde{\omega}_{jk}\tilde{b}_{jk}(\mathbf{o}_1) \\ &= \frac{\tilde{\alpha}_1(j)}{C_a C_{\omega b}}.\end{aligned} \tag{7.106}$$

- Induction

$$\begin{aligned}\tilde{\alpha}_t(j) &= \tilde{q}(\mathbf{o}_1, \cdots, \mathbf{o}_t, s_t = j|M) \\ &= \tilde{q}(\mathbf{o}_t|s_t = j, M) \sum_{i=1}^{J} \tilde{q}(s_t = j|s_{t-1} = i, M)\tilde{q}(\mathbf{o}_1, \cdots, \mathbf{o}_{t-1}, s_{t-1} = i|M) \\ &= \left(C_a \sum_{i=1}^{J} \tilde{\alpha}_{t-1}(i)\tilde{a}_{ij}\right) C_{\omega b} \sum_{k=1}^{K} \tilde{\omega}_{jk}\tilde{b}_{jk}(\mathbf{o}_t) \\ &= \left(C_a (C_a C_{\omega b})^{t-1} \sum_{i=1}^{J} \tilde{\tilde{\alpha}}_{t-1}(i)\tilde{a}_{ij}\right) C_{\omega b} \sum_{k=1}^{K} \tilde{\omega}_{jk}\tilde{b}_{jk}(\mathbf{o}_t) \\ &= (C_a C_{\omega b})^t \left(\sum_{i=1}^{J} \tilde{\tilde{\alpha}}_{t-1}(i)\tilde{a}_{ij}\right) \sum_{k=1}^{K} \tilde{\omega}_{jk}\tilde{b}_{jk}(\mathbf{o}_t), \quad \begin{array}{c} 2 \leq t \leq T \\ 1 \leq j \leq J, \end{array}\end{aligned} \tag{7.107}$$

where the unnormalized forward variable $\tilde{\tilde{\alpha}}_t(j)$ is represented as

$$\tilde{\tilde{\alpha}}_t(j) = \left(\sum_{i=1}^{J} \tilde{\tilde{\alpha}}_{t-1}(i)\tilde{a}_{ij}\right) \sum_{k=1}^{K} \tilde{\omega}_{jk}\tilde{b}_{jk}(\mathbf{o}_t), \quad \begin{array}{c} 2 \leq t \leq T \\ 1 \leq j \leq J. \end{array} \tag{7.108}$$

- Termination

$$\begin{aligned}\tilde{q}(\mathbf{O}|M) &= \sum_{j=1}^{J} \tilde{\alpha}_T(j) \\ &= (C_a C_{\omega b})^T \sum_{j=1}^{J} \tilde{\tilde{\alpha}}_T(j).\end{aligned} \tag{7.109}$$

The VB forward variable $\tilde{\alpha}_t(j)$ is obtained with the unnormalized forward variable $\tilde{\tilde{\alpha}}_t(j)$ and normalization constants C_a and $C_{\omega b}$. From this algorithm, we can compute the unnormalized forward variable $\tilde{\tilde{\alpha}}_t(j)$ similarly to the original forward algorithm, but we should be careful that the unnormalized forward variable is not a probability, and probabilistic calculation (sum and product rules etc.) must be performed via the normalized VB forward variable $\tilde{\alpha}_t(j)$.

Similarly, the VB backward variable $\tilde{\beta}_t(j)$ is computed by using the following equations:

- Initialization

$$\tilde{\beta}_T(j) = 1, \quad 1 \leq j \leq J. \tag{7.110}$$

- Induction

$$\begin{aligned}
\widetilde{\beta}_t(i) &= \widetilde{q}(\mathbf{o}_{t+1},\cdots,\mathbf{o}_T|s_t=i,M)\\
&= \sum_{j=1}^{J}\widetilde{q}(\mathbf{o}_{t+2},\cdots,\mathbf{o}_T|s_{t+1}=j,M)\widetilde{q}(\mathbf{o}_{t+1}|s_{t+1}=j,M)\widetilde{q}(s_{t+1}=j|s_t=i,M)\\
&= \sum_{j=1}^{J}C_a\widetilde{a}_{ij}\sum_{k=1}^{K}C_{\omega b}\widetilde{\omega}_{jk}\widetilde{b}_{jk}(\mathbf{o}_{t+1})\widetilde{\beta}_{t+1}(j)\\
&= (C_aC_{\omega b})^{T-t}\sum_{j=1}^{J}\widetilde{a}_{ij}\sum_{k=1}^{K}\widetilde{\omega}_{jk}\widetilde{b}_{jk}(\mathbf{o}_{t+1})\tilde{\widetilde{\beta}}_{t+1}(j),\\
&\quad t=T-1,T-2,\cdots,1,\qquad 1\le i\le J,
\end{aligned}\tag{7.111}$$

where the unnormalized backward variable $\tilde{\widetilde{\beta}}_t(i)$ is represented as

$$\tilde{\widetilde{\beta}}_t(i)=\sum_{j=1}^{J}\widetilde{a}_{ij}\sum_{k=1}^{K}\widetilde{\omega}_{jk}\widetilde{b}_{jk}(\mathbf{o}_{t+1})\tilde{\widetilde{\beta}}_{t+1}(j).\tag{7.112}$$

- Termination

$$\begin{aligned}
\beta_0 &\triangleq \widetilde{q}(\mathbf{O}|M)\\
&= \sum_{j=1}^{J}\widetilde{a}_j\sum_{k=1}^{K}\widetilde{\omega}_{jk}\widetilde{b}_{jk}(\mathbf{o}_1)\widetilde{\beta}_1(j)\\
&= (C_aC_{\omega b})^{T}\sum_{j=1}^{J}\widetilde{a}_j\sum_{k=1}^{K}\widetilde{\omega}_{jk}\widetilde{b}_{jk}(\mathbf{o}_1)\tilde{\widetilde{\beta}}_1(j).
\end{aligned}\tag{7.113}$$

Therefore, based on the VB forward and backward variables, we can compute the posterior probabilities as follows:

$$\begin{aligned}
&\widetilde{\xi}_t(i,j)\\
&= \frac{\widetilde{\alpha}_t(i)\widetilde{a}_{ij}\left(\sum_{k=1}^{K}\widetilde{\omega}_{jk}\widetilde{b}_{jk}(\mathbf{o}_t)\right)\widetilde{\beta}_{t+1}(j)}{\sum_{i'=1}^{J}\sum_{j'=1}^{J}\widetilde{\alpha}_t(i')\widetilde{a}_{i'j'}\left(\sum_{k=1}^{K}\widetilde{\omega}_{j'k}\widetilde{b}_{j'k}(\mathbf{o}_t)\right)\widetilde{\beta}_{t+1}(j')}\\
&= \frac{(C_aC_{\omega b})^t\tilde{\widetilde{\alpha}}_t(i)C_a\widetilde{a}_{ij}C_{\omega b}\left(\sum_{k=1}^{K}\widetilde{\omega}_{jk}\widetilde{b}_{jk}(\mathbf{o}_t)\right)(C_aC_{\omega b})^{T-t-1}\tilde{\widetilde{\beta}}_{t+1}(j)}{\sum_{i'=1}^{J}\sum_{j'=1}^{J}(C_aC_{\omega b})^t\tilde{\widetilde{\alpha}}_t(i')C_a\widetilde{a}_{i'j'}C_{\omega b}\left(\sum_{k=1}^{K}\widetilde{\omega}_{j'k}\widetilde{b}_{j'k}(\mathbf{o}_t)\right)(C_aC_{\omega b})^{T-t-1}\tilde{\widetilde{\beta}}_{t+1}(j')}\\
&= \frac{\tilde{\widetilde{\alpha}}_t(i)\widetilde{a}_{ij}\left(\sum_{k=1}^{K}\widetilde{\omega}_{jk}\widetilde{b}_{jk}(\mathbf{o}_t)\right)\tilde{\widetilde{\beta}}_{t+1}(j)}{\sum_{i'=1}^{J}\sum_{j'=1}^{J}\tilde{\widetilde{\alpha}}_t(i')\widetilde{a}_{i'j'}\left(\sum_{k=1}^{K}\widetilde{\omega}_{j'k}\widetilde{b}_{j'k}(\mathbf{o}_t)\right)\tilde{\widetilde{\beta}}_{t+1}(j')}.
\end{aligned}\tag{7.114}$$

Thus, we can compute the transition probability with unnormalized VB variables $\widetilde{a}_{ij}, \widetilde{\omega}_{jk}$, and $\widetilde{b}_{jk}(\mathbf{o}_t)$, and unnormalized forward and backward variables $\tilde{\widetilde{\alpha}}_t(i)$ and $\tilde{\widetilde{\beta}}_{t+1}(j)$, where the normalization constants C_a and $C_{\omega b}$ are canceled out. This is based on a well-known scaling property of the HMM forward–backward algorithm (Rabiner & Juang 1993). Similarly the occupation probability is also calculated as:

$$\widetilde{\gamma}_t(j,k) = \frac{\tilde{\tilde{\alpha}}_t(j)\tilde{\tilde{\beta}}_t(j)}{\sum_{j'=1}^{J} \tilde{\tilde{\alpha}}_t(j')\tilde{\tilde{\beta}}_t(j')} \cdot \frac{\widetilde{\omega}_{jk}\widetilde{b}_{jk}(\mathbf{o}_t)}{\sum_{k'=1}^{K} \widetilde{\omega}_{jk'}\widetilde{b}_{jk'}(\mathbf{o}_t)}. \tag{7.115}$$

These probabilities are obtained similarly to the ML cases in Eqs. (3.126) and (3.122), and the MAP case in Eqs. (4.91) and (4.92) with the VB-based variables obtained by Eq. (7.101). Thus, $\widetilde{\xi}_t(i,j)$ and $\widetilde{\gamma}_t(j,k)$ are calculated efficiently by using a probabilistic assignment via the familiar *forward–backward algorithm*. This algorithm is called the VB forward–backward algorithm.

Similarly to the VB forward–backward algorithm, the *Viterbi algorithm* is also derived within the VB approach by exchanging the summation over i for the maximization over i in the calculation of the unnormalized forward probability $\tilde{\tilde{\alpha}}_t(j)$. This algorithm is called the VB Viterbi algorithm.

Thus, VB posteriors can be calculated iteratively in the same way as the Baum–Welch algorithm, even for a complicated sequential model that includes latent variables such as HMM and GMM for acoustic models. These calculations are referred to as a VB Baum–Welch algorithm, as proposed in MacKay (1997), Watanabe *et al.* (2002), Beal (2003) and Watanabe *et al.* (2004).

7.3.4 Variational lower bound

This section discusses the VB objective function $\mathcal{F}^M$ for a whole acoustic model topology, i.e., the variational lower bound, and provides general calculation results. The variational lower bound is a criterion for both posterior distribution estimation, and model topology optimization in acoustic model construction. This section begins by focusing on one phoneme category. By substituting the VB posterior distribution obtained in Section 7.3.3, we obtain analytical results for $\mathcal{F}^M$, and therefore, this calculation also requires a VB iterative calculation based on the VB Baum–Welch algorithm used in the VB posterior calculation. We can separate $\mathcal{F}^M$ into two components: one is composed solely of $\widetilde{q}(S, V|\mathbf{O}, M)$, whereas the other is mainly composed of $\widetilde{q}(\Theta|\mathbf{O}, M)$. Therefore, we define $\mathcal{F}^M_\Theta$ and $\mathcal{F}^M_{S,V}$, and represent $\mathcal{F}^M$ as follows:

$$\begin{aligned}
\mathcal{F}^M &= \mathbb{E}_{(\Theta,S,V)}\left[\log \frac{p(\mathbf{O}, S, V|\Theta, M)p(\Theta|M)}{\widetilde{q}(\Theta|\mathbf{O}, M)}\right] - \mathbb{E}_{(S,V)}\left[\log \widetilde{q}(S, V|\mathbf{O}, M)\right] \\
&= \mathcal{F}^M_\Theta - \mathcal{F}^M_{S,V},
\end{aligned} \tag{7.116}$$

where

$$\begin{aligned}
\mathcal{F}^M_\Theta &\triangleq \mathbb{E}_{(\Theta,S,V)}\left[\log \frac{p(\mathbf{O}, S, V|\Theta, M)p(\Theta|M)}{\widetilde{q}(\Theta|\mathbf{O}, M)}\right], \\
\mathcal{F}^M_{S,V} &\triangleq \mathbb{E}_{(S,V)}\left[\log \widetilde{q}(S, V|\mathbf{O}, M)\right].
\end{aligned} \tag{7.117}$$

First, we focus on $\mathcal{F}^M_\Theta$. Based on the variational solution of $\tilde{q}(S, V|\mathbf{O}, M)$ in Eq. (7.42), $\mathcal{F}^M_\Theta$ is rewritten as follows:

$$\begin{aligned}
\mathcal{F}^M_\Theta &\triangleq \mathbb{E}_{(\Theta,S,V)}\left[\log \frac{p(\mathbf{O}, S, V|\Theta, M)p(\Theta|M)}{\frac{1}{Z}p(\Theta|M)\exp\left(\mathbb{E}_{(S,V)}\left[\log p(\mathbf{O}, S, V|\Theta, M)\right]\right)}\right] \\
&= \mathbb{E}_{(\Theta,S,V)}\left[\log p(\mathbf{O}, S, V|\Theta, M)\right] + \mathbb{E}_{(\Theta)}\left[\log p(\Theta|M)\right] \\
&\quad - \mathbb{E}_{(\Theta,S,V)}\left[\log p(\mathbf{O}, S, V|\Theta, M)\right] - \mathbb{E}_{(\Theta)}\left[\log p(\Theta|M)\right] + \log Z \\
&= \log Z,
\end{aligned} \tag{7.118}$$

where Z is a normalization constant. This equation means that $\mathcal{F}^M_\Theta$ is represented by the logarithmic function of the normalization constant Z. By using the definition of the VB auxiliary function $\tilde{q}(\Theta)$ in Eq. (7.60), Z can be rewritten as

$$\begin{aligned}
Z &\triangleq \int p(\Theta|M)\exp\left(\mathbb{E}_{(S,V)}\left[\log p(\mathbf{O}, S, V|\Theta, M)\right]\right) d\Theta \\
&= \int \exp\left(\tilde{Q}(\Theta)\right) d\Theta.
\end{aligned} \tag{7.119}$$

Here, from the similarity of the VB and MAP auxiliary functions, as discussed in Section 7.3.3, $\tilde{Q}(\Theta)$ can be obtained by using the analytical results of the MAP auxiliary function. From Eq. (4.38), $\tilde{Q}(\Theta)$ is decomposed into the following auxiliary functions:

$$\tilde{Q}(\Theta) = \tilde{Q}(\mathbf{A}) + \tilde{Q}(\boldsymbol{\omega}) + \tilde{Q}(\boldsymbol{\mu}, \boldsymbol{\Sigma}), \tag{7.120}$$

where $\tilde{Q}(\mathbf{A})$, $\tilde{Q}(\boldsymbol{\omega})$, and $\tilde{Q}(\boldsymbol{\mu}, \boldsymbol{\Sigma})$ are obtained from the analytical solutions of the corresponding MAP auxiliary functions in Eqs. (4.51), (4.54), and (4.73), as follows:

$$\tilde{Q}(\mathbf{A}) = \sum_{i=1}^{J} \log\left(\mathrm{Dir}(\{a_{ij}\}_{j=1}^J | \{\tilde{\phi}^a_{ij}\}_{j=1}^J)\right) + \sum_{i=1}^{J} \log \frac{C_{\mathrm{Dir}}(\{\phi^a_{ij}\}_{j=1}^J)}{C_{\mathrm{Dir}}(\{\tilde{\phi}^a_{ij}\}_{j=1}^J)}, \tag{7.121}$$

$$\tilde{Q}(\boldsymbol{\omega}) = \sum_{j=1}^{J} \log\left(\mathrm{Dir}(\{\omega_{jk}\}_{k=1}^K | \{\tilde{\phi}^\omega_{jk}\}_{k=1}^K)\right) + \sum_{j=1}^{J} \log \frac{C_{\mathrm{Dir}}(\{\phi^\omega_{jk}\}_{k=1}^K)}{C_{\mathrm{Dir}}(\{\tilde{\phi}^\omega_{jk}\}_{k=1}^K)}, \tag{7.122}$$

$$\begin{aligned}
&\tilde{Q}(\boldsymbol{\mu}, \mathbf{R}) \\
&= \sum_{j=1}^{J}\sum_{k=1}^{K} \log\left(\mathcal{N}(\boldsymbol{\mu}_{jk}|\tilde{\boldsymbol{\mu}}_{jk}, (\tilde{\phi}^{\boldsymbol{\mu}}_{jk}\mathbf{R}_{jk})^{-1})\mathcal{W}(\mathbf{R}_{jk}|\tilde{\mathbf{R}}_{jk}, \tilde{\phi}^{\mathbf{R}}_{jk})\right) \\
&+ \sum_{j=1}^{J}\sum_{k=1}^{K}\left(-\sum_{t=1}^{T}\frac{\tilde{\gamma}_t(j,k)D}{2}\log(2\pi) + \frac{D}{2}\log\frac{\phi^{\mu}}{\tilde{\phi}^{\mu}} + \log\frac{C_{\mathcal{W}}(\mathbf{R}^0_{jk}, \phi^{\mathbf{R}}_{jk})}{C_{\mathcal{W}}(\tilde{\mathbf{R}}_{jk}, \tilde{\phi}^{\mathbf{R}}_{jk})}\right).
\end{aligned} \tag{7.123}$$

If we consider the diagonal covariance matrix, Eq. (7.123) is modified by using the gamma distribution as follows:

$$
\begin{aligned}
&\tilde{Q}(\boldsymbol{\mu}, \mathbf{R}) \\
&= \sum_{j=1}^{J}\sum_{k=1}^{K}\sum_{d=1}^{D} \log\left(\mathcal{N}(\mu_{jkd}|\tilde{\mu}_{jkd}, (\tilde{\phi}^{\mu}_{jk} r_{jkd})^{-1})\mathrm{Gam}_2(r_{jkd}|\tilde{r}_{jkd}, \tilde{\phi}^{r}_{jk})\right) \\
&+ \sum_{j=1}^{J}\sum_{k=1}^{K}\left(-\sum_{t=1}^{T}\frac{\tilde{\gamma}_t(j,k)D}{2}\log(2\pi) + \frac{D}{2}\log\frac{\phi^{\mu}_{jk}}{\tilde{\phi}^{\mu}_{jk}} + \log\frac{C_{\mathrm{Gam}_2}(r^0_{jkd}, \phi^r_{jk})}{C_{\mathrm{Gam}_2}(\tilde{r}_{jkd}, \tilde{\phi}^r_{jk})}\right).
\end{aligned}
\tag{7.124}
$$

Here C_{Gam_2} in Eq. (7.124) and C_{Dir} in Eqs. (7.121) and (7.122) are normalization constants of Gamma and Dirichlet distributions, respectively, and are defined as follows:

$$
\begin{cases}
C_{\mathrm{Dir}}(\{\phi_j\}_{j=1}^{J}) &= \dfrac{\Gamma\left(\sum_{j=1}^{J}\phi_j\right)}{\prod_{j=1}^{J}\Gamma(\phi_j)}, \\
C_{\mathrm{Gam}_2}(\phi, r^0) &= \dfrac{\left(\frac{r^0}{2}\right)^{\frac{\phi}{2}}}{\Gamma\left(\frac{\phi}{2}\right)}.
\end{cases}
\tag{7.125}
$$

Therefore, by substituting Eqs. (7.121), (7.122), and (7.124) into Eq. (7.119), and by using the definition of the normalization constant of the Dirichlet and gamma distributions in Eq. (7.125), the integral in Z is performed with the normalization of Θ, and Z is simply obtained as follows:

$$
\begin{aligned}
\mathcal{F}^M_{\Theta} &= \log Z \\
&= \log\left(\int\prod_{i,j}\exp\left(\tilde{Q}(\mathbf{A})\right)da_{ij}\right) + \log\left(\int\prod_{j,k}\exp\left(\tilde{Q}(\boldsymbol{\omega})\right)d\omega_{jk}\right) \\
&+ \log\left(\int\prod_{j,k}\exp\left(\tilde{Q}(\boldsymbol{\mu}, \mathbf{R})\right)d\boldsymbol{\mu}_{jk}d\mathbf{R}_{jk}\right) \\
&= \sum_i \log\frac{\Gamma(\sum_j \phi^a_{ij})\prod_j\Gamma(\tilde{\phi}^a_{ij})}{\Gamma(\sum_{j'}\tilde{\phi}^a_{ij})\prod_j\Gamma(\phi^a_{ij})} + \sum_j \log\frac{\Gamma(\sum_k \phi^{\omega}_{jk})\prod_k\Gamma(\tilde{\phi}^{\omega}_{jk})}{\Gamma(\sum_k\tilde{\phi}^{\omega}_{jk})\prod_k\Gamma(\phi^{\omega}_{jk})} \\
&+ \sum_{j,k}\log\left((2\pi)^{-\frac{\tilde{\gamma}_{jk}D}{2}}\left(\frac{\phi^{\mu}_{jk}}{\tilde{\phi}^{\mu}_{jk}}\right)^{\frac{D}{2}}\frac{\left(\Gamma\left(\frac{\tilde{\phi}^r_{jk}}{2}\right)\right)^D\prod_d\left(\frac{r^0_{jkd}}{2}\right)^{\frac{\phi^r_{jk}}{2}}}{\left(\Gamma\left(\frac{\phi^r_{jk}}{2}\right)\right)^D\prod_d\left(\frac{\tilde{r}_{jkd}}{2}\right)^{\frac{\tilde{\phi}^r_{jk}}{2}}}\right).
\end{aligned}
\tag{7.126}
$$

From Eq. (7.126), $\mathcal{F}^M_{\Theta}$ can be calculated by using the statistics of the posterior distribution parameters $\widetilde{\Phi}$ given in Eq. (7.66). This part is equivalent to the objective function for model selection based on Akaike's Bayesian information criterion (Akaike 1980). The whole $\mathcal{F}^M$ for all categories is obtained by simply summing up the $\mathcal{F}^M$ results obtained in this section for all categories as in Eq. (7.36).

Now we focus on $\mathcal{F}^M_{S,V}$. From the definition in Eq. (7.117), $-\mathcal{F}^M_{S,V}$ can be represented as follows:

$$
\begin{aligned}
-\mathcal{F}^M_{S,V} &= -\mathbb{E}_{(S,V)}\left[\log \widetilde{q}(S, V|\mathbf{O}, M)\right] \\
&= -\sum_{S,V} \widetilde{q}(S, V|\mathbf{O}, M) \log \widetilde{q}(S, V|\mathbf{O}, M).
\end{aligned}
\tag{7.127}
$$

Therefore, $\mathcal{F}^M_{S,V}$ denotes the entropy of the posterior distribution $\widetilde{q}(S, V|\mathbf{O}, M)$. As we discussed in Section 7.69, it is difficult to obtain the analytical form of $\widetilde{q}(S, V|\mathbf{O}, M)$ due to the normalization constant, and direct calculation of the above entropy is also difficult. Instead, we focus on the variational complete data likelihood form $\widetilde{q}(\mathbf{O}, S, V|M)$, which is obtained by the Bayes theorem as follows:

$$
\widetilde{q}(S, V|\mathbf{O}, M) = \frac{\widetilde{q}(\mathbf{O}, S, V|M)}{\sum_{S,V} \widetilde{q}(\mathbf{O}, S, V|M)}.
\tag{7.128}
$$

Based on the discussion in Eq. (7.73), $\widetilde{q}(\mathbf{O}, S, V|M)$ is represented with the unnormalized function $\widetilde{u}(\mathbf{O}, S, V|M)$ as follows:

$$
\begin{aligned}
\widetilde{q}(\mathbf{O}, S, V|M) &= C\widetilde{u}(\mathbf{O}, S, V|M) \\
&= C \prod_{t=1}^{T} \tilde{a}_{s_{t-1}s_t} \tilde{\omega}_{s_t v_t} \tilde{b}_{s_t v_t}(\mathbf{o}_t),
\end{aligned}
\tag{7.129}
$$

where C is a normalization constant, defined as follows:

$$
C = \int \sum_{S,V} \widetilde{u}(\mathbf{O}, S, V|M) d\mathbf{O}.
\tag{7.130}
$$

$\tilde{a}_{ij}$, $\tilde{\omega}_{jk}$, $\tilde{b}_{jk}(\mathbf{o}_t)$ are analytically calculated in the VB-E step (Eq. (7.101)). Therefore, by substituting (7.129) into (7.128), the normalization constant is canceled out, and we can obtain the following equation:

$$
\begin{aligned}
\widetilde{q}(S, V|\mathbf{O}, M) &= \frac{C\widetilde{u}(\mathbf{O}, S, V|M)}{\sum_{S',V'} C\widetilde{u}(\mathbf{O}, S', V'|M)} \\
&= \frac{\widetilde{u}(\mathbf{O}, S, V|M)}{\sum_{S',V'} \widetilde{u}(\mathbf{O}, S', V'|M)}.
\end{aligned}
\tag{7.131}
$$

Therefore, $\mathcal{F}^M_{S,V}$ can be rewritten as follows:

$$
\begin{aligned}
\mathcal{F}^M_{S,V} &= \sum_{S,V} \widetilde{q}(S, V|\mathbf{O}, M) \log \widetilde{q}(S, V|\mathbf{O}, M) \\
&= \sum_{S,V} \widetilde{q}(S, V|\mathbf{O}, M) \log \left(\widetilde{u}(\mathbf{O}, S, V|M)\right) \\
&\quad - \log \left(\sum_{S,V} \widetilde{u}(\mathbf{O}, S, V|M) \right).
\end{aligned}
\tag{7.132}
$$

Note that the second term corresponds to the summation of all possible S and V for unnormalized function $\widetilde{u}(\mathbf{O}, S, V|M)$, which can be computed in the VB forward algorithm in Eq. (7.109) as follows:

$$\sum_{S,V} \widetilde{u}(\mathbf{O}, S, V|M) = \sum_{j=1}^{J} \tilde{\tilde{\alpha}}_T(j). \tag{7.133}$$

Now we focus on the first term in Eq. (7.132). From the discussions in Section 3.4.2, we can convert summation over sequences S, V to a summation over HMM states i and j, and mixture component k in this term. Therefore, this term is represented as follows:

$$\begin{aligned}
&\sum_{S,V} \widetilde{q}(S, V|\mathbf{O}, M) \log (\widetilde{u}(\mathbf{O}, S, V|M)) \\
&= \sum_{S,V} \widetilde{q}(S, V|\mathbf{O}, M) \sum_{t=1}^{T} \left(\log \tilde{a}_{s_{t-1}s_t} + \log \tilde{\omega}_{s_t v_t} + \log \tilde{b}_{s_t v_t}(\mathbf{o}_t)\right) \\
&= \sum_{i,j,t} \tilde{\xi}_t(i,j) \log \tilde{a}_{ij} + \sum_{j,k,t} \tilde{\gamma}_t(j,k) \left(\log \tilde{\omega}_{jk} + \log \tilde{b}_{jk}(\mathbf{o}_t)\right).
\end{aligned} \tag{7.134}$$

Thus, we obtain the term without computing the summation over S and V.

Finally, we summarize calculation of $\mathcal{F}^M_{S,V}$ by using the definitions of $\tilde{a}_{ij}$, $\tilde{\omega}_{jk}$, $\tilde{b}_{jk}(\mathbf{o}_t)$ in Eq. (7.101), as follows:

$$\begin{aligned}
\mathcal{F}^M_{S,V} &= \sum_{i,j} \widetilde{\xi}_{ij} \left(\Psi\left(\widetilde{\phi}^a_{ij}\right) - \Psi\left(\sum\nolimits_{j'} \widetilde{\phi}^a_{ij'}\right)\right) + \sum_{j,k} \widetilde{\gamma}_{jk} \left(\Psi\left(\widetilde{\phi}^\omega_{jk}\right) - \Psi\left(\sum\nolimits_{k'} \widetilde{\phi}^\omega_{jk'}\right)\right) \\
&- \frac{1}{2} \sum_{j,k} \widetilde{\gamma}_{jk} \left(D\left(\log(2\pi) + \frac{1}{\widetilde{\phi}^\mu_{jk}} - \Psi\left(\frac{\widetilde{\phi}^r_{jk}}{2}\right)\right) + \sum_d \log \frac{\widetilde{r}_{jkd}}{2}\right) \\
&- \frac{1}{2} \sum_{j,k} \left(\widetilde{\phi}^r_{jk} \sum_{t,d} \frac{\widetilde{\gamma}_t(j,k)(o_{td} - \widetilde{\mu}_{jkd})^2}{\widetilde{r}_{jkd}}\right) - \log\left(\sum_j \tilde{\tilde{\alpha}}_T(j)\right).
\end{aligned} \tag{7.135}$$

Thus, we also obtain the analytical result for $\mathcal{F}^M_{S,V}$, which corresponds to the latent variable effect for the variational lower bound.

The analytical result for the variational lower bound $\mathcal{F}^M$ for the CDHMM is determined using $\mathcal{F}^M_\Theta$ in Eq. (7.126) and $\mathcal{F}^M_{S,V}$ in Eq. (7.135). Although the analytical result looks complicated, all variables are already computed in the VB expectation and maximization steps. We also want to emphasize that the computation is quite feasible since it is carried out without a summation over all possible latent variable sequences S and V. The variational lower bound is derived analytically so that it retains the effects of the dependence between model parameters and of the latent variables, defined in the generative model distribution in Eq. (7.53), unlike the conventional Bayesian information criterion and minimum description length (BIC/MDL) approaches, as discussed in Section 6.5. Therefore, the variational lower bound can compare any acoustic models with respect to all topological aspects and their combinations, e.g., contextual and temporal topologies in HMMs, the number of components per GMM in an HMM state, and the dimensional size of feature vectors, based on the following equation:

$$\widetilde{M} = \arg \max_{M \in (\mathbb{T} \times \mathbb{S} \times \mathbb{G} \times \mathbb{D})} \mathcal{F}^M. \tag{7.136}$$

Here $\mathbb{T}$, $\mathbb{S}$, $\mathbb{G}$, and $\mathbb{D}$ denote search spaces of HMM-temporal, HMM-contextual, GMM and feature vector topologies, respectively.

Based on the discussion in Section 7.3, the seven steps in Algorithm 11 provide a VB training algorithm for acoustic modeling. Here, τ denotes an iteration count, and ε denotes a threshold that checks whether $\mathcal{F}^M$ converges. Thus, the posterior distribution estimation in the VB framework can be effectively constructed based on the VB Baum–Welch algorithm, which is analogous to the ML Baum–Welch algorithm (Algorithm 4). In addition, VB can realize the model selection using the VB objective function as shown in **Step 9**. Thus, VB can construct an acoustic model consistently based on the Bayesian approach.

Algorithm 11 Variational Bayesian Baum–Welch algorithm for CDHMMs with model selection

Require: Set posterior parameter $\tilde{\Phi}[\tau = 0]$ from initialized transition probability $\tilde{\xi}[\tau = 0]$, occupation probability $\tilde{\gamma}[\tau = 0]$, and model structure M (prior parameter Φ^0 is included) for each category

1: **repeat**
2: Compute $\tilde{a}[\tau + 1]$, $\tilde{\omega}[\tau + 1]$, and $\tilde{b}(\mathbf{O})[\tau + 1]$ using $\tilde{\Phi}[\tau]$. (By Eq. (7.101))
3: Update $\tilde{\xi}[\tau + 1]$ and $\tilde{\gamma}[\tau + 1]$ via the Viterbi algorithm or forward–backward algorithm. (By Eqs. (7.114) and (7.115))
4: Accumulate the sufficient statistics $\tilde{\xi}[\tau + 1]$, $\tilde{\gamma}[\tau + 1]\tilde{\gamma}^{(1)}[\tau + 1]$, $\tilde{\gamma}^{(2)}[\tau + 1]$ (by Eq. (7.67)
5: Compute $\tilde{\Phi}[\tau + 1]$ using $\tilde{\Xi}[\tau + 1]$ and Φ^0. (By Eq. (7.66))
6: Calculate total $\mathcal{F}^M[\tau + 1]$ for all categories. (By using Eqs. (7.126) and (7.135) and summing up all categories' $\mathcal{F}^M$)
7: Calculate $\Delta = |(\mathcal{F}^M[\tau + 1] - \mathcal{F}^M[\tau])/\mathcal{F}^M[\tau + 1]|$, $\tau \leftarrow \tau + 1$
8: **until** $\Delta \leq \varepsilon$
Calculate $\mathcal{F}^M$ for all possible M and find $\tilde{M}(= \arg\max_M \mathcal{F}^M)$

Note that if we change $\tilde{} \to \hat{}$ (a value with $\hat{}$ attached indicates an ML estimate), $\Phi \to \Theta$ and $\mathcal{F}^M \to Ł^M$ (where $Ł^M$ means the log-likelihood for a model M), this algorithm becomes an ML-based framework, except for the model selection. Therefore, in the implementation phase, the VB framework can be realized in the conventional systems of acoustic model construction by adding the prior distribution setting and by changing the estimation procedure and objective function calculation.

7.3.5 VB posterior for Bayesian predictive classification

This subsection deals with the Bayes decision rule based on the VB approach. It is related to the Bayesian predictive classification, as discussed in Section 6.3.1 with the Laplace approximation, but this section deals with the same issue with VB. In this section, we use the following notation to clearly distinguish the training and recognition data, similar to Section 6.3.1:

$$\begin{aligned}&\mathbf{O} : \text{future data},\\ &\mathcal{O} : \text{training data}.\end{aligned} \tag{7.137}$$

In recognition, $\mathbf{O} = \{\mathbf{O}_t \in \mathbb{R}^D | t = 1, \cdots, T\}$ denotes the feature vector sequence of input speech, and $S = \{s_t \in \{1, \cdots, J\} | t = 1, \cdots, T\}$ denotes the corresponding HMM state sequence. Although our target application is ASR, which outputs the word sequence W, this section simplifies the decoding rule for the explanation. That is, the decoding needs to handle the word sequence W in addition to the state sequence S, but it can be combined with the LVCSR decoder if we can build the Viterbi algorithm. Therefore, this section focuses on formulating the Viterbi algorithm within the VB framework, similarly to that within the ML framework, as discussed in Section 3.3.2.

The Viterbi algorithm can achieve the optimal state sequence $\widetilde{S}$ by using a conditional probability function $p(S|\mathbf{O}, \mathcal{O})$ given input data $\mathbf{O}$ and training data $\mathcal{O}$, as follows:

$$\begin{aligned}\bar{S} = \arg\max_S p(S|\mathbf{O}, \mathcal{O}) &= \arg\max_S \frac{p(\mathbf{O}, S|\mathcal{O})}{p(\mathbf{O}|\mathcal{O})}\\ &= \arg\max_S p(\mathbf{O}, S|\mathcal{O}).\end{aligned} \tag{7.138}$$

$p(\mathbf{O}, S|\mathcal{O})$ is a variant of *predictive distribution* (Berger 1985, Bernardo & Smith 2009), because this distribution predicts the probability of unknown data $\mathbf{O}$ conditioned by training data $\mathcal{O}$. Note that Eq. (7.138) does not depend on parameters Θ and model M, and these can be explicitly involved by considering the following sum rule:

$$p(\mathbf{O}, S|\mathcal{O}) = \sum_M \int p(\mathbf{O}, S|\Theta, \mathcal{O}, M) p(\Theta|\mathcal{O}, M) p(M|\mathcal{O}) d\Theta. \tag{7.139}$$

This predictive distribution based approach involves considering the integrals and true posterior distributions, an approach which is also applied to speech recognition (Huo & Lee 2000, Jiang *et al.* 1999, Lee & Huo 2000, Chien & Liao 2001), as discussed in Section 6.3.1, with the Laplace approximation.

After VB-based acoustic modeling in Section 7.3.4, an appropriate model structure $\widetilde{M}$ is selected based on the VB model selection Eq. (7.136), and the optimal VB posterior distributions are obtained $\widetilde{q}(\Theta|\mathcal{O}, \widetilde{M})$. Therefore, the true posterior distributions can be approximated by the VB posteriors, and Eq. (7.139) is approximated as:

$$\begin{aligned}p(\mathbf{O}, S|\mathcal{O}) &\approx \sum_M \int p(\mathbf{O}, S|\Theta, M) \widetilde{q}(\Theta|\mathcal{O}, M) \delta(M, \widetilde{M}) d\Theta\\ &= \int p(\mathbf{O}, S|\Theta, \widetilde{M}) \widetilde{q}(\Theta|\mathcal{O}, \widetilde{M}) d\Theta\\ &= \mathbb{E}_{(\Theta)} \left[p(\mathbf{O}, S|\Theta, \widetilde{M}) \right].\end{aligned} \tag{7.140}$$

Thus, we can build the Viterbi algorithm for the expectation over the model parameter Θ by using the VB posterior. In the following section we omit the models structure index $\widetilde{M}$ for simplicity.

Similarly to Section 3.3.2, we first define the following *expected* highest probability along a single path, at time t, which accounts for the first t observations and ends in state j:

$$\widetilde{\delta}_t(j) \triangleq \max_{s_1,\cdots,s_{t-1}} \mathbb{E}_{(\Theta)} \left[p(s_1, \cdots, s_t = j, \mathbf{o}_1, \cdots, \mathbf{o}_t | \Theta) \right]. \tag{7.141}$$

By using $\widetilde{\delta}_t(j)$ recursively, we can obtain the most probable state sequence as follows:

- Initialization

$$\widetilde{\delta}_1(i) = \mathbb{E}_{(\boldsymbol{\pi})} [\pi_i] \sum_{k=1}^{K} \mathbb{E}_{(\boldsymbol{\omega})} [\omega_{ik}] \, \mathbb{E}_{(\boldsymbol{\mu},\boldsymbol{\Sigma})} \left[\mathcal{N}(\mathbf{o}_1 | \boldsymbol{\mu}_{ik}, \boldsymbol{\Sigma}_{ik}) \right],$$
$$\psi_1(i) = 0, \quad 1 \le i \le J. \tag{7.142}$$

- Recursion

$$\widetilde{\delta}_t(j) = \left(\max_{1 \le i \le J} \widetilde{\delta}_{t-1}(i) \mathbb{E}_{(\mathbf{A})} \left[a_{ij} \right] \right) \sum_{k=1}^{K} \mathbb{E}_{(\boldsymbol{\omega})} [\omega_{ik}] \, \mathbb{E}_{(\boldsymbol{\mu},\boldsymbol{\Sigma})} \left[\mathcal{N}(\mathbf{o}_1 | \boldsymbol{\mu}_{jk}, \boldsymbol{\Sigma}_{jk}) \right],$$
$$\psi_t(j) = \left(\arg\max_{1 \le i \le J} \widetilde{\delta}_{t-1}(i) \mathbb{E}_{(\mathbf{A})} \left[a_{ij} \right] \right), \quad \begin{array}{l} 2 \le t \le T \\ 1 \le j \le J. \end{array} \tag{7.143}$$

- Termination

$$p(\widetilde{S}, \mathbf{O} | \mathcal{O}) = \max_{1 \le j \le J} \widetilde{\delta}_T(i),$$
$$\widetilde{s}_T = \arg\max_{1 \le j \le J} \widetilde{\delta}_T(i). \tag{7.144}$$

- State sequence backtracking

$$\widetilde{s}_t = \psi_{t+1}(\widetilde{s}_{t+1}), \quad t = T-1, T-2, \cdots, 1. \tag{7.145}$$

Thus, we can perform the Viterbi algorithm for the predictive distribution based on the VB posteriors. To realize the Viterbi algorithm, we need to consider the following expectation:

$$\mathbb{E}_{(\mathbf{A})} \left[a_{ij} \right],$$
$$\mathbb{E}_{(\boldsymbol{\omega})} [\omega_{ik}],$$
$$\mathbb{E}_{(\boldsymbol{\mu},\boldsymbol{\Sigma})} \left[\mathcal{N}(\mathbf{o}_t | \boldsymbol{\mu}_{jk}, \boldsymbol{\Sigma}_{jk}) \right]. \tag{7.146}$$

We provide the solution for each of the expected variables. Note that these are different from the expected variables of the CDHMM parameters in the VB E-step, as discussed in Eq. (7.73),[4] i.e.,

[4] Again we omit the initial transition parameters for simplicity.

$$\widetilde{a}_{ij} = \exp\left(\mathbb{E}_{(a_{ij})}\left[\log(a_{ij})\right]\right),$$
$$\widetilde{\omega}_{jk} = \exp\left(\mathbb{E}_{(\omega_{jk})}\left[\log(\omega_{jk})\right]\right),$$
$$\widetilde{b}_{jk}(\mathbf{o}_t) = \exp\left(\mathbb{E}_{(\boldsymbol{\mu}_{jk},\boldsymbol{\Sigma}_{jk})}\left[\log(\mathcal{N}(\mathbf{o}_t|\boldsymbol{\mu}_{jk},\boldsymbol{\Sigma}_{jk}))\right]\right). \tag{7.147}$$

This is because the VB E-step is a training step, and we need to optimize these values based on the variational method, which necessitates considering the expectation of the parameters in the logarithmic domain. In the prediction case, since we already have the posterior distributions in the training step, Eq. (7.146) simply performs the expectation determination for the CDHMM parameters directly.

Expected state transition $\mathbb{E}_{(\mathbf{A})}\left[a_{ij}\right]$

We first focus on calculation of the expected state transition $\widetilde{a}_{ij}$. Although this can be obtained as the mean result of the Dirichlet distribution in Appendix C.4, we provide the derivation for its educational value. Based on the definition of the Dirichlet distribution in Eq. (7.62), we can obtain the following equation:

$$\begin{aligned}\mathbb{E}_{(\mathbf{A})}\left[a_{ij}\right] &= \int a_{ij}\mathrm{Dir}(\{a_{ij'}\}_{j'=1}^{J}|\{\widetilde{\phi}^a_{ij'}\}_{j'=1}^{J})\prod_{j'=1}^{J} da_{ij'} \\ &= C_{\mathrm{Dir}}(\{\widetilde{\phi}^a_{ij}\}_{j=1}^{J})\int a_{ij}\prod_{j'=1}^{J}(a_{ij'})^{\widetilde{\phi}^a_{ij'}-1}da_{ij'}.\end{aligned} \tag{7.148}$$

Now we define the following variable:

$$\hat{\phi}^a_{ij'} \triangleq \begin{cases}\widetilde{\phi}^a_{ij'}+1 & j'=j \\ \widetilde{\phi}^a_{ij'} & j'\neq j.\end{cases} \tag{7.149}$$

By using $\hat{\phi}^a_{ij'}$, the integral in Eq. (7.148) is solved as:

$$\begin{aligned}\mathbb{E}_{(\mathbf{A})}\left[a_{ij}\right] &= C_{\mathrm{Dir}}(\{\widetilde{\phi}^a_{ij}\}_{j=1}^{J})\int\prod_{j'=1}^{J}(a_{ij'})^{\hat{\phi}^a_{ij'}-1}da_{ij'} \\ &= \frac{C_{\mathrm{Dir}}(\{\widetilde{\phi}^a_{ij}\}_{j=1}^{J})}{C_{\mathrm{Dir}}(\{\hat{\phi}^a_{ij}\}_{j=1}^{J})}.\end{aligned} \tag{7.150}$$

The normalization constant of the Dirichlet distribution is defined (Appendix C.4) as

$$C_{\mathrm{Dir}}(\{\phi_j\}_{j=1}^{J}) \triangleq \frac{\Gamma(\sum_{j=1}^{J}\phi_j)}{\prod_{j=1}^{J}\Gamma(\phi_j)}. \tag{7.151}$$

Therefore, by substituting the concrete form of the normalization constant into Eq. (7.150) and by using the definition of $\hat{\phi}^a_{ij'}$, Eq. (7.150) can be represented as follows:

$$\begin{aligned}\mathbb{E}_{(\mathbf{A})}\left[a_{ij}\right] &= \frac{\prod_{j'=1}^{J}\Gamma(\hat{\phi}^a_{ij'})}{\prod_{j'=1}^{J}\Gamma(\widetilde{\phi}^a_{ij'})}\frac{\Gamma(\sum_{j'=1}^{J}\widetilde{\phi}^a_{ij'})}{\Gamma(\sum_{j'=1}^{J}\hat{\phi}^a_{ij'})} \\ &= \frac{\prod_{j'\neq j}^{J}\Gamma(\widetilde{\phi}^a_{ij'})}{\prod_{j'\neq j}^{J}\Gamma(\widetilde{\phi}^a_{ij'})}\frac{\Gamma(\widetilde{\phi}^a_{ij}+1)}{\Gamma(\widetilde{\phi}^a_{ij})}\frac{\Gamma(\sum_{j'=1}^{J}\widetilde{\phi}^a_{ij'})}{\Gamma(1+\sum_{j'=1}^{J}\widetilde{\phi}^a_{ij'})} \\ &= \frac{\Gamma(\widetilde{\phi}^a_{ij}+1)}{\Gamma(\widetilde{\phi}^a_{ij})}\frac{\Gamma(\sum_{j'=1}^{J}\widetilde{\phi}^a_{ij'})}{\Gamma(1+\sum_{j'=1}^{J}\widetilde{\phi}^a_{ij'})}. \end{aligned} \tag{7.152}$$

Finally, we use the following formula for the gamma function:

$$\Gamma(x+1) = x\Gamma(x). \tag{7.153}$$

Then, Eq. (7.152) is analytically obtained as the following simple equation:

$$\begin{aligned}\mathbb{E}_{(\mathbf{A})}\left[a_{ij}\right] &= \frac{\widetilde{\phi}^a_{ij}\Gamma(\widetilde{\phi}^a_{ij})}{\Gamma(\widetilde{\phi}^a_{ij})}\frac{\Gamma(\sum_{j'=1}^{J}\widetilde{\phi}^a_{ij'})}{\sum_{j'=1}^{J}\widetilde{\phi}^a_{ij'}\Gamma(\sum_{j'=1}^{J}\widetilde{\phi}^a_{ij'})} \\ &= \frac{\widetilde{\phi}^a_{ij}}{\sum_{j'=1}^{J}\widetilde{\phi}^a_{ij'}}. \end{aligned} \tag{7.154}$$

Note that the state transition probability is obtained from the normalized weight, which is proportional to the posterior hyperparameter $\widetilde{\phi}^a_{ij}$, and the result is very intuitive.

Expected mixture weight $\mathbb{E}_{(\omega)}\left[\omega_{jk}\right]$

Since the mixture weight ω_{jk} is represented by a multinomial distribution, it is similar to the state transition a_{ij}. Similarly to $\mathbb{E}_{(\mathbf{A})}\left[a_{ij}\right]$, the expected state transition $\widetilde{\omega}_{jk}$ is calculated as follows:

$$\mathbb{E}_{(\omega)}\left[\omega_{jk}\right] = \frac{\widetilde{\phi}^\omega_{jk}}{\sum_{k'=1}^{K}\widetilde{\phi}^\omega_{jk'}}. \tag{7.155}$$

Again, the mixture weight probability is obtained using the normalized weight of the posterior hyperparameter $\widetilde{\phi}^\omega_{jk}$.

Expected Gaussian distribution $\mathbb{E}_{(\boldsymbol{\mu},\boldsymbol{\Sigma})}\left[\mathcal{N}(\mathbf{o}_t|\boldsymbol{\mu}_{jk},\boldsymbol{\Sigma}_{jk})\right]$

Finally, we calculate the expected value of the Gaussian distribution with VB posteriors for Gaussian parameters. First the expectation is factorized for each dimension when we use the diagonal covariance as follows:

$$\mathbb{E}_{(\boldsymbol{\mu},\boldsymbol{\Sigma})}\left[\mathcal{N}(\mathbf{o}_t|\boldsymbol{\mu}_{jk},\boldsymbol{\Sigma}_{jk})\right] = \prod_{d=1}^{D}\mathbb{E}_{(\mu_{jkd},r_{jkd})}\left[\mathcal{N}(o_{td}|\mu_{jkd},(r_{jkd})^{-1})\right]. \tag{7.156}$$

The indexes of state ij, mixture component k, frame t, and dimension d are removed to simplify the derivation.

Based on the definition of the Gaussian and gamma distributions in Eq. (7.62), we can obtain the following equation:

$$\begin{aligned}
&\mathbb{E}_{(\mu,r)}\left[\mathcal{N}(o|\mu,r^{-1})\right] \\
&= \int \mathcal{N}(\mu|\widetilde{\mu},(\widetilde{\phi}^{\mu}r)^{-1})\mathrm{Gam}_2(r|\widetilde{\phi}^{r},\widetilde{r})\mathcal{N}(o|\mu,r^{-1})d\mu dr \\
&= C_{\mathcal{N}}(\widetilde{\phi}^{\mu})C_{\mathrm{Gam}_2}(\widetilde{\phi}^{r},\widetilde{r})C_{\mathcal{N}} \\
&\times \int r^{\frac{1}{2}}\exp\left(-\frac{\widetilde{\phi}^{\mu}r(\mu-\widetilde{\mu})^2}{2}\right)r^{\frac{\widetilde{\phi}^{r}}{2}-1}\exp\left(-\frac{\widetilde{r}r}{2}\right)r^{\frac{1}{2}}\exp\left(-\frac{r(o-\mu)^2}{2}\right)d\mu dr \\
&\propto \int r^{\frac{\widetilde{\phi}^{r}}{2}}\exp\left(-\frac{\widetilde{r}r}{2}\right)\exp\left(-\frac{\widetilde{\phi}^{\mu}r(\mu-\widetilde{\mu})^2}{2}\right)\exp\left(-\frac{r(o-\mu)^2}{2}\right)d\mu dr.
\end{aligned} \tag{7.157}$$

First, we focus on the integration with respect to μ, and completing the square with respect to μ. Then, by integrating with respect to μ, and arranging the equation, the following equation is obtained:

$$\begin{aligned}
&\int \exp\left(-\frac{r}{2}\left((o-\mu)^2+\widetilde{\phi}^{\mu}(\mu-\widetilde{\mu})^2\right)\right)d\mu \\
&= \int \exp\left(-\frac{r}{2}\left((1+\widetilde{\phi}^{\mu})\left(\mu-\frac{o+\widetilde{\phi}^{\mu}\widetilde{\mu}}{1+\widetilde{\phi}^{\mu}}\right)^2-\frac{(o+\widetilde{\phi}^{\mu}\widetilde{\mu})^2}{1+\widetilde{\phi}^{\mu}}+o^2+\widetilde{\phi}^{\mu}\widetilde{\mu}^2\right)\right)d\mu \\
&\propto r^{-\frac{1}{2}}\exp\left(-\frac{r}{2}\left(-\frac{(o+\widetilde{\phi}^{\mu}\widetilde{\mu})^2}{1+\widetilde{\phi}^{\mu}}+o^2+\widetilde{\phi}^{\mu}\widetilde{\mu}^2\right)\right) \\
&= r^{-\frac{1}{2}}\exp\left(-\frac{r}{2(1+\widetilde{\phi}^{\mu})}\left(-(o+\widetilde{\phi}^{\mu}\widetilde{\mu})^2+(1+\widetilde{\phi}^{\mu})(o^2+\widetilde{\phi}^{\mu}\widetilde{\mu}^2)\right)\right) \\
&= r^{-\frac{1}{2}}\exp\left(-r\frac{\widetilde{\phi}^{\mu}(o-\widetilde{\mu})^2}{2(1+\widetilde{\phi}^{\mu})}\right).
\end{aligned} \tag{7.158}$$

Here we discuss the case when the VB posterior for r is the Dirac delta function around the MAP value of r, and the argument of its Dirac delta function is the maximum value of the VB posterior. Then, the result of the integration with respect to r is obtained by changing r to the MAP value $(\widetilde{\phi}^{r}-2)\widetilde{r}^{-1}$ in Eq. (7.158) in Appendix C.11. Therefore, the following equation is obtained:

$$\begin{aligned}
\mathbb{E}_{(\mu)}\left[\mathcal{N}(o|\mu,r^{-1})\right] &\propto \exp\left(-\frac{\widetilde{\phi}^{r}-2}{\widetilde{r}}\frac{\widetilde{\phi}^{\mu}(o-\widetilde{\mu})^2}{2(1+\widetilde{\phi}^{\mu})}\right) \\
&= \mathcal{N}\left(o\left|\widetilde{\mu},\frac{1+\widetilde{\phi}^{\mu}}{(\widetilde{\phi}^{r}-2)\widetilde{\phi}^{\mu}}\widetilde{r}\right.\right).
\end{aligned} \tag{7.159}$$

Thus, by recovering the omitted indexes, we can obtain

$$\mathbb{E}_{(\boldsymbol{\mu})}\left[\mathcal{N}(\mathbf{o}_t|\boldsymbol{\mu},\boldsymbol{\Sigma})\right] = \prod_{d=1}^{D}\mathcal{N}\left(o_{td}\left|\widetilde{\mu}_{jkd},\frac{1+\widetilde{\phi}^{\mu}_{jk}}{(\widetilde{\phi}^{r}_{jk}-2)\widetilde{\phi}^{\mu}_{jk}}\widetilde{r}_{jkd}\right.\right). \tag{7.160}$$

This is the analytical result of the expected function of a Gaussian distribution with expectation only over the mean parameter μ.

By substituting Eq. (7.158) into Eq. (7.157), we can obtain the following integral:

$$\int r^{\frac{\widetilde{\phi}^r+1}{2}-1} \exp\left(-r\frac{\widetilde{\phi}^\mu(o-\widetilde{\mu})^2+(1+\widetilde{\phi}^\mu)\widetilde{r}}{2(1+\widetilde{\phi}^\mu)}\right) dr. \tag{7.161}$$

First we use the following notation to simplify the integral:

$$\alpha \triangleq \frac{\widetilde{\phi}^r+1}{2},$$
$$\beta \triangleq \frac{\widetilde{\phi}^\mu(o-\widetilde{\mu})^2+(1+\widetilde{\phi}^\mu)\widetilde{r}}{2(1+\widetilde{\phi}^\mu)}. \tag{7.162}$$

Note that β depends on the observation o. Then, the integral with the explicit range of r is rewritten as follows:

$$\int_0^\infty r^{\alpha-1}e^{-\beta r}dr. \tag{7.163}$$

Now, we convert r with the following variable x:

$$r = \frac{x}{\beta},$$
$$dr = \frac{1}{\beta}dx,$$
$$r \in [0,\infty] \to x \in [0,\infty]. \tag{7.164}$$

Now $\widetilde{\phi}^\mu$ is a hyperparameter of the Dirichlet distribution, and $\widetilde{\phi}^\mu > 0$, therefore $\beta > 0$ from Eq. (7.162), and the range of x becomes $[0,\infty]$. Then, the integral can be rewritten as:

$$\begin{aligned}\int_0^\infty r^{\alpha-1}e^{-\beta r}dr &= \int_0^\infty \left(\frac{x}{\beta}\right)^{\alpha-1} e^{-x}\frac{1}{\beta}dx \\ &= \left(\frac{1}{\beta}\right)^\alpha \int_0^\infty x^{\alpha-1}e^{-x}dx.\end{aligned} \tag{7.165}$$

Here, from the formula of the gamma function, we can further rewrite the above integral as:

$$\begin{aligned}\int_0^\infty r^{\alpha-1}e^{-\beta r}dr &= \left(\frac{1}{\beta}\right)^\alpha \Gamma(\alpha) = \left(\frac{1}{\beta}\right)^\alpha (\alpha-1)! \\ &\propto \left(\frac{1}{\beta}\right)^\alpha.\end{aligned} \tag{7.166}$$

Since $(\alpha-1)!$ does not depend on the observation o, we can disregard it as a constant value. Finally, by recovering the variables of α and β from Eq. (7.162), Eq. (7.161) is obtained as the following equation:

$$\begin{aligned}&\int r^{\frac{\widetilde{\phi}^r+1}{2}-1} \exp\left(-r\frac{\widetilde{\phi}^\mu(o-\widetilde{\mu})^2+(1+\widetilde{\phi}^\mu)\widetilde{r}}{2(1+\widetilde{\phi}^\mu)}\right) dr \\ &\propto \left(\frac{\widetilde{\phi}^\mu(o-\widetilde{\mu})^2+(1+\widetilde{\phi}^\mu)\widetilde{r}}{2(1+\widetilde{\phi}^\mu)}\right)^{-\frac{\widetilde{\phi}^r+1}{2}} \\ &\propto \left(1+\frac{\widetilde{\phi}^\mu}{(1+\widetilde{\phi}^\mu)\widetilde{r}}(o-\widetilde{\mu})^2\right)^{-\frac{\widetilde{\phi}^r+1}{2}}.\end{aligned} \tag{7.167}$$

Here we refer to the concrete form of the Student's t-distribution given in Appendix C.16:

$$\mathrm{St}(x|\mu, \lambda, \kappa) \triangleq C_{\mathrm{St}} \left(1 + \frac{1}{\kappa\lambda}(x-\mu)^2\right)^{-\frac{\kappa+1}{2}}. \tag{7.168}$$

The parameters μ, κ, and λ of the Student's t-distribution correspond to those of the above equation as follows:

$$\begin{cases} \mu & = \widetilde{\mu}, \\ \lambda & = \frac{(1+\widetilde{\phi}^{\mu} \;)\widetilde{r}}{\widetilde{\phi}^{\mu} \; \widetilde{\phi}^{r}}, \\ \kappa & = \widetilde{\phi}^{r}. \end{cases} \tag{7.169}$$

Thus, the result of the integral with respect to μ and r (Eq. (7.157)) is represented as the Student's t-distribution:

$$\mathrm{St}\left(o \left| \widetilde{\mu}, \frac{(1+\widetilde{\phi}^{\mu})\widetilde{r}}{\widetilde{\phi}^{\mu}\widetilde{\phi}^{r}}, \widetilde{\phi}^{r}\right.\right). \tag{7.170}$$

The third parameter in the t-distribution is called the degree of freedom, and if this value is large, the distribution approaches the Gaussian distribution theoretically.

Thus, by recovering the omitted indexes, we can obtain

$$\mathbb{E}_{(\boldsymbol{\mu},\boldsymbol{\Sigma})}\left[\mathcal{N}(\mathbf{o}_t|\boldsymbol{\mu}, \boldsymbol{\Sigma})\right] = \prod_{d=1}^{D} \mathrm{St}\left(o_{td} \left| \widetilde{\mu}_{jkd}, \frac{(1+\widetilde{\phi}^{\mu}_{jk})\widetilde{r}_{jkd}}{\widetilde{\phi}^{\mu}_{jk}\widetilde{\phi}^{r}_{jk}}, \widetilde{\phi}^{r}_{jk}\right.\right). \tag{7.171}$$

This is the analytical result of the expected Gaussian distribution with marginalization of both mean and precision parameters μ and r. Compared with Eq. (7.160), the marginalization of both parameters changes the distribution from the Gaussian distribution to the Student's t-distribution. The latter is called a long tail distribution since it is a power law function, and it provides a robust classification in general, when the amount of training data is small.

Since the degree of freedom in this solution is the posterior hyperparameter of the precision parameter $\widetilde{\phi}^{r}$, and it is proportional to the amount of data, as shown in Eq. (7.67), this solution approaches the Gaussian distribution. Then, the variance parameters in Eq. (7.171) also approximately approach the following value:

$$\frac{(1+\widetilde{\phi}^{\mu}_{jk})\widetilde{r}_{jkd}}{\widetilde{\phi}^{\mu}_{jk}\widetilde{\phi}^{r}_{jk}} \approx \frac{\widetilde{r}_{jkd}}{\widetilde{\phi}^{r}_{jk}}. \tag{7.172}$$

Thus, Eq. (7.171) is approximated by the following Gaussian distribution when the amount of data $\mathbf{O}$ is large:

$$\mathbb{E}_{(\boldsymbol{\mu},\boldsymbol{\Sigma})}\left[\mathcal{N}(\mathbf{o}_t|\boldsymbol{\mu}, \boldsymbol{\Sigma})\right] \approx \prod_{d=1}^{D} \mathcal{N}\left(o_{td} \left| \widetilde{\mu}_{jkd}, \frac{\widetilde{r}_{jkd}}{\widetilde{\phi}^{r}_{jk}}\right.\right). \tag{7.173}$$

This solution corresponds to the MAP estimation result of the Gaussian distribution in CDHMM, as discussed in Section 4.3.5.

In Watanabe & Nakamura (2006), experimental results were reported to show the effectiveness of the Bayesian predictive classification without marginalization (it corresponds to the MAP estimation in Section 4.3), with marginalization of only mean parameters (corresponds to Eq. (7.160)), and with marginalization of both mean and covariance parameters (corresponds to Eq. (7.171)). Speaker adaptation experiments for LVCSR (30 000 vocabulary size) show that the Student's t-distribution-based Bayesian predictive classification improved the performance from the MAP estimation and the marginalization results, reducing the WERs by 2.3 % and 1.2 %, respectively, when we only used one utterance (3.3 seconds on average) for the adaptation data. Since all the results use the same prior hyperparameter values, the improvement purely comes from the marginalization effect. In addition, if the amount of adaptation data increased, the performance of these three methods converged to the same value, which is also expected, based on the discussion of analytical results of the Student's t-distribution in Eq. (7.173).

The use of VB-based Bayesian predictive classification makes acoustic modeling in speech recognition a totally Bayesian framework that follows a consistent concept, whereby all acoustic procedures (model parameter estimation, model selection, and speech classification) are carried out based on posterior distributions. For example, compare the variational Bayesian speech recognition framework with a conventional ML-BIC approach: the model parameter estimation, model selection and speech classification are based on ML (Chapter 3) and BIC (Chapter 6). BIC is an asymptotic criterion that is theoretically effective only when the amount of training data is sufficiently large. Therefore, for a small amount of training data, model selection does not perform well because of the uncertainty of the ML estimates. The next section aims at solving the problem caused by a small amount of training data by using VB.

7.3.6 Decision tree clustering

This section revisits decision tree clustering of the context-dependent HMM states, as we discussed in Section 6.5, based on the VB framework (Watanabe *et al.* 2004, Hashimoto, Zen, Nankaku *et al.* 2008, Shiota, Hashimoto, Nankaku *et al.* 2009). Similarly to Eq. (6.74), we approximate the Bayes factor in Section 6.2 by selecting an appropriate question at each split, chosen to increase the variational lower bound/VB objective function $\mathcal{F}^M$ in the VB framework, as discussed in Section 7.3.4. When node n is split into a Yes node (n_Y^Q) and No node (n_N^Q) by question Q (we use $M_{Q(n)}$ with this hypothesized model, obtained from question Q, and M_n with the original model), the appropriate question $\widetilde{Q}(n)$ is chosen from a set of questions as follows:

$$\begin{aligned}\widetilde{Q}(n) &= \arg\max_{Q} \log\left(\frac{p(M_{Q(n)}|\mathbf{O})}{p(M_n|\mathbf{O})}\right)\\ &\approx \arg\max_{Q} \Delta\mathcal{F}_{Q(n)},\end{aligned} \tag{7.174}$$

where $\Delta\mathcal{F}^{Q(n)}$ is the gain in the VB objective function when node n is split by Q, which is defined as:

$$\Delta\mathcal{F}_{Q(n)} = \arg\max_{Q} \mathcal{F}_{\Omega(n_Y^Q)} + \mathcal{F}_{\Omega(n_N^Q)} - \mathcal{F}_{\Omega(n)}. \tag{7.175}$$

$\Omega(n)$ denotes a set of (non-shared) context-dependent HMM states at node n in a decision tree. The question is chosen to maximize the gain in $\mathcal{F}^M$ by splitting. The VB objective function for decision tree construction is also simply calculated under the following same constraints as the ML approach:

- Data alignments $\widetilde{\gamma}_t(j,k)$ and $\widetilde{\xi}_t(i,j)$ for each state are fixed while splitting.
- Emission probability distribution in a state is represented by a single Gaussian distribution (i.e., $K=1$).
- Covariance matrices have only diagonal elements.
- A contribution of state transitions a_{ij} and initial weights π_j for likelihood is disregarded.

By using these conditions, the objective function is obtained without iterative calculations, which reduces the calculation time. Under conditions of fixed data assignment and single Gaussian assumptions, the latent variable part of $\mathcal{F}^M$ can be disregarded, i.e., Eq. (7.116) is approximated as

$$\mathcal{F}^M \approx \mathcal{F}^M_\Theta. \tag{7.176}$$

In the VB objective function of model parameter $\mathcal{F}^M_\Theta$ (Eq. (7.126)), the factors of posterior parameters of state transition $\widetilde{\phi}^a$ and mixture component $\widetilde{\phi}^\omega$ can also be disregarded under the above conditions. Therefore, the objective function $\mathcal{F}_\Omega$ in node n (we omit the index n for simplicity) for assigned data set $\mathbf{O}_\Omega = \{\mathbf{O}(i)|i\in\Omega\}$, where $\mathbf{O}(i) = \{\mathbf{o}_t(i)\in\mathbb{R}^D|t=1,\cdots,T(i)\}$ can be obtained from the modification of $\mathcal{F}^M_\Theta$ in Eq. (7.126) as follows:

$$\mathcal{F}_\Omega = \log\left((2\pi)^{-\frac{\widetilde{\gamma}_\Omega D}{2}}\left(\frac{\phi^\mu_\Omega}{\widetilde{\phi}^\mu_\Omega}\right)^{\frac{D}{2}}\frac{2^{\frac{\widetilde{\phi}^r_\Omega D}{2}}\left(\Gamma\left(\frac{\widetilde{\phi}^r_\Omega}{2}\right)\right)^D\prod_{d=1}^D\left(r^0_{\Omega d}\right)^{\frac{\phi^r_\Omega}{2}}}{2^{\frac{\phi^r_\Omega D}{2}}\left(\Gamma\left(\frac{\phi^r_\Omega}{2}\right)\right)^D\prod_{d=1}^D\left(\widetilde{r}_{\Omega d}\right)^{\frac{\widetilde{\phi}^r_\Omega}{2}}}\right), \tag{7.177}$$

where $\{\widetilde{\phi}^\mu_\Omega, \widetilde{\boldsymbol{\mu}}_\Omega, \widetilde{\phi}^r_\Omega, \{\widetilde{r}_{\Omega d}\}_{d=1}^D\}(\triangleq \widetilde{\Psi}_\Omega)$ is a subset of the posterior parameters in Eq. (7.66), and is represented by:

$$\begin{cases} \widetilde{\phi}^\mu_\Omega &= \phi^\mu_\Omega + \widetilde{\gamma}_\Omega, \\ \widetilde{\boldsymbol{\mu}}_\Omega &= \frac{\phi^\mu_\Omega \boldsymbol{\mu}^0_\Omega + \widetilde{\boldsymbol{\gamma}}^{(1)}_\Omega}{\phi^\mu_\Omega + \widetilde{\gamma}_\Omega}, \\ \widetilde{\phi}^r_\Omega &= \phi^r_\Omega + \widetilde{\gamma}_\Omega, \\ \widetilde{r}_{\Omega d} &= \widetilde{\gamma}^{(2)}_{\Omega d} + \phi^\mu_\Omega(\mu^0_{\Omega d})^2 - \widetilde{\phi}^\mu_\Omega(\widetilde{\mu}_{\Omega d})^2 + r^0_{\Omega d}. \end{cases} \tag{7.178}$$

$\widetilde{\gamma}_\Omega, \widetilde{\boldsymbol{\gamma}}^{(1)}_\Omega$ and $\widetilde{\gamma}^{(2)}_{\Omega d}$ are the sufficient statistics of a set of states in node n, as defined as follows.

$$\begin{cases} \widetilde{\gamma}_\Omega &\triangleq \sum_{i\in\Omega}\sum_{t=1}^{T(i)}, \\ \widetilde{\boldsymbol{\gamma}}^{(1)}_\Omega &\triangleq \sum_{i\in\Omega}\sum_{t=1}^{T(i)} \mathbf{o}_t(i), \\ \widetilde{\gamma}^{(2)}_{\Omega d} &\triangleq \sum_{i\in\Omega}\sum_{t=1}^{T(i)} (o_{td}(i))^2. \end{cases} \tag{7.179}$$

Note that since we use the hard aligned data $\mathbf{O}(i)$ (based on the Viterbi algorithm), the assignment information $\widetilde{\gamma}_t(i)$ is included in this representation. Here,

$\{\phi^{\mu}_{\Omega}, \boldsymbol{\mu}^0_{\Omega}, \phi^r_{\Omega}, \{r^0_{\Omega d}\}^D_{d=1}\}(\triangleq \Psi_{\Omega})$ is a set of prior parameters. One choice of setting the prior hyperparameters $\boldsymbol{\mu}^0_{\Omega}$ and $r^0_{\Omega d}$ would be to set them by using monophone (root node) HMM state statistics ($\widetilde{\gamma}_{\text{root}}$, $\widetilde{\gamma}^{(1)}_{\text{root}}$ and $\widetilde{\gamma}^{(2)}_{\text{root}}$) as follows:

$$\boldsymbol{\mu}^0_{\Omega} = \frac{\widetilde{\boldsymbol{\gamma}}^{(1)}_{\text{root}}}{\widetilde{\gamma}_{\text{root}}},$$
$$r^0_{\Omega d} = \phi^r_{\Omega}\left(\frac{\widetilde{\gamma}^{(2)}_{\text{root, d}}}{\widetilde{\gamma}_{\text{root}}} - \left(\mu^0_{\text{root},d}\right)^2\right). \tag{7.180}$$

The other parameters ϕ^{μ}_{Ω} and ϕ^r_{Ω} are set manually. By substituting Eq. (7.178) into Eq. (7.177), the gain $\Delta\mathcal{F}_{Q(n)}$ can be obtained when n is split into n^Q_Y, n^Q_N by question Q:

$$\Delta\mathcal{F}_{Q(n)} = f(\widetilde{\Psi}_{\Omega(n^Q_Y)}) + f(\widetilde{\Psi}_{\Omega(n^Q_N)}) - f(\widetilde{\Psi}_{\Omega(n)}) - f(\Psi_{\Omega(n)}). \tag{7.181}$$

Here, $f(\Psi)$ is defined by:

$$f(\Psi) \triangleq -\frac{D}{2}\log\phi^{\mu} - \frac{\phi^r}{2}\sum_{d=1}^{D}\log r_d + D\log\Gamma\left(\frac{\phi^r}{2}\right). \tag{7.182}$$

The terms that do not contribute to $\Delta\mathcal{F}^{Q(n)}$ are disregarded. The final term in Eq. (7.181) is only computed from the prior hyperparameter Ψ. Similarly to the BIC criterion in Eq. (6.92), node splitting stops when the condition

$$\Delta\mathcal{F}_{Q(n)} \leq 0 \tag{7.183}$$

is satisfied. A model structure based on the VB framework can be obtained by executing this construction for all trees, resulting in the maximization of total $\mathcal{F}^M$. This implementation based on the decision tree method does not require iterative calculations, and can construct clustered-state HMMs efficiently. There is another major method for the construction of clustered-state HMMs that uses a successive state splitting algorithm, and which does not remove latent variables in HMMs (Takami & Sagayama 1992, Ostendorf & Singer 1997). Therefore, this requires the VB Baum–Welch algorithm and calculation of the latent variable part of the lower bound/VB objective function for each splitting. This is realized as the VB SSS algorithm by Jitsuhiro & Nakamura (2004).

The relationship between VB model selection and the conventional BIC model selection, based on Eqs. (7.181) and (6.89), respectively, is discussed below. Based on the condition of a sufficiently large amount of data, the posterior hyperparameters in Eq. (7.178) are approximated as follows:

$$\widetilde{\phi}^{\mu}_{\Omega}, \widetilde{\phi}^r_{\Omega} \to \widetilde{\gamma}_{\Omega},$$
$$\widetilde{\boldsymbol{\mu}}_{\Omega} \to \frac{\widetilde{\boldsymbol{\gamma}}^{(1)}_{\Omega}}{\widetilde{\gamma}_{\Omega}},$$
$$\widetilde{r}_{\Omega d} \to \widetilde{\gamma}^{(2)}_{\Omega d} - \frac{\left(\widetilde{\gamma}^{(1)}_{\Omega d}\right)^2}{\widetilde{\gamma}_{\Omega}}. \tag{7.184}$$

In addition, from Stirling's approximation, the logarithmic gamma function has the following relationship:

$$\log \Gamma\left(\frac{x}{2}\right) \to \frac{x}{2}\log\left(\frac{x}{2}\right) - \frac{x}{2} - \frac{1}{2}\log\left(\frac{x}{2\pi}\right), \tag{7.185}$$

when $|x| \to \infty$. By substituting Eq. (7.184) into Eq. (7.182) and using Eq. (7.185), $f(\widetilde{\Psi})$ is approximated as

$$\begin{aligned} f(\widetilde{\Psi}) &\to -\frac{D}{2}\log\widetilde{\gamma}_\Omega - \frac{\widetilde{\gamma}_\Omega}{2}\sum_{d=1}^{D}\log\left(\widetilde{\gamma}_{\Omega d}^{(2)} - \frac{\left(\widetilde{\gamma}_{\Omega d}^{(1)}\right)^2}{\widetilde{\gamma}_\Omega}\right) \\ &\quad + D\left(\frac{\widetilde{\gamma}_\Omega}{2}\log\left(\frac{\widetilde{\gamma}_\Omega}{2}\right) - \frac{\widetilde{\gamma}_\Omega}{2} - \frac{1}{2}\log\left(\frac{\widetilde{\gamma}_\Omega}{2\pi}\right)\right) \\ &= -\frac{\widetilde{\gamma}_\Omega}{2}\left(D\left(1+\log(2\pi)\right) + \underbrace{\sum_{d=1}^{D}\log\left(\frac{\widetilde{\gamma}_{\Omega d}^{(2)}}{\widetilde{\gamma}_\Omega} - \frac{\left(\widetilde{\gamma}_{\Omega d}^{(1)}\right)^2}{(\widetilde{\gamma}_\Omega)^2}\right)}_{\approx \log|\boldsymbol{\Sigma}_\Omega^{\mathrm{ML}}|}\right) - D\log\widetilde{\gamma}_\Omega. \end{aligned} \tag{7.186}$$

Then, an asymptotic form of Eq. (7.182) is composed of a log-likelihood gain term and a penalty term depending on the number of free parameters ($2D$ in this diagonal covariance Gaussian case) and the amount of training data, i.e., the asymptotic form becomes the BIC-type objective function form, as shown in Eq. (6.88). Therefore, VB theoretically involves the BIC objective function, and so BIC model selection is asymptotically equivalent to VB model selection, which demonstrates the advantages of VB, especially for small amounts of training data.

7.3.7 Determination of HMM topology

Once a clustered-state model structure is obtained, acoustic model selection is completed by determining the number of mixture components per state. GMMs include latent variables, and their determination requires the VB Baum–Welch algorithm and computation of the latent variable part of the variational lower bound, unlike the clustering triphone HMM states in Section 7.4.5. Therefore, this section deals with determination of the number of GMM components per state by considering the latent variable effects. Then, the effectiveness of VB model selection in latent variable models is confirmed (Jitsuhiro & Nakamura 2004) for the successive state splitting algorithm, and the effectiveness of VB model selection for GMMs is re-confirmed (Valente & Wellekens 2003). In general, there are two methods for determining the number of mixture components. With the first method, the number of mixture components per state is the same for all states. The objective function $\mathcal{F}^M$ is calculated for each number of mixture components, and the number of mixture components that maximizes the total $\mathcal{F}^M$ is determined as being the appropriate one (fixed-number GMM method). With the second method, the number of mixture components per state can vary by state; here, Gaussians are split and merged to increase $\mathcal{F}^M$ and determine the number of mixture components in each state (varying-number GMM method). A model obtained by the varying-number GMM method is expected to be more accurate than one obtained by the fixed-number GMM method, although the varying-number GMM method requires more computation time.

We require the variational lower bound for each state to determine the number of mixture components. In this case, the state alignments vary and states are expressed as GMMs. Therefore, the model includes latent variables and the component $\mathcal{F}^M_{S,V}$ cannot be disregarded, unlike the case of triphone HMM state clustering. However, since the number of mixture components is determined for each state and the state alignments do not change greatly, the contribution of the state transitions to the objective function is expected to be small, and can be ignored. Therefore, the objective function $\mathcal{F}^M$ for a particular state j is represented from Eqs. (7.126) and (7.135) as follows:

$$(\mathcal{F}^M)_j = (\mathcal{F}^M_\Theta)_j - (\mathcal{F}^M_V)_j, \tag{7.187}$$

where $(\mathcal{F}^M_\Theta)_j$ is represented by removing the HMM terms in Eq. (7.126) as follows:

$$\begin{aligned}(\mathcal{F}^M_\Theta)_j &= \log \frac{\Gamma(\sum_k \phi^\omega_{jk}) \prod_k \Gamma(\tilde{\phi}^\omega_{jk})}{\Gamma(\sum_k \tilde{\phi}^\omega_{jk}) \prod_k \Gamma(\phi^\omega_{jk})} \\ &\quad + \sum_k \log \left((2\pi)^{-\frac{\tilde{\gamma}_{jk} D}{2}} \left(\frac{\phi^\mu_{jk}}{\tilde{\phi}^\mu_{jk}} \right)^{\frac{D}{2}} \frac{\left(\Gamma\left(\frac{\tilde{\phi}^r_{jk}}{2}\right)\right)^D \prod_d \left(\frac{r^0_{jkd}}{2}\right)^{\frac{\phi^r_{jk}}{2}}}{\left(\Gamma\left(\frac{\phi^r_{jk}}{2}\right)\right)^D \prod_d \left(\frac{\tilde{r}_{jkd}}{2}\right)^{\frac{\tilde{\phi}^r_{jk}}{2}}} \right).\end{aligned} \tag{7.188}$$

Similarly, $(\mathcal{F}^M_V)_j$ is also represented as follows:

$$\begin{aligned}(\mathcal{F}^M_V)_j &= \sum_k \tilde{\gamma}_{jk} \left(\Psi\left(\tilde{\phi}^\omega_{jk}\right) - \Psi\left(\sum_{k'} \tilde{\phi}^\omega_{jk'}\right)\right) \\ &\quad - \frac{1}{2} \sum_k \tilde{\gamma}_{jk} \left(D \left(\log(2\pi) + \frac{1}{\tilde{\phi}^\mu_{jk}} - \Psi\left(\frac{\tilde{\phi}^r_{jk}}{2}\right)\right) + \sum_d \log \frac{\tilde{r}_{jkd}}{2} \right) \\ &\quad - \frac{1}{2} \sum_k \left(\tilde{\phi}^r_{jk} \sum_{t,d} \frac{\tilde{\gamma}_t(j,k)(o_{td} - \tilde{\mu}_{jkd})^2}{\tilde{r}_{jkd}} \right) - \log \left(\sum_V \tilde{u}(\mathbf{O}, V|M) \right).\end{aligned} \tag{7.189}$$

Therefore, with the fixed-number GMM method, the total $\mathcal{F}^M$ is obtained by summing up all states' $(\mathcal{F}^M)_j$, which determines the number of mixture components per state. With the varying-number GMM method, the change of $(\mathcal{F}^M)_j$ per state is compared after merging or splitting the Gaussians, which also determines the number of mixture components. The number of mixture components is also automatically determined by using the BIC/MDL objective function (Chen & Gopinath 1999, Shinoda & Iso 2001). However, the BIC/MDL objective function is based on the asymptotic condition and cannot be applied to latent models in principle. On the other hand, the variational lower bound derived by VB does not need the asymptotic condition and can determine an appropriate model structure with latent variables.

Table 7.2 Automatic determination of acoustic model topology.

	Read speech (JNAS)	Read speech (WSJ)	Isolated word (JEIDA)	Lecture (CSJ)
VB	91.7 %	91.3 %	97.9 %	74.5 %
# states	912	2504	254	1986
# components	40	32	35	32
ML + Dev. Set	91.4 %	91.3 %	98.1 %	74.2 %
# states	1000	7500	1000	3000
# components	30	32	15	32

Table 7.2 shows experimental results for automatic determination of the acoustic model topology by using VB and the conventional heuristic approach that determines the model topology by evaluating ASR performance on development sets. Note that VB was only used for the model topology determination, and the other procedures (e.g., training and decoding) were performed by using the conventional (ML) approaches. Therefore, Table 7.2 simply shows the effectiveness of the model selection. We used two tasks based on read speech recognition of news articles, *JNAS* (Shikano, Kawahara, Kobayashi *et al.* 1999) and *WSJ* (Paul & Baker 1992), an isolated word speech recognition task (JEIDA 100 city name recognition), and a lecture speech recognition task, CSJ (Furui, Maekawa & Isahara 2000). Table 7.2 provides the ASR performance of the determined model topology with the number of total HMM states and a mixture component in an HMM state, where we used the same number of mixture component for all states. In the various ASR tasks, VB achieved comparable performance to the conventional method by selecting appropriate model topologies without using a development set. Thus, these experiments proved that the VB model selection method can *automatically* determine an appropriate acoustic model topology with a comparable performance to that obtained by using a development set.

7.4 Structural Bayesian linear regression for hidden Markov model

As discussed in Section 3.5, a Bayesian treatment of the affine transformation parameters of CDHMM is an important issue to improve the generalization capability of the model adaptation. While the regression tree used in the conventional maximum likelihood linear regression (MLLR) can be considered one form of prior knowledge, i.e., how various Gaussian distributions are related, another approach is to explicitly construct and use *prior knowledge of regression parameters* in an approximated Bayesian paradigm.

For example, maximum a-posteriori linear regression (MAPLR) (Chesta, Siohan & Lee 1999) replaces the ML criterion with the MAP criterion introduced in Chapter 4 in the estimation of regression parameters. Quasi-Bayes linear regression (Chien 2002) also replaces the ML/MAP criterion with a quasi-Bayes criterion. With the explicit prior knowledge acting as a regularization term, MAPLR appears to be less susceptible to the overfitting problem. The MAPLR is extended to the structural MAP (SMAP) (Shinoda & Lee 2001) and the structural MAPLR (SMAPLR) (Siohan, Myrvoll & Lee 2002), both of which fully utilize the Gaussian tree structure used in the model selection approach to efficiently set the hyperparameters in prior distributions. In SMAP and SMAPLR, the hyperparameters in the prior distribution in a target node are obtained from the statistics in its parent node. Since the total number of speech frames assigned to a set of Gaussians in the parent node is always larger than that in the target node, the statistics obtained in the parent node are more reliable than those in the target node, and these can be good prior knowledge for transformation parameter estimation in the target node.

Another extension of MAPLR is to replace MAP approximation by a fully Bayesian treatment of latent models, using VB. This section employs VB for the linear regression problem (Watanabe & Nakamura 2004, Yu & Gales 2006, Watanabe, Nakamura & Juang 2013), but we focus on model selection and efficient prior utilization at the same time, in addition to estimation of the linear transformation parameters of HMMs proposed in previous work (Watanabe & Nakamura 2004, Yu & Gales 2006). In particular, we consistently use the variational lower bound as the optimization criterion for the model structure and hyperparameters, in addition to the posterior distributions of the transformation parameters and the latent variables. As we discussed in Section 7.2, since this optimization leads the approximated variational posterior distributions to the true posterior distributions theoretically in the sense of minimizing Kullback–Leibler divergence between them, the above consistent approach leads to improved generalization capability (Neal & Hinton 1998, Attias 1999, Ueda & Ghahramani 2002).

This section provides an analytical solution to the variational lower bound by marginalizing all possible transformation parameters and latent variables introduced in the linear regression problem. The solution is based on a variance-normalized representation of Gaussian mean vectors to simplify the solution as normalized domain MLLR. As a result of variational calculation, we can marginalize the transformation parameters in all nodes used in the structural prior setting. This is a part of the solution of the variational message passing algorithm (Winn & Bishop 2006), which is a general framework of variational inference in a graphical model. Furthermore, the optimization of the model topology and hyperparameters in the proposed approach yields an additional benefit in the improvement of the generalization capability. For example, this approach infers the linear regression without controlling the Gaussian cluster topology and hyperparameters as the tuning parameters. Thus linear regression for HMM parameters is accomplished without excessive parameterization in a Bayesian sense.

7.4.1 Variational Bayesian linear regression

This section provides an analytical solution for Bayesian linear regression by using a variational lower bound. The previous section only considers a regression matrix in leaf node $j \in \mathcal{J}_M$, but we also consider a regression matrix in leaf or non-leaf node $i \in \mathcal{I}_M$ in the Gaussian tree given model structure M. Then we focus on a set of regression matrices in all nodes $\mathbf{\Lambda}_{\mathcal{I}_M} = \{\mathbf{W}_i | i = 1, \cdots, |\mathcal{I}_M|\}$, instead of $\mathbf{\Lambda}_{\mathcal{J}_M}$, and marginalize $\mathbf{\Lambda}_{\mathcal{I}_M}$ in a Bayesian manner. This extension involves the structural prior setting as proposed in

SMAP and SMAPLR (Shinoda & Lee 2001, Siohan *et al.* 2002, Yamagishi, Kobayashi, Nakano *et al.* 2009).

In this section, we mainly deal with:

- prior distribution of model parameters $p(\mathbf{\Lambda}_{\mathcal{I}_M}; M, \Psi)$;
- true posterior distribution of model parameters and latent variables $p(\mathbf{\Lambda}_{\mathcal{I}_M}, Z|\mathbf{O}; M, \Psi)$;
- variational posterior distribution of model parameters and latent variables $q(\mathbf{\Lambda}_{\mathcal{I}_M}, Z|\mathbf{O}; M, \Psi)$;
- generative model distribution $p(\mathbf{O}, Z|\mathbf{\Lambda}_{\mathcal{I}_M}; \Theta)$.

Note that the prior and generative model distributions are given, as shown in the generative process of Algorithm 12, and we obtain the variational posterior distribution, which is an approximation of the true posterior distribution.

7.4.2 Generative model

As discussed in Section 3.5, the generative model distribution with the expectation with respect to the posterior distributions of latent variables is represented as follows:

$$\mathbb{E}_{(Z)}\left[\log p(\mathbf{O}, Z|\mathbf{\Lambda}_{\mathcal{I}_M}; \Theta)\right] = \sum_{k=1}^{K}\sum_{t=1}^{T} \gamma_t(k) \log \mathcal{N}(\mathbf{o}_t|\boldsymbol{\mu}_k^{ad}, \mathbf{\Sigma}_k), \tag{7.190}$$

where $p(\mathbf{O}, Z|\mathbf{\Lambda}_{\mathcal{I}_M}; \Theta)$ is the generative model distribution of the transformed HMM parameters with transformed mean vectors $\boldsymbol{\mu}_k^{ad}$. We use $\boldsymbol{\mu}_k^{ad}$ based on the following variance normalized representation:

$$\boldsymbol{\mu}_k^{ad} = \mathbf{C}_k \mathbf{W}_j \boldsymbol{\xi}_k. \tag{7.191}$$

$\mathbf{C}_k$ is the Cholesky decomposition matrix of $\mathbf{\Sigma}_k$, and $\boldsymbol{\xi}_k = [1, ((\mathbf{C}_k)^{-1}\boldsymbol{\mu}_k^{ini})^\intercal]^\intercal$ is obtained based on the initial mean vector $\boldsymbol{\mu}_k^{ini}$. This representation makes the calculation simple.[5]

7.4.3 Variational lower bound

With regard to variational Bayesian approaches, we first focus on the following marginal log-likelihood $p(\mathbf{O}; \Theta, M, \Psi)$ with a set of HMM parameters Θ, a set of hyperparameters Ψ, and a model structure:[6,7]

$$\begin{aligned}&\log p(\mathbf{O}; \Theta, M, \Psi)\\&= \log\left(\int \sum_{Z} p(\mathbf{O}, Z|\mathbf{\Lambda}_{\mathcal{I}_M}; \Theta) p(\mathbf{\Lambda}_{\mathcal{I}_M}; M, \Psi) d\mathbf{\Lambda}_{\mathcal{I}_M}\right).\end{aligned} \tag{7.192}$$

[5] Hahm, Ogawa, Fujimoto *et al.* (2012) discuss the use of conventional MLLR estimation without the variance normalization in the VB framework and its application to feature-space MLLR (fVBLR).

[6] Ψ and M can also be marginalized by setting their distributions. This section point-estimates Ψ and M by a MAP approach, similar to the evidence approximation in Chapter 5.

[7] We can also marginalize the HMM parameters Θ. This corresponds to jointly optimizing HMM and linear regression parameters.

$p(\mathbf{\Lambda}_{\mathcal{I}_M}; M, \Psi)$ is a prior distribution of transformation matrices $\mathbf{\Lambda}_{\mathcal{I}_M}$. In the following explanation, we omit Θ, M, and Ψ in the prior distribution and generative model distribution for simplicity, i.e., $p(\mathbf{\Lambda}_{\mathcal{I}_M}; M, \Psi) \to p(\mathbf{\Lambda}_{\mathcal{I}_M})$, and $p(\mathbf{O}, Z|\mathbf{\Lambda}_{\mathcal{I}_M}; \Theta) \to p(\mathbf{O}, Z|\mathbf{\Lambda}_{\mathcal{I}_M})$.

Similarly to Eq. (7.15), since the variational Bayesian approach focuses on the variational lower bound of the marginal log likelihood $\mathcal{F}(M, \Psi)$ with a set of hyperparameters Ψ and a model structure M, Eq. (7.192) is represented as follows:

$$\begin{aligned}
&\log p(\mathbf{O}; \Theta, M, \Psi) \\
&= \log \left(\int \sum_{Z} \frac{p(\mathbf{O}, Z|\mathbf{\Lambda}_{\mathcal{I}_M}) p(\mathbf{\Lambda}_{\mathcal{I}_M})}{q(\mathbf{\Lambda}_{\mathcal{I}_M}, Z)} q(\mathbf{\Lambda}_{\mathcal{I}_M}, Z) d\mathbf{\Lambda}_{\mathcal{I}_M} \right) \\
&\geq \underbrace{\mathbb{E}_{(\mathbf{\Lambda}_{\mathcal{I}_M}, Z)} \left[\log \frac{p(\mathbf{O}, Z|\mathbf{\Lambda}_{\mathcal{I}_M}) p(\mathbf{\Lambda}_{\mathcal{I}_M})}{q(\mathbf{\Lambda}_{\mathcal{I}_M}, Z)} \right]}_{\triangleq \mathcal{F}(M, \Psi)}.
\end{aligned} \tag{7.193}$$

The inequality in Eq. (7.193) is supported by the Jensen's inequality in Eq. (7.10). $q(\mathbf{\Lambda}_{\mathcal{I}_M}, Z)$ is an arbitrary distribution, and is optimized by using a variational method to be discussed later. For simplicity, we omit M, Ψ, and $\mathbf{O}$ from the distributions. As discussed in Section 7.1, the variational lower bound is a better approximation of the marginal log likelihood than the auxiliary functions of maximum likelihood EM and maximum a-posteriori EM algorithms that point-estimate model parameters, especially for small amount of training data. Therefore, the variational Bayes can mitigate the sparse data problem that the conventional approaches must resolve.

The variational Bayes regards the variational lower bound $\mathcal{F}(M, \Psi)$ as an objective function for the model structure and hyperparameter, and an objective functional for the joint posterior distribution of the transformation parameters and latent variables (Attias 1999, Ueda & Ghahramani 2002). In particular, if we consider the true posterior distribution $p(\mathbf{\Lambda}_{\mathcal{I}_M}, Z|\mathbf{O})$ (we omit conditional variables M and Ψ for simplicity), we obtain the following relationship:

$$\mathrm{KL}\left(q(\mathbf{\Lambda}_{\mathcal{I}_M}, Z) \| p(\mathbf{\Lambda}_{\mathcal{I}_M}, Z|\mathbf{O})\right) = \log p(\mathbf{O}; \Theta, M, \Psi) - \mathcal{F}(M, \Psi). \tag{7.194}$$

This equation means that maximizing the variational lower bound $\mathcal{F}(M, \Psi)$ with respect to $q(\mathbf{\Lambda}_{\mathcal{I}_M}, Z)$ corresponds to minimizing the KL divergence between $q(\mathbf{\Lambda}_{\mathcal{I}_M}, Z)$ and $p(\mathbf{\Lambda}_{\mathcal{I}_M}, Z|\mathbf{O})$ indirectly. Therefore, this optimization leads to finding $q(\mathbf{\Lambda}_{\mathcal{I}_M}, Z)$, which approaches the true posterior distribution.[8]

[8] The following sections assume factorization forms of $q(\mathbf{\Lambda}_{\mathcal{I}_M}, Z)$ to make solutions mathematically tractable. However, this factorization assumption weakens the relationship between the KL divergence and the variational lower bound. For example, if we assume $q(\mathbf{\Lambda}_{\mathcal{I}_M}, Z) = q(\mathbf{\Lambda}_{\mathcal{I}_M})q(Z)$, and focus on the KL divergence between $q(\mathbf{\Lambda}_{\mathcal{I}_M})$ and $p(\mathbf{\Lambda}_{\mathcal{I}_M}|\mathbf{O})$, we obtain the following inequality:

$$\mathrm{KL}\left(q(\mathbf{\Lambda}_{\mathcal{I}_M}) \| p(\mathbf{\Lambda}_{\mathcal{I}_M}|\mathbf{O})\right) \leq \log p(\mathbf{O}; \Theta, M, \Psi) - \mathcal{F}(M, \Psi). \tag{7.195}$$

Compared with Eq. (7.194), the relationship between the KL divergence and the variational lower bound is less direct due to the inequality relationship. In general, the factorization assumption distances optimal variational posteriors from the true posterior within the VB framework.

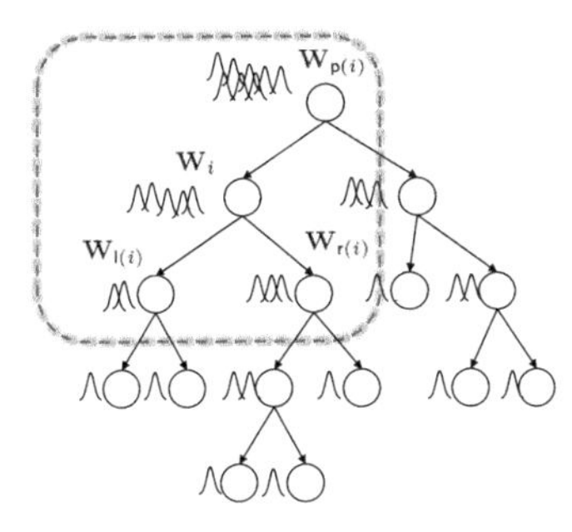

Figure 7.1 Binary tree structure with transformation matrices. If we focus on node i, the transformation matrices in the parent node, left child node, and right child node are represented as $\mathbf{W}_{\mathsf{p}(i)}$, $\mathbf{W}_{\mathsf{l}(i)}$, and $\mathbf{W}_{\mathsf{r}(i)}$, respectively.

Thus, in principle, we can straightforwardly obtain the (sub) optimal model structure, hyperparameters, and posterior distribution, as follows:

$$\begin{aligned}\tilde{m} &= \arg\max_{M} \mathcal{F}(M, \Psi),\\ \tilde{\Psi} &= \arg\max_{\Psi} \mathcal{F}(M, \Psi),\\ \tilde{q}(\mathbf{\Lambda}_{\mathcal{I}_M}, Z) &= \arg\max_{q(\mathbf{\Lambda}_{\mathcal{I}_M}, Z)} \mathcal{F}(M, \Psi).\end{aligned} \tag{7.196}$$

These optimization steps are performed alternately, and finally lead to local optimum solutions, similar to the EM algorithm. However, it is difficult to deal with the joint distribution $q(\mathbf{\Lambda}_{\mathcal{I}_M}, Z)$ directly, and we propose factorizing them by utilizing a Gaussian tree structure. In addition, we also set a conjugate form of the prior distribution $p(\mathbf{\Lambda}_{\mathcal{I}_M})$. This procedure is a typical recipe of VB to make a solution mathematically tractable similarly to that of the classical Bayesian adaptation approach.

Structural prior distribution setting in a binary tree

We utilize a Gaussian tree structure to factorize the prior distribution $p(\mathbf{\Lambda}_{\mathcal{I}_M})$. We consider a binary tree structure, but the formulation is applicable to a general non-binary tree. We define the parent node of i as $\mathsf{p}(i)$, the left child node of i as $\mathsf{l}(i)$, and the right child node of i as $\mathsf{r}(i)$, as shown in Figure 7.1, where a transformation matrix is prepared for each corresponding node i. If we define $\mathbf{W}_1$ as the transformation matrix in the root node, we assume the following factorization for the hierarchical prior distribution $p(\mathbf{\Lambda}_{\mathcal{I}_M})$:

$$\begin{aligned}p(\mathbf{\Lambda}_{\mathcal{I}_M}) &= p(\mathbf{W}_1, \cdots, \mathbf{W}_{|\mathcal{I}_M|})\\ &= p(\mathbf{W}_1)p(\mathbf{W}_{\mathsf{l}(1)}|\mathbf{W}_1)p(\mathbf{W}_{\mathsf{r}(1)}|\mathbf{W}_1)\\ &\quad p(\mathbf{W}_{\mathsf{l}(\mathsf{l}(1))}|\mathbf{W}_{\mathsf{l}(1)})p(\mathbf{W}_{\mathsf{r}(\mathsf{l}(1))}|\mathbf{W}_{\mathsf{l}(1)})\\ &\quad p(\mathbf{W}_{\mathsf{l}(\mathsf{r}(1))}|\mathbf{W}_{\mathsf{r}(1)})p(\mathbf{W}_{\mathsf{r}(\mathsf{r}(1))}|\mathbf{W}_{\mathsf{r}(1)})\cdots\\ &= \prod_{i\in\mathcal{I}_M} p(\mathbf{W}_i|\mathbf{W}_{\mathsf{p}(i)}).\end{aligned} \tag{7.197}$$

To make the prior distribution a product form in the last line of Eq. (7.197), we define $p(\mathbf{W}_1) \triangleq p(\mathbf{W}_1|\mathbf{W}_{\mathrm{p}(1)})$. As seen, the effect of the transformation matrix in a target node propagates to its child nodes.

This hierarchical prior setting is based on an intuitive assumption that the statistics in a target node are highly correlated with the statistics in its parent node. In addition, since the total number of speech frames assigned to a set of Gaussians in the parent node is always larger than that in the target node, the statistics obtained in the parent node are more reliable than in the target node, and these can be good prior knowledge for the transformation parameter estimation in the target node.

With a Bayesian approach, we need to set a practical form of the above prior distributions. A conjugate distribution in Section 2.1.4 is preferable as far as obtaining an analytical solution is concerned, and we set a matrix variate Gaussian distribution similar to maximum a-posteriori linear regression (MAPLR (Chesta *et al.* 1999)). A matrix variate Gaussian distribution is defined in Appendix C.9 as follows:

$$\begin{aligned} p(\mathbf{W}_i) &= \mathcal{N}(\mathbf{W}_i|\mathbf{M}_i, \boldsymbol{\Phi}_i, \boldsymbol{\Omega}_i) \\ &\triangleq C_{\mathcal{N}}(\boldsymbol{\Phi}_i, \boldsymbol{\Omega}_i) \exp\left(-\frac{1}{2}\mathrm{tr}\left[(\mathbf{W}_i - \mathbf{M}_i)^{\intercal}\boldsymbol{\Phi}_i^{-1}(\mathbf{W}_i - \mathbf{M}_i)\boldsymbol{\Omega}_i^{-1}\right]\right), \end{aligned} \tag{7.198}$$

where $C_{\mathcal{N}}(\boldsymbol{\Phi}_i, \boldsymbol{\Omega}_i)$ is a normalization constant defined as:

$$C_{\mathcal{N}}(\boldsymbol{\Phi}_i, \boldsymbol{\Omega}_i) \triangleq (2\pi)^{-\frac{D(D+1)}{2}} |\boldsymbol{\Omega}_i|^{-\frac{D}{2}} |\boldsymbol{\Phi}_i|^{-\frac{D+1}{2}}. \tag{7.199}$$

$\mathbf{M}_i$ is a $D \times (D+1)$ location matrix, $\boldsymbol{\Omega}_i$ is a $(D+1) \times (D+1)$ symmetric scale matrix, and $\boldsymbol{\Phi}_i$ is a $D \times D$ symmetric scale matrix. The term $\boldsymbol{\Omega}_i$ represents correlation of column vectors, and $\boldsymbol{\Phi}_i$ represents correlation of raw vectors. These are hyperparameters of the matrix variate Gaussian distribution. There are many hyperparameters to be set, and this makes the implementation complicated. In this section, we try to find another conjugate distribution with fewer hyperparameters than Eq. (7.198). To obtain a simple solution for the final analytical results, we use a spherical Gaussian distribution that has the following constraints on $\boldsymbol{\Omega}_i$ and $\boldsymbol{\Phi}_i$:

$$\begin{aligned} \boldsymbol{\Phi}_i &\approx \mathbf{I}_D, \\ \boldsymbol{\Omega}_i &\approx \rho_i^{-1}\mathbf{I}_{D+1}, \end{aligned} \tag{7.200}$$

where $\mathbf{I}_D$ is the $D \times D$ identity matrix and ρ_i indicates a precision parameter. Then Eq. (7.198) can be rewritten as follows:

$$\begin{aligned} &\mathcal{N}(\mathbf{W}_i|\mathbf{M}_i, \mathbf{I}_D, \rho_i^{-1}\mathbf{I}_{D+1}) \\ &\quad = C_{\mathcal{N}}(\mathbf{I}_D, \rho_i^{-1}\mathbf{I}_{D+1}) \exp\left(-\frac{1}{2}\mathrm{tr}\left[\rho_i(\mathbf{W}_i - \mathbf{M}_i)^{\intercal}(\mathbf{W}_i - \mathbf{M}_i)\right]\right), \end{aligned} \tag{7.201}$$

where $C_{\mathcal{N}}(\mathbf{I}_D, \rho_i^{-1}\mathbf{I}_{D+1})$ is a normalization factor, and is defined as

$$C_{\mathcal{N}}(\mathbf{I}_D, \rho_i^{-1}\mathbf{I}_{D+1}) \triangleq \left(\frac{\rho_i}{2\pi}\right)^{\frac{D(D+1)}{2}}. \tag{7.202}$$

This approximation means that matrix elements do not have any correlation with each other. This can produce simple solutions for Bayesian linear regression.[9]

Based on the spherical matrix variate Gaussian distribution, the conditional prior distribution $p(\mathbf{W}_i|\mathbf{W}_{\mathsf{p}(i)})$ in Eq. (7.197) is obtaining by setting the location matrix as the transformation matrix $\mathbf{W}_{\mathsf{p}(i)}$ in the parent node with the precision parameter ρ_i as follows:

$$p(\mathbf{W}_i|\mathbf{W}_{\mathsf{p}(i)}) = \mathcal{N}(\mathbf{W}_i|\mathbf{W}_{\mathsf{p}(i)}, \mathbf{I}_D, \rho_i^{-1}\mathbf{I}_{D+1}). \tag{7.204}$$

Note that in the following sections $\mathbf{W}_i$ and $\mathbf{W}_{\mathsf{p}(i)}$ are marginalized. In addition, we set the location matrix in the root node as the deterministic value of $\mathbf{W}_{\mathsf{p}(1)} = [\mathbf{0}, \mathbf{I}_\mathbf{D}]$. Since $\boldsymbol{\mu}_k^{ad} = \mathbf{C}_k\mathbf{W}_{\mathsf{p}(1)}\boldsymbol{\xi}_k = \boldsymbol{\mu}_k^{ini}$ from Eq. (7.191), this hyperparameter setting means that the initial mean vectors are not changed if we only use the prior knowledge. This makes sense in the case of a small amount of data by fixing the HMM parameters as their initial values; this in a sense also inherits the philosophical background of Bayesian adaptation, although the objective function has been changed from a-posteriori probability to a lower bound of the marginal likelihood. Therefore, we just have $\{\rho_i|i = 1, \cdots, |\mathcal{I}_M|\}$ as a set of hyperparameters Ψ, which will also be optimized in our framework.

Algorithm 12 Generative process of structural Bayesian transformation of CDHMM

Require: Ψ and Θ
1: Draw $\boldsymbol{\Lambda}_{\mathcal{I}_M}$ from $p(\boldsymbol{\Lambda}_{\mathcal{I}_M})$
2: Update Θ^{ad} from transformation matrices in leaf nodes $\boldsymbol{\Lambda}_{\mathcal{J}_M}$
3: Draw $\mathbf{O}$ from CDHMM with Θ^{ad}

Variational calculus

In VB, we also assume the following factorization form for the posterior distribution $q(Z, \boldsymbol{\Lambda}_{\mathcal{I}_M})$:

$$q(Z, \boldsymbol{\Lambda}_{\mathcal{I}_M}) = q(Z)q(\boldsymbol{\Lambda}_{\mathcal{I}_M}) = q(Z) \prod_{i \in \mathcal{I}_M} q(\mathbf{W}_i). \tag{7.205}$$

Then, from the general variational calculation for $\mathcal{F}(M, \Psi)$ with respect to $q(\mathbf{W}_i)$ based on Eq. (7.25), we obtain the following (sub) optimal solution for $q(\mathbf{W}_i)$:

[9] A matrix variate Gaussian distribution in Eq. (7.198) is also represented by the following multivariate Gaussian distribution (Dawid 1981):

$$\begin{aligned}&\mathcal{N}(\mathbf{W}_i|\mathbf{M}_i, \boldsymbol{\Phi}_i, \boldsymbol{\Omega}_i)\\ &\propto \exp\left(-\frac{1}{2}\,\mathrm{vec}(\mathbf{W}_i - \mathbf{M}_i)^\mathsf{T}(\boldsymbol{\Omega}_i \otimes \boldsymbol{\Phi}_i)^{-1}\,\mathrm{vec}(\mathbf{W}_i - \mathbf{M}_i)^{-1}\right),\end{aligned} \tag{7.203}$$

where $\mathrm{vec}(\mathbf{W}_i - \mathbf{M}_i)$ is a vector formed by the concatenation of the columns of $(\mathbf{W}_i - \mathbf{M}_i)$, and $\otimes$ denotes the Kronecker product. Based on this form, a VB solution in this section could be extended without considering the variance normalized representation used (Chien 2002).

$$
\begin{aligned}
&\log \tilde{q}(\mathbf{W}_i) \\
&\quad \propto \mathbb{E}_{(Z,\mathbf{W}_{\setminus i})}\left[\log p(\mathbf{O}, Z, \mathbf{\Lambda}_{\mathcal{I}_M})\right] \\
&\quad \propto \mathbb{E}_{(Z,\mathbf{W}_{\setminus i})}\left[\log p(\mathbf{O}, Z|\mathbf{\Lambda}_{\mathcal{I}_M})\right] + \mathbb{E}_{(\mathbf{W}_{\setminus i})}\left[\log p(\mathbf{\Lambda}_{\mathcal{I}_M})\right].
\end{aligned}
\tag{7.206}
$$

$\mathbf{W}_{\setminus i}$ means a set of transformation matrices at a set of nodes for $\mathcal{I}_M$ that does not include $\mathbf{W}_i$, i.e.,

$$
\mathbf{W}_{\setminus i} = \{\mathbf{W}_{i'} | i' \in \mathcal{I}_M \setminus i\}. \tag{7.207}
$$

Then, by using Eqs. (7.197) for $p(\mathbf{\Lambda}_{\mathcal{I}_M})$ and (7.205) for $q(\mathbf{W}_{\setminus i})$, we can rewrite the equation, as follows:

$$
\begin{aligned}
&\log \tilde{q}(\mathbf{W}_i) \\
&\propto \mathbb{E}_{(\mathbf{W}_{\setminus i})}\left[\log \prod_{i' \in \mathcal{I}_M} p(\mathbf{W}_{i'} | \mathbf{W}_{\mathsf{p}(i')})\right] + \mathbb{E}_{(Z,\mathbf{W}_{\setminus i})}\left[\log p(\mathbf{O}, Z|\mathbf{\Lambda}_{\mathcal{I}_M})\right] \\
&\propto \sum_{i' \in \mathcal{I}_M} \mathbb{E}_{(\mathbf{W}_{\setminus i})}\left[\log p(\mathbf{W}_{i'} | \mathbf{W}_{\mathsf{p}(i')})\right] + \mathbb{E}_{(Z,\mathbf{W}_{\setminus i})}\left[\log p(\mathbf{O}, Z|\mathbf{\Lambda}_{\mathcal{I}_M})\right].
\end{aligned}
\tag{7.208}
$$

Note that the second term depends on two nodes i' and $\mathsf{p}(i')$, and the expectation over i' is not trivial. In this expectation, we can consider the following two cases of variational posterior distributions:

1) Leaf node

We first focus on the initial term of Eq. (7.208). If i is a leaf node, we can disregard the expectation with respect to $\prod_{i' \neq i \in \mathcal{I}_M} q(\mathbf{W}_{i'})$ in the nodes other than the parent node $\mathsf{p}(i)$ of the target leaf node. Thus, we obtain the following simple solution:

$$
\log \tilde{q}(\mathbf{W}_i) \propto \mathbb{E}_{(\mathbf{W}_{\mathsf{p}(i)})}\left[\log p(\mathbf{W}_i | \mathbf{W}_{\mathsf{p}(i)})\right] + \mathbb{E}_{(Z,\mathbf{W}_{\setminus i})}\left[\log p(\mathbf{O}, Z|\mathbf{\Lambda}_{\mathcal{I}_M})\right]. \tag{7.209}
$$

2) Non-leaf node (with child nodes)

Similarly, if i is a non-leaf node, in addition to the parent node $\mathsf{p}(i)$ of the target node, we also have to consider the child nodes $\mathsf{l}(i)$ and $\mathsf{r}(i)$ of the target node for the expectation, as follows:

$$
\begin{aligned}
\log \tilde{q}(\mathbf{W}_i) \propto \quad & \mathbb{E}_{(\mathbf{W}_{\mathsf{p}(i)})}\left[\log p(\mathbf{W}_i | \mathbf{W}_{\mathsf{p}(i)})\right] && (7.210) \\
& + \mathbb{E}_{(\mathbf{W}_{\mathsf{l}(i)})}\left[\log p(\mathbf{W}_{\mathsf{l}(i)} | \mathbf{W}_i)\right] && (7.211) \\
& + \mathbb{E}_{(\mathbf{W}_{\mathsf{r}(i)})}\left[\log p(\mathbf{W}_{\mathsf{r}(i)} | \mathbf{W}_i)\right] && (7.212) \\
& + \mathbb{E}_{(Z,\mathbf{W}_{\setminus i})}\left[\log p(\mathbf{O}, Z|\mathbf{\Lambda}_{\mathcal{I}_M})\right]. && (7.213)
\end{aligned}
$$

In both cases, the posterior distribution of the transformation matrix in the target node depends on those in the parent and child nodes. Therefore, the posterior distributions are iteratively calculated. This inference is known as a variational message passing algorithm (Winn & Bishop 2006), and Eqs. (7.209)–(7.213) are specific solutions of the variational message passing algorithm to a binary tree structure. The next section provides a concrete form of the posterior distribution of the transformation matrix.

Posterior distribution of transformation matrix

We first focus on Eq. (7.213), which is a general form of Eq. (7.209) that has additional terms based on child nodes to Eq. (7.209). Equation (7.213) is based on the expectation with respect to $\prod_{i' \neq i \in \mathcal{I}_M} q(\mathbf{W}_{i'})$ and $q(Z)$. The term with $q(Z)$ is represented as the following expression similar to Eqs. (3.168) and (3.161):

$$
\begin{aligned}
&\mathbb{E}_{(Z)}\left[\log p(\mathbf{O}, Z|\mathbf{\Lambda}_{\mathcal{I}_M})\right] \\
&= \sum_{k=1}^{K}\sum_{t=1}^{T} \gamma_t(k) \log \mathcal{N}(\mathbf{o}_t|\boldsymbol{\mu}_k^{ad}, \mathbf{\Sigma}_k) \\
&= \sum_{i \in \mathcal{I}_M}\left(\sum_{k \in \mathcal{K}_i} \gamma_k \log C_{\mathcal{N}}(\mathbf{\Sigma}_k) - \frac{1}{2}\mathrm{tr}\left[\mathbf{W}_i^{\mathsf{T}}\mathbf{W}_i\mathbf{\Xi}_i - 2\mathbf{W}_i^{\mathsf{T}}\mathbf{Z}_i + \sum_{k \in \mathcal{K}_i} \mathbf{\Sigma}_k^{-1}\mathbf{\Gamma}_k\right]\right).
\end{aligned} \tag{7.214}
$$

This equation is calculated from the sufficient statistics (γ_k, $\mathbf{S}_k$, $\mathbf{\Xi}_i$, and $\mathbf{Z}_i$ in Eqs. (3.169) and (3.167)), that is

$$
\begin{cases}
\gamma_k = \sum_{t=1}^{T} \gamma_t(k), \\
\boldsymbol{\gamma}_k = \sum_{t=1}^{T} \gamma_t(k)\mathbf{o}_t, \\
\mathbf{\Gamma}_k = \sum_{t=1}^{T} \gamma_t(k)\mathbf{o}_t\mathbf{o}_t^{\mathsf{T}},
\end{cases} \tag{7.215}
$$

$$
\begin{cases}
\mathbf{\Xi}_i \triangleq \sum_{k \in \mathcal{K}_i} \boldsymbol{\xi}_k\boldsymbol{\xi}_k^{\mathsf{T}}\gamma_k, \\
\mathbf{Z}_i \triangleq \sum_{k \in \mathcal{K}_i} (\mathbf{C}_k)^{-1}\boldsymbol{\gamma}_k\boldsymbol{\xi}_k^{\mathsf{T}}.
\end{cases} \tag{7.216}
$$

These are computed by the VB-E step (e.g., $\gamma_t(k) = q(v_t = k)$), which is described in the next section. This equation form means that the term can be factorized by node i. This factorization property is important for the following analytic solutions and algorithm. Actually, by considering the expectation with respect to $\prod_{i' \neq i \in \mathcal{I}_M} q(\mathbf{W}_{i'})$, we can integrate out the terms that do not depend on $\mathbf{W}_i$, as follows:

$$
\mathbb{E}_{(\mathbf{W}_{\backslash i})}\left[\mathbb{E}_{(Z)}\left[\log p(\mathbf{O}, Z|\mathbf{\Lambda}_{\mathcal{I}_M})\right]\right] \propto -\frac{1}{2}\mathrm{tr}\left[\mathbf{W}_i^{\mathsf{T}}\mathbf{W}_i\mathbf{\Xi}_i - 2\mathbf{W}_i^{\mathsf{T}}\mathbf{Z}_i\right]. \tag{7.217}
$$

Thus, we can obtain the simple quadratic form for this expectation.

Next, we consider Eq. (7.210). Since we use a conjugate prior distribution, $q(\mathbf{W}_{\mathrm{p}(i)})$ is also represented by the following matrix variate Gaussian distribution as the same distribution family with the prior distribution:

$$
q(\mathbf{W}_{\mathrm{p}(i)}) = \mathcal{N}(\mathbf{W}_{\mathrm{p}(i)}|\mathbf{M}_{\mathrm{p}(i)}, \mathbf{I}_D, \mathbf{\Omega}_{\mathrm{p}(i)}). \tag{7.218}
$$

Note that the posterior distribution has a unique form in that the first covariance matrix is an identity matrix while the second is a symmetric matrix. We discuss this form with the analytical solution, later.

By substituting Eqs. (7.197) and (7.218) into Eq. (7.210), Eq. (7.210) is represented as follows:

$$\begin{aligned}
&\mathbb{E}_{(\mathbf{W}_{\mathrm{p}(i)})}\left[\log p(\mathbf{W}_i|\mathbf{W}_{\mathrm{p}(i)})\right] \\
&= \int \mathcal{N}(\mathbf{W}_{\mathrm{p}(i)}|\mathbf{M}_{\mathrm{p}(i)}, \mathbf{I}_D, \mathbf{\Omega}_{\mathrm{p}(i)}) \log \mathcal{N}(\mathbf{W}_i|\mathbf{W}_{\mathrm{p}(i)}, \mathbf{I}_D, \rho_i^{-1}\mathbf{I}_{D+1}) d\mathbf{W}_{\mathrm{p}(i)}. \qquad (7.219)
\end{aligned}$$

To solve the integral, we use the following matrix distribution formula:

$$\begin{aligned}
\int \mathcal{N}(\mathbf{W}_{\mathrm{p}(i)}|\mathbf{M}_{\mathrm{p}(i)}, \mathbf{I}_D, \mathbf{\Omega}_{\mathrm{p}(i)}) d\mathbf{W}_{\mathrm{p}(i)} &= 1, \\
\int \mathbf{W}_{\mathrm{p}(i)} \mathcal{N}(\mathbf{W}_{\mathrm{p}(i)}|\mathbf{M}_{\mathrm{p}(i)}, \mathbf{I}_D, \mathbf{\Omega}_{\mathrm{p}(i)}) d\mathbf{W}_{\mathrm{p}(i)} &= \mathbf{M}_{\mathrm{p}(i)}. \qquad (7.220)
\end{aligned}$$

Then, by using the concrete form of the prior distribution in Eq. (7.201) and by disregarding the terms that do not depend on $\mathbf{W}_i$, Eq. (7.219) can be solved as the logarithmic function of the matrix variate Gaussian distribution that has the posterior distribution parameter $\mathbf{M}_{\mathrm{p}(i)}$ as a hyperparameter:

$$\begin{aligned}
&\mathbb{E}_{(\mathbf{W}_{\mathrm{p}(i)})}\left[\log p(\mathbf{W}_i|\mathbf{W}_{\mathrm{p}(i)})\right] \\
&\propto \rho_i \int \mathrm{tr}\left[\mathbf{W}_i^{\mathsf{T}} \mathbf{W}_{\mathrm{p}(i)}\right] \mathcal{N}(\mathbf{W}_{\mathrm{p}(i)}|\mathbf{M}_{\mathrm{p}(i)}, \mathbf{I}_D, \mathbf{\Omega}_{\mathrm{p}(i)}) d\mathbf{W}_{\mathrm{p}(i)} \\
&\quad - \frac{\rho_i}{2} \int \mathrm{tr}\left[\mathbf{W}_i^{\mathsf{T}} \mathbf{W}_i\right] \mathcal{N}(\mathbf{W}_{\mathrm{p}(i)}|\mathbf{M}_{\mathrm{p}(i)}, \mathbf{I}_D, \mathbf{\Omega}_{\mathrm{p}(i)}) d\mathbf{W}_{\mathrm{p}(i)} \\
&\propto \rho_i \mathrm{tr}\left[\mathbf{W}_i^{\mathsf{T}} \mathbf{M}_{\mathrm{p}(i)}\right] - \frac{\rho_i}{2} \mathrm{tr}\left[\mathbf{W}_i^{\mathsf{T}} \mathbf{W}_i\right] \\
&\propto \log \mathcal{N}(\mathbf{W}_i|\mathbf{M}_{\mathrm{p}(i)}, \mathbf{I}_D, \rho_i^{-1}\mathbf{I}_{D+1}). \qquad (7.221)
\end{aligned}$$

Similarly, Eqs. (7.211) and (7.212) are solved as follows:

$$\begin{aligned}
\mathbb{E}_{(\mathbf{W}_{\mathrm{l}(i)})}\left[\log p(\mathbf{W}_{\mathrm{l}(i)}|\mathbf{W}_i)\right] &\propto \log \mathcal{N}(\mathbf{W}_i|\mathbf{M}_{\mathrm{l}(i)}, \mathbf{I}_D, \rho_{\mathrm{l}(i)}^{-1}\mathbf{I}_{D+1}), \\
\mathbb{E}_{(\mathbf{W}_{\mathrm{r}(i)})}\left[\log p(\mathbf{W}_{\mathrm{r}(i)}|\mathbf{W}_i)\right] &\propto \log \mathcal{N}(\mathbf{W}_i|\mathbf{M}_{\mathrm{r}(i)}, \mathbf{I}_D, \rho_{\mathrm{r}(i)}^{-1}\mathbf{I}_{D+1}). \qquad (7.222)
\end{aligned}$$

Thus, the expected value terms of the three prior distributions in Eq. (7.210) are represented as the following matrix variate Gaussian distribution:

$$\begin{aligned}
&\mathbb{E}_{(\mathbf{W}_{\mathrm{p}(i)})}\left[\log p(\mathbf{W}_i|\mathbf{W}_{\mathrm{p}(i)})\right] + \mathbb{E}_{(\mathbf{W}_{\mathrm{l}(i)})}\left[\log p(\mathbf{W}_{\mathrm{l}(i)}|\mathbf{W}_i)\right] \\
&\quad + \mathbb{E}_{(\mathbf{W}_{\mathrm{r}(i)})}\left[\log p(\mathbf{W}_{\mathrm{r}(i)}|\mathbf{W}_i)\right] \\
&\propto \log \mathcal{N}(\mathbf{W}_i|\mathbf{M}_{\mathrm{p}(i)}, \mathbf{I}_D, \rho_i^{-1}\mathbf{I}_{D+1}) + \log \mathcal{N}(\mathbf{W}_i|\mathbf{M}_{\mathrm{l}(i)}, \mathbf{I}_D, \rho_{\mathrm{l}(i)}^{-1}\mathbf{I}_{D+1}) \\
&\quad + \log \mathcal{N}(\mathbf{W}_i|\mathbf{M}_{\mathrm{r}(i)}, \mathbf{I}_D, \rho_{\mathrm{r}(i)}^{-1}\mathbf{I}_{D+1}) \\
&\propto \rho_i \mathrm{tr}\left[\mathbf{W}_i^{\mathsf{T}} \mathbf{M}_{\mathrm{p}(i)}\right] - \frac{\rho_i}{2} \mathrm{tr}\left[\mathbf{W}_i^{\mathsf{T}} \mathbf{W}_i\right] + \rho_{\mathrm{l}(i)} \mathrm{tr}\left[\mathbf{W}_i^{\mathsf{T}} \mathbf{M}_{\mathrm{l}(i)}\right] - \frac{\rho_{\mathrm{l}(i)}}{2} \mathrm{tr}\left[\mathbf{W}_i^{\mathsf{T}} \mathbf{W}_i\right] \\
&\quad + \rho_{\mathrm{r}(i)} \mathrm{tr}\left[\mathbf{W}_i^{\mathsf{T}} \mathbf{M}_{\mathrm{r}(i)}\right] - \frac{\rho_{\mathrm{r}(i)}}{2} \mathrm{tr}\left[\mathbf{W}_i^{\mathsf{T}} \mathbf{W}_i\right]
\end{aligned}$$

$$\propto -\frac{\rho_i + \rho_{\mathsf{l}(i)} + \rho_{\mathsf{r}(i)}}{2} \mathrm{tr}\left[\mathbf{W}_i^{\mathsf{T}} \mathbf{W}_i\right] + \mathrm{tr}\left[\mathbf{W}_i^{\mathsf{T}} \left(\rho_i \mathbf{M}_{\mathsf{p}(i)} + \rho_{\mathsf{l}(i)} \mathbf{M}_{\mathsf{l}(i)} + \rho_{\mathsf{r}(i)} \mathbf{M}_{\mathsf{r}(i)}\right)\right]$$
$$\propto \log \mathcal{N}\left(\mathbf{W}_i \middle| \frac{\rho_i \mathbf{M}_{\mathsf{p}(i)} + \rho_{\mathsf{l}(i)} \mathbf{M}_{\mathsf{l}(i)} + \rho_{\mathsf{r}(i)} \mathbf{M}_{\mathsf{r}(i)}}{\rho_i + \rho_{\mathsf{l}(i)} + \rho_{\mathsf{r}(i)}}, \mathbf{I}_D, (\rho_i + \rho_{\mathsf{l}(i)} + \rho_{\mathsf{r}(i)})^{-1} \mathbf{I}_{D+1}\right). \tag{7.223}$$

It is an intuitive solution, since the location parameter $\mathbf{W}_i$ is represented as a linear interpolation of the location values of the posterior distributions in the parent and child nodes. The precision parameters control the linear interpolation ratio.

Similarly, we can also obtain the expected value term of the prior term in Eq. (7.209), and we summarize the prior terms of the non-leaf and leaf node cases as follows:

$$\hat{q}(\mathbf{W}_i) = \mathcal{N}(\mathbf{W}_i | \hat{\mathbf{M}}_i, \mathbf{I}_D, \hat{\rho}_i^{-1} \mathbf{I}_{D+1}), \tag{7.224}$$

where

$$\hat{\mathbf{M}}_i = \begin{cases} \frac{\rho_i \mathbf{M}_{\mathsf{p}(i)} + \rho_{\mathsf{l}(i)} \mathbf{M}_{\mathsf{l}(i)} + \rho_{\mathsf{r}(i)} \mathbf{M}_{\mathsf{r}(i)}}{\rho_i + \rho_{\mathsf{l}(i)} + \rho_{\mathsf{r}(i)}} & \text{Non-leaf node,} \\ \mathbf{M}_{\mathsf{p}(i)} & \text{Leaf node,} \end{cases}$$
$$\hat{\rho}_i = \begin{cases} \rho_i + \rho_{\mathsf{l}(i)} + \rho_{\mathsf{r}(i)} & \text{Non-leaf node,} \\ \rho_i & \text{Leaf node.} \end{cases} \tag{7.225}$$

Thus, the effect of prior distributions becomes different depending on whether the target node is a non-leaf node or leaf node. The solution is different from our previous solution (Watanabe, Nakamura & Juang 2011), since the previous solution does not marginalize the transformation parameters in non-leaf nodes. In the Bayesian sense, this solution is stricter than the previous solution.

Based on Eqs. (7.214) and (7.224), we can finally derive the quadratic form of $\mathbf{W}_i$ as follows:

$$\log \tilde{q}(\mathbf{W}_i) \propto -\frac{1}{2} \mathrm{tr}\left[\hat{\rho}_i \mathbf{W}_i^{\mathsf{T}} \mathbf{W}_i + \mathbf{W}_i^{\mathsf{T}} \mathbf{W}_i \boldsymbol{\Xi}_i - 2\hat{\rho}_i \mathbf{W}_i^{\mathsf{T}} \hat{\mathbf{M}}_i - 2\mathbf{W}_i^{\mathsf{T}} \mathbf{Z}_i\right]$$
$$\propto -\frac{1}{2} \mathrm{tr}\left[\mathbf{W}_i^{\mathsf{T}} \mathbf{W}_i (\hat{\rho}_i \mathbf{I}_{D+1} + \boldsymbol{\Xi}_i) - 2\mathbf{W}_i^{\mathsf{T}} (\hat{\rho}_i \hat{\mathbf{M}}_i + \mathbf{Z}_i)\right], \tag{7.226}$$

where we disregard the terms that do not depend on $\mathbf{W}_i$. Thus, by defining the following matrix variables:

$$\tilde{\boldsymbol{\Omega}}_i = \left(\hat{\rho}_i \mathbf{I}_{D+1} + \boldsymbol{\Xi}_i\right)^{-1}$$
$$= \begin{cases} \left((\rho_i + \rho_{\mathsf{l}(i)} + \rho_{\mathsf{r}(i)}) \mathbf{I}_{D+1} + \boldsymbol{\Xi}_i\right)^{-1} & \text{Non-leaf node,} \\ (\rho_i \mathbf{I}_{D+1} + \boldsymbol{\Xi}_i)^{-1} & \text{Leaf node,} \end{cases}$$
$$\tilde{\mathbf{M}}_i = \left(\hat{\rho}_i \hat{\mathbf{M}}_i + \mathbf{Z}_i\right) \tilde{\boldsymbol{\Omega}}$$
$$= \begin{cases} \left(\rho_i \mathbf{M}_{\mathsf{p}(i)} + \rho_{\mathsf{l}(i)} \mathbf{M}_{\mathsf{l}(i)} + \rho_{\mathsf{r}(i)} \mathbf{M}_{\mathsf{r}(i)} + \mathbf{Z}_i\right) \tilde{\boldsymbol{\Omega}} & \text{Non-leaf node,} \\ \left(\rho_i \mathbf{M}_{\mathsf{p}(i)} + \mathbf{Z}_i\right) \tilde{\boldsymbol{\Omega}} & \text{Leaf node,} \end{cases} \tag{7.227}$$

we can derive the posterior distribution of $\mathbf{W}_i$ analytically. The analytical solution is expressed as

$$\tilde{q}(\mathbf{W}_i) = \mathcal{N}(\mathbf{W}_i|\tilde{\mathbf{M}}_i, \mathbf{I}_D, \tilde{\boldsymbol{\Omega}}_i)$$
$$= C_{\mathcal{N}}(\mathbf{I}_D, \tilde{\boldsymbol{\Omega}}_i) \exp\left(-\frac{1}{2}\mathrm{tr}\left[(\mathbf{W}_i - \tilde{\mathbf{M}}_i)^\intercal(\mathbf{W}_i - \tilde{\mathbf{M}}_i)\tilde{\boldsymbol{\Omega}}_i^{-1}\right]\right), \quad (7.228)$$

where

$$C_{\mathcal{N}}(\mathbf{I}_D, \tilde{\boldsymbol{\Omega}}_i) = (2\pi)^{-\frac{D(D+1)}{2}} |\tilde{\boldsymbol{\Omega}}_i|^{-\frac{D}{2}}. \quad (7.229)$$

The posterior distribution also becomes a matrix variate Gaussian distribution, since we use a conjugate prior distribution for $\mathbf{W}_i$. From Eq. (7.227), $\tilde{\mathbf{M}}_i$ are linearly interpolated by hyperparameter $\hat{\mathbf{M}}_i$ and the first-order statistics of the linear regression matrix $\mathbf{Z}_i$. $\hat{\rho}_i$ controls the balance between the effects of the prior distribution and adaptation data. This solution is the M-step of the VB–EM algorithm, and corresponds to that of the ML–EM algorithm in Section 3.5.

Compared with Eq. (7.201), Eq. (7.228) keeps the first covariance matrix as a diagonal matrix, while the second covariance matrix $\tilde{\boldsymbol{\Omega}}$ has off-diagonal elements. This means that the posterior distribution only considers the correlation between column vectors in $\mathbf{W}$. This unique property comes from the variance normalized representation introduced in Section 3.5, which makes multivariate Gaussian distributions in HMMs uncorrelated, and this relationship is taken over to the VB solutions.

Although the solution for a non-leaf node would make the prior distribution robust by taking account of the child node hyperparameters, this structure makes the dependency of the target node on the other linked nodes complex. Therefore, in the implementation step, we approximate the hyperparameters of the posterior distribution for a non-leaf node to those for a leaf node by $\hat{\mathbf{M}}_i \approx \mathbf{M}_{\mathrm{p}(i)}$ and $\hat{\rho}_i \approx \rho_i$ in the Eq. (7.225), and this makes an algorithm simple.

The next section explains the E-step of the VB–EM algorithm, which computes sufficient statistics γ_k, $\boldsymbol{\Gamma}_k$, $\boldsymbol{\Xi}_i$, and $\mathbf{Z}_i$ in Eqs. (3.169) and (3.167). These are obtained by using $\tilde{q}(\mathbf{W}_i)$, of which mode $\tilde{\mathbf{M}}_i$ is used for the Gaussian mean vector transformation.

Posterior distribution of latent variables

From the variational calculation of $\mathcal{F}(M, \Psi)$ with respect to $q(Z)$ based on Eq. (7.25), we also obtain the following posterior distribution:

$$\log \tilde{q}(Z) \propto \mathbb{E}_{(\boldsymbol{\Lambda}_{\mathcal{I}_M})}\left[\log p(\mathbf{O}, Z|\boldsymbol{\Lambda}_{\mathcal{I}_M})\right]. \quad (7.230)$$

By using the factorization form of the variational posterior (Eq. (7.205)), we can disregard the expectation with respect to the variational posteriors other than that of the target node i. Therefore, to obtain the above VB posteriors of latent variables, we have to consider the following integral:

$$\int \tilde{q}(\mathbf{W}_i) \log \mathcal{N}(\mathbf{o}_t|\mathbf{C}_k\mathbf{W}_i\boldsymbol{\xi}_k, \boldsymbol{\Sigma}_k) d\mathbf{W}_i. \quad (7.231)$$

Since the Gaussian mean vectors are only updated in the leaf nodes, node i in this section is regarded as a leaf node. By substituting Eqs. (7.228) and (3.162) into Eq. (7.231), the equation can be represented as:

$$\int \tilde{q}(\mathbf{W}_i) \log \mathcal{N}(\mathbf{o}_t | \mathbf{C}_k \mathbf{W}_i \boldsymbol{\xi}_k, \boldsymbol{\Sigma}_k) d\mathbf{W}_i$$
$$= \log \mathcal{N}(\mathbf{o}_t | \tilde{\boldsymbol{\mu}}_k, \boldsymbol{\Sigma}_k) - \frac{1}{2} \mathrm{tr}\left[\boldsymbol{\xi}_k \boldsymbol{\xi}_k^\mathsf{T} \tilde{\boldsymbol{\Omega}}_i \right]. \tag{7.232}$$

where

$$\tilde{\boldsymbol{\mu}}_k = \mathbf{C}_k \tilde{\mathbf{M}}_i \boldsymbol{\xi}_k. \tag{7.233}$$

The analytical result is almost equivalent to the E-step of conventional MLLR, which means that the computation time is almost the same as that of the conventional MLLR E-step.

We derive the posterior distribution of latent variables $\tilde{q}(Z)$, introduced in Section 7.4.3, based on the VB framework. In this derivation, we omit indexes i, k, and t for simplicity. By substituting the concrete form (Eq. (3.162)) of the multivariate Gaussian distribution into Eq. (7.231), the equation can be represented as:

$$\int \tilde{q}(\mathbf{W}) \log \mathcal{N}(\mathbf{o} | \mathbf{C}\mathbf{W}\boldsymbol{\xi}, \boldsymbol{\Sigma}) d\mathbf{W}$$
$$= -\frac{D}{2} \log(2\pi |\boldsymbol{\Sigma}|) - \frac{1}{2} \int \tilde{q}(\mathbf{W}) \underbrace{\left((\mathbf{o} - \mathbf{C}\mathbf{W}\boldsymbol{\xi})^\mathsf{T} \boldsymbol{\Sigma}^{-1} (\mathbf{o} - \mathbf{C}\mathbf{W}\boldsymbol{\xi}) \right)}_{(*1)} d\mathbf{W}, \tag{7.234}$$

where we use the following equation for the normalization term:

$$\int \tilde{q}(\mathbf{W}) d\mathbf{W} = 1. \tag{7.235}$$

Let us now focus on the quadratic form $(*1)$ of Eq. (7.234). By considering $\boldsymbol{\Sigma} = \mathbf{C}(\mathbf{C})^\mathsf{T}$ in Eq. (3.164), $(*1)$ can be rewritten as follows:

$$(*1) = (\mathbf{C}^{-1}\mathbf{o} - \mathbf{W}\boldsymbol{\xi})^\mathsf{T} (\mathbf{C}^{-1}\mathbf{o} - \mathbf{W}\boldsymbol{\xi})$$
$$= \mathrm{tr}\left[(\mathbf{C}^{-1}\mathbf{o} - \mathbf{W}\boldsymbol{\xi})(\mathbf{C}^{-1}\mathbf{o} - \mathbf{W}\boldsymbol{\xi})^\mathsf{T} \right]$$
$$= \mathrm{tr}\left[\mathbf{R}\mathbf{W}^\mathsf{T}\mathbf{W} - 2\mathbf{W}\mathbf{Y}^\mathsf{T} + \mathbf{U} \right], \tag{7.236}$$

where we use the fact that the trace of the scalar value is equal to the original scalar value and the cyclic property of the trace in Appendix B:

$$a = \mathrm{tr}[a], \tag{7.237}$$

$$\mathrm{tr}[\mathbf{ABC}] = \mathrm{tr}[\mathbf{BCA}]. \tag{7.238}$$

We also define $(D+1) \times (D+1)$ matrix $\mathbf{R}$, $D \times (D+1)$ matrix $\mathbf{Y}$, and $D \times D$ matrix $\mathbf{U}$ in Eq. (7.236) as follows:

$$\mathbf{R} \triangleq \boldsymbol{\xi}\boldsymbol{\xi}^\mathsf{T},$$
$$\mathbf{Y} \triangleq \mathbf{C}^{-1}\mathbf{o}\boldsymbol{\xi}^\mathsf{T},$$
$$\mathbf{U} \triangleq \boldsymbol{\Sigma}^{-1}\mathbf{o}\mathbf{o}^\mathsf{T}. \tag{7.239}$$

The integral of Eq. (7.236) over $\mathbf{W}$ can be decomposed into the following three terms:

$$\int \tilde{q}(\mathbf{W})\mathrm{tr}\left[\mathbf{R}\mathbf{W}^{\mathsf{T}}\mathbf{W} - 2\mathbf{W}\mathbf{Y}^{\mathsf{T}} + \mathbf{U}\right]d\mathbf{W}$$
$$= \underbrace{\int \tilde{q}(\mathbf{W})\mathrm{tr}\left[\mathbf{R}\mathbf{W}^{\mathsf{T}}\mathbf{W}\right]d\mathbf{W}}_{(*2)} - 2\underbrace{\int \tilde{q}(\mathbf{W})\mathrm{tr}\left[\mathbf{W}\mathbf{Y}^{\mathsf{T}}\right]d\mathbf{W}}_{(*3)} + \mathrm{tr}\left[\mathbf{U}\right], \tag{7.240}$$

where we use the distributive property of the trace in Appendix B:

$$\mathrm{tr}[\mathbf{A}(\mathbf{B}+\mathbf{C})] = \mathrm{tr}[\mathbf{A}\mathbf{B}+\mathbf{A}\mathbf{C}], \tag{7.241}$$

and use Eq. (7.235) in the third term of the second line in Eq. (7.240).

We focus on the integrals $(*2)$ and $(*3)$. Since $\tilde{q}(\mathbf{W})$ is a scalar value, $(*3)$ can be rewritten as follows:

$$(*3) = \int \mathrm{tr}\left[\tilde{q}(\mathbf{W})\mathbf{W}\mathbf{Y}^{\mathsf{T}}\right]d\mathbf{W}$$
$$= \mathrm{tr}\left[\int \tilde{q}(\mathbf{W})\mathbf{W}\mathbf{Y}^{\mathsf{T}}d\mathbf{W}\right]. \tag{7.242}$$

Here, we use the following matrix properties:

$$\mathrm{tr}[a\mathbf{A}] = a\,\mathrm{tr}[\mathbf{A}], \tag{7.243}$$

$$\int \mathrm{tr}[f(\mathbf{A})]d\mathbf{A} = \mathrm{tr}\left[\int f(\mathbf{A})d\mathbf{A}\right]. \tag{7.244}$$

Thus, the integral is finally solved as

$$(*3) = \mathrm{tr}\left[\left(\int \tilde{q}(\mathbf{W})\mathbf{W}d\mathbf{W}\right)\mathbf{Y}^{\mathsf{T}}\right]$$
$$= \mathrm{tr}\left[\tilde{\mathbf{M}}\mathbf{Y}^{\mathsf{T}}\right], \tag{7.245}$$

where we use

$$\int \tilde{q}(\mathbf{W})\mathbf{W}d\mathbf{W} = \tilde{\mathbf{M}}. \tag{7.246}$$

Similarly, we also rewrite $(*2)$ in Eq. (7.240) based on Eqs. (7.243) and (7.244) as follows:

$$(*2) = \int \mathrm{tr}\left[\tilde{q}(\mathbf{W})\mathbf{R}\mathbf{W}^{\mathsf{T}}\mathbf{W}\right]d\mathbf{W}$$
$$= \mathrm{tr}\left[\int \tilde{q}(\mathbf{W})\mathbf{R}\mathbf{W}^{\mathsf{T}}\mathbf{W}d\mathbf{W}\right]$$
$$= \mathrm{tr}\left[\mathbf{R}\int \tilde{q}(\mathbf{W})\mathbf{W}^{\mathsf{T}}\mathbf{W}d\mathbf{W}\right]. \tag{7.247}$$

Thus, the integral is finally solved as

$$(*2) = \mathrm{tr}\left[\mathbf{R}\left(\tilde{\boldsymbol{\Omega}} + \tilde{\mathbf{M}}^{\mathsf{T}}\tilde{\mathbf{M}}\right)\right], \tag{7.248}$$

where we use

$$\int \tilde{q}(\mathbf{W})\mathbf{W}^\mathsf{T}\mathbf{W}d\mathbf{W} = \tilde{\mathbf{\Omega}} + \tilde{\mathbf{M}}^\mathsf{T}\tilde{\mathbf{M}}. \tag{7.249}$$

Thus, we have solved all the integrals in Eq. (7.240).

Finally, we substitute the integral results of (∗2) and (∗3) (i.e., Eqs. (7.248) and (7.245)) into Eq. (7.240), and rewrite Eq. (7.240) based on the concrete forms of **R**, **Y**, and **U** defined in Eq. (7.239) as follows:

$$\begin{aligned} &\text{Eq. (7.240)} \\ &= \mathrm{tr}\left[\mathbf{R}\left(\tilde{\mathbf{\Omega}} + \tilde{\mathbf{M}}^\mathsf{T}\tilde{\mathbf{M}}\right) - 2\tilde{\mathbf{M}}\mathbf{Y}^\mathsf{T} + \mathbf{U}\right] \\ &= \mathrm{tr}\left[\boldsymbol{\xi}\boldsymbol{\xi}^\mathsf{T}(\tilde{\mathbf{\Omega}} + \tilde{\mathbf{M}}^\mathsf{T}\tilde{\mathbf{M}}) - 2\tilde{\mathbf{M}}\boldsymbol{\xi}\mathbf{o}^\mathsf{T}(\mathbf{C}^{-1})^\mathsf{T} + \mathbf{\Sigma}^{-1}\mathbf{o}\mathbf{o}^\mathsf{T}\right]. \end{aligned} \tag{7.250}$$

Then, by using the cyclic property in Eq. (7.238) and $\mathbf{\Sigma} = \mathbf{C}(\mathbf{C})^\mathsf{T}$ in Eq. (3.164), we can further rewrite Eq. (7.240) as follows:

$$\begin{aligned} &\text{Eq. (7.240)} \\ &= \mathrm{tr}\left[\boldsymbol{\xi}\boldsymbol{\xi}^\mathsf{T}\tilde{\mathbf{\Omega}} + \mathbf{\Sigma}^{-1}\left(\mathbf{\Sigma}\tilde{\mathbf{M}}\boldsymbol{\xi}\boldsymbol{\xi}^\mathsf{T}\tilde{\mathbf{M}}^\mathsf{T} - 2\mathbf{C}\tilde{\mathbf{M}}\boldsymbol{\xi}\mathbf{o}^\mathsf{T} + \mathbf{o}\mathbf{o}^\mathsf{T}\right)\right] \\ &= \mathrm{tr}\left[\boldsymbol{\xi}\boldsymbol{\xi}^\mathsf{T}\tilde{\mathbf{\Omega}} + \mathbf{\Sigma}^{-1}\left(\mathbf{o} - \mathbf{C}\tilde{\mathbf{M}}\boldsymbol{\xi}\right)\left(\mathbf{o} - \mathbf{C}\tilde{\mathbf{M}}\boldsymbol{\xi}\right)^\mathsf{T}\right]. \end{aligned} \tag{7.251}$$

Thus, we obtain the quadratic form with respect to **o**, which becomes a multivariate Gaussian distribution form. By recovering the omitted indexes i, k, and t, and substituting the integral result in Eq. (7.251) into Eq. (7.234), we finally solve Eq. (7.231) as:

$$\begin{aligned} &\int \tilde{q}(\mathbf{W}_i)\log\mathcal{N}(\mathbf{o}_t|\mathbf{C}_k\mathbf{W}_i\boldsymbol{\xi}_k, \mathbf{\Sigma}_k)d\mathbf{W}_i \\ &= -\frac{D}{2}\log(2\pi|\mathbf{\Sigma}_k|) - \frac{1}{2}\mathrm{tr}\left[\boldsymbol{\xi}_k\boldsymbol{\xi}_k^\mathsf{T}\tilde{\mathbf{\Omega}}_i + (\mathbf{\Sigma}_k)^{-1}\left(\mathbf{o}_t - \mathbf{C}_k\tilde{\mathbf{M}}_i\boldsymbol{\xi}_k\right)\left(\mathbf{o}_t - \mathbf{C}_k\tilde{\mathbf{M}}_i\boldsymbol{\xi}_k\right)^\mathsf{T}\right] \\ &= \log\mathcal{N}(\mathbf{o}_t|\mathbf{C}_k\tilde{\mathbf{M}}_i\boldsymbol{\xi}_k, \mathbf{\Sigma}_k) - \frac{1}{2}\mathrm{tr}\left[\boldsymbol{\xi}_k\boldsymbol{\xi}_k^\mathsf{T}\tilde{\mathbf{\Omega}}_i\right]. \end{aligned} \tag{7.252}$$

Here, we use the concrete form of the multivariate Gaussian distribution in Eq. (3.162).

Note that the Gaussian mean vectors are updated in the leaf nodes in this result, while the posterior distributions of the transformation parameters are updated for all nodes.

Variational lower bound

By using the factorization form (Eq. (7.205)) of the variational posterior distribution, the variational lower bound defined in Eq. (7.193) is decomposed as follows:

$$\begin{aligned} \mathcal{F}(M,\Psi) &= \mathbb{E}_{(Z,\mathbf{\Lambda}_{\mathcal{I}_M})}\left[\log\frac{p(\mathbf{O},Z|\mathbf{\Lambda}_{\mathcal{I}_M})p(\mathbf{\Lambda}_{\mathcal{I}_M})}{q(Z)\prod_{i\in\mathcal{I}_M}q(\mathbf{W}_i)}\right] \\ &= \underbrace{\mathbb{E}_{(Z,\mathbf{\Lambda}_{\mathcal{I}_M})}\left[\log\frac{p(\mathbf{O},Z|\mathbf{\Lambda}_{\mathcal{I}_M})p(\mathbf{\Lambda}_{\mathcal{I}_M})}{\prod_{i\in\mathcal{I}_M}q(\mathbf{W}_i)}\right]}_{\triangleq\mathcal{L}(M,\Psi)} - \mathbb{E}_{(Z)}\left[\log q(Z)\right]. \end{aligned} \tag{7.253}$$

The second term, which contains $q(Z)$, is an entropy value and is calculated at the E-step in the VB–EM algorithm. The first term ($\mathcal{L}(M, \Psi)$) is a logarithmic evidence term for M and $\Psi = \{\rho_i | i = 1, \cdots |\mathcal{I}_M|\}$, and we can obtain an analytical solution of $\mathcal{L}(M, \Psi)$. Because of the factorization forms in Eqs. (7.205), (7.197), and (7.214), $\mathcal{L}(M, \Psi)$ can be represented as the summation over i, as follows:

$$\begin{aligned}\mathcal{L}(M, \Psi) &= \mathbb{E}_{(Z,\mathbf{\Lambda}_{\mathcal{I}_M})}\left[\log \frac{\prod_{i\in\mathcal{I}_M} p(\mathbf{O}, Z|\mathbf{W}_i)p(\mathbf{W}_i|\mathbf{W}_{\mathsf{p}(i)})}{\prod_{i\in\mathcal{I}_M} q(\mathbf{W}_i)}\right] \\ &= \sum_{i\in\mathcal{I}_M} \mathcal{L}_i(\rho_i, \rho_{\mathsf{l}(i)}, \rho_{\mathsf{r}(i)}),\end{aligned} \tag{7.254}$$

where

$$\mathcal{L}_i(\rho_i, \rho_{\mathsf{l}(i)}, \rho_{\mathsf{r}(i)}) \triangleq \sum_{i\in\mathcal{I}_M} \mathbb{E}_{(Z,\mathbf{\Lambda}_{\mathcal{I}_M})}\left[\log \frac{p(\mathbf{O}, Z|\mathbf{W}_i)p(\mathbf{W}_i|\mathbf{W}_{\mathsf{p}(i)})}{q(\mathbf{W}_i)}\right]. \tag{7.255}$$

Note that this factorization form has some dependencies from parent and child node parameters through Eqs. (7.225) and (7.227). To derive an analytical solution, we first consider the expectation with respect to $q(Z)$ only for cluster i. By substituting Eqs. (3.168), (7.201), and (7.228) into $\mathcal{L}_i(\rho_i, \rho_{\mathsf{l}(i)}, \rho_{\mathsf{r}(i)})$, the expectation can be rewritten, as follows:

$$\begin{aligned}&\mathbb{E}_{(Z)}\left[\log \frac{p(\mathbf{O}, Z|\mathbf{W}_i)p(\mathbf{W}_i|\mathbf{W}_{\mathsf{p}(i)})}{q(\mathbf{W}_i)}\right] \\ &= \sum_{k\in\mathcal{K}_i} \gamma_k \log C_{\mathcal{N}}(\mathbf{\Sigma}_k) - \frac{1}{2}\mathrm{tr}\left[\mathbf{W}_i^{\mathsf{T}}\mathbf{W}_i\mathbf{\Xi}_i - 2\mathbf{W}_i^{\mathsf{T}}\mathbf{Z}_i + \sum_{k\in\mathcal{K}_i} \mathbf{\Sigma}_k^{-1}\mathbf{\Gamma}_k\right] \\ &\quad + \log C_{\mathcal{N}}(\mathbf{I}_D, \rho_i^{-1}\mathbf{I}_{D+1}) - \frac{1}{2}\mathrm{tr}\left[\rho_i(\mathbf{W}_i - \mathbf{W}_{\mathsf{p}(i)})^{\mathsf{T}}(\mathbf{W}_i - \mathbf{W}_{\mathsf{p}(i)})\right] \\ &\quad - \log C_{\mathcal{N}}(\mathbf{I}_D, \tilde{\mathbf{\Omega}}_i) + \frac{1}{2}\mathrm{tr}\left[(\mathbf{W}_i - \tilde{\mathbf{M}}_i)^{\mathsf{T}}(\mathbf{W}_i - \tilde{\mathbf{M}}_i)\tilde{\mathbf{\Omega}}_i^{-1}\right] \\ &= \sum_{k\in\mathcal{K}_i} \gamma_k \log C_{\mathcal{N}}(\mathbf{\Sigma}_k) + \log \frac{C_{\mathcal{N}}(\mathbf{I}_D, \hat{\rho}_i^{-1}\mathbf{I}_{D+1})}{C_{\mathcal{N}}(\mathbf{I}_D, \tilde{\mathbf{\Omega}}_i)} + (*).\end{aligned} \tag{7.256}$$

If we consider only the leaf node case, by using Eq. (7.227), the expectation of $(*)$ part can be rewritten as:

$$\begin{aligned}&(*) \text{ in Eq. (7.256)} \\ &\quad = -\frac{1}{2}\mathrm{tr}\left[\hat{\rho}_i\hat{\mathbf{M}}_i^{\mathsf{T}}\hat{\mathbf{M}}_i - \tilde{\mathbf{M}}_i^{\mathsf{T}}\tilde{\mathbf{M}}_i\tilde{\mathbf{\Omega}}_i^{-1} + \sum_{k\in\mathcal{K}_i} \mathbf{\Sigma}_k^{-1}\mathbf{\Gamma}_k\right].\end{aligned} \tag{7.257}$$

The result obtained does not depend on $\mathbf{W}_i$. Therefore, the expectation with respect to $q(\mathbf{W}_i)$ can be disregarded in $\mathcal{L}_i(\rho_i, \rho_{\mathsf{l}(i)}, \rho_{\mathsf{r}(i)})$. Consequently, we can obtain the following analytical result for the lower bound:

$$
\begin{aligned}
&\mathcal{L}_i(\rho_i, \rho_{\mathsf{l}(i)}, \rho_{\mathsf{r}(i)}) \\
&= -\frac{D}{2}\log(2\pi)\sum_{k\in\mathcal{K}_i}\gamma_k - \frac{1}{2}\sum_{k\in\mathcal{K}_i}\gamma_k\log|\boldsymbol{\Sigma}_k| \\
&\quad + \frac{D(D+1)}{2}\log\hat{\rho}_i + \frac{D}{2}\log|\tilde{\boldsymbol{\Omega}}_i| \\
&\quad - \frac{1}{2}\mathrm{tr}\left[\hat{\rho}_i\hat{\mathbf{M}}_i'\hat{\mathbf{M}}_i - \tilde{\mathbf{M}}_i'\tilde{\mathbf{M}}_i\tilde{\boldsymbol{\Omega}}_i^{-1} + \sum_{k\in\mathcal{K}_i}\boldsymbol{\Sigma}_k^{-1}\boldsymbol{\Gamma}_k\right].
\end{aligned}
\tag{7.258}
$$

The first line of the result obtained corresponds to the likelihood value given the amount of data and the covariance matrices of the Gaussians. The other terms consider the effect of the prior and posterior distributions of the model parameters. This result is used as an optimization criterion with respect to the model structure M and the hyperparameters Ψ.

Note that the objective function can be represented as a summation over i because of the factorization form of the prior and posterior distributions. This representation property is used for our model structure optimization in Section 7.4.5 for a binary tree structure representing a set of Gaussians used in the conventional MLLR.

7.4.4 Optimization of hyperparameters and model structure

In this section, we describe how to optimize hyperparameters Ψ and model structure M by using the variational lower bound as an objective function. Once we obtain the variational lower bound, we can obtain an appropriate model structure and hyperparameters that maximize the lower bound at the same time as follows:

$$
\{\widetilde{\Psi}, \widetilde{M}\} = \arg\max_{M,\Psi} \mathcal{F}(M, \Psi). \tag{7.259}
$$

We use two approximations for the variational lower bound to make the inference algorithm practical. First, we fix latent variables Z during the above optimization. Then, $\mathbb{E}_{(Z)}\left[\log q(Z)\right]$ in Eq. (7.253) is also fixed for M and Ψ, and can be disregarded in the objective function. Thus, we can only focus on $\mathcal{L}(M, \Psi)$ in the optimization step, which reduces computational cost greatly, as follows:

$$
\{\widetilde{\Psi}, \widetilde{M}\} \approx \arg\max_{M,\Psi} \mathcal{L}(M, \Psi). \tag{7.260}
$$

This approximation is widely used in acoustic model selection (likelihood criterion (Odell 1995) and Bayesian criterion (Watanabe *et al.* 2004)). Second, as we discussed in Section 7.4.3, the solution for a non-leaf node (Eq. (7.224)) makes the dependency of the target node on the other linked nodes complex. Therefore, we approximate $\mathcal{L}_i(\rho_i, \rho_{\mathsf{l}(i)}, \rho_{\mathsf{r}(i)}) \approx \mathcal{L}_i(\rho_i)$ by $\hat{\rho}_i \approx \rho_i$ and so on, where $\mathcal{L}_i(\rho_i)$ is defined in the next section. Therefore, in the implementation step, we approximate the posterior distribution for a non-leaf node to that for a leaf node to make the algorithm simple.

7.4.5 Hyperparameter optimization

Even though we marginalize all of transformation matrix ($\mathbf{W}_i$), we still have to set the precision hyperparameters ρ_i for all nodes. Since we can derive the variational lower bound, we can optimize the precision hyperparameter, and can remove the manual tuning of the hyperparameters with the proposed approach. This is an advantage of the proposed approach with regard to SMAPLR (Siohan *et al.* 2002), since SMAPLR has to hand-tune its hyperparameters corresponding to $\{\rho_i\}_i$.

Based on the leaf node approximation for variational posterior distributions, in addition to the fixed latent variable approximation ($\mathcal{F}(M, \Psi) \approx \mathcal{L}(M, \Psi)$), in this section the method we implement approximately optimizes the precision hyperparameter as follows:

$$\begin{aligned}
\tilde{\rho}_i &= \arg\max_{\rho_i} \mathcal{L}(M, \Psi) \\
&= \begin{cases} \arg\max_{\rho_i} \left(\mathcal{L}_i(\rho_i, \rho_{\mathsf{l}(i)}, \rho_{\mathsf{r}(i)}) + \mathcal{L}_{\mathsf{p}(i)}(\rho_{\mathsf{p}(i)}, \rho_i, \rho_{\mathsf{r}(\mathsf{p}(i))})\right) \\ \qquad i \text{ is a left child node of } \mathsf{p}(i) \\ \arg\max_{\rho_i} \left(\mathcal{L}_i(\rho_i, \rho_{\mathsf{l}(i)}, \rho_{\mathsf{r}(i)}) + \mathcal{L}_{\mathsf{p}(i)}(\rho_{\mathsf{p}(i)}, \rho_{\mathsf{l}(\mathsf{p}(i))}, \rho_i)\right) \\ \qquad i \text{ is a right child node of } \mathsf{p}(i) \end{cases} \\
&\approx \arg\max_{\rho_i} \mathcal{L}_i(\rho_i),
\end{aligned} \tag{7.261}$$

where

$$\begin{aligned}
\mathcal{L}_i(\rho_i) \triangleq\; & \frac{D(D+1)}{2} \log \rho_i + \frac{D}{2} \log |\tilde{\mathbf{\Omega}}_i| \\
& - \frac{1}{2} \text{tr}\left[\rho_i \mathbf{M}_{\mathsf{p}(i)}^{\mathsf{T}} \mathbf{M}_{\mathsf{p}(i)} - \tilde{\mathbf{M}}_i^{\mathsf{T}} \tilde{\mathbf{M}}_i \tilde{\mathbf{\Omega}}_i^{-1}\right].
\end{aligned} \tag{7.262}$$

This approximation makes the algorithm simple because we can optimize the precision hyperparameter within the target and parent nodes, and do not have to consider the child nodes. Since we only have one scalar parameter for this optimization step, we simply used a line search algorithm to obtain the optimal precision hyperparameter. If we consider a more complex precision structure (e.g., a precision matrix instead of a scalar precision parameter in the prior distribution setting Eq. (7.200)), the line search algorithm may not be adequate. In that case, we need to update hyperparameters by using some other optimization technique (e.g., gradient ascent).

Model selection

The remaining tuning parameter in the proposed approach is how many clusters we prepare. This is a model selection problem, and we can also automatically obtain the number of clusters by optimizing the variational lower bound. In the binary tree structure, we focus on a subtree composed of a target non-leaf node i and its child nodes $\mathsf{l}(i)$

Algorithm 13 Structural Bayesian linear regression.

1: Prepare an initial Gaussian tree with a set of nodes $\mathcal{I}$
2: Initialize $\tilde{\Psi} = \{\tilde{\rho}_i, \tilde{\mathbf{M}}_i | i = 1, \cdots |\mathcal{I}|\}$
3: **repeat**
4: VB E-step
5: $\mathcal{L}(M, \mathbf{\Phi})$ = Prune_tree(root node) // prune a tree by model selection
6: # of leaf nodes = Transform_HMM(root node) // Transform HMMs in the pruned tree
7: **until** Total lower bound is converged or a specified number of iterations has been reached.

and $\mathsf{r}(i)$. We compute the following difference based on Eq. (7.262) of the parent and that of the child nodes:[10]

$$\Delta\mathcal{L}_i(\rho_i) \triangleq \mathcal{L}_{\mathsf{l}(i)}(\rho_{\mathsf{l}(i)}) + \mathcal{L}_{\mathsf{r}(i)}(\rho_{\mathsf{r}(i)}) - \mathcal{L}_i(\rho_i). \tag{7.263}$$

This difference function is used for a stopping criterion in a top-down clustering strategy. This difference function similarly appeared in the model selection of the context-dependent CDHMM topologies in Sections 6.5 and 7.3.6. Then if the sign of $\Delta\mathcal{L}$ is negative, the target non-leaf node is regarded as a new leaf node determined by the model selection in terms of optimizing the lower bound. Next we prune the child nodes $\mathsf{l}(i)$ and $\mathsf{r}(i)$. By checking the signs of $\Delta\mathcal{L}_i$ for all possible nodes, and pruning the child nodes when $\Delta\mathcal{L}_i$ have negative signs, we can obtain the pruned tree structure, which corresponds to maximizing the variational lower bound locally. This optimization is efficiently accomplished by using a depth-first search.

Thus, by optimizing the hyperparameters and model structure, we can avoid setting any tuning parameters. We summarize this optimization in Algorithms 13, 14, and 15. Algorithm 13 prepares a large Gaussian tree with a set of nodes $\mathcal{I}$, prunes a tree based on the model selection (Algorithm 14), and transforms HMMs (Algorithm 15). Algorithm 14 first optimizes the precision hyperparameters Ψ, and then the model structure M. Algorithm 15 transforms Gaussian mean vectors in HMMs at the new root nodes in the pruned tree $\mathcal{I}_M$ obtained by Algorithm 14.

Watanabe *et al.* (2013) compare the VB linear regression method (VBLR) with MLLR and SMAPLR, as regards the *WSJ*, for various amounts of adaptation data by using LVCSR experiments for the Corpus of Spontaneous Japanese (CSJ). With a small amount of adaptation data, VBLR outperforms the conventional approaches by about 1.0% absolute accuracy improvement, while with a large amount of adaptation data, the accuracies of all approaches are comparable. This property is theoretically reasonable

[10] Since we approximate the posterior distribution for a non-leaf node to that for a leaf node, the contribution of the variational lower bounds from the non-leaf nodes to the total lower bounds can be disregarded, and Eq. (7.263) is used as a pruning criterion. If we do not use this approximation, we just compare the difference between the values $\mathcal{L}_i(\rho_i, \rho_{\mathsf{l}(i)}, \rho_{\mathsf{r}(i)})$ of the leaf and non-leaf node cases in Eq. (7.258).

Algorithm 14 Prune_tree(node i)

1: **if** First iteration **then**
2: $\tilde{\rho}_i = \arg\max_{\rho_i} \mathcal{L}_i(\rho_i)$ // These are used as
3: Update $\tilde{q}(\mathbf{W}_i)$ // hyperparameters of parent nodes
4: **end if**
5: **if** Node i has child nodes **then**
6: $\tilde{\rho}_i = \arg\max_{\rho_i} \mathcal{L}_i(\rho_i)$
7: Update $\tilde{q}(\mathbf{W}_i)$
8: $\Delta\mathcal{L}$ = Prune_tree(node *left*(i)) + Prune_tree(node *right*(i)) $-\mathcal{L}_i(\tilde{\rho}_i)$
9: **if** $\Delta\mathcal{L} < 0$ **then**
10: Prune child nodes // this node becomes a leaf node
11: **end if**
12: **return** $\mathcal{L}_i(\tilde{\rho}_i)$
13: **else**
14: $\tilde{\rho}_i = \arg\max_{\rho_i} \mathcal{L}_i(\rho_i)$
15: Update $q(\mathbf{W}_i)$
16: **return** $\mathcal{L}_i(\tilde{\rho}_i)$
17: **end if**

Algorithm 15 Transform_HMM(node i)

1: **if** Node i has child nodes **then**
2: **return** Transform_HMM(node *left*(i)) + Transform_HMM(node *right*(i))
3: **else**
4: Update $\tilde{\boldsymbol{\mu}}_k = C_k \tilde{\mathbf{M}}_i \boldsymbol{\xi}_k$
5: **return** 1
6: **end if**

since the variational lower bound would be tighter than the EM-based objective function for a small amount of data, while the lower bound would approach it for a large amount of data asymptotically. Therefore, it can be concluded that this improvement comes from the optimization of the hyperparameters and the model structure in VBLR, in addition to mitigation of the sparse data problem arising in the Bayesian approach.

7.5 Variational Bayesian speaker verification

This section describes an application of VB to speaker verification (Zhao, Dong, Zhao *et al.* 2009, Kenny 2010, Villalba & Brümmer 2011). The main goal of this approach is to obtain the feature representation that only holds speaker specific characteristics. As discussed in Section 4.6.2, state-of-the-art speaker verification systems use the super vector obtained by the GMM–UBM or MLLR techniques, as a feature. However, the super vector still includes various factors other than the speaker characteristics with very high dimensional representation. Use of factor analysis is critical in speaker verification to remove these irrelevant factors of the super vector and find the lower dimensional representation (Kenny 2010, Kinnunen & Li 2010, Dehak *et al.* 2011). In addition, a Bayesian treatment of the factor analysis yields robust modeling of the speaker verification. This section discusses a VB treatment of the factor analysis model by providing a generative model of the super vector (Section 7.5.1), prior distributions (Section 7.5.2), variational posteriors (Section 7.5.3), and variational lower bound (Section 7.5.4).

7.5.1 Generative model

Let $\mathbf{O} = \{\mathbf{o}_n \in \mathbb{R}^D | n = 1, \cdots, N\}$ be a D dimensional feature vector of n recordings. Note that $\mathbf{o}_n$ is a super vector, and it can be the Gaussian super vector, vectorized form of the MLLR matrix, or the factor vector obtained after the initial factor analysis process.[11] If we use the Gaussian super vector, the number of dimensions D would be the product of multiplying the number of mixture components in GMM (usually $K = 1024$) and the number of speech feature dimensions (usually $D_{\text{MFCC}} = 39$ when we use MFCC and delta features) when we use GMM–UBM, that is

$$\mathbf{o}_n = [\boldsymbol{\mu}_1^\top, \cdots, \boldsymbol{\mu}_k^\top, \cdots, \boldsymbol{\mu}_K^\top]^\top, \quad \boldsymbol{\mu}_k \in \mathbb{R}^{D_{\text{MFCC}}}. \tag{7.264}$$

Therefore, D would be $1024 \times 39 \approx 40$ thousands, and it is much larger than the number of speech feature dimensions that we are dealing with at CDHMMs. Note also that the feature $\mathbf{o}_n$ is obtained for each recording (utterance), and the frame level process is performed when super vectors are extracted by GMM–UBM.

The generative model is represented as follows (Kenny, Boulianne, Ouellet *et al.* 2007, Kenny 2010):

$$\mathbf{o}_n = \mathbf{m} + \mathbf{U}_1\mathbf{x}_1 + \mathbf{U}_2\mathbf{x}_{2n} + \boldsymbol{\varepsilon}_n, \tag{7.265}$$

where $\mathbf{m} \in \mathbb{R}^D$ is a global mean vector for the feature vectors and can be regarded as a bias vector of the feature vectors. Vector $\mathbf{x}_1 \in \mathbb{R}^{D_1}$ is a vector having a D_1 dimensional standard Gaussian distribution, which does not depend on the recording n, and it can represent stationary speaker characteristics across recordings. On the other hand, $\mathbf{x}_{2n} \in \mathbb{R}^{D_2}$ is a vector having a D_2 dimensional standard Gaussian distribution depending on the recording n, and it denotes channel characteristics changing over a recording. We also define $\mathbf{X}_2 \triangleq \{\mathbf{x}_{2n} | n = 1, \cdots, N\}$. $\boldsymbol{\varepsilon}_n \in \mathbb{R}^D$ as a D dimensional vector having a Gaussian distribution with $\mathbf{0}$ mean vector, and $\mathbf{R} \in \mathbb{R}^{D \times D}$ precision matrix.

In this book, $\mathbf{m}$, $\mathbf{U}_1 \in \mathbb{R}^{D \times D_1}$, $\mathbf{U}_2 \in \mathbb{R}^{D \times D_2}$, and $\mathbf{R}$ ($\Psi = \{\mathbf{m}, \mathbf{U}_1, \mathbf{U}_2, \mathbf{R}\}$) are regarded as non-probabilistic model parameters and assumed to be estimated without

[11] A Bayesian treatment of the factor analysis in a state-of-the-art speaker verification system can be performed after the first step of the factor analysis, called *i* vector analysis (Dehak *et al.* 2011). That is, Bayesian factor analysis is often performed for the first-step factor vector (*i* vector) for each utterance, instead of the Gaussian super vector. This book discusses Bayesian factor analysis for the Gaussian super vector for simplicity.

the Bayesian framework based on ML/MAP. However, Villalba & Brümmer (2011) deal with these parameters as probabilistic variables, and provide a fully Bayesian solution of the factor analysis based speaker modeling by VB. The other probabilistic variables $\mathbf{x}_1$, $\mathbf{x}_{2n}$, and $\boldsymbol{\varepsilon}_n$ are generated from the following Gaussian distributions:

$$\begin{aligned} \mathbf{x}_1 &\sim \mathcal{N}(\mathbf{0}, u_1^{-1}\mathbf{I}_{D_1}), \\ \mathbf{x}_{2n} &\sim \mathcal{N}(\mathbf{0}, u_{2n}^{-1}\mathbf{I}_{D_2}), \\ \boldsymbol{\varepsilon}_n &\sim \mathcal{N}(\mathbf{0}, v_n^{-1}\mathbf{R}^{-1}), \end{aligned} \tag{7.266}$$

where we assume a zero mean spherical Gaussian distribution for $\mathbf{x}_1$ and $\mathbf{x}_{2n}$. The model parameters $u_1 \in \mathbb{R}_{>0}$, $u_{2n} \in \mathbb{R}_{>0}$, and $v_n \in \mathbb{R}_{>0}$ are positive, and the probabilistic treatment of these parameters is discussed later.

In summary, the conditional distribution of $\mathbf{O}$ is represented as follows:

$$p(\mathbf{O}|\mathbf{x}_1, \mathbf{X}_2, \{v_n\}_{n=1}^N, \Psi) = \prod_{n=1}^N \mathcal{N}(\mathbf{o}_n|\mathbf{m} + \mathbf{U}_1\mathbf{x}_1 + \mathbf{U}_2\mathbf{x}_{2n}, v_n^{-1}\mathbf{R}^{-1}). \tag{7.267}$$

The conditional joint distribution of $\mathbf{O}$, $\mathbf{x}_1$, and $\mathbf{X}_2$ is represented as follows:

$$\begin{aligned} &p(\mathbf{O}, \mathbf{x}_1, \mathbf{X}_2|\{v_n\}_{n=1}^N, \Psi, u_1, \{u_{2n}\}_{n=1}^N) \\ &\quad = p(\mathbf{O}|\mathbf{x}_1, \mathbf{X}_2, \{v_n\}_{n=1}^N, \Psi)p(\mathbf{x}_1|u_1)\prod_{n=1}^N p(\mathbf{x}_{2n}|u_{2n}), \end{aligned} \tag{7.268}$$

where

$$\begin{aligned} p(\mathbf{x}_1|u_1) &= \mathcal{N}(\mathbf{x}_1|\mathbf{0}, u_1^{-1}\mathbf{I}_{D_1}), \\ p(\mathbf{x}_{2n}|u_{2n}) &= \mathcal{N}(\mathbf{x}_{2n}|\mathbf{0}, u_{2n}^{-1}\mathbf{I}_{D_2}). \end{aligned} \tag{7.269}$$

The following section regards u_1, u_{2n}, and v_n as probabilistic variables.

7.5.2 Prior distributions

We provide the conjugate prior distributions for u_1, u_{2n}, and v_n that are represented by a gamma distribution, as we discussed in Section 2.1.3. Kenny (2010) provides a simple hyperparameter setting for each gamma distribution in Appendix C.11 by using only one hyperparameter for each distribution, i.e., the model parameters u_1, u_{2n}, and v_n are generated from the following prior distributions:

$$\begin{aligned} u_1 &\sim \mathrm{Gam}\left(\frac{\phi_1}{2}, \frac{\phi_1}{2}\right), \\ u_{2n} &\sim \mathrm{Gam}\left(\frac{\phi_2}{2}, \frac{\phi_2}{2}\right), \\ v_n &\sim \mathrm{Gam}\left(\frac{\phi_v}{2}, \frac{\phi_v}{2}\right), \end{aligned} \tag{7.270}$$

where ϕ_1, ϕ_2, and ϕ_v ($\triangleq \Phi$) are hyperparameters in this model. Since the mean and variance of the gamma distribution $\mathrm{Gam}(y|\alpha, \beta)$ are $\frac{\alpha}{\beta}$ and $\frac{\alpha}{\beta^2}$, respectively, this parameterization means that u_1, u_{2n}, and v_n have the same mean value with 1, but the variance values are changed with $\frac{2}{\phi_1}$, $\frac{2}{\phi_2}$, and $\frac{2}{\phi_v}$, respectively. Thus, we can provide the following concrete forms of the prior distributions for u_1, u_{2n}, and v_n:

$$\begin{aligned} p(u_1|\phi_1) &= \mathrm{Gam}\left(u_1 \middle| \frac{\phi_1}{2}, \frac{\phi_1}{2}\right), \\ p(u_{2n}|\phi_2) &= \mathrm{Gam}\left(u_{2n} \middle| \frac{\phi_2}{2}, \frac{\phi_2}{2}\right), \\ p(v_n|\phi_v) &= \mathrm{Gam}\left(v_n \middle| \frac{\phi_v}{2}, \frac{\phi_v}{2}\right). \end{aligned} \tag{7.271}$$

In this model, $Z \triangleq \{\mathbf{x}_1, u_1, \{\mathbf{x}_{2n}, u_{2n}, v_n\}_{n=1}^N\}$ is a set of hidden variables, and the posterior distribution of each variable can be obtained by using variational Bayes. Thus, the joint prior distribution given hyperparameters Φ is represented as follows:

$$\begin{aligned} p(Z|\Phi) &= p(\mathbf{x}_1, u_1, \{\mathbf{x}_{2n}, u_{2n}, v_n\}_{n=1}^N|\Phi) \\ &= p(\mathbf{x}_1|u_1)p(u_1|\phi_1)\prod_{n=1}^N p(\mathbf{x}_{2n}|u_{2n})p(u_{2n}|\phi_2)p(v_n|\phi_v). \end{aligned} \tag{7.272}$$

Note that $\mathbf{x}_1$ depends on u_1, and we cannot fully factorize them. A similar discussion applies to $\mathbf{x}_{2n}$ and u_{2n}.

Now, we can provide the complete data likelihood function given hyperparameters Ψ and Φ based on Eqs. (7.267) and (7.272), and this can be used to obtain the variational posteriors:

$$\begin{aligned} & p(\mathbf{O}, Z|\Psi, \Phi) \\ &\quad = p(\mathbf{O}|\Psi, Z)p(Z|\Phi) \\ &\quad = \mathcal{N}(\mathbf{x}_1|\mathbf{0}, u_1^{-1}\mathbf{I}_{D_1})\mathrm{Gam}\left(u_1 \middle| \frac{\phi_1}{2}, \frac{\phi_1}{2}\right) \\ &\quad\quad \times \prod_{n=1}^N \mathcal{N}(\mathbf{o}_n|\mathbf{m} + \mathbf{U}_1\mathbf{x}_1 + \mathbf{U}_2\mathbf{x}_{2n}, v_n^{-1}\mathbf{R}^{-1})\mathcal{N}(\mathbf{x}_{2n}|\mathbf{0}, u_{2n}^{-1}\mathbf{I}_{D_2}) \\ &\quad\quad \times \mathrm{Gam}\left(u_{2n} \middle| \frac{\phi_2}{2}, \frac{\phi_2}{2}\right)\mathrm{Gam}\left(v_n \middle| \frac{\phi_v}{2}, \frac{\phi_v}{2}\right). \end{aligned} \tag{7.273}$$

These probabilistic distributions are represented by Gaussian and gamma distributions. In the following sections we simplify the complete data likelihood function $p(\mathbf{O}, Z|\Psi, \boldsymbol{\Phi})$ to $p(\mathbf{O}, Z)$ to avoid complicated equations. Algorithm 16 provides a generative process for the joint factor analysis speaker model with Eq. (7.273).

Algorithm 16 Generative process for joint factor analysis speaker model

Require: $\Psi = \{\mathbf{m}, \mathbf{U}_1, \mathbf{U}_2, \mathbf{R}\}$ and $\Phi = \{\phi_1, \phi_2, \phi_v\}$

1: Draw u_1 from $\text{Gam}\left(u_1 \middle| \frac{\phi_1}{2}, \frac{\phi_1}{2}\right)$
2: Draw $\mathbf{x}_1$ from $\mathcal{N}(\mathbf{x}_1|\mathbf{0}, u_1^{-1}\mathbf{I}_{D_1})$
3: **for** $n = 1, \cdots, N$ **do**
4: Draw u_{2n} from $\text{Gam}\left(u_{2n} \middle| \frac{\phi_2}{2}, \frac{\phi_2}{2}\right)$
5: Draw v_n from $\text{Gam}\left(v_n \middle| \frac{\phi_v}{2}, \frac{\phi_v}{2}\right)$
6: Draw $\mathbf{x}_{2n}$ from $\mathcal{N}(\mathbf{x}_{2n}|\mathbf{0}, u_{2n}^{-1}\mathbf{I}_{D_2})$
7: Draw $\mathbf{o}_n$ from $\mathcal{N}(\mathbf{o}_n|\mathbf{m} + \mathbf{U}_1\mathbf{x}_1 + \mathbf{U}_2\mathbf{x}_{2n}, v_n^{-1}\mathbf{R}^{-1})$
8: **end for**

7.5.3 Variational posteriors

To deal with the variational posteriors, we assume the following factorization based on the VB recipe:

$$q(Z|\mathbf{O}) = q(\mathbf{x}_1|\mathbf{O})q(u_1|\mathbf{O}) \prod_{n=1}^{N} q(\mathbf{x}_{2n}|\mathbf{O})q(u_{2n}|\mathbf{O})q(v_n|\mathbf{O}). \tag{7.274}$$

Section 7.1.3 discusses how we can obtain the following general solution for approximated variational posteriors:

$$\tilde{q}(Z_i|\mathbf{O}) \propto \exp\left(\mathbb{E}_{(Z_{\setminus i}|\mathbf{O})}\left[\log p(\mathbf{O}, Z)\right]\right). \tag{7.275}$$

We focus on the actual solutions for $q(\mathbf{x}_1|\mathbf{O})$, $q(u_1|\mathbf{O})$, $q(\mathbf{x}_{2n})$, $q(u_{2n}|\mathbf{O})$, and $q(v_n|\mathbf{O})$.

- $q(\mathbf{x}_1|\mathbf{O})$: this is calculated by substituting the factors depending on $\mathbf{x}_1$ in Eq. (7.273) into Eq. (7.275) as follows:

$$\begin{aligned}
&\log q(\mathbf{x}_1|\mathbf{O}) \\
&\quad \propto \mathbb{E}_{(Z_{\setminus \mathbf{x}_1})}[\log p(\mathbf{O}, Z)] \\
&\quad \propto \mathbb{E}_{(Z_{\setminus \mathbf{x}_1})}\Bigg[\log\Bigg(\mathcal{N}(\mathbf{x}_1|\mathbf{0}, u_1^{-1}\mathbf{I}_{D_1}) \prod_{n=1}^{N} \mathcal{N}(\mathbf{o}_n|\mathbf{m} + \mathbf{U}_1\mathbf{x}_1 \\
&\qquad\qquad + \mathbf{U}_2\mathbf{x}_{2n}, v_n^{-1}\mathbf{R}^{-1})\Bigg)\Bigg] \\
&\quad \propto \mathbb{E}_{(u_1)}[\log \mathcal{N}(\mathbf{x}_1|\mathbf{0}, u_1^{-1}\mathbf{I}_{D_1})] \\
&\qquad + \sum_{n=1}^{N} \mathbb{E}_{(\mathbf{x}_{2n}, v_n)}[\log \mathcal{N}(\mathbf{o}_n|\mathbf{m} + \mathbf{U}_1\mathbf{x}_1 + \mathbf{U}_2\mathbf{x}_{2n}, v_n^{-1}\mathbf{R}^{-1})].
\end{aligned} \tag{7.276}$$

Now let us consider the two expectations in the above equation. From the definition of the multivariate Gaussian distribution in Appendix C.6, we can obtain the following equation by disregarding the terms that do not depend on $\mathbf{x}_1$:

$$\mathbb{E}_{(u_1)}[\log \mathcal{N}(\mathbf{x}_1|\mathbf{0}, u_1^{-1}\mathbf{I}_{D_1})] \propto \mathbb{E}_{(u_1)}\left[-\frac{u_1}{2}\mathbf{x}_1^{\mathsf{T}}\mathbf{x}_1\right]$$
$$\propto -\frac{\mathbb{E}[u_1]}{2}\mathbf{x}_1^{\mathsf{T}}\mathbf{x}_1, \tag{7.277}$$

where we omit the subscript (u_1) in the expectation, as it is trivial. Similarly, the rest of the expectations in Eq. (7.276) are also represented as follows:

$$\begin{aligned}
&\mathbb{E}_{(\mathbf{x}_{2n},v_n)}\left[\log \mathcal{N}(\mathbf{o}_n|\mathbf{m} + \mathbf{U}_1\mathbf{x}_1 + \mathbf{U}_2\mathbf{x}_{2n}, v_n^{-1}\mathbf{R}^{-1})\right] \\
&\propto \mathbb{E}_{(\mathbf{x}_{2n},v_n)}\left[-\frac{v_n}{2}(\mathbf{o}_n - (\mathbf{m} + \mathbf{U}_1\mathbf{x}_1 + \mathbf{U}_2\mathbf{x}_{2n}))^{\mathsf{T}}\right. \\
&\qquad\qquad \left.\times \mathbf{R}(\mathbf{o}_n - (\mathbf{m} + \mathbf{U}_1\mathbf{x}_1 + \mathbf{U}_2\mathbf{x}_{2n}))\right] \\
&\propto \mathbb{E}_{(\mathbf{x}_{2n},v_n)}\left[-\frac{v_n}{2}\left(\mathbf{x}_1^{\mathsf{T}}\mathbf{U}_1^{\mathsf{T}}\mathbf{R}\mathbf{U}_1\mathbf{x}_1\right) + v_n(\mathbf{o}_n - \mathbf{m} - \mathbf{U}_2\mathbf{x}_{2n})^{\mathsf{T}}\mathbf{R}\mathbf{U}_1\mathbf{x}_1\right] \\
&\propto -\frac{\mathbb{E}[v_n]}{2}\mathbf{x}_1^{\mathsf{T}}\mathbf{U}_1^{\mathsf{T}}\mathbf{R}\mathbf{U}_1\mathbf{x}_1 + \mathbb{E}[v_n](\mathbf{o}_n - \mathbf{m} - \mathbf{U}_2\mathbb{E}[\mathbf{x}_{2n}])^{\mathsf{T}}\mathbf{R}\mathbf{U}_1\mathbf{x}_1.
\end{aligned} \tag{7.278}$$

Thus, by substituting Eqs. (7.277) and (7.278) into Eq. (7.276), we find that

$$\begin{aligned}
&\log q(\mathbf{x}_1|\mathbf{O}) \\
&\propto -\frac{\mathbb{E}[u_1]}{2}\mathbf{x}_1^{\mathsf{T}}\mathbf{x}_1 + \sum_{n=1}^{N} -\frac{\mathbb{E}[v_n]}{2}\mathbf{x}_1^{\mathsf{T}}\mathbf{U}_1^{\mathsf{T}}\mathbf{R}\mathbf{U}_1\mathbf{x}_1 \\
&\quad + \sum_{n=1}^{N}\mathbb{E}[v_n](\mathbf{o}_n - \mathbf{m} - \mathbf{U}_2\mathbb{E}[\mathbf{x}_{2n}])^{\mathsf{T}}\mathbf{R}\mathbf{U}_1\mathbf{x}_1 \\
&\propto -\frac{1}{2}\mathrm{tr}\left[\left(\mathbb{E}[u_1]\mathbf{I}_{D_1} + \sum_{n=1}^{N}\mathbb{E}[v_n]\mathbf{U}_1^{\mathsf{T}}\mathbf{R}\mathbf{U}_1\right)\mathbf{x}_1\mathbf{x}_1^{\mathsf{T}}\right] \\
&\quad + \mathrm{tr}\left[\mathbf{x}_1^{\mathsf{T}}\sum_{n=1}^{N}\mathbb{E}[v_n]\mathbf{U}_1^{\mathsf{T}}\mathbf{R}(\mathbf{o}_n - \mathbf{U}_2\mathbb{E}[\mathbf{x}_{2n}] - \mathbf{m})\right] \\
&\propto \log \mathcal{N}(\mathbf{x}_1|\tilde{\boldsymbol{\mu}}_{\mathbf{x}_1}, \tilde{\boldsymbol{\Sigma}}_{\mathbf{x}_1}),
\end{aligned} \tag{7.279}$$

where we use the trace form definition of the multivariate Gaussian distribution in Appendix C.6. Thus, $q(\mathbf{x}_1|\mathbf{O})$ is represented as a Gaussian distribution, and $\tilde{\boldsymbol{\mu}}_{\mathbf{x}_1}$ and $\tilde{\boldsymbol{\Sigma}}_{\mathbf{x}_1}$ are posterior hyperparameters obtained as follows:

$$\begin{aligned}
\tilde{\boldsymbol{\mu}}_{\mathbf{x}_1} &\triangleq \left(\mathbb{E}[u_1]\mathbf{I}_{D_1} + \sum_{n=1}^{N}\mathbb{E}[v_n]\mathbf{U}_1^{\mathsf{T}}\mathbf{R}\mathbf{U}_1\right)^{-1} \\
&\qquad \times \sum_{n=1}^{N}\mathbb{E}[v_n]\mathbf{U}_1^{\mathsf{T}}\mathbf{R}(\mathbf{o}_n - \mathbf{U}_2\mathbb{E}[\mathbf{x}_2] - \mathbf{m}) \\
\tilde{\boldsymbol{\Sigma}}_{\mathbf{x}_1} &\triangleq \left(\mathbb{E}[u_1]\mathbf{I}_{D_1} + \sum_{n=1}^{N}\mathbb{E}[v_n]\mathbf{U}_1^{\mathsf{T}}\mathbf{R}\mathbf{U}_1\right)^{-1}.
\end{aligned} \tag{7.280}$$

Thus, the hyperparameters obtained are represented with prior hyperparameters Ψ and the expected values of $\mathbb{E}[u_1]$ and $\mathbb{E}[v_n]$.

- $q(\mathbf{x}_{2n}|\mathbf{O})$: this is similarly calculated by substituting the factors depending on $\mathbf{x}_{2n}$ in Eq. (7.273) into Eq. (7.275) as follows:

$$\begin{aligned}
&\log q(\mathbf{x}_{2n}|\mathbf{O}) \\
&\quad \propto \mathbb{E}_{(Z_{\backslash \mathbf{x}_{2n}})}[\log p(\mathbf{O}, Z)] \\
&\quad \propto \mathbb{E}_{(Z_{\backslash \mathbf{x}_{2n}})}\left[\prod_{n'=1}^{N} \log \left(\mathcal{N}(\mathbf{o}_{n'}|\mathbf{m} + \mathbf{U}_1\mathbf{x}_1 \right.\right. \\
&\qquad\qquad \left.\left. + \mathbf{U}_2\mathbf{x}_{2n'}, \nu_{n'}^{-1}\mathbf{R}^{-1})\mathcal{N}(\mathbf{x}_{2n'}|\mathbf{0}, u_{2n'}^{-1}\mathbf{I}_{D_2})\right)\right] \\
&\quad \propto \mathbb{E}_{(\mathbf{x}_1, \nu_n)}\left[\mathcal{N}(\mathbf{o}_n|\mathbf{m} + \mathbf{U}_1\mathbf{x}_1 + \mathbf{U}_2\mathbf{x}_{2n}, \nu_n^{-1}\mathbf{R}^{-1})\right] \\
&\qquad + \mathbb{E}_{(u_{2n})}\left[\mathcal{N}(\mathbf{x}_{2n}|\mathbf{0}, u_{2n}^{-1}\mathbf{I}_{D_2})\right].
\end{aligned} \tag{7.281}$$

The expectations are rewritten as follows:

$$\begin{aligned}
&\log q(\mathbf{x}_{2n}|\mathbf{O}) \\
&\quad \propto -\frac{1}{2}\mathbb{E}[\nu_n]\mathrm{tr}\left[\mathbf{U}_2^{\mathsf{T}}\mathbf{R}\mathbf{U}_2\mathbf{x}_{2n}\mathbf{x}_{2n}^{\mathsf{T}}\right] \\
&\qquad + 2\mathbb{E}[\nu_n]\mathrm{tr}\left[\mathbf{x}_{2n}^{\mathsf{T}}\mathbf{U}_2^{\mathsf{T}}\mathbf{R}(\mathbf{o}_n - \mathbf{U}_2\mathbb{E}[\mathbf{x}_1] - \mathbf{m})\right] \\
&\qquad - \frac{1}{2}\mathbb{E}[u_{2n}]\mathrm{tr}\left[\mathbf{x}_{2n}\mathbf{x}_{2n}^{\mathsf{T}}\right] \\
&\quad \propto -\frac{1}{2}\mathrm{tr}\left[\left(\mathbb{E}[u_{2n}]\mathbf{I}_{D_2} + \mathbb{E}[\nu_n]\mathbf{U}_2^{\mathsf{T}}\mathbf{R}\mathbf{U}_2\right)\mathbf{x}_{2n}\mathbf{x}_{2n}^{\mathsf{T}}\right] \\
&\qquad + \mathrm{tr}\left[\mathbf{x}_{2n}^{\mathsf{T}}\mathbb{E}[\nu_n]\mathbf{U}_2^{\mathsf{T}}\mathbf{R}(\mathbf{o}_n - \mathbf{U}_2\mathbb{E}[\mathbf{x}_1] - \mathbf{m})\right] \\
&\quad \propto \log \mathcal{N}(\mathbf{x}_{2n}|\tilde{\boldsymbol{\mu}}_{\mathbf{x}_{2n}}, \tilde{\boldsymbol{\Sigma}}_{\mathbf{x}_{2n}}).
\end{aligned} \tag{7.282}$$

Thus, $q(\mathbf{x}_{2n}|\mathbf{O})$ is also represented as a Gaussian distribution, and $\tilde{\boldsymbol{\mu}}_{\mathbf{x}_{2n}}$ and $\tilde{\boldsymbol{\Sigma}}_{\mathbf{x}_{2n}}$ are posterior hyperparameters obtained as follows:

$$\begin{aligned}
\tilde{\boldsymbol{\mu}}_{\mathbf{x}_{2n}} &\triangleq \left(\mathbb{E}[u_{2n}]\mathbf{I}_{D_2} + \mathbb{E}[\nu_n]\mathbf{U}_2^{\mathsf{T}}\mathbf{R}\mathbf{U}_2\right)^{-1}\mathbb{E}[\nu_n]\mathbf{U}_2^{\mathsf{T}}\mathbf{R}(\mathbf{o}_n - \mathbf{U}_2\mathbb{E}[\mathbf{x}_1] - \mathbf{m}), \\
\tilde{\boldsymbol{\Sigma}}_{\mathbf{x}_{2n}} &\triangleq \left(\mathbb{E}[u_{2n}]\mathbf{I}_{D_2} + \mathbb{E}[\nu_n]\mathbf{U}_2^{\mathsf{T}}\mathbf{R}\mathbf{U}_2\right)^{-1}.
\end{aligned} \tag{7.283}$$

Note that the hyperparameters obtained are represented with prior hyperparameters Ψ and the expected values of $\mathbb{E}[u_{2n}]$, $\mathbb{E}[\nu_n]$, and $\mathbb{E}[\mathbf{x}_1]$. Compared with Eq. (7.280), Eq. (7.283) has a similar functional form, but it is computed recording by recording, while Eq. (7.280) is computed with accumulation over every recording n.

- $q(u_1|\mathbf{O})$: this is also similarly calculated by substituting the factors depending on u_1 in Eq. (7.273) into Eq. (7.275). However, compared with the previous two cases, the gamma and Gaussian distributions appear in the formulation as follows:

$$\begin{aligned}
&\log q(u_1|\mathbf{O}) \\
&\quad \propto \mathbb{E}_{(Z_{\backslash u_1})}[\log p(\mathbf{O}, Z)] \\
&\quad \propto \mathbb{E}_{(Z_{\backslash u_1})}\left[\mathcal{N}(\mathbf{x}_1|\mathbf{0}, u_1^{-1}\mathbf{I}_{D_1})\mathrm{Gam}\left(u_1 \middle| \frac{\phi_1}{2}, \frac{\phi_1}{2}\right)\right] \\
&\quad \propto \mathbb{E}_{(\mathbf{x}_1)}[\log \mathcal{N}(\mathbf{x}_1|\mathbf{0}, u_1^{-1}\mathbf{I}_{D_1})] + \log\left(\mathrm{Gam}\left(u_1 \middle| \frac{\phi_1}{2}, \frac{\phi_1}{2}\right)\right).
\end{aligned} \tag{7.284}$$

By using the definition of the gamma distribution in Appendix C.11, this equation can be rewritten as follows:

$$\begin{aligned}
&\log q(u_1|\mathbf{O}) \\
&\propto \log\left(|u_1 \mathbf{I}_{D_1}|^{\frac{1}{2}}\right) - \frac{\mathbb{E}[\mathbf{x}_1^\mathsf{T}\mathbf{x}_1]}{2}u_1 + \log\left(u_1^{\frac{\phi_1}{2}-1}\right) - \frac{\phi_1}{2}u_1 \\
&\propto \log\left(u_1^{\frac{\phi_1+D_1}{2}-1}\right) - \frac{\phi_1 + \mathbb{E}[\mathbf{x}_1^\mathsf{T}\mathbf{x}_1]}{2}u_1 \\
&\propto \log\left(\mathrm{Gam}\left(u_1 \middle| \frac{\tilde{\phi}_1}{2}, \frac{\tilde{r}_1}{2}\right)\right).
\end{aligned} \tag{7.285}$$

Thus, $q(u_1|\mathbf{O})$ is represented as a gamma distribution, and $\tilde{\phi}_1$ and $\tilde{r}_1$ are posterior hyperparameters obtained as follows:

$$\begin{aligned}
\tilde{\phi}_1 &\triangleq \phi_1 + D_1, \\
\tilde{r}_1 &\triangleq \phi_1 + \mathbb{E}[\mathbf{x}_1^\mathsf{T}\mathbf{x}_1].
\end{aligned} \tag{7.286}$$

These posterior hyperparameters are obtained with their original prior hyperparameter ϕ_1 and the second-order expectation value of $\mathbb{E}[\mathbf{x}_1^\mathsf{T}\mathbf{x}_1]$.

- $q(u_{2n}|\mathbf{O})$: this is similarly calculated by substituting the factors depending on u_{2n} in Eq. (7.273) into Eq. (7.275):

$$\begin{aligned}
&\log q(u_{2n}|\mathbf{O}) \\
&\propto \mathbb{E}_{(Z_{\backslash u_{2n}})}[\log p(\mathbf{O}, Z)] \\
&\propto \mathbb{E}_{(Z_{\backslash u_{2n}})}\left[\prod_{n'=1}^{N} \mathcal{N}(\mathbf{x}_{2n'}|\mathbf{0}, u_{2n'}^{-1}\mathbf{I}_{D_2})\mathrm{Gam}\left(u_{2n'} \middle| \frac{\phi_2}{2}, \frac{\phi_2}{2}\right)\right] \\
&\propto \mathbb{E}_{(\mathbf{x}_{2n})}[\log \mathcal{N}(\mathbf{x}_{2n}|\mathbf{0}, u_{2n}^{-1}\mathbf{I}_{D_2})] + \log\left(\mathrm{Gam}\left(u_{2n} \middle| \frac{\phi_2}{2}, \frac{\phi_2}{2}\right)\right) \\
&\propto \log\left(\mathrm{Gam}\left(u_{2n} \middle| \frac{\tilde{r}_{2n}}{2}, \frac{\tilde{\phi}_{2n}}{2}\right)\right).
\end{aligned} \tag{7.287}$$

$q(u_{2n}|\mathbf{O})$ is represented as a gamma distribution, and $\tilde{\phi}_{2n}$ and $\tilde{r}_{2n}$ are posterior hyperparameters, obtained as follows:

$$\begin{aligned}
\tilde{\phi}_{2n} &\triangleq \phi_2 + D_2, \\
\tilde{r}_{2n} &\triangleq \phi_2 + \mathbb{E}[\mathbf{x}_{2n}^\mathsf{T}\mathbf{x}_{2n}].
\end{aligned} \tag{7.288}$$

These posterior hyperparameters are also obtained with their original prior hyperparameter ϕ_2 and the second-order expectation value of $\mathbb{E}[\mathbf{x}_{2n}^\mathsf{T}\mathbf{x}_{2n}]$, and these are very similar to the posterior hyperparameters of $q(u_1|\mathbf{O})$ in Eq. (7.286).

- $q(v_n|\mathbf{O})$: finally, this is also calculated by substituting the factors depending on v_n in Eq. (7.273) into Eq. (7.275) as follows:

$$\begin{aligned}
&\log q(v_n|\mathbf{O}) \\
&\propto \mathbb{E}_{(Z_{\backslash v_n})}[\log p(\mathbf{O}, Z)]
\end{aligned}$$

$$
\begin{aligned}
&\propto \mathbb{E}_{(Z_{\setminus v_n})}\left[\prod_{n'=1}^{N} \mathcal{N}(\mathbf{o}_{n'}|\mathbf{m} + \mathbf{U}_1\mathbf{x}_1 + \mathbf{U}_2\mathbf{x}_{2n'}, v_{n'}^{-1}\mathbf{R}^{-1}) \right.\\
&\qquad\qquad \left. \times \mathrm{Gam}\left(v_{n'}\middle|\frac{\phi_v}{2}, \frac{\phi_v}{2}\right)\right]\\
&\propto \mathbb{E}_{(\mathbf{x}_1,\mathbf{x}_{2n})}\left[\mathcal{N}(\mathbf{o}_n|\mathbf{m} + \mathbf{U}_1\mathbf{x}_1 + \mathbf{U}_2\mathbf{x}_{2n}, v_n^{-1}\mathbf{R}^{-1})\right]\\
&\quad + \log\left(\mathrm{Gam}\left(v_n\middle|\frac{\phi_v}{2}, \frac{\phi_v}{2}\right)\right)\\
&\propto \log\left(\mathrm{Gam}\left(v_n\middle|\frac{\tilde{r}_{vn}}{2}, \frac{\tilde{\phi}_{vn}}{2}\right)\right).
\end{aligned} \tag{7.289}
$$

$q(v_n|\mathbf{O})$ is represented as a gamma distribution, and $\tilde{\phi}_{vn}$ and $\tilde{r}_{vn}$ are posterior hyperparameters obtained as follows:

$$
\begin{aligned}
\tilde{\phi}_{vn} &\triangleq \phi_v + D,\\
\tilde{r}_{vn} &\triangleq \phi_v + \mathbb{E}[\boldsymbol{\varepsilon}_n^{\mathsf{T}}\mathbf{R}\boldsymbol{\varepsilon}_n],
\end{aligned} \tag{7.290}
$$

where $\boldsymbol{\varepsilon}_n$ is a residual vector appearing in the basic equation of the joint factor analysis in Eq. (7.265), and is represented as:

$$
\boldsymbol{\varepsilon}_n = \mathbf{o}_n - (\mathbf{m} + \mathbf{U}_1\mathbf{x}_1 + \mathbf{U}_2\mathbf{x}_{2n}). \tag{7.291}
$$

$\mathbb{E}[\boldsymbol{\varepsilon}_n^{\mathsf{T}}\mathbf{R}\boldsymbol{\varepsilon}_n]$ is an expectation over both $\mathbf{x}_1$ and $\mathbf{x}_{2n}$, and defined as follows:

$$
\begin{aligned}
&\mathbb{E}[\boldsymbol{\varepsilon}_n^{\mathsf{T}}\mathbf{R}\boldsymbol{\varepsilon}_n]\\
&\quad \triangleq \mathbb{E}_{\mathbf{x}_1,\mathbf{x}_{2n}}[(\mathbf{o}_n - \mathbf{m} - \mathbf{U}_1\mathbf{x}_1 - \mathbf{U}_2\mathbf{x}_{2n})^{\mathsf{T}}\mathbf{R}(\mathbf{o}_n - \mathbf{m} - \mathbf{U}_1\mathbf{x}_1 - \mathbf{U}_2\mathbf{x}_{2n})]\\
&\quad = (\mathbf{o}_n - \mathbf{m})^{\mathsf{T}}\mathbf{R}(\mathbf{o}_n - \mathbf{m}) + \mathrm{tr}\left[\mathbf{U}_1^{\mathsf{T}}\mathbf{R}\mathbf{U}_1\mathbb{E}\left[\mathbf{x}_1\mathbf{x}_1^{\mathsf{T}}\right]\right] + \mathrm{tr}\left[\mathbf{U}_2^{\mathsf{T}}\mathbf{R}\mathbf{U}_2\mathbb{E}\left[\mathbf{x}_{2n}\mathbf{x}_{2n}^{\mathsf{T}}\right]\right]\\
&\quad\quad - 2(\mathbf{o}_n - \mathbf{m})^{\mathsf{T}}\mathbf{R}\mathbf{U}_1\mathbb{E}[\mathbf{x}_1] - 2(\mathbf{o}_n - \mathbf{m})^{\mathsf{T}}\mathbf{R}\mathbf{U}_2\mathbb{E}[\mathbf{x}_{2n}]\\
&\quad\quad + 2\mathrm{tr}\left[\mathbf{U}_1^{\mathsf{T}}\mathbf{R}\mathbf{U}_2\mathbb{E}[\mathbf{x}_{2n}]\,\mathbb{E}\left[\mathbf{x}_1^{\mathsf{T}}\right]\right].
\end{aligned} \tag{7.292}
$$

This value is computed by the first- and second-order expectation of $\mathbf{x}_1$ and $\mathbf{x}_{2n}$.

Thus, we can provide the VB posterior distributions of all hidden variables analytically. Note that these equations are iteratively performed to obtain the sub-optimal posterior distributions. That is, all the posterior distribution calculations need the expectation values of hidden variables Z, which can be computed using the posterior distributions obtained with the previous iteration. We provide the expectation values of $\mathbf{x}_1$ and $\mathbf{x}_{2n}$, which are easily obtained by reference to the expectation formulas of a multivariate Gaussian distribution in Appendix C.6 and posterior hyperparameters in Eqs. (7.280) and (7.283):

$$
\begin{aligned}
\mathbb{E}[\mathbf{x}_1] &= \int \mathbf{x}_1 q(\mathbf{x}_1|\mathbf{O})d\mathbf{x}_1 = \tilde{\boldsymbol{\mu}}_{\mathbf{x}_1},\\
\mathbb{E}\left[\mathbf{x}_1\mathbf{x}_1^{\mathsf{T}}\right] &= \int \mathbf{x}_1\mathbf{x}_1^{\mathsf{T}} q(\mathbf{x}_1|\mathbf{O})d\mathbf{x}_1 = \tilde{\boldsymbol{\Sigma}}_{\mathbf{x}_1},
\end{aligned}
$$

$$\mathbb{E}[\mathbf{x}_{2n}] = \int \mathbf{x}_{2n} q(\mathbf{x}_{2n}|\mathbf{O}) d\mathbf{x}_{2n} = \widetilde{\boldsymbol{\mu}}_{\mathbf{x}_{2n}},$$
$$\mathbb{E}\left[\mathbf{x}_{2n}\mathbf{x}_{2n}^{\mathsf{T}}\right] = \int \mathbf{x}_{2n}\mathbf{x}_{2n}^{\mathsf{T}} q(\mathbf{x}_{2n}|\mathbf{O}) d\mathbf{x}_{2n} = \widetilde{\boldsymbol{\Sigma}}_{\mathbf{x}_{2n}}. \tag{7.293}$$

Similarly, we provide the expectation values of u_1, u_{2n}, and v_n, which are also easily obtained by reference to the expectation formulas of gamma distribution in Appendix C.11 and posterior hyperparameters in Eqs. (7.286), (7.288), and (7.290):

$$\mathbb{E}[u_1] = \int u_1 q(u_1|\mathbf{O}) du_1 = \frac{\widetilde{\phi}_1}{\widetilde{r}_1},$$
$$\mathbb{E}[u_{2n}] = \int u_{2n} q(u_{2n}|\mathbf{O}) du_{2n} = \frac{\widetilde{\phi}_{2n}}{\widetilde{r}_{2n}},$$
$$\mathbb{E}[v_n] = \int v_n q(v_n|\mathbf{O}) dv_n = \frac{\widetilde{\phi}_{vn}}{\widetilde{r}_{vn}}. \tag{7.294}$$

Finally, we summarize the analytical results of the posterior distributions $q(Z|\mathbf{O})$, as follows:

$$\begin{aligned} q(Z|\mathbf{O}) & \\ &= q(\mathbf{x}_1|\mathbf{O})q(u_1|\mathbf{O}) \prod_{n=1}^{N} q(\mathbf{x}_{2n}|\mathbf{O})q(u_{2n}|\mathbf{O})q(v_n|\mathbf{O}) \\ &= \mathcal{N}(\mathbf{x}_1|\widetilde{\boldsymbol{\mu}}_{\mathbf{x}_1}, \widetilde{\boldsymbol{\Sigma}}_{\mathbf{x}_1})\mathrm{Gam}\left(u_1 \middle| \frac{\widetilde{\phi}_1}{2}, \frac{\widetilde{r}_1}{2}\right) \\ &\times \prod_{n=1}^{N} \mathcal{N}(\mathbf{x}_{2n}|\widetilde{\boldsymbol{\mu}}_{\mathbf{x}_{2n}}, \widetilde{\boldsymbol{\Sigma}}_{\mathbf{x}_{2n}})\mathrm{Gam}\left(u_{2n} \middle| \frac{\widetilde{r}_{2n}}{2}, \frac{\widetilde{\phi}_{2n}}{2}\right)\mathrm{Gam}\left(v_n \middle| \frac{\widetilde{r}_{vn}}{2}, \frac{\widetilde{\phi}_{vn}}{2}\right), \end{aligned} \tag{7.295}$$

where posterior hyperparameters are represented as:

$$\begin{cases} \widetilde{\boldsymbol{\mu}}_{\mathbf{x}_1} & \triangleq \left(\mathbb{E}[u_1]\mathbf{I}_{D_1} + \sum_{n=1}^{N} \mathbb{E}[v_n]\mathbf{U}_1^{\mathsf{T}}\mathbf{R}\mathbf{U}_1\right)^{-1}, \\ & \quad \times \sum_{n=1}^{N} \mathbb{E}[v_n]\mathbf{U}_1^{\mathsf{T}}\mathbf{R}(\mathbf{o}_n - \mathbf{U}_2\mathbb{E}[\mathbf{x}_2] - \mathbf{m}), \\ \widetilde{\boldsymbol{\Sigma}}_{\mathbf{x}_1} & \triangleq \left(\mathbb{E}[u_1]\mathbf{I}_{D_1} + \sum_{n=1}^{N} \mathbb{E}[v_n]\mathbf{U}_1^{\mathsf{T}}\mathbf{R}\mathbf{U}_1\right)^{-1}, \\ \widetilde{\boldsymbol{\mu}}_{\mathbf{x}_{2n}} & \triangleq \left(\mathbb{E}[u_{2n}]\mathbf{I}_{D_2} + \mathbb{E}[v_n]\mathbf{U}_2^{\mathsf{T}}\mathbf{R}\mathbf{U}_2\right)^{-1} \mathbb{E}[v_n]\mathbf{U}_2^{\mathsf{T}}\mathbf{R}(\mathbf{o}_n - \mathbf{U}_2\mathbb{E}[\mathbf{x}_1] - \mathbf{m}), \\ \widetilde{\boldsymbol{\Sigma}}_{\mathbf{x}_{2n}} & \triangleq \left(\mathbb{E}[u_{2n}]\mathbf{I}_{D_2} + \mathbb{E}[v_n]\mathbf{U}_2^{\mathsf{T}}\mathbf{R}\mathbf{U}_2\right)^{-1}, \end{cases} \tag{7.296}$$

$$\begin{cases} \widetilde{\phi}_1 & \triangleq \phi_1 + D_1, \\ \widetilde{r}_1 & \triangleq \phi_1 + \mathbb{E}[\mathbf{x}_1^{\mathsf{T}}\mathbf{x}_1], \\ \widetilde{\phi}_{2n} & \triangleq \phi_2 + D_2, \\ \widetilde{r}_{2n} & \triangleq \phi_2 + \mathbb{E}[\mathbf{x}_{2n}^{\mathsf{T}}\mathbf{x}_{2n}], \\ \widetilde{\phi}_{vn} & \triangleq \phi_v + D, \\ \widetilde{r}_{vn} & \triangleq \phi_v + \mathbb{E}[\boldsymbol{\varepsilon}_n^{\mathsf{T}}\mathbf{R}\boldsymbol{\varepsilon}_n]. \end{cases} \tag{7.297}$$

Once we obtain the VB posterior distributions, we can also calculate the variational lower bound, which is discussed in the next section.

7.5.4 Variational lower bound

As discussed in Section 4.6, speaker verification can be performed by using the likelihood ratio (Eq. (4.129)):

$$\frac{p(\mathbf{O}|H_0)}{p(\mathbf{O}|H_1)}, \tag{7.298}$$

where H_0 means that $\mathbf{O}$ is from the hypothesized speaker, while H_1 means that $\mathbf{O}$ is *not* from the hypothesized speaker. Instead of using the likelihood of $p(\mathbf{O})$ where we neglect the hypothesis index H, using the lower bound, we can treat speaker verification in a Bayesian sense. That is, we use the variational lower bound equation (in Eq. (7.5)):

$$p(\mathbf{O}) = \int \log p(\mathbf{O}, Z) dZ \geq \mathcal{F}[q(Z|\mathbf{O})]. \tag{7.299}$$

Then we use $\mathcal{F}[q(Z|\mathbf{O})]$ instead of $p(\mathbf{O})$. From the definition of the variational lower bound in Eq. (7.4), Eq. (7.299) can be decomposed into the following terms as follows:

$$\begin{aligned}\mathcal{F}[q(Z|\mathbf{O})] &\triangleq \mathbb{E}_{(Z)}\left[\log \frac{p(\mathbf{O}, Z)}{q(Z|\mathbf{O})}\right] = \mathbb{E}_{(Z)}\left[\log \frac{p(\mathbf{O}|Z)p(Z)}{q(Z|\mathbf{O})}\right] \\ &= \mathbb{E}_{(Z)}\left[\log p(\mathbf{O}|Z)\right] - \mathrm{KL}(q(Z|\mathbf{O})\|p(Z)),\end{aligned} \tag{7.300}$$

where the second term is the Kullback–Leibler divergence between the variational posterior and prior distributions.

Let us focus on the first term of the variational lower bound. The conditional likelihood $p(\mathbf{O}|Z)$ is represented as the following Gaussian by using Eq. (7.267):

$$p(\mathbf{O}|Z) = \prod_{n=1}^{N} \mathcal{N}(\mathbf{o}_n|\mathbf{m} + \mathbf{U}_1\mathbf{x}_1 + \mathbf{U}_2\mathbf{x}_{2n}, \nu_n^{-1}\mathbf{R}^{-1}). \tag{7.301}$$

By using Eq. (7.301), $\mathbb{E}_{(Z)}\left[\log p(\mathbf{O}|Z)\right]$ is represented as follows:

$$\begin{aligned}&\mathbb{E}_{(Z)}\left[\log p(\mathbf{O}|Z)\right] \\ &= \mathbb{E}_{(Z)}\left[\log \prod_{n=1}^{N} \mathcal{N}(\mathbf{o}_n|\mathbf{m} + \mathbf{U}_1\mathbf{x}_1 + \mathbf{U}_2\mathbf{x}_{2n}, \nu_n^{-1}\mathbf{R}^{-1})\right] \\ &= \sum_{n=1}^{N} \mathbb{E}_{(\mathbf{x}_1, \mathbf{x}_{2n}, \nu_n)}\left[\log \mathcal{N}(\mathbf{o}_n|\mathbf{m} + \mathbf{U}_1\mathbf{x}_1 + \mathbf{U}_2\mathbf{x}_{2n}, \nu_n^{-1}\mathbf{R}^{-1})\right].\end{aligned} \tag{7.302}$$

This expectation is calculated by using the definition of multivariate Gaussian distribution in Appendix C.6 and the residual vector $\boldsymbol{\varepsilon}_n$ in Eq. (7.291), as follows:

$$\begin{aligned}&\mathbb{E}_{(\mathbf{x}_1, \mathbf{x}_{2n}, \nu_n)}\left[\log \mathcal{N}(\mathbf{o}_n|\mathbf{m} + \mathbf{U}_1\mathbf{x}_1 + \mathbf{U}_2\mathbf{x}_{2n}, \nu_n^{-1}\mathbf{R}^{-1})\right] \\ &= \mathbb{E}_{(\mathbf{x}_1, \mathbf{x}_{2n}, \nu_n)}\left[\log C_{\mathcal{N}}(\nu_n^{-1}\mathbf{R}^{-1})) - \frac{\nu_n}{2}\boldsymbol{\varepsilon}_n^{\mathsf{T}}\mathbf{R}\boldsymbol{\varepsilon}_n\right] \\ &= \frac{D}{2}\mathbb{E}[\log \nu_n] - \frac{D}{2}\log(2\pi) + \frac{1}{2}\log|\mathbf{R}| - \frac{1}{2}\mathbb{E}[\nu_n]\mathbb{E}[\boldsymbol{\varepsilon}_n^{\mathsf{T}}\mathbf{R}\boldsymbol{\varepsilon}_n].\end{aligned} \tag{7.303}$$

Therefore, Eq. (7.302) is calculated as:

$$\mathbb{E}_{(Z)}\left[\log p(\mathbf{O}|Z)\right] = \sum_{n=1}^{N}\left(\frac{D}{2}\mathbb{E}[\log v_n] - \frac{D}{2}\log(2\pi) + \frac{1}{2}\log|\mathbf{R}| - \frac{1}{2}\mathbb{E}[v_n]\mathbb{E}[\boldsymbol{\varepsilon}_n^{\mathsf{T}}\mathbf{R}\boldsymbol{\varepsilon}_n]\right). \tag{7.304}$$

Now, let us focus on the KL divergence in Eq. (7.301), which is decomposed into the following terms based on the factorization forms of Eqs. (7.272) and (7.274):

$$\mathrm{KL}(q(Z|\mathbf{O})\|p(Z)) = \mathrm{KL}(q(\mathbf{x}_1, u_1|\mathbf{O})\|p(\mathbf{x}_1, u_1)) + \sum_{n=1}^{N}\mathrm{KL}(q(\mathbf{x}_{2n}, u_{2n}|\mathbf{O})\|p(\mathbf{x}_{2n}, u_{2n})) + \sum_{n=1}^{N}\mathrm{KL}(q(v_n|\mathbf{O})\|p(v_n)). \tag{7.305}$$

The KL divergence is decomposed into the three KL divergence terms. Now, consider $\mathrm{KL}(q(\mathbf{x}_1, u_1|\mathbf{O})\|p(\mathbf{x}_1, u_1))$, which can be further factorized as follows:

$$\mathrm{KL}(q(\mathbf{x}_1, u_1|\mathbf{O})\|p(\mathbf{x}_1, u_1)) = \mathbb{E}_{q(u_1|\mathbf{O})}\left[\mathrm{KL}(q(\mathbf{x}_1|\mathbf{O})\|p(\mathbf{x}_1|u_1))\right] + \mathrm{KL}(q(u_1|\mathbf{O})\|p(u_1)). \tag{7.306}$$

Thus, we need to compute the KL divergences of Gaussian and gamma distributions, respectively. We use the following formulas, which are the analytical result of the KL divergence between Gaussian and gamma distributions:

$$\mathrm{KL}(\mathcal{N}(\mathbf{x}|\tilde{\boldsymbol{\mu}}, \tilde{\boldsymbol{\Sigma}})\|\mathcal{N}(\mathbf{x}|\boldsymbol{\mu}, \boldsymbol{\Sigma})) = -\frac{D}{2} - \frac{1}{2}\log|\boldsymbol{\Sigma}^{-1}\tilde{\boldsymbol{\Sigma}}| + \mathrm{tr}\left[\boldsymbol{\Sigma}^{-1}\left(\boldsymbol{\Sigma} + (\tilde{\boldsymbol{\mu}} - \boldsymbol{\mu})(\tilde{\boldsymbol{\mu}} - \boldsymbol{\mu})^{\mathsf{T}}\right)\right], \tag{7.307}$$

and

$$\mathrm{KL}(\mathrm{Gam}(y|\tilde{\alpha}, \tilde{\beta})\|\mathrm{Gam}(y|\alpha, \beta)) = \log\frac{\Gamma(\alpha)}{\Gamma(\tilde{\alpha})} + \tilde{\alpha}\log\tilde{\beta} - \alpha\log\beta + (\tilde{\alpha} - \alpha)(\psi(\tilde{\alpha}) - \log\tilde{\beta}) + \tilde{\alpha}\frac{\beta - \tilde{\beta}}{\tilde{\beta}}. \tag{7.308}$$

Therefore, by using Eq. (7.307), $\mathbb{E}_{q(u_1|\mathbf{O})}\left[\mathrm{KL}(q(\mathbf{x}_1|\mathbf{O})\|p(\mathbf{x}_1|u_1))\right]$ can be rewritten as follows:

$$\begin{aligned}
&\mathbb{E}_{q(u_1|\mathbf{O})}\left[\mathrm{KL}(q(\mathbf{x}_1|\mathbf{O})\|p(\mathbf{x}_1|u_1))\right] \\
&= \mathbb{E}_{q(u_1|\mathbf{O})}\left[\mathrm{KL}(\mathcal{N}(\mathbf{x}_1|\tilde{\boldsymbol{\mu}}_{\mathbf{x}_1}, \tilde{\boldsymbol{\Sigma}}_{\mathbf{x}_1}\|\mathcal{N}(\mathbf{x}_1|\mathbf{0}, u_1^{-1}\mathbf{I}_{D_1}))\right] \\
&= \mathbb{E}_{q(u_1|\mathbf{O})}\left[-\frac{D_1}{2} - \frac{1}{2}\log|u_1\tilde{\boldsymbol{\Sigma}}_{\mathbf{x}_1}| + \mathrm{tr}\left[u_1\left(u_1^{-1}\mathbf{I}_{D_1} + (\tilde{\boldsymbol{\mu}}_{\mathbf{x}_1} - \boldsymbol{\mu})(\tilde{\boldsymbol{\mu}}_{\mathbf{x}_1} - \boldsymbol{\mu})^{\mathsf{T}}\right)\right]\right] \\
&= -\frac{D_1}{2} - \frac{D_1}{2}\mathbb{E}[\log u_1] - \frac{1}{2}\log|\tilde{\boldsymbol{\Sigma}}_{\mathbf{x}_1}| + \frac{1}{2}\mathbb{E}[\log u_1]\mathbb{E}[\mathbf{x}_1^{\mathsf{T}}\mathbf{x}_1].
\end{aligned} \tag{7.309}$$

Similarly, other terms can be obtained by using VB analytically.

Thus, we can obtain the variational lower bound for the joint factor analysis. This can be used as an objective function of the likelihood test, as discussed before, and is also used to optimize the hyperparameters Ψ and Φ based on the evidence approximation, as discussed in Chapter 5.

7.6 Latent Dirichlet allocation

Latent Dirichlet allocation (LDA) (Blei *et al.* 2003) is known as the popular machine learning approach which was proposed to build the *latent topic model* for information retrieval, document modeling, text categorization, and collaborative filtering. LDA has been successfully applied for image modeling, music retrieval, speech recognition, and many others. In general, LDA is an extended study from a topic model based on probabilistic latent semantic analysis (PLSA) (Hofmann 1999*b*, Hofmann 2001), which was addressed in Section 3.7.3. PLSA extracts the latent semantic information and estimates the topic parameters according to maximum likelihood (ML) estimation, which suffers from the over-trained problem. In addition, PLSA could not represent the unseen words and documents. The number of PLSA parameters increases remarkably with the number of collected documents. To compensate these weaknesses, LDA incorporates the Dirichlet priors to characterize the topic mixture probabilities. The marginal likelihood over all possible values of topic mixture probabilities is calculated and maximized so as to construct the LDA-based topic model for document representation. Unseen documents are generalized by using the LDA parameters, which are estimated according to the variational Bayes inference procedure. The model complexity is controlled as the training documents become larger. PLSA and LDA extract the topic information at document level, and this could be combined in a language model for speech recognition. In what follows, we first address the construction of an LDA model from a set of training documents. The optimization objective is formulated for model training. Then the variational Bayes (VB) inference is detailed. The variational distribution over multiple latent variables is introduced to find a VB solution to the LDA model.

7.6.1 Model construction

LDA provides a powerful mechanism to discover latent topic structure from a bag of M documents with a bag of words $\mathbf{w} = \{w_1, \cdots, w_N\}$, where $w_n \in \mathcal{V}$. The text corpus is denoted by $\mathcal{D} = \{\mathbf{w}_1, \cdots, \mathbf{w}_M\}$. Figure 7.2 is a graphical representation of LDA. Using LDA, each of the n words w_n is represented by a multinomial distribution conditioned on the topic z_n:

$$w_n | z_n, \boldsymbol{\beta} \sim \text{Multi}(\boldsymbol{\beta}), \tag{7.310}$$

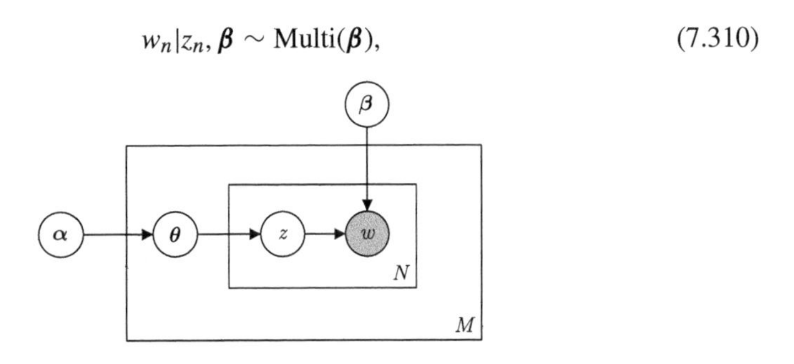

Figure 7.2 Representation of latent Dirichlet allocation.

where $\boldsymbol{\beta} \in \mathbb{R}^{|\mathcal{V}|\times K}$ denotes the multinomial parameters. The topic z_n of word w_n is also generated by a multinomial distribution given parameters $\boldsymbol{\theta} \in \mathbb{R}^K$:

$$z_n|\boldsymbol{\theta} \sim \text{Multi}(\boldsymbol{\theta}), \tag{7.311}$$

where $\theta_k \geq 0$, $\sum_{k=1}^K \theta_k = 1$ and K topics are assumed. Importantly, the latent topics of each document are treated as random variables. We assume that the multinomial parameters $\boldsymbol{\theta}$ of K topics are drawn from a Dirichlet distribution,

$$\begin{aligned}\boldsymbol{\theta}|\boldsymbol{\alpha} &\sim \text{Dir}(\boldsymbol{\alpha})\\ &= \frac{\Gamma(\sum_{k=1}^K \alpha_k)}{\prod_{k=1}^K \Gamma(\alpha_k)}\theta_1^{\alpha_1-1}\cdots\theta_K^{\alpha_K-1},\end{aligned} \tag{7.312}$$

where $\Gamma(\cdot)$ is the gamma function and $\boldsymbol{\alpha}$ is a K-vector parameter with component $\alpha_k > 0$. This K-dimensional Dirichlet random variable $\boldsymbol{\theta}$ lies in the $(K-1)$-simplex. Thus, LDA parameters $\{\boldsymbol{\alpha}, \boldsymbol{\beta}\}$ contain the K-dimensional Dirichlet parameters $\boldsymbol{\alpha} = [\alpha_1, \cdots, \alpha_K]^\top$ for K topic mixtures $\boldsymbol{\theta}$ and the topic-dependent unigram parameters $\boldsymbol{\beta} = \{\beta_{kv}\} = \{p(w = v|k)\}$. The parameters $\boldsymbol{\alpha}$ and $\boldsymbol{\beta}$ are estimated by maximizing the marginal likelihood of a text corpus $\mathcal{D}$ or an N-word document $\mathbf{w}$ over the topic mixtures $\boldsymbol{\theta}$ and the topic labels $\mathbf{z} = \{z_n\}$:

$$\{\boldsymbol{\alpha}^{\text{ML2}}, \boldsymbol{\beta}^{\text{ML2}}\} = \arg\max_{\{\boldsymbol{\alpha},\boldsymbol{\beta}\}} p(\mathbf{w}|\boldsymbol{\alpha}, \boldsymbol{\beta}), \tag{7.313}$$

where

$$p(\mathbf{w}|\boldsymbol{\alpha}, \boldsymbol{\beta}) = \int p(\boldsymbol{\theta}|\boldsymbol{\alpha})\left(\prod_{n=1}^N \sum_{k=1}^K p(z_n = k|\boldsymbol{\theta}) \times p(w_n = v|z_n = k, \boldsymbol{\beta})\right)d\boldsymbol{\theta}. \tag{7.314}$$

The marginal likelihood is calculated by integrating over continuous variable $\boldsymbol{\theta}$ and discrete variable $\mathbf{z}$. Note that LDA parameters $\{\boldsymbol{\alpha}^{\text{ML2}}, \boldsymbol{\beta}^{\text{ML2}}\}$ are estimated according to the type-2 maximum likelihood method, as addressed in Section 5.1.2, because the likelihood function considers all possible values of topic mixtures in different topics. Latent variables in LDA include topic mixtures and topic labels $\{\boldsymbol{\theta}, \mathbf{z}\}$. Strictly speaking, the LDA model parameters Θ contain $\{\boldsymbol{\theta}, \boldsymbol{\beta}\}$ while the hyperparameters Ψ should contain $\boldsymbol{\alpha}$. But, using LDA, only the parameters $\boldsymbol{\theta}$ are integrated out. Without loss of generality, we treat both $\boldsymbol{\alpha}$ and $\boldsymbol{\beta}$ as LDA parameters to be optimized from training data $\mathbf{w}$.

However, in model inference, we should apply the EM algorithm which involves a calculation of posterior distribution of latent variables $\{\boldsymbol{\theta}, \mathbf{z}\}$ given a document $\mathbf{w}$:

$$p(\boldsymbol{\theta}, \mathbf{z}|\mathbf{w}, \boldsymbol{\alpha}, \boldsymbol{\beta}) = \frac{p(\boldsymbol{\theta}, \mathbf{z}, \mathbf{w}|\boldsymbol{\alpha}, \boldsymbol{\beta})}{p(\mathbf{w}|\boldsymbol{\alpha}, \boldsymbol{\beta})}. \tag{7.315}$$

Unfortunately, this distribution is intractable because the normalization term

$$p(\mathbf{w}|\boldsymbol{\alpha}, \boldsymbol{\beta}) = \frac{\Gamma(\sum_{k=1}^K \alpha_k)}{\prod_{k=1}^K \Gamma(\alpha_k)}\int\left(\prod_{k=1}^K \theta_k^{\alpha_k-1}\right) \times \left(\prod_{n=1}^N \sum_{k=1}^K \prod_{v=1}^{|\mathcal{V}|} (\theta_k\beta_{kv})^{w_n^v}\right)d\boldsymbol{\theta} \tag{7.316}$$

is an intractable function due to the *coupling* between $\boldsymbol{\theta}$ and $\boldsymbol{\beta}$ in the summation over latent topics. Here, the superscript v in w_n^v denotes the component index, i.e., $w_n^v = 1$

and $w_n^j = 0$ for $j \neq v$. Next, we address the variational Bayes (VB) inference procedure, which is divided into three steps, namely finding the lower bound, finding the variational parameters, and finding the model parameters.

7.6.2 VB inference: lower bound

Although the posterior distribution is intractable for exact inference, a variety of approximate inference algorithms can be used for LDA including a Laplace approximation, variational approximation, and Markov chain Monte Carlo. In this section, a simple convexity-based variational inference is introduced to implement an LDA model by using Jensen's inequality. A *variational distribution*,

$$q(\boldsymbol{\theta}, \mathbf{z}|\boldsymbol{\gamma}, \boldsymbol{\phi}) = q(\boldsymbol{\theta}|\boldsymbol{\gamma}) \prod_{n=1}^{N} q(z_n|\phi_n), \tag{7.317}$$

is used as a surrogate for the posterior distribution $p(\boldsymbol{\theta}, \mathbf{z}|\mathbf{w}, \boldsymbol{\alpha}, \boldsymbol{\beta})$, where the *variational parameters* $\boldsymbol{\gamma}$ and $\boldsymbol{\phi}$ are estimated via an optimization. According to Jensen's inequality using the convex function $-\log(\cdot)$, we have

$$\begin{aligned}
\log p(\mathbf{w}|\boldsymbol{\alpha}, \boldsymbol{\beta}) &= \log \int \sum_{\mathbf{z}} p(\boldsymbol{\theta}, \mathbf{z}, \mathbf{w}|\boldsymbol{\alpha}, \boldsymbol{\beta}) d\boldsymbol{\theta} \\
&= \log \int \sum_{\mathbf{z}} \frac{p(\boldsymbol{\theta}, \mathbf{z}, \mathbf{w}|\boldsymbol{\alpha}, \boldsymbol{\beta}) q(\boldsymbol{\theta}, \mathbf{z})}{q(\boldsymbol{\theta}, \mathbf{z})} d\boldsymbol{\theta} \\
&\geq \int \sum_{\mathbf{z}} q(\boldsymbol{\theta}, \mathbf{z}) \log p(\boldsymbol{\theta}, \mathbf{z}, \mathbf{w}|\boldsymbol{\alpha}, \boldsymbol{\beta}) d\boldsymbol{\theta} \\
&\quad - \int \sum_{\mathbf{z}} q(\boldsymbol{\theta}, \mathbf{z}) \log q(\boldsymbol{\theta}, \mathbf{z}) d\boldsymbol{\theta} \\
&= \mathbb{E}_{(\boldsymbol{\theta}, \mathbf{z})}[\log p(\boldsymbol{\theta}, \mathbf{z}, \mathbf{w}|\boldsymbol{\alpha}, \boldsymbol{\beta})] - \mathbb{E}_{(\boldsymbol{\theta}, \mathbf{z})}[\log q(\boldsymbol{\theta}, \mathbf{z})] \\
&\triangleq \mathcal{L}[\boldsymbol{\gamma}, \boldsymbol{\phi}; \boldsymbol{\alpha}, \boldsymbol{\beta}] \triangleq \mathcal{F}[q(\boldsymbol{\theta}|\boldsymbol{\gamma}), q(\mathbf{z}|\boldsymbol{\phi})],
\end{aligned} \tag{7.318}$$

where $\mathbb{E}[\cdot]$ denotes the expectation operation and variational parameters $\boldsymbol{\gamma}$ and $\boldsymbol{\phi}$ are omitted for simplicity. Jensen's inequality provides us with a lower bound $\mathcal{L}[\boldsymbol{\gamma}, \boldsymbol{\phi}; \boldsymbol{\alpha}, \boldsymbol{\beta}]$ on the logarithm of marginal likelihood, given an arbitrary variational distribution $q(\boldsymbol{\theta}, \mathbf{z}|\boldsymbol{\gamma}, \boldsymbol{\phi})$. It can be easily verified that the difference between the left-hand-side and the right-hand-side of Eq. (7.318) is the KL divergence between the variational posterior distribution and the true posterior distribution. We have

$$\log p(\mathbf{w}|\boldsymbol{\alpha}, \boldsymbol{\beta}) = \mathcal{L}(\boldsymbol{\gamma}, \boldsymbol{\phi}; \boldsymbol{\alpha}, \boldsymbol{\beta}) + \mathrm{KL}(q(\boldsymbol{\theta}, \mathbf{z}|\boldsymbol{\gamma}, \boldsymbol{\phi}) \| p(\boldsymbol{\theta}, \mathbf{z}|\mathbf{w}, \boldsymbol{\alpha}, \boldsymbol{\beta})). \tag{7.319}$$

Therefore, maximizing the lower bound $\mathcal{L}[\boldsymbol{\gamma}, \boldsymbol{\phi}; \boldsymbol{\alpha}, \boldsymbol{\beta}]$ with respect to $\boldsymbol{\gamma}$ and $\boldsymbol{\phi}$ is equivalent to finding the optimal variational distribution $q(\boldsymbol{\theta}, \mathbf{z}|\hat{\boldsymbol{\gamma}}, \hat{\boldsymbol{\phi}})$ with variational parameters $\hat{\boldsymbol{\gamma}}$ and $\hat{\boldsymbol{\phi}}$, which is closest to the true posterior distribution $p(\boldsymbol{\theta}, \mathbf{z}|\mathbf{w}, \boldsymbol{\alpha}, \boldsymbol{\beta})$. To do so, the lower bound is expanded by

$$\begin{aligned}
\mathcal{L}(\boldsymbol{\gamma}, \boldsymbol{\phi}; \boldsymbol{\alpha}, \boldsymbol{\beta}) &= \mathbb{E}_{(\boldsymbol{\theta})}[\log p(\boldsymbol{\theta}|\boldsymbol{\alpha})] + \mathbb{E}_{(\boldsymbol{\theta}, \mathbf{z})}[\log p(\mathbf{z}|\boldsymbol{\theta})] \\
&\quad + \mathbb{E}_{(\mathbf{z})}[\log p(\mathbf{w}|\mathbf{z}, \boldsymbol{\beta})] - \mathbb{E}_{(\boldsymbol{\theta})}[\log q(\boldsymbol{\theta})] - \mathbb{E}_{(\mathbf{z})}[\log q(\mathbf{z})].
\end{aligned} \tag{7.320}$$

Using the fact that the expectation of the sufficient statistics is equivalent to the derivative of the log normalization factor with respect to the natural parameter, we obtain (Blei *et al.* 2003)

$$\mathbb{E}_{(\boldsymbol{\theta})}[\log \theta_k|\boldsymbol{\alpha}] = \Psi(\alpha_k) - \Psi\left(\sum_{i=1}^{K} \alpha_i\right), \tag{7.321}$$

where Ψ is the first derivative of the log gamma function, also called the di-gamma function, as used in Eqs. (5.82) and (7.81). The lower bound is further expanded in terms of variational parameters $\{\boldsymbol{\gamma}, \boldsymbol{\phi}\}$ and model parameters $\{\boldsymbol{\alpha}, \boldsymbol{\beta}\}$ by

$$\begin{aligned}
\mathcal{L}(\boldsymbol{\gamma}, \boldsymbol{\phi}; \boldsymbol{\alpha}, \boldsymbol{\beta}) = {} & \log \Gamma\left(\sum_{k=1}^{K} \alpha_k\right) - \sum_{k=1}^{K} \log \Gamma(\alpha_k) \\
& + \sum_{k=1}^{K} (\alpha_k - 1)\left(\Psi(\gamma_k) - \Psi\left(\sum_{i=1}^{K} \gamma_i\right)\right) \\
& + \sum_{n=1}^{N} \sum_{k=1}^{K} \phi_{nk}\left(\Psi(\gamma_k) - \Psi\left(\sum_{i=1}^{K} \gamma_i\right)\right) \\
& + \sum_{n=1}^{N} \sum_{k=1}^{K} \sum_{v=1}^{|\mathcal{V}|} \phi_{nk} w_n^v \log \beta_{kv} - \log \Gamma\left(\sum_{k=1}^{K} \gamma_k\right) \\
& + \sum_{k=1}^{K} \log \Gamma(\gamma_k) - \sum_{k=1}^{K} (\gamma_k - 1)\left(\Psi(\gamma_k) - \Psi\left(\sum_{i=1}^{K} \gamma_i\right)\right) \\
& - \sum_{n=1}^{N} \sum_{k=1}^{K} \phi_{nk} \log \phi_{nk}.
\end{aligned} \tag{7.322}$$

Typically, finding the lower bound of marginal likelihood in VB inference is equivalent to performing the VB E-step. However, we need to estimate the optimal variational parameters $\boldsymbol{\gamma}$ and $\boldsymbol{\phi}$ to finalize the VB E-step.

7.6.3 VB inference: variational parameters

The variational Dirichlet parameters $\boldsymbol{\gamma}$ and variational multinomial parameters $\boldsymbol{\phi}$ are estimated by maximizing the lower bound in Eq. (7.322). The terms related to $\boldsymbol{\gamma}$ are collected and arranged thus:

$$\begin{aligned}
\mathcal{L}(\boldsymbol{\gamma}) = {} & \sum_{k=1}^{K} \left(\Psi(\gamma_k) - \Psi\left(\sum_{i=1}^{K} \gamma_i\right)\right)\left(\alpha_k + \sum_{n=1}^{N} \phi_{nk} - \gamma_k\right) \\
& - \log \Gamma\left(\sum_{k=1}^{K} \gamma_k\right) + \sum_{k=1}^{K} \log \Gamma(\gamma_k).
\end{aligned} \tag{7.323}$$

Differentiating with respect to the individual parameter γ_k, we have

$$\frac{\partial \mathcal{L}(\boldsymbol{\gamma})}{\partial \gamma_k} = \Psi'(\gamma_k)\left(\alpha_k + \sum_{n=1}^{N} \phi_{nk} - \gamma_k\right) - \Psi'\left(\sum_{i=1}^{K} \gamma_i\right) \sum_{i=1}^{K} \left(\alpha_i + \sum_{n=1}^{N} \phi_{ni} - \gamma_i\right). \tag{7.324}$$

Setting this equation to zero yields the optimal variational parameters

$$\hat{\gamma}_k = \alpha_k + \sum_{n=1}^{N} \phi_{nk} \quad 1 \le k \le K. \tag{7.325}$$

Note that the variational Dirichlet parameters $\hat{\boldsymbol{\gamma}} = \{\hat{\gamma}_k\}$ are seen as the surrogate of the Dirichlet model parameters $\boldsymbol{\alpha}$, which sufficiently reflect the topic mixture probabilities $\boldsymbol{\theta}$.

On the other hand, when optimizing the lower bound $\mathcal{L}(\boldsymbol{\gamma}, \boldsymbol{\phi}; \boldsymbol{\alpha}, \boldsymbol{\beta})$ with respect to variational multinomial parameters $\boldsymbol{\phi}$, a constrained maximization problem should be tackled under the constraint

$$\sum_{k=1}^{K} \phi_{nk} = 1. \tag{7.326}$$

Therefore, we collect the terms in the lower bound related to the individual variational parameter ϕ_{nk} and form the Lagrangian with the Lagrange multiplier λ_n:

$$\mathcal{L}(\phi_{nk}) = \phi_{nk}\left(\Psi(\gamma_k) - \Psi\left(\sum_{i=1}^{K} \gamma_i\right)\right) + \phi_{nk} \log \beta_{kv} - \phi_{nk} \log \phi_{nk} + \lambda_n \left(\sum_{i=1}^{K} \phi_{ni} - 1\right), \tag{7.327}$$

where the unique word v is selected for w_n such that $w_n^v = 1$ and $w_n^j = 0$ for $j \neq v$. Differentiating this Lagrangian with respect to ϕ_{nk} yields

$$\frac{\partial \mathcal{L}(\phi_{nk})}{\partial \phi_{nk}} = \Psi(\gamma_k) - \Psi\left(\sum_{i=1}^{K} \gamma_i\right) + \log \beta_{kv} - \log \phi_{nk} - 1 + \lambda_n. \tag{7.328}$$

We set this differentiation to zero and derive the variational parameter ϕ_{nk} which is written as a function of Lagrange multiplier λ_n. By substituting the derived variational parameter ϕ_{nk} into the constraint given in Eq. (7.326), we can estimate the multiplier and then obtain the optimal variational parameters:

$$\hat{\phi}_{nk} = \frac{\beta_{kv} \exp\left(\Psi(\gamma_k) - \Psi\left(\sum_{i=1}^{K} \gamma_i\right)\right)}{\sum_{j=1}^{K} \beta_{jv} \exp\left(\Psi(\gamma_j) - \Psi\left(\sum_{i=1}^{K} \gamma_i\right)\right)} \quad 1 \le n \le N, \quad 1 \le k \le K. \tag{7.329}$$

7.6.4 VB inference: model parameters

After finding the optimal variational parameters $\{\boldsymbol{\gamma}, \boldsymbol{\phi}\}$, we fix these parameters and substitute the optimal variational distribution $q(\boldsymbol{\theta}, \mathbf{z}|\hat{\boldsymbol{\gamma}}, \hat{\boldsymbol{\phi}})$ to calculate the updated lower bound $\mathcal{L}(\hat{\boldsymbol{\gamma}}, \hat{\boldsymbol{\phi}}; \boldsymbol{\alpha}, \boldsymbol{\beta})$. The VB M-step treats the variational lower bound $\mathcal{L}(\hat{\boldsymbol{\gamma}}, \hat{\boldsymbol{\phi}}; \boldsymbol{\alpha}, \boldsymbol{\beta})$ as a surrogate of the intractable marginal log likelihood $\log p(\mathbf{w}|\boldsymbol{\alpha}, \boldsymbol{\beta})$, and estimates the LDA parameters by

$$\{\hat{\boldsymbol{\alpha}}, \hat{\boldsymbol{\beta}}\} = \arg \max_{\{\boldsymbol{\alpha}, \boldsymbol{\beta}\}} \mathcal{L}(\hat{\boldsymbol{\gamma}}, \hat{\boldsymbol{\phi}}; \boldsymbol{\alpha}, \boldsymbol{\beta}). \tag{7.330}$$

First, we deal with the optimization over the conditional multinomial distributions $\boldsymbol{\beta} = \{\beta_{kv}\} = \{p(w_n = v|z_n = k)\}$. The lower bound is hereafter calculated from a text corpus $\mathcal{D} = \{\mathbf{w}_1, \cdots, \mathbf{w}_M\}$. The terms containing model parameters $\boldsymbol{\beta}$ are collected and the constraints

$$\sum_{v=1}^{|\mathcal{V}|} \beta_{kv} = 1 \tag{7.331}$$

are imposed, so as to form the Lagrangian with Lagrange multipliers $\{\lambda_k\}$:

$$\begin{aligned} \mathcal{L}(\boldsymbol{\beta}) = & \sum_{d=1}^{M} \sum_{n=1}^{N_d} \sum_{k=1}^{K} \sum_{v=1}^{|\mathcal{V}|} \hat{\phi}_{dnk} w_{dn}^{v} \log \beta_{kv} \\ & + \sum_{k=1}^{K} \lambda_k \left(\sum_{v=1}^{|\mathcal{V}|} \beta_{kv} - 1 \right). \end{aligned} \tag{7.332}$$

We differentiate this Lagrangian with respect to individual β_{kv} and set it to zero to find the optimal conditional multinomial distributions:

$$\begin{aligned} \hat{\beta}_{kv} = & \frac{\sum_{d=1}^{M} \sum_{n=1}^{N_d} \hat{\phi}_{dnk} w_{dn}^{v}}{\sum_{m=1}^{|\mathcal{V}|} \sum_{d=1}^{M} \sum_{n=1}^{N_d} \hat{\phi}_{dnk} w_{dn}^{m}} \\ & 1 \le k \le K, \quad 1 \le v \le |\mathcal{V}|. \end{aligned} \tag{7.333}$$

To deal with the optimization over the Dirichlet parameters $\boldsymbol{\alpha} = \{\alpha_k\}$, we collect the terms in the lower bound which contain $\boldsymbol{\alpha}$ and give

$$\begin{aligned} \mathcal{L}(\boldsymbol{\alpha}) = \sum_{d=1}^{M} \Bigg[& \log \Gamma \left(\sum_{k=1}^{K} \alpha_k \right) - \sum_{k=1}^{K} \log \Gamma(\alpha_k) \\ & + \sum_{k=1}^{K} (\alpha_k - 1) \left(\Psi(\hat{\gamma}_{dk}) - \Psi \left(\sum_{i=1}^{K} \hat{\gamma}_{di} \right) \right) \Bigg]. \end{aligned} \tag{7.334}$$

Differentiating with respect to individual α_k gives

$$\begin{aligned} \frac{\partial \mathcal{L}(\boldsymbol{\alpha})}{\partial \alpha_k} = & M \left(\Psi \left(\sum_{i=1}^{K} \alpha_i \right) - \Psi(\alpha_k) \right) \\ & + \sum_{d=1}^{M} \left(\Psi(\hat{\gamma}_{dk}) - \Psi \left(\sum_{i=1}^{K} \hat{\gamma}_{di} \right) \right). \end{aligned} \tag{7.335}$$

There is no closed-form solution to the optimal Dirichlet parameter α_k, since the right-hand-side of Eq. (7.335) depends on α_i where $i \neq k$. We should use an iterative algorithm to find the $K \times 1$ optimal parameter vector $\boldsymbol{\alpha}$. Here, the Newton–Raphson optimization is applied to find the optimal parameters by iteratively performing

$$\boldsymbol{\alpha}^{(\tau+1)} = \boldsymbol{\alpha}^{(\tau)} - (\mathbf{H}(\boldsymbol{\alpha}^{(\tau)}))^{-1}\nabla\mathcal{L}(\boldsymbol{\alpha}^{(\tau)}), \tag{7.336}$$

where τ is the iteration index, $\nabla\mathcal{L}$ is the $K \times 1$ gradient vector and $\mathbf{H}$ is the $K \times K$ Hessian matrix consisting of the second-order differentiations in different entries:

$$\frac{\partial\mathcal{L}(\boldsymbol{\alpha})}{\partial\alpha_k\alpha_j} = \delta(k,j)M\Psi'(\alpha_k) - \Psi'\left(\sum_{i=1}^{K}\alpha_i\right), \tag{7.337}$$

where $\delta(k,j)$ denotes a Kronecker delta function. The inverse of Hessian matrix $(\mathbf{H}(\boldsymbol{\alpha}^{(\tau)}))^{-1}$ can be obtained by applying the Woodbury matrix inversion. As a result, LDA model parameters $\{\hat{\boldsymbol{\alpha}}, \hat{\boldsymbol{\beta}}\}$ are estimated in a VB M-step. The VB inference based on an EM algorithm is accordingly completed by maximizing the variational lower bound of marginal likelihood $\mathcal{L}(\boldsymbol{\gamma}, \boldsymbol{\phi}; \boldsymbol{\alpha}, \boldsymbol{\beta})$ through performing a VB E-step for updating variational parameters $\{\boldsymbol{\gamma}, \boldsymbol{\phi}\}$ and a VB M-step for estimating model parameters $\{\boldsymbol{\alpha}, \boldsymbol{\beta}\}$. The increase of the lower bound is assured by VB–EM iterations.

7.7 Latent topic language model

LDA is established as a latent topic model which is designed for document representation by using a bag of words in the form of document $\mathbf{w}$ or text corpus $\mathcal{D}$. LDA has been successfully extended for language modeling and applied for continuous speech recognition in Tam & Schultz (2005) and Chien & Chueh (2011). However, before the LDA language model, the topic model based on PLSA was constructed and merged in the n-gram language model (Gildea & Hofmann 1999), as addressed in Section 3.7.3. But, the PLSA-based topic model suffers from the weaknesses of poor generalization and redundant model complexity. The performance of the resulting PLSA language model is limited. Compared to the PLSA language model, we are more interested in the LDA language model and its application in speech recognition. In the literature, the LDA-based topic model is incorporated into the n-gram language model based on an indirect method (Tam & Schultz 2005) and a direct method (Chien & Chueh 2011), which are introduced in Section 7.7.1 and Section 7.7.2, respectively.

7.7.1 LDA language model

LDA is generally *indirect* for characterizing the n-gram regularities of a current word w_i given its history words w_{i-n+1}^{i-1}. The word index n in the document model differs from i in the language model. A document $\mathbf{w}$ has N words while a sentence has T words. The hierarchical Dirichlet language model in Section 5.3 was presented as an alternative to language model smoothing (MacKay & Peto 1995). In Yaman, Chien & Lee (2007), the hierarchical Dirichlet priors were estimated by a maximum a-posteriori method for language model adaptation. These two methods have adopted the Dirichlet priors to explore the structure of a language model from lower-order n-gram to higher-order n-gram. There was no topic-based language model involved. In Wallach (2006), the LDA bi-gram was proposed by considering a bag of bi-gram events from the collected documents in construction of an LDA. This LDA bi-gram was neither derived nor specifically employed for speech recognition. The basic LDA model ignores the word order and is not in accordance with sentence generation in speech recognition. In Tam & Schultz (2005, 2006), the LDA model parameters $\{\hat{\boldsymbol{\alpha}}, \hat{\boldsymbol{\beta}}\}$ were calculated from training documents $\mathcal{D}$ and then used to estimate the online topic probabilities or the variational Dirichlet parameters $\hat{\boldsymbol{\gamma}} = \{\hat{\gamma}_k\}$. The online parameters $\hat{\boldsymbol{\gamma}}$ were estimated by treating all history words in a sentence w_1^{i-1} (Tam & Schultz 2005), or even the transcription of a whole sentence w_1^T (Tam & Schultz 2006) as a single *document* $\mathbf{w}$.

Similarly to the PLSA n-gram as illustrated in Section 3.7.3, the LDA n-gram is formed as a soft-clustering model or a topic mixture model, which is calculated by combining the topic probabilities $p(z_i = k|w_1^{i-1})$ driven by history words w_1^{i-1} and the topic-dependent unigrams $\boldsymbol{\beta} = \{\beta_{kv}\} = \{p(w_i = v|z_i = k)\}$ of current word w_i. The combination is marginalized over different topics:

$$\begin{aligned} p_{\text{LDA}}(w_i = v|w_1^{i-1}) &= \sum_{k=1}^{K} p(w_i = v|z_i = k)p(z_i = k|w_1^{i-1}) \\ &\approx \sum_{k=1}^{K} \beta_{kv} \frac{\hat{\gamma}_k}{\sum_{j=1}^{K} \hat{\gamma}_j}. \end{aligned} \tag{7.338}$$

Here, the topic multinomial distributions can be driven either by the history words $p(z_i = k|w_1^{i-1})$ or by the whole sentence words $p(z_i = k|w_1^T)$. These multinomial distributions are approximated and proportional to the variational Dirichlet parameters $\hat{\boldsymbol{\gamma}}$ as calculated in Eq. (7.325). In this implementation, the pre-trained model parameters $\{\hat{\boldsymbol{\alpha}}, \hat{\boldsymbol{\beta}}\}$ and the online estimated variational parameters $\{\hat{\boldsymbol{\gamma}}, \hat{\boldsymbol{\phi}}\}$ should be calculated to determine the LDA n-gram $p_{\text{LDA}}(w_i|w_1^{i-1})$ in an online fashion. In Tam & Schultz (2005), the LDA language model was improved by further adapting the standard ML-based n-gram model using the LDA n-gram according to a linear interpolation scheme with a parameter $0 < \lambda < 1$:

$$\hat{p}(w_i|w_1^{i-1}) = \lambda p_{\text{ML}}(w_i|w_{i-n+1}^{i-1}) + (1 - \lambda)p_{\text{LDA}}(w_i|w_1^{i-1}). \tag{7.339}$$

In Tam & Schultz (2006), the language model adaptation based on the unigram rescaling was implemented by

$$\hat{p}(w_i|w_1^{i-1}) = p_{\text{ML}}(w_i|w_{i-n+1}^{i-1}) \frac{p_{\text{LDA}}(w_i|w_1^{i-1})}{p_{\text{ML}}(w_i)}. \tag{7.340}$$

Typically, the model parameters $\{\hat{\boldsymbol{\alpha}}, \hat{\boldsymbol{\beta}}\}$ in the LDA language model are inferred at the document level, catching the long-distance topic information but only *indirectly* representing the n-gram events.

7.7.2 Dirichlet class language model

LDA (Blei *et al.* 2003) builds a hierarchical Bayesian topic model and extracts the latent topics or clusters from a collection of M documents $\mathcal{D} = \{\mathbf{w}_m\}$. The bag-of-words scheme is adopted without considering the word sequence, and so it is not directly designed for speech recognition where the performance is seriously affected by the prior probability of a word string $W = w_1^T \triangleq \{w_1, \cdots, w_T\}$ or the language model $p(W)$. In Bengio, Ducharme, Vincent *et al.* (2003), the neural network language model was proposed to deal with the data sparseness problem in language modeling by projecting the ordered history vector into a continuous space and then calculating the n-gram language model based on the multilayer perceptron. MLP involves the error back-propagation training algorithm based on the least-squares estimation, which is vulnerable to the overfitting problem (Bishop 2006). Considering the topic modeling in LDA and the continuous representation of history word sequence w_{i-n+1}^{i-1} in a neural network language model, the Dirichlet class language model (DCLM) (Chien & Chueh 2011) is presented to build a *direct* LDA language model for speech recognition. For a vocabulary with $|\mathcal{V}|$ words, the $n-1$ history words w_{i-n+1}^{i-1} are first represented by a $(n-1)|\mathcal{V}| \times 1$ vector $\mathbf{h}_{i-n+1}^{i-1}$ consisting of $n-1$ block subvectors. Each block is represented by the 1-to-$|\mathcal{V}|$ coding scheme with a $|\mathcal{V}|$ dimensional vector where the vth word of vocabulary is encoded by setting the vth entry of the vector to be one and all the other entries to be zero. The order of history words is considered in $\mathbf{h}_{i-n+1}^{i-1}$. Figure 7.3 shows the system architecture of calculating a DCLM $p_{\text{DC}}(w_i = v|w_{i-n+1}^{i-1})$. A global projection is involved to project the ordered history vector $\mathbf{h}_{i-n+1}^{i-1}$ into a latent topic or class space where the projection $\mathbf{g}(\mathbf{h}_{i-n+1}^{i-1})$ could be either a linear function or a non-linear function. In the case of a linear function, a projection matrix $\mathbf{A} = [\mathbf{a}_1, \cdots, \mathbf{a}_C]$ consisting of

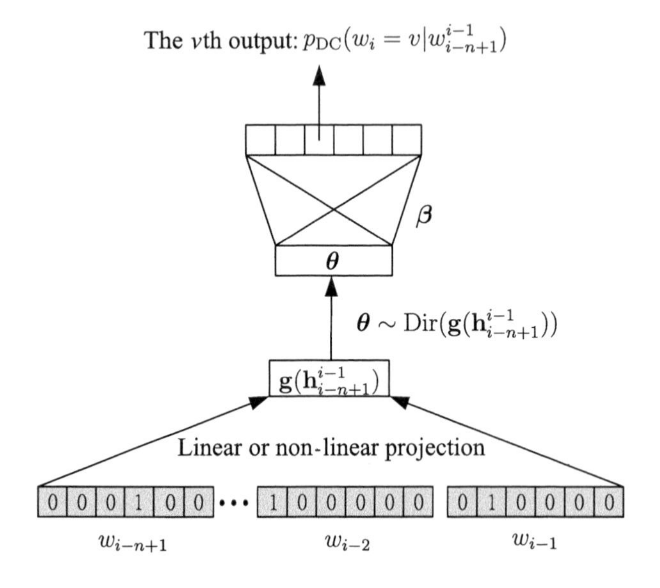

Figure 7.3 System architecture for Dirichlet class language model.

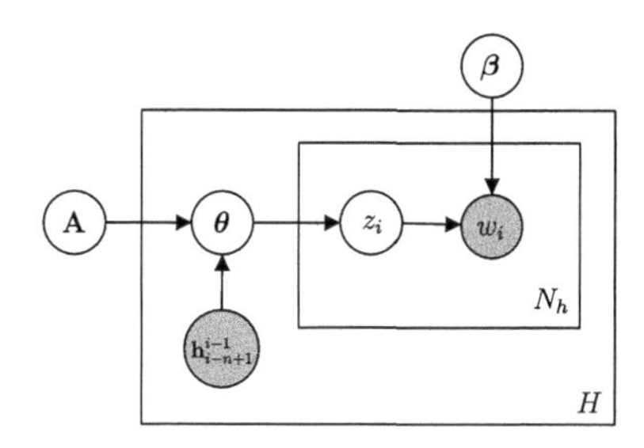

Figure 7.4 Representation of a Dirichlet class language model.

C basis vectors $\{\mathbf{a}_c\}$ is involved. The class structure of all n-gram events from training corpus $\mathcal{D} = \{w_i, w_{i-n+1}^{i-1}\}$ is represented by Dirichlet distributions. The class uncertainty is compensated by marginalizing the likelihood function over the Dirichlet priors. The latent structure in the DCLM reflects the class of an n-gram event rather than the topic in an LDA model. A DCLM is regarded as a kind of class-based language model, which is inferred by the variational Bayes (VB) procedure by maximizing the variational lower bound of a marginal likelihood of training data.

7.7.3 Model construction

A DCLM acts as a Bayesian class-based language model, which involves the prior distribution of the class variable. Figure 7.4 is a graphical representation of a DCLM from a training set of current words and history words $\mathcal{D} = \{w_i, w_{i-n+1}^{i-1}\} = \{w_i, \mathbf{h}_{i-n+1}^{i-1}\}$. The training corpus has H history events in $\{w_{i-n+1}^{i-1}\}$. Each history event $h = w_{i-n+1}^{i-1}$ has N_h possible predicted words $\{w_i\}$. Note that the $(n-1)|\mathcal{V}|$-dimensional discrete history vector is projected to a C-dimensional continuous class space using the class-dependent linear discriminant function

$$g_c(\mathbf{h}_{i-n+1}^{i-1}) = \mathbf{a}_c^\mathsf{T} \mathbf{h}_{i-n+1}^{i-1}. \tag{7.341}$$

This function is used as the hyperparameter of a Dirichlet prior for the class mixture probability θ_c or equivalently the class posterior probability of latent variable $z_i = c$ given history vector $\mathbf{h}_{i-n+1}^{i-1}$:

$$\theta_c \triangleq p(z_i = c|\boldsymbol{\theta}, \mathbf{h}_{i-n+1}^{i-1}). \tag{7.342}$$

Thus, the uncertainty of class posterior probabilities $\mathbf{g}(\mathbf{h}_{i-n+1}^{i-1}) = \{g_c(\mathbf{h}_{i-n+1}^{i-1})\}$ is characterized by a Dirichlet prior distribution:

$$\begin{aligned}\boldsymbol{\theta} = [\theta_1, \cdots, \theta_C]^\mathsf{T} | \mathbf{h}_{i-n+1}^{i-1}, \mathbf{A} &\sim \mathrm{Dir}(\mathbf{g}(\mathbf{h}_{i-n+1}^{i-1})) \\ &= \mathrm{Dir}(\mathbf{A}^\mathsf{T} \mathbf{h}_{i-n+1}^{i-1}),\end{aligned} \tag{7.343}$$

subject to the constraint $g_c(\mathbf{h}_{i-n+1}^{i-1}) > 0$ or $\mathbf{a}_c > 0$. Each word $w_i = v$ is generated by the conditional probability $\beta_{cv} \triangleq p(w_i = v|z_i = c)$ given a latent class $z_i = c$.

The n-gram probability obtained using DCLM is calculated by marginalizing the joint likelihood over the latent variables including class mixtures $\boldsymbol{\theta}$ and C latent classes:

$$\begin{aligned}
p_{\text{DC}}(w_i = v|\mathbf{h}_{i-n+1}^{i-1}, \mathbf{A}, \boldsymbol{\beta}) &= \sum_{c=1}^{C} p(w_i = v|z_i = c, \boldsymbol{\beta}) p(z_i = c|\mathbf{h}_{i-n+1}^{i-1}, \mathbf{A}) \\
&= \sum_{c=1}^{C} p(w_i = v|z_i = c, \boldsymbol{\beta}) \\
&\quad \times \int p(\boldsymbol{\theta}|\mathbf{h}_{i-n+1}^{i-1}, \mathbf{A}) p(z_i = c|\boldsymbol{\theta}) d\boldsymbol{\theta} \\
&= \sum_{c=1}^{C} \beta_{cv} \mathbb{E}_{(\boldsymbol{\theta})}[p(z_i = c|\boldsymbol{\theta}, \mathbf{h}_{i-n+1}^{i-1}, \mathbf{A})] \\
&= \sum_{c=1}^{C} \beta_{cv} \mathbb{E}_{(\boldsymbol{\theta})}[\theta_c|\mathbf{h}_{i-n+1}^{i-1}, \mathbf{A}] \\
&= \sum_{c=1}^{C} \beta_{cv} \frac{\mathbf{a}_c^{\mathsf{T}} \mathbf{h}_{i-n+1}^{i-1}}{\sum_{j=1}^{C} \mathbf{a}_j^{\mathsf{T}} \mathbf{h}_{i-n+1}^{i-1}}.
\end{aligned} \tag{7.344}$$

In Eq. (7.344), the integral is calculated over a multinomial variable $p(z_i = c|\boldsymbol{\theta}, \mathbf{h}_{i-n+1}^{i-1}, \mathbf{A}))$ with Dirichlet prior distribution $p(\boldsymbol{\theta}|\mathbf{h}_{i-n+1}^{i-1}, \mathbf{A})$ and is equivalent to the distribution mean. This integral is seen as a marginalization over different classes c, and is also obtained as a class mixture model with the class-dependent unigram probabilities $\{\beta_{cv}|1 \leq c \leq C\}$ weighted by the class mixture probabilities $\{p(z_i = c|\mathbf{h}_{i-n+1}^{i-1}, \mathbf{A})|1 \leq c \leq C\}$ driven by the ordered history vector $\mathbf{h}_{i-n+1}^{i-1}$. However, we should estimate DCLM parameters $\{\mathbf{A}, \boldsymbol{\beta}\}$ and substitute them into Eq. (7.344) to calculate the Dirichlet class (DC) n-gram probabilities.

7.7.4 VB inference: lower bound

Similarly to LDA, DCLM parameters $\{\mathbf{A}, \boldsymbol{\beta}\}$ are estimated according to the type-2 maximum likelihood method where the marginal likelihood over latent variables is maximized. As seen in Figure 7.4, the latent variables in DCLM include the continuous values of class mixture probabilities $\boldsymbol{\theta} = \{\theta_c\}$ and the discrete values of class labels $\mathbf{z} = \{z_i\}$. The optimization problem is formulated as

$$\{\mathbf{A}^{\text{ML2}}, \boldsymbol{\beta}^{\text{ML2}}\} = \arg\max_{\mathbf{A},\boldsymbol{\beta}} \log p(\mathcal{D}|\mathbf{A}, \boldsymbol{\beta}), \tag{7.345}$$

where

$$\begin{aligned}
\log p(\mathcal{D}|\mathbf{A}, \boldsymbol{\beta}) &= \sum_{\{w_i, \mathbf{h}_{i-n+1}^{i-1}\} \in \mathcal{D}} \log p(w_i|\mathbf{h}_{i-n+1}^{i-1}, \mathbf{A}, \boldsymbol{\beta}) \\
&= \sum_{\mathbf{h}_{i-n+1}^{i-1}} \log p(\mathbf{w}_h|\mathbf{h}_{i-n+1}^{i-1}, \mathbf{A}, \boldsymbol{\beta})
\end{aligned}$$

$$= \sum_{\mathbf{h}_{i-n+1}^{i-1}} \log \left(\prod_{i=1}^{N_h} p(w_i|\mathbf{h}_{i-n+1}^{i-1}, \mathbf{A}, \boldsymbol{\beta}) \right)$$

$$= \sum_{\mathbf{h}_{i-n+1}^{i-1}} \log \Bigg(\int p(\theta|\mathbf{h}_{i-n+1}^{i-1}, \mathbf{A})$$

$$\times \left(\prod_{i=1}^{N_h} \sum_{c=1}^{C} p(w_i = v|z_i = c, \boldsymbol{\beta}) p(z_i = c|\boldsymbol{\theta}) \right) d\boldsymbol{\theta} \Bigg). \tag{7.346}$$

The log marginal likelihood is calculated by integrating over the continuous $\boldsymbol{\theta}$ and summing the discrete $\{z_i = c|1 \leq c \leq C\}$. However, directly optimizing Eq. (7.346) is intractable because of the coupling between $\boldsymbol{\theta}$ and $\boldsymbol{\beta}$ in the summation over latent classes. The posterior distribution

$$p(\boldsymbol{\theta}, \mathbf{z}|\mathbf{w}_h, \mathbf{h}_{i-n+1}^{i-1}, \mathbf{A}, \boldsymbol{\beta}) = \frac{p(\boldsymbol{\theta}, \mathbf{z}, \mathbf{w}_h|\mathbf{h}_{i-n+1}^{i-1}, \mathbf{A}, \boldsymbol{\beta})}{p(\mathbf{w}_h|\mathbf{h}_{i-n+1}^{i-1}, \mathbf{A}, \boldsymbol{\beta})} \tag{7.347}$$

becomes intractable for model inference. This posterior distribution is calculated from the n-gram events with the N_h predicted words occurring after the fixed history words $h \triangleq w_{i-n+1}^{i-1}$,

$$\mathbf{w}_h \triangleq \{w_i|c(hw_i) > 0\}, \tag{7.348}$$

where $c(\cdot)$ denotes the count of an n-gram event. The variational Bayes (VB) inference procedure is involved to construct the variational DCLM where the lower bound of marginal likelihood in Eq. (7.346) serves as the surrogate to be maximized to find the optimal $\{\mathbf{A}^{\text{ML2}}, \boldsymbol{\beta}^{\text{ML2}}\}$.

To do so, a decomposed variational distribution,

$$q(\boldsymbol{\theta}, \mathbf{z}|\boldsymbol{\gamma}_h, \boldsymbol{\phi}_h) = q(\boldsymbol{\theta}|\boldsymbol{\gamma}_h) \prod_{i=1}^{N_h} q(z_i|\boldsymbol{\phi}_{h,i}), \tag{7.349}$$

is introduced to approximate the true posterior distribution $p(\boldsymbol{\theta}, \mathbf{z}|\mathbf{w}_h, \mathbf{h}_{i-n+1}^{i-1}, \mathbf{A}, \boldsymbol{\beta})$. A graphical representation of the variational DCLM is illustrated in Figure 7.5. Here, $\boldsymbol{\gamma}_h = \{\gamma_{h,c}\}$ and $\boldsymbol{\phi}_h = \{\phi_{h,ic}\}$ denote the history-dependent variational Dirichlet and multinomial parameters, respectively. By referring to Section 7.6.2, the lower bound on the log marginal likelihood is derived by applying the Jensen's inequality and is expanded by

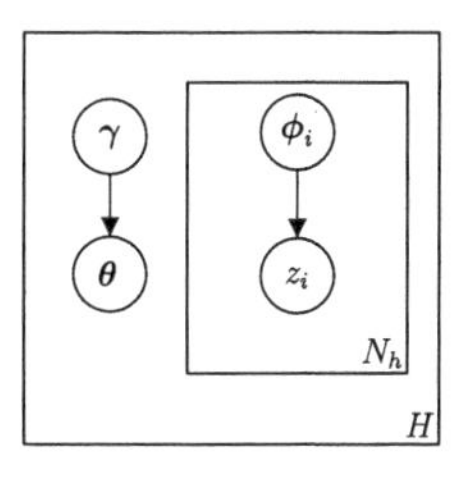

Figure 7.5 Representation of a variational Dirichlet class language model.

$$
\begin{aligned}
\mathcal{L}(\boldsymbol{\gamma}_h, \boldsymbol{\phi}_h; \mathbf{A}, \boldsymbol{\beta}) = \sum_{\mathbf{h}_{i-n+1}^{i-1}} \Bigg\{ & \mathbb{E}_{(\boldsymbol{\theta})}[\log p(\boldsymbol{\theta} | \mathbf{h}_{i-n+1}^{i-1}, \mathbf{A})] + \mathbb{E}_{(\boldsymbol{\theta}, \mathbf{z})}[\log p(\mathbf{z} | \boldsymbol{\theta})] \\
& + \mathbb{E}_{(\mathbf{z})}[\log p(\mathbf{w}_h | \mathbf{z}, \boldsymbol{\beta})] - \mathbb{E}_{(\boldsymbol{\theta})}[\log q(\boldsymbol{\theta} | \boldsymbol{\gamma}_h)] - \mathbb{E}_{(\mathbf{z})}[\log q(\mathbf{z} | \boldsymbol{\phi}_h)] \Bigg\} \\
= \sum_{\mathbf{h}_{i-n+1}^{i-1}} \Bigg\{ & \log \Gamma \left(\sum_{c=1}^{C} \mathbf{a}_c^\top \mathbf{h}_{i-n+1}^{i-1} \right) - \sum_{c=1}^{C} \log \Gamma \left(\mathbf{a}_c^\top \mathbf{h}_{i-n+1}^{i-1} \right) \\
& + \sum_{c=1}^{C} \left(\mathbf{a}_c^\top \mathbf{h}_{i-n+1}^{i-1} - 1 \right) \left(\Psi(\gamma_{h,c}) - \Psi \left(\sum_{j=1}^{C} \gamma_{h,j} \right) \right) \\
& + \sum_{i=1}^{N_h} \sum_{c=1}^{C} \phi_{h,ic} \left(\Psi(\gamma_{h,c}) - \Psi \left(\sum_{j=1}^{C} \gamma_{h,j} \right) \right) \\
& + \sum_{i=1}^{N_h} \sum_{c=1}^{C} \sum_{v=1}^{|\mathcal{V}|} \phi_{h,ic} w_i^v \log \beta_{cv} - \log \Gamma \left(\sum_{c=1}^{C} \gamma_{h,c} \right) + \sum_{c=1}^{C} \log \Gamma(\gamma_{h,c}) \\
& - \sum_{c=1}^{C} (\gamma_{h,c} - 1) \left(\Psi(\gamma_{h,c}) - \Psi \left(\sum_{j=1}^{C} \gamma_{h,j} \right) \right) - \sum_{i=1}^{N_h} \sum_{c=1}^{C} \phi_{h,ic} \log \phi_{h,ic} \Bigg\}.
\end{aligned}
\tag{7.350}
$$

We have applied Eq. (7.321) to find the variational lower bound for DCLM.

7.7.5 VB inference: parameter estimation

A VB–EM algorithm has been developed for inference of DCLM parameters. In this VB–EM procedure, we first conduct the VB E-step to find the optimal expectation function or lower bound $\mathcal{L}(\hat{\boldsymbol{\gamma}}_h, \hat{\boldsymbol{\phi}}_h; \mathbf{A}, \boldsymbol{\beta})$ in Eq. (7.350), or equivalently to estimate the optimal variational Dirichlet parameters $\hat{\boldsymbol{\gamma}}_h$ and variational multinomial parameters $\hat{\boldsymbol{\phi}}_h$. Similarly to Section 7.6.3, we collect the terms in Eq. (7.350) that are related to $\boldsymbol{\gamma}_h$ and maximize these terms, expressed by $\mathcal{L}(\boldsymbol{\gamma}_h)$, with respect to $\boldsymbol{\gamma}_h$ to find the optimal variational Dirichlet parameters:

$$
\hat{\gamma}_{h,c} = \mathbf{a}_c^\top \mathbf{h}_{i-n+1}^{i-1} + \sum_{i=1}^{N_h} \phi_{h,ic} \quad 1 \le c \le C. \tag{7.351}
$$

In addition, when maximizing the lower bound $\mathcal{L}(\boldsymbol{\gamma}_h, \boldsymbol{\phi}_h; \mathbf{A}, \boldsymbol{\beta})$ with respect to the variational multinomial parameters $\boldsymbol{\phi}_h$, we need to collect all terms related to $\boldsymbol{\phi}_h$ and solve a constrained optimization problem subject to a constraint for multinomial distributions:

$$
\sum_{c=1}^{C} \phi_{h,ic} = 1. \tag{7.352}
$$

The Lagrangian $\mathcal{L}(\boldsymbol{\phi})$ given the history-dependent Lagrange multipliers $\{\lambda_{h,i}\}$ is arranged and maximized so as to find the optimal variational multinomial distributions:

$$\hat{\phi}_{h,ic} = \frac{\beta_{cv} \exp\left(\Psi(\gamma_{h,c}) - \Psi\left(\sum_{j=1}^{C} \gamma_{h,j}\right)\right)}{\sum_{l=1}^{C} \beta_{lv} \exp\left(\Psi(\gamma_{h,l}) - \Psi\left(\sum_{j=1}^{C} \gamma_{h,j}\right)\right)}, \quad (7.353)$$
$$1 \leq i \leq T, \quad 1 \leq c \leq C$$

where the unique word v is selected for w_i such that $w_i^v = 1$ and $w_i^l = 0$ for $l \neq v$. The variational lower bound is updated with the optimal variational parameters $\mathcal{L}(\hat{\boldsymbol{\gamma}}_h, \hat{\boldsymbol{\phi}}_h; \mathbf{A}, \boldsymbol{\beta})$.

On the other hand, in a VB M-step, we fix the variational parameters $\hat{\boldsymbol{\gamma}}_h, \hat{\boldsymbol{\phi}}_h$ and optimize the updated lower bound to estimate the DCLM model parameters:

$$\{\hat{\mathbf{A}}, \hat{\boldsymbol{\beta}}\} = \arg\max_{\{\mathbf{A},\boldsymbol{\beta}\}} \mathcal{L}(\hat{\boldsymbol{\gamma}}_h, \hat{\boldsymbol{\phi}}_h; \mathbf{A}, \boldsymbol{\beta}). \quad (7.354)$$

In estimating the conditional multinomial distributions $\boldsymbol{\beta} = \{\beta_{cv}\} = \{p(w_i = v|z_i = c)\}$, the terms containing model parameters $\boldsymbol{\beta}$ are collected and the constraints

$$\sum_{v=1}^{|\mathcal{V}|} \beta_{cv} = 1 \quad (7.355)$$

are imposed to arrange the Lagrangian $\mathcal{L}(\boldsymbol{\beta})$ with C Lagrange multipliers $\{\lambda_c\}$. By solving

$$\frac{\partial \mathcal{L}(\boldsymbol{\beta})}{\partial \beta_{cv}} = 0 \quad (7.356)$$

and considering the constraints, we estimate the optimal conditional multinomial distributions

$$\hat{\beta}_{cv} = \frac{\sum_{\mathbf{h}_{i-n+1}^{i-1}} \sum_{i=1}^{N_h} \hat{\phi}_{h,ic} w_i^v}{\sum_{l=1}^{|\mathcal{V}|} \sum_{\mathbf{h}_{i-n+1}^{i-1}} \sum_{i=1}^{N_h} \hat{\phi}_{h,ic} w_i^l} \quad (7.357)$$
$$1 \leq c \leq C, \quad 1 \leq v \leq |\mathcal{V}|$$

by substituting the updated variational multinomial distributions $\hat{\boldsymbol{\phi}}_h = \{\hat{\phi}_{h,ic}\}$. However, there is no closed-form solution to optimal DCLM parameter $\mathbf{A} = [\mathbf{a}_1, \cdots, \mathbf{a}_C]$. This parameter is used to project the ordered history vector $\mathbf{h}_{i-n+1}^{i-1}$ into latent class space. The projected parameter is treated as the Dirichlet class parameter. We may apply the Newton–Raphson algorithm or simply adopt the gradient descent algorithm

$$\mathbf{a}_c^{(\tau+1)} = \mathbf{a}_c^{(\tau)} - \eta \nabla \mathcal{L}(\mathbf{a}_c^{(\tau)}) \quad 1 \leq c \leq C \quad (7.358)$$

to derive the optimal Dirichlet class parameter $\hat{\mathbf{A}}$ by using the gradient

$$\nabla_{\mathbf{a}_c}\mathcal{L}(\hat{\boldsymbol{\gamma}},\hat{\boldsymbol{\phi}};\mathbf{A},\boldsymbol{\beta}) = \sum_{\mathbf{h}_{i-n+1}^{i-1}} \left\{\Psi\left(\sum_{j=1}^{C}\mathbf{a}_j^\top \mathbf{h}_{i-n+1}^{i-1}\right)\right.$$
$$\left. - \Psi\left(\mathbf{a}_c^\top \mathbf{h}_{i-n+1}^{i-1}\right) + \Psi(\hat{\gamma}_{h,c}) - \Psi\left(\sum_{j=1}^{C}\hat{\gamma}_{h,j}\right)\right\} \cdot \mathbf{h}_{i-n+1}^{i-1}. \tag{7.359}$$

Given the estimated model parameters $\{\hat{\mathbf{A}},\hat{\boldsymbol{\beta}}\}$, the DCLM n-gram $p_{\text{DC}}(w_i = v|\mathbf{h}_{i-n+1}^{i-1},\mathbf{A},\boldsymbol{\beta})$ in Eq. (7.344) is implemented. The DCLM n-gram could be improved by interpolating with the modified Kneser–Ney (MKN) language model, as discussed in Eq. (3.251). However, the performance of DCLM is limited due to the modeling of Dirichlet classes only inside the n-gram window.

7.7.6 Cache Dirichlet class language model

In general, the long-distance information outside the n-gram window is not captured. This weakness can be compensated in the cache DCLM (Chien & Chueh 2011). The cache DCLM treats all history words w_1^{i-1} as cache memory and incorporates their class information $z_1^{i-1} = \{z_1,\cdots,z_{i-1}\}$ into language modeling as

$$\begin{aligned} & p(w_i = v|\mathbf{h}_{i-n+1}^{i-1},\mathbf{A},\boldsymbol{\beta},w_1^{i-1}) \\ & = \sum_{z_1^{i-1}} p(z_1^{i-1}|w_1^{i-1}) p(w_i|z_1^{i-1},\mathbf{h}_{i-n+1}^{i-1},\mathbf{A},\boldsymbol{\beta}) \\ & = \sum_{z_1^{i-1}} p(z_1^{i-1}|w_1^{i-1}) \sum_{c=1}^{C} p(w_i = v|z_i = c,\boldsymbol{\beta}) \\ & \times \int p(\boldsymbol{\theta}|\mathbf{h}_{i-n+1}^{i-1},\mathbf{A},z_1^{i-1}) p(z_i = c|\boldsymbol{\theta}) d\boldsymbol{\theta}, \end{aligned} \tag{7.360}$$

where the marginalization over latent classes c and class mixtures $\boldsymbol{\theta}$ is performed. We have the posterior distribution

$$p(\boldsymbol{\theta}|\mathbf{h}_{i-n+1}^{i-1},\mathbf{A},z_1^{i-1}) = \frac{p(\boldsymbol{\theta}|\mathbf{h}_{i-n+1}^{i-1},\mathbf{A}) p(z_1^{i-1}|\boldsymbol{\theta},\mathbf{h}_{i-n+1}^{i-1},\mathbf{A})}{p(z_1^{i-1}|\mathbf{h}_{i-n+1}^{i-1},\mathbf{A})}, \tag{7.361}$$

where

$$\begin{aligned} p(z_1^{i-1}|\boldsymbol{\theta},\mathbf{h}_{i-n+1}^{i-1},\mathbf{A}) &= p(z_1^{i-1}|\boldsymbol{\theta}) \\ &= \prod_{c=1}^{C} \theta_c^{\sum_{j=1}^{i-1}\delta(z_j,c)}. \end{aligned} \tag{7.362}$$

The denominator term of this posterior distribution is independent of w_i and could be ignored in calculating the cache DCLM in Eq. (7.360). For practical purposes, the summation over z_1^{i-1} is simplified by adopting a single best class sequence $\hat{z}_1^{i-1}$, where the best class $\hat{z}_{i-1}$ of word w_{i-1} is detected from the previous $n-1$ words w_{i-n}^{i-2} and the best classes $\hat{z}_1^{i-2}$ corresponding to previous $i-1$ words, namely

$$\begin{aligned}\hat{z}_{i-1} &= \arg\max_{z_{i-1}} p(w_{i-1} = v, z_{i-1} | \hat{z}_1^{i-2}, w_{i-n}^{i-2}, \mathbf{A}, \boldsymbol{\beta}) \\ &= \arg\max_{c} \beta_{cv} \frac{\mathbf{a}_c^{\top} \mathbf{h}_{i-n}^{i-2} + \sum_{j=1}^{i-2} \delta(\hat{z}_j, c)}{\sum_{l=1}^{C} \left[\mathbf{a}_l^{\top} \mathbf{h}_{i-n}^{i-2} + \sum_{j=1}^{i-2} \delta(\hat{z}_j, l) \right]}. \end{aligned} \tag{7.363}$$

A detailed derivation is given in Chien & Chueh (2011).

As a result, the recursive detection from $\hat{z}_1$ to $\hat{z}_{i-1}$ is done and applied to approximate the cache DCLM as

$$\begin{aligned} & p(w_i = v | \mathbf{h}_{i-n+1}^{i-1}, \mathbf{A}, \boldsymbol{\beta}, w_1^{i-1}) \\ & \approx \sum_{c=1}^{C} p(w_i = v | \hat{z}_i = c, \boldsymbol{\beta}) \int p(\boldsymbol{\theta} | \mathbf{h}_{i-n+1}^{i-1}, \mathbf{A}) \\ & \times \prod_{m=1}^{C} \theta_m^{\sum_{j=1}^{i-1} \delta(\hat{z}_j, m)} p(z_i = c | \boldsymbol{\theta}) d\boldsymbol{\theta} \\ & \approx \sum_{c=1}^{C} \beta_{cv} \frac{\mathbf{a}_c^{\top} \mathbf{h}_{i-n+1}^{i-1} + \rho \sum_{j=1}^{i-1} \tau^{i-j-1} \delta(\hat{z}_j, c)}{\sum_{l=1}^{C} \left[\mathbf{a}_l^{\top} \mathbf{h}_{i-n+1}^{i-1} + \rho \sum_{j=1}^{i-1} \tau^{i-j-1} \delta(\hat{z}_j, l) \right]}. \end{aligned} \tag{7.364}$$

Here, the product of the Dirichlet distribution $p(\boldsymbol{\theta} | \mathbf{h}_{i-n+1}^{i-1}, \mathbf{A})$ and the multinomial distribution $\prod_{m=1}^{C} \theta_m^{\sum_{j=1}^{i-1} \delta(\hat{z}_j, m)}$ is a new Dirichlet distribution. Taking the integral in Eq. (7.364) is equivalent to finding the mean of the new Dirichlet distribution. In this cache DCLM, we introduce a weighting factor $0 < \rho \leq 1$ to balance two terms in the numerator and the denominator and a forgetting factor $o < \tau \leq 1$ to discount the distant class information. A graphical representation of the cache DCLM is given in Figure 7.6. The best classes $\{\hat{z}_1, \cdots, \hat{z_{i-1}}\}$ corresponding to all history words $\{w_1, \cdots, w_{i-1}\}$ are recursively detected and then merged in prediction of the next word w_i.

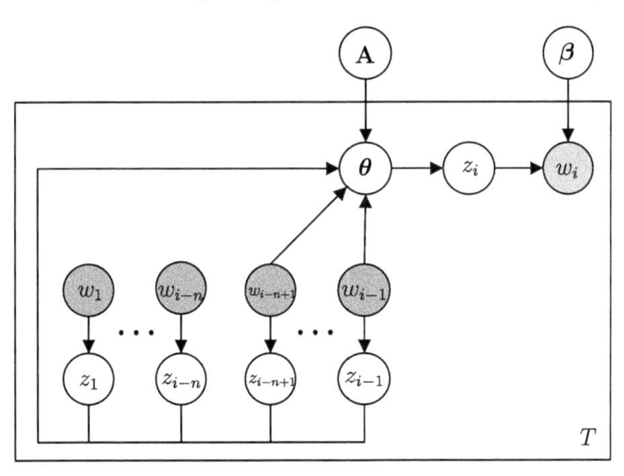

Figure 7.6 Representation of a cache Dirichlet class language model.

Table 7.3 Comparison of frequently used words of the latent variables extracted by LDA LM and DCLM.

	Topic/class	Frequently used words in latent topics or classes
LDA LM	Family	toy, kids, theater, event, season, shoe, teen, children's, plays, films, sports, magazines, Christmas, bowling, husband, anniversary, girls, festival, couple, parents,wife, friends
	Election	candidates, race, voters, challenger, democrat, state's, selection, county, front, delegates, elections, reverend, republicans, polls, conventions, label, politician, ballots
	War	troops, killed, Iraqi, attack, ship, violence, fighting, soldiers, mines, Iranian, independence, marines, revolution, died, nation, protect, armed, democracy, violent, commander
DCLM	Quantity	five, seven, two, eight, cents, six, one, nine, four, three, zero, million, point, percent, years, megabyte, minutes, milligrams, bushels, miles, marks, pounds, yen, dollars
	Business	exchange, prices, futures, index, market, sales, revenue, earnings, trading, plans, development, business, funds, organization, traders, ownership, holdings, investment
	In+	addition, the, fact, American, October, recent, contrast, Europe, June, Tokyo, July, March, turn, other, my, Washington, order, Chicago, case, China, general, which

7.7.7 System performance

The *Wall Street Journal* (*WSJ*) corpus was utilized to evaluate different language models for continuous speech recognition (CSR). The SI-84 training set was adopted to estimate the HMM parameters. The feature vector consisted of 12 MFCCs and one log energy and their first and second derivatives. Triphone models were built for 39 phones and one background silence. Each triphone model had three states with eight Gaussian components. The 1987-1989 *WSJ* corpus with 38M words was used to train the baseline backoff trigrams. A total of 86K documents and 3M trigram histories were used. The 20K non-verbalized punctuation, closed vocabularies were adopted. A total of 333 test utterances were sampled from the November 1992 ARPA CSR benchmark test data. In the implementation, the baseline tri-gram was used to generate 100-best lists. Various language models were interpolated with a baseline tri-gram using an interpolation weight, and were employed for N-best rescoring. The neural network LM (NNLM) (Bengio *et al.* 2003), class-based LM (Brown *et al.* 1992), PLSA LM (Gildea & Hofmann 1999), LDA LM (Tam & Schultz 2005, Tam & Schultz 2006), DCLM, and cache DCLM (Chien & Chueh 2011) were evaluated in the comparison.

Table 7.3 lists some examples of latent topics and classes, and the corresponding frequently used words. The three topics and classes were selected from LDA LM with $K = 100$ and DCLM with $C = 100$, respectively. The frequently used words were identified according to the likelihood of the words given target topics or classes. We can see that the frequently used words within a topic or class are semantically close to

Table 7.4 Comparison of word error rates for different methods with different sizes of training data and numbers of classes.

	Size of training data			
	6M	12M	18M	38M
Baseline LM	39.2%	21.3%	15.8%	12.9%
NNLM	35.5%	19.6%	15.0%	12.4%
Class-based LM	35.5%	19.7%	15.0%	12.4%
PLSA LM	36.0%	19.8%	15.0%	12.3%
LDA LM	35.9%	19.7%	14.7%	12.2%
DCLM (C=200)	35.9%	19.6%	14.6%	12.0%
Cache DCLM (C=200)	34.2%	19.3%	14.5%	11.9%
DCLM (C=500)	35.2%	19.2%	14.3%	11.7%
Cache DCLM (C=500)	33.9%	19.0%	14.2%	11.6%

each other. For some cases, the topically related words from LDA LM appear independently. These topics are not suitable for generating natural language. In contrast, DCLM extracts the class information from n-gram events and performs the history clustering based on sentence generation. For example, the latent class "In+" denotes the category of words that follow the preposition "in." The words "fact," "addition," and "June" usually follow the word "in," and appear as frequent words of the same class. The word order is reflected in the clustering procedure. Table 7.4 reports the word error rates for baseline LM, NNLM, class-based LM, PLSA LM, LDA LM, DCLM and cache DCLM with $C = 200$ and $C = 500$ under different sizes of training data. The issue of small sample size is examined. It is consistent that the word error rate is reduced when the amount of training data is increased. In the case of sparse training data (6M), the topic-based and class-based methods work well. In particular, the cache DCLM with $C = 200$ achieved an error rate reduction of 12.9% which outperforms the other related methods. When the number of classes is increased to $C = 500$, the improvement is obvious for the case of large training data (38M).

7.8 Summary

This chapter has introduced various applications of VB to speech and language processing, including CDHMM-based acoustic modeling, acoustic model adaptation, latent topic models, and latent topic language models. Compared with the previous inference approximations based on MAP, evidence, and asymptotic approximations, VB deals with Bayesian inference based on a distribution estimation without considering the asymptotic property, which often provides better solutions in terms of consistently using the Bayesian manner. In addition, the inference algorithm obtained can be regarded as an extension of the conventional EM type iterative algorithm. This makes implementation easier as we can utilize existing source codes based on the ML and MAP–EM algorithms. One of the difficulties in VB is that obtaining the analytical solutions of VB posterior distributions and variational lower bound is hard due to the complicated expectation and integral calculations. However, the solutions provided in this chapter (general formulas and specific solutions for acoustic and language modeling issues) would cover most of the mathematical analysis in the other VB applications in speech and language processing. Actually, there have been various other aspects of VB including speech feature extraction (Kwon, Lee & Chan 2002, Valente & Wellekens 2004*a*), voice activity detection (Cournapeau, Watanabe, Nakamura *et al.* 2010), speech/speaker GMM (Valente 2006, Pettersen 2008), speaker diarization (Valente, Motlicek & Vijayasenan 2010, Ishiguro, Yamada, Araki *et al.* 2012), and statistical speech synthesis (Hashimoto, Nankaku & Tokuda 2009, Hashimoto, Zen, Nankaku *et al.* 2009). Readers who are interested in these topics could follow these studies and develop new techniques with the solutions provided in this chapter.

Although VB provides a fully Bayesian treatment, VB still cannot overcome the problem of local optimum solutions based on the EM style algorithm, and VB also cannot provide analytical solutions for non-exponential family distributions. The next chapter deals with MCMC-based Bayesian approaches, which potentially overcome these problems with a fully Bayesian treatment.

8 Markov chain Monte Carlo

For most probabilistic models of practical interest, exact inference is intractable, and so we have to resort to some form of approximation. Markov chain Monte Carlo (MCMC) is another realization of full Bayesian treatment of practical Bayesian solutions (Neal 1993, Gilks, Richardson & Spiegelhalter 1996, Bishop 2006). MCMC is known as a *stochastic* approximation which acts differently from the *deterministic* approximation based on variational Bayes (VB) as addressed in Chapter 7. Variational inference using VB approximates the posterior distribution through factorization of the distribution over multiple latent variables and scales well to large applications. MCMC uses the numerical sampling computation rather than solving integrals and expectation analytically. Since MCMC can use any distributions in principle, it is capable of wide applications, and can be used for Bayesian nonparametric (BNP) learning, which provides highly flexible models whose complexity grows appropriately with the amount of data. Although, due to the computational cost, the application of MCMC to speech and language processing is limited to small-scale problems currently, this chapter describes promising new directions of Bayesian nonparametrics for speech and language processing by automatically growing models to deal with speaker diarization, acoustic modeling, language acquisition, and hierarchical language modeling. The strengths and weaknesses using VB and MCMC are complementary. In what follows, we first introduce the general background of sampling methods including MCMC and Gibbs sampling algorithms. Next, the Bayesian nonparametrics are calculated to build a flexible topic model based on the hierarchical Dirichlet process (HDP) (Teh *et al.* 2006). Several applications in speech and language processing areas are surveyed. GMM-based speaker clustering, CDHMM-based acoustic unit discovery by using MCMC, and the language model based on the hierarchical Pitman–Yor process (Teh *et al.* 2006) are described.

The fundamental problem in MCMC involves finding the expectation of some function $f(\boldsymbol{\theta})$ with respect to a probability distribution $p(\boldsymbol{\theta})$ where the components of $\boldsymbol{\theta}$ might comprise discrete or continuous variables, which are some factors or parameters to be inferred under a probabilistic model. In the case of continuous variables, we would like to evaluate the expectation

$$\mathbb{E}_{(\boldsymbol{\theta})}[f(\boldsymbol{\theta})] = \int f(\boldsymbol{\theta})p(\boldsymbol{\theta})d\boldsymbol{\theta}, \tag{8.1}$$

where the integral is replaced by summation in the case of discrete variables. We assume that such expectations are too complicated to be evaluated analytically. The general idea behind sampling methods is to obtain a set of samples $\{\boldsymbol{\theta}^{(l)}, l = 1, \cdots, L\}$ drawn independently from the distribution $p(\boldsymbol{\theta})$. We may approximate the integral by a sample mean of function $f(\boldsymbol{\theta})$ over these samples $\boldsymbol{\theta}^{(l)}$:

$$\widehat{f} = \frac{1}{L}\sum_{l=1}^{L} f(\boldsymbol{\theta}^{(l)}). \tag{8.2}$$

Since the samples $\boldsymbol{\theta}^{(l)}$ are drawn from the distribution $p(\boldsymbol{\theta})$, the estimator $\widehat{f}$ has the correct mean, i.e., $\widehat{f} = \mathbb{E}_{(\boldsymbol{\theta})}[f(\boldsymbol{\theta})]$. In general, ten to twenty independent samples may suffice to estimate an expectation. However, the samples $\{\boldsymbol{\theta}^{(l)}\}$ may not be drawn independently. The effective sample size might be much smaller than the apparent sample size L. This implies that a relatively large sample size is required to achieve sufficient accuracy.

8.1 Sampling methods

A simple strategy can be designed to generate random samples from a given distribution. We first consider how to generate random numbers from non-uniform distributions, assuming that we already have a source of uniformly distributed random numbers. Let θ be uniformly distributed by $p(\theta) = 1$ over the interval $(0, 1)$. We transform the values of θ using some function $f(\cdot)$ by $y = f(\theta)$. The distributions of variables θ and y are related by

$$p(y) = p(\theta)\left|\frac{d\theta}{dy}\right|. \tag{8.3}$$

Taking integrals for both sides of Eq. (8.3), we have

$$\theta = h(y) = \int_{-\infty}^{y} p(\widetilde{y})d\widetilde{y}, \tag{8.4}$$

which is the indefinite integral of $p(y)$. Thus, $y = h^{-1}(\theta)$, meaning that we have to transform the uniformly distributed random numbers θ using a function $h^{-1}(\cdot)$, which is the inverse of the indefinite integral of the desired distribution of y. Figure 8.1 depicts the geometrical interpretation of the transformation method for generating non-uniformly distributed random numbers. In addition, the generalization to multiple variables is straightforward and involves the Jacobian of the transform of variables:

$$p(y_1, \cdots, y_M) = p(\theta_1, \cdots, \theta_M)\left|\frac{\partial(\theta_1, \cdots, \theta_M)}{\partial(y_1, \cdots, y_M)}\right|. \tag{8.5}$$

A similar scheme can be applied to draw a multivariate distribution with M variables.

8.1.1 Importance sampling

The technique of *importance sampling* provides a framework for approximating the expectation in Eq. (8.1) directly but does not provide the mechanism for drawing samples from distribution $p(\boldsymbol{\theta})$. The finite sum approximation to expectation in Eq. (8.2)

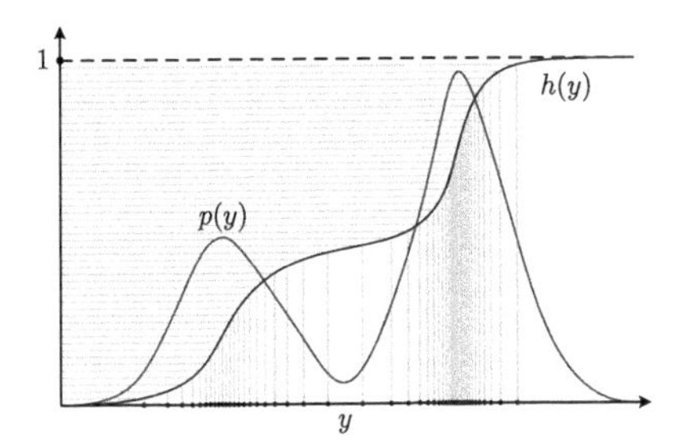

Figure 8.1 Transformation method for generating non-uniformly distributed random numbers from probability distribution $p(y)$. Function $h(y)$ represents the indefinite integral of $p(y)$. Adapted from Bishop (2006).

depends on being able to draw samples from the distribution $p(\boldsymbol{\theta})$. One simple strategy for evaluating expectation function would be to discretize $\boldsymbol{\theta}$-space into a uniform grid, and to evaluate the integrand as a sum of the form

$$\mathbb{E}_{(\boldsymbol{\theta})}[f(\boldsymbol{\theta})] \simeq \sum_{l=1}^{L} p(\boldsymbol{\theta}^{(l)}) f(\boldsymbol{\theta}^{(l)}). \tag{8.6}$$

However, the problem with this approach is that the number of terms in the summation grows exponentially with the dimensionality of $\boldsymbol{\theta}$. In many cases, the probability distributions of interest often have much of their mass confined to relatively small regions of $\boldsymbol{\theta}$ space. Accordingly, uniform sampling is very inefficient because in high-dimensional space, only a very small proportion of the samples make a significant contribution to the sum. We would like to sample the points falling in regions where $p(\boldsymbol{\theta})$ is large, or where the product $p(\boldsymbol{\theta})f(\boldsymbol{\theta})$ is large.

Suppose we wish to sample from a distribution $p(\boldsymbol{\theta})$ that is not simple or a standard distribution. Sampling directly from $p(\boldsymbol{\theta})$ is difficult. Importance sampling is based on the use of a *proposal distribution* $q(\boldsymbol{\theta})$ from which it is easy to draw samples. Figure 8.2 illustrates the proposal distribution for importance sampling. The expectation in Eq. (8.1) is expressed in the form of a finite sum over samples $\{\boldsymbol{\theta}^{(l)}\}$ drawn from $q(\boldsymbol{\theta})$:

$$\begin{aligned}
\mathbb{E}_{(\boldsymbol{\theta})}[f(\boldsymbol{\theta})] &= \int f(\boldsymbol{\theta}) p(\boldsymbol{\theta}) d\boldsymbol{\theta} \\
&= \int f(\boldsymbol{\theta}) \frac{p(\boldsymbol{\theta})}{q(\boldsymbol{\theta})} q(\boldsymbol{\theta}) d\boldsymbol{\theta} \\
&\simeq \frac{1}{L} \sum_{l=1}^{L} \frac{p(\boldsymbol{\theta}^{(l)})}{q(\boldsymbol{\theta}^{(l)})} f(\boldsymbol{\theta}^{(l)}).
\end{aligned} \tag{8.7}$$

The quantities $r_l = p(\boldsymbol{\theta}^{(l)})/q(\boldsymbol{\theta}^{(l)})$ are known as the *importance weights*, and they correct the bias introduced by sampling from the wrong distribution.

Usually, the distribution $p(\boldsymbol{\theta})$ can only be evaluated up to a normalization constant, so that $p(\boldsymbol{\theta}) = \widetilde{p}(\boldsymbol{\theta})/Z_p$, where $\widetilde{p}(\boldsymbol{\theta})$ can be evaluated easily and Z_p is unknown. We may use the importance sampling distribution $q(\boldsymbol{\theta}) = \widetilde{q}(\boldsymbol{\theta})/Z_q$ to determine the expectation function:

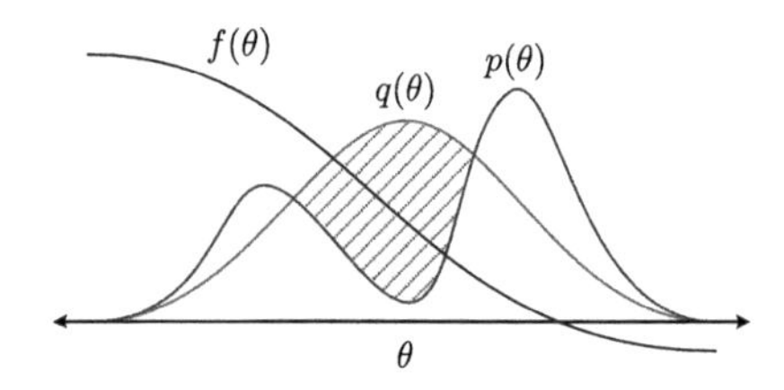

Figure 8.2 Proposal distribution $q(\theta)$ for importance sampling in estimation of expectation of function $f(\theta)$ with probability distribution $p(\theta)$. Adapted from Bishop (2006).

$$\begin{aligned}\mathbb{E}_{(\boldsymbol{\theta})}[f(\boldsymbol{\theta})] &= \frac{Z_q}{Z_p}\int f(\boldsymbol{\theta})\frac{\widetilde{p}(\boldsymbol{\theta})}{\widetilde{q}(\boldsymbol{\theta})}q(\boldsymbol{\theta})d\boldsymbol{\theta} \\ &\simeq \frac{Z_q}{Z_p}\frac{1}{L}\sum_{l=1}^{L}\widetilde{r}_l f(\boldsymbol{\theta}^{(l)}),\end{aligned} \tag{8.8}$$

where $\widetilde{r}_l = \widetilde{p}(\boldsymbol{\theta}^{(l)})/\widetilde{q}(\boldsymbol{\theta}^{(l)})$. We have

$$\begin{aligned}\frac{Z_p}{Z_q} &= \frac{1}{Z_q}\int \widetilde{p}(\boldsymbol{\theta})d\boldsymbol{\theta} = \int \frac{\widetilde{p}(\boldsymbol{\theta})}{\widetilde{q}(\boldsymbol{\theta})}q(\boldsymbol{\theta})d\boldsymbol{\theta} \\ &\simeq \frac{1}{L}\sum_{l=1}^{L}\widetilde{r}_l.\end{aligned} \tag{8.9}$$

We then obtain

$$\mathbb{E}_{(\boldsymbol{\theta})}[f(\boldsymbol{\theta})] \simeq \sum_{l=1}^{L} w_l f(\boldsymbol{\theta}^{(l)}), \tag{8.10}$$

where we define

$$w_l = \frac{\widetilde{r}_l}{\sum_m \widetilde{r}_m} = \frac{\widetilde{p}(\boldsymbol{\theta}^{(l)})/q(\boldsymbol{\theta}^{(l)})}{\sum_m \widetilde{p}(\boldsymbol{\theta}^{(m)})/q(\boldsymbol{\theta}^{(m)})}. \tag{8.11}$$

Clearly, the performance of the importance sampling method highly depends on how well the sampling distribution $q(\boldsymbol{\theta})$ matches the desired distribution $p(\boldsymbol{\theta})$.

8.1.2 Markov chain

One major weakness in evaluation of expectation function based on the importance sampling strategy is the severe limitation in spaces of high dimensionality. We accordingly turn to a very general and powerful framework called Markov chain Monte Carlo (MCMC), which allows sampling from a large class of distributions and which scales well with the dimensionality of the sample space. Before discussing MCMC methods in more detail, it is useful to study some general properties of Markov chains and investigate under what conditions a Markov chain can converge to the desired distribution. A first-order Markov chain is defined for a series of variables $\boldsymbol{\theta}^{(1)}, \cdots, \boldsymbol{\theta}^{(M)}$ such that the following conditional independence holds for $m \in \{1, \cdots, M-1\}$:

$$p(\boldsymbol{\theta}^{(m+1)}|\boldsymbol{\theta}^{(1)}, \cdots, \boldsymbol{\theta}^{(m)}) = p(\boldsymbol{\theta}^{(m+1)}|\boldsymbol{\theta}^{(m)}). \tag{8.12}$$

This Markov chain starts from the probability distribution for initial variable $p(\boldsymbol{\theta}^{(0)})$ and operates with the transition probability $p(\boldsymbol{\theta}^{(m+1)}|\boldsymbol{\theta}^{(m)})$. A Markov chain is called homogeneous if the transition probabilities are unchanged for all m. The marginal probability for a variable $\boldsymbol{\theta}^{(m+1)}$ is expressed in terms of the marginal probabilities over the previous variable $\{\boldsymbol{\theta}^{(1)}, \cdots, \boldsymbol{\theta}^{(m)}\}$ in the chain,

$$p(\boldsymbol{\theta}^{(m+1)}) = \sum_{\boldsymbol{\theta}^{(m)}} p(\boldsymbol{\theta}^{(m+1)}|\boldsymbol{\theta}^{(m)})p(\boldsymbol{\theta}^{(m)}). \tag{8.13}$$

A distribution is said to be *invariant* or stationary with respect to a Markov chain if each step in the chain keeps the distribution invariant. For a homogeneous Markov chain with transition probability $T(\boldsymbol{\theta}', \boldsymbol{\theta})$, the distribution $p^\star(\boldsymbol{\theta})$ is invariant if it has the following property:

$$p^\star(\boldsymbol{\theta}) = \sum_{\boldsymbol{\theta}'} T(\boldsymbol{\theta}', \boldsymbol{\theta})p^\star(\boldsymbol{\theta}'). \tag{8.14}$$

Our goal is to use Markov chains to sample from a given distribution. We can achieve this goal if we set up a Markov chain such that the desired distribution is invariant. It is required that for $m \to \infty$, the distribution $p(\boldsymbol{\theta}^{(m)})$ converges to the required invariant distribution $p^\star(\boldsymbol{\theta})$, which is obtained irrespective of the choice of initial distribution $p(\boldsymbol{\theta}^{(0)})$. This invariant distribution is also called the *equilibrium* distribution. A sufficient condition for an invariant distribution $p(\boldsymbol{\theta})$ is to choose the transition probabilities to satisfy the *detailed balance*, i.e.

$$p^\star(\boldsymbol{\theta})T(\boldsymbol{\theta}, \boldsymbol{\theta}') = p^\star(\boldsymbol{\theta}')T(\boldsymbol{\theta}', \boldsymbol{\theta}), \tag{8.15}$$

for a particular distribution $p^\star(\boldsymbol{\theta})$.

8.1.3 The Metropolis–Hastings algorithm

As discussed in importance sampling, we keep sampling from a proposal distribution and maintain a record of the current state $\boldsymbol{\theta}^{(\tau)}$. The proposal distribution $q(\boldsymbol{\theta}|\boldsymbol{\theta}^{(\tau)})$ depends on this current state. The sequence of samples $\boldsymbol{\theta}^{(1)}, \boldsymbol{\theta}^{(2)}, \cdots$ forms a Markov chain. The proposal distribution is chosen to be sufficiently simple to draw samples directly. At each sampling cycle, we generate a candidate sample $\boldsymbol{\theta}^\star$ from the proposal distribution and then accept the sample according to an appropriate criterion. In a basic Metropolis algorithm (Metropolis, Rosenbluth, Rosenbluth *et al.* 1953), the proposal distribution is assumed to be symmetric, namely $q(\boldsymbol{\theta}_a|\boldsymbol{\theta}_b) = q(\boldsymbol{\theta}_b|\boldsymbol{\theta}_a)$ for all values of $\boldsymbol{\theta}_a$ and $\boldsymbol{\theta}_b$. The candidate sample is accepted with the probability

$$A(\boldsymbol{\theta}^\star, \boldsymbol{\theta}^{(\tau)}) = \min\left(1, \frac{\widetilde{p}(\boldsymbol{\theta}^\star)}{\widetilde{p}(\boldsymbol{\theta}^{(\tau)})}\right), \tag{8.16}$$

where $p(\boldsymbol{\theta}) = \widetilde{p}(\boldsymbol{\theta})/Z_p$ with a readily evaluated distribution $\widetilde{p}(\boldsymbol{\theta})$ and an unknown normalization value Z_p. To fulfil this algorithm, we can choose a random number u with uniform distribution over the unit interval $(0, 1)$ and accept the sample if $A(\boldsymbol{\theta}^\star, \boldsymbol{\theta}^{(\tau)}) > u$. Definitely, if the step from $\boldsymbol{\theta}^{(\tau)}$ to $\boldsymbol{\theta}^\star$ causes an increase in the value of $p(\boldsymbol{\theta})$, then the candidate point is accepted. Once the candidate sample is accepted, then $\boldsymbol{\theta}^{(\tau+1)} = \boldsymbol{\theta}^\star$, otherwise the candidate sample $\boldsymbol{\theta}^\star$ is discarded, $\boldsymbol{\theta}^{(\tau+1)}$ is set to $\boldsymbol{\theta}^{(\tau)}$. The next candidate sample is drawn from the distribution $q(\boldsymbol{\theta}|\boldsymbol{\theta}^{(\tau+1)})$. This leads to multiple copies of samples in the final list of samples. As long as $q(\boldsymbol{\theta}_a|\boldsymbol{\theta}_b)$ is positive for any values of $\boldsymbol{\theta}_a$ and $\boldsymbol{\theta}_b$, the distribution of $\boldsymbol{\theta}^{(\tau)}$ approaches $p(\boldsymbol{\theta})$ as $\tau \to \infty$.

The basic Metropolis algorithm is further generalized to the Metropolis–Hastings algorithm (Hastings 1970) which is widely adopted in MCMC inference. This generalization is developed by relaxing the assumption in the Metropolis algorithm that the proposal distribution is no longer a symmetric function of its arguments. Using this algorithm, at step τ with current state $\boldsymbol{\theta}^{(\tau)}$, we draw a sample $\boldsymbol{\theta}^\star$ from the proposal distribution $q_k(\boldsymbol{\theta}|\boldsymbol{\theta}^{(\tau)})$ and then accept it with the probability

$$A_k(\boldsymbol{\theta}^\star, \boldsymbol{\theta}^{(\tau)}) = \min\left(1, \frac{\widetilde{p}(\boldsymbol{\theta}^\star) q_k(\boldsymbol{\theta}^{(\tau)}|\boldsymbol{\theta}^\star)}{\widetilde{p}(\boldsymbol{\theta}^{(\tau)}) q_k(\boldsymbol{\theta}^\star|\boldsymbol{\theta}^{(\tau)})}\right), \tag{8.17}$$

where k denotes the members of the set of possible transitions. For the case of a symmetric proposal distribution, the Metropolis–Hastings criterion in Eq. (8.17) is reduced to the Metropolis criterion in Eq. (8.16). We can show that $p(\boldsymbol{\theta})$ is an invariant distribution of the Markov chain generated by the Metropolis–Hastings algorithm by investigating the property of the detailed balance in Eq. (8.15). We find that

$$\begin{aligned} p(\boldsymbol{\theta}) q_k(\boldsymbol{\theta}|\boldsymbol{\theta}') A_k(\boldsymbol{\theta}', \boldsymbol{\theta}) &= \min(p(\boldsymbol{\theta}) q_k(\boldsymbol{\theta}|\boldsymbol{\theta}'), p(\boldsymbol{\theta}') q_k(\boldsymbol{\theta}'|\boldsymbol{\theta})) \\ &= \min(p(\boldsymbol{\theta}') q_k(\boldsymbol{\theta}'|\boldsymbol{\theta}), p(\boldsymbol{\theta}) q_k(\boldsymbol{\theta}|\boldsymbol{\theta}')) \\ &= p(\boldsymbol{\theta}') q_k(\boldsymbol{\theta}'|\boldsymbol{\theta}) A_k(\boldsymbol{\theta}, \boldsymbol{\theta}'). \end{aligned} \tag{8.18}$$

The choice of proposal distribution is important in an MCMC algorithm. For continuous state spaces, a common choice is a Gaussian centered on the current state, leading to an important trade-off in determining the variance parameter of this distribution. If the variance is small, the proportion of accepted transitions is high, but a slow random walk is taken through the state space. On the other hand, if the variance parameter is large, the rejection rate is high because the updated state has low probability $p(\boldsymbol{\theta})$. Figure 8.3 shows a schematic for selecting an isotropic Gaussian proposal distribution to sample random numbers from a correlated multivariate Gaussian distribution. In order to keep the rejection rate low, the scale ρ of the proposal distribution $q_k(\boldsymbol{\theta}|\boldsymbol{\theta}^{(\tau)})$ should be comparable to the smallest standard deviation $\sigma_{\min}$, which leads to a random walk so that the number of steps for separating states is of order $(\sigma_{\max}/\sigma_{\min})^2$ where $\sigma_{\max}$ is the largest standard deviation.

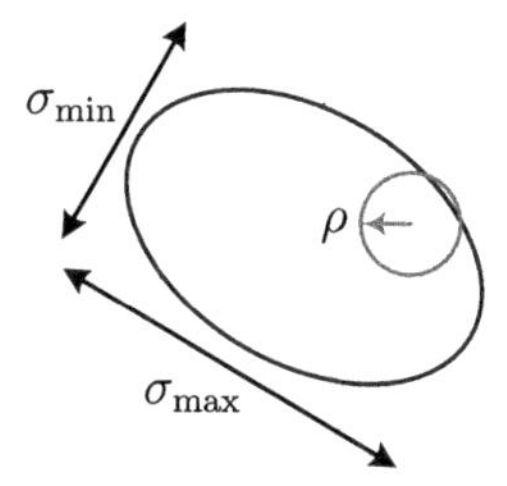

Figure 8.3 Using an isotropic Gaussian proposal distribution (circle) to sample random numbers for a correlated bivariate Gaussian distribution (ellipse). Adapted from Bishop (2006).

8.1.4 Gibbs sampling

Gibbs sampling (Geman & Geman 1984, Liu 2008) is a simple and widely applicable realization of an MCMC algorithm which is a special case of the Metropolis–Hastings algorithm. Consider the distribution of M random variables $p(\boldsymbol{\theta}) = p(\theta_1, \cdots, \theta_M)$ and suppose that we have some initial state for the Markov chain. Each step of the Gibbs sampling procedure replaces the value of one of the variables by a value drawn from the distribution of that variable conditioned on the values of the remaining states. That is to say, we replace the ith component θ_i by a value drawn from the distribution $p(\theta_i|\boldsymbol{\theta}^{-i})$, where $\boldsymbol{\theta}^{-i}$ denotes $\theta_1, \cdots, \theta_M$ but with θ_i omitted. The sampling procedure is repeated by cycling through the variables in a particular order or in a random order from some distribution. This procedure samples the required distribution $p(\boldsymbol{\theta})$, which should be invariant at each of the Gibbs sampling steps or in the whole Markov chain. This is because the marginal distribution $p(\boldsymbol{\theta}^{-i})$ is invariant and the conditional distribution $p(\theta_i|\boldsymbol{\theta}^{-i})$ is correct at each sampling step. Gibbs sampling of M variables for T steps follows this procedure:

- Initialize $\{\theta_i : i = 1, \cdots, M\}$.
- For $\tau = 1, \cdots, T$:
  - Sample $\theta_1^{(\tau+1)} \sim p(\theta_1|\theta_2^{(\tau)}, \theta_3^{(\tau)}, \cdots, \theta_M^{(\tau)})$.
  - Sample $\theta_2^{(\tau+1)} \sim p(\theta_2|\theta_1^{(\tau+1)}, \theta_3^{(\tau)}, \cdots, \theta_M^{(\tau)})$.

    $\vdots$
  - Sample $\theta_j^{(\tau+1)} \sim p(\theta_j|\theta_1^{(\tau+1)}, \cdots, \theta_{j-1}^{(\tau+1)}, \theta_{j+1}^{(\tau)}, \cdots, \theta_M^{(\tau)})$.

    $\vdots$
  - Sample $\theta_M^{(\tau+1)} \sim p(\theta_M|\theta_1^{(\tau+1)}, \theta_2^{(\tau+1)}, \cdots, \theta_{M-1}^{(\tau+1)})$.

Gibbs sampling can be shown to be a special case of the Metropolis–Hastings algorithm. Consider the Metropolis–Hastings sampling step involving the variable θ_k in which the remaining variables $\boldsymbol{\theta}^{-k}$ are fixed. The transition probability from $\boldsymbol{\theta}$ to $\boldsymbol{\theta}^\star$ is then given by $q_k(\boldsymbol{\theta}^\star|\boldsymbol{\theta}) = p(\theta_k^\star|\boldsymbol{\theta}^{-k})$, because the remaining variables are unchanged by the sampling step, and $(\boldsymbol{\theta}^\star)^{-k} = \boldsymbol{\theta}^{-k}$. By using $p(\boldsymbol{\theta}) = p(\theta_k|\boldsymbol{\theta}^{-k})p(\boldsymbol{\theta}^{-k})$, the acceptance probability in the Metropolis–Hastings algorithm is obtained by

$$A(\boldsymbol{\theta}^\star, \boldsymbol{\theta}) = \frac{p(\boldsymbol{\theta}^\star) q_k(\boldsymbol{\theta}|\boldsymbol{\theta}^\star)}{p(\boldsymbol{\theta}) q_k(\boldsymbol{\theta}^\star|\boldsymbol{\theta})}$$
$$= \frac{p(\theta_k^\star|(\boldsymbol{\theta}^\star)^{-k}) p((\boldsymbol{\theta}^\star)^{-k}) p(\theta_k|(\boldsymbol{\theta}^\star)^{-k})}{p(\theta_k|\boldsymbol{\theta}^{-k}) p(\boldsymbol{\theta}^{-k}) p(\theta_k^\star|\boldsymbol{\theta}^{-k})} = 1, \tag{8.19}$$

where $(\boldsymbol{\theta}^\star)^{-k} = \boldsymbol{\theta}^{-k}$ is applied. This result indicates that the sampling steps in the Metropolis–Hastings algorithm are always accepted.

Collapsed Gibbs sampling

Collapsed Gibbs sampling (Liu 1994, Griffiths & Steyvers 2004) is a method for using a marginal conditional distribution where some of variables are integrated out, instead of sampling. For example, suppose we have a set of variables Λ, the proposal distribution in the Gibbs sampling is represented as follows:

$$p(\theta_i|\boldsymbol{\theta}^{-i}) = \int p(\theta_i|\boldsymbol{\theta}^{-i}, \Lambda) p(\Lambda) d\Lambda. \tag{8.20}$$

$p(\Lambda)$ is a prior distribution. This integral can be analytically solved when we use a conjugate prior, and the following sections sometimes use a marginal conditional distribution.

8.1.5 Slice sampling

One weakness in the Metropolis–Hastings algorithm is the sensitivity to step size. If this is too small, the result has slow decorrelation due to random walk behavior, while if it is too large, the sampling procedure is not efficient due to a high rejection rate. The *slice sampling* approach (Neal 2003) provides an adaptive step size which is automatically adjusted to fit the characteristics of the distribution. Again, it is required that the unnormalized distribution $\widetilde{p}(\boldsymbol{\theta})$ is available to be evaluated. Consider the univariate case. Slice sampling is performed by augmenting θ with an additional variable u and drawing samples from the joint space of (θ, u). The goal is to sample uniformly from the area under the distribution given by

$$\widetilde{p}(\theta, u) = \begin{cases} 1/Z_p & \text{if } 0 \leqslant u \leqslant \widetilde{p}(\theta), \\ 0 & \text{otherwise,} \end{cases} \tag{8.21}$$

where $Z_p = \int \widetilde{p}(\theta) d\theta$. The marginal distribution of θ is given by

$$\int \widetilde{p}(\theta, u) du = \int_0^{\widetilde{p}(\theta)} \frac{1}{Z_p} du = \frac{\widetilde{p}(\theta)}{Z_p} = p(\theta). \tag{8.22}$$

To carry out this scheme, we first sample from $p(\theta)$ by sampling from $\widetilde{p}(\theta, u)$ and then neglecting the u values. Alternatively sampling θ and u can be achieved. Given the value of θ, we evaluate $\widetilde{p}(\theta)$ and then sample u uniformly in the range $0 \leqslant u \leqslant \widetilde{p}(\theta)$. We then fix u and sample θ uniformly from the "slice" through the distribution defined by $\{\theta : \widetilde{p}(\theta) > u\}$. As illustrated in Figure 8.4(a), for a given value $\theta^{(\tau)}$, a value of u is chosen uniformly in the region $0 \leqslant u \leqslant \widetilde{p}(\theta^{(\tau)})$, which defines a slice through the

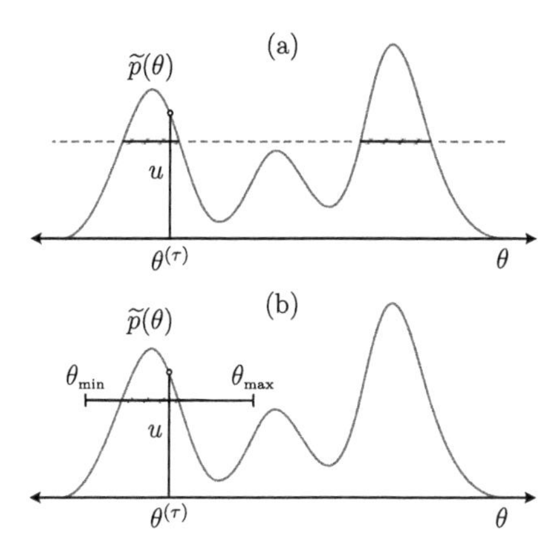

Figure 8.4 Slice sampling over a distribution $p(\theta)$ through: (a) finding the slice containing current sample $\theta^{(\tau)}$ with $\widetilde{p}(\theta) > u$; and (b) detecting the region of interest with two end points $\theta_{\min}$ and $\theta_{\max}$ so that we can uniformly draw a new sample $\theta^{(\tau+1)}$ within the slice. Adapted from Bishop (2006).

distribution, shown by solid horizontal lines. Because it is infeasible to sample directly from a slice, a new sample of θ is drawn from a region $\theta_{\min} \leqslant \theta \leqslant \theta_{\max}$, which contains the previous value $\theta^{(\tau)}$, as illustrated in Figure 8.4(b). We want the region to encompass as much of the slice as possible so as to allow large moves in θ space while having as little as possible of this region lying outside the slice, because this makes the sampling less efficient.

We start with a region of width w which contains current sample $\theta^{(\tau)}$ and then judge if both end points are within the slice. If either end point is within the slice, the region is extended in this direction by increments of value w until the end point falls outside the region. A candidate value $\theta^{\star}$ is then chosen uniformly from this region. If it lies inside the slice, it forms new sample $\theta^{(\tau+1)}$. If it lies outside the slice, the region is shrunk such that $\theta^{\star}$ forms an end point. A new candidate point is drawn uniformly from this reduced region until a value of θ is found that lies within the slice. When applying slice sampling to multivariate distributions, we repeatedly sample each variable in turn in the manner of Gibbs sampling based on a conditional distribution $p(\theta_i|\boldsymbol{\theta}^{-i})$.

8.2 Bayesian nonparametrics

The sampling methods mentioned in Section 8.1 are widely employed to infer the Bayesian nonparametrics (BNP) which are now seen as a new trend in speech and language processing areas where the data representation is a particular concern and is extensively studied. The BNP learning aims to deal with the issue that probabilistic methods are often not viewed as sufficiently expressive because of a long list of limitations and assumptions on probability distribution and fixed model complexity. It is attractive to pursue an expressive probabilistic representation with a less assumption-laden approach to inference. We would like to move beyond the simple fixed-dimensional random variables, e.g., multinomial distributions, Gaussians, and other exponential family distributions, and to mimic the flexibility in probabilistic representations. BNP methods avoid the restrictive assumptions of parametric models by defining distributions on function spaces.

Therefore, the stochastic process, containing an indexed collection of random variables, provides the flexibility to define probability distributions on spaces of probability distributions. We relax the limitation of finite parameterizations and consider the models over infinite-dimensional objects such as trees, lists, and collections of sets. The expressive data structures could be explored by computationally efficient reasoning and learning through the so-called *combinatorial stochastic processes* (Pitman 2006). BNP learning is accordingly developed from a Bayesian perspective by upgrading the priors in classic Bayesian analysis from parametric distributions to stochastic processes. Prior distributions are then replaced by the *prior process* in BNP inference. The flexible Bayesian learning and representation is conducted from the prior stochastic process to the posterior stochastic process. BNP involves the combinatorics of sums and products over prior and posterior stochastic processes and automatically learned model structure from the observed data. The model selection problem could be tackled by BNP learning.

8.2.1 Modeling via exchangeability

Some of the foundation of Bayesian inference is addressed in this section. The concept of *exchangeability* is critical in motivating BNP learning. Consider a probabilistic model with an infinite sequence of random factors or parameters $\boldsymbol{\theta} = \{\theta_1, \theta_2, \cdots\}$ to be inferred from observation data $\mathbf{x} = \{x_1, x_2, \cdots\}$, which could be either continuous such as the *speech features* $\mathbf{O}$ or discrete such as the *word labels* W. We say that such a sequence is *infinitely exchangeable* if the joint probability distribution of any finite subset of those random variables is invariant to permutation. For any N, we have

$$p(\theta_1, \theta_2, \cdots, \theta_N) = p(\theta_{\pi(1)}, \theta_{\pi(2)}, \cdots, \theta_{\pi(N)}), \tag{8.23}$$

where π denotes a permutation. The assumption of *exchangeability* is weaker than that of *independence* among random variables. This assumption often better describes the data we encounter in realization of stochastic processes for BNP learning. De Finetti's theorem states that the infinite sequence is exchangeable if and only if for any N random variables the sequence has the following property (Bernardo & Smith 2009):

$$p(\theta_1, \theta_2, \cdots, \theta_N) = \int \prod_{i=1}^{N} p(\theta_i|G) dP(G). \tag{8.24}$$

There exists an underlying random measure G and a distribution function P such that random variables θ_i are conditionally independent given G, which is not restricted to be a finite-dimensional object.

The Pólya urn model is a probability model for sequentially labeling the balls in an urn. Consider an empty urn and a countably infinite collection of colors. Randomly pick a color according to some fixed distribution G_0 and place a ball having the same color in the urn. For all subsequent balls, either choose a ball from the urn uniformly and return that ball to the urn with another ball of the same color, or choose a new color from G_0 and place a ball of that color k in the urn. We express this process mathematically by

$$p(\theta_i = k|\theta_1, \cdots, \theta_{i-1}) \propto \begin{cases} c_k & \text{if } \theta_j = k \text{ for some } j \in \{1, \cdots, i-1\}, \\ \alpha_0 & \text{otherwise,} \end{cases} \tag{8.25}$$

where $\alpha_0 > 0$ is a parameter of the process and c_k denotes the number of balls of color k. Even though we define the model by considering a particular ordering of the balls, the resulting distribution is independent of the order. It can be proved that the joint distribution $p(\theta_1, \theta_2, \cdots, \theta_N)$ is written as a product of conditionals given in Eq. (8.25) and the resulting expression is independent of the order of N random variables. The exchangeability in the Pólya urn model is confirmed. Because of this property, by De Finetti's theorem, the existence of an underlying probability measure G renders the ball colors conditionally independent. This random measure corresponds to a stochastic process known as the *Dirichlet process*, which is introduced in Section 8.2.2.

The property of exchangeability is essential for an MCMC inference procedure. Let us consider the joint distribution of $\boldsymbol{\theta}$ and $\mathbf{x}$ given by

$$p(\boldsymbol{\theta}, \mathbf{x}) = p(\theta_1, \theta_2, \cdots, \theta_N) \prod_{i=1}^{N} p(x_i|\theta_i), \tag{8.26}$$

which is viewed as the product of a prior in the first term and a likelihood function in the second term. The first term $p(\theta_1, \theta_2, \cdots, \theta_N)$ is modeled by the Pólya urn marginal distributions. In particular, our goal is to sample $\boldsymbol{\theta}$ from observation data $\mathbf{x}$ based on the Gibbs sampling. The problem is to sample a particular component θ_i while all of the other components are fixed. Because the joint distribution of $\{\theta_1, \cdots, \theta_N\}$ is invariant to permutation, we can freely permute the vector to move θ_i to the end of the list. The prior probability of the last component given all of the preceding components is given by the urn model as given in Eq. (8.25). We multiply each of the distributions by the likelihood function $p(x_i|\theta_i)$ and integrate with respect to θ_i. We assume that the prior measure G_0 and the likelihood function are conjugate so that the integral can be done in closed form. For each component, the derived result is the conditional distribution of θ_i given the other components and given x_i. This is done for different components $\{\theta_1, \cdots, \theta_N\}$ and the process iterates. This link between exchangeability and an efficient inference algorithm is important for BNP learning.

In addition, it is crucial to realize the Pólya urn model for Bayesian speech and language processing over a speech and text corpus. The ball means a word w_i or a feature vector $\mathbf{o}_t$ in the corpus and the ball color indicates the cluster label of this word or feature vector. This Pólya urn model defines a distribution of acoustic features or word labels which is not fixed in dimensionality and can be used to induce a distribution on partitions or clusterings. The distribution on partitions is known as the *Chinese restaurant process* (Aldous 1985), which is addressed in Section 8.2.4. The Chinese restaurant process and the Pólya urn model are viewed as the essentials of the BNP model for clustering where the random partition provides a prior on clusterings and the color associated with a cell can be represented by a distribution associated with a given cluster.

8.2.2 Dirichlet process

The *Dirichlet process* (DP) (Ferguson 1973) plays a crucial role in BNP learning. It is a stochastic process for a random probability measure G over a measurable space Ω,

$$G \sim \mathrm{DP}(\alpha_0, G_0), \tag{8.27}$$

such that, for any finite measurable partition $(A_1, A_2, \cdots, A_r)$, the random vector $(G(A_1, \cdots, G(A_r))$ is distributed as a finite-dimensional Dirichlet distribution with parameters $(\alpha_0 G_0(A_1), \cdots, \alpha_0 G_0(A_r))$, i.e.

$$(G(A_1, \cdots, G(A_r)) \sim \mathrm{Dir}(\alpha_0 G_0(A_1), \cdots, \alpha_0 G_0(A_r)), \tag{8.28}$$

with two parameters, a positive scaling parameter or *concentration parameter*, $\alpha_0 > 0$, and a *base probability measure*, $G_0 \in \Omega$. This process is a measure of measures over space Ω where $G(\Omega) = 1$. To realize the Dirichlet process in Eq. (8.27), an infinite sequence of points $\{\phi_k\}$ is drawn independently from the base probability measure G_0 so that the probability measure of the process is established by

$$G = \sum_{k=1}^{\infty} \beta_k \delta_{\phi_k} \tag{8.29}$$

with probability 1 (Sethuraman 1994). In Eq. (8.29), δ_{ϕ_k} is an *atom* or a unit mass at the point ϕ_k and $\{\beta_k\}$ are the random weights which depend on the concentration parameter α_0. Note that the Dirichlet process G is random in two ways. One random process is for weights β_k while the other is for locations ϕ_k. We can say that Dirichlet process G in Eq. (8.29) has the *Dirichlet marginals* as given in Eq. (8.28). According to De Finetti's theorem, the DP is seen as the De Finetti mixture distribution underlying the Pólya urn model as addressed in Section 8.2.1. DP can be presented from the perspectives of the stick-breaking construction, the Pólya urn scheme, and a limit of finite mixture models which is detailed in Sections 8.2.3, 8.2.4, and 8.2.5 respectively.

8.2.3 DP: Stick-breaking construction

The stick-breaking construction for DP was presented by Sethuraman (1994). In general, the DP and the stick-breaking process (SBP) are essential tools in BNP. Consider the stick-breaking weights $\{\beta_k\}_{k=1}^{\infty}$ on a countably infinite set. We want to find a distribution of the non-negative mixture weights $\beta_1, \beta_2, \cdots$ having the property

$$\sum_{k=1}^{\infty} \beta_k = 1. \tag{8.30}$$

One solution to this problem is provided by a procedure known as "stick-breaking." Considering a unit-length stick, we independently draw a proportion β'_k in the kth break from a Beta distribution with a concentration parameter α_0:

$$\beta'_k | \alpha_0 \sim \mathrm{Beta}(1, \alpha_0) \quad k = 1, 2, \cdots. \tag{8.31}$$

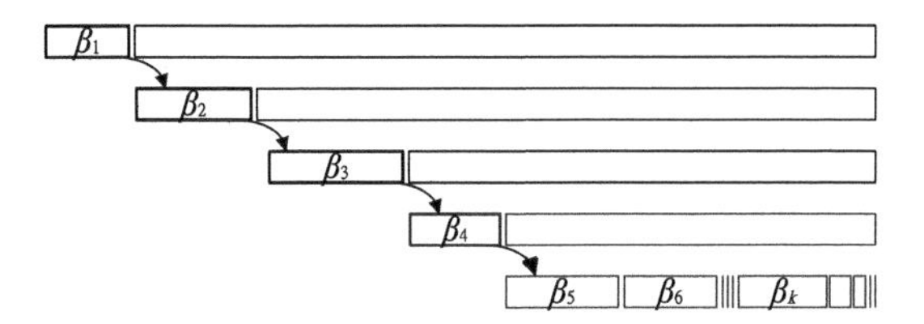

Figure 8.5 The stick-breaking process.

Each break decides a proportion β'_k while the proportion of the remaining stick is $1 - \beta'_k$. The mixture weight of the first component is $\beta_1 = \beta'_1$ and that of the kth component is determined by

$$\beta_k = \beta'_k \prod_{l=1}^{k-1}(1 - \beta'_l) \quad k = 1, 2, \cdots . \tag{8.32}$$

Under this SBP, it is straightforward to show the property of mixture weights that sum up to one with probability one. Figure 8.5 illustrates an infinite sequence of segments β_k from SBP. Let us define an infinite sequence of independent random variables $\{\beta_k\}_{k=1}^{\infty}$ and $\{\phi_k\}_{k=1}^{\infty}$ where

$$\phi_k | G_0 \sim G_0. \tag{8.33}$$

Then, the probability measure G defined in Eq. (8.29) can be shown to be the measure distributed according to DP(α_0, G_0) (Sethuraman 1994). We may interpret the sequence $\boldsymbol{\beta} = \{\beta_k\}_{k=1}^{\infty}$ as a random probability measure on the positive integers. This measure is formally denoted by the GEM distribution (Pitman 2002)

$$\boldsymbol{\beta} \sim \text{GEM}(\alpha_0). \tag{8.34}$$

8.2.4 DP: Chinese restaurant process

The second perspective on the DP is provided by the Pólya urn scheme (Blackwell & MacQueen 1973) showing that the draws from DP are both discrete and exhibit a clustering property. Let $\theta_1, \theta_2, \cdots$ be a sequence of independently and identically distributed (iid) random factors or parameters drawn from G for individual observations $x_1, x_2, \cdots$ under some distribution function. The probability model over the infinite sequence is written as

$$\begin{aligned} \theta_i | G &\sim G \\ x_i | \theta_i &\sim p(x_i|\theta_i) \qquad \text{for each } i. \end{aligned} \tag{8.35}$$

The factors $\theta_1, \theta_2, \cdots$ are conditionally independent given G, and hence are exchangeable. Consider the successive conditional distributions of θ_i of x_i given the previous factors $\theta_1, \cdots, \theta_{i-1}$ of observations $x_1, \cdots, x_{i-1}$, where G has been integrated out. It was shown that these conditional distributions have the following form (Blackwell & MacQueen 1973):

$$\theta_i|\theta_1, \cdots, \theta_{i-1}, \alpha_0, G_0 \sim \sum_{l=1}^{i-1} \frac{1}{i-1+\alpha_0}\delta_{\theta_l} + \frac{\alpha_0}{i-1+\alpha_0}G_0. \tag{8.36}$$

We can interpret the conditional distributions as a simple urn model where a ball of a distinct color is associated with each atom δ_{θ_l}. The balls are drawn equiprobably or uniformly. As seen in the second term of Eq. (8.36), a new atom is created by drawing from G_0 with probability proportional to α_0. A ball of new color is added to the urn. Equation (8.36) means that θ_i has a positive probability of being equal to one of the previous draws $\theta_1, \cdots, \theta_{i-1}$. This process results in a reinforcement effect: the more often a point is drawn, the more probable it is to be drawn in the future. To make the clustering more explicitly, we introduce a new set of variables that represent distinct values of atoms. Let $\phi_1, \cdots, \phi_K$ denote the distinct values taken from previous factors $\theta_1, \cdots, \theta_{i-1}$ and c_k be the number of customers or values $\theta_{i'}$ that are sitting at or are associated with ϕ_k of table or cluster k. Equation (8.36) is re-expressed by

$$\theta_i|\theta_1, \cdots, \theta_{i-1}, \alpha_0, G_0 \sim \sum_{k=1}^{K} \frac{c_k}{i-1+\alpha_0}\delta_{\phi_k} + \frac{\alpha_0}{i-1+\alpha_0}G_0. \tag{8.37}$$

From this re-expression, we find that the Pólya urn scheme produces a distribution on partitions which is closely related to the one using a different metaphor called the Chinese restaurant process (CRP) (Aldous 1985). The metaphor of CRP is addressed as follows. Consider a Chinese restaurant with an unbounded number of tables. Each θ_i corresponds to a customer x_i who enters the restaurant. The distinct values ϕ_k correspond to the tables at which the customers sit. The ith customer sits at the table indexed by ϕ_k with probability proportional to the number of customers c_k who are already seated there (in this case we set $\theta_i = \phi_k$), or sits at a new table with probability proportional to α_0 (in this case, we increment K, draw $\phi_K \sim G_0$, and set $\theta_i = \phi_K$). Figure 8.6 shows the scenario of the CRP where the current customer θ_{11} either chooses an occupied table from $\{\phi_1, \cdots, \phi_4\}$ or a new table ϕ_{new} according to

$$\begin{aligned} p(\text{occupied table } k|\text{previous customers}) &= \frac{c_k}{i-1+\alpha_0}, \\ p(\text{next unoccupied table}|\text{previous customers}) &= \frac{\alpha_0}{i-1+\alpha_0}. \end{aligned} \tag{8.38}$$

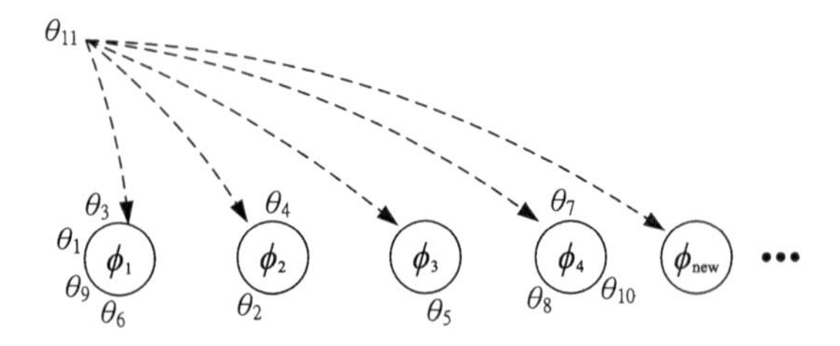

Figure 8.6 The Chinese restaurant process.

The CRP probability model in Eq. (8.38) is closely related to the Pólya urn model in Eq. (8.25).

8.2.5 Dirichlet process mixture model

One of the most important applications of the DP is to explore the nonparametric prior over the parameters of a mixture model. The resulting model is referred to as the *DP mixture model* (Antoniak 1974). Consider the probability model of an infinite sequence of observations $x_1, x_2, \cdots$ in Eq. (8.35), with graphical representation as in Figure 8.7(a). The probability measure G can be represented by using a stick-breaking construction. The factors θ_i take on values ϕ_k with probability β_k. Here, an indicator value z_i is introduced to reveal the positive integer value or cluster index for factor θ_i, and it is distributed according to $\boldsymbol{\beta}$. A DP mixture model can be represented by the following conditional distributions:

$$\begin{aligned}
\boldsymbol{\beta}|\alpha_0 &\sim \text{GEM}(\alpha_0),\\
z_i|\boldsymbol{\beta} &\sim \boldsymbol{\beta},\\
\phi_k|G_0 &\sim G_0,\\
x_i|z_i, \{\phi_k\}_{k=1}^{\infty} &\sim p(x_i|\phi_{z_i}).
\end{aligned} \tag{8.39}$$

Therefore, we have the mixture model $G = \sum_{k=1}^{\infty} \beta_k \delta_{\phi_k}$ and $\theta_i = \phi_{z_i}$.

Alternatively, the DP mixture model can be derived as the limit of a sequence of finite mixture models where the number of mixture components is taken to infinity (Neal 1992). This limiting process provides the third perspective on DP. Suppose that we have K mixture components. Let $\boldsymbol{\beta} = \{\beta_1, \cdots, \beta_K\}$ denote the mixture weights. In the limit $K \to \infty$, the vectors $\boldsymbol{\beta}$ are closely related and are equivalent up to a random size-biased permutation. We use a *Dirichlet prior* on $\boldsymbol{\beta}$ with symmetric parameters $(\alpha_0/L, \cdots, \alpha_0/L)$. We thus have the following model:

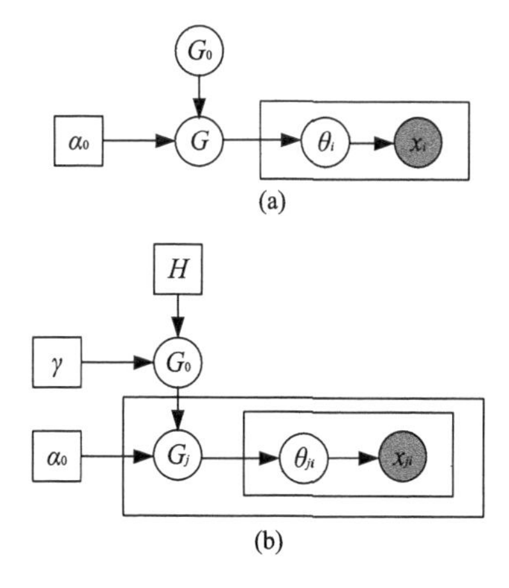

Figure 8.7 (a) Dirichlet process mixture model; (b) hierarchical Dirichlet process mixture model.

$$\begin{aligned}
\boldsymbol{\beta}|\alpha_0 &\sim \text{Dir}(\alpha_0/L, \cdots, \alpha_0/L), \\
z_i|\boldsymbol{\beta} &\sim \boldsymbol{\beta}, \\
\phi_k|G_0 &\sim G_0, \\
x_i|z_i, \{\phi_k\}_{k=1}^K &\sim p(x_i|\phi_{z_i}).
\end{aligned} \tag{8.40}$$

The corresponding mixture model $G^K = \sum_{k=1}^K \beta_k \delta_{\phi_k}$ was shown to have the property

$$\int f(\theta)dG^K(\theta) \longrightarrow \int f(\theta)dG(\theta), \tag{8.41}$$

as $K \rightarrow \infty$ for every measurable function $f(\cdot)$. The marginal distribution of the observations $x_1, \cdots, x_n$ approaches that using a DP mixture model.

8.2.6 Hierarchical Dirichlet process

The spirit of the graphical model based on directed graphs, as mentioned in Section 2.2, is mainly from that of hierarchical Bayesian modeling, while the graphical model literature has focused almost exclusively on parametric hierarchies where each of the conditionals is a finite-dimensional distribution. Nevertheless, it is possible to build hierarchies in which the components are stochastic processes. We illustrate how to do this based on the Dirichlet process.

In particular, we are interested in finding solutions to the problems in which the observations are organized into groups, where the observations are assumed to be exchangeable both within each group and across groups. For example, in document representation, the words in each document in a text corpus are associated with the data in each group from a set of grouped data. We want to discover the structures of words from a set of training documents. It is important to find the clusterings of word labels for data representation. Let j index the groups or documents and i index the observations or words within each group. We assume that $x_{j1}, x_{j2}, \cdots$ are exchangeable within each group j and also between groups. The group data $\mathbf{x}_j = \{x_{j1}, x_{j2}, \cdots\}$ in an infinite set of groups $\mathbf{x}_1, \mathbf{x}_2, \cdots$ are exchangeable.

Assuming that each observation is drawn independently from a mixture model, each observation x_{ji} is associated with a mixture component k. Let θ_{ji} denote a parameter or factor specifying the mixture component ϕ_k associated with the observation x_{ji}. The factors θ_{ji} are not generally distinct. Let $p(x_{ji}|\theta_{ji})$ denote the distribution of observation x_{ji} given the factor θ_{ji}. Let G_j denote a distribution for the factors $\boldsymbol{\theta}_j = \{\theta_{j1}, \theta_{j2}, \cdots\}$ associated with group j. Due to exchangeability, the factors are conditionally independent given G_j. The probability model for this stochastic process is expressed by

$$\begin{aligned}
\theta_{ji}|G_j &\sim G_j, \\
x_{ji}|\theta_{ji} &\sim p(x_{ji}|\theta_{ji}) \quad \text{for each } j \text{ and } i.
\end{aligned} \tag{8.42}$$

The hierarchical Dirichlet process (HDP) (Teh *et al.* 2006) was proposed to conduct BNP learning of grouped data, in which each group is associated with a mixture model and we wish to link these mixture models. An HDP is a nonparametric prior distribution over a set of random probability measures. The process defines a set of probability measures G_j, one for each group j, and a global probability measure G_0 shared for different groups. The nonparametric priors in HDP are given by the following hierarchy:

$$\begin{aligned} G_0|\gamma, H &\sim \mathrm{DP}(\gamma, H), \\ G_j|\alpha_0, G_0 &\sim \mathrm{DP}(\alpha_0, G_0) \quad \text{for each } j. \end{aligned} \tag{8.43}$$

The prior random measure of jth group G_j is a DP with a shared base measure G_0 for grouped data which is itself drawn from the DP with a base measure H. Here, the hyperparameters γ and α are the concentration parameters and H provides the prior distribution for the factors θ_{ji}. Basically, the distribution G_0 varies around the prior H with variations governed by γ while the distribution G_j over the factors in the jth group deviates from G_0 with variations controlled by α_0. Analogous to the DP mixture model mentioned in Section 8.2.5, HDP is feasible to construct the HDP mixture model as shown graphically in Figure 8.7(b). The HDP mixture model is expressed by:

$$\begin{aligned} G_0|\gamma, H &\sim \mathrm{DP}(\gamma, H), \\ G_j|\alpha_0, G_0 &\sim \mathrm{DP}(\alpha_0, G_0) \quad \text{for each } j, \\ \theta_{ji}|G_j &\sim G_j, \\ x_{ji}|\theta_{ji} &\sim p(x_{ji}|\theta_{ji}) \quad \text{for each } j \text{ and } i. \end{aligned} \tag{8.44}$$

We can see that this model achieves the goal of sharing clusters across groups by assigning the same parameters or factors to those observations. That is, if $\theta_{ji} = \theta_{ji'}$, the observations x_{ji} and $x_{ji'}$ belong to the same cluster. The equality of factors is possible because both θ_{ji} and $\theta_{ji'}$ are possibly drawn from G_j, which is a discrete measure over a measurable partition $(A_1, A_2, \cdots)$. Since G_j from different groups shares the atoms from G_0, the observations in different groups j can be assigned to the same cluster. In what follows, we present the realization of the HDP from the perspectives of a stick-breaking construction and a Chinese restaurant process.

8.2.7 HDP: Stick-breaking construction

Using a stick-breaking construction, the global measure G_0 and the individual measure G_j in HDP are expressed by the mixture models with the shared atoms $\{\phi_k\}_{k=1}^{\infty}$ but different weights $\boldsymbol{\beta} = \{\beta_k\}_{k=1}^{\infty}$ and $\boldsymbol{\pi}_j = \{\pi_{jk}\}_{k=1}^{\infty}$, respectively, as given by:

$$\begin{aligned} G_0 &= \sum_{k=1}^{\infty} \beta_k \delta_{\phi_k}, \\ G_j &= \sum_{k=1}^{\infty} \pi_{jk} \delta_{\phi_k}, \end{aligned} \tag{8.45}$$

where atom ϕ_k is drawn from base measure H and weights $\boldsymbol{\beta}$ are drawn from the GEM distribution $\boldsymbol{\beta} \sim \mathrm{GEM}(\gamma)$. Note that the weights $\boldsymbol{\pi}_j$ are independent given $\boldsymbol{\beta}$ because the G_js are independent given G_0.

Let $(A_1, \cdots, A_r)$ denote a measurable partition and $K_r = k : \phi_k \in A_l$ for $l = 1, \cdots, r$. Here, $(K_1, \cdots, K_r)$ is a finite partition of the positive integers. For each group j, we have

$$\begin{aligned}
&(G_j(A_i), \cdots, G_j(A_r)) \\
&\quad \sim \text{Dir}(\alpha_0 G_0(A_1), \cdots, \alpha_0 G_0(A_r)) \\
&\Rightarrow \left(\sum_{k \in K_1} \pi_{jk}, \cdots, \sum_{k \in K_r} \pi_{jk} \right) \\
&\quad \sim \text{Dir}\left(\alpha_0 \sum_{k \in K_1} \beta_k, \cdots, \alpha_0 \sum_{k \in K_r} \beta_k \right) \\
&\Rightarrow \boldsymbol{\pi}_j \sim \text{DP}(\alpha_0, \boldsymbol{\beta}).
\end{aligned} \tag{8.46}$$

Hence, each $\boldsymbol{\pi}_j$ is independently distributed according to $\text{DP}(\alpha_0, \boldsymbol{\beta})$. The HDP mixture model is then represented by:

$$\begin{aligned}
\boldsymbol{\beta} | \gamma &\sim \text{GEM}(\gamma), \\
\boldsymbol{\pi}_j | \alpha_0, \boldsymbol{\beta} &\sim \text{DP}(\alpha_0, \boldsymbol{\beta}), \\
z_{ji} | \boldsymbol{\pi}_j &\sim \boldsymbol{\pi}_j, \\
\phi_k | H &\sim H, \\
x_{ji} | z_{ji}, \{\phi_k\}_{k=1}^{\infty} &\sim p(x_{ji} | \phi_{z_{ji}}).
\end{aligned} \tag{8.47}$$

We may further show the explicit relationship between the elements of $\boldsymbol{\beta}$ and $\boldsymbol{\pi}_j$. The stick-breaking construction for $\text{DP}(\gamma, H)$ defines the variables β_k as

$$\begin{aligned}
\beta'_k &\sim \text{Beta}(1, \gamma), \\
\beta_k &= \beta'_k \prod_{l=1}^{k-1} (1 - \beta'_l).
\end{aligned} \tag{8.48}$$

Also, the stick-breaking construction for a probability measure $\boldsymbol{\pi}_j \sim \text{DP}(\alpha_0, \boldsymbol{\beta})$ of group j is performed by

$$\begin{aligned}
\pi'_{jk} &\sim \text{Beta}\left(\alpha_0 \beta_k, \alpha_0 \sum_{l=k+1}^{\infty} \beta_l \right) \\
&= \text{Beta}\left(\alpha_0 \beta_k, \alpha_0 \left(1 - \sum_{l=1}^{k} \beta_l \right) \right), \\
\pi_{jk} &= \pi'_{jk} \prod_{l=1}^{k-1} (1 - \pi'_{jl}).
\end{aligned} \tag{8.49}$$

This completes the stick-breaking construction for the HDP.

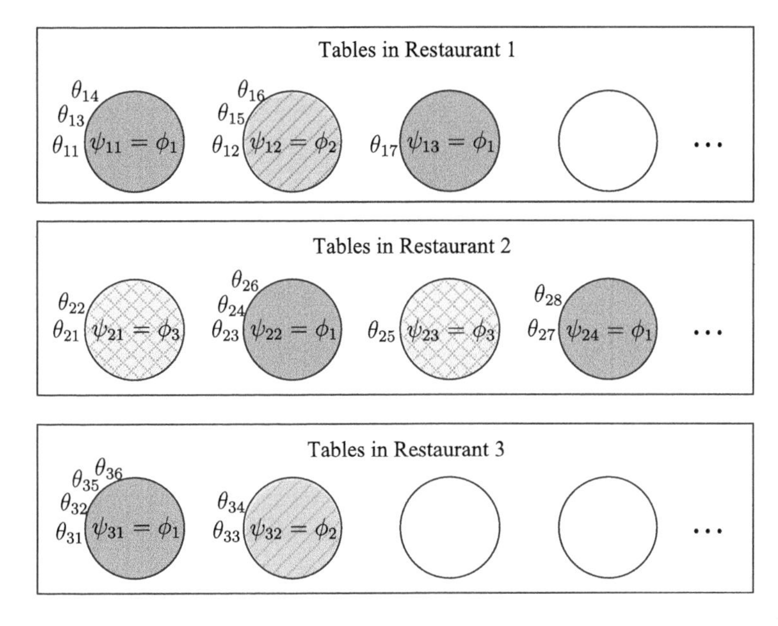

Figure 8.8 A Chinese restaurant franchise containing three restaurants (rectangles) with an infinite number of tables (circles) and dishes $\{\phi_k\}_{k=1}^{\infty}$. The term θ_{ji} denotes the ith customer in restaurant j, and ψ_{jt} is the indicator of a dish on the tth table of restaurant j. Different dishes ϕ_1, ϕ_2, and ϕ_3 are marked by different shading patterns.

8.2.8 HDP: Chinese restaurant franchise

The Chinese restaurant process for a DP is further extended to the *Chinese restaurant franchise* for an HDP, which allows for multiple restaurants sharing a set of dishes, as depicted in Figure 8.8. The metaphor is as follows. There is a restaurant franchise which shares the menu across restaurants. At each table of each restaurant, one dish is ordered from the menu by the first customer who sits there. This dish is shared among all customers sitting at that table. Different tables in different restaurants can serve the same dish. In this scenario, the restaurants correspond to the groups while the customers correspond to the factors θ_{ji} of observations x_{ji}. Let $\phi_1, \cdots, \phi_K$ denote the dishes in a global menu which are drawn from H. We introduce the variable, ψ_{jt}, that indicates the dish served at table t in restaurant j. In this restaurant franchise, we first consider the conditional distribution for a customer θ_{ji}, given the previous customers $\theta_{j1}, \cdots, \theta_{j,i-1}$ in restaurant j, and G_0, where the DP G_j is integrated out. From Eq. (8.37), we obtain

$$\theta_{ji}|\theta_{j1}, \cdots, \theta_{j,i-1}, \alpha_0, G_0 \sim \sum_{t=1}^{m_{j\cdot}} \frac{c_{jt\cdot}}{i-1+\alpha_0}\delta_{\psi_{jt}} + \frac{\alpha_0}{i-1+\alpha_0}G_0. \tag{8.50}$$

The notation c_{jtk} denotes the number of customers in restaurant j at table t eating dish k. Marginal count is represented by dot. Thus, $c_{jt\cdot}$ denotes the number of customers in restaurant j at table t. The notation m_{jk} means the number of tables in restaurant j serving dish k. Thus, $m_{j\cdot}$ represents the number of tables in restaurant j. This conditional is a mixture or a draw, which can be obtained by drawing from the terms on the right-hand-side of Eq. (8.50) with the probabilities given by the corresponding mixture weights for the occupied tables and an unoccupied table. If a term in the first summation is chosen, we increment $c_{jt\cdot}$ and set $\theta_{ji} = \psi_{jt}$. If the second term is chosen, we increment $m_{j\cdot}$, draw $\psi_{jm_{j\cdot}} \sim G_0$ and set $\theta_{ji} = \psi_{jm_{j\cdot}}$.

Next, we proceed to integrate out G_0, which is a DP, and apply Eq. (8.37) again to find the conditional distribution of a factor ψ_{jt} given the previous factors in the different restaurants:

$$\psi_{jt}|\psi_{11}, \psi_{12}, \cdots, \psi_{21}, \cdots, \psi_{j,t-1}, \gamma, H \sim \sum_{k=1}^{K} \frac{m_{\cdot k}}{m_{\cdot\cdot} + \gamma} \delta_{\phi_k} + \frac{\gamma}{m_{\cdot\cdot} + \gamma} H. \tag{8.51}$$

If we draw ψ_{jt} by choosing the term in the summation on the right-hand-side of Eq. (8.51), we set $\psi_{jt} = \phi_k$. If we choose the second term, we increment K, draw a new $\phi_K \sim H$ and set $\psi_{jk} = \phi_K$. It is meaningful that the mixture probability of the first term is proportional to $m_{\cdot k}$, which represents the number of tables serving dish k, while the probability of the second term is proportional to the concentration parameter γ of DP G_0. This completes the HDP implementation based on the Chinese restaurant franchise. To obtain samples θ_{ji} for each j and i, we first draw θ_{ji} from Eq. (8.50). If a new sample from G_0 is needed, we use Eq. (8.51) to obtain a new sample ψ_{jt} and set $\theta_{ji} = \psi_{jt}$.

Moreover, the HDP can be derived from the perspective of infinite limit over the finite mixture models. This is done by considering the collection of finite mixture models where L-dimensional $\boldsymbol{\beta}$ and $\boldsymbol{\pi}_j$ are used as the global and group-dependent mixing probabilities, respectively. The HDP finite mixture model is accordingly expressed by

$$\begin{aligned} \boldsymbol{\beta}|\gamma &\sim \text{Dir}(\gamma/L, \cdots, \gamma/L), \\ \boldsymbol{\pi}_j|\alpha_0, \boldsymbol{\beta} &\sim \text{Dir}(\alpha_0\boldsymbol{\beta}), \\ z_{ji}\boldsymbol{\pi}_j &\sim \boldsymbol{\pi}_j, \\ \phi_k|H &\sim H, \\ x_{ji}|z_{ji}, \{\phi_k\}_{k=1}^{L} &\sim p(x_{ji}|\phi_{z_{ji}}). \end{aligned} \tag{8.52}$$

It can be shown that the limit of this model as $L \to \infty$ approaches the HDP mixture model. The parametric hierarchical prior for $\boldsymbol{\beta}$ and $\boldsymbol{\pi}$ in Eq. (8.52) is also seen as the hierarchical Dirichlet model, which has been applied for language modeling (MacKay & Peto 1995), as addressed in Section 5.3.

8.2.9 MCMC inference by Chinese restaurant franchise

The MCMC sampling schemes have been developed for inference of the HDP mixture model. A straightforward Gibbs sampler based on the Chinese restaurant franchise can be implemented. To do so, the posterior probabilities for drawing tables $\mathbf{t} = \{t_{ji}\}$ and dishes $\mathbf{k} = \{k_{jt}\}$ should be determined. Let t_{ji} be the index of the factor ψ_{jt} associated with θ_{ji}, and let k_{jt} be the index of ϕ_k associated with ψ_{jt}. In the Chinese restaurant franchise, the customer i in restaurant j sits at table t_{ji}, whereas table t in restaurant j

serves dish k_{jt}. Let $\mathbf{x} = \{x_{ji}\}$, $\mathbf{x}_{jt} = \{x_{ji}, \text{all } i \text{ with } t_{ji} = t\}$, $\mathbf{z}_{z_{ji}}$, $\mathbf{m} = m_{jk}$ and $\boldsymbol{\phi} = \{\phi_1, \cdots, \phi_K\}$. Notations $\mathbf{k}^{-jt} = \mathbf{k}\backslash k_{jt}$ and $\mathbf{t}^{-ji} = \mathbf{t}\backslash t_{ji}$ denote the vectors $\mathbf{k}$ and $\mathbf{t}$ with the exception of components k_{jt} and t_{ji}, respectively, and $c_{jt\cdot}^{-ji}$ denotes the number of customers in restaurant j who enjoy the dish ψ_{jt}, but leaving out customer x_{ji}. Similarly, we have $\mathbf{x}^{-ji} = \mathbf{x}\backslash x_{ji}$ and $\mathbf{x}^{-jt} = \mathbf{x}\backslash \mathbf{x}_{jt}$. The concentration parameters γ and α_0 of DPs are assumed to be fixed. When implementing HDP using MCMC sampling, we need to calculate the conditional distribution of x_{ji} under mixture component k given all data samples except x_{ji},

$$p(x_{ji}|\mathbf{x}^{-ji}, k) = \frac{\int p(x_{ji}|\phi_k) \prod_{j'i' \neq ji, z_{j'i'}=k} p(x_{j'i'}|\phi_k) h(\phi_k) d\phi_k}{\int \prod_{j'i' \neq ji, z_{j'i'}=k} p(x_{j'i'}|\phi_k) h(\phi_k) d\phi_k}. \tag{8.53}$$

Here, $h(\phi_k)$ is a prior distribution from H and is conjugate to the likelihood $p(x|\phi_k)$. We can integrate out the mixture component parameter ϕ_k in closed form. Similarly to Eq. (8.53), the conditional distribution $p(\mathbf{x}_{jt}|\mathbf{x}^{-jt}, k)$ can be calculated by using $\mathbf{x}_{jt}$, the customers at restaurant j sitting at table t. In what follows, we derive the posterior probabilities for sampling tables t_{ji} and dishes k_{jt} which are then used to reconstruct θ_{ji}s and ψ_{jt}s based on the ϕ_ks. The property of exchangeability is employed in Eqs. (8.50) and (8.51) for θ_{ji}s and ψ_{jt}s, or equivalently for t_{ji}s and k_{jt}s, respectively.

Sampling $\mathbf{t}$: We want to calculate the conditional distribution of t_{ji} given the rest of variables $\mathbf{t}^{-ji}$. This conditional posterior for t_{ji} is obtained by combining the conditional prior for t_{ji} with the likelihood of generating x_{ji}. Basically, the likelihood function due to x_{ji} given previously occupied table t is expressed by $p(x_{ji}|\mathbf{x}^{-ji}, k)$, as shown in Eq. (8.53). The likelihood function for a new table $t_{ji} = t_{\text{new}}$ can be calculated by integrating out all possible values of $k_{jt_{\text{new}}}$ using Eq. (8.51):

$$\begin{aligned} & p(x_{ji}|\mathbf{t}^{-ji}, t_{ji} = t_{\text{new}}, \mathbf{k}) \\ & \quad = \sum_{k=1}^{K} \frac{m_{\cdot k}}{m_{\cdot\cdot} + \gamma} p(x_{ji}|\mathbf{x}^{-ji}, k) + \frac{\gamma}{m_{\cdot\cdot} + \gamma} p(x_{ji}|\mathbf{x}^{-ji}, k_{\text{new}}), \end{aligned} \tag{8.54}$$

where $p(x_{ji}|\mathbf{x}^{-ji}, k_{\text{new}})$ is a prior density of x_{ji} given by

$$p(x_{ji}) = \int p(x_{ji}|\phi) h(\phi) d\phi. \tag{8.55}$$

According to Eq. (8.50), the conditional prior probability of taking a previously occupied table $t_{ji} = t$ is proportional to $c_{jt\cdot}^{-ji}$, while the probability of taking a new table $t = t_{\text{new}} = m_{j\cdot} + 1$ is proportional to α_0. Thus, the conditional posterior probability for sampling table t_{ji} is derived from

$$\begin{aligned} & p(t_{ji} = t|\mathbf{t}^{-ji}, \mathbf{x}, \mathbf{k}) \\ & \quad \propto \begin{cases} c_{jt\cdot}^{-ji} \cdot p(x_{ji}|\mathbf{x}^{-ji}, k_{jt}) & \text{if } t \text{ previously occupied,} \\ \alpha_0 \cdot p(x_{ji}|\mathbf{t}^{-ji}, t_{ji} = t_{\text{new}}, \mathbf{k}) & \text{if } t = t_{\text{new}}. \end{cases} \end{aligned} \tag{8.56}$$

Sampling $\mathbf{k}$: When the sampled table is a new one, $t_{ji} = t_{\text{new}}$, we need to further order a new dish $k_{jt_{\text{new}}}$ from the global menu. This is done according to the conditional posterior probability, which is combined from a conditional prior for k_{jt} and a likelihood function due to x_{ji}. The conditional posterior probability is then derived from

$$\begin{aligned} p(k_{jt_{\text{new}}} &= k|\mathbf{t}, \mathbf{x}, \mathbf{k}^{-jt_{\text{new}}}) \\ &\propto \begin{cases} m_{\cdot k} \cdot p(x_{ji}|\mathbf{x}^{-ji}, k) & \text{if } k \text{ previously ordered,} \\ \gamma \cdot p(x_{ji}|\mathbf{x}^{-ji}, k_{\text{new}}) & \text{if } k = k_{\text{new}}. \end{cases} \end{aligned} \quad (8.57)$$

Because changing k_{jt} actually changes the component membership of all customers $\mathbf{x}_{jt}$ at table t, we can calculate the conditional posterior probability of ordering dish $k_{jt} = k$ from

$$\begin{aligned} p(k_{jt} &= k|\mathbf{t}, \mathbf{x}, \mathbf{k}^{-jt}) \\ &\propto \begin{cases} m_{\cdot k}^{-jt} \cdot p(\mathbf{x}_{jt}|\mathbf{x}^{-jt}, k) & \text{if } k \text{ previously ordered,} \\ \gamma \cdot p(\mathbf{x}_{jt}|\mathbf{x}^{-jt}, k_{\text{new}}) & \text{if } k = k_{\text{new}}. \end{cases} \end{aligned} \quad (8.58)$$

where the likelihood function $p(\mathbf{x}_{jt}|\mathbf{x}^{-jt}, k)$ due to the customers $\mathbf{x}_{jt}$ at table t of restaurant j is used.

8.2.10 MCMC inference by direct assignment

In the first MCMC procedure based on the Chinese restaurant franchise representation, the observations are first assigned to some table t_{ji}, and the tables are then assigned to some mixture component k_{jt}. This indirect association with mixture components could be realized by directly assigning mixture components through a new variable z_{ji} which is the same as $k_{jt_{ji}}$. The variable z_{ji} indicates the index of a mixture component corresponding to the customer x_{ji}. Using this direct assignment, the tables are represented only in terms of the number of tables m_{jk}. A bookkeeping scheme is involved. We would like to instantiate and sample from G_0 by using the *factorized posterior* on G_0 across groups. To do so, an explicit construction for $G_0 \sim \text{DP}(\gamma, H)$ is expressed in the form of

$$\begin{aligned} \boldsymbol{\beta} &= (\beta_1, \cdots, \beta_K, \beta_u) \sim \text{Dir}(m_{\cdot 1}, \cdots, m_{\cdot K}, \gamma), \\ G_u &\sim \text{DP}(\gamma, H), \\ p(\phi_k|\mathbf{z}) &\propto h(\phi_k) \prod_{ji:z_{ji}=k} p(x_{ji}|\phi_k, z_{ji}), \\ G_0 &= \sum_{k=1}^{K} \beta_k \delta_{\phi_k} + \beta_u G_u, \end{aligned} \quad (8.59)$$

which can be also expressed as

$$G_0 \sim \text{DP}\left(\gamma + m_{\cdot\cdot}, \frac{\gamma H + \sum_{k=1}^{K} m_{\cdot k} \delta_{\phi_k}}{\gamma + m_{\cdot\cdot}}\right). \quad (8.60)$$

From the expressions in Eq. (8.59), we can see that the values $m_{\cdot k}$ and γ in the first MCMC inference based on the *Chinese restaurant franchise* are replaced by β_k and β_u in the second MCMC inference based on the respective *direct assignments*. In the MCMC inference based on direct assignment, we need to sample the indicators of mixture components $\mathbf{z} = \{z_{ij}\}$ and the numbers of tables $\mathbf{m} = \{m_{jk}\}$, according to the corresponding posterior probabilities which are provided in what follows.

Sampling $\mathbf{z}$: The idea of sampling $\mathbf{z}$ is to group together the terms associated with each k in Eqs. (8.54) and (8.56) so as to calculate the conditional posterior probability:

$$p(z_{ji} = k|\mathbf{z}^{-ji}, \mathbf{x}, \mathbf{m}, \boldsymbol{\beta}) = \begin{cases} (c_{j\cdot k}^{-ji} + \alpha_0 \beta_k) p(x_{ji}|\mathbf{x}^{-ji}, k) & \text{if } k \text{ previously occupied,} \\ \alpha_0 \beta_u p(x_{ji}|\mathbf{x}^{-ji}, k_{\text{new}}) & \text{if } k = k_{\text{new}}. \end{cases} \tag{8.61}$$

Note that we have combined Eqs. (8.54) and (8.56) based on a new variable z_{ji} and have replaced $m_{\cdot k}$ with β_k and γ with β_u. To fulfil the sampling of $\mathbf{z}$ in Eq. (8.61), we have to further sample $\mathbf{m}$ and $\boldsymbol{\beta}$.

Sampling $\mathbf{m}$: According to the direct assignment of data items to mixture components $\mathbf{z}$, it is sufficient to sample $\mathbf{m}$ and $\boldsymbol{\beta}$ in place of $\mathbf{t}$ and $\mathbf{k}$. To find the conditional distribution of m_{jk}, we consider the conditional distribution of t_{ji} under the condition $k_{jt_{ji}} = z_{ji}$. From Eq. (8.50), the prior probability that data item x_{ji} is assigned to some table t such that $k_{jt} = k$ is

$$p(t_{ji} = t|k_{jt} = k, \mathbf{t}^{-ji}, \mathbf{k}, \boldsymbol{\beta}) \propto c_{jt\cdot}^{-ji}, \tag{8.62}$$

while the probability that is assigned to a new table under mixture component k is

$$p(t_{ji} = t_{\text{new}}|k_{jt_{\text{new}}} = k, \mathbf{t}^{-ji}, \mathbf{k}, \boldsymbol{\beta}) \propto \alpha_0 \beta_k. \tag{8.63}$$

These probabilities in a Gibbs sampler have the equilibrium distribution, which is the prior probability over the assignment of $c_{j\cdot k}$ observations to components in a DP with concentration parameter $\alpha_0 \beta_k$. The corresponding distribution over the number of mixture components is the desired conditional distribution of m_{jk} which is written as (Antoniak 1974, Teh *et al.* 2006):

$$p(m_{jk} = m|\mathbf{z}, \mathbf{m}^{-jk}, \boldsymbol{\beta}) = \frac{\Gamma(\alpha_0 \beta_k)}{\Gamma(\alpha_0 \beta_k + c_{j\cdot k})} s(c_{j\cdot k}, m)(\alpha_0 \beta_k)^m. \tag{8.64}$$

Here, $s(c, m)$ denotes the unsigned Stirling numbers of the first kind, which are calculated from

$$s(c + 1, m) = s(c, m - 1) + cs(c, m), \tag{8.65}$$

with initial conditions $s(0, 0) = s(1, 1) = 1$, $s(c, 0) = 0$ for $c > 0$ and $s(c, m) = 0$ for $m > c$.

Sampling $\boldsymbol{\beta}$: Having the samples $\mathbf{m}$ and the fixed hyperparameter γ, the β_k parameters are sampled from a Dirichlet distribution,

$$(\beta_1, \cdots, \beta_K, \beta_u)|\mathbf{m}, \gamma \sim \text{Dir}(m_{\cdot 1}, \cdots, m_{\cdot K}, \gamma). \tag{8.66}$$

This completes the MCMC inference for the HDP, based on the scheme of direct assignment of mixture components. It was shown that MCMC inference with the scheme of direct assignment is better than the scheme using the Chinese restaurant franchise in terms of convergence speed.

8.2.11 Relation of HDP to other methods

In general, HDP can be seen as a building block for a variety of speech and language processing applications. An instance of application is the latent Dirichlet allocation (LDA) model (Blei *et al.* 2003), where each entity is associated not with a single cluster but with a set of clusters. In LDA terminology, each document is associated with a set of topics. As described in Section 7.6, an LDA model is constructed as a Bayesian parametric model with a fixed number of topics. HDP relaxes the limitation of a finite dimension in latent topic space in LDA and builds the flexible topic model with infinite clusters or topics. Multiple DPs are used to capture the uncertainty regarding the number of mixture components. HDP is viewed as a BNP version of an LDA model. The topics or clusters ϕ_k for the jth document are drawn from the nonparametric prior G_j, while the measures G_j are drawn from a DP with a base measure G_0. This allows the same topics to appear in multiple documents.

There are some other ways to connect multiple DPs. One idea is based on the *nested Dirichlet process* (NDP) (Rodriguez, Dunson & Gelfand 2008), which was proposed to model a collection of dependent distributions by using random variables as atoms at the higher level and random distributions as atoms at the lower level. This combinatorial process borrows information across DPs while also allowing DPs to be clustered. The simultaneous multilevel clustering can be achieved by such a nested setting. NDP is characterized by

$$\begin{aligned} G_j &\sim Q \quad \text{for each } j, \\ Q &\sim \text{DP}(\alpha_0, \text{DP}(\gamma, H)). \end{aligned} \tag{8.67}$$

Using HDP, the distributions $\{G_j\}$ of different j share the same atoms but assign them with different weights. However, using NDP, these distributions may have completely different atoms and weights. By marginalizing over the DPs, the resulting urn model is closely related to the *nested Chinese restaurant process* (nCRP) (Blei, Griffiths & Jordan 2010), which is known as a tree model of Chinese restaurants. The scenario of nCRP is addressed as follows. A customer enters the tree at a root Chinese restaurant and sits at a table. This table points to another Chinese restaurant, where the customer goes to dine the next evening. The construction is done recursively. Thus, a given customer follows a path through the tree of restaurants. Through BNP inference of topic hierarchies, a hierarchical topic model is established (Blei, Griffiths, Jordan *et al.* 2004). Each document or customer chooses a tree path of topics while each word in the document is represented by a mixture model of hierarchical topics along this tree path. In the following sections, we address some BNP methods which have been successfully applied for building the speaker diarization system and for developing the solutions of acoustic and language models to speech recognition systems.

8.3 Gibbs sampling-based speaker clustering

This section describes an application of MCMC techniques for speech features. We focus on speaker clustering, as discussed in Section 6.6.2 based on a Bayesian information criterion (BIC). Section 6.6.2 models a speaker cluster with a single Gaussian where we assume a stationary property for speech features within a segment. However, this is actually not correct since it has various temporal patterns based on linguistic variations, speaking styles, and noises. Therefore, we model these short-term variations with a GMM, while speaker cluster characteristics are modeled by a mixture of GMMs to consider the multi-scale properties of speech dynamics (Moraru, Meignier, Besacier *et al.* 2003, Valente & Wellekens 2004*b*, Wooters, Fung, Peskin *et al.* 2004, Meignier, Moraru, Fredouille *et al.* 2006).

An important aspect of this speech modeling technique is the consideration of the multi-scale property in dynamics within a probabilistic framework. For example, PLSA (in Section 3.7.3) and LDA (in Section 7.6) are successful approaches in terms of the multi-scale property. They deal accurately with two types of scales, namely, word-level and document-level scales (i and m in the complete data likelihood function of the latent topic model in Eq. (3.293)), based on a latent topic model (Hofmann 1999*a*). The approach discussed in this section is inspired by these successful approaches, and aims to apply a fully Bayesian treatment to the multi-scale properties of speech dynamics.

There have been several studies on Bayesian speech modeling, e.g., by using maximum a-posteriori (MAP) in Chapter 4 or variational Bayesian (VB) approaches in Chapter 7 for speech recognition (Gauvain & Lee 1994, Watanabe *et al.* 2004), speaker verification (Reynolds *et al.* 2000), and speaker clustering (Valente & Wellekens 2004*b*). While all of these approaches are based on the EM-type deterministic algorithm, this section focuses on another method of realizing fully Bayesian treatment, namely *sampling approaches* based on MCMC. The main advantage of the sampling approaches is that they can avoid local optimum problems in addition to providing other Bayesian advantages (mitigation of data sparseness problems and capability of model structure optimization). While their heavy computational cost could be a problem in realization, recent improvements in computational power and the development of theoretical and practical aspects related to the sampling approaches allow us to apply them to practical problems (e.g., Griffiths & Steyvers (2004), Goldwater & Griffiths (2007), Porteous, Newman, Ihler *et al.* (2008) in natural language processing). Therefore, the aim of this work is to apply a sampling approach to speech modeling considering the multi-scale properties of speech dynamics.

The following sections formulate the multi-scale GMM by utilizing a Gibbs sampling approach. In this section, for its educational value, we first describe Gibbs sampling for a standard GMM. Section 8.3.2 derives the marginal likelihood of the GMM, which is used for deriving GMM Gibbs samplers in Section 8.3.3. Section 8.3.4 provides the generative process and a graphical model of multi-scale GMM for speaker clustering. Based on the analytical results of GMM Gibbs sampling, Section 8.3.5 derives the marginal likelihood of the multi-scale GMM, which is used for deriving Gibbs samplers in Section 8.3.6.

8.3.1 Generative model

This section considers the two types of observation vector sequences. One is an utterance- (or segment-) level sequence and the other is a frame-level sequence. Then, a

D dimensional observation vector (e.g., MFCC) at frame t in utterance u is represented as $\mathbf{o}_{ut} \in \mathbb{R}^D$. A set of observation vectors in utterance u is represented as

$$\mathbf{O}_u \triangleq \{\mathbf{o}_{ut} \in \mathbb{R}^D | t = 1, \cdots, T_u\}. \tag{8.68}$$

T_u denotes the number of frames at an utterance u.

Next, we assume that the frame-level sequence is modeled by a GMM as usual, and the utterance-level sequence is modeled by a mixture of these GMMs. Two kinds of latent variables are involved in multi-scale GMM for each sequence: utterance-level latent variable z_u and frame-level latent variable v_{ut}. Utterance-level latent variables may represent emotion, topic, and speaking style as well as speakers, depending on the speech variation. The joint likelihood function of U observation vectors ($\mathbf{O} \triangleq \{\mathbf{O}_u | u = 1, \cdots, U\}$) with the latent variable sequences ($Z \triangleq \{z_u \in \{1, \cdots, S\} | u = 1, \cdots, U\}$ and $V \triangleq \{v_{ut} \in \{1, \cdots, K\} | t = 1, \cdots, T_u, u = 1, \cdots, U\}$) can be expressed as follows:

$$p(\mathbf{O}, Z, V|\Theta) = \prod_{u=1}^{U} h_{z_u} \prod_{t=1}^{T_u} w_{z_u v_{ut}} \mathcal{N}(\mathbf{o}_{ut} | \boldsymbol{\mu}_{z_u v_{ut}}, \mathbf{R}^{-1}_{z_u v_{ut}}), \tag{8.69}$$

where $h_s \in [0, 1]$ denotes the utterance-level mixture weight, and $w_{sk} \in [0, 1]$ denotes the frame-level mixture weight. The terms $\boldsymbol{\mu}_{sk} \in \mathbb{R}^D$ and $\mathbf{R}_{sk} \in \mathbb{R}^{D \times D}$ denote the mean vector and precision matrix parameters of the Gaussian distribution, s and k denote utterance-level and frame-level mixture indexes, respectively, and S and K denote the number of speakers and the number of mixture components, respectively. Therefore, a set of model parameters Θ is defined as

$$\Theta \triangleq \{h_s, w_{sk}, \boldsymbol{\mu}_{sk}, \mathbf{R}_{sk} | s = 1, \cdots, S, k = 1, \cdots, K\}. \tag{8.70}$$

Note that this is almost equivalent to the following pdf of the GMM in Section 3.2.4:

$$p(\mathbf{O}, V|\Theta) = \prod_{t=1}^{T} w_{v_t} \mathcal{N}(\mathbf{o}_t | \boldsymbol{\mu}_{v_t}, \mathbf{R}^{-1}_{v_t}). \tag{8.71}$$

The next section first describes Gibbs sampling for the standard GMM.

8.3.2 GMM marginal likelihood for complete data

We assume a diagonal covariance matrix for the Gaussian distributions as usual, where the d-d diagonal element of the precision/covariance matrix is expressed as r_d/Σ_d. The following conjugate distributions are used as the prior distributions of the model parameters:

$$p(\Theta|\Psi^0): \begin{cases} \mathbf{w} \sim \mathrm{Dir}(\boldsymbol{\phi}^w), \\ \boldsymbol{\mu}_k \sim \mathcal{N}(\boldsymbol{\mu}^0_k, (\phi^\mu)^{-1}\mathbf{R}^{-1}_k), \\ r_{kd} \sim \mathrm{Gam}(\phi^r, r^0_{kd}), \end{cases} \tag{8.72}$$

where $\boldsymbol{\phi}^w, \boldsymbol{\mu}^0_k, \phi^\mu, r^0_{kd}, \phi^r (\triangleq \Psi^0)$ are the hyperparameters. $\mathrm{Dir}(\cdot)$ and $\mathrm{Gam}(\cdot)$ denote Dirichlet and gamma distributions in Appendices C.4 and C.11, respectively.

In a Bayesian inference framework, we focus on the marginal likelihood for the complete data. In the complete data case, all of the latent variables are treated as observations, i.e., the assignments of all the latent variables are hypothesized to be given in advance. Then, $p(v_u = k|\cdot) \triangleq \delta(v_u, k)$ return 0 or 1 based on the assignment information, and the sufficient statistics of the GMM given all latent variables V can be represented as follows:

$$\begin{cases} \gamma_{V,k} &= \sum_t \delta(v_t, k), \\ \boldsymbol{\gamma}^{(1)}_{V,k} &= \sum_t \delta(v_t, k)\mathbf{o}_t, \\ \gamma^{(2)}_{V,kd} &= \sum_t \delta(v_t, k)(o_{td})^2. \end{cases} \tag{8.73}$$

The quantity $\gamma_{V,k} \in \mathbb{Z}^+$ is a count of frames assigned to k, and $\boldsymbol{\gamma}^{(1)}_{V,k}$ and $\gamma^{(2)}_{V,kd}$ are first-order and second-order sufficient statistics, respectively. We define a set of these sufficient statistics as

$$\Xi_{V,k} \triangleq \{\gamma_{V,k}, \boldsymbol{\gamma}^{(1)}_{V,k}, \gamma^{(2)}_{V,kd} | k = 1, \cdots, K, d = 1, \cdots, D\}. \tag{8.74}$$

The subscript V would be omitted when it is obvious. Note that the statistics γ_k is a positive discrete number while those appearing in the EM algorithm (ML in Eq. (3.153), MAP in Eq. (4.93), and VB in Eq. (7.67)) are positive continuous values since the occurrences of latent variables in the EM algorithm are expectation values based on the posterior probabilities of latent variables.

Based on the sufficient statistics representation in Eq. (8.73), the complete data likelihood of Eq. (8.71) can be represented by using the product formula of the Kronecker delta function in Eq. (A.4) ($f_b = \prod_a f_a^{\delta(a,b)}$) as follows:

$$\begin{aligned} p(\mathbf{O}, V|\Theta) &= \prod_{k=1}^{K}\prod_{t=1}^{T}\left(w_k^{\delta(v_t,k)}\mathcal{N}(\mathbf{o}_t|\boldsymbol{\mu}_k, \mathbf{R}_k^{-1})\right)^{\delta(v_t,k)} \\ &= \prod_{k=1}^{K}(w_k)^{\gamma_k}\prod_{t=1}^{T}\left(\mathcal{N}(\mathbf{o}_t|\boldsymbol{\mu}_k, \mathbf{R}_k^{-1})\right)^{\delta(v_t,k)}. \end{aligned} \tag{8.75}$$

Since the likelihood function is represented by the exponential distributions (multinomial and Gaussian distributions), these parameters integrate out, and we can use collapsed Gibbs sampling, as discussed in Section 8.1.4. The marginal likelihood for the complete data, $p(\mathbf{O}, V|\Psi^0)$, is represented by the following expectations by substituting Eqs. (8.72) and (8.75) into the following integration:

$$\begin{aligned} &p(\mathbf{O}, V|\Psi^0) \\ &= \int p(\mathbf{O}, V|\Theta)p(\Theta|\Psi^0)d\Theta \\ &= \left(\mathbb{E}_{(\mathbf{w})}\left[\prod_{k=1}^{K}(w_k)^{\gamma_k}\right]\right)\left(\prod_{k=1}^{K}\mathbb{E}_{(\boldsymbol{\mu}_k,\mathbf{R}_k)}\left[\prod_{t=1}^{T}\left(\mathcal{N}(\mathbf{o}_t|\boldsymbol{\mu}_k, \mathbf{R}_k^{-1})\right)^{\delta(v_t,k)}\right]\right). \end{aligned} \tag{8.76}$$

Note that this calculation is similar to those of the MAP auxiliary function in Section 4.3.5 and variational lower bound in Section 7.3.4. For example, $\mathbb{E}_{(\mathbf{w})}\left[\prod_{k=1}^{K}(w_k)^{\gamma_k}\right]$ can be rewritten as

$$
\begin{aligned}
&\mathbb{E}_{(\mathbf{w})}\left[\prod_{k=1}^{K}(w_k)^{\gamma_k}\right] \\
&= \int p(\{w_k\}_{k=1}^{K})\prod_{k=1}^{K}(w_k)^{\gamma_k} dw_k \\
&= \int \exp\left(\log\left(\mathrm{Dir}(\{w_k\}_{k=1}^{K}|\{\phi_k^w\}_{k=1}^{K})\right) + \sum_{k=1}^{K}\gamma_k \log w_k\right)\prod_{k=1}^{K} dw_k. \qquad (8.77)
\end{aligned}
$$

This is the same as the MAP auxiliary function of Eq. (4.54), and by analogy, Eq. (8.77) can be rewritten with the posterior distribution based equation in Eq. (4.54), as follows:

$$
\begin{aligned}
&\mathbb{E}_{(\mathbf{w})}\left[\prod_{k=1}^{K}(w_k)^{\gamma_k}\right] \\
&= \int \exp\left(\log\left(\mathrm{Dir}(\{w_k\}_{k=1}^{K}|\{\tilde{\phi}_k^w\}_{k=1}^{K})\right) + \log\left(\frac{C_{\mathrm{Dir}}(\{\phi_k^w\}_{k=1}^{K})}{C_{\mathrm{Dir}}(\{\tilde{\phi}_k^w\}_{k=1}^{K})}\right)\right)\prod_{k=1}^{K} dw_k, \qquad (8.78)
\end{aligned}
$$

where $\tilde{\phi}_s^w$ is a posterior hyperparameter, which is defined as:

$$
\tilde{\phi}_k^w \triangleq \phi_k^w + \gamma_k. \qquad (8.79)
$$

Therefore, the integral is finally solved as follows:

$$
\begin{aligned}
\mathbb{E}_{(\mathbf{w})}\left[\prod_{k=1}^{K}(w_k)^{\gamma_k}\right] &= \frac{C_{\mathrm{Dir}}(\{\phi_k^w\}_{k=1}^{K})}{C_{\mathrm{Dir}}(\{\tilde{\phi}_k^w\}_{k=1}^{K})}\int \mathrm{Dir}(\{w_k\}_{k=1}^{K}|\{\tilde{\phi}_k^w\}_{k=1}^{K})\prod_{k=1}^{K} dw_k \\
&= \frac{C_{\mathrm{Dir}}(\{\phi_k^w\}_{k=1}^{K})}{C_{\mathrm{Dir}}(\{\tilde{\phi}_k^w\}_{k=1}^{K})}. \qquad (8.80)
\end{aligned}
$$

Thus, the integration is calculated analytically, similarly to the calculation of the variational lower bound. The integral with respect to $\boldsymbol{\mu}_k$ and $\mathbf{R}_k$ in Eq. (8.76) is also analytically calculated by using Eq. (7.124). First, the expectation is rewritten as follows:

$$
\begin{aligned}
&\mathbb{E}_{(\boldsymbol{\mu}_k,\mathbf{R}_k)}\left[\prod_{t=1}^{T}\left(\mathcal{N}(\mathbf{o}_t|\boldsymbol{\mu}_k,\mathbf{R}_k^{-1})\right)^{\delta(v_t,k)}\right] \\
&= \int \exp\Bigg(\sum_{d=1}^{D}\log\left(\mathcal{N}(\mu_{kd}|\tilde{\mu}_{kd},(\tilde{\phi}_k^{\mu} r_{kd})^{-1})\mathrm{Gam}_2(r_{kd}|\tilde{r}_{kd},\tilde{\phi}_k^r)\right) \\
&\quad + \left(-\frac{\gamma_k D}{2}\log(2\pi) + \frac{D}{2}\log\frac{\phi^{\mu}}{\tilde{\phi}_k^{\mu}} + \log\frac{C_{\mathrm{Gam}_2}(r_{kd}^0,\phi^r)}{C_{\mathrm{Gam}_2}(\tilde{r}_{kd},\tilde{\phi}_k^r)}\right)\Bigg)d\boldsymbol{\mu}_k d\mathbf{R}_k \\
&= \exp\left(-\frac{\gamma_k D}{2}\log(2\pi) + \frac{D}{2}\log\frac{\phi^{\mu}}{\tilde{\phi}_k^{\mu}} + \log\frac{C_{\mathrm{Gam}_2}(r_{kd}^0,\phi^r)}{C_{\mathrm{Gam}_2}(\tilde{r}_{kd},\tilde{\phi}_k^r)}\right), \qquad (8.81)
\end{aligned}
$$

where posterior hyperparameters $\tilde{\phi}_k^\mu$, $\tilde{\mu}_k$, $\tilde{\phi}_k^r$, and $\tilde{r}_{kd}$ are defined as follows:

$$\begin{cases} \tilde{\phi}_k^\mu & \triangleq \phi^\mu + \gamma_k, \\ \tilde{\mu}_k & \triangleq \frac{\phi^\mu \mu_k^0 + \gamma_k^{(1)}}{\tilde{\phi}_k^\mu}, \\ \tilde{\phi}_k^r & \triangleq \phi^r + \gamma_k, \\ \tilde{r}_{kd} & \triangleq r_{kd}^0 + \gamma_{kd}^{(2)} + \phi^\mu (\mu_{kd}^0)^2 - \tilde{\phi}_k^\mu (\tilde{\mu}_{kd})^2. \end{cases} \tag{8.82}$$

Thus, we finally solve all integrals in Eq. (8.76) with use of concrete forms of the normalization constants of the Dirichlet and gamma distributions in Appendices C.4 and C.11, as follows:

$$p(\mathbf{O}, V|\Psi^0)$$

$$= \frac{\Gamma(\sum_k \phi_k^w) \prod_k \Gamma(\tilde{\phi}_k^w)}{\prod_k \Gamma(\phi_k^w)\, \Gamma(\sum_k \tilde{\phi}_k^w)} \prod_k (2\pi)^{-\frac{\gamma_k D}{2}} \frac{(\phi^\mu)^{\frac{D}{2}} \left(\Gamma\left(\frac{\phi^r}{2}\right)\right)^{-D} \left(\prod_d \frac{r_{kd}^0}{2}\right)^{\frac{\phi^r}{2}}}{(\tilde{\phi}_k^\mu)^{\frac{D}{2}} \left(\Gamma\left(\frac{\tilde{\phi}_k^r}{2}\right)\right)^{-D} \left(\prod_d \frac{\tilde{r}_{kd}}{2}\right)^{\frac{\tilde{\phi}_k^r}{2}}}. \tag{8.83}$$

Below we summarize the posterior hyperparameters, which are obtained from the hyperparameters of the prior distributions (Ψ^0) and sufficient statistics (Eq. (8.73)) as follows:

$$\begin{cases} \tilde{\phi}_k^w & = \phi_k^w + \gamma_k, \\ \tilde{\phi}_k^\mu & = \phi^\mu + \gamma_k, \\ \tilde{\boldsymbol{\mu}}_k & = \frac{\phi^\mu \boldsymbol{\mu}_k^0 + \boldsymbol{\gamma}_k^{(1)}}{\tilde{\phi}_k^\mu}, \\ \tilde{\phi}_k^r & = \phi^r + \gamma_k, \\ \tilde{r}_{kd} & = r_{kd}^0 + \gamma_{kd}^{(2)} + \phi^\mu (\mu_{kd}^0)^2 - \tilde{\phi}_k^\mu (\tilde{\mu}_{kd})^2. \end{cases} \tag{8.84}$$

The marginal likelihood obtained is quite similar to the model parameter part of the variational lower bound in Eq. (7.126), since both functions are obtained by integrating out the Gaussian parameters for complete data likelihood. Based on the marginal likelihood for these complete data, we can calculate the marginal conditional distribution of v_t, as shown below.

8.3.3 GMM Gibbs sampler

As discussed in Section 8.1.4, a collapsed Gibbs sampler can assign latent variables by using the marginal conditional distribution. First, from the sum and product rules, the marginal conditional distribution $p(v_t = k|\mathbf{O}, V_{\backslash t})$ is represented as follows:

$$p(v_t = k|\mathbf{O}, V_{\backslash t}) = \frac{p(\mathbf{O}, V_{\backslash t}, v_t = k)}{p(\mathbf{O}, V_{\backslash t})} \propto p(\mathbf{O}, V_{\backslash t}, v_t = k|\Psi^0). \tag{8.85}$$

Here, $V_{\backslash t}$ indicates a set that does not include the tth frame element. Therefore, by considering the normalization constant, the posterior probability can be obtained by using Eq. (8.83) as follows:

$$p(v_t = k|\mathbf{O}, V_{\backslash t}) = \frac{p(\mathbf{O}, V_{\backslash t}, v_t = k|\Psi^0)}{\sum_{k'=1}^{K} p(\mathbf{O}, V_{\backslash t}, v_t = k'|\Psi^0)}$$
$$= \frac{g(\tilde{\Psi}_{V_{\backslash t}, v_t = k})}{\sum_{k'=1}^{K} g(\tilde{\Psi}_{V_{\backslash t}, v_t = k'})}, \tag{8.86}$$

where $\tilde{\Psi}_{V_{\backslash t}, v_t=k}$ is a set of posterior hyperparameters computed given latent variables of $V_{\backslash t}, v_t = k$. Note that the denominators of $\tilde{\Psi}_{V_{\backslash t}, v_t=k'}$ have the same latent variable for all frames except t. To compute $\tilde{\Psi}_{V_{\backslash t}, v_t=k'}$, we first need to compute $\Xi_{V_{\backslash t},k}$, which is a set of sufficient statistics for all frames except t as follows:

$$\Xi_{V_{\backslash t},k} : \begin{cases} \gamma_{V_{\backslash t},k} & = \sum_{t'=\{1,\cdots,T\}\backslash t} \delta(v_{t'}, k), \\ \boldsymbol{\gamma}^{(1)}_{V_{\backslash t},k} & = \sum_{t'=\{1,\cdots,T\}\backslash t} \delta(v_{t'}, k)\mathbf{o}_t, \\ \gamma^{(2)}_{V_{\backslash t},kd} & = \sum_{t'=\{1,\cdots,T\}\backslash t} \delta(v_{t'}, k)(o_{td})^2. \end{cases} \tag{8.87}$$

From Eq. (8.73), $\Xi_{V_{\backslash t},k}$ can be computed by simply subtracting the zeroth-, first-, and second-order values of $v_t = k$ as:

$$\Xi_{V_{\backslash t},k} : \begin{cases} \gamma_{V_{\backslash t},k} & = \gamma_k - \delta(v_t, k), \\ \boldsymbol{\gamma}^{(1)}_{V_{\backslash t},k} & = \boldsymbol{\gamma}^{(1)}_k - \delta(v_t, k)\mathbf{o}_t, \\ \gamma^{(2)}_{V_{\backslash t},kd} & = \gamma^{(2)}_{kd} - \delta(v_{t'}, k)(o_{td})^2. \end{cases} \tag{8.88}$$

Thus, $\Xi_{V_{\backslash t}, v_t=k'}$ can be obtained by simply adding the zeroth-, first-, and second-order values of $v_t = k'$ for all k' as:

$$\Xi_{V_{\backslash t}, v_t=k'} : \begin{cases} \gamma_{V_{\backslash t}, v_t=k'} & = \gamma_{V_{\backslash t},k'} + \delta(v_t, k'), \\ \boldsymbol{\gamma}^{(1)}_{V_{\backslash t}, v_t=k'} & = \boldsymbol{\gamma}^{(1)}_{V_{\backslash t},k'} + \delta(v_t, k')\mathbf{o}_t, \qquad \text{for all } k' \\ \gamma^{(2)}_{V_{\backslash t}, v_t=k',d} & = \gamma^{(2)}_{V_{\backslash t},k'd} + \delta(v_{t'}, k')(o_{td})^2. \end{cases} \tag{8.89}$$

$g(\cdot)$ in Eq. (8.86) is defined as follows:

$$g(\tilde{\Psi}_k) \triangleq \Gamma(\tilde{\phi}^w_k)(\tilde{\phi}^\mu_k)^{-\frac{D}{2}} \left(\Gamma\left(\frac{\tilde{\phi}^r_k}{2}\right)\right)^D \left(\prod_d \frac{\tilde{r}_{kd}}{2}\right)^{-\frac{\tilde{\phi}^r_k}{2}}. \tag{8.90}$$

This equation is obtained by canceling out the factors in the numerator and denominator of Eq. (8.83). Thus, we obtain the Gibbs sampler, which assigns mixture component k at frame t.

Note that if we use multinomial distributions in LDA instead of Gaussian distributions, the numerator and denominator of the Gibbs sampler are further canceled out (see Griffiths & Steyvers (2004)) based on the formula of the gamma function in Appendix A.4. Actually, $\Gamma(\tilde{\phi}^w_k)$ in Eq. (8.90) can be similarly canceled out, while the other factors cannot be canceled out. Therefore, the computational cost of the Gaussian-based Gibbs sampler is large compared with LDA, since we need to compute Eq. (8.90) for every k and every frame t.

8.3.4 Generative process and graphical model of multi-scale GMM

Based on solution of GMM Gibbs sampling in the previous section, we consider the Bayesian treatment of this multi-scale GMM for speaker clustering, which is an extension of GMM. For the conditional likelihood equation in Eq. (8.69), we again assume a diagonal covariance matrix for the Gaussian distributions. We also assume that the prior hyperparameters of the GMM parameters $\{w_{sk}\}_{sk}$, $\{\boldsymbol{\mu}_{sk}\}_{sk}$, and $\{\mathbf{R}_{sk}\}_{sk}$ for each s are shared with the parameters of one GMM (universal background model assumption (Reynolds *et al.* 2000)), which is used for speaker and speech recognition (subspace GMM (Povey, Burget, Agarwal *et al.* 2010)). Then the following conjugate distributions are used as the prior distributions of the model parameters:

$$p(\Theta|\Psi^0): \begin{cases} \mathbf{h} & \sim \mathrm{Dir}(\boldsymbol{\phi}^h), \\ \mathbf{w}_s & \sim \mathrm{Dir}(\boldsymbol{\phi}^w), \\ \boldsymbol{\mu}_{sk} & \sim \mathcal{N}(\boldsymbol{\mu}_k^0, (\phi^\mu)^{-1}\mathbf{R}_{sk}^{-1}), \\ r_{skd} & \sim \mathrm{Gam}(\phi^r, r_{kd}^0), \end{cases} \tag{8.91}$$

where $\boldsymbol{\phi}^h, \boldsymbol{\phi}^w, \boldsymbol{\mu}_k^0, \phi^\mu, r_{kd}^0, \phi^r (\triangleq \Psi^0)$ are the hyperparameters.

Based on the likelihood function and prior distributions, the generative process of multi-scale GMM can be expressed in Algorithm 17. The corresponding graphical model is shown in Figure 8.9. Now we have introduced multi-scale GMM, the following sections derive a solution for multi-scale GMM based on Gibbs sampling.

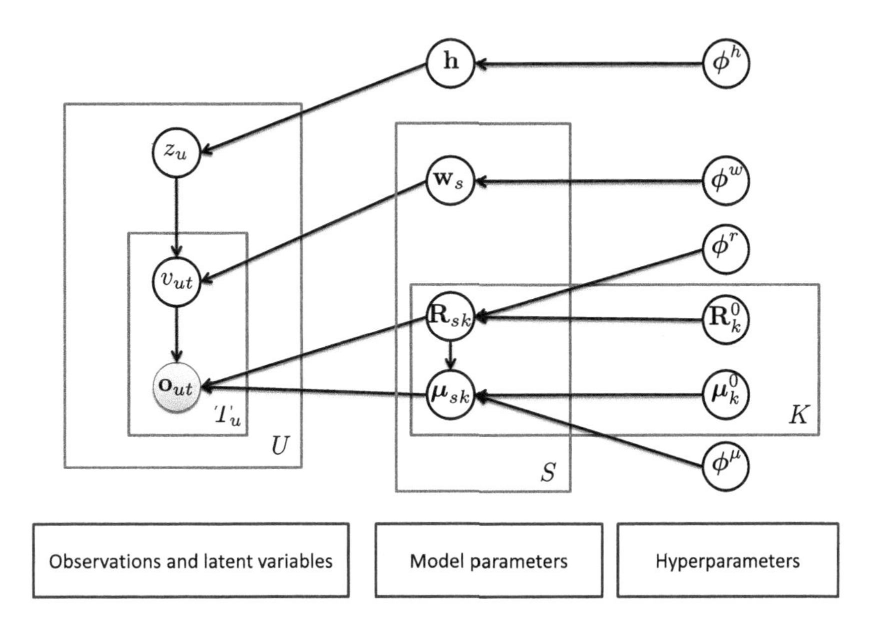

Figure 8.9 Model of multi-scale Gaussian mixture model. The model deals with two time scales based on frame t and utterance u.

Algorithm 17 Generative process of multi-scale GMM

Require: Ψ^0

1: Draw $\mathbf{h}$ from $\text{Dir}(\boldsymbol{\phi}^h)$
2: **for** each utterance-level mixture component $s = 1, \cdots, S$ **do**
3: Draw $\mathbf{w}_s$ from $\text{Dir}(\boldsymbol{\phi}^w)$
4: **for** each frame-level mixture component $k = 1, \cdots, K$ **do**
5: **for** each dimension $d = 1, \cdots, D$ **do**
6: Draw r_{skd} from $\text{Gam}(\phi^r, r^0_{kd})$
7: **end for**
8: Draw $\boldsymbol{\mu}_{sk}$ from $\mathcal{N}(\boldsymbol{\mu}^0_k, (\phi^\mu)^{-1}\mathbf{R}^{-1}_{sk})$
9: **end for**
10: **end for**
11: **for** each utterance $u = 1, \cdots, U$ **do**
12: Draw z_u from Mult $(\mathbf{h})$
13: **for** each frame $t = 1, \cdots, T_u$ **do**
14: Draw v_{ut} from Mult $(\mathbf{w}_{z_u})$
15: Draw $\mathbf{o}_{ut}$ from $\mathcal{N}(\boldsymbol{\mu}_{z_u v_{ut}}, \mathbf{R}^{-1}_{z_u v_{ut}})$
16: **end for**
17: **end for**

8.3.5 Marginal likelihood for the complete data

Similarly to the GMM Gibbs sampler, we prepare $p(v_{ut} = k|\cdot) \triangleq \delta(v_{ut}, k)$ given utterance u, which returns 0 or 1 based on the assignment information. In addition, we also prepare the assignment information of an utterance-level mixture with $p(z_u = s|\cdot) \triangleq \delta(z_u, s)$. The sufficient statistics of multi-scale GMM can be represented as follows:

$$\begin{cases} \xi_s &= \sum_u \delta(z_u, s), \\ \gamma_{sk} &= \sum_{u,t} \delta(z_u, s)\delta(v_{ut}, k), \\ \boldsymbol{\gamma}^{(1)}_{sk} &= \sum_{u,t} \delta(z_u, s)\delta(v_{ut}, k)\mathbf{o}_{ut}, \\ \gamma^{(2)}_{skd} &= \sum_{u,t} \delta(z_u, s)\delta(v_{ut}, k)(o_{utd})^2. \end{cases} \tag{8.92}$$

Here $\xi_s \in \mathbb{Z}^+$ is a count of utterances assigned to speaker cluster s, and $\gamma_{sk} \in \mathbb{Z}^+$ is a count of frames assigned to mixture component k in s. $\boldsymbol{\gamma}^{(1)}_{sk}$ and $\gamma^{(2)}_{skd}$ are first-order and second-order sufficient statistics, respectively.

Based on the sufficient statistics representation in Eq. (8.92), the complete data likelihood of Eq. (8.69) can be represented by using the product formula of the Kronecker delta function in Eq. (A.4) ($f_b = \prod_a f_a^{\delta(a,b)}$) as follows:

$$\begin{aligned} &p(\mathbf{O}, Z, V|\Theta) \\ &= \prod_{s=1}^{S}\prod_{u=1}^{U}\left(h_{z_u}\prod_{k=1}^{K}\prod_{t=1}^{T_u}\left(w_{z_u v_{ut}}\mathcal{N}(\mathbf{o}_{ut}|\boldsymbol{\mu}_{z_u v_{ut}}, \mathbf{R}^{-1}_{z_u v_{ut}})\right)^{\delta(v_{ut},k)}\right)^{\delta(z_u,s)} \\ &= \prod_{s=1}^{S}(h_s)^{\xi_s}\prod_{k=1}^{K}(w_{sk})^{\gamma_{sk}}\prod_{u=1}^{U}\prod_{t=1}^{T_u}\left(\mathcal{N}(\mathbf{o}_{ut}|\boldsymbol{\mu}_{sk}, \mathbf{R}^{-1}_{sk})\right)^{\delta(z_u,s)\delta(v_{ut},k)}. \end{aligned} \tag{8.93}$$

The marginal likelihood for the complete data, $p(\mathbf{O}, Z, V|\Psi^0)$, is represented by the following expectations by substituting Eqs. (8.91) and (8.93) into the following integration:

$$
\begin{aligned}
&p(\mathbf{O}, Z, V|\Psi^0) \\
&= \int p(\mathbf{O}, Z, V|\Theta)p(\Theta|\Psi^0)d\Theta \\
&= \mathbb{E}_{(\mathbf{h})}\left[\prod_{s=1}^{S}(h_s)^{\xi_s}\right]\left(\prod_{s=1}^{S}\mathbb{E}_{(\mathbf{w}_s)}\left[\prod_{k=1}^{K}(w_{sk})^{\gamma_{sk}}\right]\right) \\
&\times\left(\prod_{s=1}^{S}\prod_{k=1}^{K}\mathbb{E}_{(\boldsymbol{\mu}_{sk},\mathbf{R}_{sk})}\left[\prod_{u=1}^{U}\prod_{t=1}^{T_u}\left(\mathcal{N}(\mathbf{o}_{ut}|\boldsymbol{\mu}_{sk}, \mathbf{R}_{sk}^{-1})\right)^{\delta(z_u,s)\delta(v_{ut},k)}\right]\right).
\end{aligned} \tag{8.94}
$$

By following the derivations in Section 8.3.2, the expectations are calculated analytically. This section simply provides the analytical results of the expectations. By using Eq. (4.45), $\mathbb{E}_{(\mathbf{h})}\left[\prod_{s=1}^{S}(h_s)^{\xi_s}\right]$ is solved as:

$$
\mathbb{E}_{(\mathbf{h})}\left[\prod_{s=1}^{S}(h_s)^{\xi_s}\right] = \frac{C_{\mathrm{Dir}}(\{\phi_s^h\}_{s=1}^S)}{C_{\mathrm{Dir}}(\{\tilde{\phi}_s^h\}_{s=1}^S)}, \tag{8.95}
$$

where $\tilde{\phi}_s^h$ is a posterior hyperparameter, which is defined as:

$$
\tilde{\phi}_s^h \triangleq \phi_s^h + \xi_s. \tag{8.96}
$$

The other integral with respect to $\mathbf{w}_s$ in Eq. (8.94) is also calculated as follows:

$$
\mathbb{E}_{(\mathbf{w}_s)}\left[\prod_{k=1}^{K}(w_{sk})^{\gamma_{sk}}\right] = \frac{C_{\mathrm{Dir}}(\{\phi_k^w\}_{k=1}^K)}{C_{\mathrm{Dir}}(\{\tilde{\phi}_{sk}^w\}_{k=1}^K)}, \tag{8.97}
$$

where $\tilde{\phi}_{sk}^w$ is a posterior hyperparameter, which is defined as:

$$
\tilde{\phi}_{sk}^w \triangleq \phi_k^w + \gamma_{sk}. \tag{8.98}
$$

The integral with respect to $\boldsymbol{\mu}_{sk}$ and $\mathbf{R}_{sk}$ in Eq. (8.94) is also analytically calculated as follows:

$$
\begin{aligned}
&\mathbb{E}_{(\boldsymbol{\mu}_{sk},\mathbf{R}_{sk})}\left[\prod_{u=1}^{U}\prod_{t=1}^{T_u}\left(\mathcal{N}(\mathbf{o}_{ut}|\boldsymbol{\mu}_{sk}, \mathbf{R}_{sk}^{-1})\right)^{\delta(z_u,s)\delta(v_{ut},k)}\right] \\
&= \exp\left(-\frac{\gamma_{sk}D}{2}\log(2\pi) + \frac{D}{2}\log\frac{\phi^\mu}{\tilde{\phi}_{sk}^\mu} + \log\frac{C_{\mathrm{Gam}_2}(r_{kd}^0, \phi^r)}{C_{\mathrm{Gam}_2}(\tilde{r}_{skd}, \tilde{\phi}_{sk}^r)}\right),
\end{aligned} \tag{8.99}
$$

where posterior hyperparameters $\tilde{\phi}_{sk}^\mu$, $\tilde{\mu}_{sk}$, $\tilde{\phi}_{sk}^r$, and $\tilde{r}_{skd}$ are defined as follows:

$$
\begin{cases}
\tilde{\phi}_{sk}^\mu & \triangleq \phi^\mu + \gamma_{sk}, \\
\tilde{\mu}_{sk} & \triangleq \frac{\phi^\mu \boldsymbol{\mu}_k^0 + \boldsymbol{\gamma}_{sk}^{(1)}}{\tilde{\phi}_{sk}^\mu}, \\
\tilde{\phi}_{sk}^r & \triangleq \phi^r + \gamma_{sk}, \\
\tilde{r}_{skd} & \triangleq r_{kd}^0 + \gamma_{skd}^{(2)} + \phi^\mu(\mu_{kd}^0)^2 - \tilde{\phi}_{sk}^\mu(\tilde{\mu}_{skd})^2.
\end{cases} \tag{8.100}
$$

Thus, we finally solve all the integrals in Eq. (8.94) as follows:

$$p(\mathbf{O}, Z, V|\Psi^0) = \frac{\Gamma(\sum_s \phi_s^h) \prod_s \Gamma(\tilde{\phi}_s^h)}{\prod_s \Gamma(\phi_s^h) \Gamma(\sum_s \tilde{\phi}_s^h)} \prod_s \frac{\Gamma(\sum_k \phi_k^w) \prod_k \Gamma(\tilde{\phi}_{sk}^w)}{\prod_k \Gamma(\phi_k^w) \Gamma(\sum_k \tilde{\phi}_{sk}^w)}$$

$$\times \prod_{s,k} (2\pi)^{-\frac{\gamma_{sk} D}{2}} \frac{(\phi^\mu)^{\frac{D}{2}} \left(\Gamma\left(\frac{\phi^r}{2}\right)\right)^{-D} \left(\prod_d \frac{r_{kd}^0}{2}\right)^{\frac{\phi^r}{2}}}{(\tilde{\phi}_{sk}^\mu)^{\frac{D}{2}} \left(\Gamma\left(\frac{\tilde{\phi}_{sk}^r}{2}\right)\right)^{-D} \left(\prod_d \frac{\tilde{r}_{skd}}{2}\right)^{\frac{\tilde{\phi}_{sk}^r}{2}}}. \tag{8.101}$$

We summarize below the posterior hyperparameters, which are obtained from the hyperparameters of the prior distributions (Ψ^0) and sufficient statistics (Eq. (8.92)) as follows:

$$\begin{cases} \tilde{\phi}_s^h &= \phi_s^h + \xi_s, \\ \tilde{\phi}_{sk}^w &= \phi_k^w + \gamma_{sk}, \\ \tilde{\phi}_{sk}^\mu &= \phi^\mu + \gamma_{sk}, \\ \tilde{\boldsymbol{\mu}}_{sk} &= \frac{\phi^\mu \boldsymbol{\mu}_k^0 + \boldsymbol{\gamma}_{sk}^{(1)}}{\tilde{\phi}_{sk}^\mu}, \\ \tilde{\phi}_{sk}^r &= \phi^r + \gamma_{sk}, \\ \tilde{r}_{skd} &= r_{kd}^0 + \gamma_{skd}^{(2)} + \phi^\mu (\mu_{kd}^0)^2 - \tilde{\phi}_{sk}^\mu (\tilde{\mu}_{skd})^2. \end{cases} \tag{8.102}$$

Based on the marginal likelihood for these complete data, we can calculate the marginal conditional distribution of v_{ut} and z_u, as shown below.

8.3.6 Gibbs sampler

We provide a collapsed Gibbs sampler $p(v_{ut} = k|\mathbf{O}, V_{\backslash t}, Z_{\backslash u}, z_u = s)$ for a frame-level mixture component k, which has similarities with the GMM Gibbs sampler in Section 8.3.3. In addition, we also provide a collapsed Gibbs sampler $p(v_{u,t} = k'|\mathbf{O}, V_{\backslash t}, Z_{\backslash u}, z_u = s)$ for utterance-level mixture component s, which is a result of speaker clustering.

Frame-level mixture component

The Gibbs sampler assigns frame-level mixture component k at frame t by using the following equation:

$$\begin{aligned} p(v_{ut} = k|\mathbf{O}, V_{\backslash t}, Z_{\backslash u}, z_u = s) &\propto p(\mathbf{O}, V_{\backslash t}, v_{ut} = k, Z_{\backslash u}, z_u = s) \\ &\propto g(\tilde{\Psi}_{V_{\backslash t}, v_{ut}=k, Z_{\backslash u}, z_u = s}), \end{aligned} \tag{8.103}$$

where $g(\cdot)$ is defined as follows:

$$g(\tilde{\Psi}_{sk}) \triangleq \Gamma(\tilde{\phi}_{sk}^w) \left(\tilde{\phi}_{sk}^\mu\right)^{-\frac{D}{2}} \left(\Gamma\left(\frac{\tilde{\phi}_{sk}^r}{2}\right)\right)^D \left(\prod_d \frac{\tilde{r}_{sd}}{2}\right)^{-\frac{\tilde{\phi}_{sk}^r}{2}}. \tag{8.104}$$

Algorithm 18 Gibbs sampling-based multi-scale mixture model.

1: Initialize $\mathbf{\Phi}^0$
2: **repeat**
3: **for** u = shuffle $(1 \cdots U)$ **do**
4: **for** t = shuffle $(1 \cdots T_u)$ **do**
5: Sample $v_{u,t}$ by using Eq. (8.105)
6: **end for**
7: **end for**
8: **for** u = shuffle $(1 \cdots U)$ **do**
9: Sample z_u by using Eq. (8.107)
10: **end for**
11: **until** some condition is met

Therefore, by considering the normalization constant, the posterior probability can be obtained as follows:

$$p(v_{u,t} = k|\mathbf{O}, V_{\backslash t}, Z_{\backslash u}, z_u = s) = \frac{g(\tilde{\Psi}_{V_{\backslash t}, v_{ut}=k, Z_{\backslash u}, z_u=s})}{\sum_{k=1^K} g(\tilde{\Psi}_{V_{\backslash t}, v_{ut}=k', Z_{\backslash u}, z_u=s})}. \tag{8.105}$$

This equation is analytically derived by using the marginal likelihood for complete data (Eq. (8.101)).

Utterance-level mixture component

As with the frame-level mixture component case, the Gibbs sampler assigns utterance-level mixture s at utterance u by using the following equation:

$$\begin{aligned} p(z_u = s|\mathbf{O}, V, Z_{\backslash u}) &\propto p(\mathbf{O}, V, Z_{\backslash u}, z_u = s) \\ &\propto \frac{\Gamma(\sum_k \tilde{w}_{s\backslash u,k})}{\Gamma(\sum_k \tilde{w}_{s,k})} \prod_k g(\tilde{\Psi}_{sk}). \end{aligned} \tag{8.106}$$

The value of $\tilde{\Psi}_{s\backslash u,k}$ is computed by the sufficient statistics using $\mathbf{O}_{\backslash u}$ and $V_{\backslash u}$. Therefore, the posterior probability can be obtained as follows:

$$\begin{aligned} &p(z_u = s'|\mathbf{O}, V, Z_{\backslash u}) \\ &= \frac{\exp\left(\log \frac{\Gamma(\sum_k \tilde{w}_{s'\backslash u,k})}{\Gamma(\sum_k \tilde{w}_{s',k})} + \sum_k g_{s',k}(\tilde{\Psi}_{s',k}) - g_{s',k}(\tilde{\Psi}_{s'\backslash u,k})\right)}{\sum_{s,k} \exp\left(\log \frac{\Gamma(\sum_k \tilde{w}_{s\backslash u,k})}{\Gamma(\sum_k \tilde{w}_{s,k})} + g_{s,k}(\tilde{\Psi}_{s,k}) - g_{s,k}(\tilde{\Psi}_{s\backslash u,k})\right)}. \end{aligned} \tag{8.107}$$

Thus, we can derive a solution for the multi-scale mixture model based on Gibbs sampling, which jointly infers the latent variables by interleaving frame-level and utterance-level samples. Algorithm 18 provides a sample code for the proposed approach.

Table 8.1 Comparison of MCMC and VB for speaker clustering. ACP: average cluster purity, ASP: average speaker purity, and K value: geometric mean of ACP and ASP.

Evaluation data	Method	ACP	ASP	K value
CSJ-1	MCMC	0.808	0.898	0.851
(# spkr10, # utt 50)	VB	0.704	0.860	0.777
CSJ-2	MCMC	0.852	0.892	0.871
(# spkr10, # utt 100)	VB	0.695	0.846	0.782
CSJ-3	MCMC	0.866	0.892	0.879
(# spkr10, # utt 200)	VB	0.780	0.870	0.823
CSJ-4	MCMC	0.784	0.694	0.738
(# spkr10, # utt 2,491)	VB	0.773	0.673	0.721
CSJ-5	MCMC	0.740	0.627	0.681
(# spkr10, # utt 2,321)	VB	0.693	0.676	0.684

MCMC-based acoustic modeling for speaker clustering was investigated with respect to the difference in the MCMC and VB estimation methods by Tawara, Ogawa, Watanabe *et al.* (2012). Table 8.1 shows speaker clustering results in terms of the average cluster purity (ACP), average speaker purity (ASP), and geometric mean of those values (K value) with respect to the evaluation criteria in speaker clustering. We used the Corpus of Spontaneous Japanese (CSJ) dataset (Furui *et al.* 2000) and investigated the speaker clustering performance for MCMC and VB for various amounts of data. Table 8.1 shows that the MCMC-based method outperformed the VB method by avoiding local optimum solutions, especially when only a few utterances could be used. These results also supported the importance of the Gibbs-based Bayesian properties.

Since the mixture of GMM is trained by MCMC, it is a straightforward extension to deal with a Dirichlet process mixture model, as discussed in Section 8.2.5, for speaker clustering, where the number of speaker clusters is jointly optimized based on this model. There are several studies of applying the Dirichlet process mixture model to speaker clustering (Fox *et al.* 2008, Tawara, Ogawa, Watanabe *et al.* 2012*b*). The next section introduces the application of an MCMC-based Dirichlet process mixture model to cluster HMMs (mixture of HMMs).

8.4 Nonparametric Bayesian HMMs to acoustic unit discovery

This section describes an application of Bayesian nonparametrics in Section 8.2 to acoustic unit discovery based on HMMs (Lee & Glass 2012, Lee, Zhang & Glass 2013, Torbati, Picone & Sobel 2013, Lee 2014). Acoustic unit discovery aims to automatically find an acoustic unit (e.g., phoneme) from speech data *without transcriptions*, and this is used to build ASR or spoken term detection systems with limited language resources (Schultz & Waibel 2001, Lamel, Gauvain & Adda 2002, Jansen, Dupoux, Goldwater *et al.* 2013). One of the powerful advantages of Bayesian nonparametrics is to find the model structure appropriately, and it is successfully applied to word unit discovery in natural language processing (Goldwater, Griffiths & Johnson 2009, Mochihashi, Yamada & Ueda 2009) and in spoken language processing (Neubig, Mimura, Mori & Kawahara 2010). This section regards sub-word units as latent variables in one nonparametric Bayesian model. More specifically, it formulates a Dirichlet process mixture model where each mixture is an HMM used to model a sub-word unit and to generate observed segments of that unit. This model seeks the set of sub-word units, segmentation, clustering, and HMMs that best represents the observed data through an iterative inference process. This inference process can be performed by using Gibbs sampling.

To realize this acoustic unit discovery, the approach deals with the following variables:

- D dimensional speech feature $\mathbf{o}_t^n \in \mathbb{R}^D$ at frame t in utterance n.
- Binary boundary variable $b_t^n \in \{0, 1\}$ that has value 1 when the speech frame t is at the end point of a segment, and 0 otherwise.
- Boundary index $g_q^n = \{1, \cdots, t, \cdots, T^n\}$ returns the frame index of the qth boundary in utterance n. The initial boundary $g_0^n = 0$.
- Segment of features $\mathbf{O}_{t:t'}^n = \{\mathbf{o}_t^n, \cdots, \mathbf{o}_{t'}^n\}$.
- Unit label $c_{t:t'}^n \in \{1, \cdots, u, \cdots, U\}$ to specify the unit label of $\mathbf{O}_{t:t'}^n$. U is the number of the cluster, and u is a unit index. In addition, $c_t^n \in \{1, \cdots, u, \cdots, U\}$ indicates a unit label at frame t and utterance n.
- HMM Θ_u that represents one cluster unit u with a standard continuous density HMM (Section 3.2.3) that has state transition $a_{uij} \in [0, 1]$ from HMM state i to j, mixture weight $\omega_{ujk} \in [0, 1]$ at mixture component k in state j, and the mean vector $\boldsymbol{\mu}_{ujk} \in \mathbb{R}^D$ and (diagonal) precision matrix $\mathbf{R}_{ujk} \in \mathbb{R}^{D \times D}$.
- HMM state and GMM component $s_t^n \in \{1, \cdots, j, \cdots, J\}$ and $v_t^n \in \{1, \cdots, k, \cdots, K\}$.

The difference between this HMM and the conventional CDHMM in Section 3.2.3 given a phoneme unit is that this approach regards unit label $c_{t:t'}^n$ and the number of units U as a latent variable, which is obtained by a Dirichlet process. Therefore, the notation is similar to that in Section 3.2.3 except that it includes the cluster index u explicitly. A similar model is used in Gish, Siu, Chan *et al.* (2009) and Siu, Gish, Chan *et al.* (2014) based on an ML-style iterative procedure instead of Bayesian nonparametrics. In this section, the numbers of HMM states J and mixture components K are assumed to be the same fixed values for all units, but they can also be optimized by using Bayesian nonparametrics (Rasmussen 2000, Beal, Ghahramani & Rasmussen 2002, Griffiths & Ghahramani 2005).

8.4.1 Generative model and generative process

Since we use a fully Bayesian approach, the variables introduced in this model are regarded as probabilistic variables. For simplicity we consider that the boundary variable b_t^n is given in this formulation. The Bayesian approach first provides a generative process for complete data. We define latent variable Z as

$$Z \triangleq \{C, S, V, \{\gamma_u\}_{u=1}^{\infty}, \{\Theta_u\}_{u=1}^{\infty}\}. \tag{8.108}$$

Here, we assume the number of units is infinite, i.e., $U = \infty$, and latent variables will be generated based on the Dirichlet process. The other variables are defined as

$$\begin{aligned} C &\triangleq \{c_t^n | t = 1, \cdots, T_n, n = 1, \cdots, N\} \\ &= \{c_{(g_q+1):g_{q+1}}^n | q = 0, \cdots, Q_n, n = 1, \cdots, N\}, \\ S &\triangleq \{s_t^n | t = 1, \cdots, T_n, n = 1, \cdots, N\}, \\ V &\triangleq \{v_t^n | t = 1, \cdots, T_n, n = 1, \cdots, N\}. \end{aligned} \tag{8.109}$$

Q_n is the number of units appearing in utterance n, and $\zeta_u \in [0, 1]$ is a weight parameter of unit u. Then, the conditional distribution is represented as follows:

$$\begin{aligned} &p(\mathbf{O}, C, S, V | \{\zeta_u\}_{u=1}^{\infty}, \{\Theta_u\}_{u=1}^{\infty}) \\ &= \prod_{q=0}^{Q^n} p(c_{(g_q+1):g_{q+1}}^n = u | \{\zeta_u\}_{u=1}^{\infty}) p(\mathbf{o}_{g_q+1}^n, s_{g_q+1}^n, v_{g_q+1}^n | u, \Theta_u) \\ &\times \prod_{t=g_q+2}^{g_{q+1}} p(\mathbf{o}_t^n, s_{t-1}^n, s_t^n, v_t^n | u, \Theta_u), \end{aligned} \tag{8.110}$$

where each likelihood function can be represented as follows:

$$p(c_{(g_q+1):g_{q+1}}^n = u | \{\zeta_u\}_{u=1}^{\infty}) = \zeta_u, \tag{8.111}$$

and

$$\begin{cases} p(\mathbf{o}_t^n, s_t^n = j, v_t^n = k | u, \Theta_u) = a_{uj} \omega_{ujk} \mathcal{N}(\mathbf{o}_t^n | \boldsymbol{\mu}_{ujk} \boldsymbol{\Sigma}_{ujk}) & (t = g_q + 1), \\ p(\mathbf{o}_t^n, s_{t-1}^n = i, s_t^n = j, v_t^n = k | u, \Theta_u) = a_{uij} \omega_{ujk} \mathcal{N}(\mathbf{o}_t^n | \boldsymbol{\mu}_{ujk} \boldsymbol{\Sigma}_{ujk}) & \text{Otherwise.} \end{cases} \tag{8.112}$$

a_{uj} is an initial weight in an HMM. For $p(\{\zeta_u\}_{u=1}^{\infty}, \{\Theta_u\}_{u=1}^{\infty})$, we use the Dirichlet process mixture model, described in Section 8.2.5.

The model parameters are sampled from a base distribution with hyperparameter Θ_0, and we use a conjugate prior distribution of CDHMM $p(\Theta_u | \Theta_0)$ with diagonal covariance matrices, as we discussed in Section 4.3.2, which is represented as

$$\begin{aligned} &p(\Theta_u) \\ &= p(\{a_{uj}\}_{j=1}^J) \left(\prod_{i=1}^J p(\{a_{uij}\}_{j=1}^J) \right) \left(\prod_{j=1}^J p(\{\omega_{ujk}\}_{k=1}^K) \right) \left(\prod_{j=1}^J \prod_{k=1}^K p(\boldsymbol{\mu}_{ujk}, \boldsymbol{\Sigma}_{ujk}) \right) \\ &= \mathrm{Dir}(\{a_{uj}\}_{j=1}^J | \phi^{\pi}) \left(\prod_{i=1}^J \mathrm{Dir}(\{a_{uij}\}_{j=1}^J | \phi^a) \right) \left(\prod_{j=1}^J \mathrm{Dir}(\{\omega_{ujk}\}_{k=1}^K | \phi^{\omega}) \right) \\ &\times \left(\prod_{j=1}^J \prod_{k=1}^K \prod_{d=1}^D \mathcal{N}(\mu_{jkd} | \mu_d^0, (\phi^{\mu} r_{jkd})^{-1}) \mathrm{Gam}(r_{jkd} | r_d^0, \phi^r) \right). \end{aligned} \tag{8.113}$$

ϕ^a, ϕ^ω, ϕ^μ, ϕ^r, $\boldsymbol{\mu}^0$, and $\mathbf{r}^0$ are the prior hyperparameter ($\triangleq \Theta_0$). $\boldsymbol{\mu}^0$ and $\mathbf{r}^0$ can be obtained using the Gaussian mean and precision parameters of all data.

Algorithm 19 provides a generative process of a nonparametric Bayesian HMM. A Dirichlet process mixture model (DPM) can sample the acoustic unit u from existing clusters or a new cluster for every speech segment. Thus, the model finally generates a sequence of speech features without fixing an acoustic unit explicitly.

Algorithm 19 Generative process of a nonparametric Bayesian HMM

Require: Concentration parameter γ, Base distribution of DP Θ_0, Boundary b_t^n.

1: **for** every utterance $n = 1, \cdots, N$ **do**
2:   **for** every segment $q = 1, \cdots, Q_n$ **do**
3:     Draw u and Θ_u from DPM($c^n_{(g_q+1):g_{q+1}}, \Theta_u|\gamma, \Theta_0$) (from existing clusters or a new one)
4:     Draw $\mathbf{a}_u$ from Dir($\boldsymbol{\phi}^\pi_u$)
5:     Draw $\mathbf{a}_{ui}$ from Dir($\boldsymbol{\phi}^a_{ui}$)
6:     Draw $\boldsymbol{\omega}_{uj}$ from Dir($\boldsymbol{\phi}^\omega_{uj}$)
7:     **for** every feature dimension $d = 1, \cdots, D$ **do**
8:       Draw r_{ujkd} from Gam(ϕ^r, r^0_d)
9:       Draw μ_{ujkd} from $\mathcal{N}(\mu^0_d, (\phi^\mu r_{ujkd})^{-1})$
10:     **end for**
11:     Draw j from Mult($s^n_{g_q+1}|\{a_{uj'}\}^J_{j'=1}$)
12:     Draw k from Mult($v^n_{g_q+1}|\{\omega_{ujk'}\}^K_{k'=1}$)
13:     Draw $\mathbf{o}$ from $\mathcal{N}(\mathbf{o}^n_{g_q+1}|\boldsymbol{\mu}_{ujk}, \boldsymbol{\Sigma}_{ujk})$
14:     **for** every frame $t = g_q + 2, \cdots, g_{q+1}$ **do**
15:       Draw j from Mult($s^n_t|\{a_{us_{t-1}j'}\}^J_{j'=1}$)
16:       Draw k from Mult($v^n_t|\{\omega_{ujk'}\}^K_{k'=1}$)
17:       Draw $\mathbf{o}$ from $\mathcal{N}(\mathbf{o}^n_t|\boldsymbol{\mu}_{ujk}, \boldsymbol{\Sigma}_{ujk})$
18:     **end for**
19:   **end for**
20: **end for**

8.4.2 Inference

The approach infers all latent variables by using Gibbs sampling, as discussed in Section 8.1.4, which samples a target latent variable z from the following conditional posterior distribution:

$$z \sim p(z|\mathcal{Z}_{\backslash z}, \mathbf{O}), \tag{8.114}$$

where $\mathcal{Z}_{\backslash z}$ denotes a set of all hidden variables in $\mathcal{Z}$ except for z. In this section we provide conditional distributions for cluster label $c^n_{t:t'}$, HMM state sequence $S^n_{t:t'}$, GMM component sequence $V^n_{t:t'}$, and HMM parameters Θ_u so that we can perform the Gibbs sampling of these latent variables.

- Cluster label $c^n_{t:t'}$:
  Let $\mathcal{U}$ be the set of distinctive cluster units. The conditional posterior distribution of $c_{t:t'} = u \in \mathcal{U}$ is represented as follows:

$$\begin{aligned} p(c^n_{t:t'} = u|\cdots) &\propto p(c^n_{t:t'} = u|\mathcal{U}_{\backslash u}, \gamma)p(\mathbf{O}^n_{t:t'}|\Theta_u) \\ &= \frac{n^u}{N_u - 1 + \gamma} p(\mathbf{O}^n_{t:t'}|\Theta_u), \end{aligned} \tag{8.115}$$

  where γ is a hyperparameter of the DP prior, n^u represents the number of cluster labels in $\mathcal{U}_{\backslash u}$ taking the value u, and N_u is the number of speech segments. In this formulation, we do not marginalize Θ_u unlike Section 8.3, but use the sampled values, as discussed below.

  If $c^n_{t:t'}$ belongs to a new cluster that has not existed before, the conditional posterior distribution for this new cluster is represented as

$$p(c^n_{t:t'} \neq u, u \in \mathcal{U}|\cdots) \propto \frac{\gamma}{N_u - 1 + \gamma} \int p(\mathbf{O}^n_{t:t'}|\Theta)G(\Theta|\Theta_0)d\Theta, \tag{8.116}$$

  where the integral is approximated by a Monte Carlo estimation (Rasmussen 1999, Neal 2000, Tawara, Ogawa, Watanabe *et al.* 2012*b*). Note that Eqs. (8.115) and (8.116) are a typical solution of the Dirichlet process in Eq. (8.38). The Gibbs sampler for existing clusters $p(c^n_{t:t'} = u|\cdots)$ depends on the number of occurrences n^u and their likelihood, while that for a new cluster $p(c^n_{t:t'} \neq u, u \in \mathcal{U}|\cdots)$ depends on the concentration parameter γ and the marginalized likelihood.

  The likelihood values of $p(\mathbf{O}^n_{t:t'}|\Theta_u)$ and $p(\mathbf{O}^n_{t:t'}|\Theta)$ can be computed by considering all possible HMM states $S^n_{t:t'}$ and mixture components $V^n_{t:t'}$ based on the forward algorithm in Section 3.3.1. However, since the following Gibbs samplers can sample $S^n_{t:t'}$ and $V^n_{t:t'}$, we can use the following conditional likelihood values:

$$p(\mathbf{O}^n_{t:t'}|\Theta) \approx p(\mathbf{O}^n_{t:t'}|S^n_{t:t'}, V^n_{t:t'}, \Theta). \tag{8.117}$$

  This is easily computed by accumulating all Gaussian likelihood values given $S^n_{t:t'}$ and $V^n_{t:t'}$.
- HMM state s^n_t:
  The conditional posterior distribution of HMM state s^n_t is obtained from the following distribution, given the previous state s^n_{t-1} and the succeeding state s^n_{t+1}:

$$\begin{aligned} &p(s^n_t = j|\cdots) \\ &\quad \propto p(s^n_t = j|s^n_{t-1})p(\mathbf{o}^n_t|\Theta_u, s^n_t = j)p(s^n_{t+1}|s^n_t = j) \\ &\quad = \begin{cases} a_{uj}\left(\sum_{k=1}^K \omega_{ujk}\mathcal{N}(\mathbf{o}^n_t|\boldsymbol{\mu}_{ujk}\boldsymbol{\Sigma}_{ujk})\right) a_{ujs_{t+1}} & (t = g_q + 1) \\ a_{us_{t-1}j}\left(\sum_{k=1}^K \omega_{ujk}\mathcal{N}(\mathbf{o}^n_t|\boldsymbol{\mu}_{ujk}\boldsymbol{\Sigma}_{ujk})\right) a_{ujs_{t+1}} & \text{otherwise.} \end{cases} \end{aligned} \tag{8.118}$$

  Note that this algorithm does not require the forward–backward algorithm in Section 3.3.1 to obtain the occupation probability, compared with the conventional HMM. There is an alternative algorithm to sample a state sequence similar to the forward-backward algorithm, called the forward filtering backward sampling algorithm of an HMM (Scott 2002, Mochihashi *et al.* 2009) in the MCMC framework.

Given $v_t^n = k$, which is also obtained by the Gibbs sampler, we can also approximate Eq. (8.118) as

$$\begin{aligned} p(s_t^n = j | \cdots) &\approx p(s_t^n = j | v_t^n = k, \cdots) \\ &\propto \begin{cases} a_{uj} \omega_{ujk} \mathcal{N}(\mathbf{o}_t^n | \boldsymbol{\mu}_{ujk} \boldsymbol{\Sigma}_{ujk}) a_{u,j,s_{t+1}} & (t = g_q + 1) \\ a_{us_{t-1}j} \omega_{ujk} \mathcal{N}(\mathbf{o}_t^n | \boldsymbol{\mu}_{ujk} \boldsymbol{\Sigma}_{ujk}) a_{u,j,s_{t+1}} & \text{otherwise.} \end{cases} \end{aligned} \tag{8.119}$$

This avoids computing the Gaussian likelihoods for all mixture components.

- GMM component v_t^n:
  The conditional posterior distribution of GMM component $v_t^n = k$ at cluster u and state $s_t^n = j$ is obtained from the following distribution:

$$\begin{aligned} p(v_t^n = k | \cdots) &\propto p(v_t^n = k | \Theta_u, s_t^n = j) p(\mathbf{o}_t^n | \Theta_u, s_t^n = j, v_t^n = k) \\ &= \omega_{ujk} \mathcal{N}(\mathbf{o}_t^n | \boldsymbol{\mu}_{ujk}, \boldsymbol{\Sigma}_{ujk}). \end{aligned} \tag{8.120}$$

Thus, the Gibbs samplers of $p(c_{t:t'}^n | \cdots)$, $p(s_t^n | \cdots)$, and $p(v_t^n | \cdots)$ can provide the latent variable sequences of C, S, and V. Once we have C, S, and V, we can compute the sufficient statistics for the CDHMM, as follows:

$$\begin{cases} \xi_{ui} &= \sum_{n,q} \delta(c^n_{(g_q+1):g_{q+1}}, u) \delta(s^n_{g_q+1}, i), \\ \xi_{uij} &= \sum_{n,t} \delta(c_t^n, u) \delta(s_{t-1}^n, i) \delta(s_t^n, j), \\ \gamma_{ujk} &= \sum_{n,t} \delta(c_t^n, u) \delta(s_t^n, j) \delta(v_t^n, k), \\ \boldsymbol{\gamma}_{ujk}^{(1)} &= \sum_{n,t} \delta(c_t^n, u) \delta(s_t^n, j) \delta(v_t^n, k) \mathbf{o}_t^n, \\ \gamma_{ujkd}^{(2)} &= \sum_{n,t} \delta(c_t^n, u) \delta(s_t^n, j) \delta(v_t^n, k) (o_{td}^n)^2. \end{cases} \tag{8.121}$$

Thus, we can obtain the posterior distribution analytically based on the conjugate analysis, as we have shown in Sections 2.1.4 and 4.3. This section only provides the analytical solutions for the posterior distributions of CDHMM parameters, which are used to sample CDHMM parameters.

- HMM parameters Θ_u:
  - Initial weight

$$p(\mathbf{a}_u | \cdots) \propto \mathrm{Dir}(\mathbf{a}_u | \boldsymbol{\phi}_u^{\pi}), \tag{8.122}$$

where

$$\tilde{\phi}_{ui}^{\pi} = \phi^{\pi} + \xi_{ui}. \tag{8.123}$$

  - State transition

$$p(\mathbf{a}_{ui} | \cdots) \propto \mathrm{Dir}(\mathbf{a}_{ui} | \boldsymbol{\phi}_{ui}^{a}), \tag{8.124}$$

where

$$\tilde{\phi}_{uij}^{a} = \phi^{a} + \xi_{uij}. \tag{8.125}$$

  - Mixture weight

$$p(\boldsymbol{\omega}_{uj} | \cdots) \propto \mathrm{Dir}(\boldsymbol{\omega}_{uj} | \boldsymbol{\phi}_{uj}^{\omega}), \tag{8.126}$$

where

$$\tilde{\phi}^{\omega}_{ujk} = \phi^{\omega} + \gamma_{ujk}. \tag{8.127}$$

– Mean vector and covariance matrix at dimension d

$$\begin{aligned} &p(\mu_{ujkd}, r_{ujkd}|\cdots) \\ &\quad \propto \mathcal{N}(\mu_{ujkd}|\tilde{\mu}_{ujkd}, (\tilde{\phi}^{\mu}_{ujk} r_{ujkd})^{-1})\mathrm{Gam}(r_{ujkd})|\tilde{\phi}^{r}_{ujk}, \tilde{r}_{ujkd}), \end{aligned} \tag{8.128}$$

where

$$\begin{cases} \tilde{\phi}^{\mu}_{ujk} &= \phi^{\mu} + \gamma_{ujk}, \\ \tilde{\mu}_{ujk} &= \frac{\phi^{\mu}\mu^0 + \gamma^{(1)}_{ujk}}{\phi^{\mu} + \gamma_{ujk}}, \\ \tilde{\phi}^{r}_{ujk} &= \phi^{r} + \gamma_{ujk}, \\ \tilde{r}_{ujkd} &= \gamma^{(2)}_{ujkd} + \phi^{\mu}(\mu^0_d)^2 - \tilde{\phi}^{\mu}_{ujk}(\tilde{\mu}_{ujkd})^2 + r^0_d. \end{cases} \tag{8.129}$$

Note that compared with the collapsed Gibbs sampling solutions in Section 8.3, this samples the Gaussian parameters, as well as the other variables. Therefore, the Gibbs samplers for latent variables become rather simple equations.

Thus, we can obtain the nonparametric Bayesian HMMs for acoustic unit discovery given cluster boundaries. The MCMC is performed by sampling latent variables C, S, and V, and model parameters $\{\Theta_u\}_{u=1}^{U}$, iteratively. Note that the number of clusters U is changing according to the Dirichlet process, and finally becomes a fixed number. This clustering corresponds to automatically obtaining the acoustic unit in a Bayesian nonparametric manner.

Lee & Glass (2012) also considered cluster boundaries as latent variables which can be obtained by Gibbs sampling. In addition, to provide an appropriate initialization of the boundaries, the approach uses a pre-segmentation technique based on Glass (2003). The model obtained was compared with the other acoustic unit discovery method based on the dynamic time warping technique using the distance between GMM posteriors (Zhang & Glass 2009). The nonparametric Bayesian HMM achieved comparable results to the dynamic time warping technique in terms of the spoken term detection measures, which shows the effectiveness of the nonparametric Bayesian acoustic unit discovery. The approach carries a huge computational cost compared with the other EM type approaches (based on ML, MAP, and VB), and its scalability for the amount of data and the difficulty of parallelization remain serious problems. However, this is one of a few successful approaches using Bayesian nonparametrics for speech data, and it potentially has various advantages over the conventional EM type approaches (e.g., it could mitigate the local optimum problem and use any other distributions than conjugate distributions).

8.5 Hierarchical Pitman–Yor language model

In an LVCSR system, in addition to an acoustic model based on HMMs, the other key component is a language model based on n-grams. In recent years, BNP learning has been substantially developed for language modeling (Teh *et al.* 2006) and successfully applied for several LVCSR tasks in Huang & Renals (2008). In this section, we introduce a hierarchical Bayesian interpretation for language modeling, based on a nonparametric prior called the Pitman–Yor (PY) process. The motivation of conducting Bayesian learning of n-gram language models is to tackle the limitations such as overfitting of maximum likelihood estimation and the lack of rich contextual knowledge sources. The PY process offers a principled approach to language model smoothing which produces a power-law distribution for natural language. The Bayesian language model based on Bayesian nonparametrics is a realization of a full Bayesian solution according to the MCMC inference procedure. Such a model is a *distribution estimate*, which is different from the Bayesian language model based on maximum a-posteriori (MAP) estimation as addressed in Section 4.7. A MAP-based language model is known as a *point estimate* of language model. A BNP-based language model integrates the values of parameters into a marginalized language model. It is interesting to note that the resulting hierarchical PY language model is a direct generalization of the *hierarchical Dirichlet language model.*

In what follows, we first survey the PY process and explain the importance of the power-law property in language modeling. Then we revisit language model smoothing based on Kneser–Ney smoothing and find the clue for connection to BNP learning. Next the hierarchical PY process is constructed to estimate a language model which provides the hierarchical Bayesian interpretation for a Kneser–Ney smoothed language model. The relation to the hierarchical Dirichlet language model will be illustrated. Lastly, the MCMC inference for the hierarchical PY language model is addressed.

8.5.1 Pitman–Yor process

The PY process (Pitman & Yor 1997) is known as the two-parameter Poisson–Dirichlet process $\mathrm{PY}(d, \alpha_0, G_0)$, which is expressed as a three-parameter distribution over distributions where $0 \leq d < 1$ is a discount parameter, α_0 is a strength parameter and G_0 is a base distribution. Base distribution G_0 can be understood as the mean of draws from the PY process. This process can be used to draw the unigram language model. Let $G(w)$ denote the unigram probability of a word $w \in \mathcal{V}$ and $G = [G(w)]_{w \in \mathcal{V}}$ denote the vector of unigram probabilities, which is drawn from a PY process:

$$G \sim \mathrm{PY}(d, \alpha_0, G_0). \tag{8.130}$$

Here, $G_0 = [G_0(w)]_{w \in \mathcal{V}}$ is a mean vector where $G_0(w)$ is the a-priori probability of word w. In practice, this base measure is usually set to be uniform $G_0 = 1/|\mathcal{V}|$ for all $w \in \mathcal{V}$. The parameters d and α_0 both control the degree of variability around G_0 in different ways. When $d = 0$, the PY process reverts to the DP, which is denoted by $\mathrm{DP}(\alpha_0, G_0)$. The PY process is seen as a generalization of the DP.

Basically, there is no analytic form for distribution of the PY process. We would like to work out the nonparametric distribution over sequences of words from the PY process. Let $\{w_1, w_2, \cdots\}$ be a sequence of words drawn independently and identically from G given the mean distribution G_0. The procedure of generating draws from a PY process that can be described according to the metaphor of the "Chinese restaurant process" (Pitman 2006). As introduced in Section 8.2.4, imagine a Chinese restaurant containing an infinite number of tables, each with infinite seating capacity. Customers enter the restaurant and seat themselves. The first customer sits at the first table while each subsequent customer sits at an occupied table with probability proportional to the number of customers who are already sitting there $c_k - d$, or at a new unoccupied table with probability proportional to $\alpha_0 + dm.$ where $m.$ is the current number of occupied tables. That is, if z_i is the index of the table chosen by the ith customer, this customer sits at the table k given the seating arrangement of the previous $i-1$ customers, $\mathbf{z}^{-i} = \{z_1, \cdots, z_{i-1}\}$, with probability

$$p(z_i = k|\mathbf{z}^{-i}, d, \alpha_0) = \begin{cases} \frac{c_k - d}{\alpha_0 + c.} & 1 \leq k \leq m., \\ \frac{\alpha_0 + dm.}{\alpha_0 + c.} & k = m. + 1, \end{cases} \tag{8.131}$$

where c_k denotes the number of customers sitting at table k and $c. = \sum_k c_k$ is the total number of customers. The above generative procedure produces a sequence of words drawn independently from G with G marginalized out.

It is important to investigate the behaviors of drawing the sequence of words from the PY process. Firstly, the *rich-gets-richer* clustering property can be observed. That is, the more words have been assigned to a draw from G_0, the more likely subsequent words will be assigned to the draw. Secondly, the more we draw from G_0, the more likely a new word will be assigned to a new draw from G_0. Combining these two behaviors produces the so-called *power-law distribution* where many unique or distinct words are observed, most of them rarely. This distribution resembles the distribution of words which is seen in natural language. The power-law distribution has been found to be one of the most striking statistical properties of word frequencies in natural language. Figure 8.10 demonstrates the power-law behavior of the PY process which is controlled by parameters d and α_0, showing the average number of unique words among 25 sequences of words drawn from G, as a function of the number of words, for various values of α_0 and d. We find that α_0 controls the total number of unique words while d adjusts the asymptotic growth of the number of unique words. Figure 8.11 displays the proportion of words appearing only once among the unique words. These figures indicate the proportion of words that occur rarely. We can see that larger d and α_0 produce more rare words. This phenomenon is reflected by the probability of producing a new unoccupied table

$$p(z_i = k_{\text{new}}|\mathbf{z}^{-i}, d, \alpha_0) = \frac{\alpha_0 + dm.}{\alpha_0 + c.}. \tag{8.132}$$

8.5.2 Language model smoothing revisited

A key issue in a language model is to handle the sparseness of training data for training n-gram parameters under the conditions of a large n-gram window size n and large dictionary size $|\mathcal{V}|$. As addressed in Section 3.6, a series of language model smoothing methods has been developed to tackle the data sparseness issue in a statistical n-gram

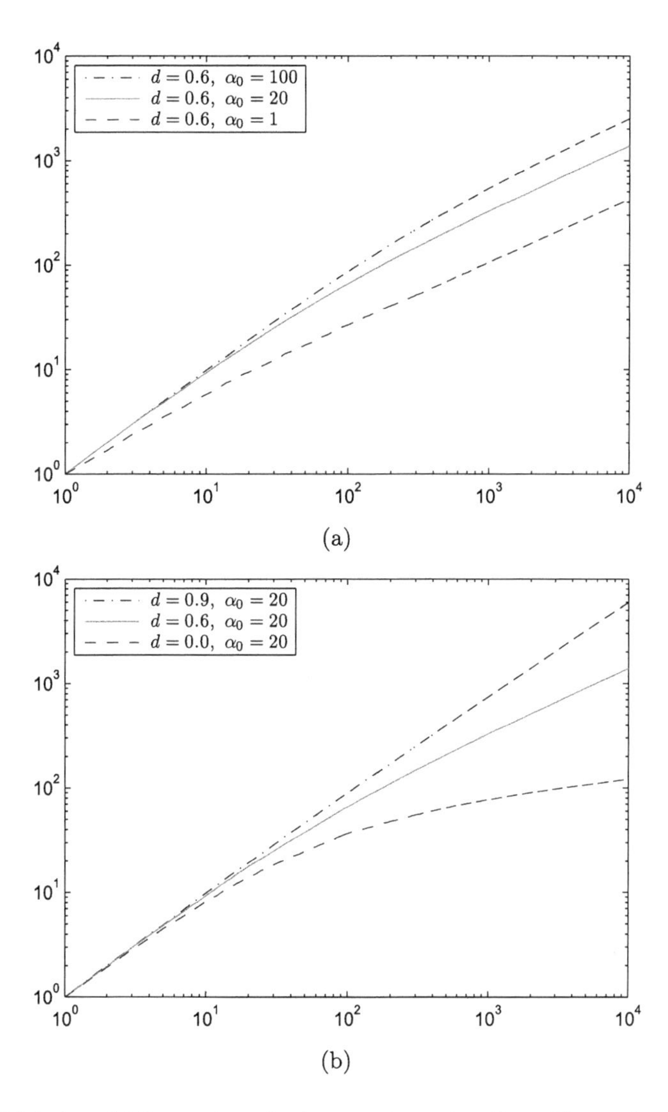

Figure 8.10 Power-law distribution: (a) the number of unique words as a function of the number of words, drawn on a log–log scale with $d = 0.6$ and $\alpha_0 = 100$ (dashdot line), 20 (solid line), and 1 (dashed line); (b) the same as (a) with $\alpha_0 = 20$ and $d = 0.9$ (dashdot line), 0.6 (solid line), and 0 (dashed line).

model. One important trick in these methods is to incorporate an absolute discount parameter d in the count of an observed n-gram event $w_{i-n+1}^{i} = \{w_i\, w_{i-n+1}^{i-1}\} \triangleq \{w\, \mathbf{u}\}$ of a word $w = w_i$ and its preceding history words $\mathbf{u} = w_{i-n+1}^{i-1}$. Owing to this discount, we modify the counts for lower order n-gram probabilities so as to construct the interpolated Kneser–Ney smoothing (Kneser & Ney 1995):

$$p_{\mathsf{KN}}(w|\mathbf{u}) = \frac{\max\{c_{\mathbf{u}w} - d_{|\mathbf{u}|}, 0\}}{c_{\mathbf{u}\cdot}} + \frac{d_{|\mathbf{u}|} m_{\mathbf{u}\cdot}}{c_{\mathbf{u}\cdot}} p_{\mathsf{KN}}(w|\pi(\mathbf{u})), \tag{8.133}$$

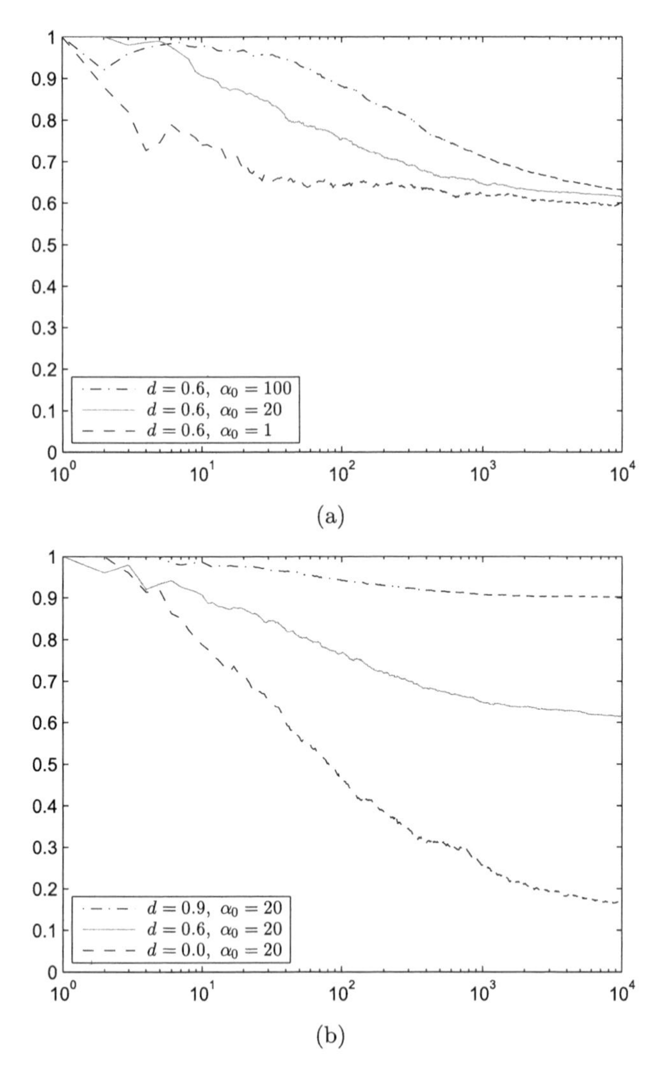

Figure 8.11 Power-law distribution: (a) the proportion of words appearing only once, as a function of the number of words drawn with $d = 0.6$ and $\alpha_0 = 100$ (dashdot line), 20 (solid line), and 1 (dashed line); (b) the same as (a) with $\alpha_0 = 20$ and $d = 0.9$ (dashdot line), 0.6 (solid line), and 0 (dashed line).

which is rewritten from Eq. (3.232) with an order-dependent discount parameter $d \to d_{|\mathbf{u}|}$ and a new notation $N_{1+}(w_{i-n+1}^{i-1}, \bullet) \triangleq m_{\mathbf{u}\cdot} = |\{w' | c_{\mathbf{u}w'} > 0\}|$, expressing the number of unique words that follow the history words $\mathbf{u}$. The term $\pi(\mathbf{u})$ denotes the backoff context of $\mathbf{u}$. If $\mathbf{u} = w_{i-n+1}^{i-1}$, then $\pi(\mathbf{u}) = w_{i-n+2}^{i-1}$.

We would like to investigate the relation between language model smoothing and Bayesian learning. Such a relation is crucial to developing a BNP-based language model. In a standard Bayesian framework for language modeling, a prior distribution is placed over the predictive distribution for the language model. The predictive distribution is estimated by marginalizing out the latent variables from the posterior distribution. The problem of zero-probability can be avoided by taking advantage of knowledge expressed by the priors. A smoothed language model can be obtained. In Yaman *et al.* (2007), the n-gram smoothing was conducted under a framework called the structural maximum a-posteriori (MAP) adaptation where the interpolation of n-gram statistics with an $n-1$-gram model was performed in a hierarchical and recursive fashion. As mentioned in Chapter 4, such a MAP approximation only finds a point estimate of a language model without considering the predictive distribution. To pursue a full Bayesian language model, the hierarchical Dirichlet language model (MacKay & Peto 1995) in Section 5.3 calculates the hierarchical predictive distribution of an n-gram by marginalizing the Dirichlet posterior over the Dirichlet priors.

However, the PY process was shown to be more fitted as a prior distribution than a Dirichlet distribution to the applications in natural language processing (Goldwater, Griffiths & Johnson 2006). It is because the *power-law* distributions of word frequencies produced by the PY process are more likely to be close to the *heavy-tailed distributions* observed in natural language. But, the PY process, addressed in Section 8.5.1, is only designed for a unigram language model. In what follows, we extend the hierarchical Dirichlet language model, which adopts the Dirichlet prior densities, to the hierarchical PY language model, which utilizes the PY process as nonparametric priors and integrates out these prior measures.

8.5.3 Hierarchical Pitman–Yor language model

Similarly to the extension from the Dirichlet process to the hierarchical Dirichlet process for representation of multiple documents, in this section, we address the extension from the PY process to the hierarchical PY (HPY) process which is developed to realize the probability measure for the smoothed n-gram language model. Let $G_\emptyset = [G_\emptyset(w)]_{w \in \mathcal{V}}$ represent the vector of word probability estimates for unigrams of all words w in a vocabulary $\mathcal{V}$. A PY process prior for unigram probabilities is expressed by

$$G_\emptyset \sim \text{PY}(d_0, \alpha_0, G_0), \tag{8.134}$$

where G_0 is a global mean vector in the form of a noninformative base distribution or uniform distribution:

$$G_0(w) = \frac{1}{|\mathcal{V}|} \quad \text{for all } w \in \mathcal{V}. \tag{8.135}$$

This PY process can be used to calculate the predictive unigram of a word w according to the metaphor of the Chinese restaurant process. The customers or word tokens enter the restaurant and seat themselves at either an occupied table or a new table with the probabilities in Eq. (8.131). Each table is labeled by a word $w \in \mathcal{V}$ initialized by the first customer sitting on it. The next customer can only sit on those tables with the same label w. That is, those c_w customers corresponding to the same word label w can sit at different tables, with m_w being the number of tables with label w. In our notation, $m_\cdot = \sum_k m_k$ means the total number of tables. The number of customers at table k is denoted by c_k, and the total number of customers is expressed by $c_\cdot = \sum_k c_k$. Given the seating arrangement of customers S, the discount parameter d_0, and the strength parameter α_0, the predictive unigram probability of a new word w is given by

$$\begin{aligned} p(w|S, d_0, \alpha_0) &= \sum_{k=1}^{m.} \frac{c_k - d_0}{\alpha_0 + c.} \delta(k, w) + \frac{\alpha_0 + d_0 m.}{\alpha_0 + c.} G_0(w) \\ &= \frac{c_w - d_0 m_w}{\alpha_0 + c.} + \frac{\alpha_0 + d_0 m.}{\alpha_0 + c.} G_0(w), \end{aligned} \tag{8.136}$$

where $\delta(k, w) = 1$ if table k has the label w, $\delta(k, w) = 0$ otherwise. This equation is derived similarly to the metaphor of the Chinese restaurant process designed for the Dirichlet process, as mentioned in Section 8.2.4. The key difference of the PY process compared to the Dirichlet process is the discount parameter d_0 which is used to adjust the power-law distribution of unigrams based on the PY process. By averaging over seating arrangements and hyperparameters (S, d_0, α_0), we obtain the probability $p(w)$ for a unigram language model.

Interestingly, if we set $d_0 = 0$, the PY process is reduced to a DP which produces the Dirichlet distribution $\text{Dir}(\alpha_0 G_0)$. In this case, the predictive unigram probability based on the PY process prior in Eq. (8.136) is accordingly reduced to

$$\begin{aligned} p(w|S, \alpha_0) &= \frac{c_w}{\alpha_0 + c.} + \frac{\alpha_0 G_0(w)}{\alpha_0 + c.} \\ &= \frac{c_w + \frac{\alpha_0}{|\mathcal{V}|}}{\sum_{w \in \mathcal{V}} [c_w + \frac{\alpha_0}{|\mathcal{V}|}]}, \end{aligned} \tag{8.137}$$

where $G_0(w) = 1/|\mathcal{V}|$ is used for all w. This equation is equivalent to the hierarchical Dirichlet unigram model as given in Eq. (5.70). The only difference here is to adopt a single shared hyperparameter α_0 for all word labels $w \in \mathcal{V}$.

Next, we extend the unigram language model to the n-gram language model based on the HPY process. An n-gram language model is defined as a probability measure over the current word $w = w_i$ given a history context $\mathbf{u} = w_{i-n+1}^{i-1}$. Let $G_\mathbf{u} = [G_\mathbf{u}(w)]_{w \in \mathcal{V}}$ be the vector of the target probability distributions of all vocabulary words $w \in \mathcal{V}$ given the history context $\mathbf{u}$. We use a PY process as the prior for $G_\mathbf{u}[G_\mathbf{u}(w)]_{w \in \mathcal{V}}$ in the form of

$$G_\mathbf{u} \sim \text{PY}(d_{|\mathbf{u}|}, \alpha_{|\mathbf{u}|}, G_{\pi(\mathbf{u})}), \tag{8.138}$$

with the hyperparameters $d_{|\mathbf{u}|}$ and $\alpha_{|\mathbf{u}|}$ specific to the length of context $|\pi(\mathbf{u})|$. However, $G_{\pi(\mathbf{u})}$ is still an unknown base measure. A PY process is recursively placed over it:

$$G_{\pi(\mathbf{u})} \sim \text{PY}(d_{|\pi(\mathbf{u})|}, \alpha_{|\pi(\mathbf{u})|}, G_{\pi(\pi(\mathbf{u}))}), \tag{8.139}$$

with parameters which are functions of $|\pi(\pi(\mathbf{u}))|$. This is repeated until we reach the PY process prior $G_\emptyset$ for a unigram model. Figure 8.12 is a schematic diagram showing the hierarchical priors for the smoothed language model based on the HPY process. This process enables us to generalize from the unigram language model to the n-gram language model. The resulting probability distribution is called the hierarchical Pitman–Yor language model (HPYLM).

A hierarchical Chinese restaurant process can be used to develop a generative procedure for drawing words from the HPYLM with all $G_\mathbf{u}$ marginalized out. The context

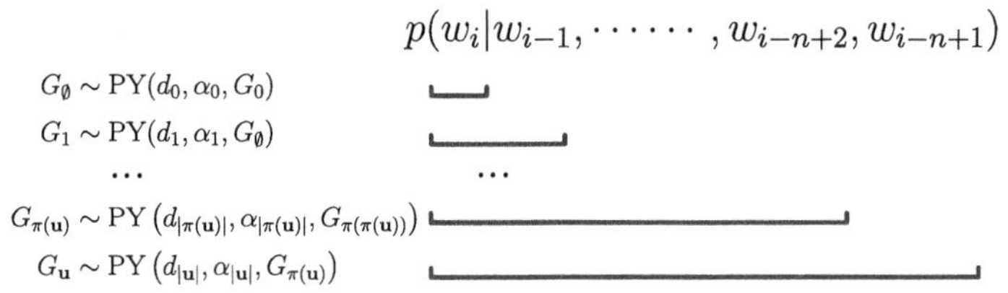

$$p(w_i|w_{i-1},\cdots\cdots,w_{i-n+2},w_{i-n+1})$$

$$G_\emptyset \sim \text{PY}(d_0,\alpha_0,G_0)$$
$$G_1 \sim \text{PY}(d_1,\alpha_1,G_\emptyset)$$
$$\cdots$$
$$G_{\pi(\mathbf{u})} \sim \text{PY}\left(d_{|\pi(\mathbf{u})|},\alpha_{|\pi(\mathbf{u})|},G_{\pi(\pi(\mathbf{u}))}\right)$$
$$G_{\mathbf{u}} \sim \text{PY}\left(d_{|\mathbf{u}|},\alpha_{|\mathbf{u}|},G_{\pi(\mathbf{u})}\right)$$

Figure 8.12 Hierarchical Pitman–Yor process for n-gram language model.

$\mathbf{u}$ corresponds to a restaurant. This procedure gives us a representation of HPYLM for efficient inference using an MCMC algorithm, and easy computation of the predictive probability distribution from new test words. Through this representation, the correspondence between the Kneser–Ney language model and HPYLM can be illustrated. In the metaphor of hierarchical CRP, there are multiple hierarchical restaurants or PY processes, with each corresponding to one context. Different orders of n-gram models share information with each other through the interpolation of higher-order n-grams with lower-order n-grams. We draw words from $G_{\mathbf{u}}$ of the PY process by using the CRP as discussed in Section 8.5.1. Further, we draw words from another CRP to sample the parent distribution $G_{\pi(\mathbf{u})}$, which is itself sampled according to a PY process. This is recursively applied until we need draws from the global mean distributions G_0. By referring to the predictive unigram probability from the PY process as shown in Eq. (8.136), the predictive n-gram probability $p(w|\mathbf{u},S,d_{|\mathbf{u}|},\alpha_{|\mathbf{u}|})$ under a particular combination of seating arrangement S and the hyperparameters $d_{|\mathbf{u}|}$ and $\alpha_{|\mathbf{u}|}$ can be obtained from

$$p(w|\mathbf{u},S,d_{|\mathbf{u}|},\alpha_{|\mathbf{u}|}) = \frac{c_{\mathbf{u}w\cdot} - d_{|\mathbf{u}|}m_{\mathbf{u}w}}{\alpha_{|\mathbf{u}|}+c_{\mathbf{u}\cdot\cdot}} + \frac{\alpha_{|\mathbf{u}|}+d_{|\mathbf{u}|}m_{\mathbf{u}\cdot}}{\alpha_{|\mathbf{u}|}+c_{\mathbf{u}\cdot\cdot}} p(w|\pi(\mathbf{u}),S,d_{|\pi(\mathbf{u})|},\alpha_{|\pi(\mathbf{u})|}), \tag{8.140}$$

where $c_{\mathbf{u}wk}$ is the number of customers sitting at table k with label w, $c_{\mathbf{u}w\cdot} = \sum_k c_{\mathbf{u}wk}$ and $m_{\mathbf{u}w}$ is the number of occupied tables with label w. It is interesting to see that the Kneser–Ney language model (KN–LM) in Eq. (8.133) is closely related to the HPYLM in Eq. (8.140). HPYLM is a generalized realization of KN–LM with an additional concentration parameter $\alpha_{|\mathbf{u}|}$. We can interpret the interpolated KN–LM as an approximate inference scheme for the HPYLM.

8.5.4 MCMC inference for HPYLM

An MCMC algorithm can be used to infer the posterior probability of seating arrangement S. Given some training data $\mathcal{D}$, we count the number of occurrences $c_{\mathbf{u}w\cdot}$ of each word w appearing after each context $\mathbf{u}$ of length $n-1$. This means that there are $c_{\mathbf{u}w\cdot}$ samples drawn from the PY process $G_{\mathbf{u}}$. We are interested in the posterior probability over the latent variables $\mathcal{G} = \{G_{\mathbf{u}} \mid \text{all contexts } \mathbf{u}\}$ and the parameters $\Theta = \{d_m,\alpha_m\}$ of all lower-order models with $0 \le m \le n-1$. Because the hierarchical Chinese restaurant process marginalizes out each $G_{\mathbf{u}}$, we can replace $\mathcal{G}$ by the seating arrangement in the corresponding restaurant using $S = \{S_{\mathbf{u}} \mid \text{all contexts } \mathbf{u}\}$. The posterior probability is then obtained from

$$p(S, \Theta|\mathcal{D}) = \frac{p(S, \Theta, \mathcal{D})}{p(\mathcal{D})}. \tag{8.141}$$

Given this posterior probability, we can calculate the predictive n-gram probability of a test word w after observing a context $\mathbf{u}$:

$$\begin{aligned} p(w|\mathbf{u}, \mathcal{D}) &= \int p(w|\mathbf{u}, S, \Theta) p(S, \Theta|\mathcal{D}) d(S, \Theta) \\ &= \mathbb{E}_{(S,\Theta)}[p(w|\mathbf{u}, S, \Theta)] \\ &\approx \frac{1}{L} \sum_{l=1}^{L} p(w|\mathbf{u}, S^{(l)}, \Theta^{(l)}), \end{aligned} \tag{8.142}$$

which is an expectation of predictive probability under a particular set of seating arrangements S and PY parameters Θ. The overall predictive probability in the integral of Eq. (8.142) is approximated by using the L samples $\{S^{(l)}, \Theta^{(l)}\}$ drawn from posterior probability $p(S, \Theta|\mathcal{D})$.

In the implementation of Teh *et al.* (2006) and Huang & Renals (2008), the discount parameter and concentration parameter were drawn by a beta distribution and a gamma distribution, respectively. Here, we address the approach to sampling the seating arrangement $S_{\mathbf{u}}$ corresponding to $G_{\mathbf{u}}$. We employ Gibbs sampling to keep track of the current state of each variable of interest in the model, and iteratively re-sample the state of each variable given the current states of all other variables. After a sufficient number of iterations, the states of variables in the seating arrangement S converge to the required samples from the posterior probability. The variables consist of, for each context (restaurant) $\mathbf{u}$ and each word (customer) $x_{\mathbf{u}l}$ drawn from $G_{\mathbf{u}}$, the index $k_{\mathbf{u}l}$ of the draw from $G_{\pi(\mathbf{u})}$ assigned $x_{\mathbf{u}l}$. In the Chinese restaurant metaphor, this is the table index of where the lth customer sat at the restaurant corresponding to $G_{\mathbf{u}}$. If $x_{\mathbf{u}l}$ has value w, it can only be assigned to draws from $G_{\pi(\mathbf{u})}$ that have the same value w. The posterior probability of drawing a table $k_{\mathbf{u}l}$ for the last word $x_{\mathbf{u}l}$ from $G_{\mathbf{u}}$ is given below:

$$p(k_{\mathbf{u}l} = k|S^{-\mathbf{u}l}, \Theta) \propto \frac{\max(0, c^{-\mathbf{u}l}_{\mathbf{u}x_{\mathbf{u}l}k} - d)}{\alpha + c^{-\mathbf{u}l}_{\mathbf{u}\cdot\cdot}}, \tag{8.143}$$

$$p(k_{\mathbf{u}l} = k_{\text{new}}|S^{-\mathbf{u}l}, \Theta) \propto \frac{\alpha + d m^{-\mathbf{u}l}_{\mathbf{u}\cdot}}{\alpha + c^{-\mathbf{u}l}_{\mathbf{u}\cdot\cdot}} p(x_{\mathbf{u}l}|\pi(\mathbf{u}), S^{-\mathbf{u}l}, \Theta), \tag{8.144}$$

where the superscript $-\mathbf{u}l$ means the corresponding set of variables or counts with $x_{\mathbf{u}l}$ excluded.

One other key difference between the KN–LM in Eq. (8.133) and the HPYLM in Eq. (8.140) is that the KN–LM adopts a fixed discount $d_{|\mathbf{u}|}$ while HPYLM produces different discounts $d_{|\mathbf{u}|} m_{\mathbf{u}w}$ for different words w due to the values of $m_{\mathbf{u}w}$, which counts the number of occupied tables labeled by w. In general, $m_{\mathbf{u}w}$ is on average larger if $c_{\mathbf{u}w}$ is larger. The actual amount of discount grows gradually as the count $c_{\mathbf{u}w}$ grows. The physical meaning of discounting in the smoothed n-grams based on HPYLM is consistent with the discount scheme in the modified Kneser–Ney language model (MKN–LM), which specifies three discounts d_1, d_2, and d_{3+} for those n-grams with one ($c = 1$), two ($c = 2$), and three or more ($c \geq 3$) counts, as addressed in Section 3.6.6. The discounts $\{d_1, d_2, d_{3+}\}$ in MKN–LM are empirically determined while the discounts $d_{|\mathbf{u}|} m_{\mathbf{u}w}$ in HPYLM are automatically associated with each word w. In particular, if we restrict $m_{\mathbf{u}w\cdot}$ to be at most 1,

$$m_{\mathbf{u}w} = \min(1, c_{\mathbf{u}w\cdot}), \tag{8.145}$$

we get the same discount value so long as $c_{\mathbf{u}w\cdot} > 0$, or equivalently we conduct absolute discounting. We also have the relationships among the $c_{\mathbf{u}w}$s and $m_{\mathbf{u}w}$:

$$c_{\mathbf{u}w} = \sum_{\mathbf{u}':\pi(\mathbf{u}')=\mathbf{u}} m_{\mathbf{u}'w\cdot}. \tag{8.146}$$

If we further assume $\alpha_{|\mathbf{u}|} = \alpha_m = 0$ for all model orders $0 \leq m \leq n-1$, the predictive probability of HPYLM in Eq. (8.140) is directly reduced to that given by the interpolated Kneser–Ney model.

8.6 Summary

Due to the high computational cost and lack of scalability for data and model sizes, MCMC approaches are still in a development stage for speech and language processing applications that essentially need to deal with large-scale data. However, the recent advance in computational powers (CPU speed, memory size, many cores, and GPU) and algorithm development make it possible to realize middle-scale data processing of MCMC, especially for language processing based on multinomial and Dirichlet distributions, as discussed in this chapter. One of the most attractive features of MCMC is that we can handle *any types of expectation calculation* by the Monte Carlo approximations, which can realize all Bayesian approaches in principle. Therefore, we expect further development of computational powers and algorithms to widen the applications of MCMC to large-scale problems, and enable fully Bayesian speech and language processing based on MCMC in the near future.

Appendix A Basic formulas

A.1 Expectation

$$\mathbb{E}_{(a)}\left[f(a)+g(a)\right]=\mathbb{E}_{(a)}\left[f(a)\right]+\mathbb{E}_{(a)}\left[g(a)\right], \tag{A.1}$$

$$\mathbb{E}_{(a)}\left[bf(a)\right]=b\mathbb{E}_{(a)}\left[f(a)\right]. \tag{A.2}$$

A.2 Delta function

A.2.1 Kronecker delta function

$$f_b=\sum_a \delta(a,b)f_a. \tag{A.3}$$

A.2.2 Product formula of Kronecker delta function

$$f_b=\prod_a f_a^{\delta(a,b)}. \tag{A.4}$$

(Proof)

$$\begin{aligned}\prod_a f_a^{\delta(a,b)}&=\exp\left(\log\left(\prod_a f_a^{\delta(a,b)}\right)\right)\\&=\exp\left(\sum_a \delta(a,b)\log\left(f_a\right)\right)\\&=\exp\left(\log\left(f_b\right)\right)=f_b.\end{aligned} \tag{A.5}$$

A.2.3 Dirac delta function

$$f(y)=\int \delta(x-y)f(x)dx. \tag{A.6}$$

A.3 Jensen's inequality

For a concave function f, where X is a probabilistic random variable sampled from a distribution function, and an arbitrary function $g(X)$, we have the following inequality:

$$f\left(\mathbb{E}_{(X)}[g(X)]\right) \geq \mathbb{E}_{(X)}[f(g(X))]. \tag{A.7}$$

In the special case of $f(\cdot) = \log(\cdot)$, (A.7) is rewritten as follows:

$$\log\left(\mathbb{E}_{(X)}[g(X)]\right) \geq \mathbb{E}_{(X)}[\log(g(X))]. \tag{A.8}$$

A.4 Gamma function

$$\Gamma(x+1) = x\Gamma(x), \tag{A.9}$$

$$\Gamma\left(\frac{1}{2}\right) = \pi^{\frac{1}{2}}. \tag{A.10}$$

A.4.1 Stirling's approximation

$$\log\Gamma\left(\frac{x}{2}\right) \to \frac{x}{2}\log\left(\frac{x}{2}\right) - \frac{x}{2} - \frac{1}{2}\log\left(\frac{x}{2\pi}\right), \qquad x \to \infty. \tag{A.11}$$

A.4.2 Di-gamma function

$$\begin{aligned}\Psi(x) &\triangleq \frac{\partial}{\partial x}\log\Gamma(x) \\ &= \frac{\frac{\partial\Gamma(x)}{\partial x}}{\Gamma(x)}.\end{aligned} \tag{A.12}$$

Appendix B Vector and matrix formulas

This appendix lists some useful vector and matrix formulas. Note that these formulas are selected for the purpose of deriving equations in this book in Bayesian speech and language processing, and do not cover the whole field of vector and matrix formulas.

B.1 Trace

$$\mathrm{tr}[a] = a, \tag{B.1}$$

$$\mathrm{tr}[\mathbf{ABC}] = \mathrm{tr}[\mathbf{BCA}] = \mathrm{tr}[\mathbf{CAB}], \tag{B.2}$$

$$\mathrm{tr}[\mathbf{A} + \mathbf{B}] = \mathrm{tr}[\mathbf{A}] + \mathrm{tr}[\mathbf{B}], \tag{B.3}$$

$$\mathrm{tr}[\mathbf{A}^\top] = \mathrm{tr}[\mathbf{A}], \tag{B.4}$$

$$\mathrm{tr}[\mathbf{A}(\mathbf{B} + \mathbf{C})] = \mathrm{tr}[\mathbf{AB} + \mathbf{AC}]. \tag{B.5}$$

B.2 Transpose

$$(\mathbf{ABC})^\top = \mathbf{C}^\top \mathbf{B}^\top \mathbf{A}^\top, \tag{B.6}$$

$$(\mathbf{A} + \mathbf{B})^\top = \mathbf{A}^\top + \mathbf{B}^\top. \tag{B.7}$$

B.3 Derivative

$$\frac{\partial \log |\mathbf{A}|}{\partial \mathbf{A}} = (\mathbf{A}^\top)^{-1}, \tag{B.8}$$

$$\frac{\partial \mathbf{a}^\top \mathbf{b}}{\partial \mathbf{a}} = \frac{\partial \mathbf{b}^\top \mathbf{a}}{\partial \mathbf{a}} = \mathbf{b}, \tag{B.9}$$

$$\frac{\partial \mathbf{a}^\top \mathbf{C} \mathbf{b}}{\partial \mathbf{C}} = \mathbf{a}\mathbf{b}^\top, \tag{B.10}$$

$$\frac{\partial \mathrm{tr}(\mathbf{AB})}{\partial \mathbf{A}} = \mathbf{B}^\top, \tag{B.11}$$

$$\frac{\partial \mathrm{tr}(\mathbf{ACB})}{\partial \mathbf{C}} = \mathbf{A}^\top \mathbf{B}^\top, \tag{B.12}$$

$$\frac{\partial \mathrm{tr}(\mathbf{AC}^{-1}\mathbf{B})}{\partial \mathbf{C}} = -(\mathbf{C}^{-1}\mathbf{BAC}^{-1})^\top, \tag{B.13}$$

$$\frac{\partial \mathbf{a}^\top \mathbf{Ca}}{\partial \mathbf{a}} = (\mathbf{C} + \mathbf{C}^\top)\mathbf{a}, \tag{B.14}$$

$$\frac{\partial \mathbf{b}^\top \mathbf{A}^\top \mathbf{DAc}}{\partial \mathbf{A}} = \mathbf{D}^\top \mathbf{Abc}^\top + \mathbf{DAcb}^\top. \tag{B.15}$$

B.4 Complete square

When $\mathbf{A}$ is a symmetric matrix,

$$\mathbf{x}^\top \mathbf{Ax} - 2\mathbf{x}^\top \mathbf{b} + c = (\mathbf{x} - \mathbf{u})^\top \mathbf{A}\,(\mathbf{x} - \mathbf{u}) + v, \tag{B.16}$$

where

$$\begin{aligned} \mathbf{u} &\triangleq \mathbf{A}^{-1}\mathbf{b} \\ v &\triangleq c - \mathbf{b}^\top \mathbf{A}^{-1}\mathbf{b}. \end{aligned} \tag{B.17}$$

By using the above complete square formula, we can also derive the following formula when matrices $\mathbf{A}_1$ and $\mathbf{A}_2$ are symmetric:

$$\begin{aligned} &(\mathbf{x} - \mathbf{b}_1)^\top \mathbf{A}_1(\mathbf{x} - \mathbf{b}_1) + (\mathbf{x} - \mathbf{b}_2)^\top \mathbf{A}_2(\mathbf{x} - \mathbf{b}_2) \\ &= \mathbf{x}^\top \underbrace{(\mathbf{A}_1 + \mathbf{A}_2)}_{\triangleq \mathbf{A}} \mathbf{x} - 2\mathbf{x}^\top \underbrace{(\mathbf{A}_1\mathbf{b}_1 + \mathbf{A}_2\mathbf{b}_2)}_{\triangleq \mathbf{b}} + \underbrace{\mathbf{b}_1^\top \mathbf{A}_1\mathbf{b}_1 + \mathbf{b}_2^\top \mathbf{A}_2\mathbf{b}_2}_{\triangleq c} \\ &= (\mathbf{x} - \mathbf{u})^\top (\mathbf{A}_1 + \mathbf{A}_2)\,(\mathbf{x} - \mathbf{u}) + v, \end{aligned} \tag{B.18}$$

where

$$\begin{aligned} \mathbf{u} &= (\mathbf{A}_1 + \mathbf{A}_2)^{-1}(\mathbf{A}_1\mathbf{b}_1 + \mathbf{A}_2\mathbf{b}_2), \\ v &= \mathbf{b}_1^\top \mathbf{A}_1\mathbf{b}_1 + \mathbf{b}_2^\top \mathbf{A}_2\mathbf{b}_2 - (\mathbf{A}_1\mathbf{b}_1 + \mathbf{A}_2\mathbf{b}_2)^\top (\mathbf{A}_1 + \mathbf{A}_2)^{-1}(\mathbf{A}_1\mathbf{b}_1 + \mathbf{A}_2\mathbf{b}_2). \end{aligned} \tag{B.19}$$

B.5 Woodbury matrix inversion

$$(\mathbf{A} + \mathbf{UCV})^{-1} = \mathbf{A}^{-1} - \mathbf{A}^{-1}\mathbf{U}(\mathbf{C}^{-1} + \mathbf{VA}^{-1}\mathbf{U})^{-1}\mathbf{VA}^{-1}. \tag{B.20}$$

Appendix C Probabilistic distribution functions

This appendix lists the probabilistic distribution functions (pdfs) used in the main discussions in this book. We basically use the definitions of these functions followed by Bernardo & Smith (2009). Each pdf section also provides values of mean, mode, variance, etc., which are used in the book, although it does not include a complete set of distribution values to avoid complicated descriptions in this appendix. The sections also provide some derivations of these values if these derivations are not complicated.

C.1 Discrete uniform distribution

When we have a discrete variable $a \in \{a_1, a_2, \cdots, a_n, \cdots, a_N\}$, the discrete uniform distribution is defined as:

- pdf:

$$\mathrm{Unif}(a) \triangleq \frac{1}{|\{a_n\}|} = \frac{1}{N}, \tag{C.1}$$

where N is the number of distinct elements.

C.2 Multinomial distribution

- pdf:

$$\mathrm{Mult}(\{x_j\}_{j=1}^J | \{\omega_j\}_{j=1}^J) \triangleq \frac{(\sum_{j=1}^J x_j)!}{\prod_{j=1}^J x_j!} \prod_{j=1}^J \omega_j^{x_j}, \tag{C.2}$$

where x_j is a nonnegative integer, and has the following constraint:

$$x_j \in \mathbb{Z}^+, \quad \sum_j^J x_j = N. \tag{C.3}$$

The parameter $\{\omega_1, \cdots, \omega_J\}$ has the following constraint:

$$\sum_j^J \omega_j = 1, \quad 0 \le \omega_j \le 1 \quad \forall j. \tag{C.4}$$

- pdf ($x_j = 1$ and $x_{j'} = 0 \; \forall j' \neq j$):

$$\mathrm{Mult}(x_j | \{\omega_j\}_{j=1}^J) \triangleq \omega_j. \tag{C.5}$$

C.3 Beta distribution

Beta distribution is a special case of the Dirichlet distribution with two probabilistic variables x and $y = 1 - x$.

- pdf:

$$\mathrm{Beta}(x|\alpha, \beta) \triangleq C_{\mathrm{Beta}}(\alpha, \beta) x^{\alpha-1} (1 - x)^{\beta-1}, \tag{C.6}$$

  where

$$x \in [0, 1], \tag{C.7}$$

  and

$$\alpha, \beta > 0. \tag{C.8}$$

- Normalization constant:

$$C_{\mathrm{Beta}}(\alpha, \beta) \triangleq \frac{\Gamma(\alpha + \beta)}{\Gamma(\alpha)\Gamma(\beta)}, \tag{C.9}$$

  where $\Gamma(\cdot)$ is a gamma function.
- Mean:

$$\mathbb{E}_{(x)}[x] = \frac{\alpha}{\alpha + \beta}. \tag{C.10}$$

  This is derived as follows:

$$\begin{aligned}
\mathbb{E}_{(x)}[x] &= \int x \mathrm{Beta}(x|\alpha, \beta) dx \\
&= C_{\mathrm{Beta}}(\alpha, \beta) \int x^{\alpha} (1 - x)^{\beta-1} dx \\
&= \frac{C_{\mathrm{Beta}}(\alpha, \beta)}{C_{\mathrm{Beta}}(\alpha + 1, \beta)} \\
&= \frac{\frac{\Gamma(\alpha+\beta)}{\Gamma(\alpha)\Gamma(\beta)}}{\frac{\Gamma(\alpha+\beta+1)}{\Gamma(\alpha+1)\Gamma(\beta)}} = \frac{\Gamma(\alpha + \beta)}{\Gamma(\alpha + \beta + 1)} \frac{\Gamma(\alpha + 1)}{\Gamma(\alpha)} \\
&= \frac{\alpha}{\alpha + \beta},
\end{aligned} \tag{C.11}$$

  where we use the following property of the gamma function:

$$\Gamma(x + 1) = x\Gamma(x). \tag{C.12}$$

- Mode:

$$\frac{\alpha - 1}{\alpha + \beta - 2}, \quad \alpha, \beta > 1. \tag{C.13}$$

It is derived as follows:

$$\begin{aligned}
&\frac{d}{dx}\text{Beta}(x|\alpha,\beta) \\
&= C_{\text{Beta}}(\alpha,\beta)\left((\alpha-1)x^{\alpha-2}(1-x)^{\beta-1}-(\beta-1)x^{\alpha-1}(1-x)^{\beta-2}\right) \\
&= C_{\text{Beta}}(\alpha,\beta)x^{\alpha-2}(1-x)^{\beta-2}\left((\alpha-1)(1-x)-(\beta-1)x\right) \\
&= C_{\text{Beta}}(\alpha,\beta)x^{\alpha-2}(1-x)^{\beta-2}\left(\alpha-1-(\alpha+\beta-2)x\right).
\end{aligned} \tag{C.14}$$

Therefore,

$$\begin{aligned}
\frac{d}{dx}\text{Beta}(x|\alpha,\beta) = 0 \\
\Rightarrow x^{\text{mode}} = \frac{\alpha-1}{\alpha+\beta-2}.
\end{aligned} \tag{C.15}$$

C.4 Dirichlet distribution

A Dirichlet distribution is a generalized beta distribution with J probabilistic variables.

- pdf:

$$\text{Dir}(\{\omega_j\}_{j=1}^J|\{\phi_j\}_{j=1}^J) \triangleq C_{\text{Dir}}(\{\phi_j\}_{j=1}^J)\prod_j(\omega_j)^{\phi_j-1}, \tag{C.16}$$

where

$$\sum_j^J \omega_j = 1, \quad 0 \leq \omega_j \leq 1, \quad 0 < \phi_j. \tag{C.17}$$

- Normalization constant:

$$C_{\text{Dir}}(\{\phi_j\}_{j=1}^J) \triangleq \frac{\Gamma\left(\sum_{j=1}^J \phi_j\right)}{\prod_{j=1}^J \Gamma(\phi_j)}. \tag{C.18}$$

- Mean:

$$\mathbb{E}_{(\omega_j)}[\omega_j] = \frac{\phi_j}{\sum_{j'=1}^J \phi_{j'}}, \tag{C.19}$$

$$\begin{aligned}
\mathbb{E}_{(\omega_j)}\left[\prod_j \omega_j^{\gamma_j}\right] &= \frac{C_{\text{Dir}}(\{\phi_j\}_{j=1}^J)}{C_{\text{Dir}}(\{\phi_j+\gamma_j\}_{j=1}^J)} \\
&= \frac{\Gamma\left(\sum_{j=1}^J \phi_j\right)}{\Gamma\left(\sum_{j=1}^J \phi_j+\gamma_j\right)}\frac{\prod_{j=1}^J \Gamma(\phi_j+\gamma_j)}{\prod_{j=1}^J \Gamma(\phi_j)},
\end{aligned} \tag{C.20}$$

$$\mathbb{E}_{(\omega_j)}[\log\omega_j] = \exp\left(\Psi(\omega_j) - \Psi\left(\sum_{j'=1}^J \omega_{j'}\right)\right). \tag{C.21}$$

- Mode:

$$
\begin{aligned}
&\frac{d}{d\omega_{j'}}\left(\log\left(\mathrm{Dir}(\{\omega_j\}_{j=1}^J|\{\phi_j\}_{j=1}^J)\right) + \lambda\left(1 - \sum_{j=1}^J \omega_j\right)\right) \\
&= \frac{\phi_j - 1}{\omega_j} + \lambda = 0, \\
&\Rightarrow \omega_j^{\text{mode}} \propto \phi_j - 1 \\
&\Rightarrow \frac{\phi_j - 1}{\sum_{j'=1}^J(\phi_{j'} - 1)}, \quad \phi_j > 1.
\end{aligned} \tag{C.22}
$$

C.5 Gaussian distribution

Note that this book uses Σ as a variance. The standard deviation σ is represented as $\sigma = \sqrt{\Sigma}$. We also list the pdf with precision scale parameter $r \triangleq 1/\Sigma$, which can also be used in this book.

- pdf (with variance Σ):

$$
\mathcal{N}(x|\mu, \Sigma) \triangleq C_{\mathcal{N}}(\Sigma)\exp\left(-\frac{(x-\mu)^2}{2\Sigma}\right), \tag{C.23}
$$

where

$$
x \in \mathbb{R} \tag{C.24}
$$

and

$$
\mu \in \mathbb{R}, \Sigma > 0. \tag{C.25}
$$

- Normalization constant (with variance Σ):

$$
C_{\mathcal{N}}(\Sigma) \triangleq (2\pi)^{-\frac{1}{2}}(\Sigma)^{-\frac{1}{2}}. \tag{C.26}
$$

- pdf (with precision r):

$$
\mathcal{N}(x|\mu, r^{-1}) \triangleq C_{\mathcal{N}}(r^{-1})\exp\left(-\frac{r(x-\mu)^2}{2}\right), \tag{C.27}
$$

where

$$
r > 0. \tag{C.28}
$$

- Normalization constant (with precision r):

$$
C_{\mathcal{N}}(r^{-1}) \triangleq (2\pi)^{-\frac{1}{2}}(r)^{\frac{1}{2}}. \tag{C.29}
$$

- Mean:

$$
\mathbb{E}_{(x)}[x] = \mu. \tag{C.30}
$$

- Variance:

$$\mathbb{E}_{(x)}[x^2] = \Sigma = r^{-1}. \tag{C.31}$$

- Mode:

$$\mu. \tag{C.32}$$

C.6 Multivariate Gaussian distribution

- pdf (with covariance matrix $\mathbf{\Sigma}$):

$$\begin{aligned}\mathcal{N}(\mathbf{x}|\boldsymbol{\mu}, \mathbf{\Sigma}) &\triangleq C_{\mathcal{N}}(\mathbf{\Sigma}) \exp\left(-\frac{1}{2}(\mathbf{x}-\boldsymbol{\mu})^{\mathsf{T}}\mathbf{\Sigma}^{-1}(\mathbf{x}-\boldsymbol{\mu})\right) \\ &= C_{\mathcal{N}}(\mathbf{\Sigma}) \exp\left(-\frac{1}{2}\mathrm{tr}\left[\mathbf{\Sigma}^{-1}(\mathbf{x}-\boldsymbol{\mu})(\mathbf{x}-\boldsymbol{\mu})^{\mathsf{T}}\right]\right),\end{aligned} \tag{C.33}$$

where

$$\mathbf{x} \in \mathbb{R}^D, \tag{C.34}$$

and

$$\boldsymbol{\mu} \in \mathbb{R}^D, \quad \mathbf{\Sigma} \in \mathbb{R}^{D\times D}. \tag{C.35}$$

$\mathbf{\Sigma}$ is positive definite.

- Normalization constant (with covariance matrix $\mathbf{\Sigma}$):

$$C_{\mathcal{N}}(\mathbf{\Sigma}) \triangleq (2\pi)^{-\frac{D}{2}}|\mathbf{\Sigma}|^{-\frac{1}{2}}. \tag{C.36}$$

- pdf (with precision matrix $\mathbf{R}$):

$$\begin{aligned}\mathcal{N}(\mathbf{x}|\boldsymbol{\mu}, \mathbf{R}^{-1}) &\triangleq C_{\mathcal{N}}(\mathbf{R}^{-1}) \exp\left(-\frac{1}{2}(\mathbf{x}-\boldsymbol{\mu})^{\mathsf{T}}\mathbf{R}(\mathbf{x}-\boldsymbol{\mu})\right) \\ &= C_{\mathcal{N}}(\mathbf{R}^{-1}) \exp\left(-\frac{1}{2}\mathrm{tr}\left[\mathbf{R}(\mathbf{x}-\boldsymbol{\mu})(\mathbf{x}-\boldsymbol{\mu})^{\mathsf{T}}\right]\right).\end{aligned} \tag{C.37}$$

- Normalization constant (with precision matrix$\mathbf{R}$):

$$C_{\mathcal{N}}(\mathbf{R}^{-1}) \triangleq (2\pi)^{-\frac{D}{2}}|\mathbf{R}|^{\frac{1}{2}}. \tag{C.38}$$

- Mean:

$$\mathbb{E}_{(\mathbf{x})}[\mathbf{x}] = \boldsymbol{\mu}. \tag{C.39}$$

- Variance:

$$\mathbb{E}_{(\mathbf{x})}[\mathbf{x}\mathbf{x}^{\mathsf{T}}] = \mathbf{\Sigma} = \mathbf{R}^{-1}. \tag{C.40}$$

- Mode:

$$\boldsymbol{\mu}. \tag{C.41}$$

C.7 Multivariate Gaussian distribution (diagonal covariance matrix)

This is a special case of the multivariate Gaussian distribution. The diagonal covariance matrix reduces the number of parameters and makes the analytical treatment simple (it is simply represented as a product of scalar Gaussian distributions).

- pdf (with covariance matrix $\mathbf{\Sigma} = \text{diag}(\boldsymbol{\sigma})$, where $\boldsymbol{\sigma} = [\Sigma_1, \cdots, \Sigma_D]^\mathsf{T}$):

$$\begin{aligned}\mathcal{N}(\mathbf{x}|\boldsymbol{\mu}, \mathbf{\Sigma}) &\triangleq \prod_{d=1}^{D} \mathcal{N}(x_d|\mu_d, \Sigma_d) \\ &= \prod_{d=1}^{D} C_\mathcal{N}(\Sigma_d) \exp\left(-\frac{1}{2}\frac{(x_d - \mu_d)^2}{\Sigma_d}\right),\end{aligned} \tag{C.42}$$

where

$$\mathbf{x} \in \mathbb{R}^D, \tag{C.43}$$

and

$$\boldsymbol{\mu} \in \mathbb{R}^D, \quad \boldsymbol{\sigma} \in \mathbb{R}^D_{>0}. \tag{C.44}$$

- Normalization constant (with variance Σ_d):

$$C_\mathcal{N}(\Sigma_d) \triangleq (2\pi)^{-\frac{1}{2}} \Sigma_d^{-\frac{1}{2}}. \tag{C.45}$$

- pdf (with precision matrix $\mathbf{R} = \text{diag}(\mathbf{r})$, where $\mathbf{r} = [r_1, \cdots, r_D]^\mathsf{T}$):

$$\begin{aligned}\mathcal{N}(\mathbf{x}|\boldsymbol{\mu}, \mathbf{R}^{-1}) &\triangleq \prod_{d=1}^{D} \mathcal{N}(x_d|\mu_d, r_d^{-1}) \\ &= \prod_{d=1}^{D} C_\mathcal{N}(r_d^{-1}) \exp\left(-\frac{r_d}{2}(x_d - \mu_d)^2\right).\end{aligned} \tag{C.46}$$

- Normalization constant (with precision r_d):

$$C_\mathcal{N}(r_d^{-1}) \triangleq (2\pi)^{-\frac{1}{2}} r_d^{\frac{1}{2}}. \tag{C.47}$$

- Mean:

$$\mathbb{E}_{(\mathbf{x})}[\mathbf{x}] = \boldsymbol{\mu}. \tag{C.48}$$

- Variance:

$$\mathbb{E}_{(x_d)}[x_d^2] = \Sigma_d = r_d^{-1}. \tag{C.49}$$

- Mode:

$$\boldsymbol{\mu}. \tag{C.50}$$

C.8 Spherical Gaussian distribution

This is another special case of the multivariate Gaussian distribution where the diagonal covariance matrix is represented with the identity matrix with a single scaling parameter.

- pdf (with covariance matrix $\mathbf{\Sigma} = \Sigma \mathbf{I}_D$):

$$\begin{aligned}\mathcal{N}(\mathbf{x}|\boldsymbol{\mu}, \mathbf{\Sigma}) &\triangleq \prod_{d=1}^{D} \mathcal{N}(x_d|\mu_d, \Sigma) \\ &= (C_{\mathcal{N}}(\Sigma))^D \prod_{d=1}^{D} \exp\left(-\frac{1}{2}\frac{(x_d - \mu_d)^2}{\Sigma}\right),\end{aligned} \tag{C.51}$$

where

$$\mathbf{x} \in \mathbb{R}^D, \tag{C.52}$$

and

$$\boldsymbol{\mu} \in \mathbb{R}^D, \quad \Sigma \in \mathbb{R}_{>0}. \tag{C.53}$$

- Normalization constant (with variance Σ):

$$C_{\mathcal{N}}(\Sigma) \triangleq (2\pi)^{-\frac{1}{2}} \Sigma^{-\frac{1}{2}}. \tag{C.54}$$

- pdf (with precision matrix $\mathbf{R} = r^{-1}\mathbf{I}_D$):

$$\begin{aligned}\mathcal{N}(\mathbf{x}|\boldsymbol{\mu}, \mathbf{R}^{-1}) &\triangleq \prod_{d=1}^{D} \mathcal{N}(x_d|\mu_d, r^{-1}) \\ &= \left(C_{\mathcal{N}}(r^{-1})\right)^D \prod_{d=1}^{D} \exp\left(-\frac{r}{2}(x_d - \mu_d)^2\right).\end{aligned} \tag{C.55}$$

- Normalization constant (with precision r):

$$C_{\mathcal{N}}(r^{-1}) \triangleq (2\pi)^{-\frac{1}{2}} r^{\frac{1}{2}}. \tag{C.56}$$

- Mean:

$$\mathbb{E}_{(\mathbf{x})}[\mathbf{x}] = \boldsymbol{\mu}. \tag{C.57}$$

- Variance:

$$\mathbb{E}_{(x_d)}[x_d^2] = \Sigma = r^{-1}. \tag{C.58}$$

- Mode:

$$\boldsymbol{\mu}. \tag{C.59}$$

C.9 Matrix variate Gaussian distribution

- pdf:

$$p(\mathbf{X}) = \mathcal{N}(\mathbf{X}|\mathbf{M}, \mathbf{\Phi}, \mathbf{\Omega}) \triangleq C_{\mathcal{N}}(\mathbf{\Phi}, \mathbf{\Omega}) \exp\left(-\frac{1}{2}\mathrm{tr}\left[(\mathbf{X}-\mathbf{M})^{\mathsf{T}}\mathbf{\Phi}^{-1}(\mathbf{X}-\mathbf{M})\mathbf{\Omega}^{-1}\right]\right), \tag{C.60}$$

where

$$\mathbf{X} \in \mathbb{R}^{D\times D'}, \tag{C.61}$$

and

$$\mathbf{M} \in \mathbb{R}^{D\times D'}, \quad \mathbf{\Phi} \in \mathbb{R}^{D\times D}, \quad \mathbf{\Omega} \in \mathbb{R}^{D'\times D'}. \tag{C.62}$$

$\mathbf{\Phi}$ and $\mathbf{\Omega}$ are correlation matrices and are positive definite. When $D' = 1$, it becomes the multivariate Gaussian distribution.

- Normalization constant:

$$C_{\mathcal{N}}(\mathbf{\Phi}, \mathbf{\Omega}) \triangleq (2\pi)^{-\frac{DD'}{2}} |\mathbf{\Omega}|^{-\frac{D}{2}} |\mathbf{\Phi}|^{-\frac{D'}{2}}. \tag{C.63}$$

- Mean:

$$\mathbb{E}_{(\mathbf{X})}[\mathbf{X}] = \mathbf{M}. \tag{C.64}$$

- Variance:

$$\mathbb{E}_{(\mathbf{X})}[\mathbf{X}\mathbf{X}^{\mathsf{T}}] = \mathbf{\Phi}, \tag{C.65}$$

$$\mathbb{E}_{(\mathbf{X})}[\mathbf{X}^{\mathsf{T}}\mathbf{X}] = \mathbf{\Omega}. \tag{C.66}$$

- Mode:

$$\mathbf{M}. \tag{C.67}$$

C.10 Laplace distribution

- pdf:

$$\mathrm{Lap}(x|\mu, \beta) \triangleq C_{\mathrm{Lap}}(\beta) \exp\left(-\frac{|x-\mu|}{\beta}\right) = C_{\mathrm{Lap}}(\beta) \begin{cases} \exp\left(-\frac{x-\mu}{\beta}\right) & \text{if } x \geq u, \\ \exp\left(-\frac{\mu-x}{\beta}\right) & \text{if } x < u, \end{cases} \tag{C.68}$$

where

$$x \in \mathbb{R}, \tag{C.69}$$

and

$$\mu \in \mathbb{R}, \quad \beta > 0. \tag{C.70}$$

- Normalization constant:

$$C_{\text{Lap}}(\beta) \triangleq \frac{1}{2\beta}. \tag{C.71}$$

- Mean:

$$\mathbb{E}_{(x)}[x] = \mu. \tag{C.72}$$

- Mode:

$$\mu. \tag{C.73}$$

Note that the probabilistic distribution function is not continuous at μ, and it is not differentiable.

C.11 Gamma distribution

A gamma distribution is used as a prior/posterior distribution of precision parameter r of a Gaussian distribution.

- pdf:

$$\text{Gam}(y|\alpha, \beta) \triangleq C_{\text{Gam}}(\alpha, \beta) y^{\alpha-1} \exp(-\beta y), \tag{C.74}$$

  where

$$y > 0, \tag{C.75}$$

  and

$$\alpha, \beta > 0. \tag{C.76}$$

- Normalization constant:

$$C_{\text{Gam}}(\alpha, \beta) \triangleq \frac{\beta^{\alpha}}{\Gamma(\alpha)}. \tag{C.77}$$

- Mean:

$$\mathbb{E}_{(y)}[y] = \frac{\alpha}{\beta}. \tag{C.78}$$

- Variance:

$$\mathbb{E}_{(y)}[y^2] = \frac{\alpha}{\beta^2}. \tag{C.79}$$

- Mode:

$$\begin{aligned} \frac{d}{dy}\text{Gam}(y|\alpha, \beta) &= C_{\text{Gam}}(\alpha, \beta)\,(\alpha - 1 - \beta y)\, y^{\alpha-2} \exp(-\beta y) = 0 \\ &\Rightarrow \frac{\alpha - 1}{\beta}, \quad \alpha > 1. \end{aligned} \tag{C.80}$$

The shape of the gamma distribution is not symmetric, and the mode and mean values are different.

To make the notation consistent with the Wishart distribution in Appendix C.14, we also use the following definition for the gamma distribution, with $\alpha \rightarrow \frac{\phi}{2}$ and $\beta \rightarrow \frac{r^0}{2}$ in the original gamma distribution defined in Eq. (C.74):

- pdf (with $\frac{1}{2}$ factor):

$$\begin{aligned} \mathrm{Gam}_2(y|\phi, r^0) &\triangleq \mathrm{Gam}\left(y \middle| \frac{\phi}{2}, \frac{r^0}{2}\right) \\ &= C_{\mathrm{Gam}_2}\left(\phi, r^0\right) y^{\frac{\phi}{2}-1} \exp\left(-\frac{r^0 y}{2}\right). \end{aligned} \tag{C.81}$$

- Normalization constant (with $\frac{1}{2}$ factor):

$$C_{\mathrm{Gam}_2}(\phi, r^0) \triangleq \frac{\left(\frac{r^0}{2}\right)^{\frac{\phi}{2}}}{\Gamma\left(\frac{\phi}{2}\right)}. \tag{C.82}$$

- Mean (with $\frac{1}{2}$ factor):

$$\mathbb{E}_{(y)}[y] = \frac{\frac{\phi}{2}}{\frac{r^0}{2}} = \frac{\phi}{r^0}. \tag{C.83}$$

- Variance (with $\frac{1}{2}$ factor):

$$\mathbb{E}_{(y)}[y^2] = \frac{\frac{\phi}{2}}{\left(\frac{r^0}{2}\right)^2} = \frac{2\phi}{\left(r^0\right)^2}. \tag{C.84}$$

- Mode (with $\frac{1}{2}$ factor):

$$\frac{\frac{\phi}{2} - 1}{\frac{r^0}{2}} = \frac{\phi - 2}{r^0}, \quad \phi > 2. \tag{C.85}$$

It is shown in Appendix C.14 that this gamma distribution ($\mathrm{Gam}_2(\cdot)$) with $\frac{1}{2}$ factor is equivalent to the Wishart distribution with the number of dimensions 1 ($D = 1$).

C.12 Inverse gamma distribution

An inverse gamma distribution is used as a prior/posterior distribution of variance parameter Σ. It can be obtained by simply replacing y in a gamma distribution by $\frac{1}{y}$ with $f(y) = g(y') \left|\frac{dy'}{dy}\right|$.

- pdf:

$$\begin{aligned} \mathrm{IGam}(y|\alpha, \beta) &= \mathrm{Gam}\left(\frac{1}{y} \middle| \alpha, \beta\right) \left|\frac{d(y^{-1})}{dy}\right| \\ &= C_{\mathrm{Gam}}(\alpha, \beta) \left(\frac{1}{y}\right)^{\alpha-1} \exp\left(-\frac{\beta}{y}\right) y^{-2} \\ &\triangleq C_{\mathrm{IGam}}(\alpha, \beta) y^{-\alpha-1} \exp\left(-\frac{\beta}{y}\right), \end{aligned} \tag{C.86}$$

where

$$y > 0, \tag{C.87}$$

and

$$\alpha, \beta > 0. \tag{C.88}$$

- Normalization constant:

$$C_{\mathrm{IGam}}(\alpha, \beta) = C_{\mathrm{Gam}}(\alpha, \beta) \triangleq \frac{\beta^\alpha}{\Gamma(\alpha)}. \tag{C.89}$$

- Mean:

$$\mathbb{E}_{(y)}[y] = \frac{\beta}{\alpha - 1}, \quad \alpha > 1. \tag{C.90}$$

- Mode:

$$\frac{\beta}{\alpha + 1}. \tag{C.91}$$

C.13 Gaussian–gamma distribution

We provide a Gaussian–gamma distribution, which is used as a conjugate prior distribution of a Gaussian distribution with mean μ and precision r. This is also known as the normal-gamma distribution. Note that we use the gamma distribution defined in Eq. (C.81) instead of the original definition of the gamma distribution in Eq. (C.74).

- pdf:

$$\begin{aligned}
&\mathcal{N}\mathrm{Gam}(\mu, r|\mu^0, \phi^\mu, r^0, \phi^r) \\
&\triangleq \mathcal{N}(\mu|\mu^0, (r\phi^\mu)^{-1})\mathrm{Gam}_2\left(r \middle| \phi^r, r^0\right) \\
&\triangleq C_{\mathcal{N}\mathrm{Gam}}(\phi^\mu, r^0, \phi^r) r^{\frac{\phi^r - 1}{2}} \exp\left(-\frac{r^0 r}{2} - \frac{\phi^\mu r(\mu - \mu^0)^2}{2}\right).
\end{aligned} \tag{C.92}$$

Note that the power of r is $\frac{\phi^r - 1}{2}$, which is different from the original gamma distribution definition in Section (C.11), because it also considers an $r^{\frac{1}{2}}$ factor in the Gaussian distribution.

- Normalization constant:

$$C_{\mathcal{N}\mathrm{Gam}}(r, \alpha, \beta) \triangleq \frac{\sqrt{\phi^\mu}\left(\frac{r^0}{2}\right)^{\frac{\phi^r}{2}}}{\sqrt{2\pi}\,\Gamma\left(\frac{\phi^r}{2}\right)}. \tag{C.93}$$

- Mean:

$$\mathbb{E}_{(\mu, r)}[\{\mu, r\}] = \left\{\mu^0, \frac{\phi^r}{r^0}\right\}. \tag{C.94}$$

- Mode:

$$\left\{\mu^0, \frac{\phi^r - 2}{r^0}\right\}, \quad \phi^r > 2. \tag{C.95}$$

The means and modes of mean and precision parameters are the same as those of the Gaussian and gamma distributions, respectively.

C.14 Wishart distribution

- pdf:

$$\mathcal{W}(\mathbf{Y}|\mathbf{R}^0, \phi) \triangleq C_{\mathcal{W}}(\mathbf{R}^0, \phi)|\mathbf{Y}|^{\frac{\phi - D - 1}{2}} \exp\left(-\frac{1}{2}\mathrm{tr}\left[\mathbf{R}^0\mathbf{Y}\right]\right), \tag{C.96}$$

 where

$$\mathbf{Y} \in \mathbb{R}^{D \times D}, \tag{C.97}$$

 and

$$\mathbf{R}^0 \in \mathbb{R}^{D \times D}, \quad \phi > D - 1. \tag{C.98}$$

 $\mathbf{Y}$ and $\mathbf{R}^0$ are positive definite.
- Normalization constant:

$$C_{\mathcal{W}}(\mathbf{R}^0, \phi) \triangleq \frac{|\mathbf{R}^0|^{\frac{\phi}{2}}}{(2)^{\frac{D\phi}{2}}\Gamma_D\left(\frac{\phi}{2}\right)}, \tag{C.99}$$

 where $\Gamma_D(\cdot)$ is a multivariate Gamma function.
- Mean:

$$\mathbb{E}_{(\mathbf{Y})}[\mathbf{Y}] = \phi(\mathbf{R}^0)^{-1}. \tag{C.100}$$

- Mode:

$$(\phi - D - 1)(\mathbf{R}^0)^{-1}, \quad \phi \geq D + 1. \tag{C.101}$$

When $D \to 1$, the Wishart distribution is equivalent to the gamma distribution defined in Eq. (C.81).

C.15 Gaussian–Wishart distribution

- pdf:

$$\begin{aligned}
&\mathcal{NW}(\boldsymbol{\mu}, \mathbf{R}|\boldsymbol{\mu}^0, \phi^\mu, \mathbf{R}^0, \phi^{\mathbf{R}}) \\
&\triangleq \mathcal{N}(\boldsymbol{\mu}|\boldsymbol{\mu}^0, (\phi^\mu\mathbf{R})^{-1})\mathcal{W}(\mathbf{R}|\mathbf{R}^0, \phi^{\mathbf{R}}) \\
&\triangleq C_{\mathcal{NW}}(\phi^\mu, \mathbf{R}^0, \phi^{\mathbf{R}})|\mathbf{R}|^{\frac{\phi^{\mathbf{R}} - D}{2}} \\
&\quad \times \exp\left(-\frac{1}{2}\mathrm{tr}\left[\mathbf{R}^0\mathbf{R}\right] - \frac{\phi^\mu}{2}(\boldsymbol{\mu} - \boldsymbol{\mu}^0)^\intercal\mathbf{R}(\boldsymbol{\mu} - \boldsymbol{\mu}^0)\right).
\end{aligned} \tag{C.102}$$

Note that when $D \to 1$, Gaussian–Wishart distribution $\mathcal{NW}(\boldsymbol{\mu}, \mathbf{R}|\boldsymbol{\mu}^0, \phi^{\mu}, \mathbf{R}^0, \phi^{\mathbf{R}})$ approaches Gaussian–gamma distribution $\mathcal{N}\mathrm{Gam}(\mu, r|\mu^0, \phi^{\mu}, r^0, \phi^r)$ in Appendix C.13.

- Normalization constant:

$$C_{\mathcal{NW}}(\phi^{\mu}, \mathbf{R}^0, \phi^R) \triangleq (2\pi)^{\frac{D}{2}} (\phi^{\mu})^{-\frac{D}{2}} \frac{|\mathbf{R}^0|^{\frac{\phi^R}{2}}}{(2)^{\frac{D\phi^R}{2}} \Gamma_D\left(\frac{\phi^R}{2}\right)}. \tag{C.103}$$

- Mean:

$$\mathbb{E}_{(\boldsymbol{\mu},\mathbf{R})}[\{\boldsymbol{\mu}, \mathbf{R}\}] = \left\{\boldsymbol{\mu}^0, \phi^R(\mathbf{R}^0)^{-1}\right\}. \tag{C.104}$$

- Mode:

$$\left\{\boldsymbol{\mu}^0, \left(\phi^R - D - 1\right)(\mathbf{R}^0)^{-1}\right\}, \quad \phi^R \geq D + 1. \tag{C.105}$$

Again, when $D \to 1$, these modes are equivalent to the modes of the Gaussian–gamma distribution in Appendix C.13.

C.16 Student's *t*-distribution

- pdf:

$$\mathrm{St}(x|\mu, \lambda, \kappa) \triangleq C_{\mathrm{St}} \left(1 + \frac{1}{\kappa\lambda}(x - \mu)^2\right)^{-\frac{\kappa+1}{2}}, \tag{C.106}$$

where

$$x \in \mathbb{R}, \tag{C.107}$$

and

$$\mu \in \mathbb{R}, \quad \kappa > 0, \quad \lambda > 0. \tag{C.108}$$

- Normalization constant:

$$C_{\mathrm{St}}(\kappa, \lambda) \triangleq \frac{\Gamma\left(\frac{\kappa+1}{2}\right)}{\Gamma\left(\frac{\kappa}{2}\right)\Gamma\left(\frac{1}{2}\right)} \left(\frac{1}{\kappa\lambda}\right)^{\frac{1}{2}}, \tag{C.109}$$

- Mean:

$$\mathbb{E}_{(x)}[x] = \mu, \tag{C.110}$$

- Mode:

$$\mu. \tag{C.111}$$

Parameter κ is called the degrees of freedom, and if κ is large, the distribution approaches the Gaussian distribution. Note that Student's t-distribution is not included in the exponential family (Section 2.1.3).

References

Abu-Mostafa, Y. S. (1989), "The Vapnik–Chervonenkis dimension: information versus complexity in learning," *Neural Computation* **1**, 312–317.

Akaike, H. (1974), "A new look at the statistical model identification," *IEEE Transactions on Automatic Control* **19**(6), 716–723.

Akaike, H. (1980), "Likelihood and the Bayes procedure," *in* J. M. Bernardo, M. H. DeGroot, D. V. Lindley & A. F. M. Smith, eds, *Bayesian Statistics*, University Press, Valencia, Spain, pp. 143–166.

Akita, Y., & Kawahara, T. (2004), "Language model adaptation based on PLSA of topics and speakers," Proceedings of Annual Conference of International Speech Communication Association (INTERSPEECH), pp. 1045–1048.

Aldous, D. (1985), "Exchangeability and related topics," *École d'Été de Probabilités de Saint-Flour XIII1983*, pp. 1–198.

Anastasakos, T., McDonough, J., Schwartz, R., & Makhoul, J. (1996), "A compact model for speaker-adaptive training," Proceedings of International Conference on *Spoken Language Processing (ICSLP)*, pp. 1137–1140.

Anguera Miro, X., Bozonnet, S., Evans, N., *et al.* (2012), "Speaker diarization: A review of recent research," *IEEE Transactions on Audio, Speech, and Language Processing* **20**(2), 356–370.

Antoniak, C. E. (1974), "Mixtures of Dirichlet processes with applications to Bayesian nonparametric problems," *Annals of Statistics* **2**(6), 1152–1174.

Attias, H. (1999), "Inferring parameters and structure of latent variable models by variational Bayes," Proceedings of the Conference on *Uncertainty in Artificial Intelligence (UAI)*, pp. 21–30.

Axelrod, S., Gopinath, R., & Olsen, P. (2002), "Modeling with a subspace constraint on inverse covariance matrices," Proceedings of International Conference on *Spoken Language Processing (ICSLP)*, pp. 2177–2180.

Bahl, L. R., Brown, P. F., de Souza, P. V., & Mercer, R. L. (1986), "Maximum mutual information estimation of hidden Markov model parameters for speech recognition," Proceedings of International Conference on *Acoustics, Speech, and Signal Processing (ICASSP)*, pp. 49–52.

Barber, D. (2012), *Bayesian Reasoning and Machine Learning*, Cambridge University Press.

Barker, J., Vincent, E., Ma, N., Christensen, H., & Green, P. (2013), "The PASCAL CHiME speech separation and recognition challenge," *Computer Speech and Language* **27**, 621–633.

Baum, L. E., Petrie, T., Soules, G., & Weiss, N. (1970), "A maximization technique occurring in the statistical analysis of probabilistic functions of Markov chains," *The Annals of Mathematical Statistics*, pp. 164–171.

Beal, M. J. (2003), Variational algorithms for approximate Bayesian inference, PhD thesis, University of London.

Beal, M. J., Ghahramani, Z., & Rasmussen, C. E. (2002), "The infinite hidden Markov model," *Advances in Neural Information Processing Systems* **14**, 577–584.

Bellegarda, J. (2004), "Statistical language model adaptation: review and perspectives," *Speech Communication* **42**(1), 93–108.

Bellegarda, J. R. (2000), "Exploiting latent semantic information in statistical language modeling," *Proceedings of the IEEE* **88**(8), 1279–1296.

Bellegarda, J. R. (2002), "Fast update of latent semantic spaces using a linear transform framework," Proceedings of International Conference on *Acoustics, Speech, and Signal Processing (ICASSP)*, pp. 769–772.

Bengio, Y., Ducharme, R., Vincent, P., & Jauvin, C. (2003), "A neural probabilistic language model," *Journal of Machine Learning Research* **3**, 1137–1155.

Berger, J. O. (1985), *Statistical Decision Theory and Bayesian Analysis*, Second Edition, Springer-Verlag.

Bernardo, J. M., & Smith, A. F. M. (2009), *Bayesian Theory*, Wiley.

Berry, M. W., Dumais, S. T., & O'Brien, G. W. (1995), "Using linear algebra for intelligent information retrieval," *SIAM Review* **37**(4), 573–595.

Bilmes, J. A. (1998), A gentle tutorial of the EM algorithm and its application to parameter estimation for Gaussian mixture and hidden Markov models, Technical Report TR-97-021, International Computer Science Institute.

Bilmes, J., & Zweig, G. (2002), "The graphical models toolkit: An open source software system for speech and time-series processing," Proceedings of International Conference on *Acoustics, Speech, and Signal Processing (ICASSP)*, pp. 3916–3919.

Bishop, C. M. (2006), *Pattern Recognition and Machine Learning*, Springer.

Blackwell, D., & MacQueen, J. B. (1973), "Ferguson distribution via Pólya urn schemes," *The Annals of Statistics* **1**, 353–355.

Blei, D., Griffiths, T., & Jordan, M. (2010), "The nested Chinese restaurant process and Bayesian nonparametric inference of topic hierarchies," *Journal of the ACM* **57**(2), article 7.

Blei, D., Griffiths, T., Jordan, M., & Tenenbaum, J. (2004), "Hierarchical topic models and the nested Chinese restaurant process," *Advances in Neural Information Processing Systems* **16**, 17–24.

Blei, D. M., Ng, A. Y., & Jordan, M. I. (2003), "Latent Dirichlet allocation," *Journal of Machine Learning Research* **3**, 993–1022.

Brants, T., Popat, A. C., Xu, P., Och, F. J., & Dean, J. (2007), "Large language models in machine translation," Proceedings of the 2007 Joint Conference on *Empirical Methods in Natural Language Processing and Computational Natural Language Learning (EMNLP-CoNLL)*, Association for Computational Linguistics, pp. 858–867.

Brill, E., & Moore, R. C. (2000), "An improved error model for noisy channel spelling correction," Proceedings of the 38th Annual Meeting of Association for Computational Linguistics, Association for Computational Linguistics, pp. 286–293.

Brown, P., Desouza, P., Mercer, R., Pietra, V., & Lai, J. (1992), "Class-based n-gram models of natural language," *Computational Linguistics* **18**(4), 467–479.

Brown, P. F., Cocke, J., Pietra, S. A. D., *et al.* (1990), "A statistical approach to machine translation," *Computational Linguistics* **16**(2), 79–85.

Campbell, W. M., Sturim, D. E., & Reynolds, D. A. (2006), "Support vector machines using GMM supervectors for speaker verification," *Signal Processing Letters, IEEE* **13**(5), 308–311.

Chen, K.-T., Liau, W.-W., Wang, H.-M., & Lee, L.-S. (2000), "Fast speaker adaptation using eigenspace-based maximum likelihood linear regression," Proceedings of International Conference on *Spoken Language Processing (ICSLP)*, pp. 742–745.

Chen, S. F. (2009), "Shrinking exponential language models," in Proceedings of *Human Language Technologies*: The 2009 Annual Conference of the North American Chapter of the Association for Computational Linguistics, Association for Computational Linguistics, pp. 468–476.

Chen, S. F., & Goodman, J. (1999), "An empirical study of smoothing techniques for language modeling," *Computer Speech & Language* **13**(4), 359–393.

Chen, S., & Gopinath, R. (1999), "Model selection in acoustic modeling," Proceedings of European Conference on *Speech Communication and Technology (EUROSPEECH)*, pp. 1087–1090.

Chesta, C., Siohan, O., & Lee, C.-H. (1999), "Maximum a posteriori linear regression for hidden Markov model adaptation," Proceedings of European Conference on *Speech Communication and Technology (EUROSPEECH)*, pp. 211–214.

Chien, J.-T. (1999), "Online hierarchical transformation of hidden Markov models for speech recognition," *IEEE Transactions on Speech and Audio Processing* **7**(6), 656–667.

Chien, J.-T. (2002), "Quasi-Bayes linear regression for sequential learning of hidden Markov models," *IEEE Transactions on Speech and Audio Processing* **10**(5), 268–278.

Chien, J.-T. (2003), "Linear regression based Bayesian predictive classification for speech recognition," *IEEE Transactions on Speech and Audio Processing* **11**(1), 70–79.

Chien, J.-T., & Chueh, C.-H. (2011), "Dirichlet class language models for speech recognition," *IEEE Transactions on Audio, Speech, and Language Processing* **19**(3), 482–495.

Chien, J.-T., Huang, C.-H., Shinoda, K., & Furui, S. (2006), "Towards optimal Bayes decision for speech recognition," Proceedings of International Conference on *Acoustics, Speech, and Signal Processing (ICASSP)*, pp. 45–48.

Chien, J.-T., Lee, C.-H., & Wang, H.-C. (1997), "Improved Bayesian learning of hidden Markov models for speaker adaptation," Proceedings of International Conference on *Acoustics, Speech, and Signal Processing (ICASSP)*, pp. 1027–1030.

Chien, J. T., & Liao, G.-H. (2001), "Transformation-based Bayesian predictive classification using online prior evolution," *IEEE Transactions on Speech and Audio Processing* **9**(4), 399–410.

Chien, J.-T., & Wu, M.-S. (2008), "Adaptive Bayesian latent semantic analysis," *IEEE Transactions on Audio, Speech, and Language Processing* **16**(1), 198–207.

Chou, W., & Reichl, W., (1999), "Decision tree state tying based on penalized Bayesian information criterion," Proceedings of International Conference on *Acoustics, Speech, and Signal Processing (ICASSP)*, pp. 345–348.

Coccaro, N., & Jurafsky, D. (1998), "Towards better integration of semantic predictors in statistical language modeling," Proceedings of International Conference on *Spoken Language Processing (ICSLP)*, pp. 2403–2406.

Cournapeau, D., Watanabe, S., Nakamura, A., & Kawahara, T. (2010), "Online unsupervised classification with model comparison in the variational Bayes framework for voice activity detection," *IEEE Journal of Selected Topics in Signal Processing* **4**(6), 1071–1083.

Dahl, G. E., Yu, D., Deng, L., & Acero, A. (2012), "Context-dependent pre-trained deep neural networks for large-vocabulary speech recognition," *IEEE Transactions on Audio, Speech and Language Processing* **20**(1), 30–42.

Davis, S. B., & Mermelstein, P. (1980), "Comparison of parametric representations for monosyllabic word recognition in continuously spoken sentences," *IEEE Transactions on Acoustics, Speech, and Signal Processing* **28**(4), 357–366.

Dawid, A. P. (1981), "Some matrix-variate distribution theory: notational considerations and a Bayesian application," *Biometrika* **68**(1), 265–274.

De Bruijn, N. G. (1970), *Asymptotic Methods in Analysis*, Dover Publications.

Dehak, N., Kenny, P., Dehak, R., Dumouchel, P., & Ouellet, P. (2011), "Front-end factor analysis for speaker verification," *IEEE Transactions on Audio, Speech, and Language Processing* **19**(4), 788–798.

Delcroix, M., Nakatani, T., & Watanabe, S. (2009), "Static and dynamic variance compensation for recognition of reverberant speech with dereverberation preprocessing," *IEEE Transactions on Audio, Speech, and Language Processing* **17**(2), 324–334.

Dempster, A. P., Laird, N. M., & Rubin, D. B. (1976), "Maximum likelihood from incomplete data via the EM algorithm," *Journal of Royal Statistical Society B* **39**, 1–38.

Digalakis, V., & Neumeyer, L. (1996), "Speaker adaptation using combined transformation and Bayesian methods," *IEEE Transactions on Speech and Audio Processing* **4**, 294–300.

Digalakis, V., Ritischev, D., & Neumeyer, L. (1995), "Speaker adaptation using constrained reestimation of Gaussian mixtures," *IEEE Transactions on Speech and Audio Processing* **3**, 357–366.

Ding, N., & Ou, Z. (2010), "Variational nonparametric Bayesian hidden Markov model," Proceedings of International Conference on *Acoustics, Speech, and Signal Processing (ICASSP)*, pp. 2098–2101.

Droppo, J., Acero, A., & Deng, L. (2002), "Uncertainty decoding with SPLICE for noise robust speech recognition," Proceedings of International Conference on *Acoustics, Speech, and Signal Processing* (*ICASSP*)', Vol. 1, pp. I–57.

Federico, M. (1996), "Bayesian estimation methods of *n*-gram language model adaptation," Proceedings of International Conference on *Spoken Language Processing (ICSLP)*, pp. 240–243.

Ferguson, T. (1973), "A Bayesian analysis of some nonparametric problems," *The Annals of Statistics* **1**, 209–230.

Fosler, E., & Morris, J. (2008), "Crandem systems: Conditional random field acoustic models for hidden Markov models," Proceedings of International Conference on *Acoustics, Speech, and Signal Processing (ICASSP)*, pp. 4049–4052.

Fox, E. B., Sudderth, E. B., Jordan, M. I., & Willsky, A. S. (2008), "An HDP-HMM for systems with state persistence," Proceedings of International Conference on *Machine Learning (ICML)*, pp. 312–319.

Fukunaga, K. (1990), *Introduction to Statistical Pattern Recognition*, Academic Press.

Furui, S. (1981), "Cepstral analysis technique for automatic speaker verification," *IEEE Transactions on Acoustics, Speech and Signal Processing* **29**(2), 254–272.

Furui, S. (1986), "Speaker independent isolated word recognition using dynamic features of speech spectrum," *IEEE Transactions on Acoustics, Speech and Signal Processing* **34**, 52–59.

Furui, S. (2010), "History and development of speech recognition," in *Speech Technology*, F. Chen and K. Jokinen, eds., Springer, pp. 1–18.

Furui, S., Maekawa, K., & H. Isahara, M. (2000), "A Japanese national project on spontaneous speech corpus and processing technology," Proceedings of ASR'00, pp. 244–248.

Gales, M. (1998), "Maximum likelihood linear transformations for HMM-based speech recognition," *Computer Speech and Language* **12**, 75–98.

Gales, M., Center, I., & Heights, Y. (2000), "Cluster adaptive training of hidden Markov models," *IEEE Transactions on Speech and Audio Processing* **8**(4), 417–428.

Gales, M. J. F. (1999), "Semi-tied covariance matrices for hidden Markov models," *IEEE Transactions on Speech and Audio Processing* **7**(3), 272–281.

Gales, M. J. F., & Woodland, P. C. (1996), Variance compensation within the MLLR framework, Technical Report 242, Cambridge University Engineering Department.

Gales, M., Watanabe, S., & Fossler-Lussier, E. (2012), "Structured discriminative models for speech recognition," *IEEE Signal Processing Magazine* **29**(6), 70–81.

Ganapathiraju, A., Hamaker, J., & Picone, J. (2004), "Applications of support vector machines to speech recognition," *IEEE Transactions on Signal Processing* **52**(8), 2348–2355.

Gaussier, E., & Goutte, C. (2005), "Relation between PLSA and NMF and implications," Proceedings of the 28th Annual International ACM SIGIR Conference on *Research and Development in Information Retrieval*, ACM, pp. 601–602.

Gauvain, J.-L., & Lee, C.-H. (1991), "Bayesian learning of Gaussian mixture densities for hidden Markov models," Proceedings of DARPA *Speech and Natural Language Workshop*, pp. 272–277.

Gauvain, J.-L., & Lee, C.-H. (1994), "Maximum a posteriori estimation for multivariate Gaussian mixture observations of Markov chains," *IEEE Transactions on Speech and Audio Processing* **2**, 291–298.

Gelman, A., Carlin, J. B., Stern, H. S., *et al.* (2013), *Bayesian Data Analysis*, CRC Press.

Geman, S., & Geman, D. (1984), "Stochastic relaxation, Gibbs distributions, and the Bayesian restoration of images," *IEEE Transactions on Pattern Analysis and Machine Intelligence* **6**(1), 721–741.

Genkin, A., Lewis, D. D., & Madigan, D. (2007), "Large-scale Bayesian logistic regression for text categorization," *Technometrics* **49**(3), 291–304.

Ghahramani, Z. (1998), "Learning dynamic Bayesian networks," in *Adaptive Processing of Sequences and Data Structures*, Springer, pp. 168–197.

Ghahramani, Z. (2004), "Unsupervised learning," *Advanced Lectures on Machine Learning*, pp. 72–112.

Ghosh, J. K., Delampady, M., & Samanta, T. (2007), *An Introduction to Bayesian Analysis: Theory and Methods*, Springer.

Gildea, D., & Hofmann, T. (1999), "Topic-based language models using EM," Proceedings of European Conference on *Speech Communication and Technology (EUROSPEECH)*, pp. 2167–2170.

Gilks, W. R., Richardson, S., & Spiegelhalter, D. J. (1996), *Markov Chain Monte Carlo in Practice*, Chapman & Hall/CRC Interdisciplinary Statistics.

Gish, H., Siu, M.-h., Chan, A., & Belfield, W. (2009), "Unsupervised training of an HMM-based speech recognizer for topic classification," Proceedings of Annual Conference of International Speech Communication Association (INTERSPEECH), pp. 1935–1938.

Glass, J. (2003), "A probabilistic framework for segment-based speech recognition," *Computer Speech & Language* **17**(2-3), 137–152.

Goel, V., & Byrne, W. (2000), "Minimum Bayes-risk automatic speech recognition," *Computer Speech and Language* **14**, 115–135.

Goldwater, S. (2007), Nonparametric Bayesian models of lexical acquisition, PhD thesis, Brown University.

Goldwater, S., & Griffiths, T. (2007), "A fully Bayesian approach to unsupervised part-of-speech tagging," Proceedings of Annual Meeting of the Association of Computational Linguistics, pp. 744–751.

Goldwater, S., Griffiths, T., & Johnson, M. (2009), "A Bayesian framework for word segmentation: Exploring the effects of context," *Cognition* **112**(1), 21–54.

Goldwater, S., Griffiths, T. L., & Johnson, M. (2006), "Interpolating between types and tokens by estimating power-law generators," *Advances in Neural Information Processing Systems* **18**.

Good, I. J. (1953), "The population frequencies of species and the estimation of populations," *Biometrika* **40**, 237–264.

Grézl, F., Karafiát, M., Kontár, S., & Cernocky, J. (2007), "Probabilistic and bottle-neck features for LVCSR of meetings," Proceedings of International Conference on *Acoustics, Speech, and Signal Processing (ICASSP)*, pp. 757–760.

Griffiths, T., & Ghahramani, Z. (2005), Infinite latent feature models and the Indian buffet process, Technical Report, Gatsby Unit.

Griffiths, T., & Steyvers, M. (2004), "Finding scientific topics," in *Proceedings of the National Academy of Sciences*, **101** Suppl. 1, 5228–5235.

Gunawardana, A., Mahajan, M., Acero, A., & Platt, J. C. (2005), "Hidden conditional random fields for phone classification," Proceedings of Annual Conference of International Speech Communication Association (INTERSPEECH), pp. 1117–1120.

Haeb-Umbach, R., & Ney, H. (1992), "Linear discriminant analysis for improved large vocabulary continuous speech recognition," International Conference on *Acoustics, Speech, and Signal Processing (ICASSP)*, Vol. 1, pp. 13–16.

Hahm, S. J., Ogawa, A., Fujimoto, M., Hori, T., & Nakamura, A. (2012), "Speaker adaptation using variational Bayesian linear regression in normalized feature space," in Proceedings of Annual Conference of International Speech Communication Association (INTERSPEECH), pp. 803–806.

Hashimoto, K., Nankaku, Y., & Tokuda, K. (2009), "A Bayesian approach to hidden semi-Markov model based speech synthesis," in Proceedings of Annual Conference of International Speech Communication Association (INTERSPEECH), pp. 1751–1754.

Hashimoto, K., Zen, H., Nankaku, Y., Lee, A., & Tokuda, K. (2008), "Bayesian context clustering using cross valid prior distribution for HMM-based speech recognition," Proceedings of Annual Conference of International Speech Communication Association (INTERSPEECH), pp. 936–939.

Hashimoto, K., Zen, H., Nankaku, Y., Masuko, T., & Tokuda, K. (2009), "A Bayesian approach to HMM-based speech synthesis," Proceedings of International Conference on *Acoustics, Speech, and Signal Processing (ICASSP)* 2009, pp. 4029–4032.

Hastings, W. K. (1970), "Monte Carlo sampling methods using Markov chains and their applications," *Biometrika* **57**, 97–109.

Heigold, G., Ney, H., Schluter, R., & Wiesler, S. (2012), "Discriminative training for automatic speech recognition: Modeling, criteria, optimization, implementation, and performance," *IEEE Signal Processing Magazine* **29**(6), 58–69.

Hermansky, H. (1990), "Perceptual linear predictive (PLP) analysis of speech," *Journal of the Acoustic Society of America* **87**(4), 1738–1752.

Hermansky, H., Ellis, D., & Sharma, S. (2000), "Tandem connectionist feature extraction for conventional HMM systems," Proceedings of International Conference on *Acoustics, Speech, and Signal Processing (ICASSP)*, pp. 1635–1638.

Hinton, G., Deng, L., Yu, D., *et al.* (2012), "Deep neural networks for acoustic modeling in speech recognition," *IEEE Signal Processing Magazine* **29**(6), 82–97.

Hinton, G., Osindero, S., & Teh, Y. (2006), "A fast learning algorithm for deep belief nets," *Neural Computation* **18**, 1527–1554.

Hofmann, T. (1999*a*), "Probabilistic latent semantic analysis," *Proceedings of Uncertainty in Artificial Intelligence*, pp. 289–296.

Hofmann, T. (1999*b*), "Probabilistic latent semantic indexing," Proceedings of the Annual International ACM SIGIR Conference on *Research and Development in Information Retrieval*, pp. 50–57.

Hofmann, T. (2001), "Unsupervised learning by probabilistic latent semantic analysis," *Machine Learning* **42**(1-2), 177–196.

Hori, T., & Nakamura, A. (2013), "Speech recognition algorithms using weighted finite-state transducers," Synthesis Lectures on *Speech and Audio Processing* **9**(1), 1–162.

Hu, R., & Zhao, Y. (2007), "Knowledge-based adaptive decision tree state tying for conversational speech recognition," *IEEE Transactions on Audio, Speech, and Language Processing* **15**(7), 2160–2168.

Huang, S., & Renals, S. (2008), "Unsupervised language model adaptation based on topic and role information in multiparty meeting," Proceedings of Annual Conference of International Speech Communication Association (INTERSPEECH), pp. 833–836.

Huang, X. D., Acero, A., & Hon, H. W. (2001), *Spoken Language Processing: A Guide to Theory, Algorithm, and System Development*, Prentice Hall.

Huang, X. D., Ariki, Y., & Jack, M. A. (1990), *Hidden Markov Models for Speech Recognition*, Edinburgh University Press.

Huo, Q., & Lee, C.-H. (1997), "On-line adaptive learning of the continuous density hidden Markov model based on approximate recursive Bayes estimate," *IEEE Transactions on Speech and Audio Processing* **5**(2), 161–172.

Huo, Q., & Lee, C.-H. (2000), "A Bayesian predictive classification approach to robust speech recognition," *IEEE Transactions on Speech and Audio Processing* **8**, 200–204.

Ishiguro, K., Yamada, T., Araki, S., Nakatani, T., & Sawada, H. (2012), "Probabilistic speaker diarization with bag-of-words representations of speaker angle information," *IEEE Transactions on Audio, Speech, and Language Processing* **20**(2), 447–460.

Jansen, A., Dupoux, E., Goldwater, S., *et al.* (2013), "A summary of the 2012 JHU CLSP workshop on zero resource speech technologies and models of early language acquisition," Proceedings of International Conference on *Acoustics, Speech, and Signal Processing (ICASSP)*, pp. 8111–8115.

Jelinek, F. (1976), "Continuous speech recognition by statistical methods," *Proceedings of the IEEE* **64**(4), 532–556.

Jelinek, F. (1997), *Statistical Methods for Speech Recognition*, MIT Press.

Jelinek, F., & Mercer, R. L. (1980), "Interpolated estimation of Markov source parameters from sparse data," Proceedings of the Workshop on *Pattern Recognition in Practice*, pp. 381–397.

Ji, S., Xue, Y., & Carin, L. (2008), "Bayesian compressive sensing," *IEEE Transactions on Signal Processing* **56**(6), 2346–2356.

Jiang, H., Hirose, K., & Huo, Q. (1999), "Robust speech recognition based on a Bayesian prediction approach," *IEEE Transactions on Speech and Audio Processing* **7**, 426–440.

Jitsuhiro, T., & Nakamura, S. (2004), "Automatic generation of non-uniform HMM structures based on variational Bayesian approach," Proceedings of International Conference on *Acoustics, Speech, and Signal Processing (ICASSP)*, pp. 805–808.

Joachims, T. (2002), "Learning to classify text using support vector machines: Methods, theory, and algorithms," *Computational Linguistics* **29**(4), 656–664.

Jordan, M., Ghahramani, Z., Jaakkola, T., & Saul, L. (1999), "An introduction to variational methods for graphical models," *Machine Learning* **37**(2), 183–233.

Juang, B.-H., & Rabiner, L. (1990), "The segmental K-means algorithm for estimating parameters of hidden Markov models," *IEEE Transactions on Acoustics, Speech and Signal Processing* **38**(9), 1639–1641.

Juang, B., & Katagiri, S. (1992), "Discriminative learning for minimum error classification," *IEEE Transactions on Signal Processing* **40**(12), 3043–3054.

Jurafsky, D. (2014), "From languages to information," *http://www.stanford.edu/class/cs124/lec/languagemodeling.pdf.*

Jurafsky, D., & Martin, J. H. (2000), *Speech and Language Processing: An Introduction to Natural Language Processing, Computational Linguistics, and Speech Recognition*, Prentice Hall.

Kass, R. E., & Raftery, A. E. (1993), Bayes factors and model uncertainty, Technical Report 254, Department of Statistics, University of Washington.

Kass, R. E., & Raftery, A. E. (1995), "Bayes factors," *Journal of the American Statistical Association* **90**(430), 773–795.

Katz, S. (1987), "Estimation of probabilities from sparse data for the language model component of a speech recognizer," *IEEE Transactions on Acoustics, Speech, and Signal Processing* **35**(3), 400–401.

Kawabata, T., & Tamoto, M. (1996), "Back-off method for n-gram smoothing based on binomial posteriori distribution," Proceedings of International Conference on *Acoustics, Speech, and Signal Processing (ICASSP)*, Vol. 1, pp. 192–195.

Kenny, P. (2010), "Bayesian speaker verification with heavy tailed priors," Keynote Speech, Odyssey Speaker and Language Recognition Workshop.

Kenny, P., Boulianne, G., Ouellet, P., & Dumouchel, P. (2007), "Joint factor analysis versus eigenchannels in speaker recognition," *IEEE Transactions on Audio, Speech, and Language Processing* **15**(4), 1435–1447.

Kingsbury, B., Sainath, T. N., & Soltau, H. (2012), "Scalable minimum Bayes risk training of deep neural network acoustic models using distributed Hessian-free optimization," Proceedings of Annual Conference of International Speech Communication Association (INTERSPEECH), pp. 10–13.

Kinnunen, T., & Li, H. (2010), "An overview of text-independent speaker recognition: from features to supervectors," *Speech Communication* **52**(1), 12–40.

Kita, K. (1999), *Probabilistic Language Models*, University of Tokyo Press (in Japanese).

Kneser, R., & Ney, H. (1995), "Improved backing-off for *m*-gram language modeling," Proceedings of International Conference on *Acoustics, Speech, and Signal Processing (ICASSP)*, pp. 181–184.

Kneser, R., Peters, J., & Klakow, D. (1997), "Language model adaptation using dynamic marginals," Proceedings of European Conference on Speech Communication and Technology (EUROSPEECH), pp. 1971–1974.

Kolossa, D., & Haeb-Umbach, R. (2011), *Robust Speech Recognition of Uncertain or Missing Data: Theory and Applications*, Springer.

Kubo, Y., Watanabe, S., Nakamura, A., & Kobayashi, T. (2010), "A regularized discriminative training method of acoustic models derived by minimum relative entropy discrimination," Proceedings of Annual Conference of International Speech Communication Association (INTERSPEECH), pp. 2954–2957.

Kudo, T. (2005), "Mecab: Yet another part-of-speech and morphological analyzer," *http://mecab.sourceforge. net/.*

Kuhn, R., & De Mori, R. (1990), "A cache-based natural language model for speech recognition," *IEEE Transactions on Pattern Analysis and Machine Intelligence* **12**(6), 570–583.

Kuhn, R., Junqua, J., Ngyuen, P., & Niedzielski, N. (2000), "Rapid speaker adaptation in eigenvoice space," *IEEE Transactions on Speech and Audio Processing* **8**(6), 695–707.

Kullback, S., & Leibler, R. A. (1951), "On information and sufficiency," *Annals of Mathematical Statistics* **22**(1), 79–86.

Kwok, J. T.-Y. (2000), "The evidence framework applied to support vector machines," *IEEE Transactions on Neural Networks* **11**(5), 1162–1173.

Kwon, O., Lee, T.-W., & Chan, K. (2002), "Application of variational Bayesian PCA for speech feature extraction," Proceedings of International Conference on *Acoustics, Speech, and Signal Processing (ICASSP)*, Vol. 1, pp. 825–828.

Lafferty, J., McCallum, A., & Pereira, F. (2001), "Conditional random fields: Probabilistic models for segmenting and labeling sequence data," Proceedings of International Conference on *Machine Learning*, pp. 282–289.

Lamel, L., Gauvain, J.-L., & Adda, G. (2002), "Lightly supervised and unsupervised acoustic model training," *Computer Speech & Language* **16**(1), 115–129.

Lau, R., Rosenfeld, R., & Roukos, S. (1993), "Trigger-based language models: A maximum entropy approach," Proceedings of International Conference on *Acoustics, Speech, and Signal Processing (ICASSP)*, Vol. 2, IEEE, pp. 45–48.

Lee, C.-H., & Huo, Q. (2000), "On adaptive decision rules and decision parameter adaptation for automatic speech recognition," *Proceedings of the IEEE* **88**, 1241–1269.

Lee, C.-H., Lin, C.-H., & Juang, B.-H. (1991), "A study on speaker adaptation of the parameters of continuous density hidden Markov models," *IEEE Transactions on Acoustics, Speech, and Signal Processing* **39**, 806–814.

Lee, C.-Y. (2014), Discovering linguistic structures in speech: models and applications, PhD thesis, Massachusetts Institute of Technology.

Lee, C.-Y., & Glass., J. (2012), "A nonparametric Bayesian approach to acoustic model discovery," Proceedings of Annual Meeting of the Association for Computational Linguistics, pp. 40–49.

Lee, C.-Y., Zhang, Y., & Glass, J. (2013), "Joint learning of phonetic units and word pronunciations for ASR," Proceedings of the 2013 Conference on *Empirical Methods on Natural Language Processing (EMNLP)*, pp. 182–192.

Lee, D. D., & Seung, H. S. (1999), "Learning the parts of objects by non-negative matrix factorization," *Nature* **401**(6755), 788–791.

Leggetter, C. J., & Woodland, P. C. (1995), "Maximum likelihood linear regression for speaker adaptation of continuous density hidden Markov models," *Computer Speech and Language* **9**, 171–185.

Lewis, D. D. (1998), "Naive (Bayes) at forty: The independence assumption in information retrieval," Proceedings of the 10th European Conference on *Machine Learning*, Springer-Verlag, pp. 4–15.

Liu, J. (1994), "The collapsed Gibbs sampler in Bayesian computations with applications to a gene regulation problem," *Journal of the American Statistical Association* **89**(427).

Liu, J. S. (2008), *Monte Carlo Strategies in Scientific Computing*, Springer.

Livescu, K., Glass, J. R., & Bilmes, J. (2003), "Hidden feature models for speech recognition using dynamic Bayesian networks," Proceedings of Annual Conference of International Speech Communication Association (INTERSPEECH), pp. 2529–2532.

MacKay, D. J. C. (1992*a*), "Bayesian interpolation," *Neural Computation* **4**(3), 415–447.

MacKay, D. J. C. (1992*b*), "The evidence framework applied to classification networks," *Neural Computation* **4**(5), 720–736.

MacKay, D. J. C. (1992*c*), "A practical Bayesian framework for back-propagation networks," *Neural Computation* **4**(3), 448–472.

MacKay, D. J. C. (1995), "Probable networks and plausible predictions – a review of practical Bayesian methods for supervised neural networks," *Network: Computation in Neural Systems* **6**(3), 469–505.

MacKay, D. J. C. (1997), Ensemble learning for hidden Markov models, Technical Report, Cavendish Laboratory, University of Cambridge.

MacKay, D. J. C., & Peto, L. C. B. (1995), "A hierarchical Dirichlet language model," *Natural Language Engineering* **1**(3), 289–308.

Maekawa, T., & Watanabe, S. (2011), "Unsupervised activity recognition with user's physical characteristics data," Proceedings of International Symposium on *Wearable Computers*, pp. 89–96.

Mak, B., Kwok, J., & Ho, S. (2005), "Kernel eigenvoice speaker adaptation," *IEEE Transactions on Speech and Audio Processing* **13**(5), 984–992.

Manning, C. D., & Schütze, H. (1999), *Foundations of Statistical Natural Language Processing*, MIT Press.

Masataki, H., Sagisaka, Y., Hisaki, K., & Kawahara, T. (1997), "Task adaptation using MAP estimation in *n*-gram language modeling," Proceedings of International Conference on *Acoustics, Speech, and Signal Processing (ICASSP)*, pp. 783–786.

Matsui, T., & Furui, S. (1994), "Comparison of text-independent speaker recognition methods using VQ-distortion and discrete/continuous HMMs," *IEEE Transactions on Speech and Audio Processing* **2**(3), 456–459.

Matsumoto, Y., Kitauchi, A., Yamashita, T., *et al.* (1999), "Japanese morphological analysis system ChaSen version 2.0 manual," NAIST Technical Report.

McCallum, A., & Nigam, K. (1998), "A comparison of event models for naive Bayes text classification," in Proceedings of the Association for the Advancement of Artificial Intelligence (AAAI) Workshop on *Learning for Text Categorization*, Vol. 752, pp. 41–48.

McDermott, E., Hazen, T., Le Roux, J., Nakamura, A., & Katagiri, S. (2007), "Discriminative training for large-vocabulary speech recognition using minimum classification error," *IEEE Transactions on Audio, Speech, and Language Processing* **15**(1), 203–223.

Meignier, S., Moraru, D., Fredouille, C., Bonastre, J.-F., & Besacier, L. (2006), "Step-by-step and integrated approaches in broadcast news speaker diarization," *Computer Speech & Language* **20**(2), 303–330.

Metropolis, N., Rosenbluth, A. W., Rosenbluth, M. N., Teller, A. H., & Teller, E. (1953), "Equation of state calculations by fast computing machines," *Journal of Chemical Physics* **21**(6), 1087–1092.

Minka, T. P. (2001), "Expectation propagation for approximate Bayesian inference," Proceedings of Conference on *Uncertainty in Artificial Intelligence (UAI)*, pp. 362–369.

Mochihashi, D., Yamada, T., & Ueda, N. (2009), "Bayesian unsupervised word segmentation with nested Pitman–Yor language modeling," Proceedings of Joint Conference of Annual Meeting of the ACL and International Joint Conference on *Natural Language Processing of the AFNLP*, pp. 100–108.

Mohri, M., Pereira, F., & Riley, M. (2002), "Weighted finite-state transducers in speech recognition," *Computer Speech and Language* **16**, 69–88.

Moraru, D., Meignier, S., Besacier, L., Bonastre, J.-F., & Magrin-Chagnolleau, I. (2003), "The ELISA consortium approaches in speaker segmentation during the NIST 2002 speaker recognition evaluation," Proceedings of International Conference on *Acoustics, Speech, and Signal Processing (ICASSP)*, Vol. 2, pp. 89–92.

Mrva, D., & Woodland, P. C. (2004), "A PLSA-based language model for conversational telephone speech," in Proceedings of Annual Conference of International Speech Communication Association (INTERSPEECH), pp. 2257–2260.

Murphy, K. P. (2002), Dynamic Bayesian networks: representation, inference and learning, PhD thesis, University of California, Berkeley.

Murphy, K. P., Weiss, Y., & Jordan, M. I. (1999), "Loopy belief propagation for approximate inference: An empirical study," Proceedings of Conference on *Uncertainty in Artificial Intelligence (UAI)*, pp. 467–475.

Nadas, A. (1985), "Optimal solution of a training problem in speech recognition," *IEEE Transactions on Acoustics, Speech and Signal Processing* **33**(1), 326–329.

Nakagawa, S. (1988), *Speech Recognition by Probabilistic Model*, Institute of Electronics, Information and Communication Engineers (IEICE) (in Japanese).

Nakamura, A., McDermott, E., Watanabe, S., & Katagiri, S. (2009), "A unified view for discriminative objective functions based on negative exponential of difference measure between strings," Proceedings of International Conference on *Acoustics, Speech, and Signal Processing (ICASSP)*, pp. 1633–1636.

Neal, R., & Hinton, G. (1998), "A view of the EM algorithm that justifies incremental, sparse, and other variants," *Learning in Graphical Models*, pp. 355–368.

Neal, R. M. (1992), "Bayesian mixture modeling," Proceedings of the Workshop on *Maximum Entropy and Bayesian Methods of Statistical Analysis* **11**, 197–211.

Neal, R. M. (1993), "Probabilistic inference using Markov chain Monte Carlo methods," Technical Report CRG-TR-93-1, Dept. of Computer Science, University of Toronto.

Neal, R. M. (2000), "Markov chain sampling methods for Dirichlet process mixture models," *Journal of Computational and Graphical Statistics* **9**(2), 249–265.

Neal, R. M. (2003), "Slice sampling," *Annals of Statistics* **31**, 705–767.

Nefian, A. V., Liang, L., Pi, X., Liu, X., & Murphy, K. (2002), "Dynamic Bayesian networks for audio-visual speech recognition," *EURASIP Journal on Applied Signal Processing* **11**, 1274–1288.

Neubig, G., Mimura, M., Mori, S., & Kawahara, T. (2010), "Learning a language model from continuous speech," Proceedings of Annual Conference of International Speech Communication Association (INTERSPEECH), pp. 1053–1056.

Ney, H., Essen, U., & Kneser, R. (1994), "On structuring probabilistic dependences in stochastic language modeling," *Computer Speech and Language* **8**, 1–38.

Ney, H., Haeb-Umbach, R., Tran, B.-H., & Oerder, M. (1992), "Improvements in beam search for 10000-word continuous speech recognition," Proceedings of International Conference on *Acoustics, Speech, and Signal Processing (ICASSP)*, Vol. 1, IEEE, pp. 9–12.

Niesler, T., & Willett, D. (2002), "Unsupervised language model adaptation for lecture speech transcription," Proceedings of Annual Conference of International Speech Communication Association (INTERSPEECH), pp. 1413–1416.

Normandin, Y. (1992), "Hidden Markov models, maximum mutual information estimation, and the speech recognition problem," PhD thesis, McGill University, Montreal, Canada.

Odell, J. J. (1995), The use of context in large vocabulary speech recognition, PhD thesis, Cambridge University.

Ostendorf, M., & Singer, H. (1997), "HMM topology design using maximum likelihood successive state splitting," *Computer Speech and Language* **11**, 17–41.

Paul, D. B., & Baker, J. M. (1992), "The design for the *Wall Street Journal*-based CSR corpus," Proceedings of the Workshop on *Speech and Natural Language*, Association for Computational Linguistics, pp. 357–362.

Pettersen, S. (2008), Robust speech recognition in the presence of additive noise, PhD thesis, Norwegian University of Science and Technology.

Pitman, J. (2002), "Poisson-Dirichlet and GEM invariant distributions for split-and-merge transformation of an interval partition," *Combinatorics, Probability and Computing* **11**, 501–514.

Pitman, J. (2006), *Combinatorial Stochastic Processes*, Springer-Verlag.

Pitman, J., & Yor, M. (1997), "The two-parameter Poisson-Dirichlet distribution derived from a stable subordinator," *Annals of Probability* **25**(2), 855–900.

Porteous, I., Newman, D., Ihler, A., *et al.* (2008), "Fast collapsed Gibbs sampling for latent Dirichlet allocation," Proceedings of the 14th ACM SIGKDD International Conference on *Knowledge Discovery and Data Mining*, pp. 569–577.

Povey, D. (2003), Discriminative training for large vocabulary speech recognition, PhD thesis, Cambridge University.

Povey, D., Burget, L., Agarwal, M., *et al.* (2010), "Subspace Gaussian mixture models for speech recognition," Proceedings of International Conference on *Acoustics, Speech, and Signal Processing (ICASSP)*, pp. 4330–4333.

Povey, D., Gales, M. J. F., Kim, D., & Woodland, P. C. (2003), "MMI-MAP and MPE-MAP for acoustic model adaptation," Proceedings of European Conference on *Speech Communication and Technology (EUROSPEECH)* **8**, 1981–1984.

Povey, D., Ghoshal, A., Boulianne, G., *et al.* (2011), "The Kaldi speech recognition toolkit," Proceedings of IEEE Automatic Speech Recognition and Understanding Workshop (ASRU).

Povey, D., Kanevsky, D., Kingsbury, B., *et al.* (2008), "Boosted MMI for model and feature-space discriminative training," Proceedings of International Conference on *Acoustics, Speech, and Signal Processing (ICASSP)*, pp. 4057–4060.

Povey, D., Kingsbury, B., Mangu, L., *et al.* (2005), "fMPE: Discriminatively trained features for speech recognition," Proceedings of International Conference on *Acoustics, Speech, and Signal Processing (ICASSP)* **1**, 961–964.

Povey, D., & Woodland, P. C. (2002), "Minimum phone error and I-smoothing for improved discriminative training," Proceedings of International Conference on *Acoustics, Speech, and Signal Processing (ICASSP)*, pp. 13–17.

Povey, D., Woodland, P., & Gales, M. (2003), "Discriminative MAP for acoustic model adaptation," Proceedings of International Conference on *Acoustics, Speech, and Signal Processing (ICASSP)* **1**, I–312.

Price, P., Fisher, W., Bernstein, J., & Pallett, D. (1988), "The DARPA 1000-word resource management database for continuous speech recognition," Proceedings of International Conference on *Acoustics, Speech, and Signal Processing (ICASSP)*, pp. 651–654.

Rabiner, L. R., & Juang, B.-H. (1986), "An introduction to hidden Markov models," *IEEE ASSP Magazine* **3**(1), 4–16.

Rabiner, L. R., & Juang, B.-H. (1993), *Fundamentals of Speech Recognition*, Vol. 14, PTR Prentice Hall.

Rasmussen, C. E. (1999), "The infinite Gaussian mixture model," *Advances in Neural Information Processing Systems* **12**, 554–560.

Rasmussen, C. E., & Williams, C. K. I. (2006), *Gaussian Processes for Machine Learning, Adaptive Computation and Machine Learning*, MIT Press.

Reynolds, D., Quatieri, T., & Dunn, R. (2000), "Speaker verification using adapted Gaussian mixture models," *Digital Signal Processing* **10**(1-3), 19–41.

Rissanen, J. (1984), "Universal coding, information, prediction and estimation," *IEEE Transactions on Information Theory* **30**, 629–636.

Rodriguez, A., Dunson, D. B., & Gelfand, A. E. (2008), "The nested Dirichlet process," *Journal of the American Statistical Association* **103**(483), 1131–1154.

Rosenfeld, R. (2000), "Two decades of statistical language modeling: Where do we go from here?," *Proceedings of the IEEE* **88**(8), 1270–1278.

Sainath, T. N., Ramabhadran, B., Picheny, M., Nahamoo, D., & Kanevsky, D. (2011), "Exemplar-based sparse representation features: from TIMIT to LVCSR," *IEEE Transactions on Audio, Speech and Language Processing* **19**(8), 2598–2613.

Saito, D., Watanabe, S., Nakamura, A., & Minematsu, N. (2012), "Statistical voice conversion based on noisy channel model," *IEEE Transactions on Audio, Speech, and Language Processing* **20**(6), 1784–1794.

Salakhutdinov, R. (2009), Learning deep generative models, PhD thesis, University of Toronto.

Salton, G., & Buckley, C. (1988), "Term-weighting approaches in automatic text retrieval," *Information Processing & Management* **24**(5), 513–523.

Sanderson, C., Bengio, S., & Gao, Y. (2006), "On transforming statistical models for non-frontal face verification," *Pattern Recognition* **39**(2), 288–302.

Sankar, A., & Lee, C.-H. (1996), "A maximum-likelihood approach to stochastic matching for robust speech recognition," *IEEE Transactions on Speech and Audio Processing* **4**(3), 190–202.

Saon, G., & Chien, J.-T. (2011), "Some properties of Bayesian sensing hidden Markov models," Proceedings of IEEE Automatic Speech Recognition and Understanding Workshop (ASRU), pp. 65–70.

Saon, G., & Chien, J.-T. (2012*a*), "Bayesian sensing hidden Markov models," *IEEE Transactions on Audio, Speech, and Language Processing* **20**(1), 43–54.

Saon, G., & Chien, J.-T. (2012*b*), "Large-vocabulary continuous speech recognition systems: A look at some recent advances," *IEEE Signal Processing Magazine* **29**(6), 18–33.

Schalkwyk, J., Beeferman, D., Beaufays, F., *et al.* (2010), " 'Your word is my command': Google search by voice: A case study," in *Advances in Speech Recognition*, Springer, pp. 61–90.

Schlüter, R., Macherey, W., Müller, B., & Ney, H. (2001), "Comparison of discriminative training criteria and optimization methods for speech recognition," *Speech Communication* **34**(3), 287–310.

Schultz, T., & Waibel, A. (2001), "Language-independent and language-adaptive acoustic modeling for speech recognition," *Speech Communication* **35**(1), 31–51.

Schwarz, G. (1978), "Estimating the dimension of a model," *The Annals of Statistics* **6**(2), 461–464.

Scott, S. (2002), "Bayesian methods for hidden Markov models," *Journal of the American Statistical Association* **97**(457), 337–351.

Seide, F., Li, G., Chen, X., & Yu, D. (2011), "Conversational speech transcription using context-dependent deep neural networks," Proceedings of Annual Conference of International Speech Communication Association (INTERSPEECH), pp. 437–440.

Sethuraman, J. (1994), "A constructive definition of Dirichlet priors," *Statistica Sinica* **4**, 639–650.

Shikano, K., Kawahara, T., Kobayashi, T., *et al.* (1999), *Japanese Dictation Toolkit – Free Software Repository for Automatic Speech Recognition*, *http://www.ar.media.kyoto-u.ac.jp/dictation/*.

Shinoda, K. (2010), "Acoustic model adaptation for speech recognition," *IEICE Transactions on Information and Systems* **93**(9), 2348–2362.

Shinoda, K., & Inoue, N. (2013), "Reusing speech techniques for video semantic indexing," *IEEE Signal Processing Magazine* **30**(2), 118–122.

Shinoda, K., & Iso, K. (2001), "Efficient reduction of Gaussian components using MDL criterion for HMM-based speech recognition," Proceedings of International Conference on *Acoustics, Speech, and Signal Processing (ICASSP)*, pp. 869–872.

Shinoda, K., & Lee, C.-H. (2001), "A structural Bayes approach to speaker adaptation," *IEEE Transactions on Speech and Audio Processing* **9**, 276–287.

Shinoda, K., & Watanabe, T. (1996), "Speaker adaptation with autonomous model complexity control by MDL principle," Proceedings of International Conference on *Acoustic, Speech, and Signal Processing (ICASSP)*, pp. 717–720.

Shinoda, K., & Watanabe, T. (1997), "Acoustic modeling based on the MDL criterion for speech recognition," Proceedings of European Conference on Speech Communication and Technology (EUROSPEECH), Vol. 1, pp. 99–102.

Shinoda, K., & Watanabe, T. (2000), "MDL-based context-dependent subword modeling for speech recognition," *Journal of the Acoustical Society of Japan (E)* **21**, 79–86.

Shiota, S., Hashimoto, K., Nankaku, Y., & Tokuda, K. (2009), "Deterministic annealing based training algorithm for Bayesian speech recognition," Proceedings of Annual Conference of International Speech Communication Association (INTERSPEECH), pp. 680–683.

Siohan, O., Myrvoll, T. A., & Lee, C. H. (2002), "Structural maximum a posteriori linear regression for fast HMM adaptation," *Computer Speech and Language* **16**(1), 5–24.

Siu, M.-h., Gish, H., Chan, A., Belfield, W., & Lowe, S. (2014), "Unsupervised training of an HMM-based self-organizing unit recognizer with applications to topic classification and keyword discovery," *Computer Speech & Language* **28**(1), 210–223.

Somervuo, P. (2004), "Comparison of ML, MAP, and VB based acoustic models in large vocabulary speech recognition," Proceedings of International Conference on *Spoken Language Processing (ICSLP)*, pp. 830–833.

Spiegelhalter, D. J., & Lauritzen, S. L. (1990), "Sequential updating of conditional probabilities on directed graphical structures," *Networks* **20**(5), 579–605.

Sproat, R., Gale, W., Shih, C., & Chang, N. (1996), "A stochastic finite-state word-segmentation algorithm for Chinese," *Computational Linguistics* **22**(3), 377–404.

Stenger, B., Ramesh, V., Paragios, N., Coetzee, F., & Buhmann, J. M. (2001), "Topology free hidden Markov models: Application to background modeling," Proceedings of International Conference on *Computer Vision (ICCV)*', Vol. 1, pp. 294–301.

Stolcke, A., Ferrer, L., Kajarekar, S., Shriberg, E., & Venkataraman, A. (2005), "MLLR transforms as features in speaker recognition," Proceedings of Annual Conference of International Speech Communication Association (INTERSPEECH), pp. 2425–2428.

Stolcke, A., & Omohundro, S. (1993), "Hidden Markov model induction by Bayesian model merging," *Advances in Neural Information Processing Systems*, pp. 11–18, Morgan Kaufmann.

Takahashi, J., & Sagayama, S. (1997), "Vector-field-smoothed Bayesian learning for fast and incremental speaker/telephone-channel adaptation," *Computer Speech and Language* **11**, 127–146.

Takami, J., & Sagayama, S. (1992), "A successive state splitting algorithm for efficient allophone modeling," Proceedings of International Conference on *Acoustics, Speech, and Signal Processing (ICASSP)*, pp. 573–576.

Tam, Y.-C., & Schultz, T. (2005), "Dynamic language model adaptation using variational Bayes inference," Proceedings of Annual Conference of International Speech Communication Association (INTERSPEECH), pp. 5–8.

Tam, Y.-C., & Schultz, T. (2006), "Unsupervised language model adaptation using latent semantic marginals," Proceedings of Annual Conference of International Speech Communication Association (INTERSPEECH), pp. 2206–2209.

Tamura, M., Masuko, T., Tokuda, K., & Kobayashi, T. (2001), "Adaptation of pitch and spectrum for HMM-based speech synthesis using MLLR," Proceedings of International Conference on *Acoustics, Speech, and Signal Processing (ICASSP)*, pp. 805–808.

Tawara, N., Ogawa, T., Watanabe, S., & Kobayashi, T. (2012*a*), "Fully Bayesian inference of multi-mixture Gaussian model and its evaluation using speaker clustering," Proceedings of International Conference on *Acoustics, Speech, and Signal Processing (ICASSP)*, pp. 5253–5256.

Tawara, N., Ogawa, T., Watanabe, S., Nakamura, A., & Kobayashi, T. (2012*b*), "Fully Bayesian speaker clustering based on hierarchically structured utterance-oriented Dirichlet process mixture model," Proceedings of Annual Conference of International Speech Communication Association (INTERSPEECH), pp. 2166–2169.

Teh, Y. W. (2006), "A hierarchical Bayesian language model based on Pitman–Yor processes," Proceedings of International Conference on *Computational Linguistics* and Annual Meeting of the Association for Computational Linguistics, pp. 985–992.

Teh, Y. W., Jordan, M. I., Beal, M. J., & Blei, D. M. (2006), "Hierarchical Dirichlet processes," *Journal of the American Statistical Association* **101**(476), 1566–1581.

Tipping, M. E. (2001), "Sparse Bayesian learning and the relevance vector machine," *Journal of Machine Learning Research* **1**, 211–244.

Torbati, A. H. H. N., Picone, J., & Sobel, M. (2013), "Speech acoustic unit segmentation using hierarchical Dirichlet processes," Proceedings of Annual Conference of International Speech Communication Association (INTERSPEECH), pp. 637–641.

Ueda, N., & Ghahramani, Z. (2002), "Bayesian model search for mixture models based on optimizing variational bounds," *Neural Networks* **15**, 1223–1241.

Valente, F. (2006), "Infinite models for speaker clustering," Proceedings of Annual Conference of International Speech Communication Association (INTERSPEECH), pp. 1329–1332.

Valente, F., Motlicek, P., & Vijayasenan, D. (2010), "Variational Bayesian speaker diarization of meeting recordings," Proceedings of International Conference on *Acoustics, Speech, and Signal Processing (ICASSP)*, pp. 4954–4957.

Valente, F., & Wellekens, C. (2003), "Variational Bayesian GMM for speech recognition," Proceedings of European Conference on *Speech Communication and Technology (EUROSPEECH)*, pp. 441–444.

Valente, F., & Wellekens, C. (2004*a*), "Variational Bayesian feature selection for Gaussian mixture models," Proceedings of International Conference on *Acoustics, Speech, and Signal Processing (ICASSP)*, Vol. 1, pp. 513–516.

Valente, F., & Wellekens, C. (2004*b*), "Variational Bayesian speaker clustering," Proceedings of ODYSSEY The Speaker and Language Recognition Workshop, pp. 207–214.

Vapnik, V. (1995), *The Nature of Statistical Learning Theory*, Springer-Verlag.

Veselỳ, K., Ghoshal, A., Burget, L., & Povey, D. (2013), "Sequence-discriminative training of deep neural networks," Proceedings of Annual Conference of International Speech Communication Association (INTERSPEECH), pp. 2345–2349.

Villalba, J., & Brümmer, N. (2011), "Towards fully Bayesian speaker recognition: Integrating out the between-speaker covariance," Proceedings of Annual Conference of International Speech Communication Association (INTERSPEECH), pp. 505–508.

Vincent, E., Barker, J., Watanabe, S., *et al.* (2013), "The second 'CHiME' speech separation and recognition challenge: An overview of challenge systems and outcomes," Proceedings of IEEE *Automatic Speech Recognition and Understanding Workshop (ASRU)*, pp. 162–167.

Viterbi, A. J. (1967), "Error bounds for convolutional codes and an asymptotically optimal decoding algorithm," *IEEE Transactions on Information Theory* **IT-13**, 260–269.

Wallach, H. M. (2006), "Topic modeling: beyond bag-of-words," Proceedings of International Conference on *Machine Learning*, pp. 977–984.

Watanabe, S., & Chien, J. T. (2012), "Tutorial: Bayesian learning for speech and language processing," International Conference on *Acoustics, Speech, and Signal Processing (ICASSP).*

Watanabe, S., Minami, Y., Nakamura, A., & Ueda, N. (2002), "Application of variational Bayesian approach to speech recognition," *Advances in Neural Information Processing Systems.*

Watanabe, S., Minami, Y., Nakamura, A., & Ueda, N. (2004), "Variational Bayesian estimation and clustering for speech recognition," *IEEE Transactions on Speech and Audio Processing* **12**, 365–381.

Watanabe, S., & Nakamura, A. (2004), "Acoustic model adaptation based on coarse-fine training of transfer vectors and its application to a speaker adaptation task," Proceedings of International Conference on *Spoken Language Processing (ICSLP)*, pp. 2933–2936.

Watanabe, S., & Nakamura, A. (2006), "Speech recognition based on Student's t-distribution derived from total Bayesian framework," *IEICE Transactions on Information and Systems* **E89-D**, 970–980.

Watanabe, S., & Nakamura, A. (2009), "On-line adaptation and Bayesian detection of environmental changes based on a macroscopic time evolution system," Proceedings of IEEE International Conference on *Acoustics, Speech, and Signal Processing (ICASSP)*, pp. 4373–4376.

Watanabe, S., Nakamura, A., & Juang, B. (2011), "Bayesian linear regression for hidden Markov model based on optimizing variational bounds," Proceedings of IEEE Workshop on *Machine Learning for Signal Processing*, pp. 1–6.

Watanabe, S., Nakamura, A., & Juang, B.-H. (2013), "Structural Bayesian linear regression for hidden Markov models," *Journal of Signal Processing Systems*, 1–18.

Wegmann, S., McAllaster, D., Orloff, J., & Peskin, B. (1996), "Speaker normalization on conversational telephone speech," Proceedings of International Conference on *Acoustics, Speech, and Signal Processing (ICASSP)*, pp. 339–341.

Winn, J., & Bishop, C. (2006), "Variational message passing," *Journal of Machine Learning Research* **6**(1), 661.

Witten, I. H., & Bell, T. C. (1991), "The zero-frequency problem: estimating the probabilities of novel events in adaptive text compression," *IEEE Transactions on Information Theory* **37**, 1085–1094.

Wooters, C., Fung, J., Peskin, B., & Anguera, X. (2004), "Towards robust speaker segmentation: The ICSI-SRI fall 2004 diarization system," in RT-04F Workshop, Vol. 23.

Wooters, C., & Huijbregts, M. (2008), "The ICSI RT07s speaker diarization system," in *Multimodal Technologies for Perception of Humans*, Springer, pp. 509–519.

Yamagishi, J., Kobayashi, T., Nakano, Y., Ogata, K., & Isogai, J. (2009), "Analysis of speaker adaptation algorithms for HMM-based speech synthesis and a constrained SMAPLR adaptation algorithm," *IEEE Transactions on Audio, Speech, and Language Processing* **17**(1), 66–83.

Yaman, S., Chien, J.-T., & Lee, C.-H. (2007), "Structural Bayesian language modeling and adaptation," Proceedings of Annual Conference of International Speech Communication Association (INTERSPEECH), pp. 2365–2368.

Yedidia, J. S., Freeman, W. T., & Weiss, Y. (2003), "Understanding belief propagation and its generalizations," *Exploring Artificial Intelligence in the New Millennium* **8**, 236–239.

Young, S., Evermann, G., Gales, M., *et al.* (2006), "The HTK book (for HTK version 3.4)," Cambridge University Engineering Department.

Young, S. J., Odell, J. J., & Woodland, P. C. (1994), "Tree-based state tying for high accuracy acoustic modelling," Proceedings of the Workshop on *Human Language Technology*, pp. 307–312.

Yu, K., & Gales, M. J. F. (2006), "Incremental adaptation using Bayesian inference," Proceedings of International Conference on *Acoustics, Speech, and Signal Processing (ICASSP)*, pp. 217–220.

Zhang, Y., & Glass, J. R. (2009), "Unsupervised spoken keyword spotting via segmental DTW on Gaussian posteriorgrams," Proceedings of IEEE *Automatic Speech Recognition & Understanding Workshop (ASRU)*, pp. 398–403.

Zhang, Y., Liu, P., Chien, J.-T., & Soong, F. (2009), "An evidence framework for Bayesian learning of continuous-density hidden Markov models," Proceedings of International Conference on *Acoustics, Speech, and Signal Processing (ICASSP)*, pp. 3857–3860.

Zhao, X., Dong, Y., Zhao, J., *et al.* (2009), "Variational Bayesian joint factor analysis for speaker verification," Proceedings of IEEE International Conference on *Acoustics, Speech, and Signal Processing (ICASSP)*, pp. 4049–4052.

Zhou, B., & Hansen, J. H. (2000), "Unsupervised audio stream segmentation and clustering via the Bayesian information criterion," Proceedings of Annual Conference of International Speech Communication Association (INTERSPEECH), pp. 714–717.

Zweig, G., & Nguyen, P. (2009), "A segmental CRF approach to large vocabulary continuous speech recognition," Proceedings of IEEE *Automatic Speech Recognition and Understanding Workshop (ASRU)*, pp. 152–157.

Zweig, G., & Russell, S. (1998), "Speech recognition with dynamic Bayesian networks," Proceedings of the National Conference on *Artificial Intelligence*, pp. 173–180.

Index

Page numbers in bold type indicate a main explanatory entry.